# 8th International Conference on Inertial Fusion Sciences and Applications (IFSA 2013)



Journal of Physics: Conference Series Volume 688

Nara, Japan
8 – 13 September 2013

**Editors:**

**Hiroshi Azechi**
**Hiroyuki Shiraga**

ISBN: 978-1-5108-2279-5
ISSN: 1742-6588

**Printed from e-media with permission by:**

Curran Associates, Inc.
57 Morehouse Lane
Red Hook, NY 12571



Printed by Curran Associates, Inc. (2016)

For permission requests, please contact the Institute of Physics
at the address below.

Institute of Physics
Dirac House, Temple Back
Bristol BS1 6BE UK

Phone: 44 1 17 929 7481
Fax: 44 1 17 920 0979

techtracking@iop.org

**Additional copies of this publication are available from:**

Curran Associates, Inc.
57 Morehouse Lane
Red Hook, NY 12571 USA
Phone: 845-758-0400
Fax: 845-758-2633
Email: curran@proceedings.com
Web: www.proceedings.com

# Table of contents

## Volume 688

**March 2016**

**8th International Conference on Inertial Fusion Sciences and Applications (IFSA 2013)8–13 September 2013, Nara, Japan**

**Accepted papers received: 8 January 2016**
**Published online: 1 April 2016**

### Preface



### Papers

# Preface

The Eighth International Conference on Inertial Fusion Science and Applications was hosted on 8–13 September 2013 at the Nara Prefectural New Public Hall, organized by the Institute of Laser Engineering, Osaka University. It brought together 394 participants from 19 countries. The common goal was to share the latest advances in the fields of inertial fusion energy, laser and particle beams and fusion technology as well as high-energy density physics and applications. There were 393 presentations in total including three Keynote lectures, two Teller lectures, 17 plenary talks, 132 orals and 236 posters. Three IAEA-TM joint papers were presented during the Conference week. The proceedings collected in this volume underwent a rigorous peer-review process for which the Editors are thankful to the many referees who worked very hard to finalize the publication. They also thank the TPC members whose work made such a wonderful Conference program.

Among the tremendous achievements presented, the great progress in experiments on NIF aimed at ICF ignition was one of the most important highlights. Significant progress in high-energy density science was another important topic in this IFSA2013. Of course many other excellent papers on various subjects were presented showing the continuous growth of the related fields.

The Teller Award was bestowed during IFSA2013 on two key contributors to ICF science, J. Hammer and R. Petrasso. The student prize was awarded for six excellent papers presented by students from various countries.

On behalf of the various Committees, we are glad to thank again everyone who helped make IFSA2013 a great success.

Hiroshi Azechi

Hiroyuki Shiraga

# Members of Governing Committee

## IFSA Co-Chairs

Hiroshi Azechi, *Osaka University, Japan*

Edward Moses, *Lawrence Livermore National Laboratory, USA*

Patrick Mora, *Institut Lasers et Plasmas, France*

## Organizing Co-Chairs

Ryosuke Kodama, *Osaka University, Japan*

William Goldstein, *Lawrence Livermore National Laboratory, USA*

Sylvie Jacquemot, *Laboratoire pour l'Utilisation des Lasers Intenses, France*

## Sponsoring Organizations

*Institute of Laser Engineering, Osaka University, Japan*

*University of California, USA*

*Institut Lasers et Plasmas, France*

## Co-Organizations

*National Institute for Fusion Science, Japan*

*Kansai Photon Science Institute, Japan Atomic Energy Agency, Japan*

## Local Organizing Committee (LOC)

*Chair*

Hiroshi Azechi, *Osaka University*

*Vice Chairs*

Masataka Nishi, *Japan Atomic Energy Agency*

Akio Komori, *National Institute for Fusion Science*

*Members*

Hiroyuki Daido, *Japan Atomic Energy Agency*

Takuma Endo, *Hiroshima University*

Masayuki Fujita, *Institute for Laser Technology*

Etsuo Fujiwara, *University of Hyogo*

Kazuhiro Hanada, *Kyushu University*

Masanori Hangyo, *Osaka University*

Akira Hiruma, *Research Foundation for Opto-Science and Technology / Hamamatsu Photonics*

Hiroshi Horiike, *Osaka University*

Kazuhiko Horioka, *Tokyo Institute of Technology*

Tetsuya Ishikawa, *RIKEN*

Yasukazu Izawa, *Institute for Laser Technology*

Yoshiaki Kato, *Graduate School for Creation of New Photonics Industries*

Shigeo Kawata, *Utsunomiya University*

Mitsuru Kikuchi, *Japan Atomic Energy Agency*

Tadafumi Kishimoto, *Osaka University*

Yasuaki Kishimoto, *Kyoto University*

Ryosuke Kodama, *Osaka University*

Kiminori Kondo, *Japan Atomic Energy Agency*

Masao Matsuyama, *Toyama University*
Katsumi Midorikawa, *RIKEN*
Kunioki Mima, *Graduate School for Creation of New Photonics Industries*
Shuji Miyamoto, *University of Hyogo*
Noriaki Miyanaga, *Osaka University*
Sadao Nakai, *Laser Society of Japan / Osaka University*
Yasuyuki Nakao, *Kyushu University*
Kazuhiro Nakata, *Osaka University*
Hiroaki Nishimura, *Osaka University*
Takayoshi Norimatsu, *Osaka University*
Yuichi Ogawa, *University of Tokyo*
Shoichi Okamura, *National Institute for Fusion Science*
Kunihiko Okano, *Central Research Institute of Electric Power Industry*
Shuji Sakabe, *Kyoto University*
Hitoshi Sakagami, *National Institute for Fusion Science*
Nobuhiko Sarukura, *Osaka University*
Mamiko Sasao, *Doshisha University*
Kazunari Shibata, *Kyoto University*
Katsuya Shimizu, *Osaka University*
Hiroyuki Shiraga, *Osaka University*
Tohru Suemoto, *University of Tokyo*
Hideaki Takabe, *Osaka University*
Masayoshi Tonouchi, *Osaka University*
Kenichi Ueda, *University of Electro-Communication*
Yoshio Ueda, *Osaka University*
Hiroshi Yamada, *National Institute for Fusion Science*
Kaoru Yamauchi, *University of Tokyo*
Hitoki Yoneda, *University of Electro-Communication*
Noboru Yugami, *Utsunomiya University*

## Local Executive Committee (LEC)

*Chair*

Hiroaki Nishimura, *Osaka University*

*Members*

| | | |
|---|---|---|
| Yasunobu Arikawa | Yasushi Fujimoto | Shinsuke Fujioka |
| Hisanori Fujita | Masayuki Fujita | Masanori Hangyo |
| Takeshi Higashiguchi | Yoichiro Hironaka | Akifumi Iwamoto |
| Tomohiro Johzaki | Masaki Kando | Junji Kawanaka |
| Iwao Kawayama | Mayuko Koga | Masahiko Matsumoto |
| Hironaru Murakami | Masakatsu Murakami | Takeshi Nagashima |
| Hideo Nagatomo | Mitsuo Nakai | Yoshiki Nakata |
| Masaharu Nishikino | Takayoshi Norimatsu | Yoshihiro Ochi |
| Satoshi Orimo | Norimasa Ozaki | Satoshi Ozaki |
| Youichi Sakawa | Takayoshi Sano | Nobuhiko Sarukura |
| Keisuke Shigemori | Toshihiko Shimizu | Hiroyuki Shiraga |
| Atsushi Sunahara | Toshihiro Taguchi | Hideaki Takabe |
| Masayoshi Tonouchi | Koji Tsubakimoto | Ryusuke Tsuji |
| Hiroki Yoshida | | |

## Technical Program Committee

*Co-Chairs*

Hiroyuki Shiraga, *Osaka University, Japan*
Bruce Hammel, *Lawrence Livermore National Laboratory, USA*
Erik Lefebvre, *Commissariat à l'Energie Atomique et aux Energies Alternatives, France*

*Members*

Stefano Atzeni, *Università degli Studi di Roma "La Sapienza", Italy*
John Barnard, *Lawrence Berkeley National Laboratory, USA*
Dimitri Batani, *Centre Lasers Intenses et Applications, France*
Sophie Baton, *Laboratoire d'utilisation des Lasers Intenses, France*
Farhat Beg, *University of Calfornia, San Diego, USA*
Sergei Bulanov, *Japan Atomic Energy Agency, Japan*
Jerry Chittenden, *Imperial College London, UK*
Dan Clark, *Lawrence Livermore National Laboratory, USA*
Gilbert Collins, *Lawrence Livermore National Laboratory, USA*
Mike Cuneo, *Sandia National Laboratories, USA*
Claude Deutsch, *Laboratoire de Physique des Gaz et des Plasmas, France*
Jean-Michel Di Nicola, *Lawrence Livermore National Laboratory, USA*
Mike Dunne, *Lawrence Livermore National Laboratory, USA*
Roger Evans, *Imperial College London, UK*
Juan Fernandez, *Los Alamos National Laboratory, USA*
John Foster, *Atomic Weapons Establishment, UK*
Shinsuke Fujioka, *Osaka University, Japan*
Hisanori Fujita, *Osaka University, Japan*
Pascal Gauthier, *Commissariat à l'Energie Atomique et aux Energies Alternatives, France*
Gary Grimm, *Lawrence Livermore National Laboratory, USA*
Sergey Guskov, *Lebedev Physical Institute, Russia*
Mark Herrmann, *Sandia National Laboratories, USA*
Denise Hinkel, *Lawrence Livermore National Laboratory, USA*
Dieter Hoffman, *Technische Universität Darmstadt, Germany*
Javier Honrubia, *Universidad Politécnica de Madrid, Spain*

Stefan Hueller, *Centre de Physique Théorique de l'Ecole Polytechnique, France*
Masaki Kando, *Japan Atomic Energy Agency, Japan*
Junji Kawanaka, *Osaka University, Japan*
Joe Kilkenny, *General Atomics, USA*
Ravindra Kumar, *Tata Institute of Fundamental Research, India*
Thomas Kwan, *Los Alamos National Laboratory, USA*
Dick Lee, *Lawrence Berkeley National Laboratory, USA*
Patrick Mckenty, *University of Rochester, USA*
Nathan Meezan, *Lawrence Livermore National Laboratory, USA*
Emilio Minguez, *Universidad Politécnica de Madrid, Spain*
Masakatsu Murakami, *Osaka University, Japan*
Abbas Nikroo, *General Atomics, USA*
Philip Nilson, *University of Rochester, USA*
Takayoshi Norimatsu, *Osaka University, Japan*
Tetsuo Ozaki, *National Institute for Fusion Science, Japan*
Jose Manuel Perlado, *Universidad Politécnica de Madrid, Spain*
Tomasz Plewa, *Florida State University, USA*
André Richard, *Commissariat à l'Energie Atomique et aux Energies Alternatives, France*
Markus Roth, *Technische Universität Darmstadt, Germany*
Claude Rouyer, *Commissariat à l'Energie Atomique et aux Energies Alternatives, France*
Hitoshi Sakagami, *National Institute for Fusion Science, Japan*
Youichi Sakawa, *Osaka University, Japan*
Craig Sangster, *University of Rochester, USA*
Andy Schmitt, *Naval Research Laboratory, USA*
Zheng-Ming Sheng, *Shanghai Jiao Tong University, China*
Keisuke Shigemori, *Osaka University, Japan*
Max Tabak, *Lawrence Livermore National Laboratory, USA*
Toshihiro Taguchi, *Setsunan University, Japan*
Vladimir Tikhonchuck, *Centre Lasers Intenses et Applications, France*
Daniel Vanderhaegen, *Commissariat à l'Energie Atomique et aux Energies Alternatives, France*
Noboru Yugami, *Utsunomiya University, Japan*

Journal of Physics: Conference Series **688** (2016) 011002 doi:10.1088/1742-6596/688/1/011002

## Teller Award Chair

Robert Kauffman, *Lawrence Livermore National Laboratory, USA*

## Conference Supervisor

Saburo Aimoto, *Osaka University, Japan*

## Sponsorship

- The Fund Project of the Commemorative Organization for the Japan World Exposition '70
- Inertial Fusion Energy Forum
- Japan Society for the Promotion of Science
- Matsuo Foundation
- Nara Visitors Bureau
- Ogasawara Foundation for the Promotion of Science & Engineering
- Research Foundation for Opto-Science and Technology
- Research Foundation for the Electrotechnology of Chubu

助成 独立行政法人 日本万国博覧会記念機構
Supported by the Commemorative Organization for the Japan World Exposition('70).
この助成金は、日本万国博覧会の収益を基にしています。

IFE FORUM

JSPS 独立行政法人 日本学術振興会
Japan Society for the Promotion of Science

一般財団法人
奈良県ビジターズビューロー
Nara Visitors Bureau

Journal of Physics: Conference Series **688** (2016) 011004 doi:10.1088/1742-6596/688/1/011004

# Endorsements

- Ministry of Education, Culture, Sports, Science and Technology (MEXT)
- American Nuclear Society
- American Physical Society
- Atomic Energy Society of Japan
- European Physical Society
- Fusion Power Associates
- Japan Society of Applied Physics
- Japan Society of Plasma Science and Fusion Research
- Laser Society of Japan
- Physical Society of Japan
- Institute of Electrical Engineering of Japan
- International Atomic Energy Agency

Journal of Physics: Conference Series **688** (2016) 011005 doi:10.1088/1742-6596/688/1/011005

# Exhibitors

- INTECH CO., LTD
- HAMAMATSU PHOTONICS K.K.
- HERZ CO., LTD
- NIKON CORPORATION
- NTT Advanced Technology Corporation
- RAD Device Co., Ltd.
- Japan Laser Corporation / Amplitude Technologies
- Struck Innovative Systeme
- OptoScience, INC. / Gooch & Housego
- NIPPON ADVANCED TECHNOLOGY CO., LTD
- OKAMOTO OPTICS WORKS, INC.
- National Institute for Fusion Science
- Shin-Etsu Quartz Products Co., Ltd.
- Nippon Roper / LxRay
- HORIBA, LTD.
- NanoLabz
- High Power Laser Science and Engineering
- THALES JAPAN K.K.
- Open Advanced Research Facilities Initiative "New Opto-science industry opened by High Intensity Lasers"

Nikon

NTT AT

RAD Device

JLC
JAPAN LASER

Amplitude
TECHNOLOGIES
Nothing but ultrafast.

struck innovative systeme

OPTO SCIENCE, INC.

Gooch & Housego

NAT

OKAMOTO
optics

NIFS

Shin-Etsu QUARTZ
A JOINT VENTURE WITH Heraeus

NIPPON ROPER

LR LxRay
for the future

HORIBA
Scientific

nanolabz

HIGH POWER
LASER SCIENCE AND ENGINEERING

THALES

# Peer review statement

All papers published in this volume of *Journal of Physics: Conference Series* have been peer reviewed through processes administered by the proceedings Editors. Reviews were conducted by expert referees to the professional and scientific standards expected of a proceedings journal published by IOP Publishing.

Journal of Physics: Conference Series **688** (2016) 012001 doi:10.1088/1742-6596/688/1/012001

# Probing ultrafast dynamics in a solid-density plasma created by an intense femtosecond laser

**Amitava Adak[1], Dave Blackman[2], Gourab Chatterjee[1], Prashant Kumar Singh[1], Amit D. Lad[1], P. Brijesh[1], A. P. L. Robinson[3], John Pasley[2,3] and G. Ravindra Kumar[*1]**

[1]Tata Institute of Fundamental Research, Dr. Homi Bhabha Road, Colaba, Mumbai-400005, India
[2]York Plasma Institute, University of York, Heslington, York, YO10 5DQ, United Kingdom
[3]Central Laser Facility, RutherfordAppleton Laboratory, Chilton, Didcot, OX10 0QX, United Kingdom
*Email: grk@tifr.res.in

**Abstract.** We report a study on the dynamics of a near-solid density plasma using an ultraviolet (266 nm) femtosecond probe laser pulse, which can penetrate to densities of $\sim 10^{22}$ cm$^{-3}$, nearly an order of magnitude higher than the critical density of the 800 nm, femtosecond pump laser. Time-resolved probe-reflectivity from the plasma shows a rapid decay (picosecond-timescale) while the time-resolved reflected probe spectra show red shifts at early temporal delays and blue shifts at longer delays. This spectral behaviour of the reflected probe can be explained by a laser-driven shock moving inward and a subsequent hydrodynamic free expansion in the outward direction.

## 1. Introduction

Ultra-intense lasers can create extremely hot, high-density plasma while interacting with a solid. Extreme states of matter, achieved on a laboratory table-top by such interactions are very interesting to study, the measurement of the equation of state similar to that of an astrophysical object being a striking example[1]. Relativistic energy electron beams produced at the plasma critical surface in such interactions are crucial to the success of fast ignition[2] of the fusion pellet in inertial confinement fusion(ICF). A lot of experimental and theoretical studies have been performed in recent decades for the basic understanding of intense short-pulse-laser matter interaction[3] and for applications like particle acceleration[4, 5]. The hot dense plasma created via various laser absorption mechanisms[6] evolves very rapidly (on femtosecond and picosecond time-scales) and these dynamics can be monitored using the pump-probe technique[7, 8, 9].

In this paper, we investigate the temporal dynamics of a highly dense electron layer($n_e = 10^{22}$ cm$^{-3}$ ) inside a plasma created on an aluminium-coated BK7 glass target by a laser at relativistic light intensities. Doppler spectrometry of the reflected probe enables the observation of the ultrafast motion of its critical surface (high density layer) inside the hot dense plasma. A numerical simulation is performed which reproduces the results of the experiment.

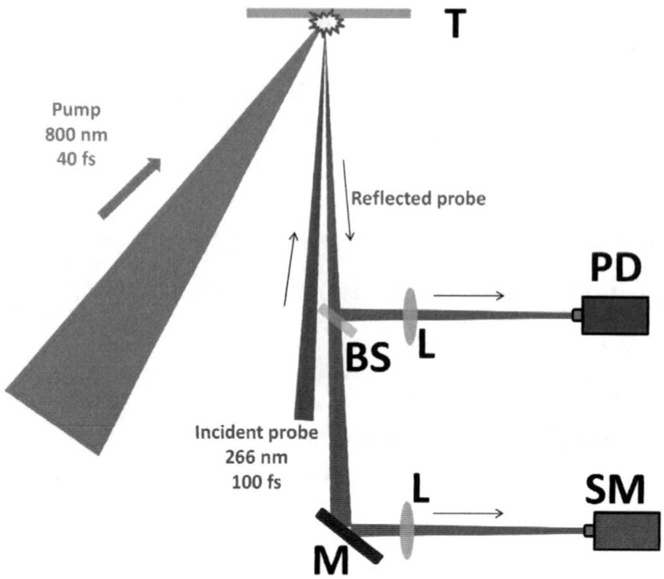


**Figure 1.** Experimental setup: T-target, BS-Beamsplitter, M-mirror, L-lens, SM-spectrometer, PD- Photo diode.

## 2. Experiment

The experiment (Fig. 1) was carried out with a chirped-pulse-amplification-based 20 TW laser system (800 nm, 40 fs, 10 Hz) at Tata Institute of Fundamental Research. An extra Pockels' cell was used to obtain a $10^{-6}$ nanosecond intensity contrast. The pump pulse was focussed on a solid slab at $45^0$ angle of incidence to a focal spot of 17 $\mu$m (FWHM) to obtain intensities of $\sim 10^{18}$ W/cm$^2$. A small portion of the laser pulse (5 %) was extracted using a beam-splitter, up-converted to 266 nm, and focussed to a spot of 60 $\mu$m at the interaction point at near-normal incidence using a fused-silica lens. The reflected probe pulse was then split into two parts and fed to a photodiode and a high-resolution ultraviolet spectrometer. A delay line was introduced in the path of the probe to change the relative temporal delay between the pump and the probe. The focussed probe intensity was $\sim 10^{11}$ W/cm$^2$. Spatial and temporal overlap was achieved by looking at the reflected probe intensity from a plasma created on a dielectric slab at relatively lower pump intensity ($10^{17}$W/cm$^2$). We define the temporal zero where the reflectivity shows a sudden spike. In this experiment we observed the probe reflectivity and spectrum from a solid-density plasma on aluminium-coated BK-7 target.

## 3. Results & Discussions

Figure 2(a) shows reflection of the probe as a function of probe delay with respect to the pump from a super critical layer ($n_e$=$10^{22}$ cm$^{-3}$) in the plasma. Target: Al-coated 5 mm BK-7 glass target. The probe reflection shows an exponential decay ($\tau$ = 5 ps) as the plasma eveolves after excitation by pump. The reflectivity of the normally incident probe can be written as [10]

$$R \propto \exp\left(-\frac{8\nu_{ei}^{*}L}{3c}\right) \tag{1}$$

where $\nu_{ei}^{*}$ is the effective electron ion collision frequency. L is the spatial scale length of plasma over which the probe gets absorbed. The collision frequency is a function of electron density

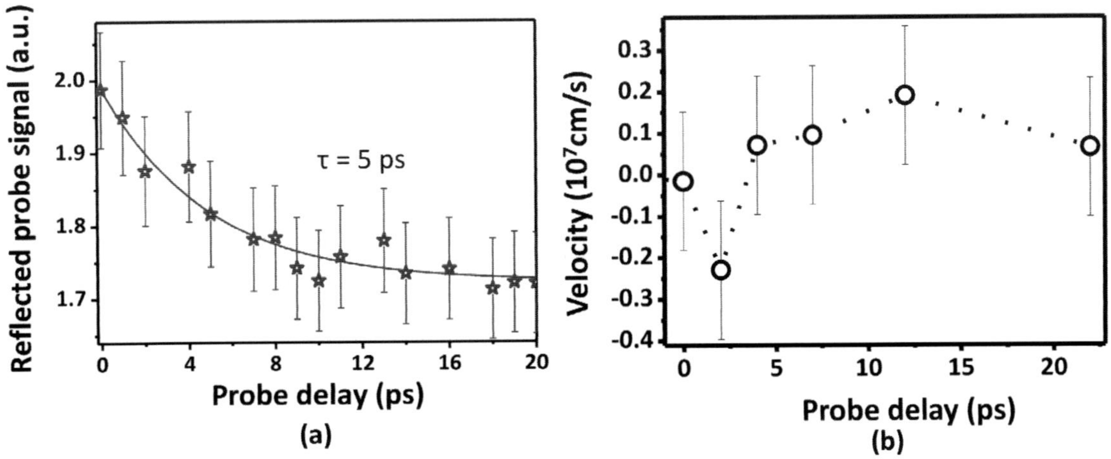


**Figure 2.** (a) Reflected probe intensity as a function of probe delay. The blue line is the exponential fit to the experimental data (blue asterisks). The decay time is measured to be 5 ps from the fit. (b) A similar plot for the velocity of the probe-critical-layer is shown. These velocities (black open circles) are calculated from the Doppler shifts of the ultraviolet probe measured from pump-probe Doppler spectrometry.

and temperature. In this context of fast time scale (few picosecond) probe reflectivity is mainly dependent on L, if we assume quasi-static values of density and temperature.

The velocity of the supercritical layer was measured by pump-probe Doppler spectrometry[8]. The velocity can be expressed as,

$$v_{exp} = -0.5c\frac{\Delta\lambda}{\lambda} \tag{2}$$

Where $\Delta\lambda$ is the Doppler shift from the experiment and $\lambda = 266$ nm in our case. Figure 2(b) shows the velocity of the probe-critical-layer, calculated from the Doppler shift at various probe delays. At initial few picoseconds, the probe critical-surface moves deeper into the plasma (negative velocity) riding on a non-relativistic shock. At subsequent times, the critical surface moves towards vacuum with the freely expanding plasma (positive velocity).

## 4. Simulation

Figure 3 shows the velocity of the probe critical-layer results from 1-D hybrid simulations. First, the HYADES code was run on a 500 $\mu$m silicate target. The output form this (ion, mass and electron density) was then interpolated onto a regular grid for use in an 1-D PIC code (ELPS) to find the hotspot formed by the pump laser. The code was run with 800 nm, 30 fs, $2 \times 10^{18}$W/cm$^2$ laser pulse. A density spike was observed and its motion was simulated by a Lagrangian hydro code. These 1-D simulations calculate a slightly higher velocity of the critical-layer of the 266 nm probe beam than the experimental observation. The sign reversal of the velocity is around 6 ps, which is close to the experimental result.

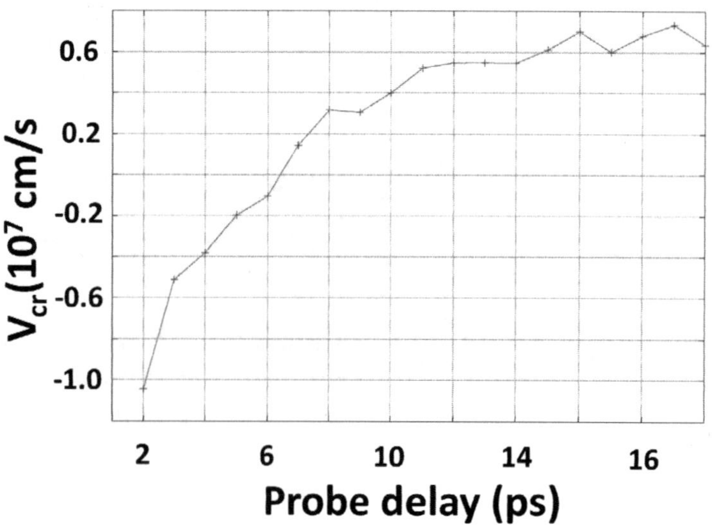


**Figure 3.** The simulated velocity of the probe-critical-layer as a function of the probe delay. The sign convention: outward motion corresponds to positive velocity.

## 5. Conclusions

In this study, we observe ultrafast motion of a supercritical ($10^{22}$ cm$^{-3}$) electron layer in a plasma created by a high-intensity, femtosecond laser pulse on a solid target. We see laser-driven density pile-up and propagation of a non-relativistic shock inside the solid. 1-D hybrid HYADES-PIC-HYDRO simulations support the experiment results.

## 6. Acknowledgments

AA and ADL thank the 'Strong Field Science' (11P-1401) programme of the DAE, Government of India for support for participation in IFSA2013. GRK acknowledges support from a J.C. Bose Felowship grant from DST, Government of India.

## References

[1] Paul McKenna*et. al.*, *Laser-Plasma Interactions and Applications* (Springer Switzerland, Heidelberg, New York, Dordrecht, London, 2013).
[2] M. Tabak *et. al.*, Phys. Plasmas **1**, 1626 (1994).
[3] P. Gibbon *et. al.*, Plasma Phys. Control. Fusion **38**, 769-793 (1996).
[4] T. Tajima *et. el.*, Phys. Rev. Lett. **43**, 267 (1979).
[5] V. Malka *et. el.*, Science **298**, 1596 (2002).
[6] S. C. Wilks *et. el.*, IEEE JOURNAL OF QUANTUM ELECTRONICS, **33**, 1954 (1997).
[7] A. S. Sandhu *et. al.*, Phys. Rev. Lett. **89**, 225002 (2002).
[8] S. Mondal *et. al.*, Phys. Rev. Lett. **105**, 105002 (2010).
[9] Y. Ping *et. al.*, Phys. Rev. Lett. **109**, 145006 (2012).
[10] W. L. Kruer, *The physics of laser plasma interactions* (Boulder, Colorado, 2003).

# A Simple Model of Hohlraum Power Balance and Mitigation of SRS

**B. J. Albright, David S. Montgomery, L. Yin, and J. L. Kline**
Los Alamos National Laboratory, Los Alamos, NM 87545, USA
E-mail: balbright@lanl.gov

**Abstract.** A simple energy balance model has been obtained for laser-plasma heating in indirect drive hohlraum plasma that allows rapid temperature scaling and evolution with parameters such as plasma density and composition. This model enables assessment of the effects on plasma temperature of, e.g., adding high-$Z$ dopant to the gas fill or magnetic fields.

## 1. Introduction
Stimulated Raman Scattering (SRS) is the resonant, three-wave coupling of a light wave into scattered light and electron plasma waves. The scaling of SRS reflectivity $R_{SRS}$ with laser intensity $I$ in a solitary laser speckle in plasma has been measured [1] and found in the electron trapping regime $k\lambda_{\rm De} \gtrsim 0.3$ ($k$ is the EPW wave number and $\lambda_{\rm De} = \sqrt{k_B T_e / 4\pi n_e e^2}$ is the Debye length for plasma of electron density $n_e$ and temperature $T_e$) to behave nonlinearly, increasing sharply at a threshold intensity $I_{\rm th}$ and saturating for $I > I_{\rm th}$. The physics in this regime is governed by the growth of large-amplitude EPW that trap "resonant" electrons with speeds along the wave propagation direction matching the wave's phase speed; this reduces local Landau damping [2] and enhances instability growth. At high intensity, trapping lowers the EPW frequency [3] and introduces variation in EPW phase velocity across the speckle and wave phase fronts bend [4, 5, 6, 7]. As EPW grow,secondary, nonlinear processes break the phase fronts into filaments of small transverse scale [4, 5, 8, 9] that further contribute to nonlinear saturation. In simulations [10], SRS saturation generates hot electrons and back- and side-scattered light waves that propagate obliquely out of hot spots and enhance SRS growth in neighboring speckles. At high gain in multi-dimensions, this coupling enables networks of speckles to self-organize [10] and exhibit emergent behavior where reflectivity exceeds that of the sum of contributions from individual, non-interacting speckles. The nonlinear nature of SRS in this regime is robust, with a threshold at modest laser intensity, $\gtrsim 10^{14}$ W/cm$^2$ for NIF laser conditions where $k\lambda_{\rm De} \approx 0.3$ and the highest levels of backscatter are found [11]. Saturated SRS reflectivity scales with electron temperature $R_{SRS} \sim (k\lambda_{\rm De})^{-4} \sim T_e^{-2}$ [12], so increasing plasma temperature (which remedies inverse bremsstrahlung losses) would likely benefit beam propagation, particularly in the NIF inner beams, which show sizable backscatter.

## 2. Simplified Model of Beam Energy Balance
Consider a cylindrical laser beam of radius $r$ and length $\ell$ with intensity $I$ propagating through optically thin plasma with electron density $n_e$; we assume the plasma is heated dominantly

by inverse bremsstrahlung absorption of laser light and it cools by thermal conduction, bremsstrahlung, and, if magnetized, cyclotron radiation. For a thin, cylindrical shell of plasma,

$$\frac{d}{dt}\left[\frac{3}{2}\left(2\pi r\, dr\right)\sum_i k_B\left(Z_i T_e + T_i\right) n_i\right] = P_{ib} + P_{br} + P_{cy} + P_{\nabla T}, \tag{1}$$

where $T_e$ and $T_i$ are electron and ion temperatures (in keV) and the terms on right are the power gained or lost through inverse bremsstrahlung, bremsstrahlung emission, cyclotron radiation, and thermal conduction, respectively. We denote the thermal gradient heat flux $\boldsymbol{q}(r)$ at radius $r$. The summation is over ions of species $i$, whose densities and ionization states are $n_i$ and $Z_i$, respectively. We shall ignore $P\,dV$ work done on or by the plasma blob, taking the plasma to be of constant density throughout the evolution, we assume no variation in ionization state, and for simplicity we assume $T_e \sim T_i \equiv T$. This leads to a simple evolution equation for the temperature $C_V dT/dt = P_{ib} + P_{br} + P_{cy} + P_{\nabla T}$, where $C_V = \frac{3}{2}k_B\left(2\pi r\, dr\right)\sum_i\left(Z_i+1\right)n_i$ is the plasma heat capacity. For light of intensity $I$ propagating through optically thin plasma, the inverse bremsstrahlung power absorbed [13] is

$$P_{ib} = 2\pi r\, dr\, I(1 - e^{-\kappa_{ib}\ell}) \approx 2\pi r\, dr\, I\ell\kappa_{ib} = 2\pi r\, dr\, I\ell\kappa_{ib0}\left(T/T_0\right)^{-3/2} \tag{2}$$

for inverse bremsstrahlung absorption coefficient [13]

$$\kappa_{ib0} = 9.8\times10^{-12}(\sum_i n_i Z_i)(\sum_i n_i Z_i^2)\omega_0^{-2}\left(1 - \frac{n_e}{n_{\rm cr}}\right)^{-1/2} T_0^{-3/2}\log\Lambda\ {\rm cm}^{-1}. \tag{3}$$

Here $\omega_0 = 2\pi c/\lambda_0$, $n_{\rm cr}$ is the critical density, $\log\Lambda$ is the Coulomb logarithm, and $T_0$ is the temperature on axis. The plasma radiates power away as x-rays at a rate $P_{br} = 5.3\times10^{-31}\left(\sum_i n_i Z_i\right)\left(\sum_i n_i Z_i^2\right)T(r)^{1/2}2\pi\ell r\, dr$ Watts. Magnetized plasma also emits cyclotron radiation $P_{cy} = 6.21\times10^{-17}\left(\pi r_{\rm beam}^2\ell\right)B^2 n_e T$ Watts, where $B$ is the magnetic field in Tesla. Generally, one can show that for indirect drive ignition hohlraum plasma, we may omit $P_{br}$ and $P_{cy}$ relative to absorbed inverse bremsstrahlung power.

The dominant loss term is thermal conduction. The heat flux is [14]

$$\begin{aligned}\boldsymbol{q}(r) &= -k\alpha\nabla T\\ &= -\frac{2}{7}k_0\alpha\hat{r}\frac{\partial}{\partial r}\left(\frac{T(r)}{T_0}\right)^{7/2}\end{aligned} \tag{4}$$

where

$$k_0 = (9.4\times10^{12}\ \text{Watts})\,\overline{S(Z)}(\sum_i n_i Z_i)T_0^{5/2}(\sum_i n_i Z_i^2)^{-1}(\log\Lambda)^{-1}, \tag{5}$$

with $\overline{S(Z)}$ representing the "Spitzer function," of order unity, and $\alpha \le 1$ is a constant that models the inhibition of thermal conductivity from magnetic insulation [15]; unmagnetized plasma has $\alpha = 1$. In general, $\alpha$ is a complicated function of temperature, magnetic field orientation and strength, and plasma density (e.g., a model interpolating between the magnetized and unmagnetized results from classical transport theory $\alpha = \left[1 + (\omega_{ce}\tau_e)^2\right]^{-1}$, where $\omega_{ce} = eB/m_e c$ and $\tau_e$ is the electron collision time, has been applied successfully in magnetized laser-plasma settings [16]). However, for our purposes here we shall assume $\alpha$ is a constant. Taylor-expanding the heat flux on the outer and inner sides of a thin, cylindrical shell of plasma of thickness $dr$ and dropping $\mathcal{O}\left(dr^2\right)$ terms results in the net conductive heat loss of the shell

$$P_{\nabla T} = -2\pi\ell\left[(r+dr)\hat{r}\cdot\boldsymbol{q}(r+dr) - rr\cdot\boldsymbol{q}(r)\right]$$

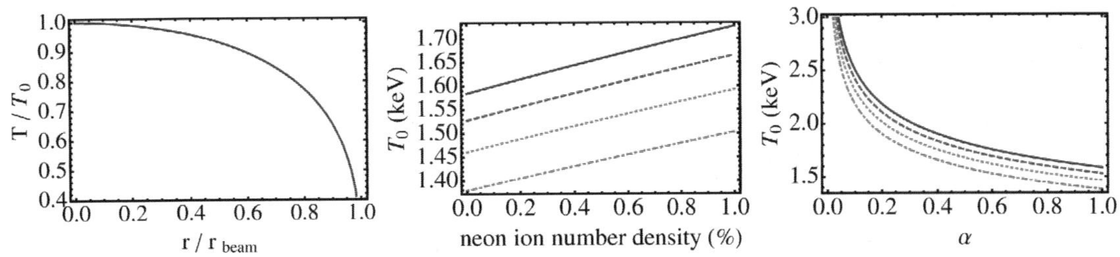


**Figure 1.** (Left) Solution to radial temperature profile equation (8) with $\eta = \eta_0 \approx 3.45$. (Center) Temperature at the center of the beam as a function of neon ion number density fraction (in percent). (Right) Temperature at the center of the beam for different values of $\alpha$, modeling thermal conductivity inhibition by magnetic fields. In the rightmost two panels are results for $\lambda_{\mu m} = 0.351$, $n_{21} = 0.5$, and $\log \Lambda \approx 5$ for four different laser beam powers: 3.0 TW (solid), 2.5 TW TW beam (dashed), 2.0 TW (dotted), and 1.5 TW (dot-dashed). From the latter two panels, we see that increasing beam power, adding neon dopant and suppressing thermal conductivity (e.g., through the introduction of magnetic fields into the hohlraum plasma) all indicate that substantial increases in plasma temperature may be possible.

$$= \frac{4\pi}{7}\ell k_0 \alpha \left[ r\frac{\partial}{\partial r}\left(\frac{T(r)}{T_0}\right)^{7/2} + dr\frac{\partial}{\partial r}\left(\frac{T(r)}{T_0}\right)^{7/2} + r\,dr\frac{\partial^2}{\partial r^2}\left(\frac{T(r)}{T_0}\right)^{7/2} - r\frac{\partial}{\partial r}\left(\frac{T(r)}{T_0}\right)^{7/2}\right]$$

$$= \frac{4\pi}{7}\ell k_0 \alpha\, dr \left[\frac{\partial}{\partial r}\left(\frac{T(r)}{T_0}\right)^{7/2} + r\frac{\partial^2}{\partial r^2}\left(\frac{T(r)}{T_0}\right)^{7/2}\right]. \qquad (6)$$

The plasma temperature profile responds rapidly (of order 10s of ps) to changing plasma or laser conditions, so the inverse bremsstrahlung heating and thermal conductive cooling nearly balance and we may solve for the radial temperature profile as a function of plasma and laser parameters. Let us define $\rho \equiv r/r_{\text{beam}}$, $\tau \equiv (T/T_0)^{7/2}$, and

$$\eta \equiv \frac{7r_{\text{beam}}^2 I\kappa_{ib0}}{2\alpha k_0}. \qquad (7)$$

Equating (2) and (6), we arrive at the nonlinear ordinary differential equation (ODE)

$$\rho\tau'' + \tau' + \eta\rho\tau^{-3/7} = 0, \qquad (8)$$

where primes denote $d/d\rho$. The boundary conditions are that $\tau(0) = 1$ and $\tau'(0) = 0$. This ODE has a singular point as $\rho \downarrow 0$, so numerical integration requires that we match to a boundary layer solution near the origin, which can be shown to be of the form $\tau(\rho) \sim 1 - (\eta\rho^2/4) + \mathcal{O}(\rho^4\eta^2)$. Numerical integration in $\rho$ obtains that for the specific value of the eigenvalue $\eta = \eta_0 \approx 3.45$, the boundary condition $\tau(1) = 0$ is satisfied (a radial power-balance boundary condition analogous to the one in Lindl's model of an ignition hot spot [14]). The resulting temperature profile as a function of radius is shown in the left panel of Fig. 1.

Setting $\eta = \eta_0$ in (7) allows us to solve for the core temperature $T_0$ and determine its dependence on physics parameters within our model. Let $P_{12}$ be the beam power in units of TW, $n_{21}$, the ion density in units of $10^{21}$ cm$^{-3}$, $\lambda_{\mu m}$, the wavelength in microns, and $\overline{Z^2} \equiv \sum_i n_i Z_i^2 / \sum_i n_i$, and let us take $\overline{S(Z)} \approx 1.5$ for a dominantly helium gas fill. Then,

$$T_0 \approx (0.67\text{ keV}) \left[P_{12} n_{21} \lambda_{\mu m}^2 (\log\Lambda)^2\, \overline{Z^2}^2 \alpha^{-1}\right]^{1/5}. \qquad (9)$$

## 3. SRS Mitigation

Let us examine (9) and consider the various ways we might increase $T_0$ and thus decrease SRS reflectivity. Let us consider nominal laser and hohlraum plasma parameters $\lambda_{\mu m} = 0.351$, $n_{21} = 0.5$, and $\log \Lambda \approx 5$, typical NIF hohlraum parameters during high SRS. An obvious way would be to increase beam power, though this is limited when near the maximum rated power on an ICF facility. A second way would be through introducing trace amounts of high-Z dopant, noting that up to around 0.5% number density fraction of neon can be added without freezing out into the cryo environment of an ignition hot spot [17]. The center panel of Fig. 1 shows the resulting increase in $T_0$ as a function of neon gas fill number density fraction. A third way would be through decreasing thermal conductivity through addition of a magnetic field, as illustrated in the rightmost panel of Fig. 1. Finally, one could consider going to green light.

Note that because we've assumed cold surrounding plasma to provide our boundary condition on the ODE, this leads to a numerically larger eigenvalue $\eta_0$ than if we had finite temperature surrounding plasma. Physically, this implies enhanced thermal conductive losses and a predicted $T_0$ is somewhat lower than in experiment. Nevertheless, we believe the power balance model a reasonable sense of the trade-space for increasing hohlraum plasma temperature. Based on the results of this model, it appears that a significant increase in $T$ may be obtained from the application of one or a combination of these approaches and that they may be fruitful avenues of further exploration.

## Acknowledgements

Work performed under the auspices of the US DOE by the Los Alamos National Security, LLC, Los Alamos National Laboratory and supported in part by the Science Campaigns and LANL LDRD. We acknowledge useful discussions with Steve Batha and Juan Fernández.

## References

[1] D. S. Montgomery, J. A. Cobble, J. C. Fernández, R. J. Focia, R. P. Johnson, R. Renard-LeGalloudec, H. A. Rose, and D. A. Russell, Phys. Plasmas **9**, 2311 (2002).

[2] T. O'Neil, Phys. Fluids **8**, 2255 (1965).

[3] G. J. Morales, and T. M. O'Neil, Phys. Rev. Lett. **28**, 417 (1972).

[4] L. Yin, B. J. Albright, K. J. Bowers, W. Daughton, and H. A. Rose, Phys. Rev. Lett., 99, 265004 (2007).

[5] L. Yin, B. J. Albright, H. A. Rose, K. J. Bowers, B. Bergen, D. S. Montgomery, J. L. Kline, and J. C. Fernández, Phys. Plasmas **16**, 113101 (2009).

[6] C. Rousseaux, S. D. Baton, D. Bénisti, L. Gremillet, J. C. Adam, A. Héron, D. J. Strozzi, and F. Amiranoff, Phys. Rev. Lett. **102**, 185003 (2009).

[7] J. W. Banks, R. L. Berger, S. Brunner, B. I. Cohen, and J. A. F. Hittinger, Phys. Plasmas **18**, 052102 (2011).

[8] H. A. Rose, Phys. Plasmas **12**, 12318 (2005); H. A. Rose and L. Yin, Phys. Plasmas **15**, 042311 (2008).

[9] P. E. Masson-Laborde, W. Rozmus, Z. Peng, D. Pesme, S. Hüller, M. Casanova, V. Yu. Bychenkov, T. Chapman, and P. Loiseau, Phys. Plasmas **17**, 092704 (2010).

[10] L. Yin, B. J. Albright, H. A. Rose, K. J. Bowers, B. Bergen and R. K. Kirkwood, Phys. Rev. Lett. **108**, 245004 (2012).

[11] R. K. Kirkwood et al., Phys. Plasmas **18**, 056311 (2011); D. E. Hinkel et al., Phys. Plasmas **18**, 056312 (2011); S. H. Glenzer et al., Phys. Rev. Lett. **106**, 085004 (2011).

[12] Yin, L., B. J. Albright, H. A. Rose, K. J. Bowers, B. Bergen, R. K. Kirkwood, D. E. Hinkel, A. B. Langdon, D. S. Montgomery, and J. L. Kline, *Phys. Plasmas* **19**, 056304 (2012); Yin, L., B. J. Albright. H. A. Rose, D. S. Montgomery, J. L. Kline, R. K. Kirkwood, P. Michel, R. London, K. J. Bowers, and B. Bergen, *Phys. Plasmas* **20**, 012702 (2013).

[13] J. D. Huba *NRL Plasma Formulary* NRL/PU/6790–00-426 (Washington, DC: Naval Research Laboratory; 2009), p. 58.

[14] John D. Lindl, *Inertial Confinement Fusion*, (Springer; New York; 1998) p. 40.

[15] D.S. Montgomery, B.J. Albright, P.Y. Chang, J.R. Davies, G. Fiksel, J.L. Kline, L. Yin, R. Betti, "Mitigation of stimulated Raman scattering in hohlraum plasmas via magnetic insulation," in preparation (2013).

[16] D. S. Montgomery, O. L. Landen, R. P. Drake, K. G. Estabrook, H. A. Baldis, S. H. Batha, K. S. Bradley, and R. J. Procassini, *Phys. Rev. Lett* **73**, 2055 (1994).

[17] B. A. Hands, *Vacuum* **26**, 11 (1975).

# Study of shock waves generation, hot electron production and role of parametric instabilities in an intensity regime relevant for the shock ignition

**L Antonelli[1,2], P Köster[3], G Folpini [2], Y Maheut[2], F Baffigi[3], G Cristoforetti[3], L Labate[3], T Levato [1,3,4], L A Gizzi[3], F Consoli[5], R De Angelis[5], Z Kalinowska[6], T Chodukowski[6], M Rosinski[6], P Parys[6], T Pisarczyk[6], P Raczka[6], L Ryc[6], J Badziak[6], J Wolowski[6], M Smid[4], O Renner[4], E Krousky[4], M Pfeifer[4], J Skala[4], J Ullschmied[7], P Nicolaï[2], X Ribeyre[2], G Shurtz[2], S Atzeni[8], A Marocchino[8], A Schiavi[8], C Spindloe[9], T O'Dell[9], Y J Rhee[10], M Richetta[1], D Batani[2]**

[1]University of Rome Tor Vergata, Rome, Italy
[2]Université Bordeaux 1, CNRS, CEA, CELIA (Centre Lasers Intenses et Applications), UMR 5107, F-33405 Talence, France
[3]Intense Laser Irradiation Laboratory, INO-CNR, Pisa, Italy
[4]Institute of Physics, Prague, ASCR, Czech Republic
[5]ENEA-Euratom, Frascati, Italy
[6]Institute of Plasma Physics and Laser Microfusion, Warsaw, Poland
[7]Institute of Plasma Physics, Prague, Czech Republic
[8]Dipartimento SBAI, Univ. di Roma La Sapienza and CNISM, Italy
[9]Scitech Precision, Rutherford Appleton Laboratory, United Kinkdom
[10]Nuclear Data Center, KAERI, Korea

E-mail: luca.antonelli.2@uniroma2.it

**Abstract.** We present experimental results at intensities relevant to Shock Ignition obtained at the sub-ns Prague Asterix Laser System in 2012 . We studied shock waves produced by laser-matter interaction in presence of a pre-plasma. We used a first beam at $1\omega$ (1315 nm) at $7 \times 10^{13}$ W/cm$^2$ to create a pre-plasma on the front side of the target and a second at $3\omega$ (438 nm) at $\sim 10^{16}$ W/cm$^2$ to create the shock wave. Multilayer targets composed of 25 (or 40 $\mu$m) of plastic (doped with Cl), 5 $\mu$m of Cu (for K$\alpha$ diagnostics) and 20 $\mu$m of Al for shock measurement were used. We used X-ray spectroscopy of Cl to evaluate the plasma temperature, K$\alpha$ imaging and spectroscopy to evaluate spatial and spectral properties of the fast electrons and a streak camera for shock breakout measurements. Parametric instabilities (Stimulated Raman Scattering, Stimulated Brillouin Scattering and Two Plasmon Decay) were studied by collecting the back scattered light and analysing its spectrum. Back scattered energy was measured with calorimeters. To evaluate the maximum pressure reached in our experiment we performed hydro simulations with CHIC and DUED codes. The maximum shock pressure generated in our experiment at the front side of the target during laser-interaction is 90 Mbar. The conversion efficiency into hot electrons was estimated to be of the order of $\sim$ 0.1% and their mean energy in the order $\sim$50 keV.

## 1. Introduction

Shock Ignition is a novel approach to ICF proposed in 2007 [1]. In this approach, the compression phase and the ignition phase are separated. Two different laser pulses are used: the first one in the ns regime to compress the target and the second one in ps regime and more intense to launch a strong convergent shock to achieve the ignition conditions. We performed an experiment at Prague Asterix Laser System [2] to investigate a regime relevant to Shock Ignition [3]. We used two different beams in planar geometry. The first one at $7 \times 10^{13}$ W/cm$^2$ and 1315 nm to create an extended plasma corona in front of the target, and the second one at $\sim 10^{16}$ W/cm$^2$ and 438 nm to lunch strong shock. We changed the delay between the beams from 0 up to 1200 ps to observe the influence of the plasma corona on the shock wave.

## 2. Experimental setup

To characterize the pre-plasma we used optical interferometry [4] at 650 nm obtaining density maps. To measure the electron plasma temperature we used a spherically bent mica crystal spectrometer configured to record the X-ray emission from Cl used as doping in the first plastic layer of the target [5]. Using a streak camera we measured the time required the shock wave to cross the target (shock breakout time). We characterized hot electrons using an X-ray CCD in single photon counting mode [6, 7] recording the K$\alpha$ emission from Cu, and a spherically bent quartz crystal for imaging of the same radiation [8]. The back-scattered light was collected within the cone of the lens used to focus the laser. We analysed this light using a calorimeter and two fiber spectrometers looking at blue frequencies (to detect back scattered laser light and Stimulated Brillouin Scattering) and red frequencies (to record the spectrum due to the Stimulated Raman Scattering). We also recorded inside the chamber signal coming from the Two Plasmon Decay. We used multilayer targets: a first layer of plastic doped with Cl to create an extend plasma corona of low Z material as in a real ICF experiment, a second layer of 5 $\mu$m Cu that we used to trace hot electrons and a third layer of 20 $\mu$m of Al as shock reference. The thickness of the first plastic layer was 0 $\mu$m, 25 $\mu$m and 40 $\mu$m to observe the attenuation of the electrons in plastic. We used on both beams Random Phase Plate (RPP) to homogenise focal spot and to have a stable interaction between laser and plasma. The focal spot diameter of the first beam was 900 $\mu$m, with a flat top energy distribution. The second beam had a gaussian spatial energy distribution with a full width half maximum of 100 $\mu$m.

## 3. Results

We reproduced the experimental shock breakout time with the hydro codes (DUED [9] and CHIC [10]). The comparison with the code CHIC is shown in Figure (1). The maximum pressure inferred is 90 Mbar using RPP on both beams. This is the highest pressure measured until now in this kinds of experiments. Nevertheless even higher values could be obtained if a larger focal spot had been used. Indeed in our setup the main beam had a focal spot of $\sim$100 $\mu$m comparable to the distance between the absorption and the ablation layers. A significant part of the absorbed energy is lost in transverse direction in the conduction (overcritical density) region. Increasing the focal spot size and keeping constant intensity, we would have obtained a pressure of the order of $\sim$120 Mbar. Such value is still low compared to simulation predictions for the nominal incident laser intensity. Probably a large part of the laser energy is lost due to the refraction at very large angles in the plasma plume in front of the target. K$\alpha$ measurements show an electron spot of $\sim$150 $\mu$m not so larger than the laser

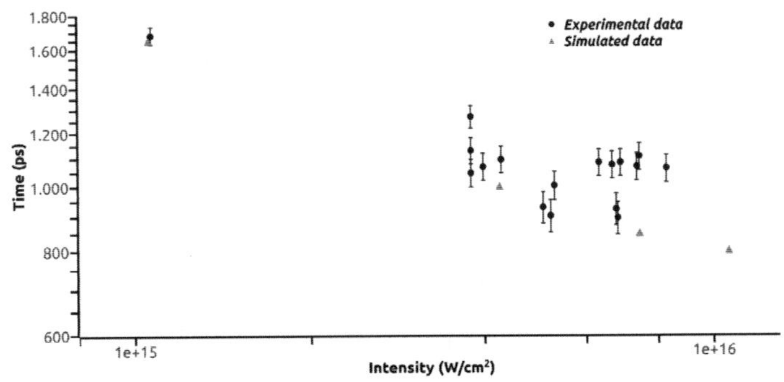


**Figure 1.** Experimental and simulated (using CHIC code) shock breakout time vs. laser intensity.

focal spot. Measuring signal coming at different depth we studied the attenuation of the K$\alpha$ signal. Considering K$\alpha$ signal proportional to the number of hot electrons generated, it is possible to estimate the hot electron range as shown in figure (2). We found a range of ∼27

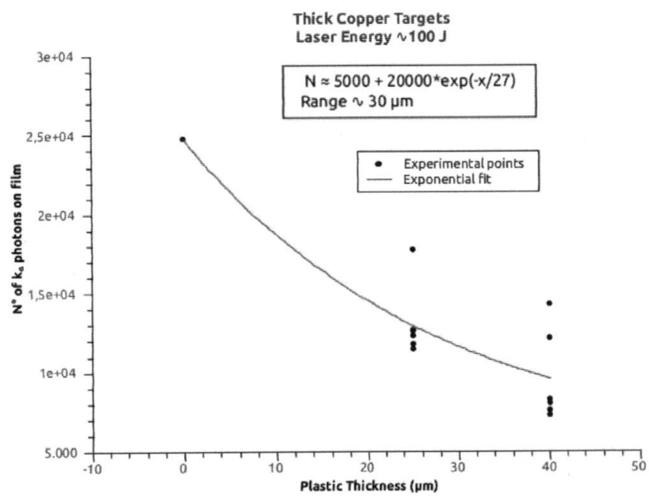


**Figure 2.** K$\alpha$ signal versus thickness of plastic over layer. Assuming an exponential penetration of hot electrons, we obtain an estimation of their range and of the associated average energy.

$\mu$m corresponding to an energy [11] of $\sim$ 45 keV with an error of 10 keV in accordance with values measured in previous experimental campain in PALS [12]. From crystal measurements we evaluate the total amount of K$\alpha$ emitted. Using the same assumption to find the range we estimated a conversion efficiency of laser energy into hot electrons in the order of 0.1 %. Back scattering measurements show an amount of back-scattered energy of $\sim$ 5-6% [13]. The energy back-scattered by Raman process was in the order of 0.1%

Journal of Physics: Conference Series **688** (2016) 012003 doi:10.1088/1742-6596/688/1/012003

## 4. Conclusion

We studied laser-matter interaction in an intensity regime relevant for Shock Ignition. The experiments were performed at the Prague Asterix Laser System. We characterized our conditions using different diagnostics, in particular Shock Chronometry, K$\alpha$ measurements, and back scattering measurements. The maximum pressure produced by the laser on the target front during interaction was inferred by reproducing the shock breakout time on target rear side. The obtained value of the order of 90 Mbar is the maximum measured until now in this kind of experiments. Hydro simulations show that, if larger focal spots were used at given intensity, pressures of about 120 Mbar could have been reached; The difference is due to lateral losses due to thermal conduction in the overcritical region, due to the distance between the absorption and ablation region being comparable to the focal spot size. Simulations also show that only about 50% of laser light is absorbed. We observed a low amount of back-scattered light within the cone of the lens ($\sim$5-6%) and we characterized the hot electrons using a spherically bent crystal, obtaining images of the K$\alpha$ emission from a Cu-tracer layer. From these images we measured the diameter of the electron beam which was compatible with our focal spot. We also measured the flux at different depths to calculate the hot electron range. From our measurements we obtained an average energy of 50 keV compatible with the assumption of hot electrons generation from Raman scattering [14] (non local effects have not,insted, been taken into account since they appear not to play a major role [15]). The total energy of the hot electrons was less than 1% laser energy.


## Acknowledgments

We acknowledge the support of HiPER project and Preparatory Phase Funding Agencies (EC, MSMT and STFC), the Italian MIUR project PRIN 2009FCC9MS and Sapienza project 2012 C26A12CZH2, the National Centre for Science (NCN), Poland under Grant No 2012/04/M/ST2/00452. The work was also supported by the European Union under the LaserLab program of the 7th FP, and by the Czech Ministry of Education, Youth and Sports, projects LC528 and LM2010014. Scitech Precision and Rutherford Appleton Laboratory Target Fabrication Group supplied targets. Finally, we thank the technical staff of PALS for help in running the experiments.


## References


[1] Betti R *et al.* (2007) *Phys. Rev. Lett.* **98** 155001
[2] Jungwirth K, *et al.* (2001) *Phys. Plasmas* **8** 2495
[3] Antonelli L, *et al.* (2011) *Acta Tecnica,* **56** T57
[4] Borodziuk S, *et al.* (2005) *Optica Apllicata* **35** 2
[5] Smid M, *et al.* (2013) *Acta Polytechnica,* **53** 233
[6] Levato T, *et al.* (2010) *Nucl. Inst. Methods A* **623**, 842
[7] Gizzi L, *et al.* (2007) *Plasma Phys. Controll. Fusion* **49** B221
[8] Morace A, *et al.* (2010) *Nuclear Instruments and Methods in Physics Research A*, 623:22, 797- 800, Elsevier BV, 11
[9] Maire P H, *et al.* (2007) *SIAM J. Sci. Comput.* **29**, 1781
[10] Atzeni S, *et al.* (2005) *Computer Phys. Commun.* **169** , 153
[11] Estardatabase: http://physics.nist.gov/PhysRefData/Star/Text/ESTAR.html
[12] Antonelli L, *et al.*,in 39th EPS Conference on Plasma Physics (Stockholm, Sweden, 2012), paper P1-097
[13] Koester P, *et al.*, in 40th EPS Conference on Plasma Physics, (2013)
[14] Klimo O, *et al.* (2010) *Plasma Phys. Control. Fusion,* **52** 055013
[15] Marocchino A, *et al.* (2013) *Physics of Plasmas* **20** 2702

# The diagnostics of the energy coupling efficiency in the Fast Ignition integrated experiment

**Y. Arikawa[1], S. Fujioka[1], A. Morace [1], Z. Zhang[1], T. Nagai[1], M. Taga[1], Y. Abe[1], S. Kojima[1], S. Sakata[1], H. Inoue[1], M. Utsugi[1], S. Hattori[1], S. H. Lee[1], T. Ikenouchi[1], T. Hosoda[1], A. Sunahara[2], T. Ozaki[3], T. Johzaki,[4] H. Nagatomo[1], K. Yamanoi[1], T. Shimizu[1], S. Tokita[1], Y. Fujimoto[1], J. Kawanaka[1], Y. Nakata[1], M. Nakai[1], H. Shiraga[1], H. Jitsuno[1], N. Sarukura[1], N. Miyanaga[1], T. Norimatsu[1], H. Nishimura[1], FIREX-group[1], and H. Azechi[1]**

[1] Institute of Laser Engineering, Osaka University, *2-6, Suita, Osaka, Japan*
[2] Institute for Laser Technology, *2-6, Suita, Osaka, Japan*
[3] National Institute for Fusion Science, 322-6 Shimoishicho, Toki, Gifu, Japan
[4] Hiroshima University, 1-1-89, Higashi Senda-cho, Naka-ku, Hiroshima City , Hiroshima,

E-mail: arikawa-y@ile.osaka-u.ac.jp

**Abstract.** The energy coupling efficiency (CE) in Fast Ignition (FI) laser fusion was studied at GEKKO XII and LFEX laser facility by using newly developed targets and plasma diagnostic instruments. The gated-liquid scintillator neutron detectors had been upgraded by using neutron collimators for intense background fluxes of $\gamma$-rays and neutrons in the FI experiment. Clear fusion neutron signal was successfully recorded in the sub-kJ heating FI experiment. Up to 5 times neutron yield enhancement was observed, and the CE of the heating laser to core plasma was estimated to be 1.6 % for cone-in-shell target implosion by 9 beams and core heating by LFEX pulse 115 ps before bang time. The laser-to-electron energy conversion efficiency was separately diagnosed using a newly developed target and resulted to be 45 %. The fast electron energy spectrum was estimated to be 2.3 MeV slope temperature by hard x-ray spectroscopy. Monte Carlo simulations demonstrate the consistency of the data set.

## 1. Introduction

The Fast Ignition (FI) laser fusion scheme is an alternative approach to Inertial Confinement Fusion which provides a more compact laser design compared to the classic Central Hot Spot ignition scheme. The goal of the FI realization experiment (FIREX) project phase-I, conducted at Institute of Laser Engineering, Osaka University is to demonstrate the fast heating of the core plasma up to 5 keV ignition threshold temperature by using LFEX laser with the designed maximum energy of 10 kJ. In order to achieve this goal more than 10 % laser-to-core plasma energy coupling efficiency is required [1]. The deposited energy in the core plasma can be measured from the increment in the ion temperature which can be measured by fusion neutron yield or neutron spectrum Doppler broadening, and the core density profiles which can be measured by x-ray radiography. The detection of the fusion neutron in the FI experiment represented a challenging subject in the past decade due to the intense background of hard x-ray and the photo-nuclear reaction neutrons. The neutron detectors have been dramatically improved by installing Oxygen-enriched liquid scintillators, gated-PMT, and neutron collimators in the neutron TOF detection systems [2,3,4]. Another fundamental parameter is represented by the laser-to-fast electron energy conversion efficiency [5,6,7]. In most of the previous studies, a short pulse laser was used, however the fast electron transport from the cone surrounded by shell plasma is of difficult study since the sheath field at the cone-outer-surface is significantly different from the one in vacuum. Moreover the electron spectrum in the core plasma has been technically difficult to be measured by using conventional magnet-based electron spectrometer, due to the significant fast electron stopping at high densities. Hard x-ray spectroscopy is an alternative scheme to diagnose the electron spectrum [8] and a specially designed target was used in the experiment. By using this target the laser-to-fast electron energy conversion efficiency was measured from absolute K-$\alpha$ yield [9], the divergence angle and the energy spectrum of the electron were measured by using five stack filter high energy x-ray spectrometers and a newly developed Compton

Journal of Physics: Conference Series **688** (2016) 012004 doi:10.1088/1742-6596/688/1/012004

γ-ray spectrometer [10]. In this work the progress on the neutron diagnostics and the measurement of the global laser to core plasma energy coupling efficiency are presented.

## 2. The progress on the neutron diagnostics

In the FI experiment the hard x-ray generated by the heating laser creates a very intense background signal preceding the fusion neutron detection [2,3,4]. The rapid decaying liquid scintillator diagnostics have been developed for this aim and they were implemented in the FI experiment. The liquid scintillators consist of oxygen gas enriched p-Xylen or TMB (Tri-Methyl benzene) having higher temperature flashing point than that of p-Xylen as a solution, BBQ (4,4'''-Bis[(2-butyloctyl)oxy]-1,1':4',1'':4'',1'''-quaterphenyl) as a dye, and Benzophenon as a additional quencher to have faster decay. Figure 2 shows the temporal profiles of the newly developed liquid scintillators with and without Benzophenon. The fall time was dramatically improved from 90 % to 10% by Benzophenon quenching while the afterglow (long time constant tail) was not significantly increased. Light output was reasonably decreased with the Benzophenon fraction. By implementing theses scintillators at GEKKO XII-LFEX facility, the neutron TOF system has been upgraded. The experimental setup in the 2012 experimental campaign is shown in the Fig. 2(a), and the comparison of the typical neutron TOF signal before and after upgrade is shown in the Fig.2 (b). After the upgrade, it was possible to detect DD fusion neutron yields equal or greater than $5\times10^5$ with 1 kJ LFEX heating shots. The detailed description of MANDALA and the liquid TOF detectors will be presented in the same conference proceedings series.

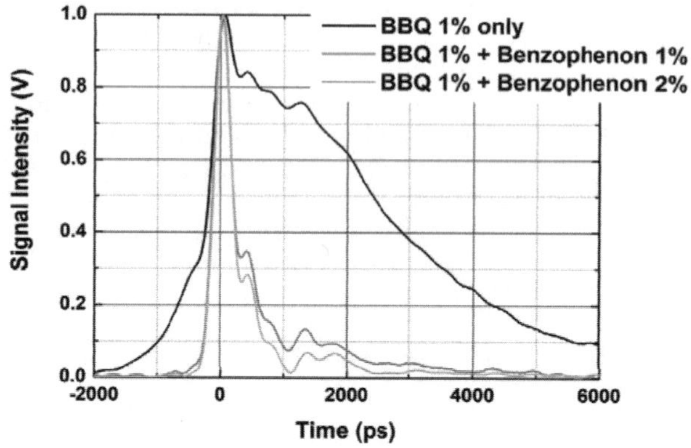


| Weight-% of the Benzophenon | Fall time (90%-10%) | Light output |
|---|---|---|
| 0% | 5400 ps | 100% |
| 1% | 875 ps | 6% |
| 2% | 625 ps | 4% |

Figure 1 The comparison of the decay curve of the Benzophenon quenched BBQ liquid scintillator.
Table 1 The comparisons of the fall time and the light output.

## 3. The energy coupling efficiency of the laser to the core plasma

The FI integrated experiment (implosion and heating joint shots) was conducted at GEKKO XII – LFEX facility. A typical FI target with a gold cone attached deuterated-polystyrene shell was irradiated by GEKKO XII and LFEX. The LFEX pulse width was 1.2 ps FWHM. The shell used was 7 μm thick and 500 μm diameter. The increment of the neutron yield by LFEX heating was clearly observed, rising from $(5.7\pm1)\times10^5$ without LFEX pulse, to $(2.2\pm1)\times10^6$ for a 612 J heating shot. Unfortunately the LFEX timing was not optimized in this shot, and it was estimated to be 115 ps before the maximum compression time by x-ray streak camera. The core density and the areal density at this time were separately measured to be respectively about 7 g/cm$^3$ and 20 mg/cm$^2$, by using a monochromatic x-ray framing backlighting image with 100-ps time resolution. The size of the core plasma was measured to be 54 μm in diameter by x-ray streak image. The ion temperature was estimated using following simple formula :

$$Yn = \frac{n^2}{2}\langle\sigma v\rangle V\tau$$

where n is the number density of the deuterium, $\langle\sigma v\rangle$ is the Maxwell averaged fusion cross section, V is the volume of the core plasma, τ is the temperature dependent confinement time. The ion temperature (Ti) without LFEX heating shot was estimated to be 0.67±0.02 keV and 0.84±0.05 keV for 612 J heating shot. A slight Doppler broadening in the neutron spectrum was observed by MANDALA, and Ti resulted to be less than 1 keV, in agreement with the estimated value.

A multi-images x-ray streak camera observed the time resolved electron temperature (Te) image around the fast heating time. At the heating time, a significant emission from the cone-tip and a small emission from the core plasma were found. Te at the core plasma resulted to be 1 keV, in agreement with the Ti estimated from neutrons while the Te at the cone-tip was around 2 keV. With these information we conclude that the DD neutrons were produced in the core plasma, and finally we conclude that the core ion temperature increase was $0.17 \pm 0.05$ keV, and the deposited energy was 9.8 J. The CE of the laser to the core plasma resulted to be $1.6 \pm 0.4$ %. If the heating time had been at maximum compression the electron stopping efficiency would much better. As shown in the next section the reason of the low CE is due to the density of the core plasama was too low and Te of the fast electron was too high.

## 4. The laser-to-electrons energy conversion efficiency

To better understand the energy transfer from the laser to the core plasma, fast electron generation must be separately studied. The electron spectrum, the divergence angle, and the laser-to-electron conversion efficiency were studied by using a newly designed target with a cone attached hemi-spherical shell coupled with a layered metal block [8]. A standard gold cone target was adopted, and the electrons spectrum was measured to be 2.3 MeV slope temperature and a 60 degree full angle angular distribution was measured by using Bremsstrahlung x-ray spectrometers. The laser-to-electron conversion efficiency was estimated to be 45 % by the absolute Kα yield spectroscopy. The detailed description of the experiment and the target will be presented in the same conference proceeding series. The electron stopping efficiency was estimated by using Monte-Carlo simulation code PHITS [11]

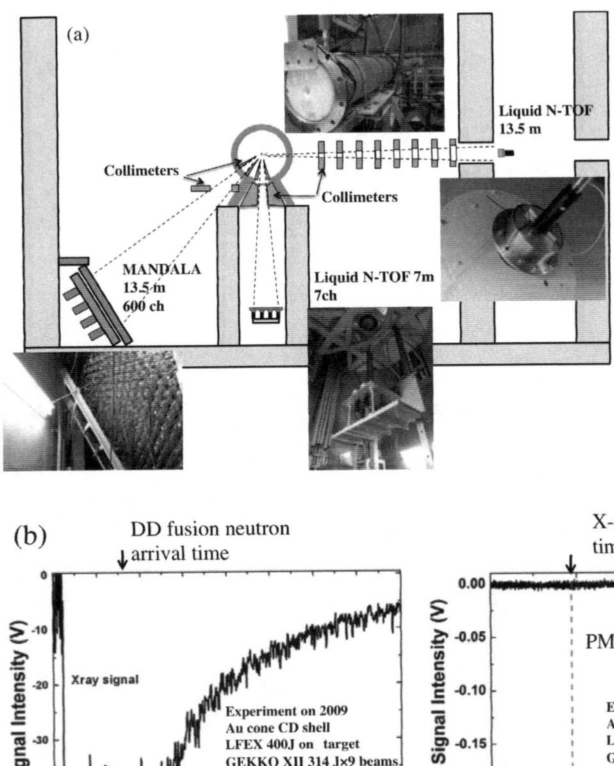


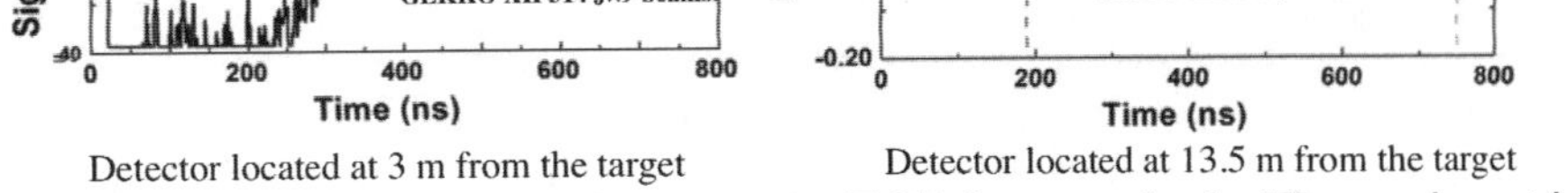


Figure 2 (a) The experimental setups of the neutron TOF detectors in the FI experiment in 2012.
(b) The comparison of the signal of neutron TOF detectors between 2009 and 2012.

assuming using a uniform CH solid core plasma with 7 g/cm$^3$ density, 54 μm core size, electron beam injection 50 μm far from the core center, beam divergence of 60 degree full angle, and 2.3 MeV slope temperature. Figure 3 shows the electron trajectories and the deposited energy map. The energy deposition efficiency on the core resulted to be 2.5 %. Since the laser-to-core energy transfer is calculated multiplying the laser-to-electron energy conversion efficiency and the electron stopping efficiency including angular distribution function, a 1.1 % energy transfer form the laser to the core was estimated. It is in good agreement with the value measured by the neutron diagnostics. If the core could be compressed to 100 g/cm$^3$ and the Te could be 1 MeV slope, the over 20 % stopping can efficiency can be obtained as shown in the Fig.3 (b). Thus a better controlling of the heating time and the suppression of the Te are crucial subjects.

## Conclusion

The laser to the core plasma CE in FI was studied by using advanced neutron detectors, x-ray spectrometers, and a new target. The laser to the core CE was measured to be 1.6±4 %. The incremental ion temperature measured by the neutron diagnostics is in agreement with the electron temperature measured by streaked x-ray image, and therefore the neutrons were considered to be originated from the core heating. The laser-to-electron conversion efficiency and the electron spectrum were separately studied by using new cone-shell-block target, and resulted respectively to be 45 %, and 2.3 MeV. The electron stopping efficiency was estimated by Monte Carlo simulations, and these values were confirmed to be well consistent with the CE of the laser to the core plasma determined by neutron diagnostics. This analysis indicates the CE can be be improved by better imploding of the core plasma and the reducing Te of fast electron. This study represents a step forward in the understanding of the Fast Ignition heating process.

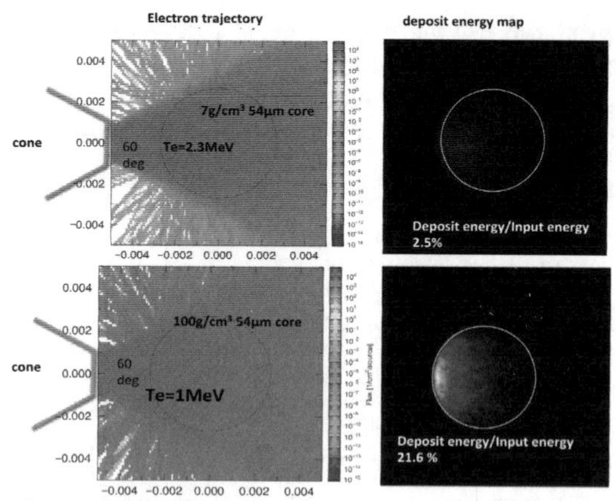


Figure 3 The Monte Carlo simulation results of the electron stopping efficiency, (a) Te 2.3 MeV electron was injected into 7 g/cm$^3$ CH solid, Te 1MeV electron was injected into 100 g/cm$^2$ CH solid, the size of the CH solid.

## Acknowledgement

The authors gratefully acknowledge the support of the LFEX development and operation group, the GEKKO XII operation group, the target fabrication group, and the plasma diagnostics operation group of the Institute of Laser Engineering, Osaka University. This work was partly supported by the Japan Society for the Promotion of Science under the contracts of Grant-in-Aid for Scientific Research (A)No. 24244095 and (B) No. 23360413, Grant-in-Aid for Young Scientists (A)No.24686103, Grant-in-Aid for Challenging Exploratory Research No.25630419, and the auspices of Japanese Ministry of Education, Culture, Sports, Science and Technology (MEXT) project on "Promotion of relativistic nuclear physics with ultra-intense laser."

## References

[1] T. Johazaki, et.al., Phys. Plasmas, **15,** 062702 (2008).
[2] C. Stoeckl, et.al., Rev. Scient. Instrum., **81**, 10, 10D302, (2010)
[3] Y. Arikawa, et.al., Rev. Scient. Instrum, **83**, 10, 10D909, (2012)
[4] T. Nagai, et.al., Jap. Journal. App.Phys. **50**, 8, 080208, (2011)
[5] W. Theobald, Phys. Plasmas, **18**, 056305, (2011)
[6] M. H. Key, et. al., Phys. Plasmas, **15**, 022701 (2008)
[7] T. Ma, et. al., Phys Rev. Letteres, **108**, 115004, (2012)
[8] Z, Zhang, et. al., High Energy Density Physics, **9**, 435, (2013)
[9] C. D. Chen, Phys, Plasmas, **16**, 082705, (2009)
[10] S. Kojima, et. al., Plasma and Fusion Research, (to be published 2014)
[11] K. Niita,et.al., JAEA-Data/Code 2010-022 (2010)

# Improved robustness study of a shock ignited target, with DUED code including non-local electron transport and 3D laser ray-tracing

**Stefano Atzeni, Alberto Marocchino and Angelo Schiavi**
Dipartimento SBAI, Università di Roma "La Sapienza" and CNISM,
Via A. Scarpa, 14–16, 00161 Roma, Italy

E-mail: stefano.atzeni@uniroma1.it

**Abstract.** Accurate descriptions of laser power coupling to the plasma and electron energy transport are crucial for designing shock-ignition targets and assessing their robustness (in particular with regard to laser and positioning errors). To this purpose, the 2D DUED laser fusion code has been improved with the inclusion of a 3D laser ray-tracing scheme and a model for non-local electron transport. 2D simulations with the upgraded code are presented; the dependence of the fusion yield vs target displacement is studied. Two different irradiation configurations are considered.

## 1. Introduction

Shock ignition [1] is a recently proposed direct-drive laser fusion scheme. The fusion target is imploded at sub-ignition velocity (250-300 km/s) by a moderate intensity laser pulse. Towards the end of the implosion, a powerful pulse (a *spike*, with intensity $5 \times 10^{15}$–$10^{16}$ W/cm$^2$, for laser with wavelength $\lambda = 0.35$ $\mu$m) drives a converging shock wave, which contributes to the creation of the central hot spot required for ignition. Shock-ignition has the potential for higher gain than standard central ignition schemes, and in principle it is less sensitive to hydrodynamic instabilities than standard direct-drive schemes. In addition, a full scale demonstration may be feasible on existing facilities (such as the NIF) without major facility modification [2, 3]. Although issues remain, mainly concerning possible deleterious effects of laser-plasma instabilities (LPI's) during the interaction of the spike with the preformed plasma corona [4], shock ignition targets are currently designed using standard fluid codes for ICF, that neglect any LPI. A widely studied target is the so-called HiPER baseline target [5], originally proposed for fast ignition, but later shown to be suitable for shock ignition too [6]. Our group performed a detailed analysis of this target [7], developed models for target scaling in energy and wavelength [8, 9] and analyzed irradiation schemes [10]. All performed studies were based on simulations with the 2D DUED radiation-hydro-nuclear code [11, 12]. We have however realized that code improvements were required for a better analysis of target robustness. In Sec. 2, discuss a new 3D laser ray-tracing scheme, which allows to study realistic irradiation geometries. In addition, we have introduced in DUED a model for non-local electron transport that replaces the scheme with flux-limited thermal conduction. Results of simulations of the shock-ignited HiPER target using the upgraded code model are reported in Sec. 3.

## 2. Code model improvements

### *2.1. Non-local electron transport*

We have introduced in DUED the model for non-local electron transport proposed by Schurtz *et al.* [13]. The model correctly reproduces features of electron transport in strong temperature gradients that are generally not captured by the commonly used flux-limited (local) conduction[14]. In addition, contrary to flux-limited conduction models, non-local electron transport does not require the introduction of any tunable parameter, such as the flux limiter.

Our simulations of shock-ignited targets with non-local transport show that the ablation front penetration during the implosion phase is in approximate agreement with simulations with flux-limited conductivity and flux limiter $f = 0.07$ (see Fig. 1), but also show some preheating in front of the heat wave and somewhat different profiles in the corona. The converging shock driven by the ignition spike, instead, is somewhat faster when simulated with non-local electron transport than with flux-limited transport. Target gain is unaffected by such differences.

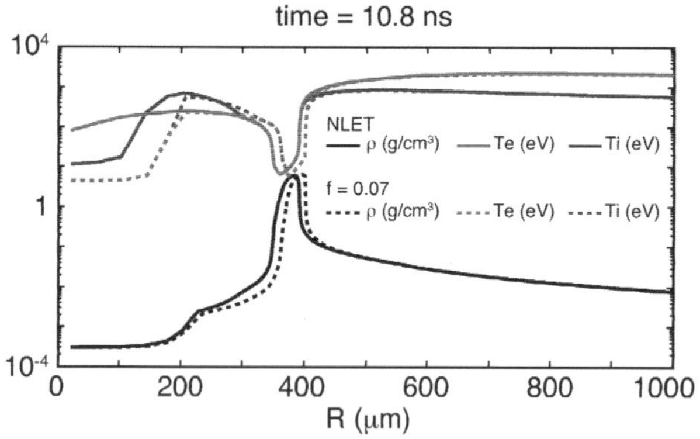


**Figure 1.** Density, electron and ion temperature radial profiles towards the end of the compression pulse, from 1D simulations of the HiPER baseline target with non-local electron transport (solid curves) and with flux-limited conductivity and $f = 0.07$ (dashed).

### *2.2. 3D laser ray-tracing*

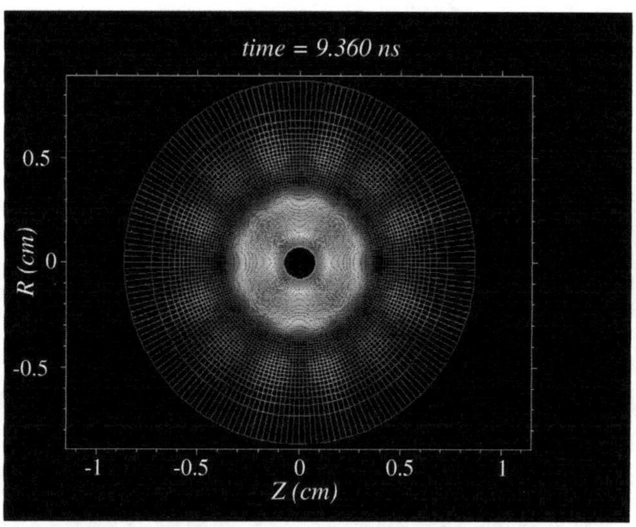


**Figure 2.** 2D hydrodynamic simulation with 3D laser ray-tracing. 2D Pattern of laser absorption clearly showing the separate beams entering the corona.

We have also introduced a 3D laser ray-tracing scheme in DUED, in order to simulate actual multi-beam irradiation geometry realistically. The 3D power absorption distribution is then mapped onto the 2-D axially symmetric mesh. An example is shown in Fig. 2. The finite number of rays representing each laser beam introduces numerical noise in the power absorption distribution, which in turn causes mass inhomogeneities, which seed hydrodynamic instabilities (Richtmyer-Meshov and Rayleigh-Taylor). To keep power deposition noise at very low level, without the use of smoothing techniques, we trace a very large number of rays at each step. The cross section of a laser beam was divided into 300x300 sectors in the radial and azimuthal direction. Each beamlet was represented by a ray uniformly sampled in a sector. With this choice of parameters, the irradiation non-uniformity $\sigma_{2D}$, projected onto the 2D mesh, was accurately reproduced, i.e. its value did not changed by increasing ten times the number of rays. Resort to massive parallelism turned out necessary to make simulation times acceptable. Good scalability was observed for the raytracing kernel with increasing number of computing cores. In order to have comparable times for both the raytracing step and the hydrodynamic step, about 800 - 1000 cores were used. A start-to-end implosion run with 3D raytracing could then be carried out in a few hours. Details of the implementation and analysis of noise levels will be presented elsewhere.

## 3. Parametric robustness studies

In this section we summarize recent studies on target robustness performed with DUED, using one or both the above improved models. All results refer to the shock-ignited HiPER baseline target. In Ref. [7] we studied gain degradation as a function of beam mispositioning (offset of the target centre with respect to the centre of the laser beams) for a target irradiated by the nominal HiPER 48-beam laser configuration. For simplicity, the simulations assumed radial irradiation, where the intensity unbalance is described via the sum of Legendre polynomials. We have now improved and extended that study, by i) making use non non-local electron transport and ii) considering different values of the power of the spike. The results are shown in Fig. 3. We see that for the nominal spike (absorbed) power of 95 TW, the allowable displacement is 10 $\mu$m (1% of the radius). Tolerance to mispositioning grows with increasing the spike power, but is in any case limited to about 2.5%.

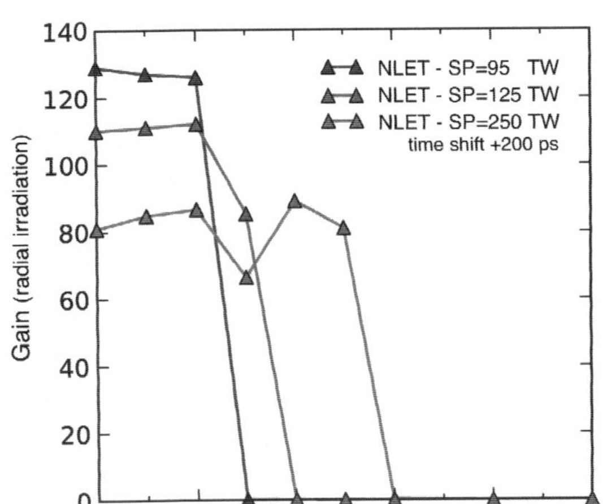


**Figure 3.** Target gain vs target centre offset for three different ignition spike powers. From 2D simulations of the HiPER baseline target with non-local electron transport, and radial irradiation. Note that, due to assumed radial irradiation the incident power is fully absorbed and then the gain reported here is about twice as large as the gain obtained from 1D simulations with nominal laser configuration and 2D ray-tracing.

The reference HiPER irradiation scheme for the compression stage (48 beams, each with intensity profile $\propto \exp(-r/w)^m$, with $m = 2.04$ and $w = 640$ $\mu$m) was chosen to minimize the initial irradiation nonuniformity in the absence of any laser or target positioning error. However, it has been shown that other schemes provide better uniformity during the implosion and are also less sensitive to errors (at the expense of larger laser energy to compensate for reduced absorption) [10, 15]. E.g., one may use wider superGaussian beams, with $m = 8$ and $w = 1300\ \mu$m [10]. First simulations with 3D raytracing confirm this prediction: using the above irradiation scheme, targets can tolerate larger displacements, and achieve nearly 1D yield; the gain is reduced by about 35%. A relevant result is shown in Fig. 4, where the density maps at stagnation are compared for a target irradiated with the reference beam configuration and displaced by somewhat less than 2% and a target irradiated by the wider beams and displaced by 3%. This second target ignites and achieves a nearly 1D yield.

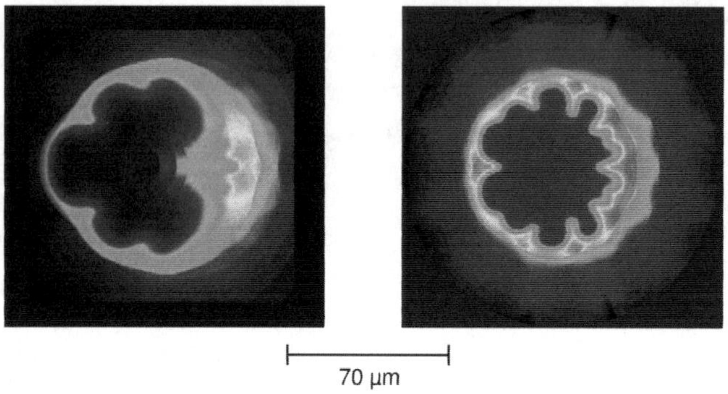


**Figure 4.** 2D simulations of the HiPER baseline target. Density maps at the shell stagnation for a) target displaced by 20 $\mu$m and irradiated by the reference HiPER beam configuration (left frame); b) target displaced by 30 $\mu$m and irradiated by beams with larger spots and flatter profiles and 50% larger power (right frame). Target (a) does not ignite; target (b) ignites and achieves nearly 1D yield.

In conclusion, we have improved DUED models for laser interaction and non-local electron transport. Simulation of shock ignited targets with the upgraded version of DUED essentially confirm prior results on target robustness. The code is now ready for systematic studies aiming at the optimization of irradiation schemes. It should be observed, however, that further model improvements, concerning LPI's and cross-beam-energy-transfer is required to tackle other critical issues for shock-ignition [4].

## Acknowledgments
Work was partially supported by the projects MIUR PRIN 2009FCC9MS and Sapienza 2012 C26A12CZH2, and by the HiPER project and Preparatory Phase Funding Agencies (EC, MSMT and STFC).

## References
[1] Betti R *et al.* 2007 *Phys. Rev. Lett.* **98** 155001
[2] Perkins L J, Betti R, LaFortune K N and Williams W H 2009 *Phys. Rev. Lett.* **103** 045004
[3] Anderson K S *et al.* 2013 *Phys. Plasmas* **20** 056312
[4] Batani D *et al.* 2014 *Nucl. Fusion* **54** 054009
[5] Atzeni S, Schiavi A and Bellei C 2007 *Phys. Plasmas* **14** 052702
[6] Ribeyre X *et al.* 2009 *Plasma Phys. Control. Fusion* **51** 015013
[7] Atzeni S, Schiavi A and Marocchino A 2011 *Plasma Phys Controll Fusion* **53** 035010
[8] Atzeni S, Marocchino A, Schiavi A and Schurtz G 2013 *New J. Phys.* **15** 045004
[9] Atzeni S, Marocchino A, Schiavi A 2012 *Phys. Plasmas* **19** 090702
[10] Schiavi A, Atzeni S and Marocchino M 2011 *EPL* **94** 35002
[11] Atzeni S 1986 *Computer Phys. Commun.* **43** 107
[12] Atzeni S *et al* 2005 *Computer Phys. Commun.* **169** 153
[13] Schurtz G P, Nicolaï P D and Busquet M 2000 *Phys. Plasmas* **7** 4238
[14] Marocchino A *et al.* 2013 *Phys. Plasmas* **20** 022702
[15] Temporal M, Canaud B and Le Garrec B 2010 *Phys. Plasmas* **17** 022701

# Direct-drive implosion physics: Results from OMEGA and the National Ignition Facility

**P. B. Radha[1], V. N. Goncharov[1], M. Hohenberger[1], T. C. Sangster[1], R. Betti[1], R. S. Craxton[1], D. H. Edgell[1], R. Epstein[1], D. H. Froula[1], J. A. Marozas[1], F. J. Marshall[1], R. L. McCrory[1], P. W. McKenty[1], D. D. Meyerhofer[1], D. T. Michel[1], S. X. Hu[1], W. Seka[1], A. Shvydky[1], S. Skupsky[1], J. A. Frenje[2], M. Gatu-Johnson[2], R. D. Petrasso[2], T. Ma[3], S. Le Pape[3], and A. J. Mackinnon[3]**

[1]Laboratory for Laser Energetics, University of Rochester, Rochester, NY, USA
[2]Plasma Science and Fusion Center, Massachusetts Institute of Technology, Cambridge, MA, USA
[3]Lawrence Livermore National Laboratory, Livermore, CA, USA

E-mail: rbah@lle.rochester.edu

**Abstract**. Direct-drive–implosion experiments from both OMEGA and the National Ignition Facility (NIF) are critical to gain confidence in ignition predictions on the NIF. Adequate performance of hydrodynamically scaled 1.8-MJ ignition designs must be obtained on OMEGA at 26 kJ. Implosions on the NIF must be used to identify and mitigate the effect of laser–plasma interactions (LPI's) on hydrodynamic parameters at the NIF scale. Results from spherically driven OMEGA cryogenic implosion experiments are described. Mitigation of nonuniformity sources and cross-beam energy transfer (CBET) is important for improving target performance on OMEGA. Initial polar-driven implosion experiments on the NIF have provided valuable measurements of trajectory and symmetry. Simulations that include the effect of CBET more closely reproduce the observed velocity.

## 1. Introduction

In the direct-drive approach to hot-spot ignition, nominally identical laser beams irradiate a capsule containing a layer of cryogenic deuterium–tritium (DT), driving the shell inward like a rocket [1]. The converging shell performs *PdV* work on the hot spot, raising the temperature. If a sufficient number of alpha particles from the DT-fusion reaction are produced and the hot spot has sufficient areal density (~300 mg/cm$^2$ at a temperature ~5 keV), the alpha particles deposit their energy, further raising the temperature and increasing the number of DT neutrons produced. Ignition is said to occur when the energy produced in the 14.1-MeV neutrons is greater than the laser energy used to irradiate the target.

One-dimensional (1-D) target designs are characterized by the fuel adiabat $\alpha$ (the ratio of the pressure to the Thomas–Fermi pressure at peak shell density), implosion velocity $V_{\text{imp}}$ (the peak mass-averaged shell velocity), and ablation pressure $P_{\text{a}}$. The ignition threshold factor $\text{ITF}_{\text{1-D}}$ [2], defined as the ratio of the shell's kinetic energy to the minimum energy for ignition [3], scales as $\text{ITF}_{\text{1-D}} = E_{\text{K}} / \min\left(E_{\text{ign}}\right) \sim \left[P_a(I)\right]^{0.8} V_{\text{imp}}^7 / I\alpha$, where $I$ is the on-target intensity. Designs that successfully ignite in multidimensional hydrodynamic simulations typically have $ITF_{1\text{-}D}$ values between 3 and 5 [2].

## 2. Results from OMEGA implosions

The OMEGA [4] laser is used to study direct-drive–implosion physics with the goal of developing and validating the models required to predict ignition. In order to be ignition relevant, cryogenic implosions on OMEGA are scaled [5] from ignition designs [6] for the NIF [7]. A typical OMEGA cryogenic implosion has a diameter of 860 $\mu$m with an ~3-ns-long shaped laser pulse at 26 kJ of laser energy. Three pickets that set the adiabat profile in the shell precede a main pulse [figure 1(a)].

Implosion energetics must be accurately modeled to simulate and predict performance. OMEGA implosions indicate that it is necessary to include the effect of nonlocal heat transport of the deposited laser energy [8] and CBET [9] to reproduce the observables related to energetics—time-resolved scattered light, trajectory, and timing of the neutron-production histories (figure 1) [9]. Nonlocal heat transport is important because of the finite mean free path of the energetic coronal electrons from the tail of their distribution. As figure 1(a) indicates, including only nonlocal transport significantly underestimates the time-resolved scattered light. Simulations, including the effect of CBET and nonlocal transport, reproduce the time-resolved scattered light very well. The trajectory [figure 1(b)] is inferred from gated framing camera images of self-emission, which peaks just outside the ablation surface [10]. Again, excellent agreement with the inferred trajectory and neutron rate timing is obtained when both CBET and nonlocal transport are included in the simulation.

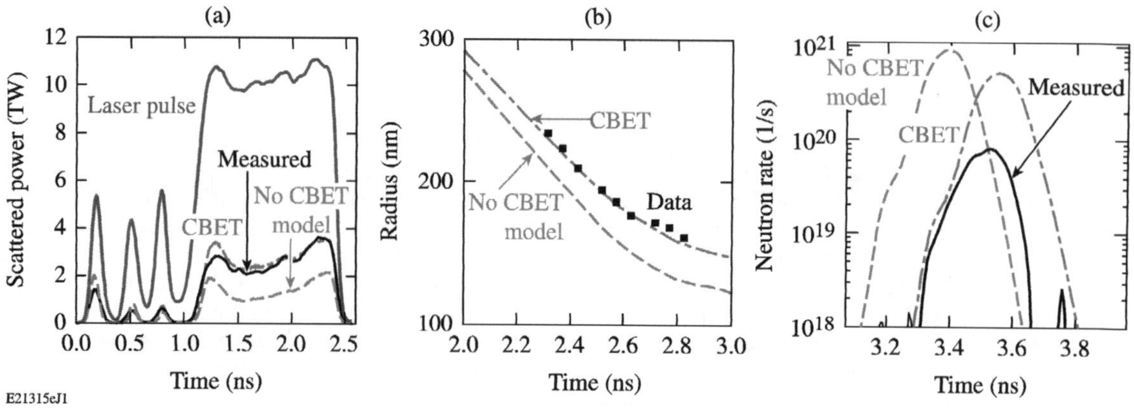


**Figure 1.** (a) Scattered light, measured laser pulse (gray); (b) trajectory of the ablation surface inferred from x-ray emission images; (c) neutron rate. Measured (black), collisional absorption with nonlocal heat conduction (dashed), collisional absorption with the effect of CBET and nonlocal heat conduction (dashed–dotted).

The triple-picket laser pulse shape has been shown to achieve high, nearly 1-D areal density in cryogenic implosions [6]. Relative to the simulation that includes only collisional absorption as the mechanism of energy deposition, the inclusion of CBET effects reduces ablation pressure by nearly 40% in OMEGA implosions and implosion velocity by nearly 15%. This can reduce $ITF_{1\text{-}D}$ by nearly a factor of 10. Options for improving $ITF_{1\text{-}D}$ include reducing the adiabat, increasing the implosion velocity by reducing shell mass or mitigating CBET, improving ablation pressure by mitigating CBET, and increasing the intensity on target.

OMEGA cryogenic implosions have been performed for a range of implosion velocities and adiabats. The adiabat is varied by primarily varying the energy in the first picket. The timing between pickets is varied to optimize the adiabat profile in the shell. Different implosion velocities are obtained by varying the shell and ablator mass. The measured yields increase with implosion velocity (figure 2), consistent with simulation, indicating that higher nonuniformity growth rates related to the higher implosion velocity do not overwhelm the 1-D effect of implosion velocity on the temperature. A fit to the data in figure 2(a) reveals, however, that while simulated yields scale as $V_{imp}^6$, measured yields scale as $V_{imp}^5$ suggesting that nonuniformities do compromise the yield.

Two-dimensional (2-D) simulations using the hydrodynamic code *DRACO* [11] have been performed for a range of cryogenic implosions [12]. These simulations include short-wavelength nonuniformities from laser imprint due to single-beam uniformity ($\ell \leq 150$) and the effect of beam smoothing through smoothing by spectral dispersion (SSD) [13], longer wavelength nonuniformities ($\ell \leq 10$) including the roughness of the inner ice layer, and beam imbalances such as mistiming and power imbalance. For the moderate adiabat ($\alpha \sim 4$) implosions, these simulations reproduce all the observables well (simulated yield is within 75% of the measured value, whereas ion temperature, burn width, hot-spot size, and areal density are within 90% of the measured values), indicating that these implosions are adequately understood. For lower-adiabat implosions, such as $\alpha = 2$, the simulations do not reproduce many of the observables including areal density and hot-spot radius. Areal density, in particular, is compromised at lower adiabats; for example, at $\alpha = 2$ the measured areal density is ~35% of the simulated value (figure 3). The decrease in areal density is correlated with increased x-ray emission from the core relative to spherically symmetric simulations, suggesting the mixing of ablator carbon into the hot spot [14]. The role of outer-surface defects has been investigated as a possible source of this degradation [15]. These simulations indicate that defects smaller than 1 $\mu$m in height can significantly compromise performance. Other possible sources of areal density degradation being investigated include higher levels of single-beam nonuniformity, void formation at the ice–ablator interface during the cooling process, early-time laser shinethrough, etc. Ongoing identification and mitigation of these sources of nonuniformity are expected to improve cryogenic target performance on OMEGA.

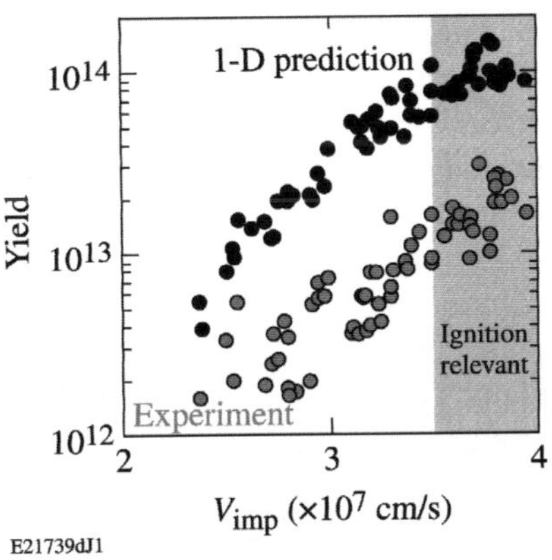


**Figure 2.** Measured yields (gray) and simulated yields (black) from cryogenic DT OMEGA implosions versus implosion velocity. Each point represents a single OMEGA shot with simulated values of adiabat between 2 and 5.

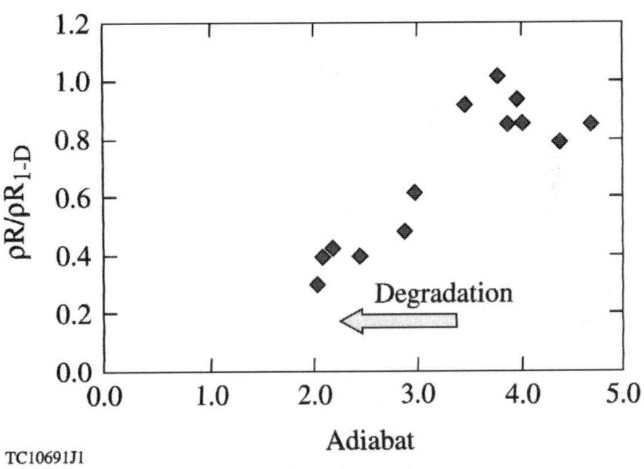


**Figure 3.** Ratio of measured areal density to simulated areal density versus the adiabat.

To improve $ITF_{1\text{-D}}$ requires a further decrease in adiabat beyond the typical values in current implosions. Given current sources of performance degradation, it is less likely that lower adiabats than 2 can lead to high-performing implosions.

Increasing ablation pressure and implosion velocity by mitigating the effect of CBET is necessary to improve OMEGA target performance and improve $ITF_{1\text{-D}}$. Targets with mid-*Z* ablators, such as Si, have been shown to mitigate the effect of CBET [16]. CBET results in the transfer of laser energy from the incoming rays to the outgoing rays mediated by ion-acoustic waves near the quarter-critical surface in the corona. The thickness of the layer is chosen such that Si stays at the quarter-critical region of the corona when the CBET gain is expected to be the highest, i.e., during the main drive of the pulse. With a higher average charge than CH, the temperature at the quarter-critical surface is larger with the mid-*Z* layer. Since the CBET gain scales inversely as the temperature [9], a Si layer reduces the extent of CBET.

Cryogenic targets with multiple layers will be investigated on OMEGA. The layers from the outside are an outer doped-plastic layer (to reduce imprint [17], thereby permitting more-stable implosions at a higher implosion velocity); a thin Si layer (<1 $\mu$m); a Be layer, with an inner cryogenic DT layer. Beryllium (Be) offers a higher ablation pressure than the historically used plastic (CH) ablator [16]. Simulations indicate that the combination of Be and a mid-*Z* material (Si) increases ablation pressure by 15% and velocity by nearly 5% through a better choice of ablator material and by mitigating CBET in the Si.

Other ways to mitigate CBET include modifications to the laser. The use of smaller phase plates during the main pulse reduces beam overlap and therefore the interactions of rays that result in the greatest CBET transfer. Improved target performance in room-temperature implosions has been demonstrated with this concept [18]. Further development involves using "zooming" phase plates [19]. There the beam size matches the target radius initially in the implosion and is reduced later during the main pulse, permitting reduced initial nonuniformity seeds while mitigating CBET later in the implosion. The use of different wavelengths on different beams causing the greatest overlap will also be investigated.

## 3. Results from NIF implosions

OMEGA implosion performance cannot be simply extrapolated to NIF scales because of the different density scale lengths in the corona (~150 $\mu$m for OMEGA-scale 26-kJ implosions compared to ~600 $\mu$m for the ignition-scale implosion at 1.8 MJ). Most laser–plasma interactions scale with increasing coronal density scale lengths.

Room-temperature experiments are being conducted in the polar-drive (PD) configuration [20] on the NIF to validate models and identify and mitigate the effect of LPI on implosions. Ideally, direct drive requires spherical drive with beams arranged symmetrically around the target chamber, like the OMEGA geometry; on the NIF, however, beams are arranged axisymmetrically. Simulations indicate that the reduced irradiation at the equator can be corrected by a combination of beam repointing, where beams at higher latitudes are repointed toward the equator, an optimal choice of pulse shapes for the different rings, and custom beam profiles. Ongoing experiments [21] use the existing NIF phase plates, defocused to improve low-order uniformity. These are not optimal, however, for PD and offer limited freedom in design. These experiments are intended to identify and demonstrate mitigation of LPI effects on gross hydrodynamic parameters, such as symmetry and implosion velocity, and not to achieve high-performance direct-drive implosions.

The target is an 80-$\mu$m-thick CH shell driven with an $\alpha \sim 3$ pulse shape comprising a low-intensity foot rising to a flat-top pulse. The average intensity at the initial target radius is ~$4 \times 10^{14}$ W/cm$^2$. The hydrodynamic code *DRACO* has been modified to include a model of CBET [22] based on the formulation of Randall *et al.* [23]. Preliminary simulations with this model indicate that CBET can have an observable effect on inferred velocity and shape. As the contour plot of the instantaneous deposited laser energy (figure 4) indicates, significant energy is transferred out of the equator into the outgoing rays of beams from the other rings relative to a simulation that includes only collisional absorption. This results in an overall reduction of time-integrated absorbed energy from 90% to 68%. The drive near the equator is also reduced, leading to a more-oblate shape.

The trajectory of the ablation surface is inferred from the time-resolved x-ray framing camera images [9]. The peak of the emission is identified using the method of matched filters and averaged in

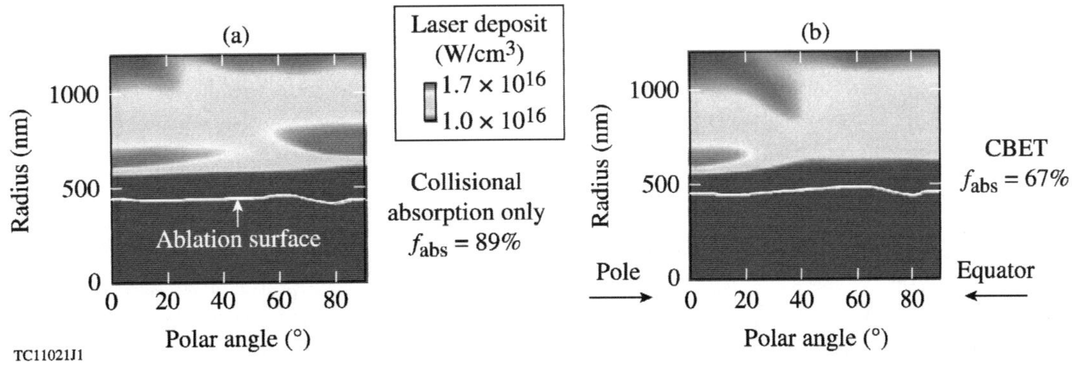


**Figure 4.** Contour plot of instantaneous laser energy deposited versus radius of the target and polar angle (a) with only collisional absorption and (b) with the inclusion of the CBET model.

angle to obtain an average radius versus time. This method has provided an important measurement for validating laser deposition and heat conduction models on OMEGA, as shown in figure 1(b). Simulations indicate that implosions at this intensity are relatively insensitive to the model of heat conduction, making this an ideal test for the CBET model. The images, including the effect of the pinhole used in the experiment, and the experimental gate width are created from the *DRACO* simulation using the code *Spect3D* [24]. The same procedure used to extract trajectories from the data is used on the simulated images. The velocity of the ablation surface inferred from the trajectories is shown in figure 5. Better agreement with experiment is obtained when CBET is included in the simulation (figure 5). The resulting difference between simulation and experiment may be due to the uncertainties in the beam profiles used in the simulation, the relatively lower resolution of the quarter-critical surface in the simulation, the effect of short-wavelength nonuniformity growth that could potentially decouple the ablation surface from the center of mass (an effect not included in the simulation), or limitations in the modeling (for example, an ion-acoustic wave damping coefficient is currently parameterized by a single value of 0.2; larger values indicate greater damping). These sources are currently being investigated.

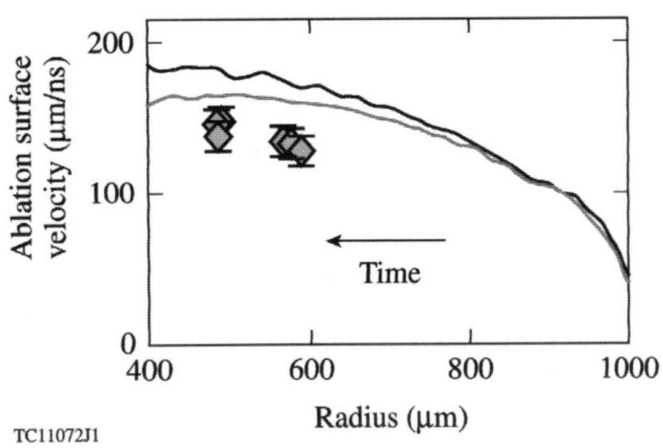


**Figure 5.** Inferred ablation surface velocity versus the radius of the peak emission region. Symbols: measured; black curve: simulated with only collisional absorption; gray curve: simulated with the CBET model.

## 4. Summary

Results from direct-drive OMEGA cryogenic implosions show that implosion energetics can be reproduced well using models of CBET and nonlocal heat conduction. Yields and ion temperatures from cryogenic implosions scale with implosion velocity, indicating that nonuniformity does not overwhelm the increase in these values from higher implosion velocity. Moderate-adiabat ($\alpha \sim 4$) target performance can be described well using two-dimensional simulations that include single-beam nonuniformity, beam imbalances, and inner ice roughness. Lower-adiabat performance cannot be explained with these sources of nonuniformity, indicating that other sources need to be accounted for. Candidates such as outer surface defects, void formation at the ice–ablator interface during the cryogenic process, and possible underestimation of laser imprint seeds are being investigated. Mitigating CBET is also important for improving cryogenic target performance.

Implosions on the NIF with the existing x-ray drive configuration are performed to study symmetry, velocity, and preheat. Simulations with the newly developed CBET model in *DRACO* indicate that velocity is significantly influenced by CBET. Future PD implosions on the NIF will be used to validate the CBET model and study various ways of mitigating CBET, including mid-*Z* layers and different wavelengths on the inner and outer cones.

**Acknowledgements**
This material is based upon work supported by the Department of Energy National Nuclear Security Administration under Award Number DE-NA0001944, the University of Rochester, and the New York State Energy Research and Development Authority. The support of DOE does not constitute an endorsement by DOE of the views expressed in this article.

**References**
[1] Nuckolls J, Wood L, Thiessen A and Zimmerman G 1972 *Nature* **239** 139 (doi: 10.1038/239139a0)
[2] Haan S W *et al* 2011 *Phys. Plasmas* **18** 051001 and references therein (doi: 10.1063/1.3592169)
[3] Herrmann M C, Tabak M and Lindl J D 2001 *Nucl. Fusion* **41** 99 (doi: 10.1088/0029-5515/41/1/308)
[4] Boehly T R *et al* 1997 *Opt. Commun.* **133** 495 (doi:10.1016/S0030-4018(96)00325-2)
[5] 1998 *LLE Review Quarterly Report* **82** 49 Laboratory for Laser Energetics University of Rochester Rochester NY LLE Document No. DOE/SF/19460-344
[6] Goncharov V N *et al* 2010 *Phys. Rev. Lett.* **104** 165001 (doi: 10.1103/PhysRevLett.104.165001)
[7] Moses E I *The National Ignition Facility: Transition to a User Facility and the Future of HED Laser Systems* presented at The Eight International Conference on Inertial Fusion Sciences and Applications, IFSA 2013, Nara, Japan, 8–13 September 2013, Paper K.Mo_A2
[8] Goncharov V N *et al* 2008 *Phys. Plasmas* **15** 056310 (doi: 10.1063/1.2856551)
[9] Michel D T *et al.* 2012 *Rev. Sci. Instrum.* **83** 10E530 (doi: 10.1063/1.4732179)
[10] Igumenshchev I V *et al* 2012 *Phys. Plasmas* **19** 056314 (doi: 10.1063/1.4718594)
[11] Radha P B *et al* 2005 *Phys. Plasmas* **12** 032702 (doi: 10.1063/1.1857530)
[12] Hu S X *et al* 2013 *Bull. Am. Phys. Soc.* **58** 324
[13] Skupsky S *et al* 1989 *J. Appl. Phys.* **66** 3456 (doi: 10.1063/1.344101)
[14] Sangster T C *et al* 2013 *Phys. Plasmas* **20** 056317 (doi: 10.1063/1.4805088)
[15] Igumenshchev I V *et al* 2013 *Phys. Plasmas* **20** 082703 (doi: 10.1063/1.4818280)
[16] Michel D T *et al Experimental Optimization of the Rocket Efficiency in Direct-Drive Implosions Using Different Ablator Materials* submitted to Physical Review Letters
[17] Hu S X *et al* 2012 *Phys. Rev. Lett.* **108** 195003 (doi: 10.1103/PhysRevLett.108.195003)
[18] Froula D H *et al* 2012 *Phys. Rev. Lett.* **108** 165003 (doi: 10.1103/PhysRevLett.108.165003)
[19] Igumenshchev I V *et al* 2013 *Phys. Rev. Lett.* **110** 145001 (doi: 10.1103/PhysRevLett.110.145001)
[20] Skupsky S *et al* 2004 *Phys. Plasmas* **11** 2763 (doi: 10.1063/1.1689665)
[21] Radha P B *et al* 2013 *Phys. Plasmas* **20** 056306 (doi: 10.1063/1.4803083)
[22] Marozas J A and Collins T J B 2012 *Bull. Am. Phys. Soc.* **57** 344
[23] Randall C J, Albritton J R and Thomson J J 1981 *Phys. Fluids* **24** 1474 (doi: 10.1063/1.863551)
[24] MacFarlane J J *et al* 2007 *High Energy Density Phys.* **3** 181 (doi: 10.1016/j.hedp.2007.02.016)

# Study on the transport of a relativistic electron beam in plasmas

**Hong-bo Cai**[1,2*], **Shao-ping Zhu**[1], **X.T. He**[1,2]
[1]Institute of Applied Physics and Computational Mathematics, Beijing, 100094, China
[2]Center for Applied Physics and Technology, Peking University, Beijing, 100871, People's Republic of China

E-mail: *cai_hongbo@iapcm.ac.cn

**Abstract.**
In the fast ignition scheme, in order to deliver enough energy (10kJ) into the pellet in 10ps, the ignition laser beam should reach the intensity of $I_0 \sim 10^{20} W/cm^2$ if the coupling from the laser to the core plasma is about 20% for a well-collimated beam. However, studies have shown that the electron divergence increases with the laser intensity. Of particular importance is the possibility of collimating the fast electron beams with the size of the compressed core. In this work, we will introduce two ways to obtain the collimated electrons: 1) collimate fast electrons in specially engineered targets with the spontaneously generated magnetic field during its transportation; 2) collimate fast electrons with imposed magnetic field.

## 1. Introduction

Fast ignition [1] is an attractive option for inertial confinement fusion because it relaxes the requirements on the symmetry of the implosion and the compression energy. One task is how to produce and transport enormous numbers of high-energy and high-quality charged particles. However, recent studies revealed that the relativistic electron beam produced from the interaction of an intense laser pulse with a solid-density target may be strongly divergent. [2, 3] If the beam divergence is large, then the coupling efficiency could not be large and the required ignitor pulse energy can rapidly become unfeasibly large. State-of-the-art integrated simulations and experiments indicate that the coupling efficiency of short-pulse energy into the compressed deuterated-plastic (CD) core is only $3.5 \pm 1.0\%$. [4] Therefore, of particular importance is the possibility of collimating the REBs with the size of the compressed core.

To date, several mitigation works have been done regarding the control of the divergence of REBs, including self-generated magnetic field at the resistivity boundary in sandwich target, [5] high-resistivity-core-low-cladding targets,[6] and switchyard target, [7] two laser pulses, [8] double cone target, [9], imposed axial magnetic field, [10], and so on. Altogether, the fast electrons are collimated by the azimuthal magnetic field or axial magnetic field. In this paper, we will introduce two ways to obtain the collimated electrons and discuss the possible influence of axial and azimuthal field on fast electrons in the fast ignition scenario.

## 2. Magnetic collimation of fast electrons in specially engineered targets

Firstly, we find two different kinds of sandwich targets can produce magnetic fields with different generation mechanisms to collimate electrons. One is the large resistivity gradient along the interfaces between different materials, the other is nonparallel density gradient and fast electron current at the interfaces. The former is consistent with the hybrid-Vlasov-Fokker-Planck's simulation results [7] and experimental results [5, 6]. The latter will be discussed with our analytical model. From the electron fluid equations (the continuity equation and force balance equation), we can get the relationship between the self-generated magnetic field and electron flow velocity [11], $\mathbf{B} = \frac{c}{e}\nabla \times \mathbf{p_e}$, where $\mathbf{p_e} = m_e\gamma_e\mathbf{v}_{e0}$ is the momentum of the background electron flow and $\gamma_e = 1/\sqrt{1 - v_{e0}^2/c^2}$ is the relativistic factor. Maxwell's equations for the self-generated electric and magnetic fields, $\mathbf{E}$, and $\mathbf{B}$, are given by $\nabla \times \mathbf{B} = -\frac{4\pi e}{c}(n_{e0}\mathbf{p}_e/m_e\gamma_e + n_h\mathbf{v}_h) + \frac{1}{c}\frac{\partial \mathbf{E}}{\partial t}$. Here, for a long beam with pulse length $l_b \gg v_h/\omega_p$, where $v_h$ and $\omega_p$ are the fast electron beam velocity and background plasma frequency, the displacement current $\frac{1}{c}\frac{\partial \mathbf{E}}{\partial t}$ is of order $(v_h/\omega_p l_b)^2 \ll 1$ compared to the electron current. In the following discussion, the displacement current $\frac{1}{c}\frac{\partial \mathbf{E}}{\partial t}$ is neglected. Therefore, we obtain the equations for self-generated magnetic field,

$$\frac{mc^2}{4\pi e^2}\nabla \times \left(\frac{\gamma_e \nabla \times \mathbf{B}}{n_{e0}}\right) + \mathbf{B} = -\frac{mc}{e}\nabla \times \left(\frac{\gamma_e n_h \mathbf{v}_h}{n_{e0}}\right), \tag{1}$$

One can see that the source term is $\nabla \times \left(\frac{\gamma_e n_h \mathbf{v}_h}{n_{e0}}\right) \approx -\frac{\gamma_e}{n_{e0}^2}\nabla n_{e0} \times \mathbf{j}_h$, which means the generation of self-generated magnetic field is due to the nonparallel density gradient and fast electron current.

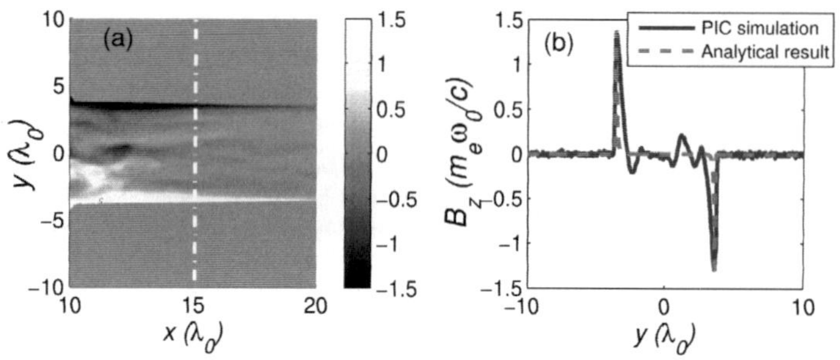


**Figure 1.** (a) The spontaneous magnetic fields at time $t = 500fs$; (b) A slice of spontaneous magnetic fields at $x = 15\lambda_0$, the solid line is for simulation result, and the dash-dotted line for the analytical result. The unit of the magnetic field is $m_e\omega_0 c/e$ (1 unit = 100MG).

Figure 1(a) shows the spontaneous magnetic field at time $t = 500fs$. Clearly, very large magnetic field has been generated at the interfaces, which plays a role in collimating fast electrons. Moreover, we observe that there are several stronger magnetic peaks in the inner target. These peaks are due to the formation of filament due to the Weibel instability. Fig. 1(b) shows a slice of spontaneous magnetic fields at $x = 15\lambda_0$, the solid line is for simulation result, and the dashed line for the analytical result. We can see that the analytical result is consistent with the simulation result.It is found that the spontaneously generated magnetic field reaches as high as $100MG$, which is large enough to collimate fast electron transport in overdense plasmas.

## 3. Effects of the imposed magnetic field on the production and transport of relativistic electron beams

The effects of the imposed uniform magnetic field, ranging from 1MG up to 50MG, on the production and transport of relativistic electron beams (REBs) in overdense plasmas irradiated by ultraintense laser pulse are investigated. This study gives clear evidence that the imposed magnetic field is capable of effectively confining the relativistic electrons in space even when the source is highly divergent since it forces the electrons moving helically. In comparison, the spontaneous magnetic fields, generated by the helically moving electrons interplaying with the current filamentation instability, are dominant in scattering the relativistic electrons. In Fig.

2 the energy density distributions of electrons with energy between $0.5 \leq E[MeV] \leq 5.0$ are plotted. It is clearly seen that the produced fast electrons are very divergent for the case without imposed magnetic field. Alternatively, in the cases with imposed magnetic field, the electrons are collimated in space. The spot of the electron beam becomes smaller in the higher imposed magnetic field case. It seems that the high energy electrons are confined much better in the higher imposed magnetic field case. Actually, when the imposed magnetic field $B_0 > 3MG$, the overall coupling from laser to the relativistic electrons which have the potential to heat the compressed core in fast ignition was found to increase from 6.9% to 21.3%.

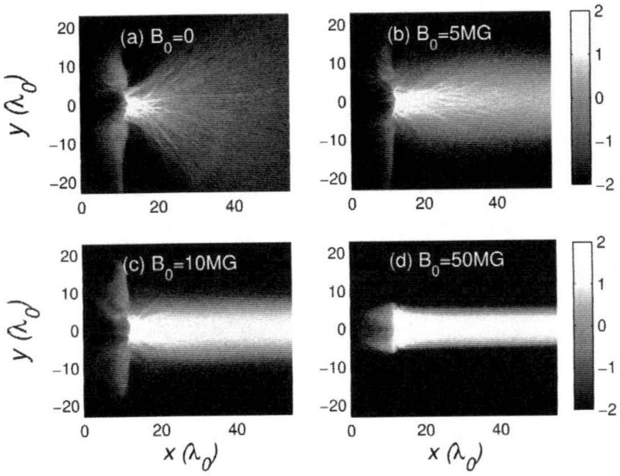


**Figure 2.** The natural logarithm of the electron energy density for different imposed magnetic fields (a) $B_0 = 0$, (b) $B_0 = 5MG$, (c) $B_0 = 10MG$, and (d) $B_0 = 50MG$ at t=1ps. Here the electron energy density is normalized by $m_e c^2 n_c$.

We next check the effects of imposed and spontaneous magnetic fields on the divergence angle of the electrons. In our simulations, the laser plasma interaction region locates at $x = 11 \sim 13\lambda_0$, while the local angular distribution function of the fast electrons was extracted at the positions $15\lambda_0$, $25\lambda_0$, and $55\lambda_0$. In order to distinguish the divergence angle of the fast electrons from that of the background electrons, only the electrons with energy higher than 200keV and propagating through the diagnostic positions at time interval t=(667, 1000)fs were considered. The local dispersion has been characterized by a Gaussian function $f(\theta) = A\exp[-\frac{(\theta-\theta_r)^2}{\Delta\theta_0^2}]$, where $\theta = \tan^{-1}[p_y/p_x]$, $\theta_r$ is the local electron mean propagation angle, and $\Delta\theta$ is the local divergence angle.

**Table 1.** Divergence angle of fast electrons diagnosed at position $x = 15\lambda_0$ for different imposed magnetic fields. Here, only the electrons in the region $y = (-8\lambda_0, 8\lambda_0)$ and $t = (667, 1000)fs$ are diagnosed.

| $B_0$ (MG) | 0 | 1 | 3 | 5 | 10 | 20 | 30 | 40 | 50 |
|---|---|---|---|---|---|---|---|---|---|
| divergence angle | $63^o$ | $64^o$ | $72^o$ | $64^o$ | $67^o$ | $60^o$ | $71^o$ | $76^o$ | $90^o$ |

Table 1 presents the divergence angle of fast electrons diagnosed at position $x = 15\lambda_0$ for different imposed magnetic fields. The divergence angle at $x = 15\lambda_0$ denotes the divergence angle in the laser plasma interaction region. It is found for the case without imposed magnetic field, the divergence angle is about $63^o$. In fact, when the imposed magnetic field is smaller than $30MG$, the divergence angle is measured as being $66(\pm 6)^o$. It means that for the cases with $B_0 \leq 30MG$, the electron divergence angle is not significantly affected by the imposed magnetic field. However, when the imposed magnetic field $B_x$ reaches as high as 50MG, which is the same order of the spontaneous magnetic field due to the current filamentation instability in the interaction region, the divergence angle rapidly increases to $90^o$. This means that the imposed magnetic field can result in larger divergence angle. Altogether, taking into consideration of the conversion efficiency and the electron divergence angle, we may conclude that $B_0 = 3 \sim 30MG$ is more suitable to collimate the fast electrons.

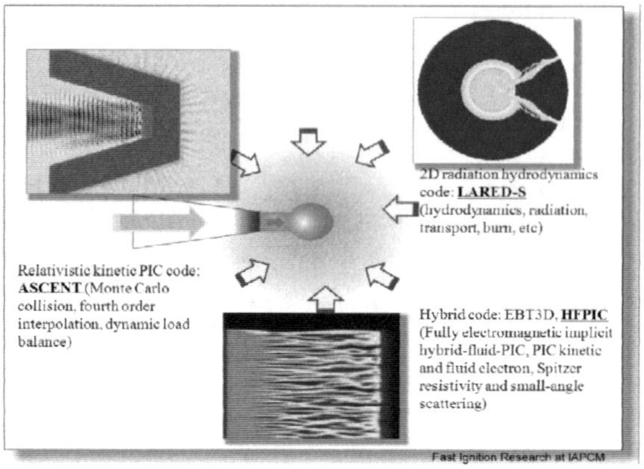


**Figure 3.** Integrated simulations coupling a hydrodynamic code, a hybrid-fluid-code, and a PIC code.

## Summary

In summary, we have shown two ways to collimate fast electrons and improve beam quality. It is found that huge surface magnetic field, which can collimate fast electrons, is generated due to the the nonparallel density gradient and fast electron current for the specially engineered low-density-core-high-density-cladding structure targets. However, this collimation scheme also bring some complexity for the target design. On the other hand, we found when the imposed magnetic field was increased from 1MG to 50MG, overall coupling from laser to the relativistic electrons which have the potential to heat the compressed core in fast ignition was found to increase from 6.9% to 21.3%. The simulations show that imposed magnetic field of the value of $3 \sim 30MG$ could be more suitable to fast-ignition inertial fusion. Further analysis of the integrated efficiency will be carried out by the our newly developed integrated simulation codes combining particle-in-cell code ASCENT, 2D radiation hydrodynamic code LARED-S, and hybrid-fluid-PIC code HFPIC (see Fig. 3) in the near future.

## ACKNOWLEDGEMENT

This work was supported by the National Natural Science Foundation of China (Grant Nos. 11275028, 11005011, and 91230205).

## References

[1] Tabak M *et al.* 1994 *Phys. Plasmas* **1** 1626
[2] Lancaster K L *et al.* 2007 *Phys. Rev. Lett.* **98** 125002
[3] Green J S *et al.* 2007 *Phys. Rev. Lett.* **100** 015003
[4] Theobald W R *et al.* 2011 *Phys. Plasmas* **18** 056305
[5] Kar S *et al.* 2009 *Phys. Rev. Lett.* **102** 055001
[6] Ramakrishna B *et al.* 2008 *Phys. Rev. Lett.* **105** 135001
[7] Robinson A P L *et al.* 2012 *Phys. Rev. Lett.* **108** 125004
[8] Scott R H H *et al.* 2012 *Phys. Rev. Lett.* **109** 015001
[9] Cai H B *et al.* 2009 *Phys. Rev. Lett.* **102** 245001
[10] Strozzi D J *et al.* 2009 *Phys. Plasmas* **19** 072711
[11] Cai H B *et al.* 2011 *Phys. Rev. E.* **83** 036408

# Convergent ablation measurements with gas-filled rugby hohlraum on OMEGA

**A. Casner, T. Jalinaud, D. Galmiche**
CEA, DAM, DIF, F-91297 Arpajon, France
E-mail: alexis.casner@cea.fr

**Abstract.** Convergent ablation experiments with gas-filled rugby hohlraum were performed for the first time on the OMEGA laser facility. A time resolved 1D streaked radiography of capsule implosion is acquired in the direction perpendicular to hohlraum axis, whereas a 2D gated radiography is acquired at the same time along the hohlraum axis on a x-ray framing camera. The implosion trajectory has been measured for various kinds of uniformly doped ablators, including germanium-doped and silicon-doped polymers (CH), at two different doping fraction (2 % and 4 % at.). Our experiments aimed also at measuring the implosion performance of laminated capsules. A laminated ablator is constituted by thin alternate layers of un-doped and doped CH. It has been previously shown in planar geometry that laminated ablators could mitigate Rayleigh Taylor growth at ablation front. Our results confirm that the implosion of a capsule constituted with a uniform or laminated ablator behaves similarly, in accordance with post-shot simulations performed with the CEA hydrocode FCI2.

## 1. Introduction

In indirect-drive inertial confinement fusion (ICF) [1, 2], a cryogenic deuterium-tritium target is imploded to high density and temperature using thermal x-ray. To ignite the fuel from the central hot spot generated at implosion stagnation an high implosion velocity $V_{fuel} \geq 370\ km.s^{-1}$ is requested. Recent experiments performed on the National Ignition Facility have achieved $V_{fuel} \geq 350\ km.s^{-1}$ at the price of a record laser shot (520 TW, 1.86 MJ laser pulse) [3]. Additional research and alternative paths are therefore actively pursued to achieve high implosion velocity at lower laser energy. Ablators more efficient than plastic provide one way to optimize implosion performances [4]. Beryllium is a well-known candidate [5] but it implies new safety rules for the facility. High Density Carbon (HDC) has been therefore tested on a few NIF shots and has performed very close to 1D predictions, approaching implosion velocity of 400 $km.s^{-1}$ [6]. An HDC capsule is the current record holder for neutron production in a non-cryogenic layered implosion with a yield of $1.6.10^{15}$ neutrons.

The other way to increase implosion velocity for a given laser energy is to improve the hohlraum coupling efficiency. Wall losses could be reduced using higher opacity wall material such as depleted uranium [7], or also by changing the hohlraum internal shape to reduce its area as suggested with rugby hohlraums [8, 9]. The baseline design for implosion experiments with the Laser MegaJoule (LMJ) relies in fact on a rugby-shaped hohlraum [10], which presents significant advantages in terms of LPI mitigation, coupling efficiency and symmetry control with a 1/2-1/2 energy balance [11]. The increased x-ray flux on capsule in rugby hohlraum compared to a classical cylindrical hohlraum was confirmed at OMEGA scale by enhanced

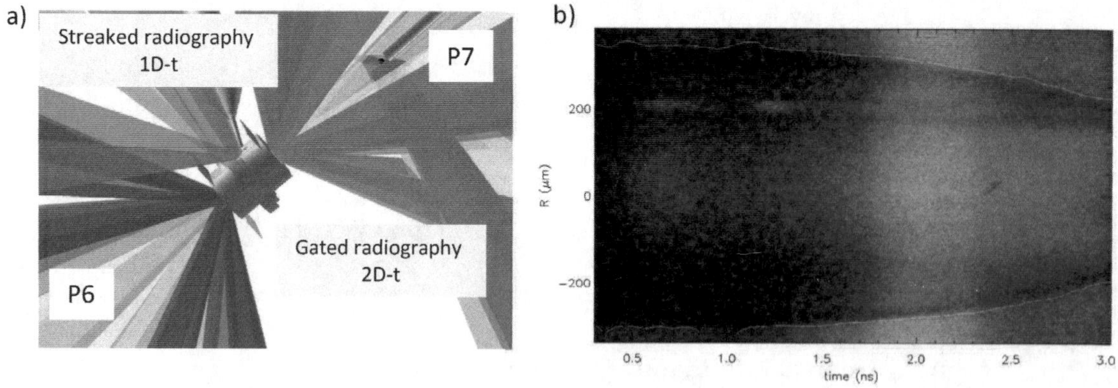


**Figure 1.** (a) Schematics of the experimental configuration. (b) Typical streaked radiography acquired on a gas-filled rugby implosion. The thin white line is the capsule limb position extracted by an automatic edge detection algorithm.

nuclear performances [12–14]. Rugby hohlraum implosions experiments are currently conducted on NIF [15, 16]. Here we present the first convergent ablation measurements [17] for gas-filled rugby hohlraum implosion experiments performed on the OMEGA laser facility.

## 2. Experimental configuration

A schematics of the experimental configuration is shown on Fig. 1a). The rugby hohlraum (spherical internal shape) is aligned along a pentagonal axis of the OMEGA target chamber and filled with 1 atm methane at room temperature. The primary diagnostic to assess the implosion velocity, key metric for implosion performance, is the SSC-A x-ray streak camera with a 20x magnification and a 10 $\mu m$ imaging slit. A typical streaked radiograph is shown on Fig. 1b). The limb of the shell is well defined and its position versus time could be extracted. Typical internal capsule diameter is 550 $\mu m$ with ablator thickness in the range from 30 to 55 $\mu m$. For radiography access two slots were made of either side of the hohlraum. One slot is covered by a 6 $\mu m$ thick vanadium backlighter foil, whereas a 50 $\mu m$ CH foil is glued on the opposite slot. Due to the presence of these lateral slots the hohlraum is driven by 36 beams (instead of 40) arranged in three cones (21°, 42° and 59°) on each Laser-Entrance-Hole. Beams were equipped with elliptical phase plates and smoothed by spectral dispersion. Polarization smoothing was not used in this experiment. Backscattering levels were measured on 2 beams belonging to 42° and 59° cones and were found to be negligible. The laser drive (shaped pulse) was adjusted by playing with the Frequency Crystal Conversion (FCC) of the 59° beams. The x-ray flux was measured with Dante, looking at one LEH at 37° of hohlraum axis. 2 radiative drive conditions have principally been tested: a low-drive case with a peak radiative temperature at 205 eV, and a high-drive case reaching 225 eV. A second radiograph was simultaneously performed along the hohlraum axis with a titanium area backlighter. Images were recorded with an x-ray framing camera coupled to a gated microchannel plate detector. We used a magnification of 8 with 10 $\mu m$ diameter pinholes for that diagnostic.

## 3. FCI2 Integrated simulations

We performed two-dimensional radiation-hydrodynamics simulations of hohlraum and capsule dynamics with FCI2 [18], taking into account M-band contributions determined with the Non Local Thermodynamical Equilibrium (NLTE) model Radiom [19]. Hohlraum wall deformations

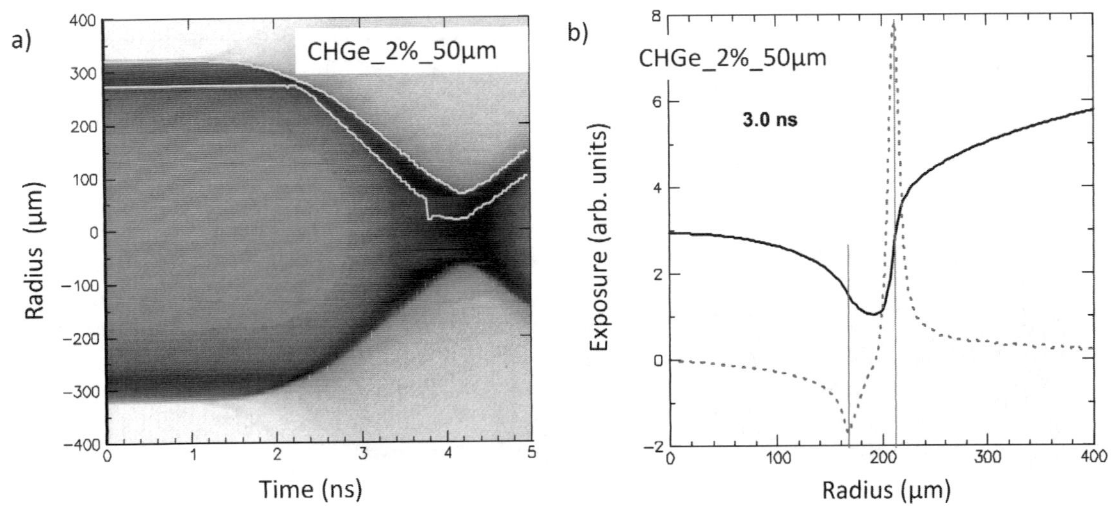


**Figure 2.** (a) FCI2 postprocessed radiograph. The external and internal radii are underlined in yellow. (b) Vertical lineout accross the previous image. The external limb radius is defined at the position of the maximum exposure gradient.

and x-ray flux asymmetry on capsule require an Adaptative Lagrangian Eulerian (ALE) mesh for the whole integrated simulation. As shown on Fig.1b) for a germanium-doped CH ablator (2 % atomic fraction and 50 $\mu m$ initial thickness) the external limb position (thin white line) of the ablator could easily be extracted from experimental radiographs by detection of the maximum slope of the exposure. Detection of the internal limb is more challenging due to backlighter profile and photocathode non-uniformities, as well as spatial resolution of the x-ray camera. The same analysis procedure is applied on postprocessed FCI2 radiographs, as explained on Fig.2. Experimental and postprocessed radiographs are therefore compared directly without Abel inversion and the subsequent hypothesis on the density and opacity profiles of the ablated plasma [17]. This is the most straightforward and robust method to compare simulations and experiments. In the remaining of the article we will refer to the limb radius and velocity ($V_{limb}$) as our implosion metrics.

## 4. Comparison of laminated and uniformly doped ablators

These experiments aimed first at comparing the implosion performance of uniformly doped and laminated capsules [20]. A laminated ablator is constituted by thin alternate layers of un-doped and (here) germanium-doped CH. The figure 3a) presents a pre-shot radiography of a laminated capsule, on which the thin layers (1 $\mu m$ each) of doped and un-doped material are clearly visible. The germanium-doped CH layers of the laminated ablator were at 3.7 ($\pm$0.1) % at. fraction. The corresponding uniformly doped CH capsules were doped at 2.4 ($\pm$0.1) % at. fraction, so that the mean doping fraction of the 55 $\mu m$ thick ablator were equivalent in both cases. It has been experimentally demonstrated in planar geometry that laminated ablators could mitigate Rayleigh Taylor instabilities at ablation front due to a transverse diffusive mechanism [20, 21]. However laminated capsules have never been fabricated and tested before those shots. Implosion trajectory for laminated and uniformly doped capsules are shown on Fig. 3b). First of all streaked and gated radiographs are in accordance demonstrating the good level of x-ray drive uniformity on capsule. Then the main result is that there is no detectable discrepancies between laminated and homogeneous ablators which implode in accordance with FCI2 simulations. This

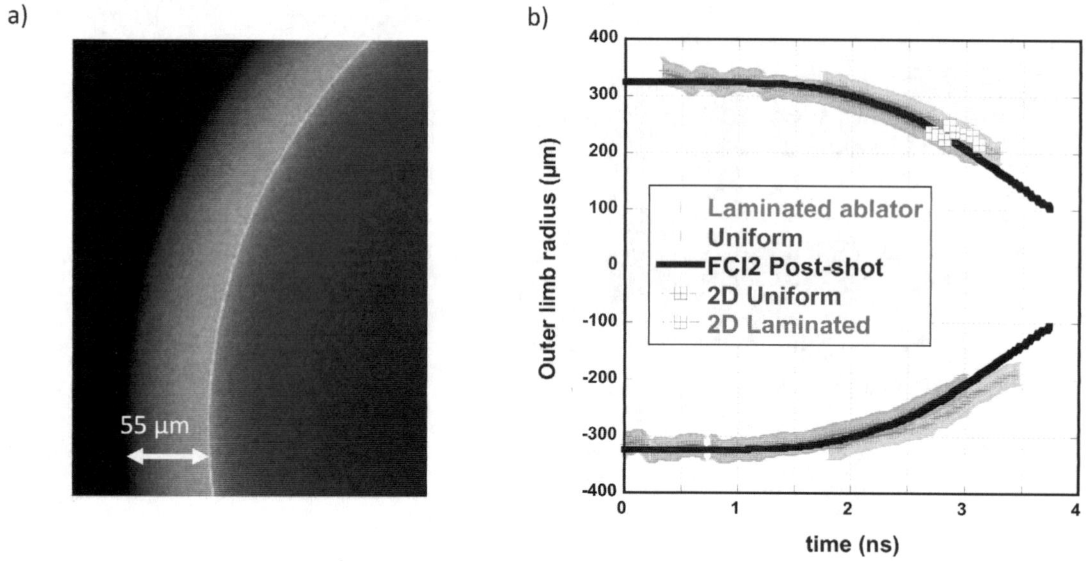


**Figure 3.** (a) Cross-section of a laminated capsule. (b) Comparison of capsule trajectory for laminated and uniformly doped ablators. The square points represents the measurements acquired on the 2D gated radiography on axis.

corroborates the trajectory measurements already acquired for planar samples [21]. A definitive answer regarding the stabilizing effect of laminated ablator on hydrodynamic instabilities in convergent geometry needs yet to be given via a serie of high nuclear performance implosions [13].

## 5. Comparison of germanium and silicon-doped polymers

We focus in that section of the relative implosion performances of CHGe and CHSi capsules. It has in fact been emphasized that Si-doped ablators are more efficient than Ge-doped ones at NIF scale [22]. A larger database has been acquired in our experiments, as parametric variations have been made under low-drive and high-drive conditions for both ablators, at two different doping fraction (2 % and 4 % at.) and for various ablator thicknesses. A detailed comparative study will be presented elsewhere [23]. We only present here the general trends. Fig.4 presents typical outer limb trajectory and velocity as a function of time in the case of a silicon-doped (4 % at.) CH capsule with an initial ablator thickness of 55 $\mu m$. The maximum drive temperature was 225 eV in that case. A reasonable good agreement is found for the experimental and postprocessed trajectories, even if simulations predict an earlier final convergence x-ray bang time [13]. The final outer limb velocities are in accordance within the error bars of the measurements, with $V_{limb}^{max} = 200 \pm 20\ km.s^{-1}$. Regarding ablator efficiency conclusions should be drawn carefully due to fluctuations in density and capsule external diameter for different production batches. Fig.5 compares in a low drive case (205 eV) the performances of CHSi and CHGe capsules (doped at 4 % at.) having the same external diameter, thus receiving the same x-ray drive. The CHSi capsule (initial thickness 55 $\mu m$) has 13 % more mass than CHGe (45 $\mu m$) thick for the same external capsule diameter. The silicon-doped capsule seems performs slighly better than the germanium-doped reaching a maximum limb velocity of 120 $km.s^{-1}$, versus 100 $km.s^{-1}$.

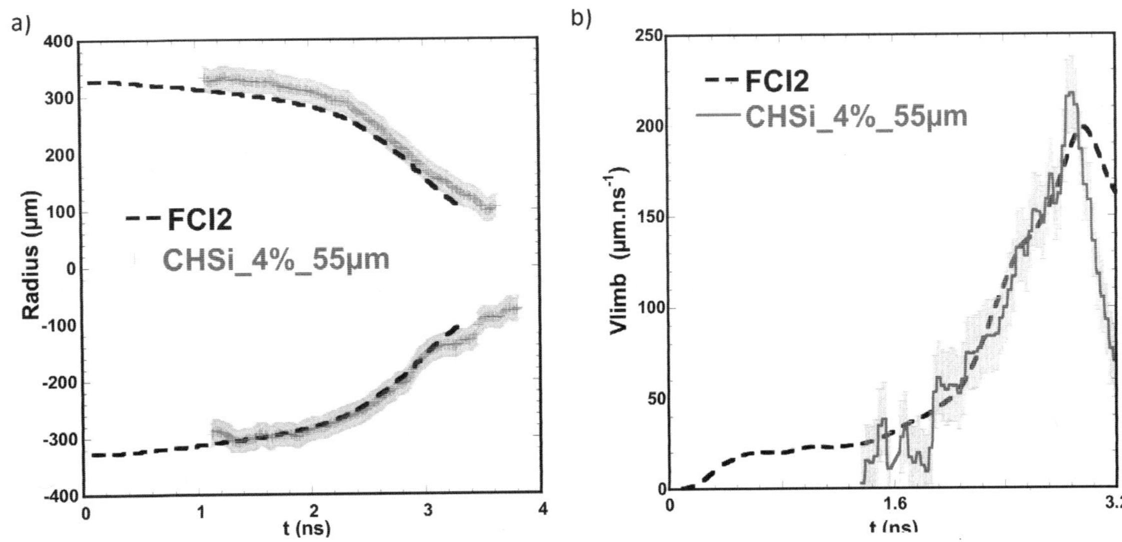


**Figure 4.** (a) Ablator outer limb radius versus time in the case of of CHSi capsule doped at 4 % at. fraction. The initial ablator thickness is 55 $\mu m$. The long dashed black lines represent the results of FCI2 postshot simulations. (b) Experimental and simulated outer limb velocities $V_{limb}$ as a function of time.

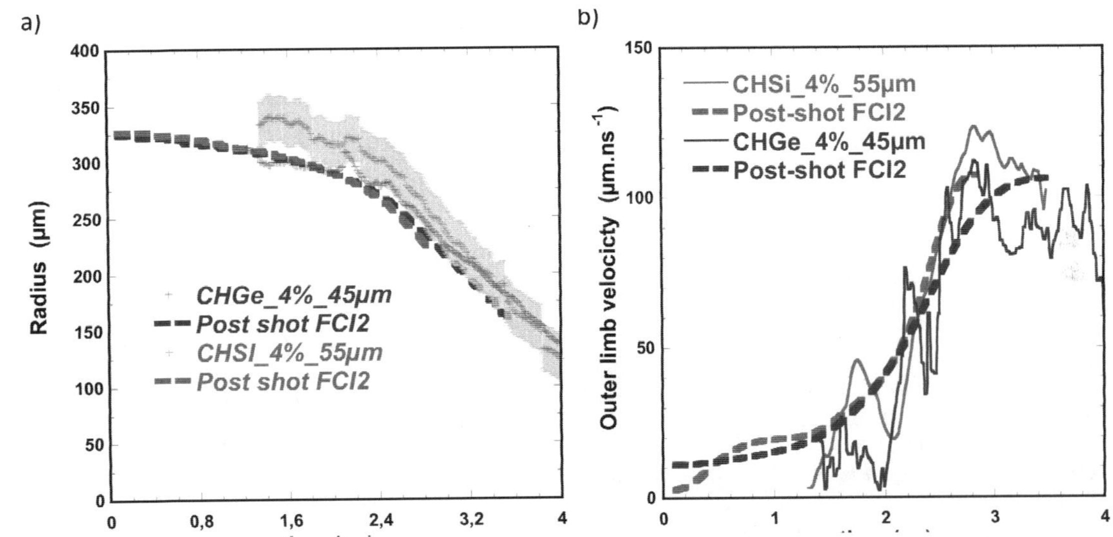


**Figure 5.** (a) Trajectories and (b) outer limb velocities $V_{limb}$ as a function of time for CHSI and CHGe capsules with the same initial external diameter.

## 6. Conclusions

Since the last IFSA [14], CEA has pursued gas-filled rugby hohlraum experiments on the OMEGA Laser Facility in preparation for LMJ and has improved the level of precision of its experiments. The first convergent ablation experiments in gas-filled rugby-hohlraum were performed in a double axis radiography configuration, with a transverse streaked radiography (aka ConA)and an axial time-resolved radiography (aka 2D ConA) recorded simultaneously on the same shot. Various kinds of ablators have been tested with this new commissioned platform.

We have verify that laminated ablator capsules performs similarly as uniformly doped capsules, as expected from homogenization principle. CHGe and CHSi ablators were compared for 2 doping fraction (2 and 4 % atomic) and 2 different drive conditions. The resulting implosion velocities span the range in between 150 and 300 $km.s^{-1}$. The general tendency is that the simulated trajectories converges slightly more early than the experimental radiographs, which is consistent with the late bang times ( 200 to 300 ps) measured in rugby implosion experiments on OMEGA. However the final limb velocity does not show an important velocity deficit as observed in NIF implosions [22]. CHSi is performing slightly better than CHGe, as found at NIF scale [22]. This rich implosion database will be used to benchmark FCI2 simulations of germanium and silicon doped ablators and to test the rocket model [24] in the high ablation regime limit.

## Acknowledgments
We would like to thank the staff of the OMEGA Laser Facility for their help in carrying these experiments, as well as the target fabrication and assembly groups at CEA Valduc and CEA CESTA.

## References
[1] Atzeni S and Meyer-Ter-Vehn J 2004 *The Physics of Inertial Confinement Fusion* (Clarendon Press, Oxford)
[2] Lindl J 1998 *Inertial Confinement Fusion* (Springer, New-York)
[3] Meezan N B *et al.* 2013 *Physics of Plasmas* **20** 056311
[4] Olson R E *et al.* 2011 *Physics of Plasmas* **18** 032706
[5] Wilson D *et al.* 1998 *Physics of Plasmas* **5**
[6] Ross J *et al.* 2013 *this conference*
[7] Schein J *et al.* 2007 *Phys. Rev. Lett.* **98** 175003
[8] Vandenboomgaerde M *et al.* 2007 *Phys. Rev. Lett.* **99**(6) 065004
[9] Amendt P *et al.* 2008 *Physics of Plasmas* **15** 012702
[10] Malinie G *et al.* 2011 *Bull. Am. Phys. Soc.* **56**
[11] Laffite S and Loiseau P 2010 *Physics of Plasmas* **17** 102704
[12] Philippe F *et al.* 2010 *Phys. Rev. Lett.* **104**(3) 035004
[13] Robey H F *et al.* 2010 *Physics of Plasmas* **17** 056313
[14] Casner A *et al.* 2012 *EPJ Web of Conference* **in press**
[15] Landen O L *et al.* 2012 *Plasma Physics and Controlled Fusion* **54** 124026
[16] Amendt P *et al.* 2013 *this conference*
[17] Hicks D G *et al.* 2010 *Physics of Plasmas* **17** 102703
[18] Schurtz G P *et al.* 2000 *Phys. Plasmas* **7** 4238–4249
[19] Busquet M 1993 *Physics of Fluids B: Plasma Physics* **5** 4191–4206
[20] Masse L 2007 *Phys. Rev. Lett.* **98**(24) 245001
[21] Masse L *et al.* 2011 *Phys. Rev. E* **83** 055401
[22] Hicks D G *et al.* 2012 *Physics of Plasmas* **19** 122702
[23] Casner A *et al.* 2013 *to be submitted*
[24] Saillard Y 2006 *Nucl. Fusion* **46** 1017

Journal of Physics: Conference Series **688** (2016) 012009 doi:10.1088/1742-6596/688/1/012009

# Simulations of the energy loss of ions at the stopping-power maximum in a laser-induced plasma

**W. Cayzac[1,2], A. Frank[3], A. Ortner[2], V. Bagnoud[3,4], M.M. Basko[5], S. Bedacht[2], A. Blažević[3,4], O. Deppert[2], D.O. Gericke[6], L. Hallo[7], A. Knetsch[2], D. Kraus[4], G. Malka[1], K. Pépitone[7], G. Schaumann[2], T. Schlegel[3], D. Schumacher[4], An. Tauschwitz[8,9], J. Vorberger[10], F. Wagner[2], and M. Roth[2]**

[1]Univ. Bordeaux-CEA-CNRS CELIA UMR 5107, Talence, France
[2]Institut für Kernphysik, Technische Universität Darmstadt, Darmstadt, Germany
[3]Helmholtzinstitut Jena, Jena, Germany
[4]GSI Helmholtzzentrum für Schwerionenforschung, Darmstadt, Germany
[5]Keldysh Institute of Applied Mathematics, Moscow, Russia
[6]University of Warwick, Coventry, United Kingdom
[7]CEA-CESTA, Le Barp, France
[8]Goethe-Universität Frankfurt, Frankfurt/Main, Germany
[9]Helmholtz International Center for FAIR (HIC for FAIR), Frankfurt/Main, Germany
[10]MPI for Physics of Complex Systems, Dresden, Germany

E-mail: w.cayzac@gsi.de

**Abstract.** Simulations have been performed to study the energy loss of carbon ions in a hot, laser-generated plasma in the velocity region of the stopping-power maximum. In this parameter range, discrepancies of up to 30 % exist between the various stopping theories and hardly any experimental data are available. The considered plasma, created by irradiating a thin carbon foil with two high-energy laser beams, is fully-ionized with a temperature of nearly 200 eV. To study the interaction at the maximum stopping power, Monte-Carlo calculations of the ion charge state in the plasma are carried out at a projectile energy of 0.5 MeV per nucleon. The predictions of various stopping-power theories are compared and experimental campaigns are planned for a first-time theory benchmarking in this low-velocity range.

## 1. Introduction

Ion stopping in plasmas is a central issue of high-energy-density physics and inertial confinement fusion. Here, not only is the precise description of the ion-plasma interaction important for all ion-based fusion concepts, as heavy-ion fusion or ion-driven fast ignition. Even more importantly, it is also essential for the understanding of alpha-particle heating in the burning DT fuel that determines the yield of a specific fusion reactor to a large extent.

When the velocity of the projectiles $v_p$ is significantly higher than the thermal velocity of the plasma electrons $v_{th}$, the stopping power (energy deposited per unit length) of the ions in the plasma can be described by perturbative approaches based on a first Born approximation. Examples are the well-known Bethe formula, or the standard stopping model (SSM) for intermediate projectile velocities [1]. Until now most stopping experiments were carried out at $v_p \gg v_{th}$ and can be interpreted within the SSM frame.

Journal of Physics: Conference Series **688** (2016) 012009 doi:10.1088/1742-6596/688/1/012009

Theoretical difficulties arise however for $v_p \approx v_{th}$, where the stopping power reaches its maximum. For such velocities, the ion-plasma interaction involves both strong collisions and collective plasma excitations, and the coupling between the projectile and the plasma is maximal. Here, large discrepancies appear between the various theories, reaching up to 30-50 % between perturbative predictions and data from nonperturbative approaches like the T-Matrix or the combined schemes [2, 3].
By now the scarce available experimental data [4] have not enabled to benchmark the theoretical predictions. In this work, simulations have been performed to study the energy loss at maximum stopping power, in a hot laser-induced plasma, as a preparation for an upcoming experimental benchmarking of the stopping theories.

## 2. Theoretical considerations

The considered setup enables to study the ion-plasma interaction in reproducible and well-controlled conditions, while simulation codes permit to access the plasma and projectile parameters at all times.

### 2.1. *Plasma target*

The plasma target is identical to the one employed at GSI in Darmstadt as described in Ref. [5]: a 100 $\mu$g/cm$^2$ planar graphite foil is irradiated from both sides by laser beams at 532 nm wavelength, 30 J energy, 7 ns pulse length at FWHM and 1 mm focus spot diameter obtained by using random phase plates for spatial beam smoothing. In this way, a hot carbon plasma with an electron density $n_e \leq 10^{21}$ cm$^{-3}$ and an electron temperature $T_e \leq 200$ eV is generated. This plasma is (i) fully-ionized by the time the laser irradiation ends, (ii) precisely reproducible and quasi-uniform in the direction transversal to the ion axis owing to the used irradiation scheme and (iii) fully characterized via the combined data from hydrodynamic simulations obtained with the two-dimensional RALEF2D code [6] and from time-resolved multi-frame interferometry

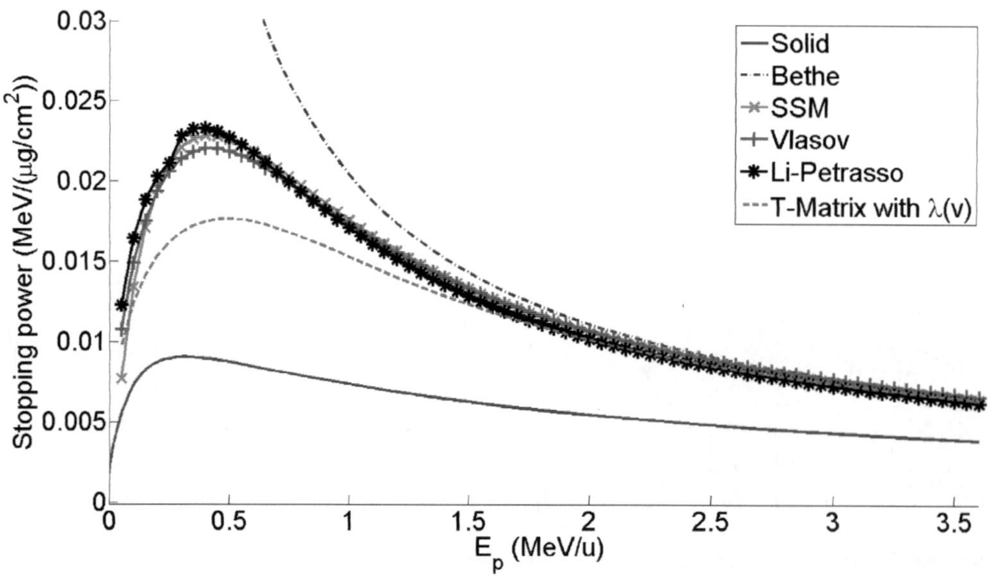


**Figure 1.** Comparison of the stopping power of $C^{6+}$ ions in a fully ionized plasma with a density of $5{\times}10^{21}$ cm$^{-3}$ and a temperature of 150 eV as predicted by different theories as a function of the projectile energy. The stopping power in solid matter is shown for comparison.

measurements of the plasma electron density [7]. Therefore, this plasma target constitutes a well-suited test bed for precise ion energy-loss measurements and studies.

### *2.2. Projectile ions*
In order to fix the charge-state of the projectiles in the plasma and, thus, to simplify the interpretation of the energy-loss results, carbon was chosen as projectile ion species as these ions are expected to be nearly fully stripped in a fully ionized carbon plasma. This was predicted by using a Monte Carlo code describing the charge transfer processes (ionization and recombination) of the projectile ions in the plasma [8] and showed that more than 90 % of the beam ions are fully ionised in the hot plasma. The required projectile energy to access $v_p \approx v_{th}$ for $T_e = 200\,\mathrm{eV}$ is $0.5\,\mathrm{MeV/u}$. The Coulomb parameter of the interaction for $C^{6+}$ ions at $0.5\,\mathrm{MeV/u}$ energy is $\eta = Z_p e^2/\hbar v_r > 1$ with $Z_p$ being the projectile charge and $v_r$ the relative projectile-electron velocity. This value suggests that the perturbative approaches may be inadequate.

### *2.3. Stopping predictions*
Stopping-power calculations are shown in Fig. 1 as a function of the projectile energy for several theories. The predictions of three perturbative theories, based on a first Born approximation, are shown: (i) the SSM, relying on a modified Bethe formula with the Bloch, Barkas and Chandrasekhar corrections [1], (ii) the dielectric theory obtained from the linearized Vlasov-Poisson equations [9], describing the stopping power resulting from large impact-parameter events in a collisionless plasma, and (iii) the Li-Petrasso stopping model, structurally similar to the SSM but including higher-order collisions [10]. Here, these perturbative approaches appear very consistent with each other. Their common high-velocity limit, the well-known Bethe formula, significantly overevaluates the stopping-power maximum. However, the nonperturbative T-Matrix approach [11], describing binary collisions with a simple model for dynamic screening, predicts a 30 % smaller stopping power than the perturbation approaches.

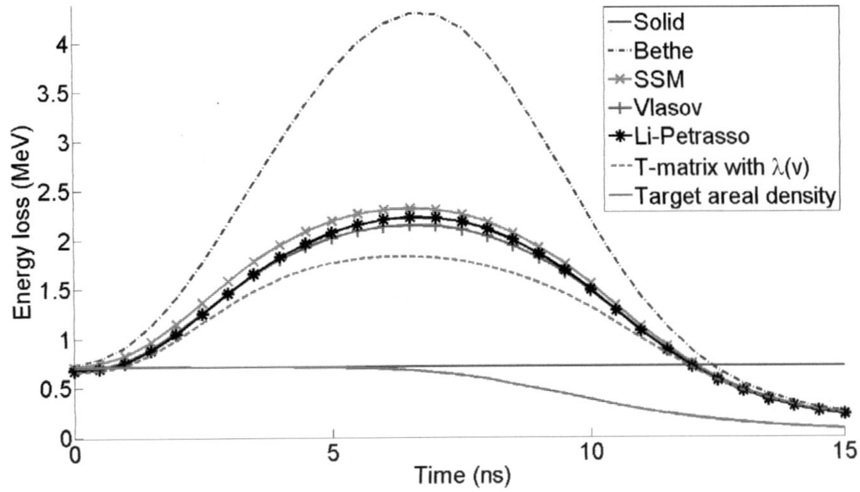


**Figure 2.** Comparison of the energy-loss predictions in the studied configuration. The time 0 marks the start of the target heating. The energy loss in solid matter is shown for comparison, as well as the target areal density, normalized to the latter, as indicator of the plasma expansion.

The resulting energy loss in the considered beam-plasma configuration was obtained by combining the simulated plasma parameters, taking both the longitudinal and transversal parameter gradient profiles into account, with the corresponding charge-state distributions

of the ion beam in plasma. It is shown in Fig. 2 over the first 15 nanoseconds of the interaction. Uncertainties on the simulated plasma parameters that arise from the comparison with interferometry data [5] lead to a 5 % error on the calculated energy loss.
As expected, the energy loss increases in plasma due to the more efficient energy transfer to free electrons and the increase in the projectile charge state. A maximum is reached when the plasma becomes fully ionized at around 7 ns. Later, the energy loss decreases due to the starting three-dimensional plasma expansion, characterized by the decrease in the target areal density. Between 6 and 13 ns, where the temperature is highest, $v_p \approx v_{th}$ and discrepancies between the theories similar to those of Fig. 1 appear. In particular, the optimal energy resolution required in the experiment to benchmark the theories can be determined as 100 keV.

## 3. Planned experiment

The experiment is planned at GSI-Darmstadt, where two high-energy lasers can be used in combination with a high-frequency pulsed ion beam. To reach a projectile energy of 0.5 MeV/u, a beam of 3.6 MeV/u carbon ions is decelerated by passing through a 45 $\mu$m thick carbon foil as shown in Fig. 3. The ions are detected after a time-of-flight (TOF) distance of 0.5 m allowing an energy resolution of 60 keV with a Chemical Vapour Deposition (CVD)-diamond device as TOF-detector [12], which enables to distinguish between different stopping theories.

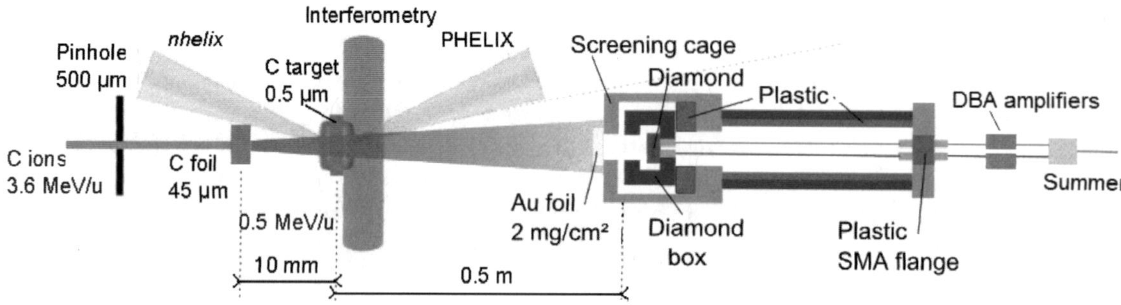


**Figure 3.** Schematics of the experimental setup.

## Acknowledgments

This work was supported by the Fonds de coopération interrégionale between the Région Aquitaine and the Land Hessen as well as by Bundesministerium für Bildung und Forschung (BMBF) and the Helmholtz International Center for FAIR (HIC4FAIR).

## References

[1] Deutsch D, Maynard G *et al.* *The open plasma physics journal* **3** 88 – 115
[2] Gericke D O and Schlanges M 1999 *Physical Review E* **60** 904–910
[3] Gericke D O and Schlanges M 2003 *Physical Review E* **67** 037401
[4] Hicks D G, Li C K *et al.* 2000 *Physics of Plasmas* **7** 5106–5117
[5] Frank A, Blažević A *et al.* 2013 *Physical Review Letters* **110** 115001
[6] Tauschwitz A, Basko M *et al.* 2013 *High Energy Density Physics* **9** 158–166
[7] Börner M, Fils J, Frank A *et al.* 2012 *Review of Scientific Instruments* **83** 043501
[8] Frank A, Blažević A *et al.* 2010 *Physical Review E* **81** 026401–+
[9] Peter T and Meyer-Ter-Vehn J 1991 *Physical Review A* **43** 1998–2030
[10] Li C K and Petrasso R D 1993 *Physical Review Letters* **70** 3059–3062
[11] Gericke D O 2002 *Laser and Particle Beams* **20** 471–474
[12] Cayzac W, Frank A *et al.* 2013 *Review of Scientific Instruments* **84** 043301

# Progress Towards a Laser Produced Relativistic Electron-Positron Pair Plasma

Hui Chen, J. Bonlie, R. Cauble, F. Fiuza, W. Goldstein, A. Hazi, C. Keane, A. Link, E. Marley, S. R. Nagel, J. Park, R. Shepherd, G. J. Williams

*Lawrence Livermore National Laboratory, Livermore, CA 94551, USA*

D. D. Meyerhofer, G. Fiksel, D. Barnak, P. Y. Chang

*LLE, University of Rochester, Rochester, NY 14623, USA*

M. Nakai, Y. Arikawa, H. Azechi, S. Fujioka, S. Kojima, N. Miyanaga, T. Morita, T. Nagai, H. Nishimura, T. Ozaki, Y. Sakawa, H. Takabe, Z. Zhang

*ILE, Osaka University, 2-6 Yamadaoka, Suita, Osaka 565-0871, Japan*

S. Kerr, R. Fedosejevs

*University of Alberta, Canada*

Y. Sentoku

*University of Nevada, Reno, Nevada 89557, USA*

M. P. Hill, D. J. Hoarty, L. M. R. Hobbs, S. F. James

*Directorate of Science and Technology, AWE plc, Reading, RG7 4PR, UK*

chen33@llnl.gov

**Abstract**. A set of experiments has been performed exploring unique characteristics of pair jets and plasmas at several energetic short-pulse laser facilities including Titan at Livermore and OMEGA EP in Rochester, as well as the Osaka LFEX and AWE Orion lasers. New results are summarized, including positron beam emittance, scaling of pair production *vs.* laser energy, and initial results on the pair jet collimation using electromagnetic fields.

## 1. Introduction

Relativistic electron-positron pair plasmas and jets are believed to be abundant in many astrophysical objects and are invoked to explain energetic phenomena related to Gamma Ray Bursts, Active Galactic Nuclei, and Black Holes [1-9]. Making a relativistic electron-positron (antimatter) plasma in the laboratory that is astrophysical-relevant has been challenging due to the difficulties associated with producing high density of the pairs, the short lifetime of positrons, and their highly relativistic energies. As a result, the experimental platforms capable of simulating astrophysical conditions have so far been absent.

In the past few years, groundbreaking steps have been made using high-energy ultra-short laser pulses to make large numbers of positrons in a small volume [10]. It was found that laser-produced positrons have several characteristics that may prove essential for making a relativistic pair plasma.

The first is that intense lasers can make a very large number of positrons ($10^{10}$ to $10^{12}$ per shot) in a short time (10 – 100 ps) [10]. This feature, in combination with the small volume (~mm$^3$) these positrons occupy, leads to a high density of positrons, even though only for a very short time. The second characteristic is that the target sheath field can accelerate these positrons to 10s of MeV, enabling positrons to be made and accelerated to the relativistic regime in one integrated process. The third characteristic is that MeV electrons and positrons produced from the laser-target interaction form overlapping jets behind the target, allowing much higher pair density to be achieved than it would be if the pairs were distributed isotropically.

Following these initial experiments that produced the high-flux jets of positrons with temperatures of MeV, we performed a new set of experiments focusing on understanding the physics of positron production, beam emittance of the jets, scaling with laser parameters and collimation. The new results show that the emittance of laser-produced positrons is 100 – 500 mm-mrad, comparable to that obtained at the Stanford Linear Collider [11]. The laser contrast was found to have a large effect on the positron yield [12]. The nonlinear scaling of positrons as a function of laser energy is evident when the laser energy is greater than 1000 J [13]. With plasma collimation, we hope to eventually use the multi-kilojoule, short-pulse laser systems worldwide, in combination with more advanced target designs, to create the first relativistic high-density pair plasmas in the laboratory - a completely novel system enabling detailed study of some of the most exotic and energetic systems in the universe [1-9].

## 2. Experiments and results

The new experiments were performed on the Titan and OMEGA EP lasers in the USA, as well as at the Osaka LFEX in Japan and AWE Orion laser in the UK. For the experiments discussed here, Titan produced 100-350 J in 1-10 ps, with 5-10 shots per day. At OMEGA EP, the laser energy was up to 1.3 kJ at pulse duration of 1-10 ps, with up to 14 shots a day. We used two beams of LFEX lasers at 1 ps, with up to 1.5 kJ per beam. One of the two ORION short pulse lasers was used with energy of 100-500 J in 0.5 ps. All experiment used lasers at fundamental frequency. The experiments were performed over a large range of laser parameters including the intensity ($10^{18}$ – $10^{21}$ Watts/cm$^2$) and contrast ($10^6$ – $10^9$). The data was grouped and compared under similar laser and target conditions.

### *2.1. Energy conversion*

Using the LFEX laser at the Institute of Laser Engineering at Osaka University, we quantitatively investigated the production of high-energy particles and photons in the relativistic regime by observing electrons, positrons, x-rays, and photoneutrons simultaneously. The diagnostics for each observable were independently calibrated to determine the fraction of the laser energy converted to electrons, gammas and positrons [12]. The laser-to-electron conversion efficiencies are obtained from Laue diagnostic measurements of gold $K_\alpha$ and $K_\beta$ lines. The total gamma number and the energy transfer efficiency were inferred. The positron numbers and energies were measured directly using a magnetic spectrometer. The conversion efficiency from laser to electrons was 10-15%, laser to gammas (energy > 1 MeV) was ~ 0.2%, and laser to positrons ~ 0.002%. These measurements are consistent with the positron production occurring through Bethe-Heitler process [14], confirming our understanding of the laser pair generation physics.

### *2.2. Positron beam emittance*

We performed the first measurements of the emittance of intense laser-produced positron beams on the Titan and OMEGA EP lasers [11]. Emittance is a basic parameter that describes the quality of particle beams and can be derived through measurements of beam divergence and source size. We measured these two parameters for different positron energies under various laser conditions. The divergence angle (full-width at half-maximum) was found to be 20 ± 6 degrees for laser energy of ~300 J at 10 ps (the positrons distribution was peaked at ~12 MeV); ~ 22 degrees for ~850 J, at 10 ps pulse duration, where the peak of the positron energy distribution shifted to ~18 MeV due to the higher laser energy. We found that both the positron beam divergence and source size varied as a function of positron beam energy at the peak of its distribution. The source size varied between 800 and 400 μm for peak energies between 6.5 MeV and 16 MeV, respectively.

In summary, the laser-produced positrons have a geometric emittance between 100 - 500 mm-mrad, comparable to the positron sources used at existing accelerators. With $10^{10}$-$10^{12}$ positrons per bunch, this low-emittance beam, which is quasi-monoenergetic in the energy range of 5 – 20 MeV, may be useful as an alternative positron source for future accelerators.

### *2.3. Scaling*

Preliminary experimental results from OMEGA EP show that the positron yield as a function of laser energy is not linear (Fig. 1) [13]. This is surprising, since, assuming that the fraction of absorbed laser energy is constant, one would expect the positron yield to be linearly proportional to the laser energy. Our result indicates that another physics parameter becomes important when the laser energy is higher than about 1 kJ. For example, simulations have shown that the electron reflux factor and the enhancement of the hot electron temperature appear to play a less important role at intensities in the range $10^{18}$ – $10^{19}$ W/cm$^2$, but can no longer be ignored when it is higher than $10^{19}$ W/cm$^2$, corresponding to laser energies higher than 1000 J.

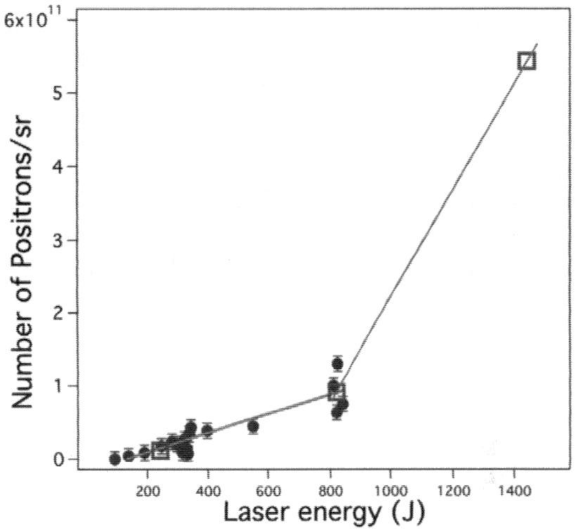


*Fig. 1: Number of positrons as a function of laser energy. Data shown are taken from two experiments (blue dots and green squares. The lines are drawn as guides. As a whole, the data cannot be well fit with a linear function.*

### *2.4. Initial pair jet collimation by external magnetic fields*

The ultimate goal is to confine the particles to make a relativistic charge-neutral electron-positron pair plasma. If successfully confined, the electron-positron plasma would offer a novel system to enable a detailed study of some of the most exotic and energetic systems in the universe. We performed initial experiments on the OMEGA EP laser system (at ~1 kJ over 10 ps), which were designed to collimate the positron jets produced in high intensity laser-plasma interactions. The targets were of 1-mm-thick gold. Previous experiments by this team have shown that quasi-monoenergetic relativistic positron jets with a beam divergence of 30° were formed during high-intensity interactions with thick gold targets [10]. The current experiments were designed to collimate the positron jet with an external magnetic lens. Using the recently developed magnetized-inertial fusion electrical delivery system (MIFEDS) [15], strong collimation of both positrons and electrons were observed. This results in a "near-pencil" beam with an equivalent beam divergence angle of 5°. The charge imbalance was reduced from ~100 (no collimation) to ~ 2.5 (with collimation), a significant step towards making a charge neutral electron-positron pair plasma in the laboratory. A jet of positrons and electrons was emitted from the rear side of the target. Between the target and the detector, MIFEDS coils produced peak magnetic field of about 7 Tesla.

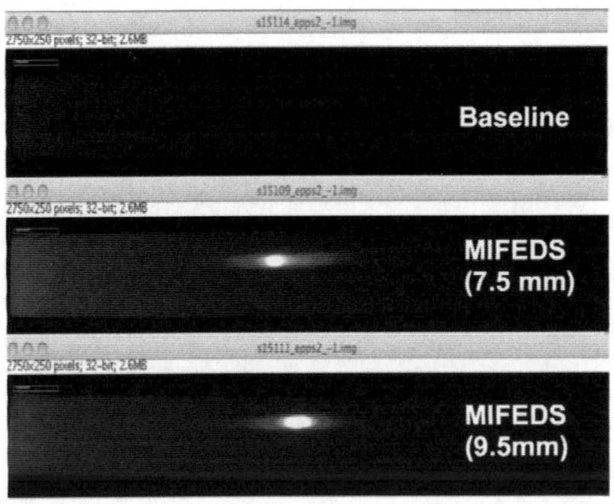


*Fig. 2: Raw image of positron data from three shots. One shot (top image) had no MIFEDS coil and the other two (middle and bottom images) with the coil at different location. The positron distribution peaked at ~13 MeV. The signal from the collimated beam is 30-40 times higher.*

The collimated beams were measured with a magnetic electron-positron spectrometer. Without the external magnetic field, the peak densities are about $10^{13}$ and $10^{15}$ $cm^{-3}$ for positrons and electrons, respectively [10]. With the external B-field applied, a factor of ~40 increase in the peak positron signal was observed (Fig. 2).

### 3. Future plans

Building upon the progress as described in this paper, our next step is to attempt trapping the particles in order to create a pair plasma. This will be a very challenging task as the particle energies are relativistic and the particle flux is orders-of-magnitudes higher than that obtained from conventional sources such as radioactive isotopes. However, this is an important scientific topic that needs to be addressed as highlighted in the recently published report on "Basic Research Needs for High Energy Density Laboratory Physics" by the US Department of Energy [17]: *"if and when we can produce them (relativistic pair plasma) in the laboratory, we would expect novel dynamics and matter-antimatter elementary thermal processes. The extreme electromagnetic fields and particle energies now accessible approach the edge of understood physics, with the forces acting on electrons exceeding past experimental and even theoretical considerations, including gravity."*

We gratefully acknowledge the facility support from Jupiter Laser Facility, Omega EP, Osaka LFEX and AWE Orion during the experiments. HC wishes to thank Henry Shaw for his encouragement and support. This work was performed under the auspices of the US DOE by Lawrence Livermore National Laboratory under Contract DE-AC52-07NA27344, and funded by the LDRD (12-ERD-062) program.

### References

[1]. E. Fermi, Phys. Rev. **75**, 1169 (1949).
[2]. A. R. Bell, Monthly Notices of the Royal Astronomical Society, **182**, 147 (1978).
[3]. R. Blandford, and J. Ostriker, Astrophysics. J, 221, L29 (1978); R. Blandford and D. Eichler, Phys. Rep. 154, 1 (1987).
[4]. P. Meszaros, Annal Review of Astronomy and Astrophysics **40**, 137 (2002).
[5]. F. Mirabel and L. F. Rodriguez, Review of Astronomy and Astrophysics **37**, 409-43 (1999).
[6]. A. Spitkovsky, ApJ Lett., 673, L39 (2008); A. Spitkovsky, AIP Conf. Proc., 801, 345 (2005); L. Sironi and A. Spitkovsky, ApJ, 698, 1523 (2009); L. Sironi and A. Spitkovsky, ApJ, 726, 75 (2011).
[7]. J. Wardle, *et al.*, Nature **395**, 457 (1998)
[8]. G. Weidenspointner, et al, Nature **451**, 159 (2008).
[9]. L. O. Silva, R. A. Fonsea, J. W. Tonge, J. M. Dawson, W. B. Mori, and M. V. Medvedev, Astrophysics. J, **596**, L121 (2003).
[10]. Hui Chen, *et al.*, Phy. Rev. Lett. **102**,105001 (2009); Chen, et al., Phys. Plasmas **16**, 122702 (2009); Chen *et al.*, Phy. Rev. Lett. **105**, 015003 (2010); Chen, *et al.*, High Energy Density Physics, **7**, 225 (2011)
[11]. Hui Chen, *et al.* Phys. Plasmas **20**, 012507 (2013)
[12]. Hui Chen, M. Nakai, Y. Sentoku, Y. Arikawa, H. Azechi, S. Fujioka, C. Keane, S. Kojima, W. Goldstein, B. R. Maddox, N. Miyanaga, T. Morita, T. Nagai, H. Nishimura, T. Ozaki, J. Park, Y. Sakawa, H. Takabe, G. Williams, Z. Zhang, New J. Phys. **15** (2013) 065010
[13]. Hui Chen, D. D. Meyerhofer, Y. Sentoku, *et al.*, in preparation.
[14]. W. Heitler, *The Quantum Theory of Radiation*, Clarendon, Oxford (1954); K. Nakashima and H. Takabe, Phys. Plasmas **9**, 1505 (2002).
[15]. O. Gotchev, *et al.*, Rev. Sci. Instrum. **80**, 043504 (2009)
[16]. R. Rosner, *et al.*, "Research Opportunities in Plasma Astrophysics" (2010); Also see Betti, R., *et al.* "Advancing the Science of High Energy Density Laboratory Plasmas" (2009)

# Progress in modelling ignition implosion experiments on the National Ignition Facility

**D S Clark[1], D C Eder, S W Haan, B A Hammel, D E Hinkel, O S Jones, M M Marinak, J L Milovich, P K Patel, J D Salmonson, S M Sepke, C A Thomas, and R P J Town**

Lawrence Livermore National Laboratory, 7000 East Ave., Livermore, CA 94550 USA

E-mail: clark90@llnl.gov

**Abstract**. The recently completed National Ignition Campaign on the National Ignition Facility showed significant discrepancies between 2-D simulations predictions of implosion performance and experimentally measured performance, particularly in thermonuclear yield. This discrepancy between simulation and observation persisted despite concerted efforts to include all of the known sources of implosion degradation within a reasonable 2-D simulation model, *e.g.*, using measured surface imperfections and radiation drives adjusted to reproduce observed implosion trajectories. Since this simulation study was undertaken, more recent experiments have brought to light several effects that can significantly impact implosion performance, in particular large inflight long-wavelength shell asymmetries and larger than expected perturbations seeded by the capsule support tent. These effects are now being included in the simulation model and show improved agreement with observation. In addition, full-capsule 3-D simulations with resolution adequate to model the dominant unstable hydrodynamic modes are being run and show further improvements in agreement with experiment.

## 1. Introduction

The National Ignition Campaign (NIC) [1, 2] on the National Ignition Facility (NIF) [3] was completed in September 2012. While substantial progress was made in experiments on NIF, the NIC ultimately fell short of its goal of demonstrating fusion ignition by indirect drive inertial confinement fusion [4, 5]. Implosions driven by record-breaking laser powers and energies on NIF and tuned with unprecedented precision and control achieved implosion velocities of 350 km/s, deuterium-tritium (DT) fuel densities of greater than 800 g/cm$^3$, and hot spot pressures greater than 100 billion atmospheres. Nevertheless, DT fusion yields did not exceed ~ $9\times10^{14}$ neutrons, well below the ~ $5\times10^{16}$ threshold of unambiguous fusion ignition. Additionally, concentrated simulation efforts, analogous to those used in designing ignition experiments for NIF, consistently over-predicted the yields from NIC experiments [6]. This discrepancy persisted despite concerted efforts to include all of the known sources of implosion degradation present during NIF implosions in sophisticated 2-D as well as 3-D simulations [7]. Despite reproducing the observed implosion characteristics reasonably

[1] To whom any correspondence should be addressed.

well (velocity, bang time, shell remaining mass) and approximately matching the measured burn-averaged ion temperatures, fuel areal densities, and hot spot x-ray image sizes, neutron yields averaged an order of magnitude higher in simulations than observed in experiment.

Since these simulation studies were completed, experiments have continued on NIF and have identified several important effects—absent in the previous simulations—that can likely resolve at least some of the large discrepancies between simulated and experimental yields. Foremost among these results were the 2-D ConA experiments that used germanium backlighting to image the shape of imploding shells inflight on NIF [8]. Time gated images from these 2-D ConA experiments revealed significantly larger low-mode asymmetries in NIF implosions at a converged radius of ~ 200 μm. In particular, larger Legendre mode four ("$P_4$") asymmetries were observed than had been expected or were included in previous simulations. The presence of these low-mode asymmetries inflight was also corroborated by the observation of large low-mode asymmetries in neutron down scattering signals from the burning hot spot. While these two indications of low-mode asymmetries are not easily connected, together they do strongly suggest the presence of larger asymmetries in the range of $P_2 - P_4$ than previously modeled.

Another important observation from the 2-D ConA experiments was the much larger role of the capsule support tent in seeding perturbation growth at the shell ablation front than had been expected. 2-D ConA images from experiments using various tent thicknesses (110 nm, 45 nm, and 15 nm) showed a clear "scar" across the face of the image that scaled weakly with tent thickness and vanished when the capsule was supported by a stalk instead of a tent. According to previous simulation-based estimates, the tent should have produced an essentially invisible perturbation in the ConA images. Such a large, coherent perturbation, again unanticipated and absent from previous simulations, can certainly be expected to impact implosion performance.

An additional effect omitted in previous systematic simulations of NIF implosion performance is the presence of deeply penetrating mix contaminating the DT hot spot with higher *Z* ablator material. While the effects of this "deep mix" have long been appreciated based on simulations [9, 10] and significant efforts have been made to measure it experimentally [11], sufficiently detailed shot-by-shot data on deep mix was only just becoming available at the time previous simulation studies were being run. Some scoping studies [7] and attempts to connect the inferred mix from NIF experiments to target quality were made [12], but only recently has a systematic analysis based on Ross pair x-ray images of NIF implosions produced a fairly comprehensive database on deep mix in NIC implosions [13]. These mix estimates, based on a model of the ratio of x-ray to neutron yield from the hot spot, still have considerable uncertainty and the origins of the large amounts of mix inferred on many NIF shots remains unexplained, but the presence of this mix is a clearly non-negligible effect in NIF implosions that must be included in any plausible simulation model.

This paper summarizes the results of a revised post-shot simulation of NIF shot N120321 incorporating the effects enumerated above. This shot was one of the best performing shots fired during the NIC featuring the highest combination of compression and yield, and as such has been the subject of much modeling work. While agreement between simulation and experiment is substantially improved by the inclusion of the effects listed above, some discrepancies remain indicating that some improvements in simulation technique are still required.

## 2. Revised 2-D post-shot simulations

With the identification of the effects described above, a sequence of 2-D simulation was undertaken to add these effect incrementally and assess their contributions to the degradation of implosion performance. Table 1 summarizes these results in the form of several simulated observables to compare with experiment. The last column in the table gives the experimental values for shot N120321. The first entry in the table lists the simulated observables of a 1-D simulation for which the x-ray drive on the capsule has been adjusted to match the measured shock timing and implosion trajectory from companion VISAR and 1-D ConA shots. The details of this technique have been described elsewhere [7]. The subsequent columns show the results of progressively adding the

Legendre $P_2$ and $P_4$ drive asymmetries tuned to match the inflight asymmetries fromß 2-D ConA experiments, adding a surrogate perturbation adjusted to model the observed effect of the tent in 2-D ConA radiographs, adding a similar surrogate perturbation to model the effect of the fill tube attached to the capsule, adding the effect of surface roughness, adding 200 ng of ablator material into the hot spot to model the effect of deep mix, and finally a Legendre $P_1$ ice layer asymmetry of 1.6 μm particular to this shot. All of the 2-D simulations used zoning sufficient to resolve Legendre mode numbers up to 100 and for the cases including roughness used the as measured ablator surface and ice surface roughness for this shot. Based on Ross pair imaging, the inferred mix mass for this shot is 409 ± 143 ng [14], while estimates based on linear regime growth factors suggest 200 – 500 ng could have been injected. Given the uncertainties in all of these estimates and the crudeness of the simulation models treatment of this effect, 200 ng was deemed a reasonable estimate.

**Table 1.** Summary of simulated observables for shot N120321 including a hierarchy of perturbation effects. Effects add sequentially from left to right from 1-D to 2-D to 3-D simulations. The right most column lists the experimentally measured values.

| | 1-D | $P_2$ & $P_4$ | tent | fill tube | rough-ness | 200 ng mix | ice $P_1$ | 3-D | expt. |
|---|---|---|---|---|---|---|---|---|---|
| bang time (ns) | 22.93 | 22.89 | 22.88 | 22.86 | 22.86 | 22.85 | 22.86 | 22.83 | 22.83 |
| burn width (ns) | 121 | 119 | 119 | 118 | 112 | 118 | 106 | 116 | 158 |
| x-ray $P_0$ (μm) | 25.8 | 24.3 | 22.9 | 22.7 | 21.9 | 19.1 | 18.8 | 18.8 | 17.6 |
| x-ray $P_2$ (μm) | — | 0.34 | 0.03 | 0.00 | - 0.05 | - 0.01 | - 0.02 | 0.07 | 0.06 |
| DSR (%) | 6.08 | 6.27 | 6.33 | 6.31 | 6.31 | 7.70 | 7.74 | 7.01 | 6.26 |
| $T_{ion}$ (keV) | 4.49 | 3.88 | 3.73 | 3.09 | 3.20 | 2.16 | 2.19 | 2.48 | 3.07 |
| neutrons ($10^{14}$) | 313 | 88.6 | 55.0 | 28.0 | 29.4 | 9.2 | 9.5 | 7.7 | 4.2 |

In progressing from a 1-D simulation to the 2-D simulation including all of the above effects, a substantial drop in yield is apparent with the presence of the tent perturbation, the fill tube perturbation, and the mix mass being leading contributors. When all of the effects are included, the simulated yield has dropped from two orders of magnitude greater than experiment to within a factor of two. The inclusion of these effects also influences several of the other observables as listed in the table. Namely, the burn-averaged ion temperature drops with increasing perturbation to a value close to experiment, but then drops below the experimental value with the addition of deep mix. Simultaneously, the neutron down-scattered ratio (DSR), a measure of compression or $\rho R$, approximately matches the data in 2-D without mix, but is greater than the experimental value with mix. Both of these effects are suggestive that higher than simulated levels of instability growth that could break up the imploding shell and decrease the DSR while increasing the ion temperature via Doppler broadening could be present.

## 3. Progress in 3-D simulations

Also included in Table 1 are the results of a 3-D full sphere simulation of N120321. This simulation is analogous to the 2-D simulations in resolution and the various perturbation sources included. Since this simulation is fully 3-D, no approximations need be made in initializing capsule surface perturbations and the actual measured surface features from pre-shot imaging of the ablator are used. Most importantly, 3-D simulations relax the artificial symmetry constraints present in 2-D axi-symmetric simulations and allow essentially arbitrary flows through the center of the capsule implosion. As such, these simulations give the most complete and faithful representation of the conditions present in the experiment.

As shown in the table, simulating in 3-D further degrades the yield toward the experimental value. The 3-D simulation also improves agreement with the measured ion temperature illustrating the importance of Doppler broadening due to flows within the 3-D hot spot. The agreement with the experimental DSR value is somewhat improved as well. These continued discrepancies, however, again suggest that greater instability growth is occurring in experiment than is currently simulated leading to greater breakup of the imploded shell than currently modeled.

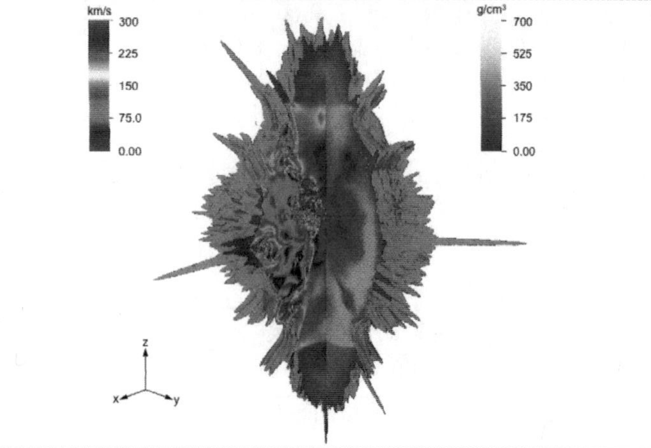


**Figure 5.** Rendering of the fuel ablator interface at bang time (22.83 ns) from a 3-D simulation of NIF shot N120321. The color scales show fluid flow velocity (left) and density (right) on corresponding slices.

As an illustration of the 3-D character of these implosions, Fig. 1 shows a rendering of the fuel ablator interface and hot spot conditions from this simulation.

## 4. Conclusions

While reasonable agreement can be shown between simulations and experimental results for the shot considered here, several caveats deserve to be borne in mind. Foremost, this analysis has considered only one of the three dozen ignition experiments conducted during the NIC. Performing a similar analysis for a significant fraction of these experiments would be necessary to validate or refute the conclusions reached for N120321. More importantly, a number of shots during the NIC showed unexplainably large amounts of ablator mix into the hot spot. Similarly, the unexpectedly large perturbation seeded by the support tent cannot be explained based on current simulations. This effect, along with the unexplained mix mass for some shots, and the lingering discrepancies in the simulations discussed here are all suggestive of hydrodynamic instability growth rates, or the sources that initiate that growth, being larger than currently modeled. This hypothesis motivates direct measurement of these growth rates as is now being done via the Hydro. Growth Radiography (HGR) experiments on NIF [15] and further simulation and theoretical work to understand these discrepancies.

**Acknowledgements**
This work was performed under the auspices of the U.S. Department of Energy by Lawrence Livermore National Laboratory under Contract DE-AC52-07NA27344.

**References**

[1] Mackinnon A J, *et al.* 2012 *Phys. Rev. Lett.* **108** 215005
[2] Glenzer S H, *et al.* 2012 *Phys. Plasmas* **19** 056318
[3] Moses E I, Boyd R N, Remington B A, Keane C J, Al-Ayat R 2009 *Phys. Plasmas* **16** 041006
[4] Atzeni S and Meyer-ter-Vehn J 2004 *The Physics of Inertial Fusion* Clarendon, Oxford
[5] Lindl J D, *et al.* 2004 *Phys. Plasmas* **11** 339
[6] Jones O S, *et al.* 2012 *Phys. Plasmas* **16** 056302
[7] Clark D S, *et al.* 2013 *Phys. Plasmas* **20** 056318
[8] Rygg R, *et al. These proceedings*
[9] Hammel B A, *et al.* 2011 *Phys. Plasmas* **18** 056310
[10] Haan S W, *et al.* 2011*Phys. Plasmas* **18** 051001
[11] Regan S P, *et al.* 2013 *Phys. Rev. Lett.* **111** 045001
[12] Hinkel D E, *et al.* 2012 *Bull. Am. Phys. Soc.* **57** 240
[13] Ma T, *et al.* 2013 *Phys. Rev. Lett.* **111** 085004
[14] Patel P K 2013 *Private communication*
[15] Smalyuk V, *et al.* 2013 *These proceedings*

# Preparation of the high power laser system PETAL for experimental studies of inertial confinement fusion and high energy density states of matter

**E. d'Humières[1*], J. Caron[1], C. Perego[1], D. Raffestin[2], J.-L. Dubois[2], J. Baggio[2], A. Compant La Fontaine[3], S. Hulin[1], J.-E. Ducret[1], F. Lubrano[2], J.C. Gommé[2], J. Gazave[2], J. Ribolzi[2], J.-L. Feugeas[1], P. Nicolai[1], E. Lefebvre[3], V.T. Tikhonchuk[1] and D. Batani[1]**

1) Univ. Bordeaux - CNRS - CEA, CELIA, Talence 33405, France
2) CEA/DAM/CESTA, BP 12, Le Barp 33405, France
3) CEA/DAM/DIF, BP 1, Arpajon 91405, France

E-mail: dhumieres@celia.u-bordeaux1.fr

**Abstract.** The paper describes the preparation of the short-pulse high-energy laser PETAL that will be coupled to the French megajoule laser (LMJ) of CEA. The LMJ/PETAL facility will be opened to academic access for the international research community. In parallel diagnostics are being developed within the PETAL project and many physical problems are being addressed ranging from the study of the problems of radiation generation and activation issues to the problem of generation of large amplitude electromagnetic pulses.

## 1. Introduction

The high power laser system PETAL (http://petal.aquitaine.fr) is funded by the Regional Council of Aquitaine and French government and constructed by the CEA/CESTA, for academic studies of high energy density states of matter and inertial confinement fusion. It will be commissioned in 2015 and will be operated together with the Laser MégaJoule (LMJ) as a joint facility [1]. Significant theoretical, numerical, experimental and technical developments are still needed during the final period of PETAL construction and the starting phase of its exploitation. This period of 4 years, from 2013 to 2016 implies a tight and efficient collaboration between the scientific and technical staff of the academic groups and laboratories and the CEA engineers in order to attain the expected laser characteristics in a short time scale of two years and to assure all necessary conditions for efficient and secure operation of this unique installation. The results presented in this article were obtained in the context of this collaboration and are related to planning and realization of the first phase of the PETAL operation so that it will be compatible with the safety requirements, the characteristics of the diagnostic equipment planned in the PETAL+ project and with the planned regular operation of PETAL together with the LMJ as an international user facility. In Section 2, new results on the expected activation in the LMJ chamber induced in high power PETAL experiments by the radionuclides generated by energetic photons and protons are presented. In the first phase of the PETAL operation, the compatibility of PETAL experiments in the context of national nuclear activity regulations has been evaluated for two regimes of operation. In Section 3, new results on the studies of strong electromagnetic fields produced during the interaction of a high intensity laser with a solid target are presented. This effect will be of prime importance for the PETAL diagnostics and experimental equipments. A model of the target electric polarization induced by a short and intense laser pulse has been developed and verified in an experiment with a sub-picosecond laser pulse.

## 2. Radiation and Activation issues

Shooting the PETAL laser pulse on targets will induce the generation of large radiation fluxes, which in turn may generate radionuclides in the LMJ chamber by energetic photons and protons, i.e. activate the materials of the diagnostics and the chamber itself, causing issues both for radioprotection and nuclear safety, and for the correct operation of diagnostics. In the first phase of the PETAL operation, the compatibility of PETAL experiments with respect to national nuclear activity regulations has been evaluated for two regimes of operation. The characteristics of X-ray photons and protons originating from typical experiments that will be conducted with the PETAL laser were studied in order to estimate the activation coming from the interaction with matter of these particles, either originating directly from the interaction of the laser or from secondary interactions with the diagnostics instruments or the materials of the chamber wall [2]. Two dimensional (2D) Particle-In-Cell simulations using the code PICLS [3] were performed to obtain the electron and proton sources produced during the interaction of a PETAL pulse with a solid target. The maximum electron density used in the simulation is $300n_c$ with a 20 $\mu$m long preplasma (with an exponential shape, as deduced from the 2D hydrodynamic simulation). The target is 10 $\mu$m thick and is composed of protons, deuterons and electrons (it is equivalent to a CH target as collisions are not treated). The laser pulse has a Gaussian profile in time with a FWHM of 0.5 ps and also in space with an FWHM of 30 $\mu$m. The pulse is linearly polarized (p-polarization). The total number of protons per MeV is obtained using the conservative estimate that the length in the third dimension is of the order of the laser FWHM. Collisions are not taken into account as the high laser energy and the small target thickness will lead to a rapid heating of the target. 2D PIC simulation results cannot be taken as precise predictions for experiments but they nevertheless provide a good idea of the involved particle energies in a reasonably majoring fashion. Laser energy was varied from 0.2 to 3.5 kJ corresponding to laser intensities of $1.4 \times 10^{19}$ W/cm$^2$ to $2.5 \times 10^{20}$ W/cm$^2$. Examples of proton angular and energy distributions for the laser energy of 1 kJ are presented in Fig. 1. Unlike electrons, the proton angular distribution is strongly collimated. The emission angle decreases with increasing proton energy and is not very sensitive to the laser energy in the conditions studied here. The energy distribution can be approximated by an exponential function characterized by a certain temperature with a sharp cut-off. These results allow to use the proton characteristics directly for activation calculations and the electron characteristics for photon conversion calculations.

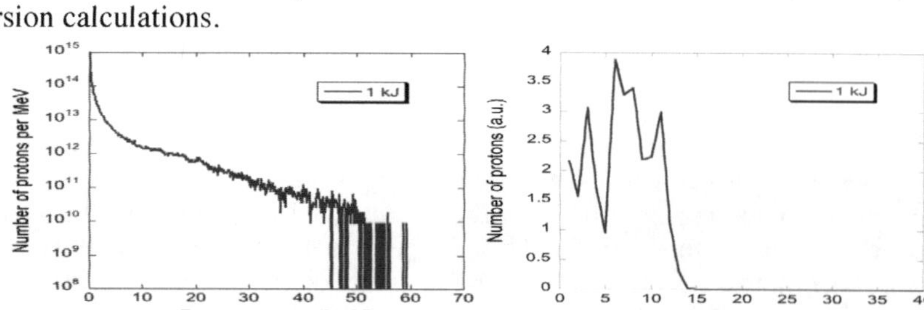


**Figure 1.** Energy (a) and angular (b) distributions of the protons simulated by the code PICLS for a total laser energy of 1 kJ.

Simulations of the electron transport and interaction with a tungsten target were performed using the Monte Carlo code GEANT4 [4]. The electrons produced with the PICLS simulations are used as initial parameters for the bremsstrahlung calculations. As the process of photon production is linear in the electron number, the input parameters of the GEANT4 simulations are the energy and angular distribution of the electrons. The characteristics of the incident electron beam obtained from the PIC simulations were the following: a Maxwellian distribution with the high energy cut-off depending on the considered laser energy and Gaussian angular and spatial distributions. In GEANT4 simulations we calculated the characteristics of exiting electrons, photons and positrons from a tungsten target. A significant fraction of incident electrons is reflected, and the majority of electrons penetrating the target are absorbed. Figure 2 shows an example of the photon energy spectra emitted from the back and the front target surface for the laser pulse energy of 1 kJ. The maximum photon energy is 43 and 59 MeV, respectively for the reflected and transmitted fractions. Similarly to the electron and the proton spectra, the photon energy distributions can be characterized by an exponential function with a sharp cut-off. The photon distributions obtained for laser energies up to 3.5 kJ have similar characteristics with more energetic photons, more photons at a given energy and more photons at a given angle being emitted when the laser energy is increased. The goal of this study was to estimate the activation induced by PETAL experiments in its early phase of operation. A complete calculation sequence consists in the simulation of the interaction of the PETAL laser with solid targets, leading to the production of energetic particles, for two types of generic experiments, TS1 for photons and TS2 for protons. This sequence takes into account the transport and the conversion of particles and radiation, and then estimates the global activities of the experiment system to determine regulatory levels expressed through the Q factor defined as the ratio between the activity computed with the simulation and its exemption value given by the French nuclear regulations. The calculation results, performed for four different levels of irradiation from 0.2 to 3.5 kJ identify two regulatory thresholds (both corresponding to Q = 1, the lower limit defining a non-nuclear facility according to French regulations) that depend on the type of experiment: a threshold of 1 kJ for TS1 experiments (x-ray generation through bremsstrahlung conversion of electrons in a massive target), and 300 J for TS2 experiments (generation of energetic protons using a thin solid foil) [2]. Although the calculations are performed specifically for the LMJ-PETAL installation, the methodology is rather general and may be applied for design of experiments with other high power laser systems.

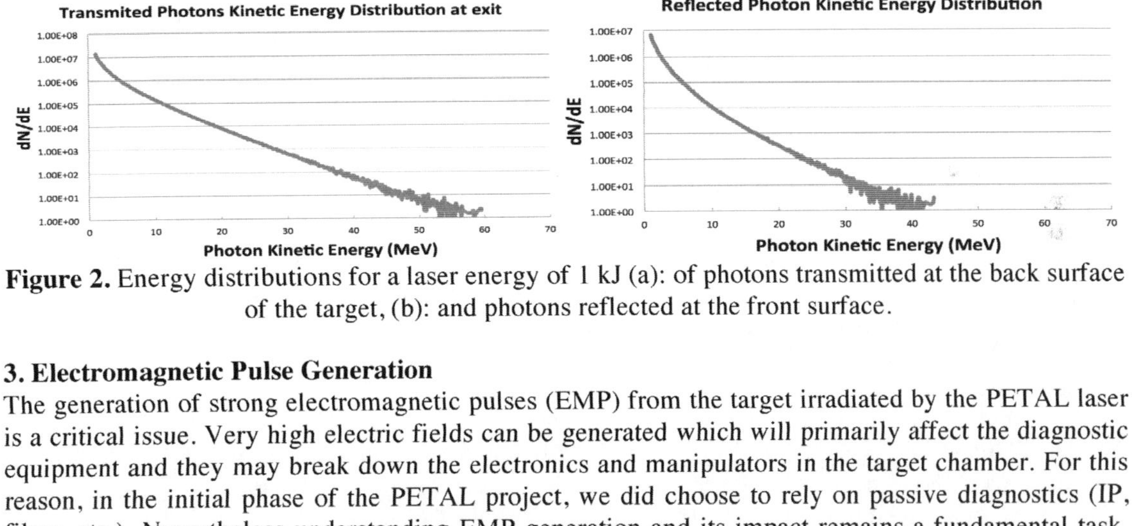


**Figure 2.** Energy distributions for a laser energy of 1 kJ (a): of photons transmitted at the back surface of the target, (b): and photons reflected at the front surface.

## 3. Electromagnetic Pulse Generation

The generation of strong electromagnetic pulses (EMP) from the target irradiated by the PETAL laser is a critical issue. Very high electric fields can be generated which will primarily affect the diagnostic equipment and they may break down the electronics and manipulators in the target chamber. For this reason, in the initial phase of the PETAL project, we did choose to rely on passive diagnostics (IP, films, etc.). Nevertheless understanding EMP generation and its impact remains a fundamental task. The origins of electromagnetic fields induced by the interaction of high intensity laser on targets are not well known. Comparison and extrapolation of the measurements from other high-intensity facilities indicate expected chamber field exceeding the MV/m level, which is of major concern with respect to vulnerability of equipments. In this context, we started a multi-year program for the comprehension of the underlying mechanisms of EMP generation. Final objectives are to produce a complete line of simulation, to provide recommendations concerning the target design and to develop mitigation procedures. A significant part of the EMP stems from the bunch of electrons ejected from the target and from the discharge current through the target support. In this context, a specially designed experiment was conducted in 2012 to provide direct measurements of the target polarization and the discharge current, the objective being to validate and enhance the accuracy of an analytical model of ejected charge [5]. The experiment was performed on the Eclipse laser system at CELIA

(Bordeaux University, France). The facility delivers a maximum energy of 100 mJ on target for a minimum pulse duration of 50 fs (800 nm) with the contrast of $10^{-7}$ and a 7.5 $\mu$m (FWHM) focal spot. The induced discharge current and the radiated field were measured in function of the laser energy, pulse duration and target size (copper, diameter 5, 10 and 15 mm). Experimental results corroborated by the model and simulation, indicate that the total charge increases approximately linearly with the laser energy, independently on the target size. In contrast, the maximum current decreases with the target diameter. According to our measurements, the charge and the maximum current depend essentially on the laser energy and much less on the laser intensity (no influence of laser pulse duration in the considered range). A suite of numerical codes integrating a PIC code (Calder) for the source term and the transport codes of particles (MCNP) and fields (Sophie) were employed to simulate the experiment. Comparison of the calculated current waveform with the experimental results shows a very good agreement and demonstrate that the target electric polarization is responsible for generation of high amplitude GHz electromagnetic pulses in the interaction chamber. However, some quantitative differences still remain and need further analysis and experiments.

## 4. Conclusion

A large effort is undergoing in France to install the short-pulse high-energy laser PETAL as a high-intensity companion to the LMJ laser. The LMJ/PETAL facility will be opened for the academic access. In parallel diagnostics are being developed within the PETAL project and many physical problems are being addressed ranging from study of the problems of radiation generation and activation issues to the problem of generation of large EMP signals. In the first phase of the PETAL operation, the compatibility of PETAL experiments in the context of national nuclear activity regulations has been evaluated for two regimes of operation. The characteristics of x-rays and protons originating from typical PETAL experiments were studied in order to estimate the activation coming from their interaction with matter. The conclusion of this study is that in order to comply with a non-nuclear classification status of facility during the first stage of operation, the PETAL energy should be limited by chamber activation at the level of 300 J for proton backlighting experiments and at around 1 kJ for X-ray backlighter experiments for a pulse duration of 500 fs. The study of strong electromagnetic fields produced during the interaction of a high intensity laser with a solid target is of prime importance for the development of PETAL diagnostics. A model of the target electric polarization induced by a short and intense laser pulse has been developed and verified in an experiment with a sub-picosecond laser pulse with energy of 100 mJ and compared with numerical simulations demonstrating that this effect is responsible for generation of high amplitude GHz electromagnetic pulses in the interaction chamber.

## 5. References

[1] D. Batani, S. Hulin, J.E. Ducret, et al. Acta Polytechnica, **53**, 103 (2013).
[2] E. d'Humières, J. Caron, J. Baggio et al., Modeling of high energy particles and radiation production and the associated Q factor during the PETAL energy rise, in preparation.
[3] Y. Sentoku, A. Kemp, J. Comput. Phys. **227**, 6846-6861 (2008).
[4] V.N. Ivanchenko, Nuclear Instruments and Methods in Physics Research **525**, 402 (2004).
[5] J.-L. Dubois, F. Lubrano-Lavaderci, D. Raffestin, et al., "Target charging in the short pulse laser-plasma experiments", Phys. Rev. E **89**, 013102 (2014).

## Acknowledgments

The PETAL project is supported by the Conseil Régional d'Aquitaine (CRA), the French Ministry of Research and the European Union and with the scientific supports of the Institut Lasers et Plasmas (ILP). The activation work was granted access to the HPC resources of CINES under allocations 2013-056129 and 2014-056129 made by GENCI and was supported by the ARIEL project funded by the CRA. The work on diagnostics takes place within the Equipex PETAL+ funded by the French National Agency for Research (ANR) and coordinated by the University of Bordeaux.

Journal of Physics: Conference Series **688** (2016) 012013 doi:10.1088/1742-6596/688/1/012013

# Ablation loading of solid target through foam absorber on ABC laser at ENEA-Frascati

R. De Angelis[1a], F. Consoli[1], S.Yu. Gus'kov[2,3], A.A. Rupasov[2], P. Andreoli[1], G, Cristofari[1], G. Di Giorgio[1], D. Giulietti[4], G. Cantono[4], M. Kalal[5]

[1]*Associazione ENEA-Euratom sulla Fusione, Frascati, Italy*
[2] *Lebedev Physical Institute, Russian Academy of Sciences, Moscow, Russia*
[3] *Lebedev Physical Institute, Russian Academy of Sciences, Moscow, Russia*
[3]*National Research Nuclear University MEPhI (Moscow Engineering Physics Institute), Russia*
[4]*Dipartimento di Fisica "E.Fermi" dell'Università di Pisa, Pisa, Italy*
[5]*Fac. of Nucl.Sciences and Phys.Engin.,Czech Techn.Univ. in Prague, Prague,Czech Rep.*

**Abstract.** This work reports an experimental characterization of the efficiency of energy transmission of porous laser absorbers as a function of their density and thickness. In this campaign the foams were deposited on different metal substrates, which finally absorbed the energy deposited by the laser on the bulk of the porous material. The dimensions of the craters produced on the substrate can be related to the energy transmitted through the foams

## Introduction

The stability of acceleration of targets irradiated by high intensity lasers is a main concern in inertial confinement fusion (ICF) research. The highest degree of both irradiation homogeneity and target surface finish will be required in ICF devices to achieve high performance implosions. The uniformity of the pressure exerted by the ablating plasma on the underlying target can be enhanced further by an even absorption of the laser light. Volume structured materials, such as foams with supercritical average density, have been proposed as random absorbers of the radiation [1,2] to mediate the interaction of the laser light with the target. In order not to limit the overall energy transfer to the target, optimization of the absorber characteristics is of primary importance. We report here a study of the energy transmitted through laser-irradiated porous absorbers in the ABC facility in Frascati at intensities ~$10^{13\text{-}14}$ W/cm$^2$. Previous experiments in similar conditions reported low levels of reflected and transmitted light [3]. The purpose of this work was to measure the energy transmitted by the absorber to a metallic holder opposite the laser beam. This energy is deduced from the volume of the craters imprinted on the substrate, knowing the properties of the substrate material. In our experiment, the dimensions and shapes of the craters were measured with a high-resolution confocal microscope. The foams used had densities in the 10 - 60 mg/cm$^3$ range and varying thickness. The experimental dependence of crater characteristics vs. the interaction parameters, the foam thickness and the foam density, are investigated. The plasma produced on the foam was characterized by 2$^{nd}$ harmonic interferometry and imaging, time resolved soft X ray

[a] riccardo.deangelis@enea.it

emission and ion-collectors measurements [4]. Density profiles of the plasma were deduced from the analysis of the complex interferometry data.

## Smoothing of laser radiation

The absorbers used in our experiments were polystyrene foams with average densities in the range 10 - 60 mg/cm$^3$, a few times higher than the critical density $\rho_c = 3$ mg/cm$^3$ at the fundamental laser wavelength (1054 nm). These can be regarded as composed of empty cells delimited by solid plastic walls a few μm thick. Laser absorption on such materials relies on a 2-stage mechanism as described in [1, 2]: in the initial phase (~1 ns) the radiation absorbed on the cell walls creates the plasma, which rapidly expands in the vacuum regions. In the following stage, extended hot plasma is created by collisional interaction of the small plasma zones. During this process the absorber's transparency length decreases and part of the laser energy is converted to the motion of a high-pressure front propagating in the plasma [5]. To optimize the efficiency of this process, the foam characteristics should be chosen in relation to the laser pulse duration, so that energy is delivered during the extended plasma phase and full absorption occurs inside the absorber thickness. In ABC for a foam of density $\rho_{foam}$ = 20 mg/cm$^3$ and typical sizes of the pores $\delta_0$ ~50 - 100 μm with wall thickness $b_0$ ~2 μm [5], typical initial transparency lengths are ~400 μm, reduced by a factor ~2 in the second phase.

## Experimental setup

ABC is a Nd-glass laser, wavelength 1054 nm, delivering 2 beams with energies ≤ 100 J each, which can be focused on opposite sides of a planar target to spots of $\varnothing_{min}$ ≃40 μm by F/1 lenses. By the use of ISI plates (Induced Spatial Incoherence) [6], intensity profiles with amplitude modulations < 1 % can be achieved on 500 μm*500 μm spots. The measurements reported in this paper were obtained with a single heating beam with $E_L$ ~50 J, pulse duration 3 ns. The laser intensity varied between $I_L$ ~ 10$^{13}$ W/cm$^2$ with ISI beam smoothing plates and ~2.5*10$^{14}$ W/cm$^2$ focusing on 100 μm spots. Multichannel ion-collectors and soft-X ray pin diodes were used for time resolved measurements of particles and radiation emission. Shadowgraphy and interferometry of the plasma were obtained by means of a 2$^{nd}$ harmonic beam, originated from the main pulse; this beam has a ≃ 0.6 ns duration and is split and delayed to probe the plasma at 4 different times. Spatial resolutions of the images are better than 50 μm. The interferometer configuration is of Nomarsky type and uses Wollaston prisms; electron density profiles are obtained by analysis of complex interferograms and Abel inversion with the 'cplxifer' code [7]. Polystyrene foams of ~10 mm$^2$ area and thicknesses 200 - 800 μm were put in close contact to metal substrates (Al, Sn), which absorbed the energy of the shock wave produced by the laser [8]. The metal substrates were previously polished with 15 μm grain size.

## Experimental results

Experimental data were collected at approximately constant laser energy. Reference shots on pure Al target were used for comparison with the foam plasma behaviour and to correct the crater volumes against shot to shot variations of the laser output. Figure 1 reports the time evolutions of the optical border of the plasma originated on pure Al targets and on a 320 μm thick foam. The pictures where obtained by transverse illumination of the target with 2$^{nd}$ harmonic probe beams delayed respectively by t = 0.3; 4 and 6 ns with respect to the heating beam. The absorption of the laser energy on the foam is accompanied by small-scale non-uniformities in the plasma, compatible with the pore sizes, and by an enhanced transvers expansion leading to profiles more flat than in pure Al targets. This is also shown in the density reconstructions obtained by complex analysis [7] of the interferograms (see Figures 2-4); marked enlargements of the density profiles have been observed in two conditions of laser Intensity: the first using an ISI smoothing plate and a 500μm*500μm focal area , the second with a sharper focus (SF) on a 100μm focal spot. Large tranverse energy propagation in the foam plasma appears as an evidence of the property of the foam of redistrubiting the absorbed energy.

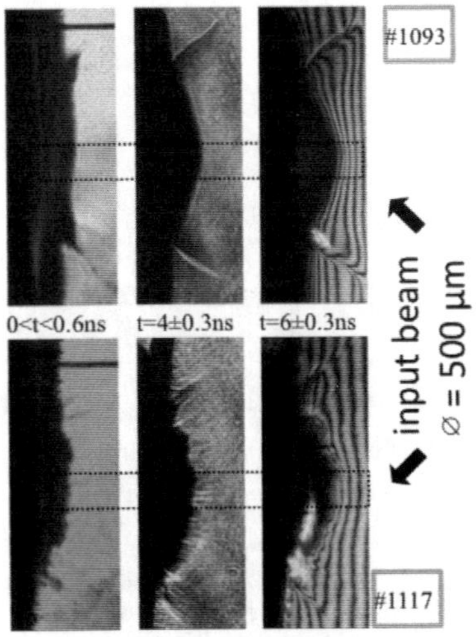


**Fig. 1** Images illustrating the plasma evolution.

- Top - Al target
- Bottom – Foam-coated Al target ($\varrho_{Foam}$=10mg/cm³, $\delta_{Foam}$=320μm)

The dense plasma appears dark.
(The first 2 columns are shadowgraphies, the last column is a portion of interferograms obtained with a Wollaston prism).

(The 'wings' on the plasma periphery are disregarded since they are know to be due to the laser profile in the focal spot).

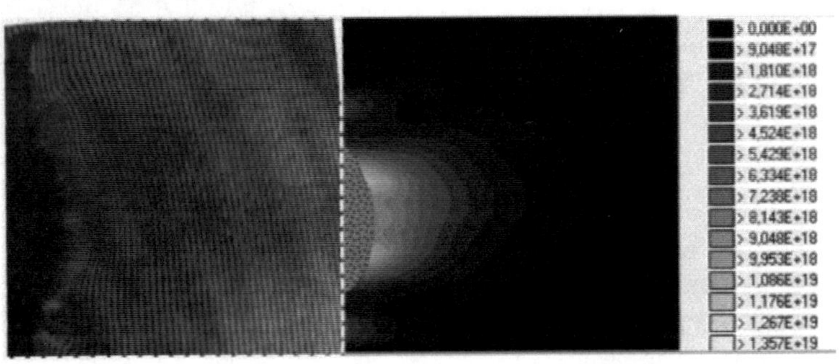


**Fig. 2 -** Image produced by the Nomarsky interferometer and density reconstruction for shot #1117 at t=4ns

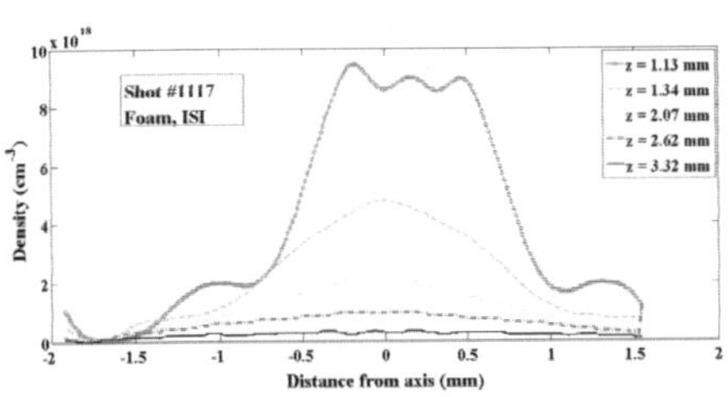


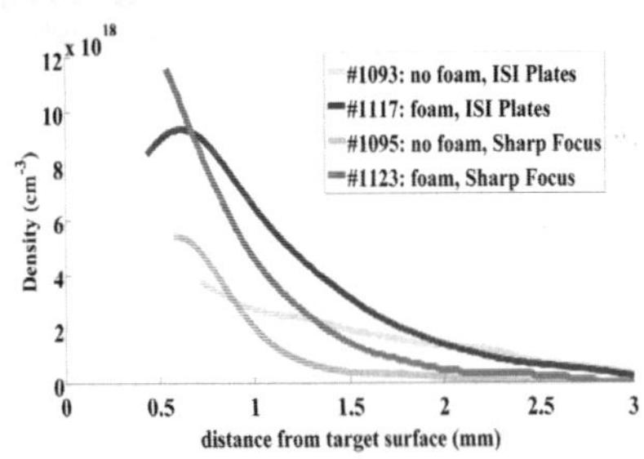


**Fig. 3** Example of a radial electron density profile (#1117 at t=4ns)

**Fig. 4 -** Density vs. distance from the target at t=4ns with foam (#1117, #1123) and on pure metal (#1093, #1095)

## Crater analysis

The analysis of craters was done with a confocal microscope LEICA DCM 3D which allowed measurements of their surface with an accuracy of ~ 1.7 μm on the planar dimensions and ~ 45 nm on the depth, an example is shown in Figure 5.The dependence of the crater volumes on foam thickness is shown in Figure 6 for various densities of the absorber. The data show a clear maximum for $\delta_{foam}$ ~ 250 μm while for $\delta_{foam}$ ~ 800 μm the crater volume reduces to about a factor 10 of the maximum.

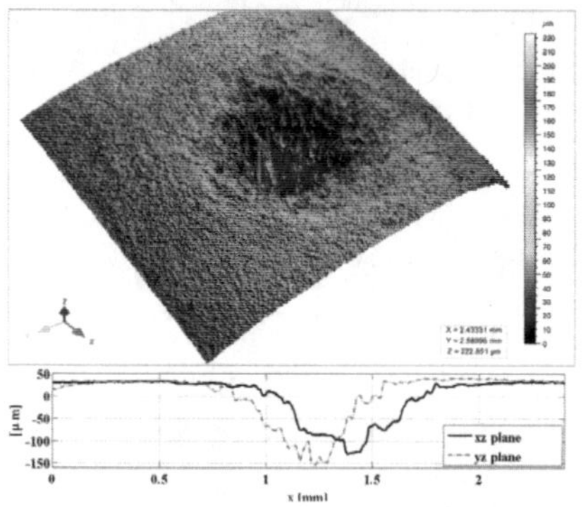


**Fig. 5** - The profile of a crater measured on the confocal microscope (shot #1117)

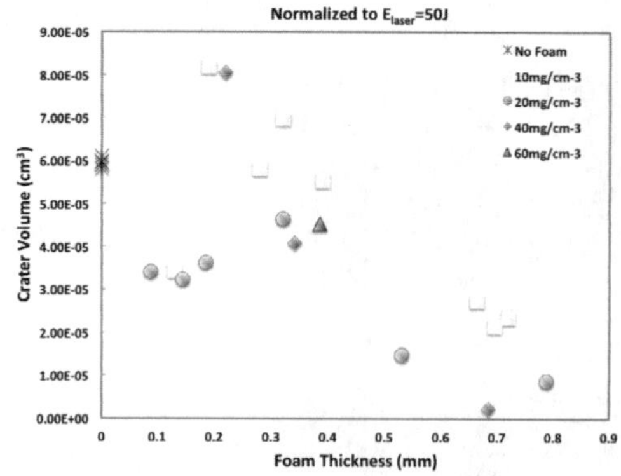


**Fig. 6** -Dependence of crater volumes on foam thickness

## Crater formation

The ablation of material from the target substrates is caused by the high-pressure shock wave traveling in the laser produced plasma. The amount of matter, which is melted and evaporated by the shock wave hitting the metal, can be related to the energy transmitted by the plasma. The volume of the crater $V_{crater}$ [$cm^3$] depends on the laser energy $E_L$, on the laser-plasma energy coupling $K_p$ and on the efficiency of conversion to shock wave $\sigma$:

$$V_{\text{crater}} = \frac{\sigma K_p E_L}{\alpha w_0 \rho_0}$$

where where $\alpha$ is the thermal/kinetic ratio in the shock adiabat, $w_0$ the specific vaporisation heat, $\rho_0$ the solid metal density. It is well known, for strong shock waves, under the approximation of negligible elastic energy, $\alpha$=2. The main part of the crater is created by the shock wave damped during several tens of nanoseconds, longer than laser pulse. So, the average factor $\alpha$ is less 2 and varies in the region 1.2-1.5 (see, for example, [9]). For $E_L = 50$ J, the efficiency of energy transformation to shock wave for pure Aluminum ($w_0=10^3$ J/g $\rho_0=2.7 g/cm^3$) is $\sigma K_p / V_{crater} \sim 10^2\ cm^{-3}$. Experimental data show that the value at the optimum thickness is $\sigma K_p \lesssim 8*10^{-3}$ which is comparable with the pure Al values (for which $\sigma K_p \lesssim 6*10^{-3)}$. The experimental data above indicate that choosing the optimal thickness of the foam the efficiency of energy transfer to the target can be similar or even better than in the case of direct irradiation, with the advantage of reduced non-uniformities.

## References

1. S.Yu. Gus'kov et al, JETP **81**, 296 (1995);
2. S.Yu. Gus'kov et al, Laser Part. Beams **17**, 287 (1999);
3. A. Caruso et al, Laser Part. Beams **18**, 25 (2000);
4. F. Consoli, R. De Angelis et al, NIMA **A 720** 149–152 (2013);
5. S.Yu. Gus'kov et al, J. Russ. Laser Res., **31**(6), 574-587 (2010)
6. C. Strangio et al. Quant. Electr. **36**(5) 424-428 (2006)
7. M. Kalal SPIE Volume 1983, Part II, pp. 686-687
8. F. Consoli, R. De Angelis et al, Proc. 40th EPS Conf. on Plasma Phys., 2013, Espoo, Finland.
9. S.Yu. Gus'kov et al, Laser and Particle Beams **32**, 177 (2014)

# Capsule Ablator Inflight Performance Measurements Via Streaked Radiography Of ICF Implosions On The NIF*

**E L Dewald[1,3], R Tommasini[1], A Mackinnon[1], A MacPhee[1], N Meezan[1], R Olson[2], D Hicks[1], S LePape[1], N Izumi[1], K Fournier[1], M A Barrios[1], S Ross[1], A Pak[1], T Döppner[1], D Kalantar[1], K Opachich[1], R Rygg[1], D Bradley[1], P Bell[1], A Hamza[1], B Dzenitis[1], O L Landen[1], B MacGowan[1], K LaFortune[1], C Widmayer[1], B Van Wonterghem[1], J Kilkenny[1], M J Edwards[1], J Atherton[1], E I Moses[1]**
[1]Lawrence Livermore National Laboratory, P.O. Box 808, Livermore, CA 94550 USA
[2]Los Alamos National Laboratory, Los Alamos, New Mexico 87545, USA

[3]E-mail: dewald3@llnl.gov

**Abstract.** Streaked 1-dimensional (slit imaging) radiography of 1.1 mm radius capsules in ignition hohlraums was recently introduced on the National Ignition Facility (NIF) and gives an inflight continuous record of capsule ablator implosion velocities, shell thickness and remaining mass in the last 3-5 ns before peak implosion time. The high quality data delivers good accuracy in implosion metrics that meets our requirements for ignition and agrees with recently introduced 2-dimensional pinhole radiography. Calculations match measured trajectory across various capsule designs and laser drives when the peak laser power is reduced by 20%. Furthermore, calculations matching measured trajectories give also good agreement in ablator shell thickness and remaining mass.

## 1. Introduction

In the indirect drive approach to inertial confinement fusion (ICF) employed on the National Ignition Facility (NIF), laser energy is converted into thermal x-rays that ablatively drive the implosion of a spherical capsule containing a deuterium-tritium (DT) fuel layer [1]. The resulting kinetic energy delivered to the fuel has to be high enough to overcome the heat lost by conduction and radiation while maintaining a relatively low fuel adiabat prior to implosion stagnation. Specifically the ignition threshold factor (ITF) [2] that quantifies the quality of an implosion relative to achieving ignition, scales as implosion velocity to the $7^{th}$ power. However, as the rocket equation and earlier data show [3.4], increasing velocity leads to decreasing remaining ablator mass and greater undesirable sensitivity to feed-through of ablation front instabilities. In our quest for ignition and high gain, as we experimentally optimize the capsule, hohlraum and laser pulse designs [5], it is hence important to measure accurately the implosion velocity and ablator remaining mass. In addition, in the absence of a direct observation of the fuel, a precise measure of the ablator thickness is required to validate the ablator drive, compression and coupling to the DT fuel.

## 2. Experimental setup

Convergent Ablator (ConA) experiments measure inflight capsule ablator performance using 1-d slit radiography as shown in Figure 1a. Typically these experiments employ gas filled (DHe3) capsule surrogates that have slightly thicker ablators to compensate for the missing DT fuel payload or use low yield THD layered capsules.

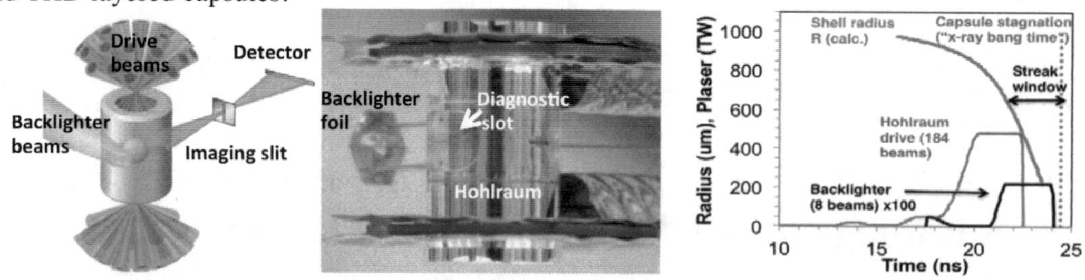


**Figure 1.** (a) ConA experimental arrangement, (b) picture of the target and (c) typical time history of a ConA measurement.

Eight laser beams (two 50° quads) of the 192 NIF beams are diverted towards a backlighter foil, mounted 6-12 mm away from the capsule center (Fig. 1b) to radiograph the capsule. Areal radiography is performed through two 0.1 mm tall and 2.4 mm wide diagnostic slots cut out in the high-Z hohlraum and filled with 100 um thick high-density carbon (HDC) that ensure minimal distortion of the hohlraum drive. The other 184 laser beams heat the hohlraum and drive the capsule implosion, whereas some beams have modified pulse shape to compensate for the missing backlighters when compared to full 192 beams DT implosions. Figure 1c superimposes the calculated time history of the ablator center of mass (CoM) trajectory on the hohlraum laser power and the backlighter pulses. The 2-5 ns long pulse backlighter beams generate He-alpha lines of the Zn or Ge foils (hv=9-10 keV), and are delayed to capture the ablator trajectory at radii ranging from 0.9 mm through stagnation (bang time, Fig. 1c). The ablator is imaged at the capsule waist using a 15 um wide slit oriented along the hohlraum axis, perpendicular to the diagnostic slots and placed 100 mm away from the capsule. Until early 2012 the measurement was performed using a 4-strip gated imager (M=9x setup) with 80 ps temporal blurring providing 4 adjustable times within a 1 ns time interval [3]. Recently we employed a x-ray streaked detector (DISC) [6] (M=12x setup) that provides reduced temporal blurring (40 ps), and at the same time captures a longer 3-5 ns continuous history of the inflight ablator.

## 3. Streaked data of NIF implosions

Figure 2 shows a target schematic as viewed by the detector and typical streaked images of imploding capsule ablators recorded with DISC.

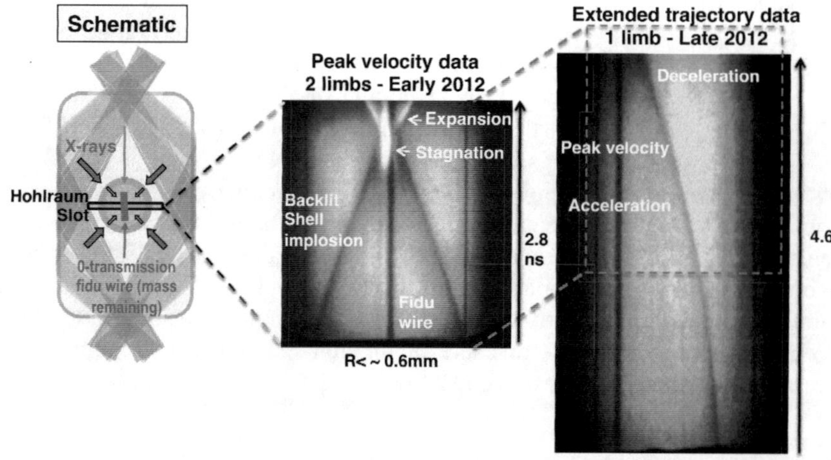


**Figure 2.** Target schematic and ConA data for peak velocity and extended trajectory of NIF ICF implosions (360 TW peak power/Au hohlraums).

Data shown here was recorded using gas filled, 210 um thick CH capsules with a embedded Si doped layer (2% Si, 50 um thick) that dominates the x-ray absorption in two distinct configurations: peak velocity and long record. Peak implosion velocity measurements, expected at a capsule radius of 0.2-0.4 mm, image both capsule limbs at the waist starting at radii > 0.6 mm, through stagnation and re-expansion. In validating our calculations we also use the expansion of the outgoing shock to constrain the energy coupled into the hot spot and remaining mass, typically in the order of 5-10 kJ [7]. Long record measurements, on the other hand, follow the trajectory of a single capsule limb over an extended range within a single experiment, at radii between 0.9 and 0.2 mm. Peak velocity data provide better statistics (2 limbs) and use a 3 ns streak record (40 ps temporal blurring), while extended record data images one limb over 5 ns to capture most of the trajectory and hence is affected by larger blurring (70 ps). The high quality streak data provides high accuracy in implosion velocity, remaining mass and shell thickness as summarized in Table 1. Besides providing a long continuous range of ablator implosion history, the accuracies of implosion metrics, inferred from data rms over the resolution element, are fairly consistent with the photon statistics, improve for streaked over gated data [3] and meet estimated requirements for ignition. Spatial blurring is reduced for streaked vs gated data due to reduced slit size and hence temporal blurring.

**Table 1.** ConA accuracies and requirements for ignition

| | Estimated req. for ignition | Gated | Streaked peak velocity | Streaked extended trajectory |
|---|---|---|---|---|
| **Radius range , mm** | | 0.2-0.4 | 0.1-0.6 | 0.1-0.9 |
| **Temporal blurring, ps** | < 50 | 80 | 40 | 70 |
| **Resolution, $\mu$m** | < 20 | 25 | 17 | 17 |
| **Signal-to-noise /20 $\mu$m** | > 10 | 80 | 35 | 45 |
| **Spatial blurring, all sources, um*** | < 20 | 34 | 18 | 23 |
| **Velocity accy., (0.3 ns av.), %** | ≤ 2 | 5 | 2 | 3 |
| **Width accy., $\mu$m*** | ≤ 4 | 6 | 3 | 3 |
| **Shell density, %*** | ≤ 10 | 15 | 10 | 9 |
| **Rem. mass/Mass(t=0) accy, %** | ≤ 2 | 1.5 | 2 | 2 |

## 4. Streaked slit vs gated 2D ConA data comparison

In addition to streaked ConA's, we have recently developed 2D ConA backlit pinhole imaging using gated detectors [8] that provide inflight shape data, typically around 0.3 mm capsule radius.

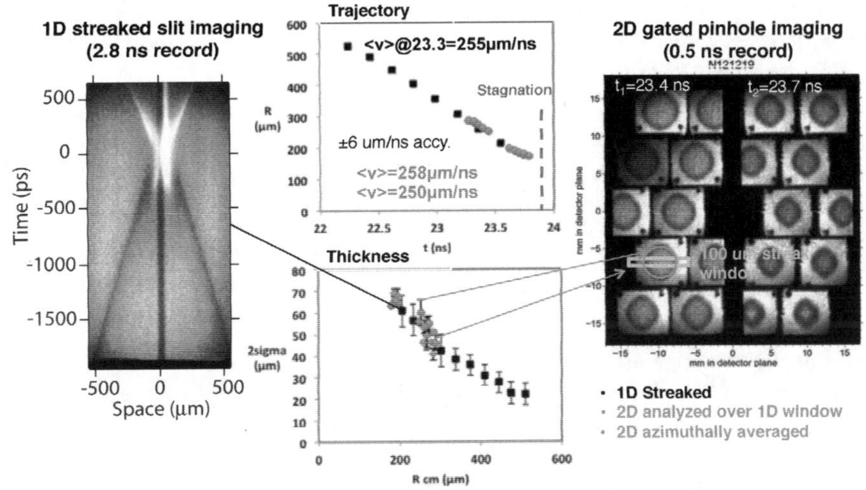


**Figure 3.** Comparison of 1D and 2D ConA data for similar implosions (360 TW peak laser power, 210 um thick graded doped (Si) CH capsules.

While 2D ConA provides implosion velocity over the entire capsule rather than the waist, it requires larger diagnostic windows (hence bigger distortion on hohlraum drive), covers a shorter temporal window (~0.5 ns) and provides lower accuracy in velocity. A comparison of ConA and 2D ConA data for similar implosions is shown in Figure 3 for 360 TW peak power and 210 $\mu$m thick graded dopant capsules. ConA and 2D ConA results show same implosion velocities and capsule thickness within measurement uncertainty, both if the 2D analysis is performed over same 100 $\mu$m window of streaked ConA or over the azimuthal average around the capsule. This cross check validates these two measurement methods and shows that, for reasonably round implosions, streaked data is representative for the overall implosion performance around the capsule.

## 5. Data comparison to simulations and discussion

We compare our data with HYDRA calculations that are employed to optimize our implosions towards ignition. Good agreement in trajectory is obtained when the peak laser power (Fig. 1c) is reduced by 20%. Figure 4 shows a comparison between a measured and a simulated streaked radiograph processed from post-shot simulations.

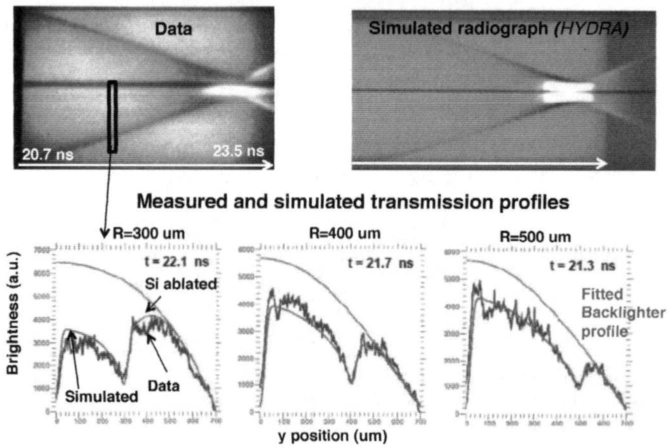


**Figure 4** Measured and simulated radiographs and corresponding spatial lineouts (360 TW power, 210 um thick CH capsule with Si doped layer).

Furthermore, once we match the measured trajectory we observe good agreement in the transmission profiles [9], even at R=0.3 mm, when ablation burns into the doped layer, and therefore in capsule thickness and remaining ablator mass. The drive reduction is due to an overestimate in the calculated radiation drive at the capsule as confirmed by View Factor experiments [10] that measure independently the x-ray drive through the hohlraum laser entrance hole and the drive as seen by the capsule. As it will be shown in future work, a similar drive reduction is required to match the data across various laser powers, capsule and ICF designs.

*This work performed under the auspices of the U.S. Department of Energy by Lawrence Livermore National Laboratory under Contract DE-AC52-07NA27344.

## 6. References

[1] Haan S et.al, *Phys. Plasmas* **18,** 051001 (2011).
[2] Edwards M J et. al., *Phys. Plasmas* **20,** 070501 (2013).
[3] Hicks D **et al**, *Phys. Plasmas* **19,** 122702 (2012).
[4] Meezan **N** et al *Phys. Plasmas* **20,** 056311 (2013).
[5] O.L. Landen et. al., *Phys. Plasmas* **18,** 051002 (2011).
[6] Opachich K et al., *Rev. Sci. Instrum.* **83**, 125105 (2012).
[7] Pak A, *Phys. Plasmas* **20,** 056315 (2013).
[8] Rygg R et al, O.Mo_A7, presented at *IFSA* 2013.
[9] Olson R et al., *Rev. Sci. Instrum.* **83**, 10D310 (2012).
[10] Schneider M et al, O.Fr_A8, presented at *IFSA* 2013.

# Experimental Test of the Polarization Persistence in Inertial Confinement Fusion

**J. P. Didelez[1] , C. Deutsch[2], M. Fujiwara[3,4], M. Nakai[5] and M. Utsuro[3]**

[1]IPN, CNRS/IN2P3 & Université Paris-Sud, ORSAY, France

[2]LPGP, Université Paris-Sud, ORSAY, France

[3]RCNP, Osaka University, IBARAKI, Japan

[4]QUBS, Japan Atomic nergy Agency, TOKAI, Japan

[5]ILE, Osaka University, OSAKA, Japan

E-Mail: didelez@ipno.in2p3.fr

**Abstract.** The complete deuteron and triton polarization in the DT fusion increases the reactivity by 50%. For Inertial Confinement Fusion (ICF), due to the dynamics of the fusion reaction process, the fusion rate could even be further increased. It has been argued that the polarization would survive as well in magnetic as in inertial confinements. Recently, we have proposed an experiment to test the persistence of the polarization in a fusion process, using a powerful laser hitting a polarized HD target. The polarized deuterons heated in the plasma induced by the laser can fuse. The corresponding reaction is: D + D → $^3$He + n. The angular distribution of the emitted neutrons and the change in the corresponding total cross section are signatures to estimate the polarization persistency. A proposal to test the persistence of the polarization in ICF has been accepted at ILE: the POLAF project (POlarization in LAser Fusion Process). It uses the polarized HD targets produced at RCNP and the powerful ILE lasers, as well as the neutron detectors existing there. Both institutions are on the same campus at Osaka University. The description of the POLAF experiment and of the corresponding set-up is given.

## 1. Introduction

The polarization of D and T nuclei increases the reactivity when they are used as fuel material in fusion processes induced either by magnetic or by inertial confinement. The fusion reaction:

$$D + T \rightarrow \alpha + n + 17.6 \text{ MeV} \tag{1}$$

goes mainly through the excitation of an $^5$He $3/2^+$ intermediate state, resulting from the coupling of the spins 1 and 1/2 of the D and T nuclei to a total spin S = 3/2. Without polarization of D and T, the

Journal of Physics: Conference Series **688** (2016) 012015 doi:10.1088/1742-6596/688/1/012015

statistical distribution of the six possible states gives four S = 3/2 and two S = 1/2 states. Only the 3/2 states can produce the intermediate 3/2 resonance. With 100% parallel polarization of D and T, all states would contribute to the fusion, increasing the reactivity by 50%. In addition, the polarization allows the control of the direction in which the reaction products are emitted, the neutron having a $\sin^2\theta$ distribution, as compared to the polarization axis. This can be very useful to reduce damage or activation of costly equipment. Theoretical considerations indicate that the polarization should persist as well in MCF (Magnetic Confinement Fusion) [1] as in ICF [2]. Recent realistic simulations show that for ICF, the required hot-spot temperature and areal density can be reduced by about 15 % for fully polarized nuclear fuel. Moreover, numerical simulations of a directly driven capsule show that the required laser power and energy to achieve high gain are significantly reduced, while the maximum achievable energy gain scales roughly as the fusion cross section [3]. It is found that for an invested energy of 1 MJ, the estimated gains are $G_A$ (1MJ) = 130 while $G_B$ (1MJ) = 300 for un-polarized and polarized DT fuel respectively. For a detailed discussion, see Ref. [3] and [4].

## 2. Experimental set-up

We have proposed to investigate the polarization persistency using the reaction:

$$D + D \rightarrow {}^3He + n + 3.3\ MeV \qquad (2)$$

induced by the fusion of polarized deuterons heated in a plasma. It is anticipated that the angular distributions of the neutrons as well as significant changes in the fusion rates can be measured and related to the persistence of the polarization. Details on the set-up and the experiment features are given in Ref. [5]. Here, we show only a sketch of the experiment.

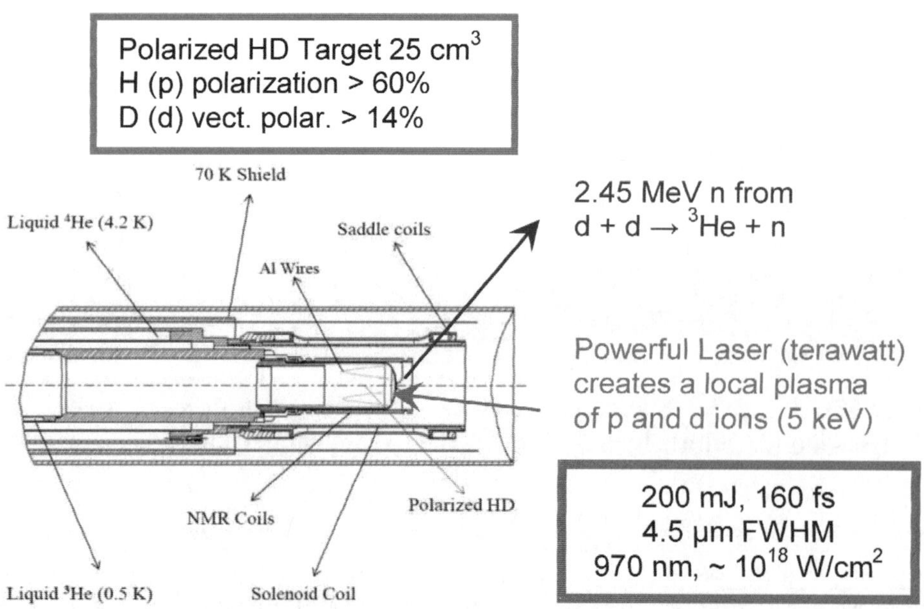


**Figure 1.** Tentative set-up showing a polarized HD target in a cryostat (temperature 1 K, holding field 1 Tesla). The target is struck by a terawatt laser producing a localized plasma. Neutrons are generated by fusion reactions between polarized deuterons.

The above described experiment uses only existing equipment: polarized HD targets are currently produced by static polarization [6] and the polarization can be accurately measured [7], on the other hand terawatt lasers and neutron detectors are easily available. A signal of 10-20% on the neutron yield due to the persistence of the polarization should be measurable. Precise data in this energy range will come from a Russian-German group which performs precise measurements of all the observables of the double polarized fusion reaction (2) at the PNPI laboratory in Gatchina [8].

## 3. Neutron detection

On the other hand, the laser fusion investigation is underway: a proposal has been made at the Institute for Laser Engineering (ILE) in Osaka, Japan and has been accepted as the POLAF project (Polarization in Laser Fusion) [9]. The experiment is planed using the ILE equipment, in particular the MANDALA neutron multi-detector as shown below.

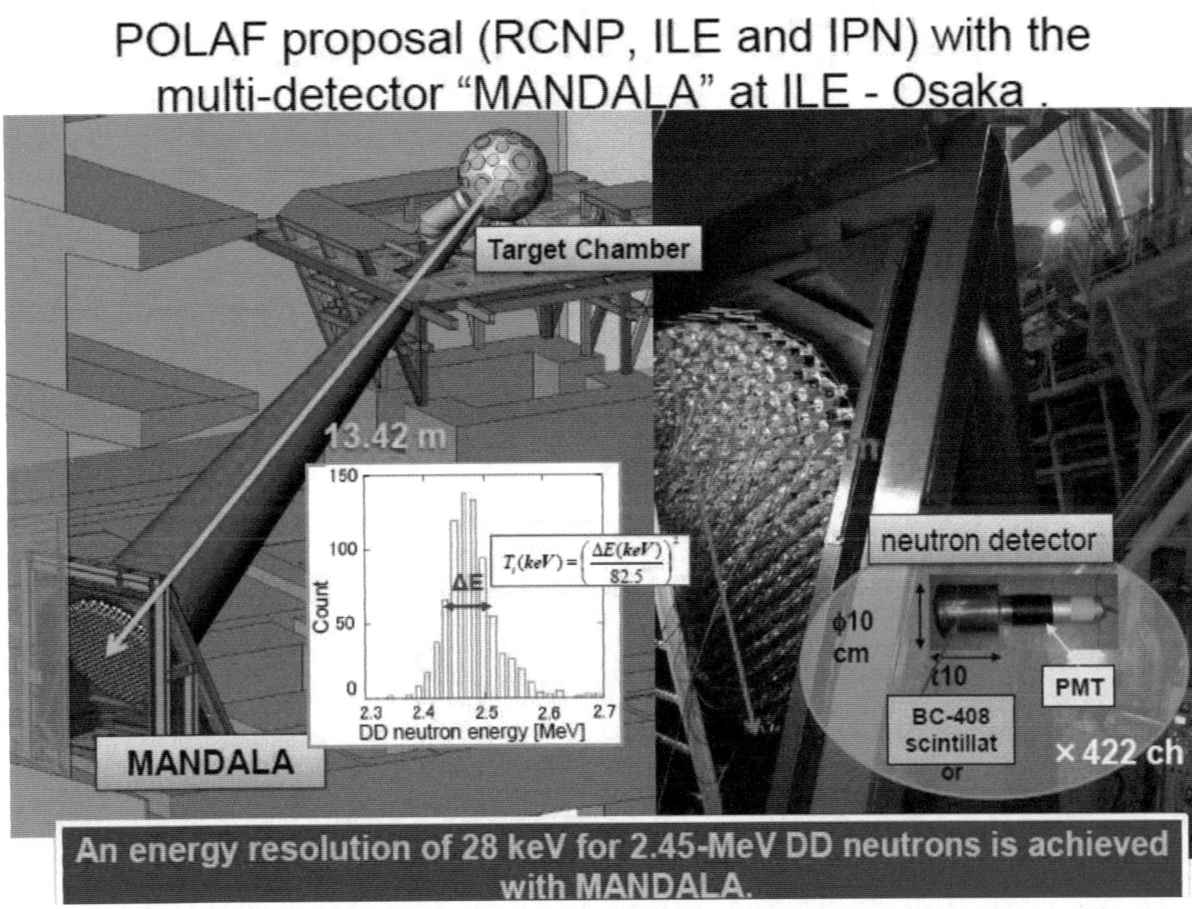


**Figure 2.** Picture showing the neutron detection set-up MANDALA at ILE.

The MANDALA neutron detection equipment comprises 422 BC-408 scintillator cylinders ($\phi$ =10 cm × 10 cm) located in the present configuration 13.42 m away from the target center, providing an energy resolution of 28 keV for the 2.45 MeV neutrons produced by the DD fusion. Scintillators are viewed by photomultipliers which can be switched off during the flash light induced by the EM particles produced by the short laser pulse interacting with the target, so that heavy shielding of the neutron detectors is not necessary. This neutron detection method is feasible, due to the relatively long delay between the impacts of the EM particles (electrons and gamma rays) and those of the slow neutrons coming from the DD fusion. The real difficulty is to bring the polarized HD target and its cryogenic support in the center of the laser target chamber. However, the complex handling of polarized HD targets for nuclear physics experiments requires various transfer and in-beam cryostats [10] which can be "easily" modified to match the ILE environment.

It should be mentioned that the present arrangement of the ILE MANDALA detector could be used to measure the Radiation Pressure Acceleration (RPA) of deuterons by lasers, at least their maximum energy, by measuring the maximum energy of emitted neutrons. Little is known concerning RPA as compared to Target Normal Sheath Acceleration (TNSA) [11]. This could be done with un-polarized $D_2$ and would be an excellent preparation towards runs with polarized HD targets.

## 4. Conclusion

A considerable effort is under way to produce energy using controlled fusion either by MCF or ICF. Polarized fusion fuel is of great interest, both to increase the fuel reactivity and to control the direction in which reaction products are emitted. The question to know if the polarization will persist in a fusion process can be answered using existing experimental equipment, and neutron detection. The corresponding investigation is underway at ILE (Osaka), through the POLAF project. The ongoing measurements at Gatchina for the double polarized DD fusion at low energy will provide the necessary experimental data to analyze the Osaka results. Finally, the investigation at ILE of the RPA by lasers using the un-polarized DD fusion neutrons to identify this acceleration mechanism as compared to TNSA, would be an excellent preparation to the future POLAF experiment.

## References

1. Kulsrud R M, Furth H P, Valeo E J and Goldhaber M, *Phys. Rev. Lett.* **49** 1248 (1982)
2. More R M, *Phys. Rev. Lett.* **51** 396 (1983)
3. Temporal M, Brandon V, Canaud B, Didelez J P, Fedosejevs R and Ramis R, *Nucl. Fusions* **52** 103011 (2012)
4. Didelez J P, Deutsch C and Guidal M, *EPJ Web of Conferences* **33** 03010 (2012), DOI 10.1051
5. Didelez J P and Deutsch C, *Laser and Particle Beams* **29** 169 (2011)
6. Bouchigny S, Didelez J P, Dubois F and Rouillé G, *Nucl. Instr. and Meth.* **A 607** 271 (2009)
7. Ohta T, Fujiwara M, Fukuda K, Kohri H, Kunimatsu T, Morisaki C, Ono S, Tanaka M, Ueda K, Utsuro M, Wang S Y and Yosoi M, *Nucl. Instr. and Meth.* **A 633** 46 (2011)
8. Grigoryev K, Chernov N, Engels R, Ivanov I, Kiselev S, Komarov E, Kotchenda L, Kravtsov P, Kroell L, Martyushov A, Marusina M, Mikirtychyats M, Nikolaev N, Rathmann F, Paetz gen Schieck H, Sherman S, Ströher H, Trofimov V, Vasilyev A and Vznuzdaev M, "Double polarized dd-fusion experiment", *Journal of Physics Conferences Series* **295** 012168 (2011), DOI: 10.1088/1742-6596/1/012168
9. Joint Research Application Form, RCNP-ILE-IPN-LPGP Collaboration, Osaka, Japan (2011)
10. Didelez J P, *Nucl. Phys. News* **4** 10 (1994)
11. Perego C, "Target Normal Sheath Acceleration for Laser-Driven Ion Generation Advances in Theoretical Modeling", *PhD Thesis, University of Milano,* January 2013 and References therein

# Study of shock waves and related phenomena motivated by astrophysics

**R P Drake[1], P A Keiter[1], C C Kuranz[1], G Malamud[1,2], M Manuel[1], C A Di Stefano[1], E J Gamboa[1], C M Krauland[1], M J MacDonald[1], W C Wan[1], R P Young[1], D S Montgomery[3], C Stoeckl[4], and D H Froula[4]**

[1] Center for Laser Experimental Astrophysics Research, University of Michigan, Ann Arbor, MI 48109 USA
[2] Nuclear Research Center, Negev, Department of Physics, Negev, Israel
[3] Los Alamos National Laboratory, Los Alamos, NM, USA
[4] Laboratory for Laser Energetics, University of Rochester, Rochester, NY, USA

E-mail: rpdrake@umich.edu

**Abstract.** This paper discusses the recent research in High-Energy-Density Physics at our Center. Our work in complex hydrodynamics is now focused on mode coupling in the Richtmyer-Meshkov process and on the supersonic Kelvin-Helmholtz instability. These processes are believed to occur in a wide range of astrophysical circumstances. In radiation hydrodynamics, we are studying radiative reverse shocks relevant to cataclysmic variable stars. Our work on magnetized flows seeks to produce magnetized jets and study their interactions. We build the targets for all these experiments, and simulate them using our CRASH code. We also conduct diagnostic research, focused primarily on imaging x-ray spectroscopy and its applications to scattering and fluorescence.

## 1. Introduction

In our Center for Laser Experimental Astrophysics Research, we explore fundamental processes in high-energy-density (HED) physics that are relevant to astrophysics, working with systems that feature shock waves or other high-Mach-number flows. This research at present is in several areas, which we discuss separately in the following. Many of our experiments are done at the Omega laser facility [?], while others are done elsewhere. The Omega experiments are our primary focus here. These experiments require complex targets, which we assemble at Michigan [?]. We also need simulations to design and analyze the experiments. We do this primarily with our CRASH code [?].

## 2. Complex Hydrodynamics

Our studies of complex hydrodynamics examine the nonlinear development of the basic instabilities that develop in dynamic media and that may evolve into turbulence. Our present focus is on the use of Omega EP, which can drive mm-scale areas for up to 30 ns while reserving one beam for diagnostics. This has enabled us to design [?] and field the first HED, Richtmyer-Meshkov experiment in which the shock wave proceeds from a denser to a less dense material,

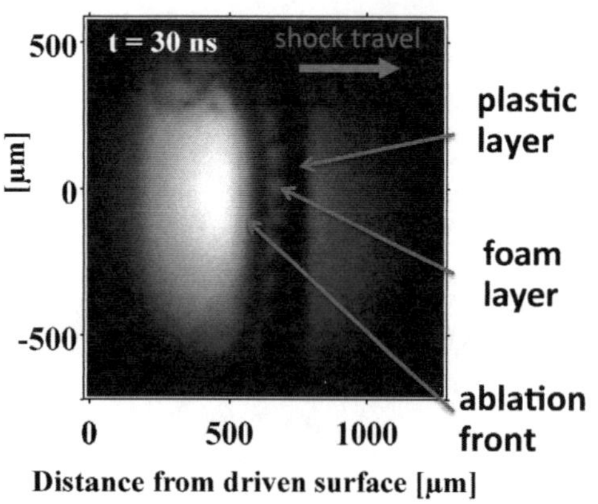


**Figure 1.** Radiographic image of nonlinear structure developed by the Richtmyer-Meshkov process. The structure involves the interpenetration of the (opaque) iodinated plastic with the (transparent) foam. Adapted from [?]

allowing a clean examination of the phenomenon of mode coupling. We drive a steady shock through an interface having a perturbation with modes of 5% amplitude at 100 $\mu$m and 50 $\mu$m wavelengths. Figure 1 shows a radiographic image obtained using a spherical crystal imager operating at the $K_\alpha$ line of Cu. One can see that the structure has grown to large amplitude and become geometrically complex. Analysis of the interface structure shows that mode coupling has become important [?]. Using a similar experimental design, we are initiating experiments on the Kelvin-Helmholtz instability [?]. In related work we are the lead group for an experiment [?] to study radiative effects on the Rayleigh-Taylor instability [?] at the National Ignition Facility [?].

## 3. Radiation Hydrodynamics

HED facilities have made it possible to undertake fundamental experiments on radiation hydrodynamics, in which the flow of radiative energy alters the hydrodynamic structure of a system. Following extensive work with driven radiative shocks, in which a "piston" of Be plasma drives a shock wave down a tube of Xe or Ar gas [?] [?] [?] [?], our emphasis has turned to the study of radiative reverse shocks, in which a supersonic flow is impeded by an obstacle. This configuration is relevant to the class of cataclysmic variable stars in which an incoming flow strikes an accretion disk at an oblique angle.

Figure ?? shows the experimental structure and some radiographic data. The 4 mm expansion distance enables the radiative shock to be sustained for tens of ns. Our experiments have included shocks formed at normal incidence and at oblique incidence [?], as shown in the figure. A future challenge is to replace the Al obstacle used now, with a flowing plasma. This system is well scaled to the astrophysical case.

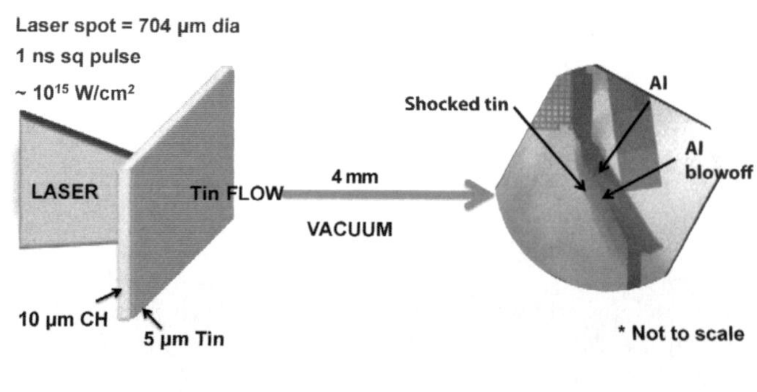


**Figure 2.** Schematic of radiative reverse shock experiment, illustrating how plasma is produced and showing radiographic data of the region where the shock forms.

## 4. Magnetized Flows

Our work with magnetized flows is aimed toward the study of interacting systems that produce magnetic field amplification within a turbulent flow, such as the flow-driven accretion disk discussed by Ryutov [?]. As a first step, we have developed a long-duration plasma jet by the irradiation of the outer surface of a conical target. This produces a supersonic flow that converges on axis, producing a jet initially confined by shocks.

By embedding this target within the equatorial plane of a cusp magnetic field we can produce a magnetized flow with the potential to amplify the frozen-in field lines as two or more jets collide. The experiments on Omega produce a jet having a velocity above 100 km/s; the plasma produced by jet collisions will have a magnetic Reynolds number of about 100. Figure ?? shows one such jet and the jet velocities of opposing jets, measured using Thomson scattering of UV light.

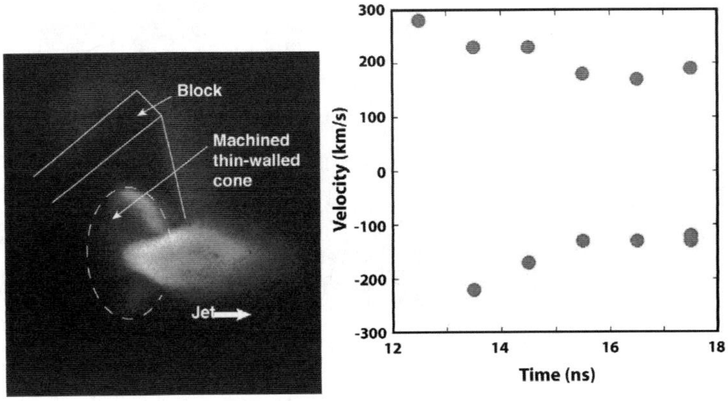


**Figure 3.** Left: An optical image of a plasma jet produced by the rear-surface irradiation of a conical target. Right: The velocity of jets produced in opposing directions.

## 5. X-ray imaging

As is evident from the above data, a key diagnostic for much of our research is x-ray imaging, which has led to our involvement in improved x-ray diagnostics. Recently, our focus has been on imaging x-ray Thomson scattering (XRTS). In collaboration with Los Alamos National Laboratory, we developed an imaging x-ray spectrometer that could be fielded on Omega [?]. Using it, we obtained the first simultaneous measurements of profiles of temperature, density, and ionization in experiments in which a blast wave was driven into foam. This provided one of the few consistency tests to date of the properties inferred from XRTS [?]. Figures ?? and ?? show a schematic of this experiment and examples of the data.

## Acknowledgments

We acknowledge contributions from S. Klein and R. Gillespie and the General Atomics technical staff in target fabrication, M. Trantham in simulations, and the operations staff at the Omega Laser Facility. This work is funded by the U.S. Department of Energy, through the NNSA-DS and SC-OFES Joint Program in High-Energy-Density Laboratory Plasmas, grant number DE-NA0001840, and the National Laser User Facility Program, grant number DE-NA0000850, and through the Laboratory for Laser Energetics, University of Rochester by the NNSA/OICF under Cooperative Agreement No. DE-FC52-08NA28302.

## References

[1] Boehly T, Craxton R, Hinterman T, Kelly J, Kessler T, Kumpman S, Letzring S, McCrory R, Morse S, Seka W, Skupsky S, Soures J and Verdon C 1995 *Rev. Sci. Intsr.* **66** 508–510
[2] Klein S R, Gamboa E J, Huntington C M, Kuranz C C, Susalla P, Chadwick S, Lairson B, Hoover D E, Elsner F, Malamud G, Di Stefano C, Gillespie R S and Drake R P 2013 *Fusion Science and Technology* **63** 305–312
[3] van der Holst B, Toth G, Sokolov I V, Torralva B R, Powell K G, Drake R P, Klapisch M, Busquet M, Fryxell B and Myra E S 2013 *High Energy Density Physics* **9** 8–16
[4] Malamud G, Di Stefano C A, Elbaz Y, Huntington C M, Kuranz C, Keiter P A and Drake R P 2013 *High Energy Density Physics* **9** 122–131
[5] Di Stefano C, Malamud G, Kuranz C, Klein S, Stoeckl C and Drake R 2015 *Appl. Phys. Lett.* **106** 114103
[6] Malamud G, Shimony A, Wan W, Di Stefano C, Elbaz Y, Kuranz C, Keiter P, Drake R and Shvarts D 2013 *High Energy Density Physics* **9** 672–686
[7] Kuranz C, Park H S, Remington B, Drake R, Miles A, Robey H, Kilkenny J, Keane C, Kalantar D, Huntington C, Krauland C, Harding E, Grosskopf M, Marion D, Doss F, Myra E, Maddox B, Young B, Kline J, Kyrala

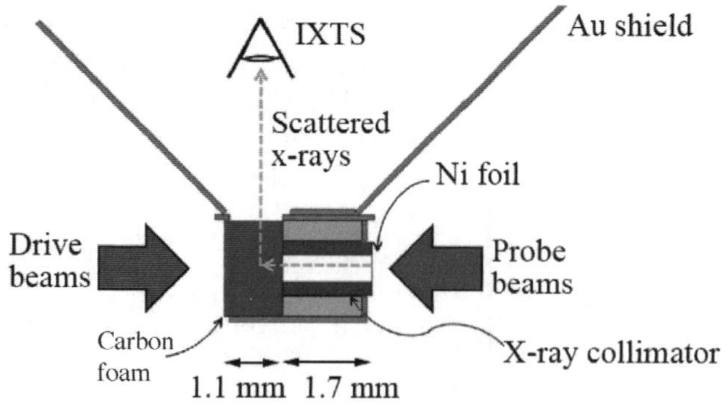


**Figure 4.** Schematic of the x-ray scattering experiment.

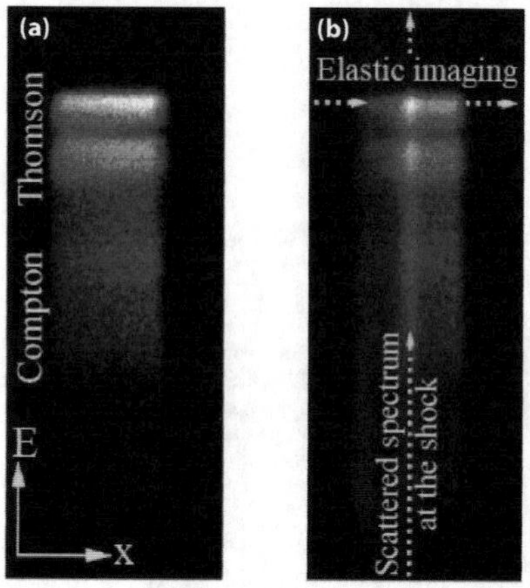


**Figure 5.** Data from two shots. Left: undriven target. Right: driven target.

G, Plewa T, Wheeler J, Arnett W, Wallace R, Giraldez E and Nikroo A 2011 *Astrophysics and Space Science* **336** 207–211

[8] Huntington C, Kuranz C, Drake R, Miles A, Prisbrey S, Park H S, Robey H and Remington B 2011 *Physics of Plasmas* **18** 112703

[9] Hogan W, Moses E, Warner B, Sorem M and Soures J 2000 *Nucl. Fusion* **41** 567–573

[10] Doss F, Drake R and Myra E 2011 *Physics of Plasmas* **18** 056901

[11] Visco A, Drake R, Glenzer S, Doeppner T, Gregori G, Froula D and Grosskopf M 2012 *Phys. Rev. Lett.* **108** 145001

[12] Kuranz C C, Drake R P, Krauland C M, Marion D C, Grosskopf M J, Rutter E, Torralva B, Holloway J P, Bingham D, Goh J, Boehly T R and Sorce A T 2013 *Physics Of Plasmas* **20**

[13] Kuranz C C, Drake R P, Huntington C M, Krauland C M, Di Stefano C A, Trantham M, Grosskopf M J, Klein S R and Marion D C 2013 *High Energy Density Physics* **9** 315–318

[14] Krauland C, Drake R, Kuranz C, Loupias B, Plewa T, Huntington C, Kaczala D, Klein S, Sweeney R, Young R, Falize E, Villette B and Keiter P 2013 *Astrophysical Journal Letters* **762** 1–4

[15] Ryutov D 2010 *Astrophysics and Space Science* DOI 10.1007/s10509–010–0558–9 - Using intense lasers to simulate aspects of accretion discs andoutflows in astrophysics - 1

[16] Gamboa E J, Montgomery D S, Hall I and Drake R P 2011 *Journal of Instrumentation* **6** P04004

[17] Gamboa E, Drake R, Falk K, Keiter P, Montgomery D and Benage J 2013 *High Energy Density Physics* **11** 75–79

# The Ignition Physics Campaign on NIF: Status and Progress

**M. J. Edwards for the Ignition Team**

Lawrence Livermore National Laboratory, Livermore CA USA 94551

E-mail: edwards39@llnl.gov

**Abstract.** We have made significant progress in ICF implosion performance on NIF since the 2011 IFSA. Employing a 3-shock, high adiabat CH ("High-Foot") design, total neutron yields have increased 10-fold to 6.3 $x10^{15}$ (a yield of ~ 17 kJ, which is greater than the energy invested in the DT fuel ~ 12kJ). At that level, the yield from alpha self-heating is essentially equivalent to the compression yield, indicating that we are close to the alpha self-heating regime. Low adiabat, 4-shock High Density Carbon (HDC) capsules have been imploded in conventional gas-filled hohlraums, and employing a 6 ns, 2-shock pulse, HDC capsules were imploded in near-vacuum hohlraums with overall coupling ~ 98%. Both the 4- and 2-shock HDC capsules had very low mix and high yield over simulated performance. Rugby holraums have demonstrated uniform x-ray drive with minimal Cross Beam Energy Transfer (CBET), and we have made good progress in measuring and modelling growth of ablation front hydro instabilities.

## 1. Introduction.

A primary goal of the Inertial Confinement Fusion (ICF) program on the National Ignition Facility (NIF) is to implode a low-Z capsule filled with ~ 0.2 mg of DT via laser indirect-drive ICF and demonstrate fusion ignition and propagating thermonuclear burn [1].

At the end of the National Ignition Campaign (NIC) in 2012, we had demonstrated implosion and compressed fuel conditions at ~ 80-90% for most ignition point design values independently, but not at the same time [1]. For low mix implosions the nuclear yield was a factor of ~ 3–10X below the simulated values and a similar factor below the alpha dominated regime. The principal reason for this appeared to be a hot spot density, and therefore pressure, that was factor of ~ 2-3 below the simulated values, although the stagnation pressure scaled with implosion velocity as expected. Ablator mix into the hot spot was also observed at lower velocities than predicted, and correlated strongly with the measured ion temperature and yield. Indications were that this mix was being driven by hydrodynamic instability at the ablation front. Angular measurements of down-scattered neutrons indicate that there could be low spatial mode asymmetries in the compressed main fuel, which may explain some of the deficit in pressure and the larger than expected mix. The NIC target is shown in Fig. 1.

In order to achieve the required symmetric capsule shape, we typically have to transfer a large fraction (~ 30%) of the energy and power from the outer to the inner beams via Cross-Beam Energy Transfer (CBET) [2, 3]. Since this transfer is dependent on the laser intensity and the plasma conditions in the region of the overlapping beams, it is not well modeled yet in an integrated fashion in the simulation codes. We believe the time and spatially varying CBET is to a large extent responsible for observed time dependent asymmetries of the imploded capsule and contributed to the decrease in implosion performance as the laser power and capsule implosion velocity increased [3, 4].

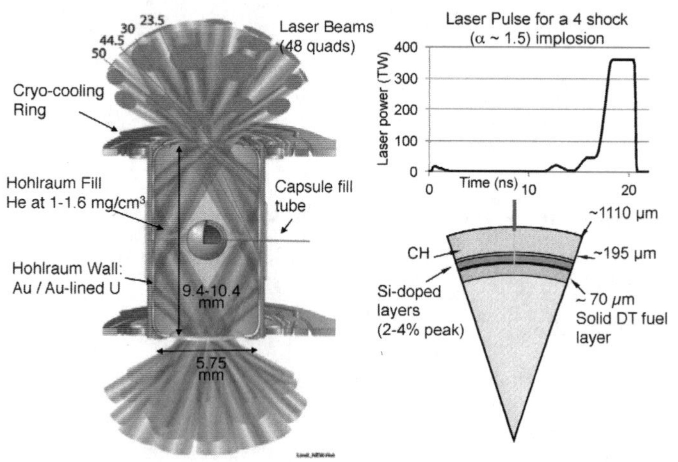


Fig 1. Schematic of NIF CH capsule in a gas-filled Au or Au-lined U hohlraum and a typical laser pulse that produces a 4-shock low adiabat (α ~1.5[1]) implosion. The laser power is stepped in time to gradually increase the X-ray drive on the capsule such that the pressure in the ice is increased in four shocks from ~1 Mbar to a value in excess of 100 Mbar during the implosion phase. The hohlraum is filled with He gas at a density ~ 0.96 mg/cm$^3$, to control wall motion for drive symmetry control.

During 2013, the theoretical and experimental effort has been focused on understanding the underlying physics issues responsible for the deviation from modeled performance. This paper provides an overview of the current progress of the X-ray drive ICF program, and in particular discusses the results of 1) implosions of CH capsules under drive conditions that are more robust to hydrodynamic mix; 2) preliminary results of implosions in Rugby-shaped hohlraums with minimal CBET; and 3) implosions with High Density Carbon (HDC) ablators both in gas-filled and Near Vacuum Hohlraums (NVH). He gas fill in NVH has so far been 0.03 mg/cm$^3$.

## 2. Higher Adiabat, High Foot, Lower Hydrogrowth Implosions with CH Capsules.

The goal of the 'High-Foot' campaign is to manipulate the laser pulse-shape and the resulting drive (see Fig.2) to create a more one-dimensional and robust implosion that is more resistant to ablation driven hydrodynamic instabilities and the resulting mix of the ablator into the DT fuel [5-8]. The high foot CH design uses the same target as the low foot NIC design. The key pulse-shape changes, as compared to the low-foot pulse are, more laser power at early time which causes the radiation temperature in the 'foot' of the pulse to be higher (hence the name 'High-Foot'), 3 shocks instead of 4 and a decrease in pulse length to maintain shock timing. The drive for the High-Foot capsule is tailored to control the ablation front instability growth. The higher foot temperature and resulting higher adiabat causes more ablative stabilization, longer scale lengths and lower In Flight Aspect Ratio (IFAR). However it also results in lower final density, reducing ideal margin for ignition in 1-D.

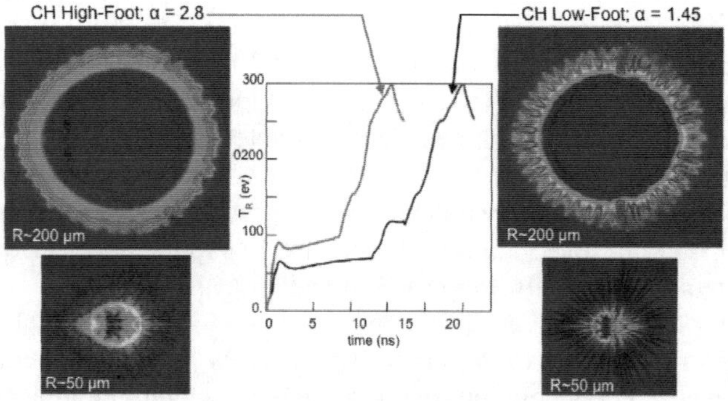


**Fig. 2** Simulations of high and low-foot implosions at two different radii. The higher foot implosions show less fine-scale structure, indicating resistance to acceleration driven instability, and lower ablator density, indicating less compression. The values of alpha shown at top are the adiabat of each implosion. The 3-shock and 4-shock radiation drives for high and low foot is shown in the center.

The reduction of growth of surface perturbations for the high foot implosions was measured in a series of hydro-growth experiments (see Fig. 3) [9, 10].

[1] Adiabat defined as $P/P_{cold}$, where $P_{cold}$ is minimum pressure at 1000g/cc from EOS of DT [17].

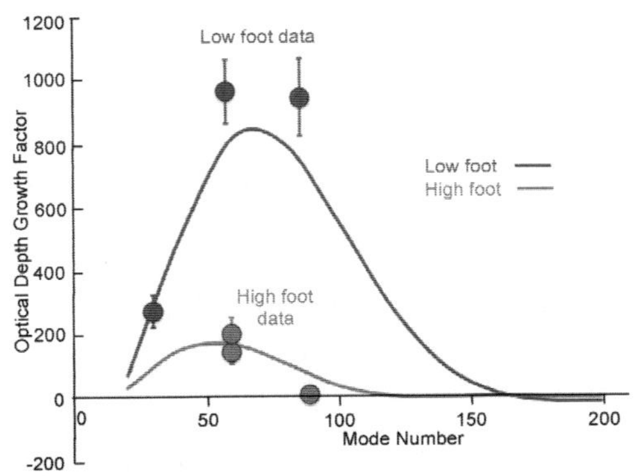


**Fig. 3.** Pre-imposed capsule surface perturbations with several different modes were probed using both the low and high foot pulse shapes. Linear growth factor is shown as a function of mode number for both low and high foot drive at the time the shell has converged to 650 $\mu$m and demonstrates reduced growth for mode 60 and significantly reduced growth of mode 90 when comparing high foot to low foot drive, as was expected due to its higher adiabat design. Simulations for growth are seen to be in reasonable agreement with the data.

A series of DT-layered high foot implosions were carried out in 2013 (see Fig. 4). They had high yield-over-simulated performance (50-70%) and have been diagnosed to have very low mix. By the end of November 2013, total DT neutron yields of 6.3 x$10^{15}$ (a yield of ~ 17 kJ) were obtained. In the highest yield shot (N131119 in Fig. 4) the yield from alpha self heating is essentially equivalent to the compression yield, meaning that NIF results are very close to alpha dominated performance. As with the low-foot NIC targets, the high-foot hohlraums also required large CBET in order to achieve adequate hot-spot symmetry [5-8]. As can be seen from the time integrated hot-spot polar and equatorial x-ray images of Fig. 4, and the corresponding equatorial neutron images in Fig. 5, achieving symmetry and high velocity at the same time remains a challenge, even for the high foot design.

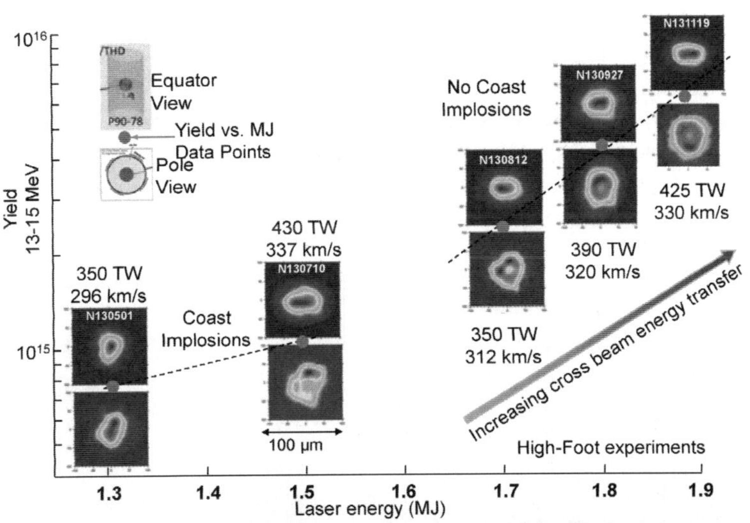


**Fig. 4**. Total neutron yield and time integrated hot-spot polar and equatorial images as a function of laser energy in MJ for the DT layered high-foot shots. Early experiments with 1.3-1.5 MJ were "coast" implosions, where the laser pulse was turned off when the ablator/fuel interface was at a radius of 600 $\mu$m. In order to increase the implosion velocity and prevent shell decompression, later "no-coast" experiments had a longer peak power pulse (and correspondingly larger total laser energy).

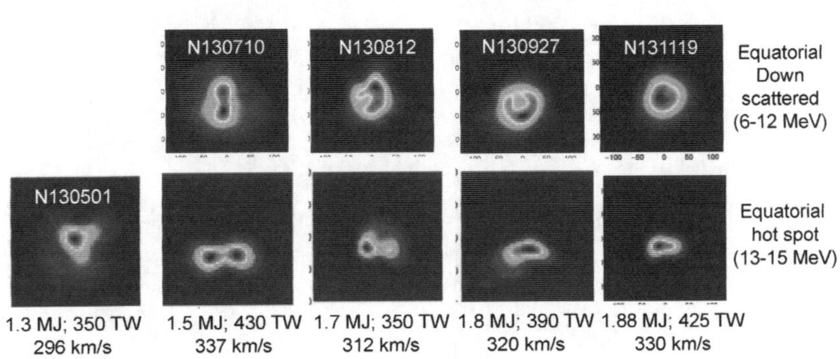


**Fig. 5.** Neutron images [11-13] for shots of Fig. 4. The upper row shows images of the down-scattered neutron from the high-density main fuel and the bottom row has images of the hot spots produced by the 13-15 MeV neutrons.

In spite of the toroidal shapes of the high-density main fuel and the asymmetric hot spots, the inferred levels of CH ablator mix into the hot-spots were low for the high-foot implosions, confirming the reduction of ablator mix for the high-foot design. Further increasing the yields will require increasing the implosion velocity ($v_{imp}$) and compression, as well as improving the hohlraum drive symmetry. Near term we will attempt to increase $v_{imp}$ by increasing laser energy and power, using thinner ablators as well as hohlraums made from Depleted Uranium (DU), and further fine tuning the CBET. Longer term, a more predictable, more efficient gas-filled hohlraum with better symmetry control is clearly desirable. Alternate ablators designs may also present more favorable conditions than the current experiments. This will be discussed in the next two sections.

## 3. Implosions in Rugby-Shaped Hohlraums.

Rugby-shaped gold hohlraums driven by low-adiabat laser pulse shapes have recently been tested on NIF [14]. The potential benefits of a rugby-shaped hohlraum (see Fig. 6) over the NIC-style cylindrical hohlraum (Fig. 1) can be exercised in two ways. The reduction in surface area over a NIC hohlraum with similar diameter and length results in a higher coupling efficiency and achieves a near 20% improvement in peak drive x-ray flux. Alternatively, a rugby hohlraum with a similar area to a NIC hohlraum will achieve a similar drive but due to the larger diameter, will have improved beam clearances (to the capsule and laser entrance holes) and higher hohlraum case-to-capsule ratio (CCR) for greater smoothing of hohlraum radiation modes, and could potentially allow uniform x-ray drive symmetry with minimal CBET.

The rugby hohlraum experiments on NIF tested the 2$^{nd}$ option. With a diameter of 7mm and a length of 10.5 mm the rugby has essentially the same surface area (to within a few %) as the current NIC hohlraum shown in Fig. 1. This results in a 22% larger CCR and nearly 30% greater beam clearance over the capsule than the NIC hohlraum. Simulations indicated that the rugby could have good inner beam propagation and x-ray drive symmetry without the large amount of CBET required to achieve adequate inner beam propagation and x-ray drive symmetry for the NIC hohlraum. For the NIC hohlraums an ~ 8Å wavelength separation between the outer and inner laser beams is used to transfer ~ 30% of the outer beam energy and power to the inner beams [3, 4].

The 7 mm rugby with a CH Symcap was fielded in May of 2013 with laser energy of 1.3 MJ and peak power of 370 TW. A 1Å separation between inner and outer beams was used, resulting in minimal CBET. The overall coupling was higher (~93% vs ~ 85%) than is typical for NIC hohlraums, and stimulated Raman scattering (SRS) backscatter from inner beams as a fraction of total energy was ~ ½ that observed for NIC hohlraums. However, the time integrated equatorial hot spot x-ray image was severely oblate with a P2/p0 of ~ 66% (see Fig. 6). Post-shot analysis done both by LLNL [14] and at the French CEA-DAM-DIF laboratory [15] indicated that Au plasma blow off from the hohlraum wall generated by the outer beams severely impeded the propagation of the inner beams.

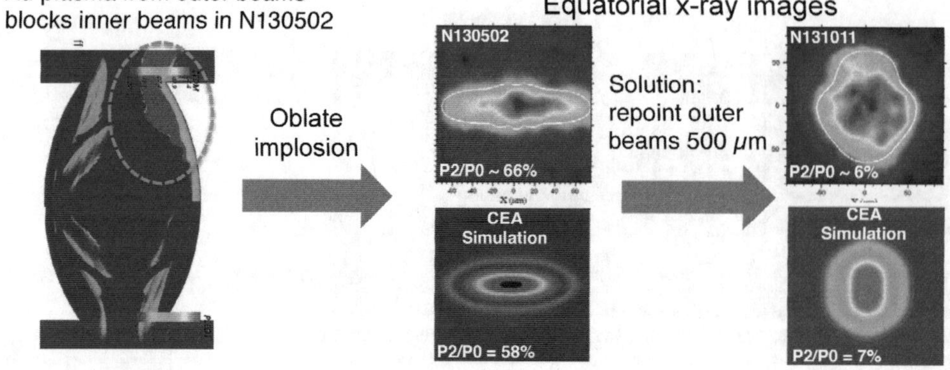


**Fig. 6.** Rugby target geometry and equatorial x-ray images from two NIF Rugby experiments. Also shown are simulated images generated by CEA-DAM-DIF.

The same post-shot simulations indicated that repointing the outer beams (moving the outer beams in by 500 $\mu$m) would eliminate the Au blow-off problem. The result from executing this repointing is also shown in Fig. 6. The laser energy and power were identical to the May shot and the same 1Å wavelength separation between inner and outer beams was used. Additionally, a small amount of Ne (2.5%) was added to the hohlraum gas fill in an attempt to reduce SRS, hot electrons generated by SRS, and 2-plasma decay. The primary objective of controlling core shape was successful and can be seen in Fig. 6. The central core emission was very close to round with a $P_2/P_0$ = 6% ± 1%. Additionally, a very low $P_2/P_0$ x-ray flux symmetry swing in time (< 10% $P_2/P_0$ per ns) was observed, in contrast to symmetry swings of ~ 30%/ns for NIC hohlraums that employ large CBET. The secondary objective of reducing SRS was marginally successful. The total measured SRS was similar to the May shot, but the hot electron signal, believed to be generated by SRS, was reduced by a factor of 4 or greater at all energies. There was also an increase in stimulated Brillion scattering which resulted in an overall coupling of ~ 88%.

It thus appears that a rugby hohlraum can indeed provide good x-ray drive symmetry with minimal CBET and offers the potential for low x-ray flux symmetry swings in time. Future high-foot experiments will take advantage of these features to increase $v_{imp}$ and capsule implosion symmetry.

## 4. Implosions with High Density Carbon Ablators.

High Density Carbon (HDC) is an interesting option as an ablator material for indirect drive ICF implosions. HDC has a higher density (3.5 g/cc) than plastic (CH ~ 1g/cc), which results in a thinner ablator with a larger inner radius for a given capsule scale. This results in more pdV work done on a larger fuel and hot-spot volume and the thinner shell leads to less absolute inward motion during shock compression. The overall result is higher x-ray absorption (and higher overall efficiency) and shorter laser pulses compared to equivalent CH designs [16, 17] (see Fig. 7 and 8). During 2013, we carried out a series of experiments to examine the feasibility of using HDC as an ablator using both gas filled hohlraums and lower density, Near Vacuum Hohlraums (NVH)[16].

A series of five experiments using a 4-shock pulse shape designed to produce an adiabat 1.5 implosion culminated in a cryogenic CH symcap target filled with DT gas [18]. Fig.7 compares the HDC capsule and its 1.3 MJ, 350 TW laser pulse shape to a CH target and the low-foot pulse used for the adiabat 1.4 CH NIC implosions. With the slightly higher picket and foot for the HDC design, the He gas fill in hohlraum was increased to 1.2 mg/cm$^3$. The measured laser to hohlraum coupling for N130628 was 91 ± 2%, compared to 87 ±3% average for gas filled hohlraums driving CH capsules [19]. The DT neutron yield of 1.6x10$^{15}$ at an ion temperature of 2.9 ± 0.1 keV is the highest gas-filled implosion yield to date and had a better than 50% Yield-over-Clean (YoC).

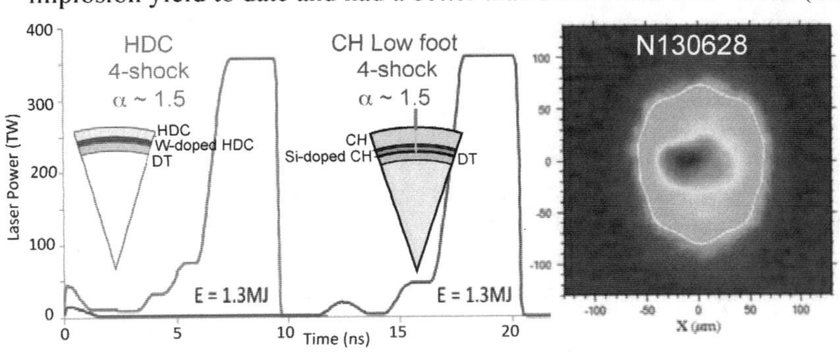


**Fig. 7a** The 4-shock, 1.5 adiabat laser pulse for HDC is 50% shorter than the 4-shock low foot 1.5 adiabat laser pulse for the NIC point design CH target.
**Fig. 7b** Time integrated x-ray image of hot spot is fairly symmetric with an average radius 57±4 $\mu$m and a P2/P0 = 0.19 ± 0.08

In addition to the 4-shock HDC design discussed above, there are also HDC designs with 2 and 3 steps (or shocks) in the x-ray drive [16]. These are compared in Fig. 9 with the 3-shock CH high-foot and the 4-shock NIC low foot, low adiabat CH designs.

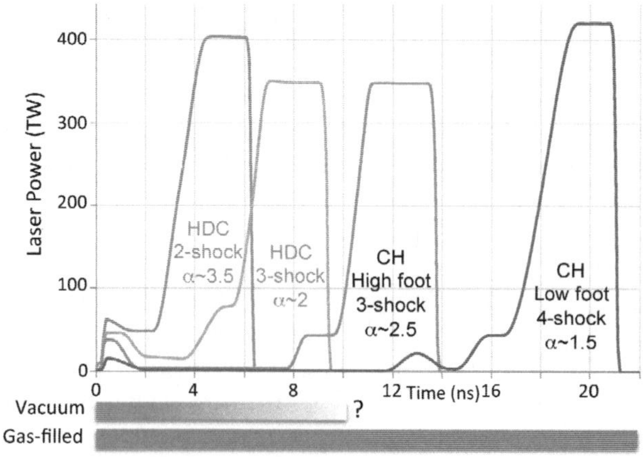


**Fig. 9.** Pulse shapes and adiabats for 2- and 3-shock HDC, and high and low foot CH designs. The bars below the plot indicates that although gas-filled hohlraums can be used for all designs, (and must be used for the 14-22 ns pulse length required for CH targets), the pulse lengths for the 2- and 3-shock HDC implosions are short enough that near vacuum hohlraums can (and have been) used for the 2-shock HDC design, and possibly could be used for the 3-shock design.

The 2-shock high adiabat ($\alpha$ ~3.5) design uses an un-doped HDC capsule to reduce the ablation front growth factors by more than a factor of 10 over the 4-shock HDC design. This 2-shock pulse is not an ignition design but it is predicted to produce ~ $2x10^{16}$ yield in 1D, which would allow valuable insight into implosion performance in the alpha heating regime while minimizing the effect of ablation front instability growth. The 6-7 ns pulse allows less time for wall motion and plasma filling and opens up the possibility of using hohlraums with very low gas fill (He = $0.03mg/cm^3$), or Near Vacuum Hohlraums (NVH). The advantage of a NVH is that it has reduced backscatter, little or no hot electron generation and very high (98-99%) hohlraum efficiency [20]. A series of shots were conducted to explore the performance of the 2-shock HDC design in a NVH [16], culminating with a 1.16 MJ, 305 TW shot on a cryogenic DT-layered shot that produced $1.8x10^{15}$ neutrons (see Fig. 10). Laser-to-hohlraum coupling was 98.5%, and post-shot simulations indicate yield-over-simulated performance of ~ 25-50%. An earlier cryogenic THD layered implosion version was consistent with a fuel velocity = 430 ±50km/s with no observed ablator mixing into the hot spot. The remaining issue with these types of hohlraums is that plasma filling in large laser spot motion. In particular, the inner beams provide the x-ray drive to the capsule equator and impaired propagation leads to asymmetry and/or time dependent symmetry swings through the peak of the x-ray drive. Future plans include the use of larger diameter cylindrical and/or shaped hohlraums to mitigate this effect.

HDC ablators have been shown to perform well in ICF implosions using both conventional gas filled and NVH. NVH may provide an alternate path for higher adiabat implosions with very high velocities and little or no hotspot mix. These high velocities, in excess of 370km/s and the high hohlraum efficiency (>98%) indicate that the NVH size can be increased for fixed capsule radius while maintaining adequate drive to achieve ignition velocities. Increasing the case to capsule ratio in this way creates more room for the inner beams to propagate. Similarly using different shape hohlraums can reduce hohlraum wall plasma blow off in the path of the inner beams, improving the drive to the capsule equator. Finally, choosing an intermediate hohlraum gas density should also reduce hohlraum wall plasma filling, further improving inner beam propagation while keeping below thresholds for laser plasma instabilities. A series of 2D simulations indicate that by using all three of these tools, time dependent symmetry should be improved for pulses as long as 9ns. If this can be realized, it opens up the possibility of extending the promising NVH to 3-shock, lower adiabat ($\alpha$ ~ 2) HDC designs. A series of experiments are planned in 2014 to examine this parameter space.

## 5. Conclusion.

Significant progress in ICF implosion performance on NIF was seen in 2013 (see Fig. 10). Options to increase $v_{imp}$ and hence the yield for the high foot target include increasing laser energy and power, using thinner ablators as well as DU hohlraums, and further fine tuning the CBET. We will also investigate CH high foot capsules in rugby hohlraums.

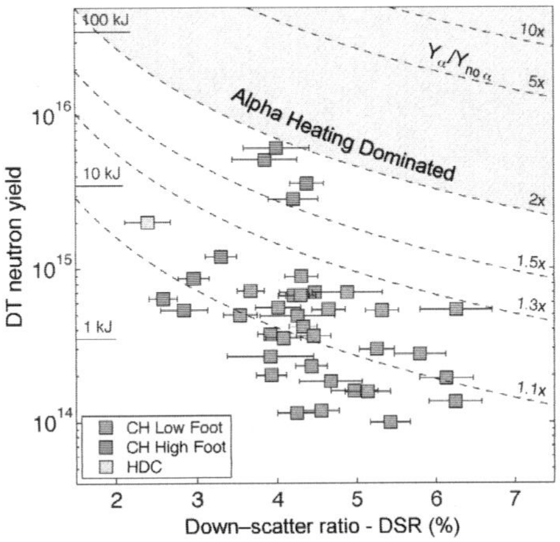


**Fig. 10.** Total neutron yield (13-15 MeV) plotted against DSR, the ratio of downscattered to unscattered neutrons. DSR is approximately proportional to the DT fuel ρr [21]. High foot, high adiabat (α~2.5) CH targets using a 3-shock design, (green squares) have demonstrated a total neutron yield of 6.3 x$10^{15}$ (a fusion yield of ~ 17 kJ) with DT layered targets. This represents a near 10-fold increase over the low-foot CH design (blue squares). At a yield of 17 kJ, the yield from alpha self-heating for the high-foot design is ~ equal to the compression yield (the threshold for alpha self-heating). A DT-layered HDC target with a 7 ns, 2-shock (α~3.5) pulse in a near vacuum hohlraum achieved a yield of 1.8 x$10^{15}$(yellow square).

HDC ablators have been shown to perform well using both gas filled and NVH. Four-shock low adiabat (α~1.5) HDC capsules were imploded in conventional gas-filled hohlraums and produced higher quality implosions than equivalent CH designs. Both the 4- and 2-shock HDC capsules had very low mix and high YoC performance. The Near Vacuum Hohlraums provide an alternate path for higher adiabat implosions with very high velocities and little or no hotspot mix.

Rugby holraums have demonstrated uniform x-ray drive with minimal CBET and will be used with high foot CH designs.

We have made good progress in measuring and modeling growth of ablation front hydro instabilities.

This work was performed under the auspices of Lawrence Livermore National Security, LLC (LLNS) under Contract DE-AC52-07NA27344.

**References**

[1] Edwards M. J *et al.* 2013 *Phys. Plasma.* **18** 050901
[2] Michel P *et al.* 2010 *Phys. Plasma.* **17** 056305
[3] Hinkel D. E. *et al.* 2014 *Proceedings of IFSA 2013*
[4] Moody J. D. *et al.* 2014 *Submitted for publication to Phys. Plasma*
[5] Hurricane O *et al.* 2014 *Nature, in print,* **DOI** 10.1038
[6] Hurricane O *et al.* 2014 *Phys. Plasma.* **21**
[7] Park, H.-S., *et. al.* 2014 *Phys. Rev. Lett.* ***112**, 055001*
[8] Dittrich, T. R. *et. al.* 2014 *Phys. Rev. Lett.* ***112**, 055002*
[9] Smalyuk V. A. *et al.* 2014 *Submitted for publication to Phys. Plasma*
[10] Casey D. T. *et al.* 2014 *Submitted for publication to Phys. Rev. Lett*
[11] Merril, F. E. *et al.* 2012 *Rev, Sci, Instrum.* ***83**, 10D317*
[12] Volegol, P. *et al.* 2014 *Rev, Sci, Instrum.* ***85**, 023508*
[13] Grim, G. *P. et al.* 2013 *Phys. Plasma.* **20**, 056320
[14] Amendt P. *et al.* 2014 *Proceedings of IFSA 2013*
[15] Leidinger J-P. 2013 *Private communication*
[16] MacKinnon A. J. *et al.* 2014 *Submitted for publication to Phys. Plasma*
[17] Haan S. W. *et al.* 2011 *Phys. Plasma.* **18** 051001
[18] Ross J. S. *et al.* 2014 *Submitted for publication to Phys. Rev. Lett*
[19] Town R. P. J. *et al.* 2011 *Phys. Plasma.* **18** 056302
[20] Le Pape S. *et al.* 2014 *Submitted for publication to Phys. Rev. Lett*
[21] Patel, P., Spears, B. K. *Private communication*

# Investigating Turbulent Mix in HEDLP Experiments

**K. A. Flippo, F. W. Doss, B. Devolder, J. R. Fincke, E. N. Loomis, J. L. Kline, L. Welser-Sherrill**
Los Alamos National Laboratory, PO Box 1663, Los Alamos, NM, USA
E-mail: kflippo@lanl.gov

**Abstract.** Mix is an important issue in High Energy Density Laboratory Plasmas (HEDLP), specifically Inertial Confinement Fusion (ICF) implosions. In ICF, shock waves traverse fuel capsule defects and material interfaces, and due to hydrodynamic instabilities transitioning into turbulence, these shocks can initiate mix between shell and fuel, degrading yield. To this end, a series of laser-driven mix experiments has been designed for the OMEGA and NIF laser facilities to investigate the turbulent mixing of materials proceeded by reshock and shear, which initiates Richtmyer-Meshkov and\or Kelvin-Helmholtz instabilities on a tracer layer. The experiments are designed to understand if the Besnard-Harlow-Rauenzahn (BHR) mix model that has been implemented in LANL's RAGE hydrodynamics code has coefficients that are properly determined for an HEDLP environment.

## 1. Introduction

Los Alamos National Laboratory has been engaged in Inertial Confinement Fusion (ICF) research in support of ignition on the National Ignition Facility (NIF). As part of this effort, we are studying the effects of material mixing inside of ICF capsules which can degrade the fusion yield. One of the areas we have focused on as part of our research is the effect of instabilities and their transition to turbulence on mix. Much of this mixing becomes relevant to ICF implosions following the shocks crossing the material boundaries in a fuel capsule, as well as in other applications like shocks in supernovae implosions. Shown in Figure 1 is a representation of two typical regions where instabilities in an implosion reside. Rayleigh-Taylor (RT) and Richtmyer-Meshkov (RM) [1, 2] instabilities can be initiated during the initial phase of the implosion. In the current point design for ignition, four shocks traverse the capsule and converge on the center of the capsule to heat and ignite the fuel. Strong shocks can cause the instabilities to progress rapidly into the turbulent regime when traversing these regions under the right conditions. Figure 1 shows two regions of interest, one subject to the RM instability labeled as the "Re-Shock Region" and another region subject to the Kelvin-Helmholtz (KH) [3, 4] instability labeled "Shear Region". The re-shock region is characterized by a shock traversing the material interface, in this case the capsule shell and ice/fuel boundary, subjecting it to RM initially; then a reflected shock hits the layer again from the opposite direction, causing the layer to be recompressed and quickly transition into a turbulent mix regime [5]. The shear region is characterized by two flows in opposite directions, one light (the fuel), and one heavy (the capsule shell), subjecting the region to KH shear instability, when one shock passes or when two shocks pass in opposite directions. This subjects the layer to intense shear forces, causing the initial instability to transition into a turbulent regime. To this end, we have developed the *Reshock/Shear* platform for the Omega laser system to study these interactions in detail and compare the conditions to one particular turbulence model, the Besnard-Harlow-Rauenzahn (BHR) mix model [6, 7], which has been implemented in RAGE [8].

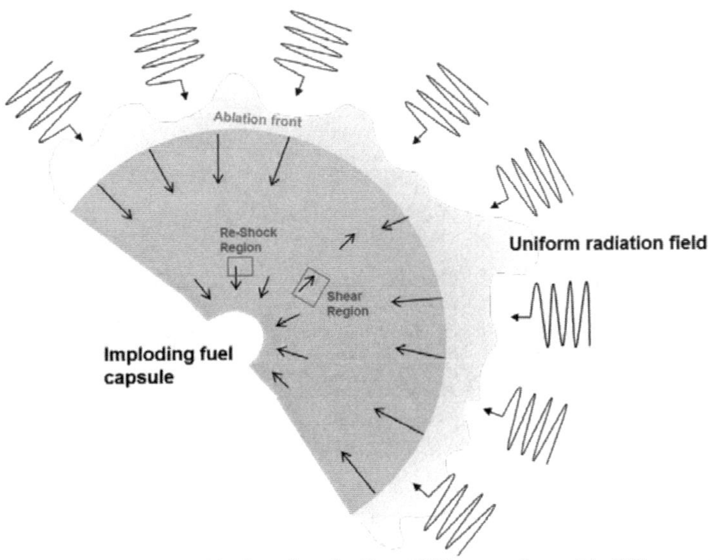


**Figure 1**. Schematic of half of an imploding ICF capsule, with DT gas at the center (white), DT ice (grey), and the capsule wall/ablator (orange).

## 2. The (Re)Shock/Shear Platforms

The Omega campaigns on reshock and shear have been successful in helping us understand the BHR mix model in the context of high energy density laboratory plasma (HEDLP) experiments, by measuring the expansion of a tracer layer as it mixes into the surrounding material. These campaigns used a single platform target exterior that is changed on the interior of the shock tube to be in any of four *Reshock* or *Shear* configurations [9]. The Omega *Reshock* campaign [9] has shown that the BHR model successfully models shock and reshock evolution in the HEDLP regime, even though the model's parameters were set from non-HEDLP hydro conditions [7]. The Omega *Shear* campaign [9, 10] has revealed similar results in the shear geometry, and revealed the dependence of the KH instability on the initial scale length parameter. Additionally, the results revealed indirect evidence of turbulent self-heating via thermal-turbulence coupling [11]. Drawing on the success of this platform, we are extending these experiments to the NIF. We have designed the platform and are currently carrying out experiments, including an ultra-large area x-ray backlighter benchmarking [12, 13].

The *Shear* experimental platforms for Omega and NIF are shown side-by-side in Figure 2. The NIF laser system is capable of delivering up to1.8 MJ, which is 45 times that of the Omega laser system. This allows for much larger experimental volumes, higher drives, and more energy deposited into the system to perform HEDLP experiments. Unlike the Omega experiments, which used direct laser-ablation to produce the shocks in the shock tubes, the NIF experiments use halfraums on the ends of the tube to indirectly drive the ablator with x-rays and produce the shocks, as this allows much longer support times for the shocks (past 10 ns) and gives greater flexibility over the shock strengths and convective Mach numbers.. This source temperature, currently designed for ~ 260 eV, can be tuned by changing the laser energy delivered to the halfraums.

Figure 3 shows the NIF setup of the shock-tube (gold) and iron backlighter (blue), with the UV laser beams in a variety of colors. Only the *outer cone* beams, those at 44.5 and 50 degrees, are used for driving the halfraums, and the *inner cone* beams, those with 23.5 degrees are used for driving the backlighter, with the exception of two *outer cone* beams (50-degree) used to flatten out the backlighter uniformity for a total of 136 kJ of laser energy on the backlighter.

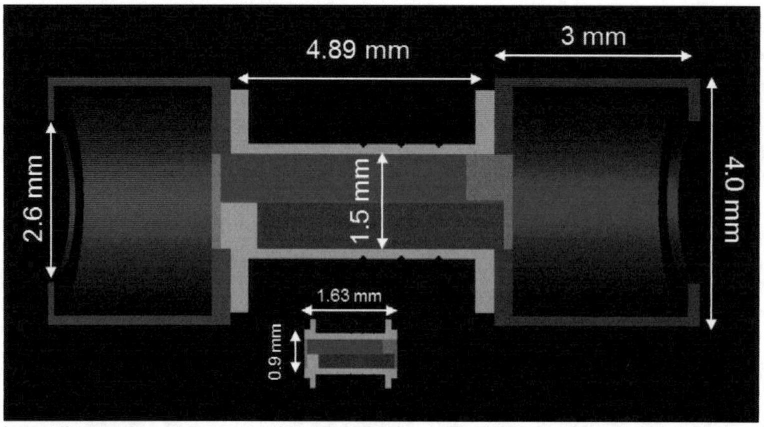


**Figure 2**. Relative dimensional comparisons of the *(Re)Shock/Shear* platforms for both Omega (bottom) and NIF (top). The NIF platform is indirectly driven by two halfraums (blue) on either end, which push the Rexolite ablators (orange and pink) with 260 eV radiation, instead of directly by the UV laser beams at Omega. The NIF platform, encased in a 250 µm thick Rexolite tube instead of a Be tube (green), is 3 times longer with a 194 times greater volume, allowing for longer experiments and fewer edge effects. The shocks are blocked from propagating on the opposite halves by dense gold plugs (red and light green); such that shocks propagate from opposite directions on either half of the tube. The interior shows the counter-propagating *Shear* geometry, with two foams (purple and dark grey) separated by a tracer layer (light grey).

## 3. NIF experiment design

The experiment was designed using the hydrocode RAGE, initially using conditions similar to the Omega experiments for comparison. The halfraum drive temperature is chosen to drive a shock of 110 km/s near the Omega experimental conditions. The Reynolds number is designed to be 6E5 (ΔΔ·) and 5E6 (ΔV), with a Zhou's [5] turbulence transition time calculated to be between 3 and 6 ns after the shocks cross. This means that in the NIF experiment, with the shock crossing time of ~17 ns into the experiment, the experiment will be fully turbulent ~23 ns into the experiment. Compared to the Omega experiment, where the shocks cross at 6 ns and the transition time was 5-10 ns after this, meaning it is not fully turbulent until >14 ns. In the Omega experiments, due to the size of the tube, the experiment ends by shocks reflecting off the end plugs back into the tube center at 20 ns. Thus, the

**Figure 3**. A visual representation of the NIF Shock/Shear setup. The main target (gold) sits at the center of the target chamber supported on a stalk from the left, with halfraums on top and bottom, which are driven by 60 top and 60 bottom *outer* cone beams (blue beams) with 310 kJ each of the halfraums. Two *outer* quads (red) and five *inner* quads (violet) are pointed at the backlighter foil (purple edge-on) driving it with ~163k J of laser energy to produce the radiograph. (*image inset*) The target as actually built.

fully turbulent data set is limited to the last 4 ns. With the NIF platform, we anticipated having at least 7 ns of data, which should be possible to extend even later with the longer delay capabilities of the backlighter beams on NIF, past 30 ns.

Figure 4 shows simulated radiographs using a 6.7 keV Fe x-ray backlighter to follow the expansion and mixing of an aluminum tracer layer initially 20-μm thick. The shocks are produced in 60 mg/cc foams inside two halves of a Rexolite cylinder from opposite directions, kept from propagating from both directions by dense gold plugs, shown in Figure 2, and as dark squares below in Figure 4. These shocks are driven by shooting 310 kJ of laser energy into each halfraum end, which is then converted into the indirect soft x-ray bath heating and ablating a Rexolite end cap, launching the shocks into the tube with a peak pressure of 143 Mbar at approximately 11 ns after the lasers arrive at the target. The shocks propagate at ~Mach 2 from opposite ends and cross around ~17 ns at the center of the tube. This produces a shear on the Al tracer layer at a rate greater than 5 $ns^{-1}$ from a $\Delta V$ of 200 μm/ns and a 40 μm initial layer width. This will cause the KH-like shear instability to transition into turbulence 5 ns later, based on the Zhou criteria. These figures of merit will be compared to the experiment to understand whether the BHR model is capturing the NIF experiment, with different conditions than those of the Omega experiment.

Figure 4(right) shows the mix width evolution for the radiographs presented on the left. One can see the evolution of the width as it starts to expand slowly due to pre-heat. Then, at 17 ns, the shocks cross and compress the layer slightly. After that, it begins to re-expand slowly due to the KH instability. Next, the layer begins to turn turbulent. Energy is then extracted from the mean flow, and converted to turbulent kinetic energy. After the transition to fully developed turbulence around 25 ns, the layer growth due to mixing proceeds at an increased rate, instead of reaching hydrodynamic equilibrium which would stop the expansion. This later time expansion is governed only by the turbulence evolution, and not by the initial conditions of the model, which is ultimately the area we would most like to explore in the NIF experiments.

## 4. Conclusions

The design and simulated results of the Shear platform on the NIF have been discussed in the context of measuring the width of a tracer layer as it undergoes shear instabilities and transitions into turbulence. This experiment will test the extension of the BHR turbulent mix model into a strongly HEDLP regime, where the turbulent energy in the tracer layer is greater than that of the surrounding material, allowing the layer to continually expand, instead of reaching hydrodynamic equilibrium with the surrounding material on the time scale of the experiment.

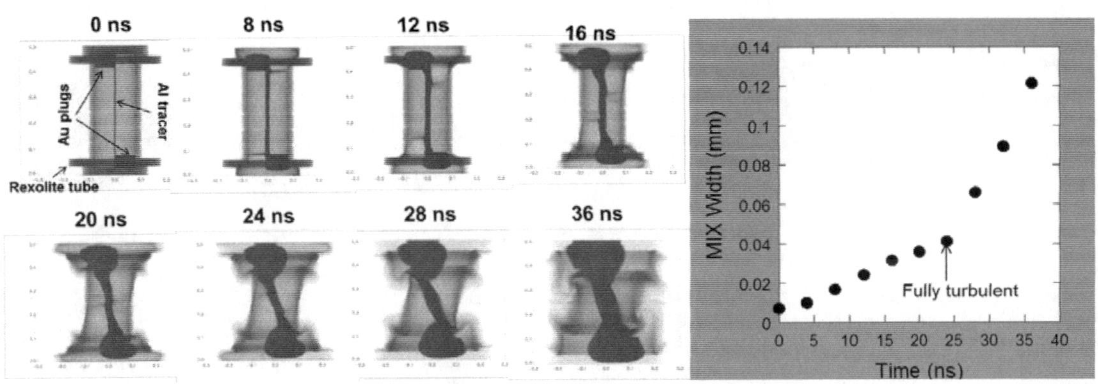


**Figure 4**. (left) Synthetic radiographs from RAGE simulations representing multiple snapshots in time for the simulated NIF Shock/Shear platform showing the expansion (initial dark central vertical line) of the mix tracer layer, which is 40 μm of Al backlit by a 6.7 keV Fe x-ray source. (right) Points from line-outs of the synthetic radiographs in Figure 4 showing the behavior of the layer, the mix width, in terms of its horizontal thickness versus time of the radiographs. An arrow represents the point where full turbulence is predicted in the simulations.

Journal of Physics: Conference Series **688** (2016) 012018 doi:10.1088/1742-6596/688/1/012018

## 5. References

[1] Meshkov EE 1969 *Sov Fluid Dyn* **4** 101
[2] Richtmyer RD 1960 *Comm on Pure and Appl Math* **13** 297
[3] Helmholtz H 1868 *Rep of Royal Prussian Acad of Phil* **23** 215
[4] Thomson WLK 1871 *Phil Mag* **42**(4) 362-77
[5] Zhou Y 2007 Unification and extension of the similarity scaling criteria and mixing transition for studying astrophysics using high energy density laboratory experiments or numerical simulations *Phys of Plasmas* **14** 082701
[6] Besnard D, Harlow F, Rauenzahn Rand Zemach C 1992 Turbulence transport equations for variable-density turbulence and their relationship to two-field models Los Alamos, Los Alamos National Laboratory LA-12303-MS
[7] Banerjee A, Gore RAand Andrews MJ 2010 *Phys Rev E* **82** 046309
[8] Gittings M, Weaver R, Clover M, Betlach T, Byrne N, Coker R, et al. 2008 The RAGE radiation-hydrodynamic code *Computational Science and Discovery* **1** 015005
[9] Welser-Sherrill L, Fincke JR, Doss FW, Loomis E, Flippo KA, Offermann D, et al. 2013 Two Laser-Driven Mix Experiments to Study Reshock and Shear *High Energy Density Physics* **9**(3) 496 – 9
[10] Doss FW, Loomis EN, Welser-Sherrill L, Fincke JR, Flippo KAand Keiter PA 2013 Instability, mixing, and transition to turbulence in a laser-driven counterflowing shear experiment *Phys of Plasmas* **20**(3) 012707
[11] Doss FW, Fincke JR, Loomis EN, Welser-Sherrill Land Flippo KA 2013 Dec The high-energy-density counterpropagating shear experiment and turbulent self-heating *Physics of Plasmas* **20**(12)
[12] Flippo KA, Kline JL, Loomis EN, Doss FW, DeVolder B, Tregillis I, et al. 2014 Development of the big area backlighter (BABL) for HED experiments *Rev Sci Inst* **85** 093501
[13] Doss FW, Loomis EN, Welser-Sherrill L, Fincke JR, Flippo KAand Keiter PA 2013 Jan Instability, mixing, and transition to turbulence in a laser-driven counterflowing shear experiment *Physics of Plasmas* **20**(1)

# Ignition calculations using a reduced coupled-mode electron-ion energy exchange model*

**W J Garbett and D A Chapman**
AWE plc, Aldermaston, Reading, Berkshire, RG7 4PR, UK

E-mail: warren.j.garbett @awe.co.uk

**Abstract.** Coupled-mode models for electron-ion energy exchange can predict large deviations from standard binary collision models in some regimes. A recently developed reduced coupled-mode model for electron-ion energy exchange, which accurately reproduces full numerical results over a wide range of density and temperature space, has been implemented in the Nym hydrocode and used to assess the impact on ICF capsule fuel assembly and performance. Simulations show a lack of sensitivity to the model, consistent with results from a range of simpler alternative models. Since the coupled-mode model is conceptually distinct to models based on binary collision theory, this result provides increased confidence that uncertainty in electron-ion energy exchange will not impact ignition attempts.

## 1. Introduction

Integrated calculations of ICF capsule implosions are generally performed using relatively simple, computationally inexpensive, plasma physics models for energy transport and exchange processes, such as those of Spitzer [1] or Brysk [2]. These models are derived using binary collision theory in ideal plasma and as such do not allow for energy transfer between electrons and ions through the interaction of collective excitations. Models based on quantum kinetic treatments which include the effects of coupled, collective modes can produce electron-ion energy exchange rates that differ significantly from Spitzer in some regimes [3,4], raising concerns about a potential lack of fidelity in ICF modelling.

During an ICF implosion, the capsule fuel accesses a range of conditions, including hot, dense, partially electron degenerate states, where coupled-mode effects may be pronounced. Although such effects are expected to be diminished as the fuel is heated to high temperatures, calculations predict electron-ion energy exchange to remain slower than the Spitzer rate even for temperatures in the kilo-electron-volt range [5]. However, it has been difficult to accurately assess the impact of coupled-mode models on capsule performance since full numerical evaluation is computationally expensive and unsuitable for direct implementation in radiation-hydrocodes.

A recently developed reduced coupled-mode model for electron-ion energy exchange [6], which accurately reproduces full numerical results over a wide range of density and temperature space, has been implemented in the Nym hydrocode [7]. The model has been applied to 1D calculations of a NIF ignition capsule to assess the impact on fuel assembly and performance.

## 2. Energy exchange model

The coupled-mode model is derived for a two-component electron-ion plasma using a quantum statistical framework, which correctly describes the particle kinetics in dense, partially-degenerate systems [4]. In the weakly-coupled limit, the model is evaluated using the random phase approximation (RPA), which yields an expression for the energy transfer rate per unit volume as a double Fourier-space integral in terms of the plasma dielectric function ε and those for the individual components $\varepsilon_{ee}$ and $\varepsilon_{ii}$.

$$Z_{ei}^{CM}(t) = \frac{\hbar}{\pi^3}\int_0^\infty dk\, k^2 \int_0^\infty d\omega\, \omega\left[n_{Bi}(\omega;t) - n_{Be}(\omega;t)\right]\frac{\operatorname{Im}\varepsilon_{ee}(k,\omega;t)\operatorname{Im}\varepsilon_{ii}(k,\omega;t)}{\left|\varepsilon(k,\omega;t)\right|^2} \quad (1)$$

Here ω and k are the frequency and wavenumber and $n_B(\omega)$ is the Bose occupancy function for the plasma modes. This approach naturally captures coupling between the collective modes of each component. However, direct numerical evaluation of the expression is computationally expensive since sharp plasmon peaks occur in the mode spectrum near the zeroes of the dielectric function.

The reduced model uses the fact that the integrand is modulated by the spectral function of the ions, which decays over much smaller frequency scales than that of the electrons, to Taylor expand the electron term. Electron screening is also approximated to first order, including both the static real term and a non-vanishing imaginary term. This latter term, representing electron damping, was found to play a critical role, such that the coupled-mode effect cannot be reproduced by alternative reduced approaches using only static screening [6]. Although sharp plasmon peaks still occur, the inclusion of electronic damping means their amplitude increases more slowly than their width and the contribution to the integral decreases and remains calculable. Moreover, the position of the peak can be estimated *a priori*, allowing high numerical resolution of the integral to be focussed over a small interval. The reduced model shows excellent agreement with full numerical calculations over a wide range of temperature and density space.

## 3. Ignition calculations

Calculations of the NIF Rev 5 Ge-doped plastic ignition design, which was the initial target used during the ignition campaign [8] were performed in 1D using Nym (figure 1). Although details of the ignition design have evolved, the basic operation and fuel conditions are essentially unchanged.

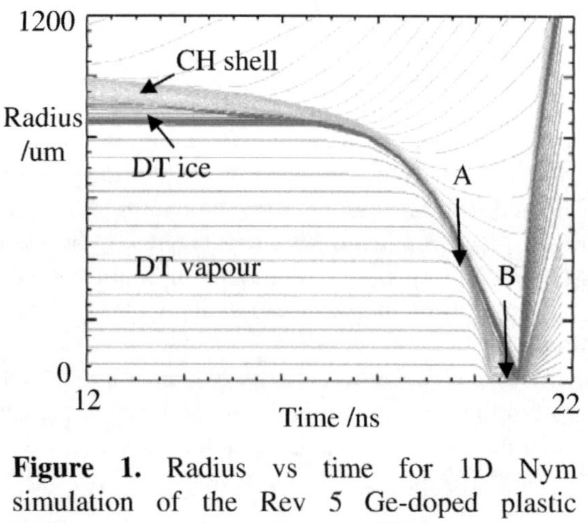


**Figure 1.** Radius vs time for 1D Nym simulation of the Rev 5 Ge-doped plastic ignition capsule. Non-equilibrium occurs following shock breakout (A) and during self-heating (B).

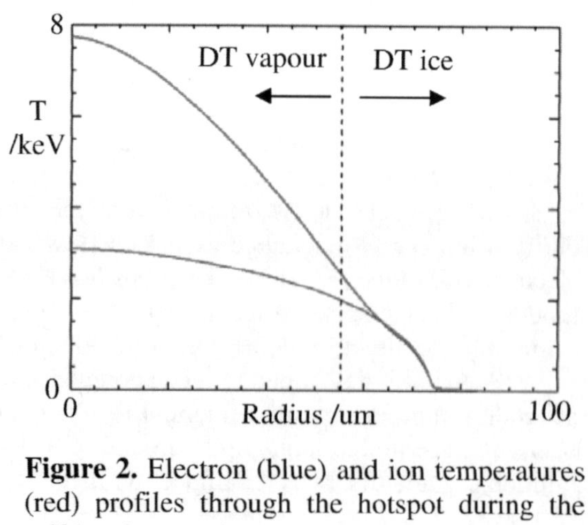


**Figure 2.** Electron (blue) and ion temperatures (red) profiles through the hotspot during the self-heating phase, showing substantial separation.

Nominal calculations suggest two significant phases of operation where the electrons and ions are out of equilibrium. Firstly, as shocks break-out from the DT ice into the low density DT vapour they coalesce into a strong shock, producing significant temperature separation ($T_i>T_e$). Thus there is potential for uncertainty in the energy exchange model to impact the calculated shock strength and subsequent energetics at stagnation. A slower exchange rate allows greater temperature separation and reduces radiative losses by maintaining a lower electron temperature. Following shock stagnation the fuel density continues to increase as the fuel shell converges, so that the temperatures tend to re-equilibrate.

The second period of separation occurs during the self-heating phase, when the ions in the DT fuel are again heated above the electrons (figure 2). This behaviour is perhaps counter-intuitive since at typical hotspot temperatures most of the alpha energy is deposited into the electrons [9]. However, despite lower energy input the ions also have much lower energy losses than the electrons, as the electrons and ions effectively become decoupled with increasing electron temperature. At these higher temperatures the hotspot is relatively weakly coupled and the electrons quickly become non-degenerate. Under these circumstances the uncertainty in the energy exchange rate may only be small.

Calculations were performed using both the Spitzer and coupled-mode energy exchange models. Integrated performance metrics, such as neutron yield, mean fuel temperatures or maximum areal density, show only small variation between the two models. Simulations show differences only during the self-heating and burn phases, and only in the central part of the fuel. The observed variation is much less than expected using a uniform multiplier on the exchange rate, but is consistent with the variation seen in a range of alternative models (Table 1). However, since the coupled-mode model is conceptually distinct from the various binary collision approaches, this result significantly strengthens confidence in those conclusions. Note also that in nearly all cases the alternative models reduce performance compared to Spitzer, albeit by only small amounts.

**Table 1.** Comparison of performance metrics for a range of models. The first line gives nominal values from simulations using the Spitzer model. Subsequent lines show deviation from nominal.

| Model | Neutron yield | Ti (keV) | Te (keV) | ρR (g/cm$^2$) |
|---|---|---|---|---|
| Spitzer | 6.1e18 | 47.4 | 36.2 | 1.46 |
| Coupled-mode | -2% | -3.5% | -2% | -1% |
| 0.5 x Spitzer | -11% | -17% | -12% | -6% |
| Brysk | -3% | -5% | -3% | -2% |
| Chang [10] | -1% | -2% | -1.5% | -0.5% |
| Zha [11] | 0 | -0.5% | 0 | 0 |
| Gericke [12] | 0 | +0.5% | +0.5% | 0 |

The lack of sensitivity to electron-ion energy exchange can be understood by detailed analysis of the simulations. The ratio of coupled-mode and Spitzer rates shows significant deviation from unity during the course of the implosion, but the largest deviations occur at early time when the fuel is cold and very strongly coupled, when neither model is strictly appropriate. The energy exchange rates have no impact at this time since the temperatures are low and the ions and electrons are in equilibrium.

When temperature separation is significant, the coupled-mode and Spitzer models tend to agree to better than 50% (figure 3). For the hotspot region the models show little variation in rate once temperature separation exceeds ~1eV. In the main fuel, the coupled-mode model can produce larger reductions relative to Spitzer (up to 70%) for temperatures separated by ~10eV, when the fuel is in the warm dense matter regime.

Furthermore, the largest differences between the coupled-mode and Spitzer rates occur not only when temperature separation is small, but also when the energy exchange rates are very high, typically for equilibration times much less than 1ps (figure 4). The coupled-mode model does show a large deviation from Spitzer for the hotspot when the equilibration time is a few picoseconds, but this corresponds to temperature differences of less than 1eV, so the impact is limited.

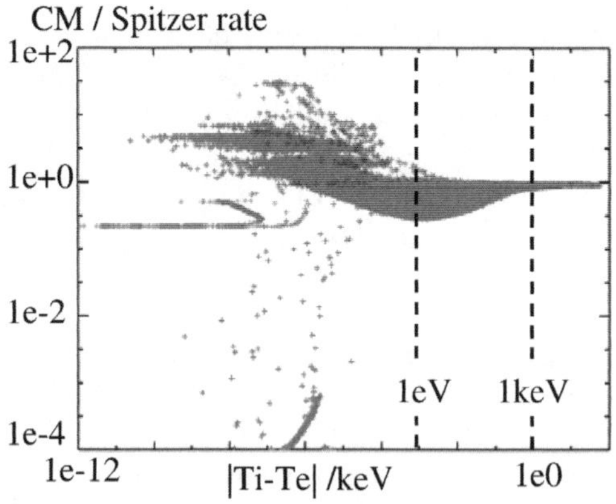


**Figure 3.** Ratio of coupled-mode to Spitzer energy exchange rate as a function of ion-electron temperature separation. Points represent values for the simulation cells in the hotspot (blue) and main fuel (red) at each time.

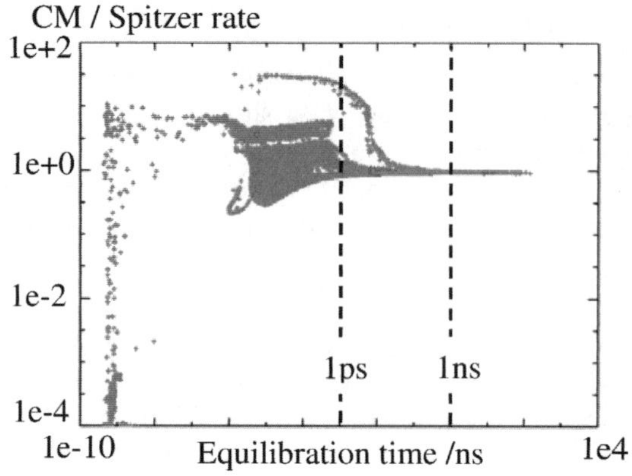


**Figure 4.** Ratio of coupled-mode to Spitzer energy exchange rate as a function of Spitzer ion-electron equilibration time, plotted for the hotspot (blue) and main fuel (red) cells at each time.

Examination of the exchange rate model ratio as a function of the (non-degenerate) plasma coupling parameter shows that the models agree to better than 25% within the weakly coupled regime ($\Gamma$<0.1). Although the coupled-mode model is derived in the RPA weak coupling limit, extension of the model to include strong coupling corrections is likely to have little impact on capsule performance, which appears insensitive to the already significant model variation in the strongly coupled regime. To have an impact on ignition capsule performance it appears that energy exchange models must deviate significantly from Spitzer in the high temperature, weakly coupled regime.

## 4. Conclusions

Coupled-mode models for electron-ion energy exchange predict large deviations from conventional binary collisions models in some regimes. A recently developed reduced coupled-mode model has been implemented in the Nym hydrocode and applied to calculations of a NIF capsule. Simulations show little sensitivity to the coupled-mode model, consistent with results from a range of alternative models. This lack of sensitivity arises because the models are in reasonable agreement whenever electron and ion temperatures separate or equilibration times become significant. Although the coupled-mode model is evaluated in the weakly-coupled RPA limit, extending the model to include strong coupling corrections is not expected to impact this conclusion. These results provide some confidence that uncertainty in electron-ion energy exchange models will not impact ignition attempts.

## 5. References


[1] Spitzer L 1967 *Physics of Fully Ionized Gases* (New York: Interscience)
[2] Brysk H 1974 *Plasma Phys.* **16** 927
[3] Dharma-Wardana M W C and Perrot F 1998 *Phys. Rev. E* **58** 3705
[4] Vorberger J and Gericke D O 2009 *Phys. Plasmas* **16** 082702
Vorberger J, Gericke D O, Bornath Th and Schlanges M 2010 *Phys. Rev. E* **81** 046404
[5] Murillo M S and Dharma-Wardana M W C 2008 *Phys. Rev. Lett* **100** 205005
[6] Chapman D A, Vorberger J and Gericke D O 2013 *Phys. Rev. E* **88** 013102
[7] Roberts P D, Rose S J, Thompson P C and Wright R J 1980 *J. Phys. D* **13** 1957
[8] Haan S W *et al.* 2011 *Phys. Plasmas* **18** 051001
[9] Fraley G S, Linnebur E J, Mason R J and Morse R L 1974 *Phys. Fluids* **17** 474
[10] Chang Y and Ordonez C A 2004 *Phys. Rev. E* **69** 037401
[11] Zha X, Han S, Xu Z and Wang Y 2006 *HEDP* **2** 70
[12] Gericke D O, Murillo M S and Schlanges M 2002 *Phys. Rev. E* **65** 036418

# High-Energy Density science at the Linac Coherent Light Source

**S H Glenzer, L B Fletcher, and J B Hastings**
SLAC National Accelerator Laboratory, 2575 Sand Hill Road, MS 19, Menlo Park, CA 94025

E-mail: glenzer@slac.stanford.edu

**Abstract.** The Matter in Extreme Conditions end station at the Linac Coherent Light Source holds great promise for novel pump-probe experiments to make new discoveries in high-energy density science. In recent experiments we have demonstrated the first spectrally-resolved measurements of plasmons using a seeded 8-keV x-ray laser beam. Forward x-ray Thomson scattering spectra from isochorically heated solid aluminum show a well-resolved plasmon feature that is down-shifted in energy by 19 eV from the incident 8 keV elastic scattering feature. In this spectral range, the simultaneously measured backscatter spectrum shows no spectral features indicating observation of collective plasmon oscillations on a scattering length comparable to the screening length. This technique is a prerequisite for Thomson scattering measurements in compressed matter where the plasmon shift is a sensitive function of the free electron density and where the plasmon intensity provides information on temperature.

## 1. Introduction

With the advent of the Linac Coherent Light Source (LCLS) x-ray laser [1] and the commissioning of the Matter of Extreme Conditions (MEC) end station [2] an unprecedented experimental capability has become available to explore extreme matter conditions with accurate pump-probe measurements [3], cf., Fig.1. Experiments that apply spectrally resolved x-ray scattering to probe plasma conditions of compressed solids offer novel techniques to determine the physical properties of matter at mega bar pressures. The ultrafast temporal resolution provided by the LCLS x-ray source and the high repetition rate of these experiments makes studies of high-pressure phase transitions [4,5], observations of novel structural properties [6], or direct measurements of material strain rates highly attractive.

For this purpose, the MEC instrument is equipped with one ultra-short pulse laser and two nanosecond laser beams that drive the material into extreme matter conditions, for example, by isochoric heating or by launching shock waves that propagate through solid foil targets. In these conditions, spectrally resolved x-ray Thomson scattering measurements in the non-collective (backward) scattering regime provide information of the microscopic physics by measuring the free electron distribution function. On the other hand, in the collective (forward) scattering regime, plasmon (Langmuir) [7-9] or ion acoustic oscillations are observed [10]. In particular, the plasmon scattering spectrum is of fundamental interest because it holds promise to determine plasma parameters and the physical properties from first principles [11]. This is particularly relevant for solid densities and above where the plasma is often closely coupled and standard theoretical approximations that have been developed for solids or ideal plasmas are not applicable. On the other hand, knowledge of dense plasma conditions are important for warm dense matter studies and potential future applications where contemporary physics pursues questions related to particle acceleration, inertial confinement fusion, and laboratory astrophysics.

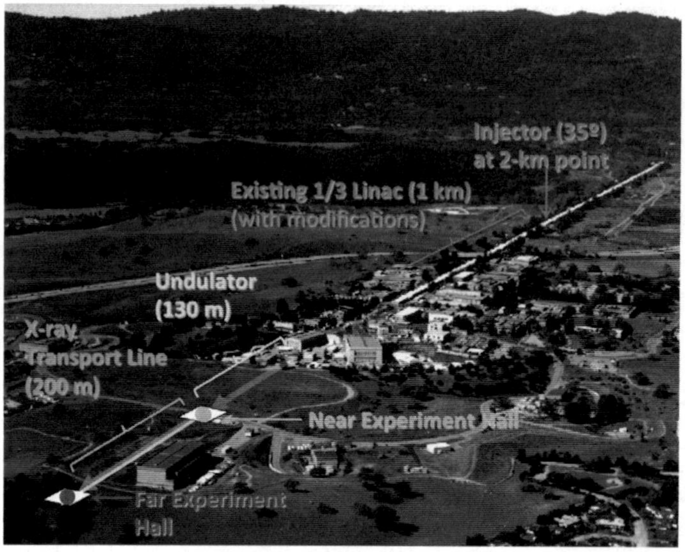


**Figure 1.** Areal view of the Linac Coherent Light Source. Electrons are injected at the 2 km point of the Linac that accelerates electron bunches over 1 km to energies ranging from 4.3 to 14 GeV. A 130 m long undulator is then producing the Free Electron Laser (FEL) x-ray beam with energies ranging from 1.5 Å to 15 Å at 120 Hz for experiments in the near and far experimental halls. The MEC is the 3rd instrument in the far experimental hall that is reached after a 200 m-long x-ray transport line. At MEC, the x-ray beam can deliver x-ray energies from 480 to 10,000 eV with pulse durations of 5 to 500 fs and up to 4 mJ FEL pulse energy.

LCLS provides a powerful penetrating x-ray beam with unique properties to probe the conditions in these short-lived hot dense states of matter. The beam delivers approximately $10^{12}$ x-ray photons in a micron-scale focal spot allowing measurements with high spectral resolution of $\Delta E/E = 10^{-4}$ (in a seeded beam mode [12]), high wave-number resolution of $\Delta k/k = 10^{-2}$, and high temporal resolution of 20-50 fs. Consequently, by employing highly efficient curved crystal spectrometers, the plasmon spectrum can be observed in single shots.

In this study, experiments have been performed in single shot mode and at high repetition rate of 115 Hz. The latter allowed us to measure quasi noise-free plasmon spectra within 7 seconds that determine the plasmon dispersion and intensity for warm solid-density aluminum. The plasmon shift of 19 eV yields $n_e = 1.8 \times 10^{23}$ cm$^{-3}$ expected for $Z = 3$ and temperatures less than 1 eV. These findings demonstrate the plasmon scattering capability at LCLS and will motivate future investigations in hot compressed matter.

## 2. X-ray Thomson scattering in dense plasmas

The x-ray scattering process is markedly different from optical scattering as the energy of the incident x-ray photon with frequency $w_0$ is large enough to give a significant Compton or plasmon shift to the frequency of the scattered radiation. During the scattering process, the incident photons transfer the momentum $h\mathbf{k}/2\pi$ and the energy $hw_C/2\pi = (h\mathbf{k}/2\pi)^2/2m_e$ to the electrons. The magnitude of the **k**-vector is given by

$$k = |\mathbf{k}| = \frac{4\pi\, hc}{E_0}\sin\left(\frac{\theta}{2}\right) \qquad (1)$$

with $E_0 = hw_0/2\pi$ being the energy of the probe x rays and $\theta$ the scattering angle. Momentum and energy is transferred to free electrons and to electrons whose binding energy is less than $hw_C/2\pi$. With the scattering vector defined by x-ray energy and scattering geometry, the Thomson scattering regime is characterized by the scattering parameter $\alpha$ that is proportional to the ratio of the x-ray probe scale-length to the plasma screening length, $\lambda_S$:

$$\alpha = \frac{1}{k\lambda_S} \tag{2}$$

For $\alpha < 1$, the non-collective regime, spectrally-resolved incoherent Thomson scattering measures the momentum distribution function for free electrons, and hence the electronic temperature. In warm dense matter, however, this spectrum blends with the bound-free scattering spectrum [13-15] that measures the momentum distribution of the bound electrons. Moreover, bound electrons with ionization energies larger than $hw_C/2\pi$ (states deep in the Fermi sphere) cannot be excited, and no energy can be transferred during the scattering process giving rise to an elastic scattering feature. For $\alpha > 1$, the collective scattering regime, the scattering is sensitive to temporal and spatial correlations between electron motion separated by a screening length, and therefore observes ion-acoustic and electron plasma wave resonances.

The total scattering spectrum results in elastic and inelastic spectral features that are calculated with the dynamic form factor. The free electron contribution is described by $S_{ee}(\mathbf{k},\omega)$, while the un-shifted elastic scattering component at $E_0$ that is commonly referred to as Rayleigh peak is described by the first term of the dynamic structure factor [16-17]

$$S(k,\omega) = \left|f_I(k) + q(k)\right|^2 S_{ii}(\mathrm{k}) + Z_f S_{ee}(k,\omega) + Z_c \int S_{CE}(k,\omega-\omega')S_S(k,\omega')d\omega' \tag{3}$$

with $Z_f$ and $Z_C$ the number of free and bound electrons, respectively. The last term of Eq. (3) calculates the bound-free scattering contribution. This expression for the dynamic structure factor allows a few simple approximations. Of interest to this study, $\alpha > 1$ results in collective scattering and the plasmon frequency shift from $E_0$ is determined by the plasmon dispersion relation and the width is determined by Landau damping and collisional damping processes. The shift can be approximated for small values of $k$ using an inversion of Fermi integrals given by Zimmerman [18] that results in a modified Bohm-Gross dispersion relation [18-19]

$$\omega_{pl}^2 = \omega_p^2 + 3k^2 v_{th}^2\left(1 + 0.088 n_e \Lambda_e^3\right) + \left(\frac{\hbar k^2}{2m_e}\right)^2 \tag{4}$$

where $\omega_p = \sqrt{n_e e^2/\varepsilon_0 m_e}$ is the plasma frequency, $v_{th} = \sqrt{k_B T_e/m_e}$ is the thermal velocity, and $\Lambda_e = h/\sqrt{2\pi m_e k_B T_e}$ is the thermal wavelength. For solid-density plasma conditions encountered in warm dense matter, Eq. (4) results in energy shifts in the range of 15 – 50 eV consequently requiring a high-energy x-ray source of $E > 3$ keV and $\Delta E/E = 10^{-3}$ - $10^{-4}$ to both penetrate through the dense plasma and to resolve the plasmon frequency shift. Furthermore, it is noteworthy to realize that simultaneous energy resolution of $\Delta E/E < 10^{-4}$ and wavenumber resolution of $\Delta k/k < 10^{-1}$ will be needed to resolve plasmon broadening determined by damping processes in the plasma.

## 3. Results and Discussions

In this study, we first validated plasmon measurements at high repetition rate with high-resolution measurements of isochorically heated targets before applying the technique in single shots on shock-compressed aluminum. Figure 4 shows the experimental scattering spectra from solid-density aluminum. Results are shown from the forward scattering and backward scattering spectrometers with the x-ray beam in seeded and SASE mode of operation. The seeded x-ray beam provides a bandwidth of 1 eV and combined with 8 eV spectrometer resolution resolves the plasmon that is downshifted from the elastic 8 keV scattering feature by 19 eV. The figure also shows the results of the backward scattering spectrometer which observes elastic scattering at 8 keV reflecting the instrument function of the spectrometer convolved with a Gaussian profile that accounts for the statistical fluctuations of the x-ray energy of a seeded beam over 700 shots.

The experiment with a seeded x-ray laser beam shows that the bandwidth is sufficient to resolve plasmons in dense matter. In forward direction, the x-ray scattering spectra with SASE operation show a slight broadening of red wing of the scattering spectrum but the SASE scattering spectrum is not suitable for inferring the dense plasma conditions.

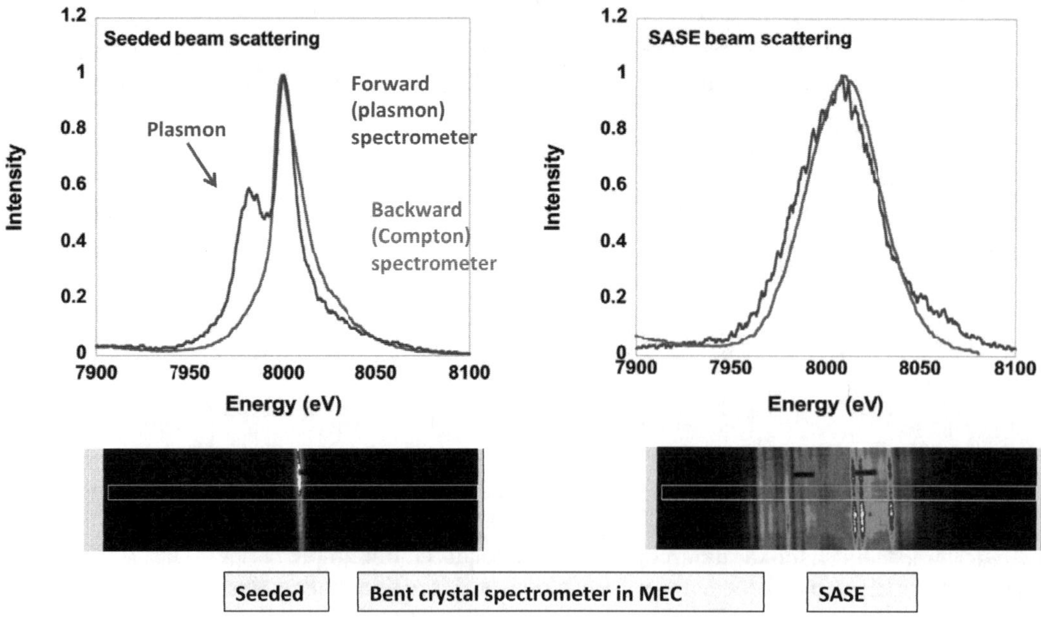


**Figure 2.** X-ray Thomson scattering data from solid density aluminum targets are shown from forward (plasmon) scattering and backscattering spectrometers for seeded and SASE x-ray beam operation. These data are accumulated over 700 shots at 120 Hz. Also shown are examples of the spectrum of a single x-ray laser shot indicating 1 eV bandwidth for seeded beam and 50 eV for SASE operations.

The backward spectrometer also measured a Compton scattering feature downshifted in energy by 250 eV (not shown). Here, we note that the plasmon resonance is only observed in forward scattering providing strong evidence of a collective phenomenon as predicted for these conditions; bound-free scattering features are predicted to be negligible in this energy range and no feature has been observed with the Compton spectrometer. For these measurements, the crystal in the forward scattering spectrometer is 40 mm thick while in backscattering we employed a 100 mm thick crystal giving rise to slightly different instrument functions and consequently slight differences in the spectral shape of the elastic scattering feature.

When fitting the dynamic structure factor, equation (3) to the experimental forward scattering spectrum we find that the shift of the plasmon peak yields the electron density of $n_e = 1.8 \times 10^{23}$ cm$^{-3}$ ± 5%. The electron temperature is 2 eV accounting for the isochoric heating of the aluminum by the x-ray laser pulse [20]. The ions are cold with temperatures of order 0.1 eV. Here, the plasmon frequency is sensitive to the electron density, but the temperatures are too small to affect the plasmon. The accuracy of the electron density measurement is very high due to the sensitivity of the plasmon resonance to the plasma frequency. With $n_e = Z/A\, 6.02 \times 10^{23} \rho$ (the electron density $n_e$ in cm$^{-3}$ and the mass density $r$ in g cm$^{-3}$), assuming ionization state of $Z$=3 and using the mass $A$=26.98 u, the measured electron density provides $r$ = 2.7 g cm$^{-3}$ as expected for solid aluminum.

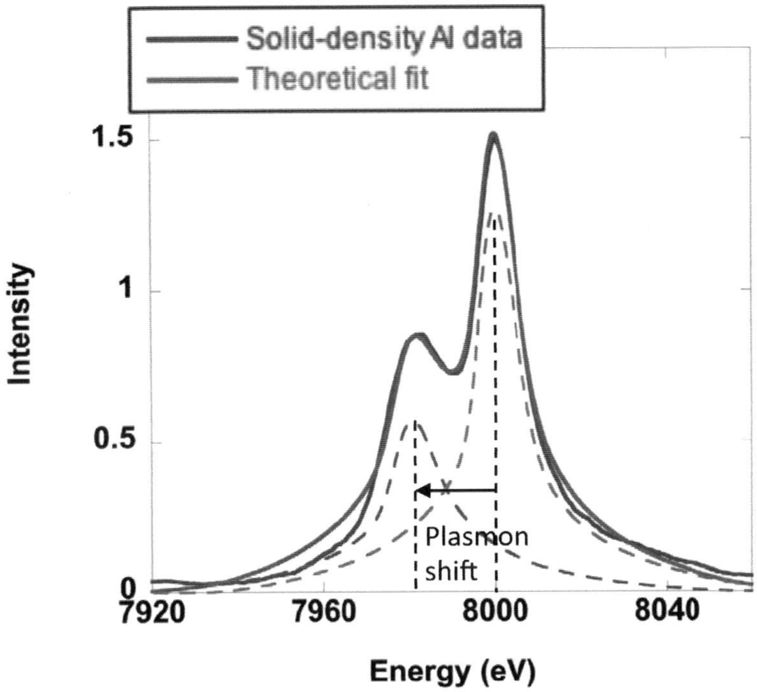


**Figure 3.** The theoretical fit to the experimental forward X-ray Thomson scattering data is shown for the electron density of $n_e$ = 1.8 x 10$^{23}$ cm$^{-3}$. The theoretical fit uses an electron temperature of 2 eV reflecting the direct heating of electrons by the x-ray laser interactions. The ions are assumed cold with temperatures of order 0.1 eV. The red dashed curve shows the elastic scattering component and the blue dashed curve represents the plasmon spectrum.

## 4. Conclusions

With the measurements of plasmon scattering we have demonstrated a highly accurate tool to measure the physical properties of dense matter. This capability is a prerequisite to test predictions of radiation hydrodynamic simulations and density function theory to understand dense matter conditions. Furthermore, the unique combination with pump-probe capability at LCLS will allow novel experiments to discover new phases and material conditions in well-characterized conditions. The high bandwidth and high photon numbers combined with high-resolution crystal spectrometers allows resolving the plasmon shift of 19 eV resulting in an electron density $n_e$ = 1.8 x 10$^{23}$ cm$^{-3}$ ± 5%. This density is consistent with Z=3 in solid aluminum. For future experiments this capability can be applied in single shots to determine, e.g., the conditions of warm dense matter that is compressed by laser-driven shock waves. Also, the technique will allow ultrafast pump-probe studies for measurements of thermodynamic equilibration. On a fundamental level, precision measurements of the plasmon dispersion and amplitude will now be possible. Future developments will include high-resolution spectrometers with sufficient efficiency for measuring the plasmon broadening through Landau and collisional damping processes.

## 5. References

[1] Emma P, et al. 2010, First lasing and operation of an ångstrom-wavelength free-electron laser, Nature Photonics **4**, 641 - 647.

[2] Lee H J, et al., to be published

[3] Glenzer S H and Redmer R 2009, X-ray Thomson scattering in high energy density plasmas, Rev. Mod. Phys. **81** 1625.

[4] Coppari F et al. 2013, Experimental evidence for a phase transition in magnesium oxide at exoplanet pressures, *Nature Geoscience* doi:10.1038/ngeo1948.

[5] Lindenberg A M et al. 2000, Time-Resolved X-Ray Diffraction from Coherent Phonons during a Laser-Induced Phase Transition, Phys. Rev. Lett. **84** 111.

[6] Fortmann C, Niemann C, and Glenzer S H 2012, Theory of x-ray scattering in high-pressure electrides, Phys. Rev. B **86** 174116.

[7] Glenzer S H et al. 2007, Observations of Plasmons in Warm Dense Matter, Phys. Rev. Lett. **98** 065002.

[8] Kritcher A L et al. 2008, Ultrafast X-ray Thomson scattering of shock-compressed matter, Science **322** 69.

[9] Neumayer P et al. 2010, Plasmons in strongly coupled shock-compressed matter, Phys. Rev. Lett. **105** 075003.

[10] Gregori G and Gericke D O 2009, Low frequency structural dynamics of warm dense matter, Physics of Plasmas **16**, 056306.

[11] Döppner T et al. 2009, Temperature measurement through detailed balance in x-ray Thomson scattering, High Energy Density Physics **5** 182–186.

[12] Amann et al. 2012, Demonstration of self-seeding in a hard-X-ray free-electron laser, Nature Photonics DOI: 10.1038/NPHOTON.2012.180

[13] Lee H J et al. 2009, X-Ray Thomson-Scattering Measurements of Density and Temperature in Shock-Compressed Beryllium, Phys. Rev. Lett. **102** 115001.

[14] Fortmann C et al. 2012, Measurement of the Adiabatic Index in Be Compressed by Counterpropagating Shocks, Phys. Rev. Lett. **108** 175006.

[15] Chihara J 1987, "Difference in x--ray scattering between metallic and non--metallic liquids due to conduction electrons", J. Phys. F: Met. Phys. **17**, 295.

[16] Gregori G et al. 2003, Theoretical model of x-ray scattering as a dense matter probe, Phys. Rev. E. **67**, 026412.

[17] Gregori G et al. 2006, Generalized x-ray scattering cross section from non-equilibrium solids and plasmas, Phys. Rev. E. **74,** 026402.

[18] Zimmerman R 1987, *Many-Particle Theory of Highly Excited Semiconductors* (Teubner, Leipzig).

[19] Thiele R et al. 2008, Plasmon resonance in warm dense matter Phys. Rev. E. **78,** 026411.

[20] Hau-Riege S et al. 2012, Ultrafast Transitions from Solid to Liquid and Plasma States of Graphite Induced by X-Ray Free-Electron Laser Pulses, Phys. Rev. Lett. **108** 217402.

### Acknowledgments

This work was performed at the Matter at Extreme Conditions (MEC) instrument of LCLS, supported by the DOE Office of Science, Fusion Energy Science under contract No. SF00515. This work was supported by DOE Office of Science, Fusion Energy Science under FWP 100182, and partially supported by DOE Office of Basic Energy Sciences, Materials Sciences and Engineering Division, under Contract DE-AC02-76SF00515

# Ion Acceleration by Ultra-intense Laser Pulse Interacting with Double-layer Near-critical Density Plasma

**Y. J. Gu[1,2†], Q. Kong[1*], S. Kawata[2], T. Izumiyama[2], T. Nagashima[2], M. Takano[2], X. F. Li[1], Q. Yu[1], D. Barada[2], Y. Y. Ma[3] and P. X. Wang[1]**
[1] Institute of Modern Physics, Fudan University, Shanghai, China
[2] Dept. of Advanced Interdisciplinary Sciences, Utsunomiya University, Utsunomiya, Japan
[3] Department of Physics, National University of Defence Technology, Changsha, China

E-mail
†: yanjun_gu@live.cn
*: qkong@fudan.edu.cn

**Abstract:** A collimated ion beam is generated through the interaction between ultra-intense laser pulse and a double layer plasma. The maximum energy is above 1GeV and the total charge of high energy protons is about several tens of nC/μm. The double layer plasma is combined with an underdense plasma and a thin overdense one. The wakefield traps and accelerates a bunch of electrons to high energy in the first underdense slab. When the well collimated electron beam accelerated by the wakefield penetrates through the second overdense slab, it enhances target normal sheath acceleration (TNSA) and breakout after-burner (BOA) regimes. The mechanism is simulated and analyzed by 2.5 dimensional Particle-in-cell code. Compared with single target TNSA or BOA, both the acceleration gradient and energy transfer efficiency are higher in the double layer regime.

## 1. Introduction

High energy ion beam has been used in cancer therapy [1], inertial confinement fusion [2], radiography of dense matter [3], fundamental particle physics [4] and other purposes. However, the size and cost of conventional accelerators are so huge that long term construction and international collaboration is required. With the development of laser technology, ultra-intense and ultra-short laser pulse are now available. Ion acceleration driven by laser-plasma interaction becomes an interesting and important topic and has been investigated intensively in these decades. Many mechanisms have been proposed including target normal sheath acceleration (TNSA) [5], breakout after-burner (BOA) [6], radiation pressure regime [7], Coulomb explosions [8] and so on. In this paper, the enhanced BOA regime through a double layer near critical density plasma has been discussed by 2.5-dimensional particle-in-cell (PIC) simulations. The leading electrons in this regime is not the hot electrons but the laser-wakefield-accelerated (LWFA) electron beam which has good beam quality. The protons are accelerated for a long distance by the LWFA electron beam. The maximum ion energy exceed 1GeV and the accelerated ion bunch is also well collimated.

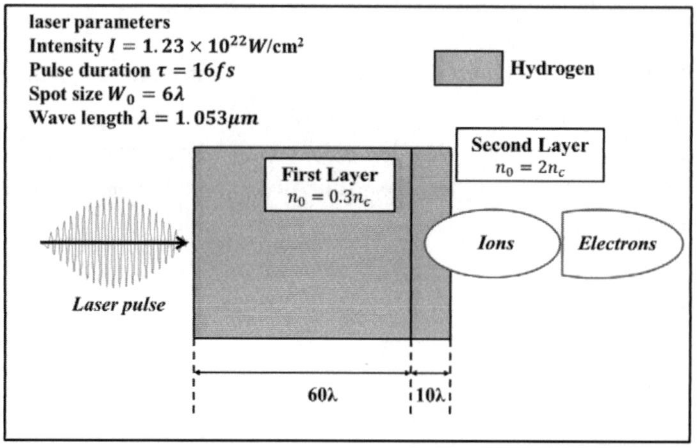


**Figure 1.** Schematic of the simulation model

## 2. Enhanced proton acceleration through the double layer target

The schematic of simulation model for double layer plasma acceleration is presented in Figure 1. The Gaussian pulse is focused at the left boundary of the target. The laser pulse is linearly polarized in y direction with intensity $I = 1.23 \times 10^{22} W/cm^2$, pulse duration 16fs and radius $W_0 = 6\lambda$, here $\lambda$=1.053μm is the pulse wavelength. The target is a hydrogen plasma which has two layers with different densities. The first layer is underdense with $0.3n_c$ locating from $20 \leq x(\lambda) \leq 80$, where $n_c = 1.005 \times 10^{21} cm^{-3}$ is the non-relativistic plasma critical density. From 80λ to 90λ, there is the slightly overdense second layer with $2n_c$ as shown in Fig. 1. The simulations are based on the electromagnetic relativistic code "ZOHAR". Free boundary conditions are used for both particles and fields. The mesh size is dx=dy=0.04λ and the total macro-particle number is about $7.3 \times 10^8$.

The ultra-intense laser pulse induces a wakefield in the first layer underdense plasma when it is propagating through the target. The wakefield is strong enough to trap and accelerate a bunch of electrons to high energy in a short distance. In Fig. 2 (a), the wakefields at three moments (150fs, 225fs and 315fs) are plotted. One can find the maximum longitudinal electric field reaches 150GV/cm. The energy spectrums of the trapped electron beam at the corresponding time are depicted in Fig. 2 (b). In about 70λ, the trapped electrons have been accelerated to hundreds MeV. The effective acceleration gradient reaches about 6.8MeV/μm. According to the angular spectrum, the electron beam is well collimated as shown in Fig. 2 (c) with the transverse emittance of $(4/N)\sqrt{\sum_{i=1}^{N}(y_i - <y>)^2} \times \sqrt{\sum_{i=1}^{N}(\theta_i - <\theta>)^2} \approx 1.35 \mathrm{mm} \cdot \mathrm{mrad}$.

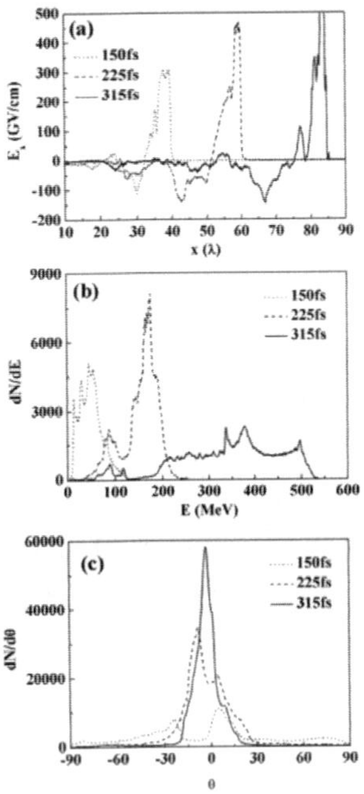


**Figure 2.** (a) The longitudinal electric field, (b) the energy spectrum and (c) the angular spectrum of LWFA electrons at 150fs, 225fs and 315fs are plotted with dotted line, dashed line and solid line, respectively.

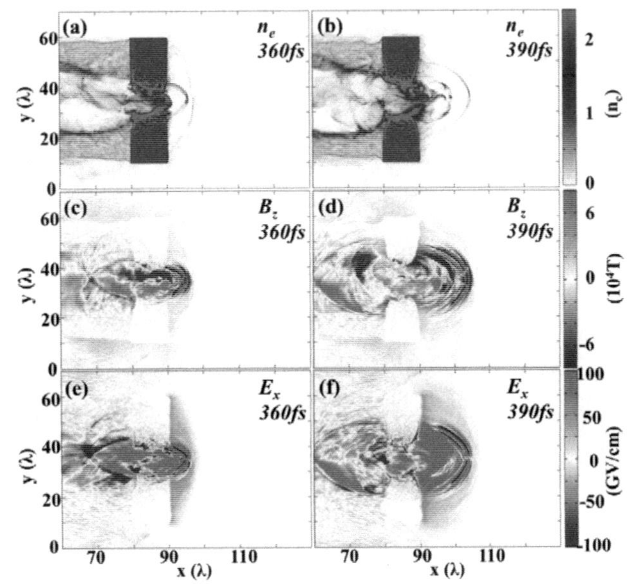


**Figure 3.** The distribution of (a, b) electron density, (c, d) z-component of magnetic field and (e, f) longitudinal electric field at 360fs and 390fs are plotted.

The accelerated electron beam induces a dipole magnetic field by its current. When the electron beam penetrates throughout the second layer target, the corresponding magnetic field expanses quickly in vacuum. Figure 3 (a) and (b) show the electron density profile at 360fs and 390fs, respectively. During this period, the LWFA electron beam is propagating through the boundary of second target and ejecting into vacuum. The magnetic field (z-components) distributions at the corresponding moments are plotted in Fig. 3 (c) and (d), respectively. The expanding magnetic field stricture is quite apparently by compared with the transverse size of $B_z$. With the expansion of magnetic field, a longitudinal electric field is induced near the boundary which accelerates the ions as shown in Fig. 3 (e) and (f). The longitudinal electric field expanses with the magnetic field and reaches about hundreds GV/cm.

The longitudinal electric field along the second layer target boundary is plotted in Fig. 4 (a). Since the second layer density is overdense with $2n_c$, the field strength exceeds 600GV/cm. It also suggests the strong field is tightly localized near the center. The ions located near the boundary are accelerated violently to high energy with large gradient. The following ion acceleration stage is breakout after-burner (BOA) regime. From the distribution of electron density, ion density and transverse electric field at 811fs shown in Fig. 4 (b), one can find the steady BOA acceleration structure is already formed. The first electron density peak represents a bunch of electrons which is directly accelerated by the laser ponderomotive force. The second electron density peak is the LWFA electrons. Both of the two electron bunches accelerate the following ion beam. The accelerated ion bunch is also collimated since the beam quality of the leading LWFA electrons. The electrons transfer their energy to the ions, while they are fed back through the pulse. This structure maintains for a long time and distance until the pulse energy is exhausted. In Fig. 4 (c) and (d), we present the proton energy spectrum and the energy angular related distribution at 946fs, respectively. The maximum proton energy is 1.03GeV. It's not the limit of the acceleration mechanism since the laser pulse still contains large energy at this moment. If the simulation area is enlarged, higher energy ions can be obtained. The proton bunch contains energy totally about $3.19\mathrm{J}/\mu\mathrm{m}$. The energy of the laser pulse in our simulation is about $10\mathrm{J}/\mu\mathrm{m}$, therefore the energy transfer efficiency from laser to the ions is about 31.9%, which is very effective. From Fig. 4 (d), one can also find most of the high energy protons are well collimated between $\pm 20°$.

## 3. Discussions and conclusions

In our work, a high energy ion bunch is obtained with the interaction between a double layer near critical density hydrogen plasma and ultra-intense ultra-short laser pulse. The leading electron beam accelerates and collimates the following ion bunch in a long distance, while the energy is supplied to the electrons by the laser pulse. The steady structure maintains for a long time. Both the TNSA and BOA are enhanced through the double layer target. Higher energy and larger charge quantity are acquired compared with the single target regime [9, 10]. This kind of interaction is one of the promising mechanisms to obtain high energy ions.

**Acknowledgements**

This work was partly supported by NSFC (No. 11175048), Shanghai Nature Science Foundation (No. 11ZR1402700), and Shanghai Scientific research innovation key projects No. 12ZZ011. Support from Shanghai Leading Academic Discipline Project B107, the JSPS and MEXT Program, and CORE of Utsunomiya University, Japan are also acknowledged.

**References**

[1] Bulanov S V and Khoroshkov V S 2002 *Plasma Phys. Rep.* **28** 453
[2] Roth M *et al* 2001 *Phys. Rev. Lett.* **86** 436
[3] Kar S *et al* 2008 *High Energy Density Phys.* **4** 26
[4] Esarey E, Schroeder C B and Leemans W P 2009 *Rev. Mod. Phys.* **81** 1229
[5] Wilks S C *et al* 2001 *Phys. Plasmas* **8** 542
[6] Yin L *et al* 2006 *Laser Part. Beams* **24** 291
[7] Esirkepov T *et al* 2004 *Phys. Rev. Lett.* **92** 175003
[8] Bulanov S V *et al* 2002 *Phys. Lett. A* **299** 240
[9] Gu Y J *et al* 2011 *Euro. Phys. Lett.* **95** 35001
[10] Gu Y J *et al* 2013 *Phys. Plasmas* **20** 070703

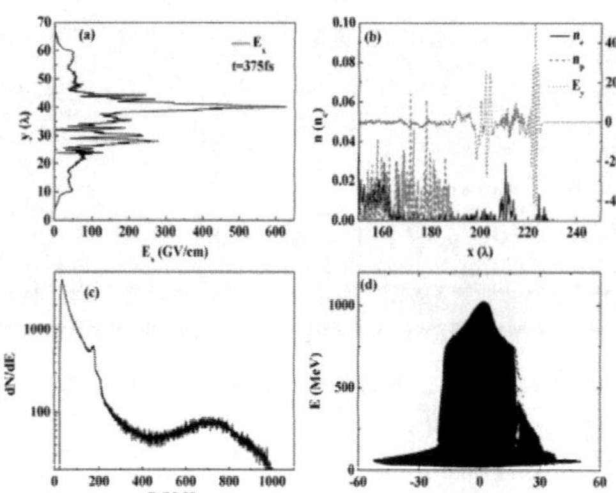

**Figure 4.** (a) The longitudinal electric field at 375fs along the target boundary. (b) The transverse electric field (blue dotted line), electron density (red dashed line) and proton density (black solid line) along y=35λ at 811fs. (c) and (d) are the energy spectrum and energy-angular related distribution of the accelerated proton bunch at 946fs.

# Ab Initio Simulation Beryllium in Solid Molecular Hydrogen: Elastic Constant

**Carlo L Guerrero and Jose M Perlado**
Instituto de Fusión Nuclear, Universidad Politécnica de Madrid.
Madrid 28006, Spain

E-mail: guerrerocarlo@yahoo.es

**Abstract.** In systems of inertial confinement fusion targets Deuterium-Tritium are manufactured with a solid layer, it must have specific properties to increase the efficiency of ignition. Currently there have been some proposals to model the phases of hydrogen isotopes and hence their high pressure, but these works do not allow explaining some of the structures present at the solid phase change effect of increased pressure. By means of simulation with first principles methods and Quantum Molecular Dynamics, we compare the structural difference of solid molecular hydrogen pure and solid molecular hydrogen with beryllium, watching beryllium inclusion in solid hydrogen matrix, we obtain several differences in mechanical properties, in particular elastic constants. For $C_{11}$ the difference between hydrogen and hydrogen with beryllium is 37.56 %. This may produce a non-uniform initial compression and decreased efficiency of ignition.

## 1. Introduction

The interest of the study of solid molecular hydrogen (SMH) includes different fields of science and technology, essentially in models of giant planets [1] and exoplanets, also in the construction of future aerospace systems where it is proposed as fuel [2]. The study of the SMH phase transitions is very important in the material compression model [3,4], which in turn are applied to systems of inertial confinement fusion (ICF). The ICF Systems are based on a target compression of deuterium tritium solid (DT ice) by irradiation with high power lasers. The DT ice content in the target must have a well-defined characteristics and properties [5]. These properties may be modified depending on the capsule container. The interest of this paper is to submit a first approximation of the effects of possible dissolution of beryllium atoms from the target capsule with DT ice. This approximation is achieved by inserting a Be atom in a molecular matrix solid H. The SMH serves as a model since this study has been more studied than T or D, and can compare some features with our simulations [4]. Our general hypothesis is that the initial conditions of the target configuration with respect to the properties of the DT ice and capsule container determine the efficiency of ICF ignition system [3]. In this paper we propose to study through the static model of phase transitions occurring in a dynamic compression. Static model called us to compression simulated by Quantum Molecular Dynamics (QMD) for a given temperature and pressure, so that it represents a pressure ramp we modeled a large area of the phase diagram of H and an approximation of dynamic compression system that occurs in ICF. We focus on the moments before or on the same ICF initial compression where low temperatures between 15 K and 300 K, which in turn can be compared to experimental results of compression systems diamond anvil cells (DAC), where the maximum pressure that can be reached is approximately 300 GPa.

The problem of segregation and dissolution of metals in contact with SMH is currently under discussion because it has presented a report which shows evidence conductivity of SMH at pressures above 200 GPa and 300 K [6]. A model that can explain this change in properties with the possible immersion of metals in the SMH crystal structure in the experimental case [6] technique they use DAC with copper coated diamonds to measure conductivity, this is where different authors have hypothesized that this copper can be dissolved into the structure and change the H properties [7]. Taking this hypothesis we have seen immersing a metallic light element as Be and changes in the structural and mechanical properties of SMH. Fuel contamination with material of the capsule to been linked to effects on the efficiency of ignition in previous works [8-10].

## 2. Methodology

Through simulation we could play a series of properties for SMH as lattice constant, crystal structure, phase changes [3] and mechanical properties such as modulus Bulk [4], in particular in this work we study the elastic constants. For which changes occur in the structure and mechanical properties of DT ice, use arrays we have obtained for a wide range of temperatures and pressures, in the case of SMH from 1 GPa to 200 GPa and temperatures of 5 K to 300 K, in this work we shows results for 300 K and pressure range 10 GPa to 25 GPa for hydrogen and hydrogen with beryllium. SMH simulations is for 576 atoms, using the methodology applied in previous works [3,4]. Every $H_2$ molecule occupies a hexagonal close-packed (HCP) lattice point, the structure consists 576 atoms in 288 H molecules. We use periodic boundary conditions with which approximates volume representation. The simulation Be in H solid is performed with 574 H atoms comprising 287 molecules, and adding a Be atom at the centre position, which should be occupied by one molecule of H in original structure H 288 molecules, this new configuration is named H_Be. With this initial structure we evolve the system to reach again the temperature and pressure at which we want to compare the structure obtained.

QMD simulations have been carried out using SIESTA code [11] which is based on Density Functional Theory (DFT) calculations with plane waves. We use exchange-correlations functional Local-Density Approximation (LDA) based on functional Cerpeley-Alder (CA) with DZP basis, MeshCutoff 100 Ry. The pseudopotential is Troullier-Martin type. For testing Generalized Gradient Approximation (GGA) functional use in revised Perdew-Burker-Ernzerhof (revPBE), these tests showed no difference with LDA calculations for pressures below 180 GPa. Our QMD exploration consists of a combination of well founded individual stages. The process is initiated by the search for more stable structural configuration, we used the method of Conjugated Gradients (CG). With configurations obtained by simulation for QMD defined temperature and pressure, we calculated the elastic constants $C_{11}$, $C_{33}$, $C_{12}$ and $C_{13}$ for H and H_Be structure in a range of 10 GPa to 25 GPa, and temperature of 300 K. The initial setting is obtained using QMD with implementation the Nose-Hoover thermostat and Parrinello-Rahman barostat. To calculate the elastic constants we realize the deformation of the structure defined in the three axes X, Y and Z, obtaining the relationship between stress and strain experienced by the structure. The relationship is $C = \sigma / \varepsilon$, where $\sigma$ is the stress and $\varepsilon$ strain. In order to estimate each constant use transverse isotropic model applied to the SMH-HCP structure, where $C_{22}= C_{11}$ and $C_{23}= C_{13}$, $C_{44}$ no show in this work.

## 3. Results

Figure 1 show a comparison between experiment SMH [12] and our results for $C_{11}$, $C_{33}$, $C_{12}$ and $C_{13}$, the experiment uses DAC, for pressure of 6 GPa to 23 GPa. We study phase I, where $H_2$ is solid state. In this pressure, figure 2, our results show a maximum difference of 18.65 %, for the constants $C_{33}$, and $C_{11}$, to $C_{12}$ and $C_{13}$ the difference is smaller. This is our starting point for interpreting the changes to add the Be atom in the crystal structure. The introduction of Be atom within a volume of 576 atoms allows us to calculate the changes in the SMH, which may well be related to changes in the properties of the DT ice. As a first step evaluate the crystal structure, to compare in figure 3 the position of SMH alone and with Be introduced. Both cases are structures obtained after thermal stabilization and pressure. For SMH pressure range 10 GPa to 25 GPa and 300 K has a hexagonal configuration recognizable, compatible with phase I for SMH [13]. In figure 3 is displaying projections of all the atoms in different planes XY, ZY and XZ. In the case of Be immersed in $H_2$ - HCP, there is no change in the crystal structure, but are observed significant changes in the mechanical properties. Figure 2 shows the comparison between the simulation results for H and H_Be, concerning the calculation of the elastic constants.

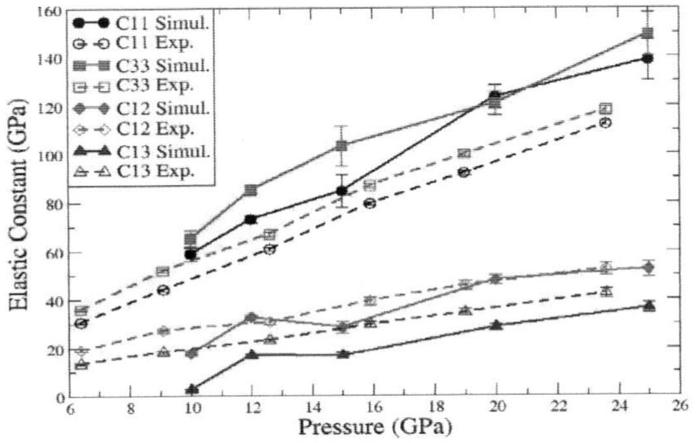


**Figure 1.** Simulated SMH [4] and experimental results [12]. The difference between the elastic constant varies according to axis. The mean difference: $C_{11}$ *16.37 %,* $C_{33}$ *18.65 %,* $C_{12}$ *3.68 % and* $C_{13}$ *9.97 %.*

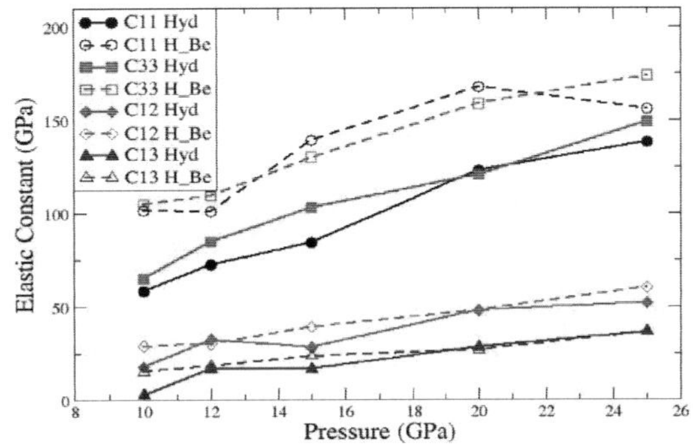


**Figure 2.** Comparison between the elastic constants calculated for H and H_Be. $C_{12}$ *and* $C_{13}$ are similar in both cases, but $C_{11}$ *and* $C_{33}$ mean difference is more than 37.56 % and 30.84 % respectively.

With a linear regression we can compare point to point each of the results. The greatest differences occur between the $C_{11}$ *and* $C_{33}$, for $C_{12}$ *and* $C_{13}$ the difference is minimal. In figure 1 $C_{11}$ mean difference is 16.37 %, $C_{33}$ 18.65 %, $C_{12}$ 3.68 % and $C_{13}$ 9.97 %, with respect to the experimental result. The inclusion of Be atom by two of H produces a marked change in the magnitude of the elastic constant, and hence in its mechanical properties, for figure 2 $C_{11}$ mean difference is 37.56 %, $C_{33}$ 30.84%, $C_{12}$ 5.65 % $C_{13}$ 4.73 %, with respect to the hydrogen simulation. Although the structure as shown in figure 3 does not change, its mechanical properties change. This study is a first step to obtain and study the mechanical properties of the phase change system, in this paper we carry only the Phase I study, eventually we see the relation between changes in the mechanical properties and crystal structure of the system to thereby characterize either the compression process in the field of DT ice in the case of ICF and turn get a good representation of the phase diagram of H isotopes. This method of study allows us to compare our results with the DAC compression systems or Laser, with additional value that our method allows a window of atomic structural information.

## 4. Conclusions

We show a comparison between the elastic constants of H and H_Be. Beryllium inclusion causes an increase in the amount of $C_{11}$ *and* $C_{33}$, this effect is equivalent to increasing the pressure of the SMH, this due to the formation of beryllium hydride, which may explain the appearance of solid hydrogen properties presented higher pressure such as metallization [7].

This method allows have a view of the early stages of compression from 15 K to 300 K and in a pressure range from 0.1 GPa to 200 GPa as we have shown in previous work [3,4]. This type of analysis can be applied to DAC compression or by laser compression, which are of interest in systems for inertial confinement fusion.

We have observed that fuel contamination with Be (Z-low) of the capsule changes the elastic properties of solid materials by 30%, then the deformation or strain induced by the initial shockwave differs at each point with Be in 30%, this may cause a non uniform initial compression. A way to reduce the contamination effect after the initial compression in ICF experiment is the so-called "high foot" method [10]. Should be studied the storing and delivering times to ICF target to know whether the SMH layer fuel contamination with material of the capsule occurs before the initial compression.

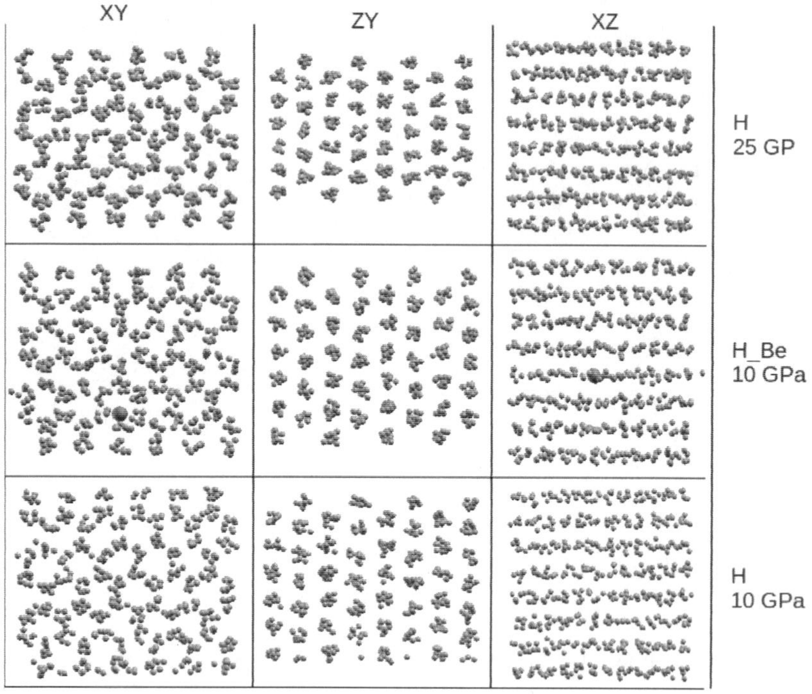


**Figure 3.** Planes XY, ZY, XZ simulation boxes for the pressures of 10 GPa at H and H_Be, and 25 GP for H. They can be seen as setting the crystal structure suffers few changes in the presence of Be atom (big sphere), in agreement with that observed experimentally for phase I of H. HCP Crystal Structure and $H_2$ molecules in lattice point.

**Acknowledgment**
To S. Cuesta. The author acknowledges the computer resources and technical assistance provided by the Centro de Supercomputación y Visualización de Madrid (CeSViMa). The authors thank the Spanish Ministry of Science and Innovation for economical support via the ACI-PROMOCIONA program 2009 (ACI2009-1040).

**References**

[1] Duffy T, Vos W, Zhang Ch, Hemley R J, and Mao H, 1994 *Science*, **263**, 5153, 1590-1593
[2] Silvera I and Cole J, 2010 *J. of Phys.: Conf. Series*, 215 012194. Bennitong S, Lovell A, Headen T, Royse D, Nathanson A and Voller S, 2013 US Patent 2013/0095307A1
[3] Guerrero C, Cuesta S and Perlado J M, 2013 *Eur. Phys. J. Web of Conferences* **59**, 16004
[4] Guerrero C, Cuesta S and Perlado J M, 2014 *Europhys. Lett.* **108**, 26001. Guerrero C and Perlado J M 2013, *25th Symposium on Fusion Engineering (SOFE)*, pp.1,4 doi: 10.1109/SOFE.2013.6635457.
[5] Aleksandrova I V et al 2007 *J. of Russian Laser Research*, **28**, 3
[6] Eremets M I and Troyan A I 2011 Conductive dense hydrogen *Nat. Mater.* **10**, 927
[7] Nellis W J, Ruoff A and Silvera I F 2012 Has Metallic Hydrogen Been Made in a Diamond Anvil Cell? *Preprint* arXiv:1201.0407
[8] Caruso A and Strangio C 2003 *J. Exp. Theor. Phys.* **97**, 5, 948-957
[9] Yi S A, et al 2014 *Phys. Plasmas* **21**, 092701
[10] Hurricane O A, et al 2014 *Nature* **506**, 343
[11] Soler J M, Artacho E, Gale J D, Garcia A, Junquera J, Ordejon P and Sanchez-Portal D 2002 *J. Phys.: Condens. Matter* **14**, 2745-2779.
[12] Zha C, Duffy T, Mao H and Hemley R J 1993 *Phys. Rev. B.* **48**, 13, 9246-9255
[13] Toledano P, Katzke H, Goncharov A F and Hemley R J 2009 *Phys. Rev. Lett.* **103**, 105301

# Laser-driven ablation through fast electrons in PALS experiment

S.Yu. Gus'kov[1,a], T. Chodukowski[2] , N. Demchenko[1], Z. Kalinowska[2], A. Kasperczuk[2], E. Krousky[3], M. Pfeifer[3], P. Pisarczyk[4], T. Pisarczyk[2], O. Renner[3], J. Skala[3], M. Smid[3], J. Ullschmied[5]

[1]P.N.Lebedev Physical Institute of RAS, 119991 Leninskii pr. 53, Moscow, Russia
[2]Institute of Plasma Physics and Laser Microfusion, 23 Hery St., 00-908 Warsaw, Poland
[3]Institute of Physics ASCR, v.v.i., Na Slovance 2, 182 21 Prague 8, Czech Republic
[4]Warsaw University of Technology, ICS, 15/19 Nowowiejska St., 00-665 Warsaw, Poland
[5]Institute of Plasma Physics ASCR, v.v.i., Za Slovankou 3, 182 00 Prague 8, Czech Republic

**Abstract.** Energy transfer to shock wave in Al and Cu targets irradiated by a laser pulse with intensity of I≈1-50 PW/cm$^2$ and duration of 250 ps was investigated at Prague Asterix Laser System (PALS). The iodine laser provided energy in the range of 100-600 J at the first and third harmonic frequencies. The focal spot radius of laser beam on the target was varied from 160 to 40 μm. The dominant contribution of fast electron energy transfer into the ablation process was found when using the first harmonic radiation, the focal spot radius of 40-100 μm, and the energy of 300-600 J. The fast electron heating results in the growth of ablation pressure from 60 Mbar at the intensity of 10 PW/cm$^2$ to 180 Mbar at the intensity of 50 PW/cm$^2$ and in the growth of the efficiency of the energy conversion into the shock wave from 2 to 7 % under the conditions of 2D ablation.

## 1 Introduction

The one of the subjects related to shock ignition method (SI) [1,2] that is influence of fast electron energy transport in laser-produced plasma on efficiency of energy coupling to shock wave in a dense target is studied. In the experiments the coupling parameter $I\lambda^2$ ($I$ and λ are the intensity and wave length of laser radiation) was varied in the range of $10^{14}$-$8\times10^{16}$ Wμm$^2$/cm$^2$ (low to high level of fast electrons effect) at the laser pulse intensity (1-50 PW/cm$^2$) and duration (250 ps) close to that at SI. In the case of relatively short plasma of PALS experiments the fast electrons are generated due to TNSA mechanism near the plasma resonance region, while in the case of the SI extended plasma the fast electrons are generated due to stimulated parametric processes in the lower density plasma. Nevertheless as shown below, the average energy of fast electrons in the high intensity PALS experiments is close to that measured in the OMEGA laser SI experiments [3]. Under the action of tightly focused laser beam the expansion of laser-produced plasma is not planar, therefore the interpretation of the results was based on 2D numerical simulations and analytical model taking into

[a] e-mail : guskov@sci.lebedev.ru

account the lateral expansion of ablated material. To determine plasma density and temperature a 3-frame interferometry and an x-ray spectroscopy were used. The laser energy transferred to the solid target was determined via measurements of the crater volume created on the target surface.

## 2 Experiment

The experiments were carried out at a normal incidence of the PALS laser radiation of the first ($\lambda_1$ = 1.315 μm) and third ($\lambda_3$ = 0.438 μm) harmonics on a surface of planar Al or Cu targets. The pulses of 120, 290, and 580 J energies (in the 3ω case only two first energies were available) with duration of 250 ps (FWHM) were applied. The values of beam radii ($R_L$) of 40, 80, 120 and 160 μm provided the variation of the incident laser intensity on the target, for example at the laser energy $E_L$=290 J, from $I_L$=1.4×10$^{15}$ ($R_L$=160 μm) to $I_L$=2.3×10$^{16}$ W/cm$^2$ ($R_L$=40 μm). The maximum value of the coupling parameter $I\lambda^2$(8 ×10$^{16}$ Wμm$^2$/cm$^2$) was achieved at the energy of 580 J for 1ω and $R_L$=40 μm while its minimum value (1.1×10$^{14}$ Wμm$^2$/cm$^2$) – at the energy of 120 J for 3ω and $R_L$=160 μm.

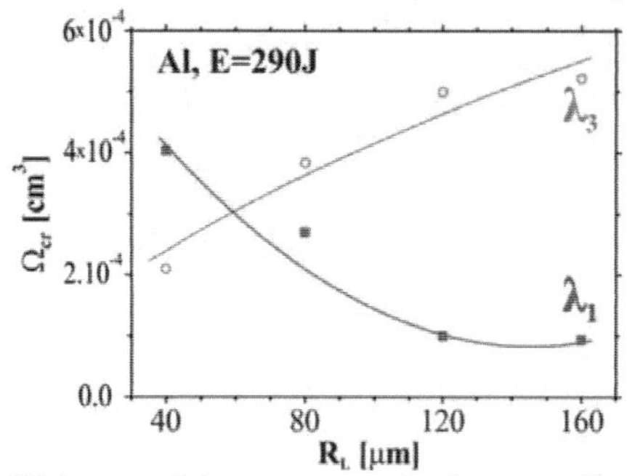


**Fig. 1**. Volume of the crater vs the beam radius at the energy of 290 J of the first and third harmonics laser radiation for Al target.

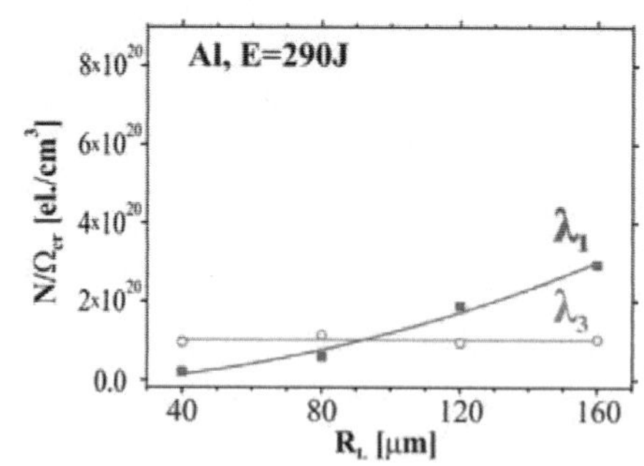


**Fig. 2**. Ratio of electron number to crater volume vs the beam radius at the energy of 290 J of the first and third harmonics of the laser radiation for Al target

The dependencies of the crater volume $\Omega_{cr}$ produced in Al target and the ratio of total number of plasma electrons $N_e$ (computed on the basis of interferometric measurements of spatial electron density distribution) to crater volume on the laser beam radius are presented in Figs. 1 and 2.

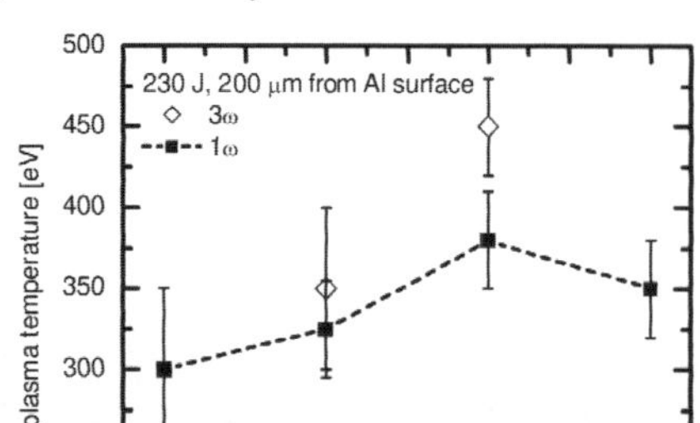


**Fig. 3**. Plasma temperatures at the distance of 200 μm from Al target surface irradiated at different laser harmonics and focal radii at the laser energy of 230 J.

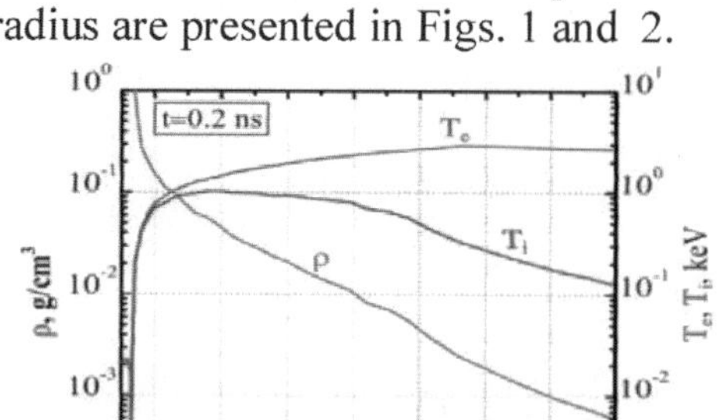


**Fig. 4**. Distributions of mass density, electron and ion temperatures on the axis of the plasma for Al target irradiated by the first harmonic beam at 40 μm radius

The data of both figures correspond to experiments with the energy of 290 J of the 1ω and 3ω laser radiation. The similar results were obtained for Cu target. In the 3ω case, the crater volume (energy transferred to the solid target) decreases with the decreasing $R_L$ despite the increasing intensity of the laser beam that is explained by lateral expansion of the laser-produced plasma at the negligible role of fast electron energy transfer. By contrast, in the 1ω case, when at the given intensity the coupling parameter $I\lambda^2$ is 9 times larger than in the 3ω case, the crater volume increases with the decreasing $R_L$ (increasing intensity), particularly fast in the range of small radii of 40-100 μm (large values of $I_L\lambda^2$). This fact points out to appear an additional factor of energy transfer, which is stronger than the lateral expansion effect, that it is natural to associate with energy transfer by fast electrons. The same conclusion follows from the data of Fig. 2, which show decreasing the number of electrons providing the creation of the unit crater volume with decreasing $R_L$ in the 1ω case.

The plasma temperatures spectroscopically measured at the distance of 200 μm from Al target surface irradiated by 230 J of the 1ω and 3ω laser radiation are presented in Fig. 3. The temperature decrease at the small radii of 1ω beam indicates the heating depth growth with increasing intensity. Temperature in the case of 3ω beam is by a factor of 1.2-1.3 higher than that in the case of 1ω beam.

## 3 Numerical simulation and model

The numerical simulations of the laser beam interaction with Al target were carried out for the energy of 290 J of the first and third harmonics laser beam with the radii 120 and 40 μm using the 2D hydrodynamic code ATLANT-HE. The code includes refraction of the laser radiation in the plasma, inverse bremsstrahlung and resonance mechanisms of the laser light absorption, fast electron generation due to the resonance absorption and its energy transfer with Coulomb collisions. The results of simulations are the following. In the 1ω case and R=120 μm ($I_L=2.6\times10^{15}$ W/cm$^2$): total absorption coefficient is $K_a=0.38$, absorption coefficient due to resonant mechanism is $K_{ra}=0.08$, fast electron temperature $T_h=14$ keV, crater volume $\Omega_{cr}=1.7\times10^{-4}$ cm$^3$; in the 1ω case and R=40 μm ($I_L=2.3\times10^{16}$ W/cm$^2$): $K_a=0.28$, $K_{ra}=0.06$, $T_h=90$ keV, $\Omega_{cr}=4.1\times10^{-4}$ cm$^3$. In the 3ω case and R=120 μm ($I_L=2.6\times10^{15}$ W/cm$^2$): $K_a=086$, $K_{ra}=0.07$, $T_h=3$ keV, $\Omega_{cr}=2.4\times10^{-4}$ cm$^3$; in the 1ω case and R=40 μm ($I_L=2.3\times10^{16}$ W/cm$^2$): $K_a=0.73$, $K_{ra}=0.1$, $T_h=19$ keV, $\Omega_{cr}=1.8\times10^{-4}$ cm$^3$. In the Fig. 4 the dependences of mass density, electron and ion temperatures on the beam radius are presented.

The analytical model is developed to explain the results of experiments and simulations. It combines the 1D and 3D self-similar solutions for isothermal expansion of a given mass of material [4] by means introducing the parameter, which is ratio of laser-produced plasma size and laser beam radius. The plasma density ($\rho_a$), temperature ($T_a$) and pressure ($P_a$) on the ablation surface as well as the efficiency of energy transfer (σ) to shock wave propagating and creating the crater in a solid target (ablation loading efficiency) determined by the ablation pressure, according this model, are

$$\rho=\frac{3\mu^{3/2}}{(2\pi)^{1/2}I^{1/2}\tau^{3/2}}\Psi^{-2},\quad T=\frac{I\tau}{2C_V\mu\left(2-\Psi^{-1}\right)},\quad P=\frac{I^{1/2}\mu^{1/2}}{(2\pi)^{1/2}\tau^{1/2}\left[\Psi\left(2\Psi-1\right)\right]} \qquad (1)$$

$$\sigma=\frac{2^{1/2}\mu^{3/4}}{(2\pi)^{3/4}(\gamma+1)^{1/2}I^{1/4}\tau^{3/4}\rho_0^{1/2}}\frac{\Psi^{1/2}}{(2\Psi-1)^{3/2}} \qquad (2)$$

where μ is the areal ablated mass, which is determined either by the depth of matter heating by thermal conductivity wave ($\mu_c$), or by the range of fast electrons ($\mu_e$)

$$\mu_c=\frac{\kappa^{1/3}I^{2/3}\tau^{2/3}}{C_V^{7/6}}\approx2.6\cdot10^{-3}\frac{A^{7/6}I_{(pw)}^{2/3}t_{(ns)}^{2/3}}{(Z+1)^{7/6}(Z+3.3)^{1/3}},\quad \mu_e=\frac{E_0^2\cdot A\cdot m_p}{4\pi e^4\cdot Z\cdot\Lambda}\approx5.2\cdot10^{-7}\frac{A}{Z}E_{0(keV)}^2\ \text{g/cm}^2 \qquad (3)$$

the electron conductivity coefficient (at the value of Coulomb logarithm $\Lambda=5$) $\kappa=8.4\times10^{19}/(Z+3.3)$ erg×cm$^{-1}$×s$^{-1}$×keV$^{-7/2}$ and $E_{0(keV)}=8(I_{L(pw)}\lambda_\mu^2)^{2/3}$ (the values $I_{(pw)}$ and $I_{L(pw)}$, $\tau_{(ns)}$, $\lambda_\mu$ are measured, respectively, in the units of $10^{15}$ W/cm$^2$, ns, μm); $I=K_aI_L$ is the absorbed radiation intensity; $C_V=Zk_B/Am_p(\gamma-1)$ is specific heat; γ is adiabatic exponent, $Z$ and $A$ are the charge and atomic number of the plasma ions, $k_B$ is the Boltzmann constant, $m_p$ is the proton mass, $\Psi=(1+2^{3/4}L_p/3R)$ is the factor of lateral plasma expansion; $L_p=(I\tau^3/\mu)^{1/2}$ is the plasma size; $\rho_0$ is the initial density of target material. Under conditions of considered experiments, according to (3), in the 3 ω case $\mu_c>\mu_e$ for all radii, while in the 1ω case $\mu_c>\mu_e$ for the radii $R$=160 and 120 μm and $\mu_c<\mu_e$ for the radii $R$ = 80 and 40μm. The results of calculations of ablation loading efficiency as well as volume of crater $\Omega_{cr}=\sigma K_a E_L/\alpha\varepsilon\rho_s$ ($\varepsilon=2.6\times10^3$ J/g is the specific energy for evaporation of unit mass of Al target, $\alpha$=1.5 is the ratio between the thermal and kinetic energy of shock wave) are presented in Fig. 5 and 6. According to (3), the fast electron transfer provides a stronger growth of ablated mass and, consequently, stronger growth of the ablation density with increasing the laser intensity than thermal conductivity wave. At the beam radii of 160 and 120 μm the ablation is provided by thermal conductivity and the lateral expansion effect is negligible for both harmonics. Therefore, $\rho_{a(c)}\propto I^{1/2}$ and for given laser intensity $\sigma_{(c)}\propto K_a^{1/4}R^{-1/2}$, $\Omega_{cr}\propto K_a^{5/4}R^{-1/2}$. Thus, for large beam radii at both harmonics crater volume depends weakly on radius, and the ratio of the crater volumes in the 3ω and 1ω cases equals approximately to the ratio of the absorption coefficients: $\Omega_{cr(3)}/\Omega_{cr(1)}\approx(K_{a(3)}/K_{a(1)})^{5/4}$. At the small beam radii of 80 and 40 μm the lateral expansion effect is strong and, therefore, $\rho/\rho_a\propto(R/L)^2\propto R^2\mu/I$. From here we get follows $\rho_{(e)}/\rho_{a(e)}\propto R^{4/3}K_a^{1/3}$ and $\rho_{(c)}/\rho_{a(c)}\propto R^{8/3}K_a^{-1/3}$. So in the 3ω case, when ablation, as before, is provided by thermal conductivity $\sigma_{(c)}\propto K_a^{1/12}R^{-5/6}$ and $V_{cr}\propto K_a^{13/12}R^{5/6}$. In the 1ω case, i.e at ablation due to fast electron heating $\sigma_{(e)}\propto K_a^{11/12}R^{-5/6}$ and $V_{cr}\propto K_a^{23/12}R^{-5/6}$.

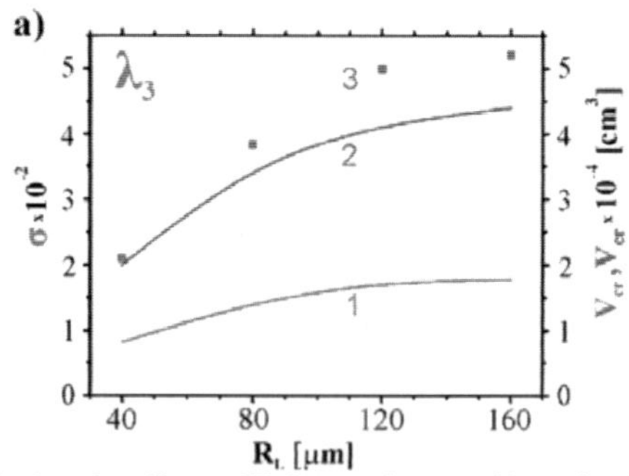


**Fig. 5**. Ablation loading efficiency (curve 1) and theoretical (curve 2) and experimental values (curve 3, green squares) of crater volume vs focal spot radius at energy of 290 J of the 3ω radiation.

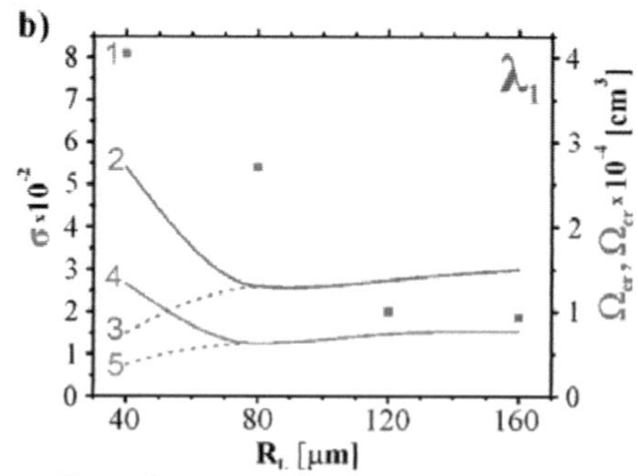


**Fig.** . Experimental (curve 1, green squares) and theoretical crater volume with (curve 2) and without (curve 3) taking into account the fast electron transport and ablation loading efficiency with (curve 4) and without (curve 5) taking into account the fast electron transport vs the beam radius at the energy of 290 J of the 1ω radiation.

## 4 Conclusion

At the laser intensity of 10-50 PW/cm$^2$ and $I\lambda^2=10^{16}$-$5\times10^{16}$ Wμm$^2$/cm$^2$ (the experiments with the laser pulse energy of 290-580 J for the first harmonic of iodine laser radiation) relevant to the laser spike in SI approach the dominant mechanism of ablation during the period of 250 ps is the matter heating by fast electrons. In spite of the lateral expansion, the ablative pressure increases from 100 Mbar for the intensity of 10 PW/cm$^2$ to 300 Mbar for the intensity of 50 PW/cm$^2$, whereas the efficiency of the laser energy transformation into the shock wave energy is on the level of 2-7%.

This work was supported by RFBR projects No 12-02-92101-JF and No 11- and 13-02-00295, by the Czech Republic AS project M100101208 and by the Czech Science Foundation project CZ.1.07/2.3.00/20.0279 and by the NCN of Poland Grant No 2012/04/M/ST2/00452.

## References

1. V.A. Scherbakov, Sov. J. Plasma Phys. , 240 (1983)
2. R. Betti, C.D. Zhou, K.S. Anderson et al, Phys. Rev. Lett. **8**, 155001 (2007)
3. W. Theobald, R. Nora, M. Lafon et all, Phys. Plasmas **1** , 102706 (2012)
4. V.S. Imshennik. Sov. Phys. Dokl. **5**, 263 (1960)

Journal of Physics: Conference Series **688** (2016) 012024 doi:10.1088/1742-6596/688/1/012024

# Habit control of deuterated potassium dihydrogen phosphate crystal for laser applications

**L A Guzman[1], M Suzuki [1], Y Fujimoto[2] and K Fujioka[2]**

[1]Dept. of Chemical and Material Engineering, National Institute of Technology, Ibaraki College, 866 Nakane, Hitachinaka, Japan

[2]Institute of Laser Engineering, Osaka University, 2-6 Yamadaoka, Osaka, Japan

E-mail: laguzman@chem.ibaraki-ct.ac.jp

**Abstract**. In this study we investigate the habit of partially deuterated potassium dihydrogen phosphate (DKDP) crystals in the presence of $Al^{3+}$ ions. We have grown single DKDP crystals in (50wt% and 80wt%) partially deuterated solutions and in solutions doped with $Al^{3+}$ ions (2 ppm) by the point-seed rapid growth technique at controlled supercooling ($\Delta T$=10°C). The growth length of each crystal face was measured and the aspect ratio was calculated. We found that crystals grown in partially deuterated solutions are similar in aspect ratio, while, crystals grown in deuterated solutions doped with $Al^{3+}$ ions showed a relative change in aspect ratio, the crystal increased in size in the pyramidal direction (vertical axis direction). Crystal characteristics were also analyzed by X-ray diffraction, FTIR and Raman spectroscopy. We have speculated that the relative habit modification is due to a probably adsorption and inclusions of $Al^{3+}$ ions in the prismatic section of the crystal.

## 1. Introduction

In recent years, high-power laser facilities require optical devises as a part of its design. KDP and DKDP crystals are recognized for its special optical properties. Those crystals can be grown sufficiently large to fulfill the requirements of the design. However for the optimization in the production of those crystals, it is crucial to control the habit which is given by the aspect ratio (ratio of crystal dimensions along its crystallographic axes, X, Y, Z and the constant angles between the faces).

In previous studies, we have reported the habit and growth kinetics of KDP crystals in presence of metallic ($Cr^{3+}$, $Fe^{3+}$, $Al^{3+}$ and $Pb^{2+}$) ions, these impurities showed the ability to control the shape and growth faces [1, 2]. Large information on KDP crystal can be found. However, limited information is available on the habit of partially deuterated DKDP crystals. A recent study on the habit of DKDP crystal in the presence of $Al^{3+}$ ions (1ppm) was reported, where $Al^{3+}$ ions cannot be used as habit modifier in the presence of EDTA solutions [3]. Another interesting study found was the habit control of KDP and DKDP crystals without impurities using the rapid growth method [4]. In these reports, high deuterium concentration levels (98%) were used. Recently, high-power laser requires 70~80% deuterated DKDP crystals. Therefore, data in partial deuterated concentrations are required.

In this work, the influence of $Al^{3+}$ ions on the habit (referred as, aspect ratio) of DKDP crystals is studied in two levels of deuterated solutions. The seed point technique and controlled supercooling conditions are applied to evaluate the possibility to control the habit and to obtain large bulk crystals.

## 2. Experimental Procedure

Two partially deuterated solutions ($D_2O$-$H_2O$, 50wt% and 80wt% in $D_2O$) were prepared using deuterium water (Heavy water) 99.8% in $D_2O$ (Isotec Inc.) and distilled water deonized. The Solutions were saturated with extra pure reagent grade KDP and Aluminum standard solution (Kanto Kagaku) was used as impurity to obtain a doped solution ($Al^{3+}$ ions, 2ppm). The doped and un-doped solutions were previously filtered, over-heated to 60 °C for 4 h and cool-down to 40°C. Seed crystals were prepared by spontaneous nucleation of KDP solution and then selected seed crystals were grown in deuterated solutions. The obtained DKDP seed crystal was glued in a crystal holder as a point seed and set into a small scale jacketed crystallizer (30ml vol.). The supercooling condition ($\Delta T$=10°C) was controlled with a programmable temperature controller at a rate of 0.5 °C /h. Before and after the growth, the crystal size was measured between the faces in the three crystallographic (X, Y and Z) axes and the aspect ratio was calculated. The characteristics of the grown crystals were also verified by X-ray diffraction, FTIR and Raman spectroscopy.

Previous to the experiments, the impurity and the impurity concentration were first evaluated between three different ions ($Pb^{2+}$, $Cr^{3+}$and $Al^{3+}$). Scanning Electron Microscope (SEM) was used to evaluate the habit change of the crystal in a wide range of impurity concentrations (up to 100 ppm).

## 3. Results

The results on the selection of impurity and impurity concentration are shown in figure 1. SEM pictures shows the habit of DKDP microcrystals nucleated from deuterated solution 50wt% doped with $Pb^{2+}$, $Cr^{3+}$ and $Al^{3+}$ ions at 2 ppm concentration. The habit of microcrystals grown in the presence of $Pb^{2+}$ and $Cr^{3+}$ ions did not show any considerable change (see, figure 1 (a) and (b), respectively). However, the crystals grown with $Al^{3+}$ ions showed a relative habit change. The crystal growth increase was clearly observed in the pyramidal direction (z-axis direction). At impurity concentration higher than 2 ppm, the crystal became slender and finally showed a needle shape (20 ppm). In the case of $Pb^{2+}$ and $Cr^{3+}$ ions, the crystals were tapered and needle shape (data not show here). From these observations, $Al^{3+}$ions and 2 ppm were selected as impurity and impurity concentration.

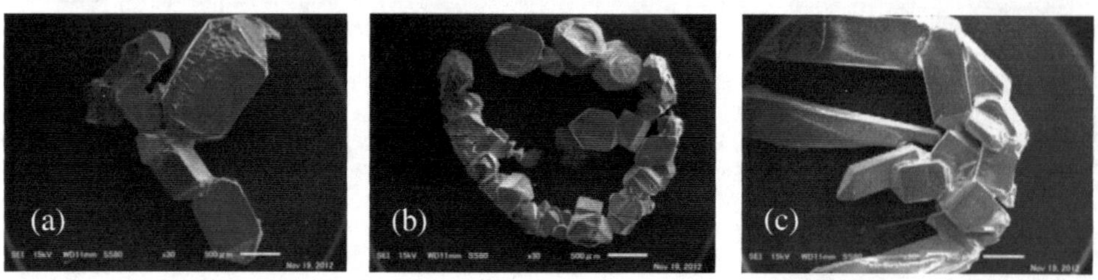


**Figure 1.** Habit of DKDP microcrystals grown from doped solutions (a)$Pb^{2+}$, (b)$Cr^{3+}$and (c)$Al^{3+}$ ions. (50wt% deuterated solutions with 2 ppm impurity concentration)

After the impurity selection, single DKDP crystals were grown in a crystallizer in two different solutions. One is, in deuterated solutions 50wt% and 80wt% and the other is in deuterated solution doped with $Al^{3+}$ ions. The habit of each crystal grown in these solutions can be seen in figure 2. Habit of crystals grown in 50wt% and 80wt% deuterated solutions free of $Al^{3+}$ ions are shown in figure 2(a) and figure 2(c), respectively. Deuterium concentration increase in the solution did not affect the crystal habit. However, in solutions doped with $Al^{3+}$ ions a relative habit change can be observed (see, figures 2(b) and 2(d)) where the crystals showed an increased growth in the pyramidal direction (z-axis) similar to those crystals showed in figure 1(c).

Size of grown crystals was measured and the aspect ratio was calculated. Average values of the aspect ratio are shown in table 1. The calculations showed that the x/y ratio of the crystals grown from deuterated solutions were similar compared to the x/y ratio of their seed crystals. In contrary, z/x and z/y ratios of crystals grown in doped $Al^{3+}$ ions exhibited an increased value indicating a habit change. In some cases, slightly tapering effect was observed on the prismatic face (100) (see, figure 2(b)).

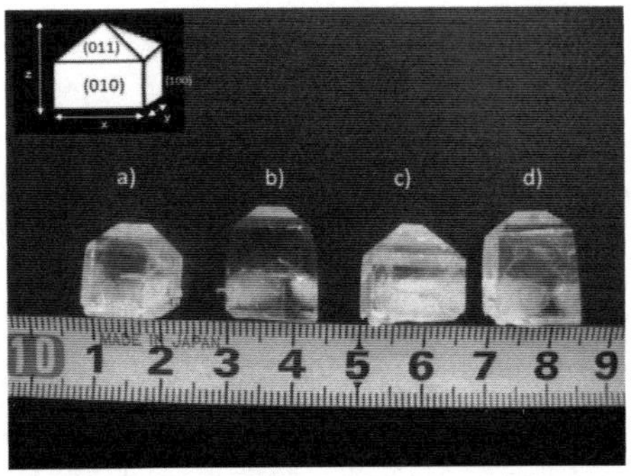


**Figure 2.** Habit of DKDP crystals grown in partial deuterated solutions and in solutions doped with $Al^{3+}$ ions.
(a) DKDP crystal in 50wt% $D_2O$,
(b) DKDP crystal in 50wt% $D_2O$ with $Al^{3+}$ 2ppm
(c) DKDP crystal in 80wt% $D_2O$
(d)DKDP crystal in 80wt% $D_2O$ with $Al^{3+}$ 2 ppm

**Table 1.** Aspect ratio change of DKDP crystals in presence of $Al^{3+}$

| Impurity content (ppm) | $D_2O$ in solution wt% | Seed crystals | | | Grown crystals | | |
|---|---|---|---|---|---|---|---|
| | | x/y | z/x | z/y | x/y | z/x | z/y |
| 0 | 50 | 1.48 | 0.89 | 1.31 | 1.48 | 1.09 | 1.61 |
| | 80 | 1.40 | 0.91 | 1.27 | 1.38 | 1.00 | 1.38 |
| 2 ppm | 50 | 1.45 | 0.87 | 1.25 | 1.47 | 1.37 | 2.01 |
| | 80 | 1.45 | 0.91 | 1.31 | 1.40 | 1.35 | 1.88 |

On the other hand, grown crystals were also analyzed to see the effect of the impurity concentration in the internal structure. The X-ray diffraction patterns of DKDP powder was used to check the DKDP crystal structure. A Rigaku X-Ray Diffractometer (RINT 2000) was used for the analysis. The X-ray pattern of DKDP crystals grown in 50wt% deuterated solution and with doped $Al^{3+}$ ion at 2 ppm, are compared and shown in figure 3. The diffraction patterns were similar in both cases. From these results can be assumed that $Al^{3+}$ ions do not affect the internal structure.

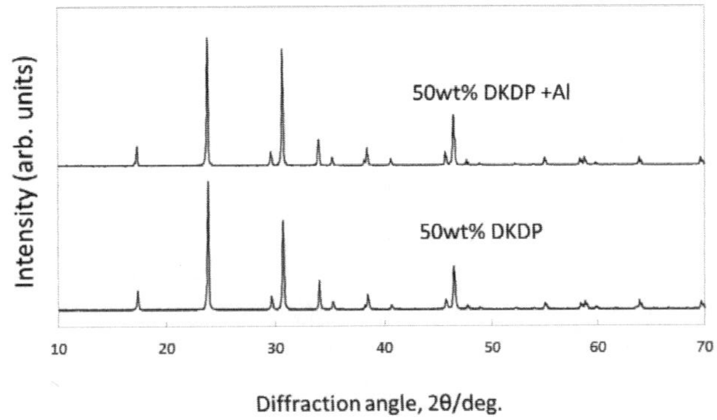


**Figure 3.** X-ray powder diffraction patterns of DKDP (50wt%) and DKDP (50wt%) doped with $Al^{3+}$ (2 ppm)

The presence of deuterium content in the crystal was examined with a Jasco Raman Spectrometer NRS-5000. Samples of DKDP crystals grown in partially deuterated solution (50wt% and 80wt%) were analyzed and for comparison a KDP crystal was also used. The Raman spectra of the samples are shown in figure 4. Here, the absorption spectra shift (930 $cm^{-1}$) increased according the deuterium concentration. The spectral shifts in this range indicate us the presence of deuterium in the crystal and the replacement of hydrogen by the deuterium.

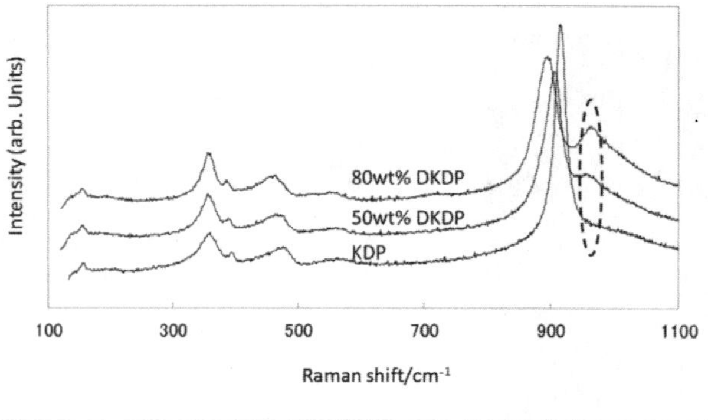


**Figure 4.** Raman spectra of pure KDP, 50wt% DKDP and 80wt% DKDP

On the other hand, the DKDP crystals grown in deuterated solution 50wt% and with $Al^{3+}$ were analyzed using an Infrared Spectrometer FT/IR-6100 (Jasco). The internal vibration modes of hydrogen was observed in a spectral range area between 4000~400($cm^{-1}$). Result of the analysis is shown in figure 5. Absorption bands were detected at relatively high-frequency region. O-H stretching appeared in a 3500 $cm^{-1}$ followed by the P-O-H asymmetric stretching at 2900 ($cm^{-1}$). And finally, O=P-OH stretching appeared at 2400 and 1600 ($cm^{-1}$) respectively. The O-H stretching adsorption band was relatively different. It is probably caused by an inclusion of $Al^{3+}$ into the crystal. The other modes remain similar.

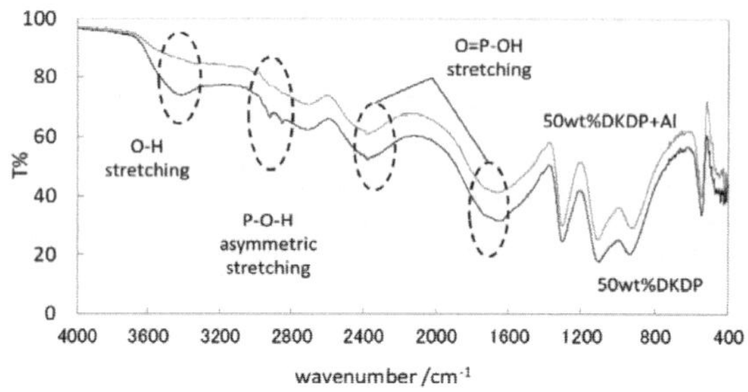


**Figure 5.** IR spectra of DKDP crystals grown in 50wt% deuterated solution and deuterated solution doped with $Al^{3+}$ ions (2 ppm).

From the results described above, we can estimate that $Al^{3+}$ ion would be a candidate to be used as a DKDP crystal habit modifier. However, to determine the factors that control the crystal habit more effectively is still necessary to understand the growth kinetics at different growth conditions (of deuterium concentration, supersaturation and temperature, etc.). As a continuation of this work, a study is under investigation and it will be reported elsewhere.

**Acknowlegements**

This work was performed under the joint research project of the Institute of Laser Engineering, Osaka University (under contract subject "B2-19").

**References**

[1] Guzman L A, Maeda K, Hirota S 1998 *Kagaku Kougaku Ronbunshu* **24-1** 143-145
[2] Guzman L A, Maeda K, Hirota S 1998 *J. Chem. Eng. of Japan* **31-2** 295-297
[3] Booth N, Land T, Erhmann P, Vekilov P 2005 *Cryst. Growth and Design* **5-1** 105-110
[4] Zaitseva N. P, Carman L, Smolsky I. L 2002 *J. Cryst. Growth* **24** 363-373

# Alternative Approaches to High Energy Density Fusion

**J Hammer**

Lawrence Livermore National Laboratory, Livermore California 94550, USA

E-mail: hammer2@llnl.gov

**Abstract.** This paper explores selected approaches to High Energy Density (HED) fusion, beginning with discussion of ignition requirements at the National Ignition Facility (NIF). The needed improvements to achieve ignition are closely tied to the ability to concentrate energy in the implosion, manifested in the stagnation pressure, $P_{stag}$. The energy that must be assembled in the imploded state to ignite varies roughly as $P_{stag}^{-2}$, so among other requirements, there is a premium on reaching higher $P_{stag}$ to achieve ignition with the available laser energy. The U.S. inertial confinement fusion program (ICF) is pursuing higher $P_{stag}$ on NIF through improvements to capsule stability and symmetry. One can argue that recent experiments place an approximate upper bound on the ultimate ignition energy requirement. Scaling the implosions consistently in spatial, temporal and energy scales shows that implosions of the demonstrated quality ignite robustly at 9-15 times the current energy of NIF. While lasers are unlikely to reach that bounding energy, it appears that pulsed-power sources could plausibly do so, giving a range of paths forward for ICF depending on success in improving energy concentration. In this paper, I show the scaling arguments then discuss topics from my own involvement in HED fusion. The recent Viewfactor experiments at NIF have shed light on both the observed capsule drive deficit and errors in the detailed modelling of hohlraums. The latter could be important factors in the inability to achieve the needed symmetry and energy concentration. The paper then recounts earlier work in Fast Ignition and the uses of pulsed-power for HED and fusion applications. It concludes with a description of a method for improving pulsed-power driven hohlraums that could potentially provide a factor of 10 in energy at NIF-like drive conditions and reach the energy bound for indirect drive ICF.

## 1. Introduction

High Energy Density (HED) fusion research is in an exciting time given the progress on the National Ignition Facility (NIF). Ignition remains elusive, yet the important milestone of alpha particle heating – doubling the yield due to self-heating – appears likely to be within reach. The basic requirements for ICF ignition are well known [1]. The temperature of the ignition region must be high enough to exceed radiative losses and give high fusion reactivity, e.g., a burn averaged value of near 4.5 keV or higher. Most of the alpha particle energy must be deposited in the ignition spot and the Lawson parameter, the product of pressure and energy confinement time, $P\tau$, must be high enough to give significant gain. The latter two conditions can usually be expressed in terms of $\rho R$, or density times radius in g/cm$^2$, with $\rho R \sim 0.3$ in the hot spot to trap alphas and the total $\rho R > 1$ to give reasonable gain.

If we focus on the energy requirements of the hot spot, then a simple relationship between energy and the stagnation pressure of the capsule exists. Since $P_{stag} \propto \rho T$ for high temperature matter, and the hot spot energy is proportional to the product of pressure and volume, we have the result:

$$E_{HotSpot} \propto \rho T R^3 = \frac{(\rho R)^3 T}{\rho^2} \propto \frac{(\rho R T)^3}{P_{stag}{}^2} \propto P_{stag}{}^{-2} \tag{1}$$

The last relationship in equation (1) follows if we take $\rho R$ and $T$ to be determined by the ignition requirements. For instance, at $\rho R = 0.3, T = 4.5keV$ we have $E_{HotSpot} \cong 70kJ * P_{stag}{}^{-2}$ for $P_{stag}$ in units of 100 GBar. The total energy invested in fuel internal energy is larger than $E_{HotSpot}$ because of the surrounding dense, cold fuel layer needed for inertial confinement. High gain hot spot targets could be dominated by the energy of the cold fuel, but ignition targets at the NIF have cold fuel energy comparable to hot spot energy. Fast ignition designs have similar scaling but require higher $\rho R$ in the ignition spot.

Equation (1) shows the importance of reaching high stagnation pressures to enable ignition at reasonable energy. At 400 GB, typical of ignition designs, the hot spot energy required is about 4.4 kJ, which should be achievable at NIF. Most experiments to date, on the other hand, have been limited to $P_{stag} \lesssim 150GB$ [2]. Many factors such as implosion symmetry, hydrodynamic instability, shock mistiming or fuel pre-heat could contribute to low $P_{stag}$, and improved understanding of these effects are among the main thrusts of the experimental program moving forward. One simple way of looking at the issue is through energy conservation. The stagnated state energy should approximately equal the $PdV$ work done by the implosion:

$$P_{Stag}V_{Stag} \propto P_{Drive}V_0 \rightarrow P_{Stag} \propto P_{Drive}\left(V_0/V_{Stag}\right) \rightarrow P_{Stag} \propto P_{Drive}C^3 \tag{2}$$

where $P_{Drive}$ is the average drive pressure, $\sim T_R^{3.5}$ for indirect drive. $V_0$ and $V_{Stag}$ are the initial and stagnation volumes and $C$ is the fuel convergence ratio. Note that $C$ defined this way differs from the usual hot spot convergence ratio. One caveat with this expression is that many NIF implosions have missing energy, i.e., kinetic energy present in the implosion is incompletely converted to internal energy [2]. The drive temperature and pressure are limited by available power or laser intensity limits within the hohlraum, so increasing $P_{stag}$ depends critically on reducing sources of asymmetry, instability, and preheat that limit $C$ or disrupt the efficient energy conversion assumed in equation (2).

The High Foot experiments [3] recently conducted at NIF, so named for the higher drive temperature and shock pressure during the foot of the pulse, are interesting not only in being the highest yield to date, but because they agree well with simulations for most measured and inferred quantities, such as yield, $\rho R$, burn temperature, and pressure. They thus represent validation of hydrodynamic and burn models for their operating regime of implosions on the NIF. With a burn-averaged $P_{stag}$ of 100 – 150 GB, and $C \sim 17$ they are below ignition requirements at the available energy, but arguably they set a bound on what energy would be needed to ignite at the values of $C$ and $P_{stag}$ that have already been demonstrated. Equation (1) suggests that bound is about one order of magnitude larger than the existing NIF implosions, however we can use simulations to better quantify the issue. Starting with a 1 dimensional radiation hydrodynamics model of the High Foot implosion experiment N130710, we can scale the implosion while preserving drive pressure, shock-timing, $C$ and $P_{stag} \cong 119GB$ using Euler scaling:

$$R \rightarrow SR, \quad t \rightarrow St, \quad P \rightarrow P, \quad E \rightarrow S^3E \tag{3}$$

where $S$ is the spatial/temporal scale factor, giving scaling as $S^3$ in energy. The calculations do not include the ablator, but use a pressure drive extracted at the fuel-ablator interface since radiation-driven ablation does not strictly follow Euler scaling. Heat conduction in the hot spot also does not follow Euler scaling, but has little effect on $C$ and $P_{stag}$. Burn-off calculations confirm that $C$ and $P_{stag}$ are preserved for $S < 3$. The results of the scaling are shown in figure 1, confirming the expectation from equation (1). Interestingly, the bounding energy for ignition (and high yield) from this argument may be within the capabilities of future laboratory drivers. The dashed lines in figure 1 show the drive energy from a hypothetical 16MJ pulsed-power source discussed below. As indicated in figure 2, the path toward ignition may come through either raising the drive energy, improving implosion quality or both. In that vein, the recent experiment N130927 achieved preliminary values of inferred pressure of ~ 150GB, suggesting a revised bounding energy ~40% lower than shown in figure 1.

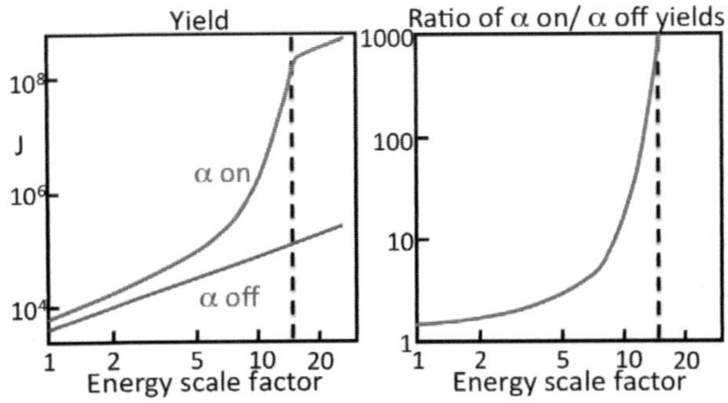


**Figure 1.** Yields with alpha deposition on/off and the ratio of alpha-on to alpha-off yield from hydrodynamically scaled High Foot capsule, versus the energy scale factor, $S^3$.

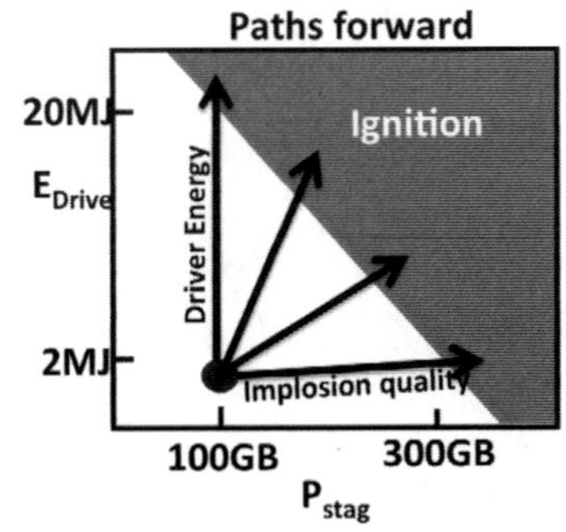


**Figure 2.** Possible paths toward ignition.

## 2. Viewfactor Experiments

A viewfactor experiment, shown in figure 3, employs an ignition-scale cryogenic hohlraum with one end cut off, exposing the interior and far laser entrance hole (LEH) to views that allow absolute measures of x-ray flux with Dante diagnostics as well as x-ray imaging. The purpose is to characterize the x-ray drive from the viewpoint of a capsule, as opposed to the usual measurement of the drive through the LEH from which the capsule drive is inferred. The experiments are designed to create a plasma and radiation environment as close as possible to the ignition target while allowing for the improved view of the hohlraum interior. The experiments are motivated by inconsistencies between the LEH-inferred drive and the implosion trajectories of capsules. As reported in greater detail in by M Schneider at another paper at this conference, viewfactor experiments are indeed largely consistent with a lower capsule drive than the LEH-inferred drive. The drive deficit of 15-25% represents much, although perhaps not all, of the discrepancy between simulations and the observed implosion trajectory. X-ray imaging diagnostics provide a wealth of data about the uniformity and spatial location of hohlraum reemission, as well as clear signatures of the effect of cross beam energy transfer on beam spot emission.

Figure 4 shows the ratio of measured to calculated drive through the open end of the hohlraum, indicating a discrepancy largely consistent with drive multipliers used to match observed capsule trajectories. In contrast, the ratio of measured to calculated drive through the LEH end is close to 1, as usually found for cryogenic hohlraums. X-ray imaging suggests the discrepancy is due to cancelling errors from over-calculating both LEH closure and the internal hohlraum drive. As shown in Figure 3, x-ray images at energies near the peak of the wall reemission spectrum show structure which can imprint asymmetries on the capsule. Variation in wall reemission is expected, as shown in the bottom image in Figure 3, but the detailed pattern differs significantly from the calculation. Images at higher x-ray energies show that the extent and location of the "gold bubble" produced by the laser spots also differ from calculations. The pattern of x-ray emission near the wall determines the time dependent implosion symmetry: if the emission is decomposed into spherical harmonics, the non-uniformities in intensity distribution at the capsule are smoothed by factors of a few at low mode number, $l$, to an order of magnitude or more at high $l$. The resulting net drive flux asymmetry is inversely proportional to the achievable convergence, $C$, so improvements in hohlraum modeling and control appear to be

Journal of Physics: Conference Series **688** (2016) 012025 doi:10.1088/1742-6596/688/1/012025

critical to increasing $P_{stag}$, as indicated by equation (2), and the chances of ignition suggested by equation (1).

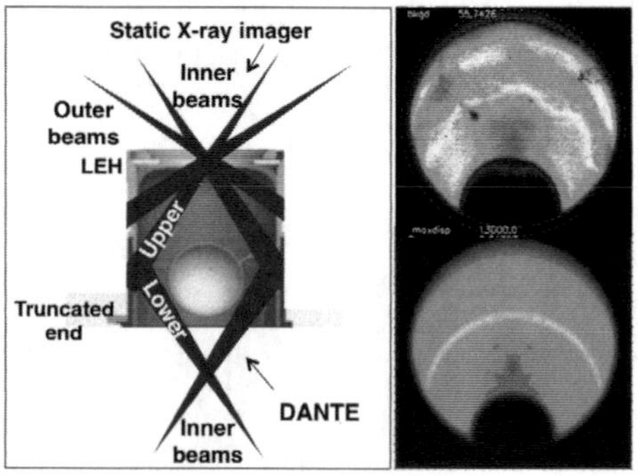


**Figure 3.** Viewfactor target and time integrated images through the open end at 870 eV for shot N120816, experiment(top) and 2D simulation(bottom). Lineouts show 20% brightness variation, differing from simulated wall reemission, especially for inner beams.

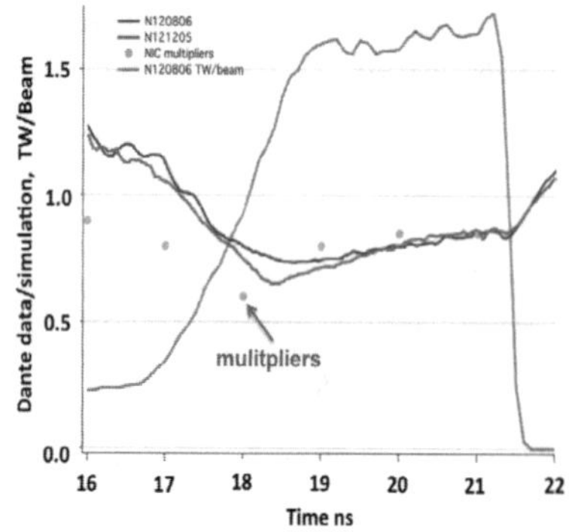


**Figure 4.** The ratio of measured to calculated open end x-ray flux for viewfactor experiments N120806 and N121205, along with drive multipliers inferred from capsule trajectories. The laser pulse is shown for reference.

## 3. Fast Ignition

Consider another chapter in the fusion story: fast ignition. Around 1990, two technological developments of interest were taking place. First, the phenomenon of radiative collapse in a cryogenic deuterium or deuterium-tritium fiber z-pinch was being studied [4], and second, the new chirped-pulse laser technology had been developed. The latter was projected to enable delivery of 10 kJ or more in a few picoseconds with a spot size of order 10 μm.

One dimensional radiation magnetohydrodynamic (MHD) models suggested that it might be possible to compress a cryogenic fiber to between a few hundred to 1000 g/cm$^3$ at nearly Fermi degenerate conditions. Theory at the time [5] suggested that the cool, collapsed pinch might be resistively stabilized against violent MHD instability (sausage and kink modes) since the magnetic Reynolds number, the ratio of diffusion time to Alfven wave time scales, could be fairly small. While the final pinch might be stabilized, the collapse could be expected (and observed) to remain unstable so that high density could only occur in very localized regions. Only high density regions have the reactivity to be relevant to ignition and burn. At a current of about 20 Megamperes, the models indicated the compressed DT could reach $\rho R \approx 1$, sufficient for isochoric ignition if a heat source could rapidly bring the fuel to fusion temperatures. Consulting equation (1), and inserting $\rho R = 1, T = 5keV$ one finds an ignition spot energy of 10kJ at a density of 500 g/cc and pressure of 1900GBar. The new short pulse laser technology seemed a good candidate as a heat source, since it could deliver the energy before the fuel could disassemble, although there were many questions about the coupling efficiency. The inward propagation of energy from the critical surface would also be complicated by the very strong magnetic field that may localize even highly relativistic electrons transverse to the field. The fiber pinch (in 1D) had the advantage of a sharp density gradient at the edge of the fuel, as shown in figure 5. The steep profile meant the critical surface for the laser was close to the pinch surface so that the ignition energy was deposited near where it was needed in the dense fuel. The sharp gradient is caused by the resistive electric field, giving a strong inward-directed Poynting flux. Low-density matter is swept inward by the associated $E \times B$ velocity. In discussions on the subject, John Nuckolls suggested replacing the pinch with laser compression. It was not clear, however, how to get through the extended corona in the laser-compressed case to deposit the energy in the dense fuel. Max Tabak independently invented the Nuckolls version of the idea and added the ideas of a hole-boring laser or a focus cone to allow coupling through the corona. The different efforts were combined and Fast Ignition at Livermore was launched [6].

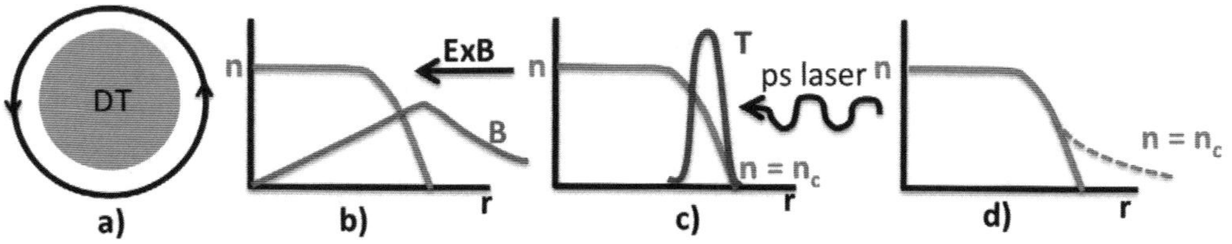


**Figure 5.** a) Collapsed pinch in cross-section b) Density and magnetic field profile. c) Heating to ignition with a short-pulse laser d) Residual low β instability draws out a corona.

The fiber pinch version of Fast Ignition fell out of favor for several reasons. There are practical difficulties in collocating the short pulse laser pulse in space and time with the radiatively collapsed regions. The more recent development of the X-pinch might ease that difficulty. Also, further work on z-pinch stability [7] showed that, while the high density portions of the pinch might be resistively stabilized, the low density edge remains unstable. It is well known from MHD theory that at low β, where β is the ratio of plasma to magnetic pressures, curvature or gravitational instabilities (e.g., sausage and Rayleigh Taylor modes) become electrostatic in nature. The electrostatic modes have negligible perturbed current and hence no resistive stabilization. The instability would likely pull out a plasma corona and complicate laser coupling, as sketched in figure 5(d).

Study of Fast Ignition led the author to a broader exploration of fusion applications of pulsed power. In the next section we will see how pulsed power leads us back to possibilities for indirect drive ICF.

## 4. Pulsed-power hohlraum-driven fusion

It has been known for many years that imploding fast pinches couple strongly to radiation and produce some of the brightest x-ray sources in existence [8]. The work done on the plasma by magnetic forces is mostly radiated at stagnation. As noted above the z-pinch is typically strongly unstable, yet it appears that the unstable dynamics actually enhance conversion of magnetic to kinetic energy by providing additional degrees of freedom for the conversion. The instability time scales are of order the Alfven transit time, which is comparable to stagnation times of the pinch. The inward $E \times B$ Poynting flux discussed above is enhanced by instability, drawing much of the magnetic energy of the system into the stagnation, and hence efficiently into x-rays. Experiments at Sandia National Laboratories on the Saturn and Z facilities showed the x-ray sources could be used to heat hohlraums [8].

Of particular interest were vacuum hohlraums where the collapsing pinch drives an enclosing hohlraum, leading to the possibility that these could be candidates for indirect drive [9]. Key parts of the concept were to place pinches in primary hohlraums, on each end of a secondary hohlraum containing the capsule, with a Faraday cage barrier between hohlraums. The Faraday cage is an array of narrow radial spokes or a metal web that is largely transparent to x-rays but prevents magnetic fields from perturbing the secondary hohlraum and capsule. The pinches reside behind shine shields so that the capsule has no direct view of the pinch and the small-scale non-uniformities of the pinch emission are highly smoothed by the hohlraum. Symmetry is tuned through hohlraum geometry, e.g. length to diameter of the secondary hohlraum and pulse shaping is achieved by adding shells to the imploding pinch. Each of these aspects of the concept was successfully tested in a remarkable series of experiments at Sandia [10]. For example, symmetry control to ~ 3% was demonstrated.

The difficulty with the vacuum hohlraum concept is in reaching high radiation temperatures. The design with 16 MJ of pinch energy reached secondary temperatures of 220 eV, sufficient to give ignition and high yield at convergence ratios comparable to the nominal NIF ignition target. If it were possible to reach higher temperatures, e.g. ~ 300 eV, then the energy advantage of the pinch could be of great value for adding margin as discussed in section 1 of this paper. Figure 6 shows such a scheme, where the imploding pinches are shifted axially by guide cones in order to flow into primary hohlraums with small wall area, and correspondingly small wall loss, compared to the original concept. The concept is similar to one proposed by Chittenden [11].

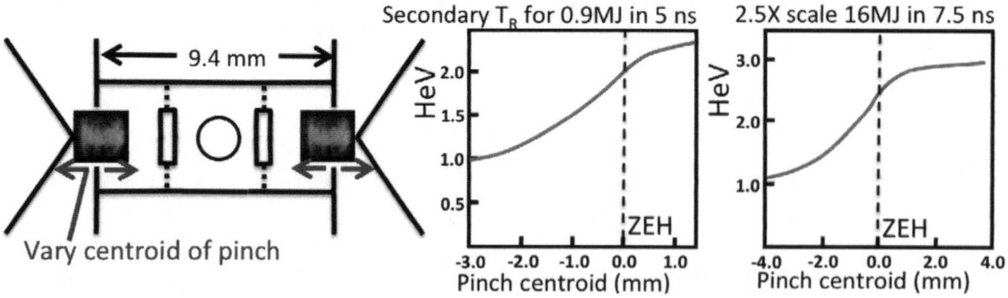


**Figure 6.** Double-ended vacuum hohlraum geometry where the pinch is projected axially through an entrance hole, the ZEH, into a small primary hohlraum. Middle image shows secondary temperatures versus pinch location for Z-like energy. The rightmost image shows secondary temperatures for a 2.5X linear scale of the Z-scale hohlraum with 16MJ of pinch energy.

The radiation temperatures in Figure 6 are derived from a simple wall loss and coupled hohlraum model [12]. The capsule in the 2.5X scale hohlraum with 16MJ pinch sources absorbs nearly 3MJ and is driven at close to 300 eV, suggesting a potential way of reaching the robust ignition regime for the Euler scaled high foot capsule shown in Figure 1.

## Conclusions

HED fusion is rich with multiple approaches, and exciting progress is occurring on the flagship NIF facility. Experimental progress points to the possible paths ahead toward ignition, as sketched in figure 2. The paths include improvements in implosion quality and hohlraum control on the NIF, but may also include other approaches such as pulsed-power driven indirect drive.

## Acknowledgements

The author thanks Omar Hurricane, Steve MacLaren, Marilyn Schneider, Paul Springer, John Lindl, Max Tabak, Mordy Rosen, Aaron Miles, Mark Herrmann, Dan Sinars and Mike Cuneo for many useful discussions. A special thanks to John Nuckolls for many years of serving as mentor and instilling "fusion by any means." This work was performed under the auspices of the U.S. Department of Energy by Lawrence Livermore National Laboratory under Contract DE-AC52-07NA27344.

## References

[1] Lindl J 1998 *Inertial Confinement Fusion* (New York: Springer-Verlag)
[2] Cerjan C, Springer P, Sepke S 2013 Phys. Plasmas 20 056319
[3] Hurricane O 2013 Bull. Am. Phys. Soc. 58 16
[4] Hammel J, Scudder D 1987 Proc. 14$^{th}$ Eur. Conf. Contrl. Fus. Plasma Phys., Madrid p.450
[5] Haines M, *et al.,* 1989 Proc. 2$^{nd}$ Intl. Conf. on Dense Z-pinches, Laguna Beach p.203
[6] Tabak M, Hammer J, et al., 1994 Phys. Plasmas 1 1626
[7] Hammer J, *et al.,* 1996 Phys. Plasmas 3 2063
[8] Matzen K, *et al.,* 2005 Phys. Plasmas 12 055503
[9] Hammer J, Tabak M, *et al.,* 1999 Phys. Plasmas 6 2129
[10] Cuneo M, *et al.,* 2006 Plasma Phys. Control. Fusion 48 R1-R35
[11] Chittenden J, *et al.,* 2002 Phys. Rev. Lett. 88 235001
[12] Rosen, M 1999 Phys. Plasmas 6 1690

# Repetitive Solid Spherical Pellet Injection and Irradiation toward the Repetitive-mode Fast-Ignition Fusion mini-Reactor CANDY.

**Ryohei HANAYAMA[1], Osamu KOMEDA[1, 2], Yasuhiko NISHIMURA[1, 3], Yoshitaka MORI[1], Katsuhiro ISHII[1], Suisei NAKAYAMA[1], Shinichiro OKIHARA[1], Kazuhisa FUJITA[1], Takashi SEKINE[4], Nakahiro SATO[4], Takashi KURITA[34], Toshiyuki KAWASHIMA [4], Hirofumi KAN [4], Naoki NAKAMURA[2], Takuya KONDO[2], Manabu FUJINE[2], Hirozumi AZUMA[5], Tatsumi HIOKI[5], Mitsutaka KAKENO[5], Tomoyoshi MOTOHIRO[5], Atsushi SUNAHARA[6], Yasuhiko SENTOKU[7], Eisuke MIURA[8] and Yoneyoshi KITAGAWA[1]**

[1] The Graduate School for the Creation of New Photonics Industries, 1955-1 Kurematsu-cho, Nishi-ku, Hamamatsu, Shizuoka 431-1202, Japan
[2] Toyota Motor Corporation, 1200 Mishuku, Susono, Shizuoka 410-1193, Japan
[3] Toyota Technical Development Corporation, 1-21 Imae, Hanamoto-cho, Toyota, Aichi 470-0334 Japan
[4] Hamamatsu Photonics K.K., 1820 Kurematsu-cho, Nishi-ku, Hamamatsu, Shizuoka 431-1202, Japan
[5] Toyota Central Research and Development Laboratories, Inc., 41-1 Yokomichi, Nagakute, Aichi 480-1192, Japan
[6] Institute for Laser Technology, 1-8-4 Utsubo-honmachi, Nishi-ku, Osaka 550-0004, Japan
[7] Department of Physics, University of Nevada, Reno 1664 N Virginia Street, Reno, NV 89557, USA
[8] The National Institute of Advanced Industrial Science and Technology, 1-1-1 Umezono, Tsukuba, Ibaraki 305-8568, Japan

E-mail: hanayama@gpi.ac.jp

**Abstract**. Pellet injection and repetitive laser illumination are key technologies for realizing inertial fusion energy[1-4]. Neutron generator using lasers also requires a repeating pellet target supplier. Here we present the first demonstration of target injection and neutron generation[5]. We injected more than 1300 spherical deuterated polystyrene($C_8D_8$) bead pellet targets during 23 minutes at 1 Hz(Fig.1). After the pellet targets fell for a distance of 18 cm, we applied the synchronized laser-diode-pumped ultra-intense laser HAMA. The laser intensity at the focal point is $5 \times 10^{18}$ W/cm$^2$, which is high enough to generate neutrons. As a result of the irradiation, we produced 2.45-MeV DD neutrons. Figure 2 shows the neutron time-of-flight signals detected by plastic scintillators coupled to photomultipliers. The neutron energy was calculated by the time-of-flight method. The maximum neutron yield was $9.5 \times 10^4/4\pi$ sr. The result is a step toward fusion power and also suggests possible industrial neutron sources.

## 1. Introduction

We have been doing research and development towards the early realization of a compact laser fusion experimental reactor (we are calling the CANDY). The CANDY is composed by the pellet injection equipment, diode-pumped ultra-high-intense laser drivers, and heat exchanger (converter) which change the heat of the fusion reaction from the reactor chamber. Pellet injection and repetitive laser illumination are key technologies for realizing inertial fusion energy[1]. Industrial neutron and X-ray generators using lasers also require a repeating pellet target supplier. Here we demonstrate flying pellet injection, high power counter laser beams' irradiation and neutron generation from the pellet.

## 2. Pellet injection system

Fig. 1 shows the pellet injector, installed in the illumination chamber, which is evacuated to $2.6 \times 10^{-3}$ Pa. The pellet is a spherical CD bead with a diameter of 970±2.7$\mu$m and a sphericity of 99%. A pellet loader stores more than 10,000 CD pellets at a time. The pellets in the loader made to free-fall in gravity onto a rotating disk, which is 110 mm in diameter and rotates at 6 rpm. Each pellet on the disk is conveyed to an exit hole and falls along a parabolic trajectory to the laser focal point 18 cm below. The exit hole is shaped as an ellipse with major and minor axes of 4.0 and 2.0 mm, respectively, and a depth of 8.0 mm. The focal point is irradiated by two counter laser beams at 1 Hz. The signals from the two photodiodes at 60 mm and 100 mm above the focal point are sequentially sent to a laser controller, which forecasts the arrival time at the focal point and sends a shooting-request signal to the laser appropriately. As soon as the laser controller receives the signal, its two laser beams irradiate the injected pellet with an appropriate delay time. Symmetric counter laser beam irradiation induces implosive fusion[3]. The laser energy, pulse duration, and wavelength are 0.63 J per beam, 104 fs, and 811 nm, respectively, and the intensity is $4.7 \times 10^{18}$ W/cm$^2$.

Fig. 2 shows the pellet irradiation. Figs.2(a) and (b) show photographic images before and after the irradiation. We can see the trajectory of the pellet. When the pellet comes to the focal point, the two counter laser beams irradiate the pellet. At that time, the velocity of the pellet is 1.8m/s. Fig.2(c) shows the shadowgraph using probe laser beam, that is perpendicular to the counter laser beam axis, and is synchronized to it. The image shows that the pellet are irradiated from both side and something comes out from the pellet.

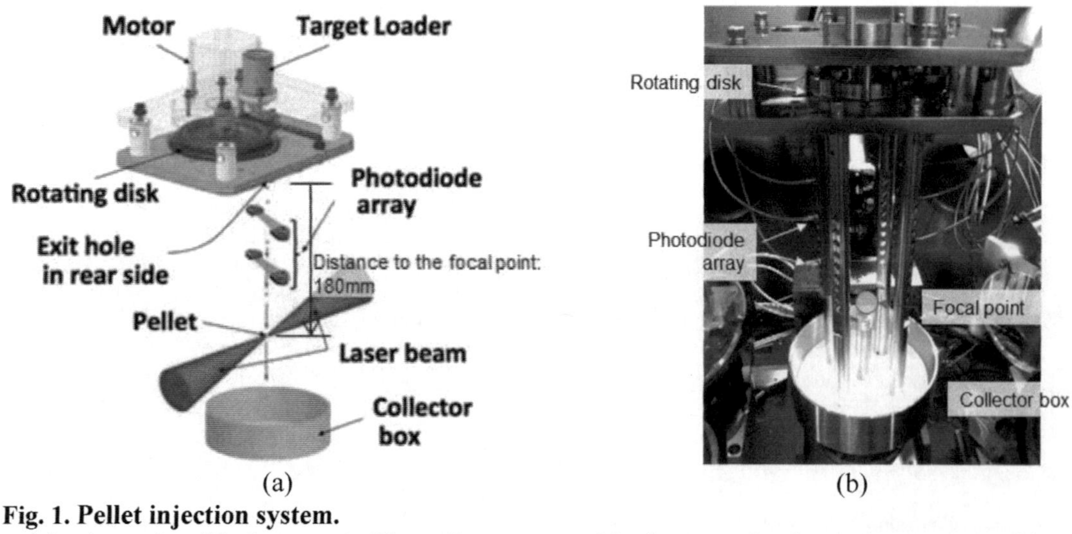


**Fig. 1. Pellet injection system.**

(a) schematics. (b) photograph. The pellets are stored in the target loader, and are conveyed by the rotating disk to the exit hole. Each pellet falls between a two-step photodiode array. These signals are sent to the laser controller. Then the laser beams are emitted at the appropriate timing.

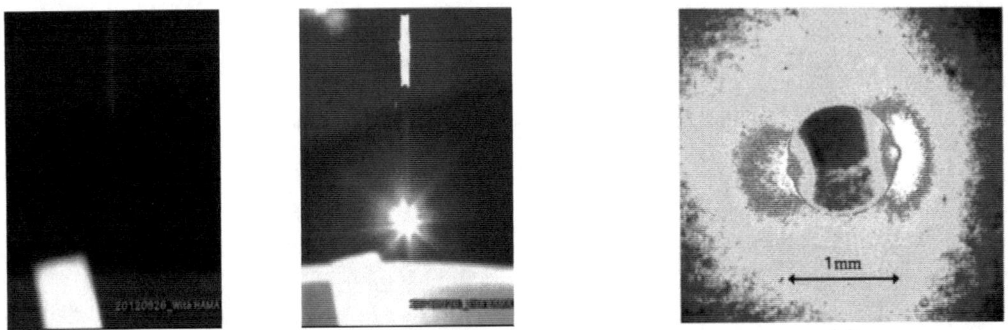


(a) (b) (c)

**Fig. 2. Snapshots of pellet irradiation.**
(a) and (b) show before and after the irradiation. The pellet trajectory can be seen in (a). When the pellet comes to the focal point, it is irradiated by the two counter laser beams. (c) shows shadowgraph using probe laser beam, that is perpendicular to the counter laser beam axis. The probe shadow and self-emissions clearly show the pellet was irradiated by the two counter laser beams.

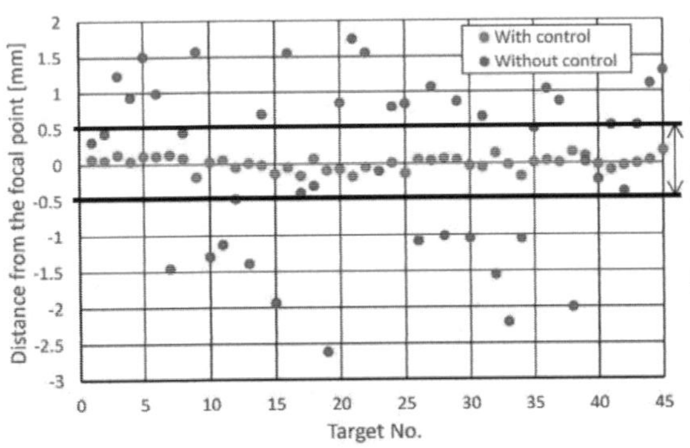


**Fig. 3. Pellet vertical placement accuracy.**
Distribution of the vertical position of the pellet. The distance from the focal point are controlled to $\pm 0.18$mm at 2 $\sigma$ .

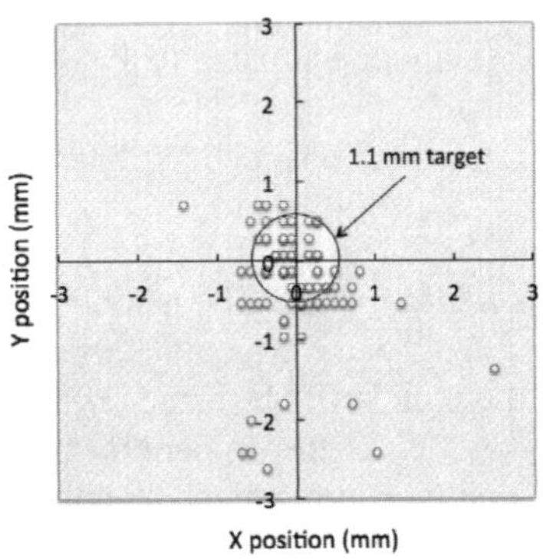


**Fig. 4. Pellet horizontal placement accuracy.**
Distribution on the plane of the laser focal point. X-axis and y-axis is perpendicular and parallel to the optical axis, respectively.

Figs. 3 and 4 shows the pellet vertical and horizontal placement accuracy, respectively. We have to control four degree of freedom of the pellet for adequate irradiation. Those are three dimensional position and time. The time is precisely controlled by the developed synchronization system. The vertical position is also well controlled to $\pm 0.18$mm at 2 $\sigma$ , which is much less than the diameter of the pellet, as shown in Fig. 3, because it is associated to the time through the vertical velocity of free fall. The distribution of the horizontal position almost falls into the shootable region, too. However it is not actively controlled. We have to control the quality of fabrication of the mechanism and manage static electricity and so on, to achieve this level of convergence.

## 3. Observation results and discussion

We could execute more than 1,000 times of repetitive pellet injection and irradiation at 1 Hz of repetitive frequency. The irradiation are succeeded at about 70% of probability out of pellet injection. The repetitive measurement system are also required for such a system. We constructed repetitive neutron detecting system using scintillation detector, oscilloscope and computer, for monitoring repetitive laser fusion experiment. We achieved about 20% and 5% of probability of detection for gamma ray and neutron, respectively. Fig. 5 shows an example of waveform observed on the oscilloscope.

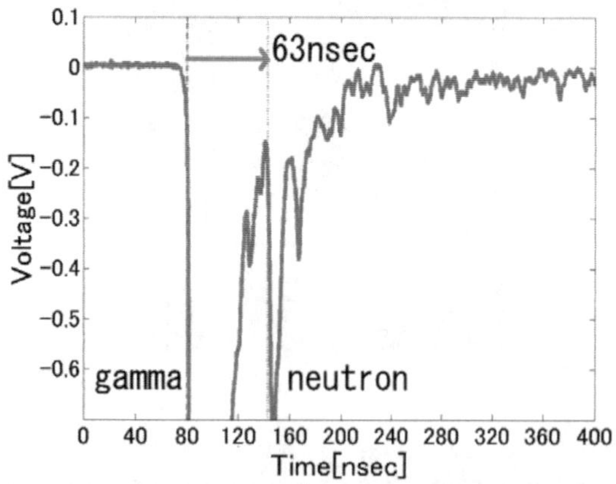


**Fig. 5. Observed waveform of neutron detector.** An obvious signal peak was observed 63nsec. delayed from gamma ray signal. It can be thought as neutron signal of 2.45MeV using TOF analysis. The yield was estimated to $1 \times 10^5/4\pi$ sr.

**Fig. 6. Bored channel along the laser axis.** These channels were observed in injected pellets. It was not observed for pellet supported using a solid SUS disk.

The laser-irradiated pellets fall down into the collecting box. Fig.6 is a scanning electron microscope (SEM) image of the cross section along the laser axis of a pellet after irradiation. The hole through the bead can be observed. The inset image shows a visible light image of the same bead. The black line is the channel. The laser light, having an intensity of $4.7 \times 10^{18}$W/cm$^2$, appears to bore from both sides through the bead and form a channel. we can see that a hole pierces through the pellet. The radius of the hole is 10 $\mu$m, which is smaller than the laser focal point size of 13 $\mu$m.

We have to make the probability of irradiation and neutron generation higher. There are two way to improve, one is improvement of injection system, and another is laser beam tracking. However our injection system has been already well optimized. And it is difficult to control the focal point position using moving mirror or deformable mirror with appropriate speed and without degradation of aberration for laser beam tracking. Moreover, we don't control the attitude of pellet, because we are using spherical target. If we will use anisotropic pellet, we have to try to control more degree of freedoms.

## 4. Summary

We succeeded in injection of spherical deuterated polystyrene bead pellets at 1 Hz and symmetrical irradiation and irradiation of them with two ultra-intense laser beams. We observed channel formation through the free-falling pellets, which might be the evidence to support a scheme for fast ignition. On an average, approximately 70% of the pellets were irradiated with the laser. Neutrons were observed for 5% of these pellets. This result represents a step toward repetitive-mode fast-ignition fusion mini-reactor, CANDY.

## References

[1] Kitagawa Y *et al* 2013 *Plasma Fusion Res.* **8** 2404000
[2] Mori Y *et al* 2013 *Nuclear Fusion* **53** 073011
[3] Kitagawa Y *et al* 2012 *Phys. Rev. Lett.* **108** 155001
[4] Kitagawa Y *et al* 2011 *Plasma Fusion Res.* **6** 1306006
[5] Komeda O *et al* 2013 *Plasma Fusion Res.* **8** 1205020
[6] Komeda O *et al* 2013 *Sci. Rep.* **3** 2561

Journal of Physics: Conference Series **688** (2016) 012027 doi:10.1088/1742-6596/688/1/012027

# Kinetic simulation of electron transport using electron magnethydrodynamic structures

**Masayasu Hata[1], Hitoshi Sakagami[2], and Amita Das[3]**

[1]Institute for Chemical Research, Kyoto University, Uji, Kyoto, 611-0011, Japan
[2]Department of Simulation Science, National Institute for Fusion Science, Toki, Gifu, 509-5292, Japan
[3]Institute for Plasma Research, Bhat, Gandhinagar 328428, India

mhata@laser.kuicr.kyoto-u.ac.jp

**Abstract**. Recent researches show the possibility that dipolar vortex structures, which are moving solutions in electron magnetohydrodynamic model, are generated by laser plasma interaction and they are applicable to electron transport in high-density plasma. However, the consideration of immobile ion and zero temperature employed in the electron magnetohydrodynamic formalism is pretty restrictive. Therefore the kinetic effects due to finite electron temperature on propagation of electron magnetohydrodynamic dipolar structures have been firstly explored using two-dimensional Particle-In-Cell simulation. Results show that the dipolar structures propagate damping the magnetic fields and its lifetime becomes longer in the case of lower temperature.

## 1. Introduction

In fast ignition scheme, various heating methods have been proposed [1,2,3]. Application of a laser-produced electron beam is one method to ignite the core, but it is difficult to produce the directional electron beam because electron beam generated by interaction between high-intensity laser and solid matter has originally large divergence angle [4]. Another method is the use of ion beam that is easy to be directive compared to the electron beam but conversion efficiency is not so good. Although it is not clarified that which method is the best way, we pay attention to electrons as the medium for core heating. It is because electron magnetohydrodynamic structures may be able to be used for electron transport [5]. In the two-dimensional electron megnetohydrodynamics (EMHD), a variety of localized solutions are permitted [6,7]. For example, there are stationary monopolar rotating electron vortex solutions in the 2-D plane. A combination of two counter-rotating electron vortices brought within a distance of the electron skin depth, forms a structure of a dipolar form, which propagates at a constant speed towards the direction of central flow if the strength of the two vortices is the same. There is the possibility that dipolar structures are generated by laser plasma interaction and applicable to electron transport in high-density plasma [5,8]. The consideration of immobile ion and zero temperature employed in the EMHD formalism, however, is pretty restrictive. Therefore, the role of kinetic effects due to finite electron temperature on EMHD structures needs to be investigated. In the previous related work, the kinetic effects on robustness of EMHD monopolar structures had been investigated [9]. As the next step, the kinetic effects on the propagation of EMHD dipolar structures have been firstly explored in this paper.

## 2. Electron magnetohydrodynamic structures

The EMHD model is relevant to phenomena for which the typical time scale of variations lies between the electron and ion gyroperiods and the length scale is in between the electron and ion gyroradii. The governing equations for EMHD can be obtained from: (i) the electron momentum equation, (ii) the expression for current in terms of the electron velocity (immobile ion), and (iii) Ampere's law, in which the displacement current is neglected. When the current and its variation are confined in two dimensions, namely the x-y plane, the equation for EMHD is simplified. Simplified equation permits exact stationary solutions. A typical solution in cylindrical coordinates $(r,\theta,z)$ can have the following form:

$$\begin{aligned} b &= \exp\left(-\frac{r^2}{R^2}\right) \\ v_\theta &= -\frac{2r}{R^2}\exp\left(-\frac{r^2}{R^2}\right) \end{aligned}, \tag{1}$$

where $\mathbf{v}_e = v_\theta \mathbf{e}_\theta$ because $\mathbf{B} = b(r)\mathbf{e}_z$, and $R$ represents a typical size of the structure. The solution consists of the vortex electron flow (currents) and its associated magnetic fields. The simplified 2-D EMHD equation also has dipolar solution that is moving solution. This solution consists of Bessel function that was derived by Ishichenko *et al* in Ref. 7 and approximate solution can be given by putting closely two-opposite monoploar solutions.

## 3. Two-dimensional Particle-In-Cell simulation of EMHD dipolar structures

### *3.1. Initial condition*

To estimate the kinetic effects on the propagation of EMHD dipolar structures, we perform kinetic simulations using 2-D Particle-In-Cell code, FISCOF-2D [10]. We suppose high-density plasma over the critical density for the Nd: glass laser. In this case, $9.92 \times 10^{20}$ electrons per $cm^3$ are the critical density for the laser wavelength of 1.06 μm. A rectangular plasma slab, whose electron density is 20 times the critical density, is placed at $-1.0 < X$ [μm] $< 4.0$ and $-1.0 < Y$ [μm] $< 1.0$ and surrounded by a vacuum, where ions are the stationary background and simulation box is defined at $-1.2 < X$ [μm] $<$ 6.2 and $-1.2 < Y$ [μm] $< 1.2$. Normally, electrons that move out over the edge of the plasma return via self-generated sheath fields and they cannot reach the boundary of the simulation box. Initial electron currents and magnetic fields of the dipolar structures are given by putting closely two-opposite monoploar solutions that follow Eq. (1). Both structure size and distance between two monopolar solutions are set to 0.15 μm. Maximum magnetic field is set to 500 mega gauss. Initial electron velocity distribution is followed by locally shifted Maxwellian, where the drift velocity is given by the fluid velocity of the EMHD solution. Three cases of the temperature of 1, 4, and 16 keV are simulated. Simulation time is 400 fs that is enough to see kinetic effects on the simulated EMHD structures.

### *3.2. Results*

Figure 1 shows 2-D spatial profiles of magnetic field in the z direction in the case of (a) 1, (b) 4, and (c) 16 keV at t = (1) 0, (2) 50, (3) 100, (4) 150, and (5) 200 fs. It is clearly shown that dipolar structures move in high-density plasma and one of them collapses in the case of high temperature, namely 16 keV. It is noted that the moving speed seems not to depend on the electron temperature. In those cases, the moving speeds are estimated at 0.005–0.006 μm/fs. Figure 2 shows two examples of electron orbits in the cases of (a) 1 and (b) 16 keV, where black and blue lines indicate orbits of electrons that are initially inside the dipolar structure and purple line indicates the orbit of electron that is initially outside the structure. Background contour is initial two-dimensional spatial profile of magnetic field in the z direction. In the case of 1 keV, electrons that are initially inside the structure do not run away outside the structure, therefore the structure does not collapse much. On the other hand,

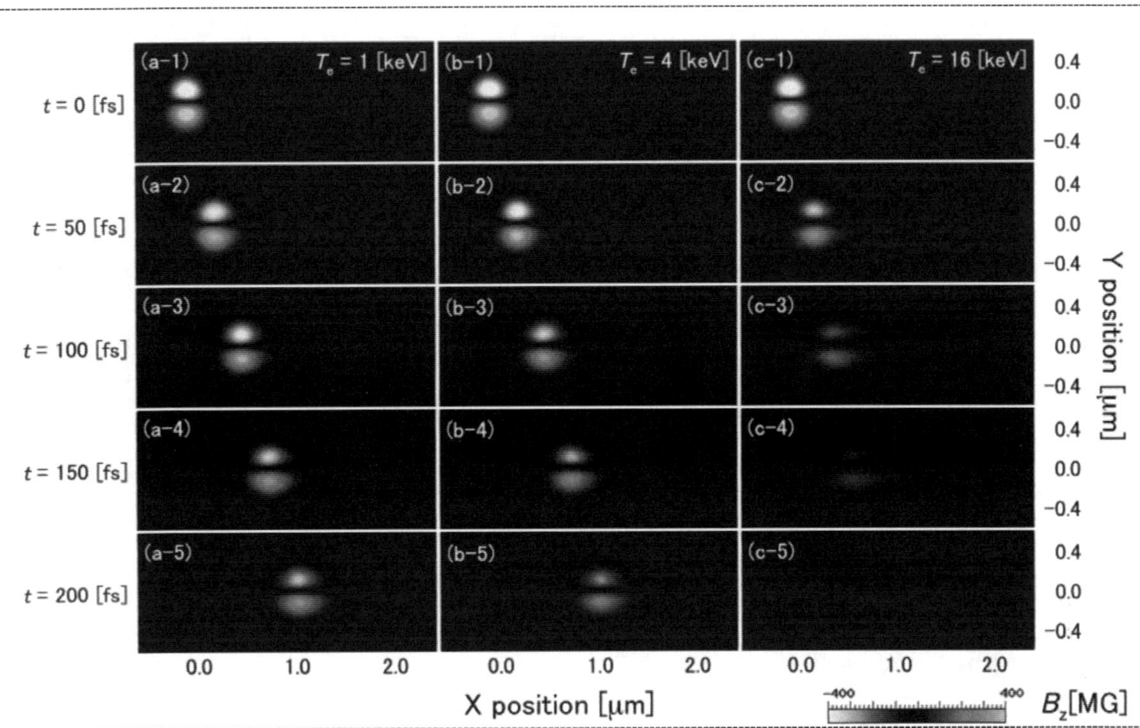


**Figure 1.** 2-D spatial profile of magnetic field in the z direction in the case of (a) 1, (b) 4, and (c) 16 keV at t = (1) 0, (2) 50, (3) 100, (4) 150, and (5) 200 fs.

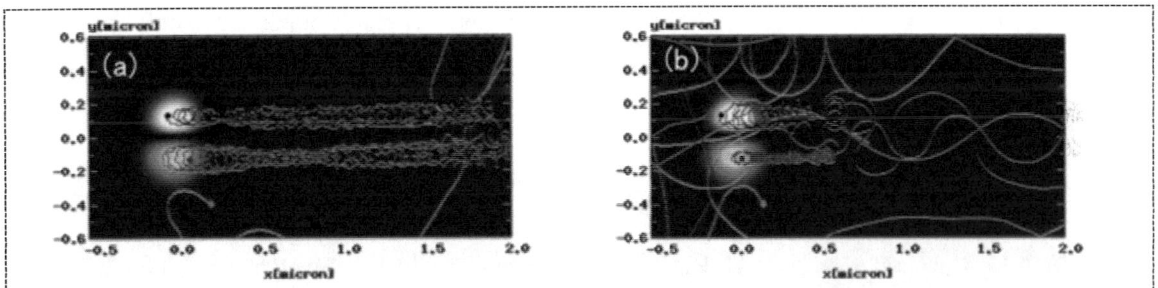


**Figure 2.** Electron orbits in the case of (a) 1 and (b) 16 keV, where black and blue orbits indicate the electrons that are initially inside the dipolar structure and purple orbit indicates the electron that is initially outside the structure.

in the case of 16 keV, electrons that are initially put inside the structure are trapped inside the structure for a while but they run away outside the structure within the simulated time, therefore the structure gradually collapses. Figure 3 shows that time evolution of magnetic energy within the circle with the radius of three and half times the structure size ($3.5R$) from the center of the structure, where red, blue, and green lines indicate the cases of 1, 4, and 16 keV, respectively. The decay times of the magnetic energy, which are estimated from Figure 3 assuming the exponential decay and summarized in Table 1, are shorter for lower temperature, but that for higher temperature is longer.

**Table 1.** Decay times of the magnetic energy assuming exponential decay

| Simulation case | 1 keV | 4 keV | 16 keV |
|---|---|---|---|
| Decay time | 315 fs | 193 fs | 65 fs |

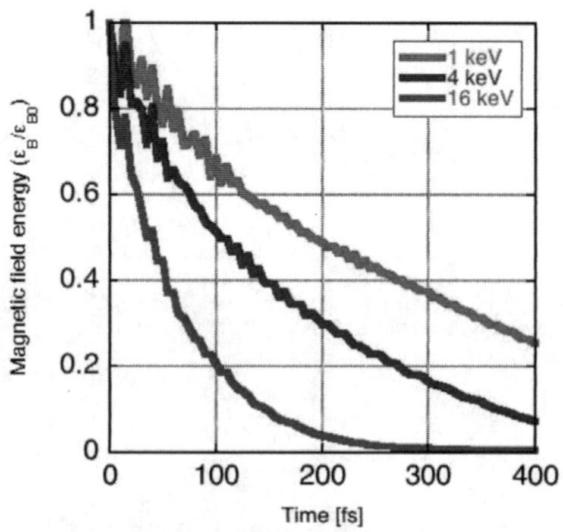


**Figure 3.** Time evolution of magnetic energy within the circle with the radius of three and half times the structure size ($3.5R$) from the center of the structure, where red, blue, and green lines indicate the cases of 1, 4, and 16 keV, respectively.

## 4. Summary and discussion

Kinetic simulations of electron transport using EMHD dipolar structures have been performed and kinetic effects due to finite electron temperature on the propagation of these structures have also been explored. Under the finite electron temperature, the dipolar structure can survive for a while and its lifetime is seemed to depend on the electron temperature. The finite temperature effects cause the damping of the magnetic field and finally its collapse. Collapsing of the dipolar structure is crucial issue for the application to the fast ignition scheme because this structure may collapse before reaching the core. Actually, simulated structures must collapse during the propagation of few microns. However, this issue can be cleared because it is considered that the larger structure has longer lifetime according to Ref. 9.

## References

[1] Tabak M, Hammer J, Glinsky M E, Kruer W L, Wilks S C, Gampbell E M, Perry M D and Mason R J 1994 *Phys. Plasmas* **1** 1626–34
[2] Roth M, Cowan T E, Key M H, Hatchett S P, Brown C, Fountain W, Johnson J, Pennington D M, Snavely R A, Wilks S C et al. 2001 *Phys. Rev. Lett.* **86** 436
[3] Hegelich B M, Jung D, Albright B J, Fernandez J C, Gautier D C, Huang C, Kwan T J, Letzring S, Palaniyappan S, Shah R C et al. 2011 *Nucl. Fusion* **51** 083011
[4] Hata M, Sakagami H, Johzaki T and Nagatomo H 2013 *Laser Part. Beams* **31** 371–7
[5] Yadav S K, Das A and Kaw P 2008 *Phys. Plasmas* **15** 062308
[6] Isichenko M B and Marnachev A M 1987 Sov. Phys. JETP **66** 702
[7] Das A 1999 *Plasma Phys. Controlled Fusion* **41** A531
[8] Nakamura T, Bulanov S V, Esirkepov T Zh and Kando M 2010 *Phys. Rev. Lett.* **105** 135002
[9] Hata M, Sakagami H and Das A 2013 *Phys. Plasmas* **20** 042303
[10] Sakagami H and Mima K 2002 *Proc. 2nd Int. Conf. on Inertial Fusion Sciences and Applications* 380–383

# Estimation on Achievable Parameter Regime of Warm Dense Matter Generated by Isochoric Heating Discharge using Intense Pulsed Power Generator

**Ryota Hayashi[1], Kenji Kashine[2], Akira Tokuchi[1,3], Fumihiro Tamura[1], Arata Watabe[1], Takahiro Kudo[1], Kazumasa Takahashi[1], Toru Sasaki[1], Takashi Kikuchi[1], Tsukasa Aso[1], Nob. Harada[1], Weihua Jiang[1]**

[1] Nagaoka University of Technology, Kamitomioka 1603-1, Nagaoka, 940-2188, Japan
[2] Yuge National College of Maritime Technology, Simoyuge 1000, Yuge, Kamijima, 794-2593, Japan
[3] Pulsed Power Japan Laboratory Ltd., Nomura 4-5-2, Kusatsu, 525-0027, Japan

E-mail: r_hayashi@stn.nagaokaut.ac.jp

**Abstract**. An evaluation method for warm dense matter (WDM) with similar timescale in inertial confinement fusion (ICF) by isochoric heating using intense pulsed power generator ETIGO-II is considered for evaluating target behavior. The temperature increase of the sample is estimated from the numerical calculation using the measured current. As a result, in the case that the shape of sample is $\phi$2 mm x 10 mm and the density is 0.01 times solid density of copper, the temperature of sample increases up to 30000 K. It is expected that the WDM is generated using the proposed method with ICF implosion timescale.

## 1. Introduction

In inertial confinement fusion (ICF) driven by heavy ion beams, the target consists of a fuel, a pusher, a tamper and an X-ray converter. The fuel target is rapidly imploded by irradiation from the energy driver. The fuel becomes dense plasma at the centre of the target, and causes thermonuclear fusion reactions.

In the implosion process, especially, regime around solid density ($10^{21}$~$10^{24}$ cm$^{-3}$) and relatively low temperature ($10^{3}$~$10^{5}$ K) is called as warm dense matter (WDM) [1, 2]. The pusher becomes from solid to plasma through WDM regime during the implosion process. To understand the implosion dynamics in ICF, numerical simulations are crucial approach. The hydrodynamics of fuel target behaviour from the numerical simulation depends on pressure and speed of sound in WDM [3]. Therefore, the reliable physical properties of matter are required for estimation on WDM regime. However, WDM is the complex regime, because of unclear theoretical model and lack of experimental evaluations. In addition, the implosion uniformity is important to obtain the nuclear fusion reactions. To suppress the implosion non-uniformity in ICF, a foamed metal is used for the pusher or the X-ray converter [4, 5].

To understand the physical properties of matter, the evaluation method for WDM by isochoric heating [6] using intense pulsed power generator ETIGO-II (~1 TW, ~50 ns) [7] with similar timescale of implosion has been considered. The features of the method are possible to generate WDM state with several-10 nanoseconds, isochoric condition and direct spectroscopic measurement using a rigid-wall capillary having transparency, and avoiding the skin effect with foamed metal sample.

In this study, the experimental apparatus is considered, and the achievable temperature of sample is estimated to understand the properties of WDM in wide temperature regime. The achievable temperature of the foamed metal sample is controlled by changing gap distance of an electron beam diode. Using the measured electron beam current, the achievable temperature is estimated by numerical calculations.

## 2. Experimental setup for measuring current

To control the current into the sample provided by ETIGO-II, an electron beam diode, which is placed at the output terminal of ETIGO-II, is used. The electron beam diode, which consists of a disk-shaped cathode ($\phi$105 mm, SUS 304) and a disk-shaped anode ($\phi$210 mm, SUS 304), is used as an adjustable resistance in the pulsed power system.

Figure 1 shows the experimental setup for measuring current of the electron beam diode. The provided current was measured in the gap distance for 10, 15, and 20 mm, respectively. Time evolution of the current $I(t)$ is measured with a Rogowski coil. To interfere the electrical breakdown, the interior pressure of the chamber was controlled to be less than 0.02 Pa. The electron beam current and the impedance are decided by the gap distance.

The sample is a foamed metal surrounding with a hollow rigid-wall capillary. The capillary is sapphire, which is possible to sustain up to ~GPa. The sample is placed on the anode side of the electron beam diode. However, the input power into the sample depends on the diode impedance because of relatively high impedance of the diode. Therefore, the current at cathode is measured without the sample, and the anode is grounding to the outer feeder.

## 3. Numerical model for temperature increase of sample

To estimate the achievable parameter regime of WDM generated by isochoric heating, the sample impedance of the foamed copper $R$ is estimated to be

$$R = \frac{l}{\sigma \cdot S}, \tag{1}$$

where, $l$ is the length of the sample with 10 mm, $\sigma$ is the electric conductivity ($10^4$ S/m is assumed from the previous experimental results [6, 8, 9]), the cross sectional area of the sample $S$ is determined with $r^2\pi$, and $r$ is the radius of the sample (1 mm).

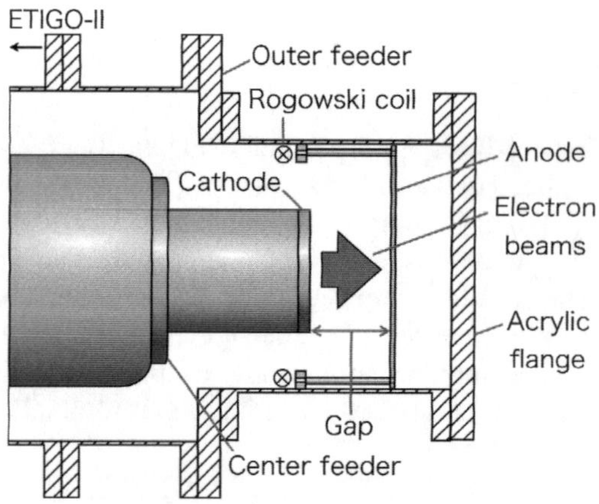


**Figure 1.** Experimental setup for measuring current.

The time evolution of the temperature increase of sample $\Delta T(t)$ is estimated to be

$$\Delta T(t) = \int_{t_1}^{t_2} \frac{I(t)^2 R dt}{S \cdot l \cdot \rho_{\mathrm{foam}} \cdot C_p(t)}, \tag{2}$$

and the temperature of sample $T(t)$ is estimated to be

$$T(t) = \Delta T(t) + 298, \tag{3}$$

where, $t_1$ is the onset time of the discharge, $t_2$ is the time after 50 ns from $t_1$, the current at the cathode $I(t)$ is given by the corresponding experimental data as shown in Fig. 2, the density of foamed copper sample is $\rho_{\mathrm{foam}} = 0.01\rho_s$ ($\rho_s$ is the solid density of copper, 8940 kg/m$^3$), and $C_p(t)$ is the heat capacity that the data of copper in solid, liquid, and gas phases are given by Refs. [10, 11]. The temperature increase is estimated after 50 ns from the onset time. The onset time to calculate the integral of the current was defined as the response of the current to rise to 10 % of its peak value. The initial temperature is set as 298 K by a room temperature.

## 4. Current measurement and estimation of achievable temperature of sample

Figure 2 shows the current waveforms at the cathode, in the case of the gap distance for 10, 15 and 20 mm, respectively. The current waveforms are the average of five times for each gap distance. The peak current increases from 10 kA to 60 kA with the decrease of the gap distance. The pulse width is ~80 ns (FWHM).

Figure 3 shows the temperature of the sample as a function of the gap distance. To estimate the temperature of sample, the experimentally obtained current waveform is used by Eqs. (2) and (3). The temperature of the sample increases with the decrease of gap distance. When the sample is 0.01 times the solid density of copper with $\phi$2 mm x 10 mm, the temperature is estimated up to 30000 K after 50 ns from the onset time. Therefore, it is expected that the WDM state is generated by the proposed method with ICF implosion timescale.

## 5. Conclusion

The experiment for WDM generation with ICF implosion timescale was proposed, and the temperature increase was estimated from the numerical calculation using the experimental data. When the sample is $\phi$2 mm x 10 mm and 0.01 times the solid density of copper, the temperature of foamed copper sample achieved up to 30000 K. The results indicated that the WDM state is produced by isochoric heating using the intense pulsed power generator with similar timescale of implosion.

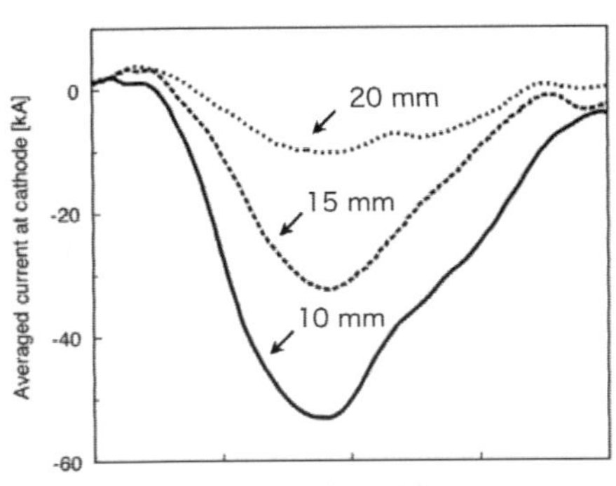


**Figure 2.** Averaged current waveforms (each 5 shot) measured experimentally at the cathode.

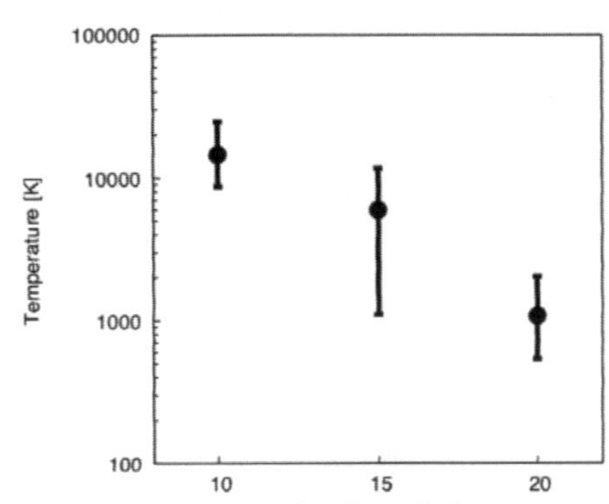


**Figure 3.** Temperature of the sample as a function of the gap distance. The density of foamed copper sample is $\rho_{\mathrm{foam}} = 0.01\rho_s$. The error bars indicate maximum and minimum values.

**Acknowledgment**
This work was supported from MEXT Grant-in-Aid for Scientific Research, and by Program for High Reliable Materials Design and Manufacturing in Nagaoka University of Technology, and by JSPS Grant-in-Aid for Challenging Exploratory Research No.25630418.

**References**

[1] Drake R P 2009 *Physics of Plasmas* **16** 055501
[2] Redmer R and Röpke G 2010 *Contributions to Plasma Physics* **50** 970
[3] Komatsu Y, Sasaki T, Kikuchi T, Harada N and Nagatomo H 2013 *EPJ Web of Conferences* **59** 04010
[4] Atzeni S and Meyer-ter-vehn J 2004 *The Physics of Inertial Fusion: Beam Plasma Interaction, Hydrodynamics, Hot Dense Matter* (Oxford Univ. Press, N.Y.) Chap. 9
[5] Iizuka Y, Kawata S, Kodera T, Ohoyski A I and Kikuchi T 2009 *Nuclear Instruments and Methods in Physics Research* A **606** pp 165-168
[6] Amano Y, Miki Y, Takahashi T, Sasaki T, Kikuchi T and Harada N 2012 *Review of Scientific Instruments* **83** 085107
[7] Jiang W, Sakagami T, Masugata K and Yatsui K 1993 *Japanese Journal of Applied Physics* **32** pp L752-L754
[8] DaSilva A W and Katsouros J D 1999 *International Journal Thermophysics* **20** 1267
[9] Sasaki T, Nakajima M, Kawamura T and Horioka K 2010 *Physics of Plasmas* **17** 084501
[10] NIST Standard Reference Database Number **69**
[11] Chase M W 1998 *Journal of Physical and Chemical Reference Data* NIST-JANAF Thermochemical Tables Fourth Edition Monograph 9

# The updated advancements of inertial confinement fusion program in China

**X. T. He[1]* with ICF teams in China**
[1]Institute of Applied Physics and Computational Mathematics, Beijing, 100094, China
E-mail: *xthe@iapcm.ac.cn

**Abstract.**
The recent achievements of ICF research in China are reviewed. The constructions of laser facilities of SG-III and SG-IIUP are completed in this year and the full energy output operation will be in 2014. The target physics studies involving numerical simulations of a new ignition scheme, which is proposed to enhance implosion velocity and suppress hydrodynamic instability and distortion at interface between hot spot and main fuel, and experimental results (a few selected examples) are presented.

## 1. Introduction

The goal of the first milestone of the inertial confinement fusion (ICF) program in China is to perform the ignition and plasma burning in laboratory around 2020. The national research institutes, such as the Research Centre of Laser Fusion (RCLF), the Institute of Applied Physics and Computational Mathematics (IAPCM), the National Laboratory on High Power Laser and Physics (NLHPLP), as well as many universities and other institutions are participating in the Program. Since 1993, based on this program, target physics studies involving theories (models and simulations) and experiments (measurement techniques), establishments of precise diagnostic systems, constructions of laser facilities (SG-series) and technology development, and precise target fabrications have been coordinately developed, and the great achievements have been gained. The pre-ignition target design and advanced target physics research, based on the SG-III laser facility, are crucial as the basis for the ignition target design in the coming years, which should identify and solve the essential science and technology issues, to reduce risks, and to scale up the design to the ignition target. Finally, we will demonstrate the ICF ignition on the designing SG-IV laser ignition facility.

## 2. Development of laser drivers

The SG-series of laser facilities were constructed and/or are being developed to serve for ICF research in China. The SG-II laser facility, which has 8 beams and energy output of $3kJ/3\omega$ with pulse duration of 1ns and is supplemented by a ninth laser beam with energy output of $5.2kJ/1\omega$ $(3kJ/3\omega)$ and pulse duration of $2-3ns$ as a backlighter, has provided over 4000 shots related to the ICF experiments since 2000. The SG-III Prototype (SG-IIIP), which has 8 beams and laser energy output of $15kJ/3\omega$ and pulse duration of 3ns, has been operating for physical experiments over 2000 shots since 2006, and a new laser beam as its backlighter is being prepared. Also, the SG-IIU is a combined laser system which is made up of the upgrading SG-II

(8 beams, $24kJ/3\omega$, 3ns) and a PW laser ($\sim 1kJ$, $3-5ps$) to be operating in 2014. The SG-III with 48 beams (6 bundles) and the output laser energy of $250kJ/3ns/3\omega$ and $400kJ/5ns/3\omega$ has been completed in 2013, see Fig.1. All bundles are now operating for experiments and full energy operation and physical shots will be in 2015. The ignition facility (SG-IV) is being designed to serve for ignition demonstration around 2020 in China.

**Figure 1.** The SG-III laser system with 48 beams (6 bundles) and the output laser energy of $250kJ/3ns/3\omega$ and $400kJ/5ns/3\omega$ has been completed in early 2013.

## 3. A hybrid indirect-direct drive ignition scheme by a target of spherical hohlraum and layered fuel capsule

For the conventional indirect-drive hotspot ignition scheme of ICF [1], the ignition occurs near stagnation (isobaric) time. For such scheme, there exist at least two issues in implosion dynamics, i.e., strong rarefaction wave due to the ablated plasma expanding towards vacuum at the ablator surface leads to severe reduction of the ablative pressure and implosion velocity; the mix from Rayleigh-Taylor instability (RTI) and asymmetry amplification (distortion) due to the implosion shock multiple-reflection at the interfaces of hotspot/main fuel and others resulting in deceleration, may lead to ignition difficulty.

To solve the issues, we proposed a new ignition scheme of indirect-direct hybrid drive for ICF: a layered fuel capsule inside a spherical hohlraum (SH) is compressed first by indirect-drive (ID) x-ray ablation with laser energy of about 0.5MJ and pulse duration of 12.7 ns, and then by intense direct-drive (DD) lasers with laser energy of about 0.85MJ and pulse duration of last $2ns$ (figure 2c, green colour). The SH, which has a radius of 3.5 mm with six laser entrance holes (LEFs) and octahedral symmetry ( figure. 2a), is designed by using plasma-filling model under conditions of the specific indirect-drive (ID) radiation temperature (figure 2c, red colour) and the filling density. The non-uniformity of about a few thousandth on the capsule surface, which has total deuterium-tritium (DT) fuel mass of 0.19 mg and an initial outer radius of $850\mu m$

(figure 2b), is obtained by view factor code simulations if the radius ratio of SH to capsule is greater than 5, the incident angles of laser beams are between $50^o - 60^o$ , the radii of the laser bundle and LEH are properly chosen.

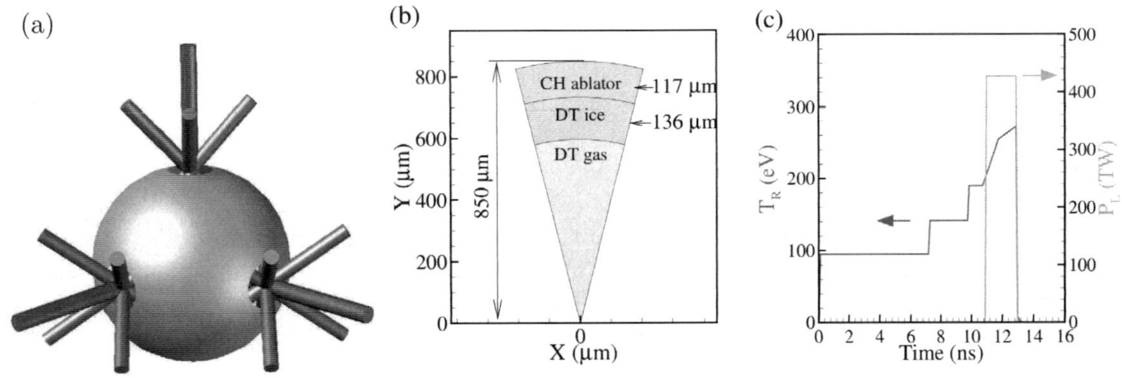


**Figure 2.** (a) The SH configuration; (b) capsule cross section; (c) indirect-drive radiation temperature (red) and direct-drive laser power (green).

The implosion dynamics of the capsule is investigated by using LARED-S code. The 1D simulation results are plotted in Figure 3. The ablator (CH material with initial density of $1.0g/cm^3$) of the capsule is ablated by 4 radiation- temperature steps with a top $Tr = 270eV$ (figure 2c, red colour), a radiation ablation front (RAF) appears in the ablated CH surface. Meanwhile four shocks are generated and propagates in the capsule that is imploded and compressed. Beginning at time $t = 10.7ns$, the DD lasers with intensity of $3.4 \times 10^{15} W/cm^2$ are imposed at a critical surface, where is $950\mu m$ away from the capsule centre, in the corona region. A supersonic electron thermal wave with an electron ablation front (EAF) is formed there and runs from the critical surface towards the RAF about $650\mu m$ away from the capsule centre. The thermal wave gradually slows down to a sonic/subsonic one due to the energy deposition in corona during propagation, followed a DD shock wave is driven. The four shocks driven by ID and the DD shock merge into a merged shock (MS) near inner wall of the main fuel. When the MS runs into the DT gas filling cavity a hot spot involving the DT gas and a part of innermost main fuel is formed and compressed by the MS, then the MS rebounds from the capsule centre and moves towards the interface of the hot spot and main fuel.

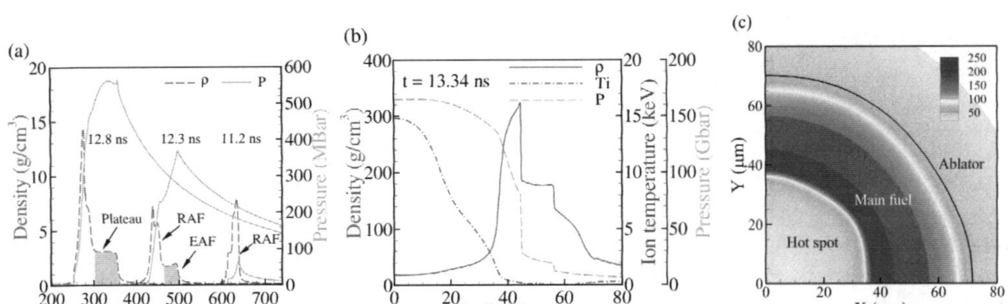


**Figure 3.** (a) and (b) are the density (red), pressure (green), ion temperature (blue) vs radii of the imploding capsule in different time, and (c) is the 2D profile (hot spot, main fuel, ablator) at ignition time.

Meanwhile a compression wave with the EAF followed the DD shock, likes a snowplow, drives the ablated CH of density $\rho \ll 1.0g/cm^3$ in the corona, which is gradually piled into a higher CH density plateau between the EAF (sonic) and RAF(subsonic). At $t = 12.3ns$, the plateau has a spatial width of $\sim 40\mu m$ (figure 3a, grey), the density $\rho \sim 2.0g/cm^3$ and the maximal radiation temperature $Tr \sim 270eV$, and thus, the peak pressure $P \sim 380$ Mbar; at t=12.8 ns, the plateau has a spatial width of $\sim 60\mu m$, the density $\rho \sim 3.0g/cm^3$ and the peak pressure $P \sim 570$ Mbar, it is far greater than ID ablation pressure of $\sim$ tens Mbar as shown at t=11.2 ns in figure 3a.

Such a plateau likes an extreme-high pressure piston to push out an enhanced shock (ES). The ES fast runs in the imploding capsule. At $t \sim 13.02$ ns the ES reaches the interface of hot spot/main fuel and collides with the MS that just arrives there from the hot spot centre, thus the MS is stopped to reflect at the hot spot interface. Then, the ES runs in the hot spot that is further compressed and heated. At $t \sim 13.34$ns, the non-isobaric ignition ( particle heating rate equals to electron cooling rate) occurs before stagnation when the ES rebounded from the hot spot centre arrives near the interface of the hot spot/main fuel in a lower convergent ratio of 25 (figure 3b). As a result, the RTI, mix, and the interface distortion are prevented due to no deceleration phase happened. Finally, the fusion energy of 17.4 MJ and a gain of 13 for total laser energy of 1.35MJ are achieved. This hybrid scheme is robust, even if the laser energy is reduced to 1.1MJ the yield still has 16MJ.

The 2D simulation results are investigated by using LARED-S. The non-uniformity from DD laser irradiation at critical surface is suppressed by the thermal smoothing effect [2] during the supersonic electron thermal wave propagation from the critical surface ($\sim 950\mu m$) to the RAF ($\sim 650\mu m$) of the implosion capsule. The growth factors (GFs) of hydrodynamic instabilities varying with mode l for hybrid scheme are shown in figure 4, and compared with the traditional 1D one. The hybrid GF at interface of hot spot/main fuel is suppressed as shown in figure 4a, while the GFs at interfaces of ablator/main fuel and RAF are about 220 and 150 (figure 4b and 4c), respectively, which are lower than that for the traditional 1D scheme as seen in figure 4 (green). The simulation results also showed that the profile of imploding core at ignition time is quite close to symmetry and 2D effects seem to be neglectable. Therefore, the yield in 2D simulations is same as 1D result.

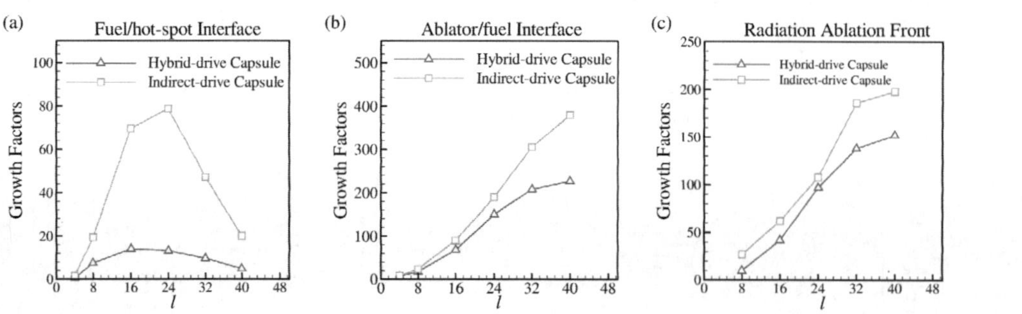


**Figure 4.** The growth factors of hydrodynamic instabilities vs mode l for hybrid scheme (red) to compare with the traditional 1D one (green). At the interface: (a) hot spot/main fuel, (b) ablator/main fuel, (c) radiation ablation front (RAF).

The expected efficiency of the laser energy coupling to the SH for ID phase is over 90% higher than the cylindrical and rugby hohlraum, and is $\sim 20\%$ for the DD lasers of intensity of $3.4 \times 10^{15} W/cm^2$ [3] with a relatively low hot electron temperature of $\sim 30keV$, which is easily stopped by the compressed surplus ablator.

## 4. Physical experiments (selected examples)

### *4.1. Hohlraum test experiments with the first two bundle beams of SG-III*

The hohlraum test experiments reported here were performed at the SG-III laser facility with the first two bundles of beams. In these experiments we compared the radiation temperature Tr of the Au hohlraums that were heated with the different incident angle lasers. We use two kinds of hohlraum in these experiments: hohlraum A is $1.4mm$ in diameter, $2.1mm$ in length, and has LEH diameters of $1.0mm$; hohlraum B is $2.4mm$ in diameter, $3.4mm$ in length, and has LEH diameters of $1.6mm$. The hohlraum A was heated by 16 beams using a 3 ns flattop laser pulse with a total energy of up to 80kJ and 45kJ, and The hohlraum B was heated with a 1ns flattop laser pulse with a total energy of up to 16kJ. Temperature Up16 (Up42, Up64) was diagnosed from upper-half-chamber with an angle of $16^o$ ($42^o$, $64^o$) between observation direction and hohlraum axis; Down20 (Down42) was diagnosed from down-half-chamber with an angle of $20^o$ ($42^o$) between observation direction and hohlraum axis. The temperature differences are shown in figure 5.

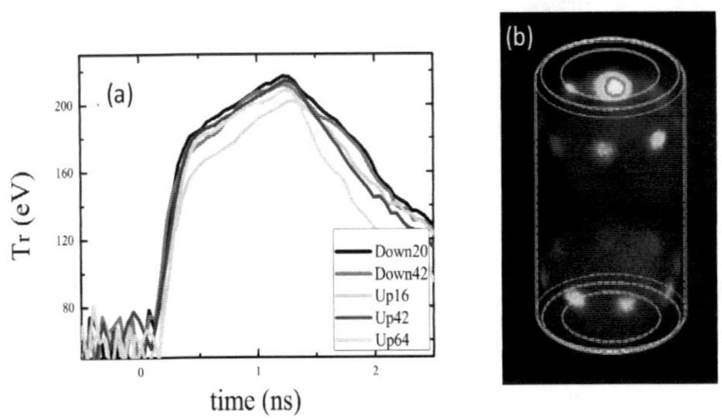


**Figure 5.** Hohlraum temperature experiment on SG-III laser system with only two bundles of beams.

### *4.2. Target physics–cryogenic capsule experiments*

Cryogenic deuterium-deuterium (DD) capsule direct-drive experiments are carried on SG-IIIP. The target assembly (figure 6a) is made up of a copper cylinder clamped by two cooling arms; the plastic capsule cooled by helium gas to below the deuterium triple point; supported and filled with D2 gas by fill tube of $10\mu m$ in diameter; ice thickness is uniformed by tuning the temperature difference between two cooling arms; inserted into target chamber on a six-dimensional adjustable platform. The experiments are carried out with a polar direct drive mode: beam pointing is optimized for best irradiation uniformity. The laser has a pulse shape of two steps with power contrast of $1 : 10$ to minimize the energy loss on the sealing film. The experiments showed that the cryogenic capsule (ice layer of $115\mu m$) implosion achieved DD neutron yield of $4.7 \times 10^6$, while warm plastic capsule implosion achieved neutron yield of $7.8 \times 10^7$. As shown in Fig. 6, the target alignment accuracy in cryogenic condition is demonstrated, the technique for fuel areal density is validated, and the implosion dynamics is measured.

### *4.3. Monochromatic imaging for implosion measurement*

Monochromatic imaging (MI) based on spherically bent crystals (BCs) was developed and used to explore implosion dynamics. Imploding ablator performance was measured on SGII using this MI system. Here, 5th order of reflection of the mica spherical crystal was used and the measured photon energy $E_{photon} = 3.14keV$ (Pd). The detrimental long-range spatial structure from backlighter non-uniformities are absent using monochromatic imaging system. The experimental results showed that crystal resolution is smaller than $5\mu m$, system resolution (involving streak

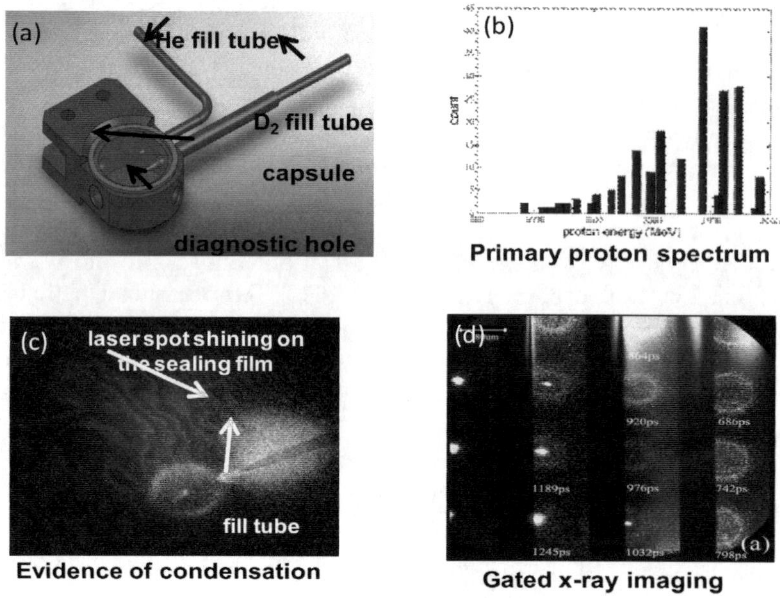


**Figure 6.** Preliminary experimental results of cryogenic deuterium-deuterium (DD) capsule direct-drive experiments on SG-IIIP: (a) cryogenic capsule involving He and D2 filling tubes, diagnostic hole; (b) primary proton spectrum; (c) evidence of condensation; (d) gated x-ray imaging.

camera) is about $10\mu m$. The uncertainty in this experiment is that the implosion velocity is about 10% and the residual mass is about 15%, see Fig. 7.

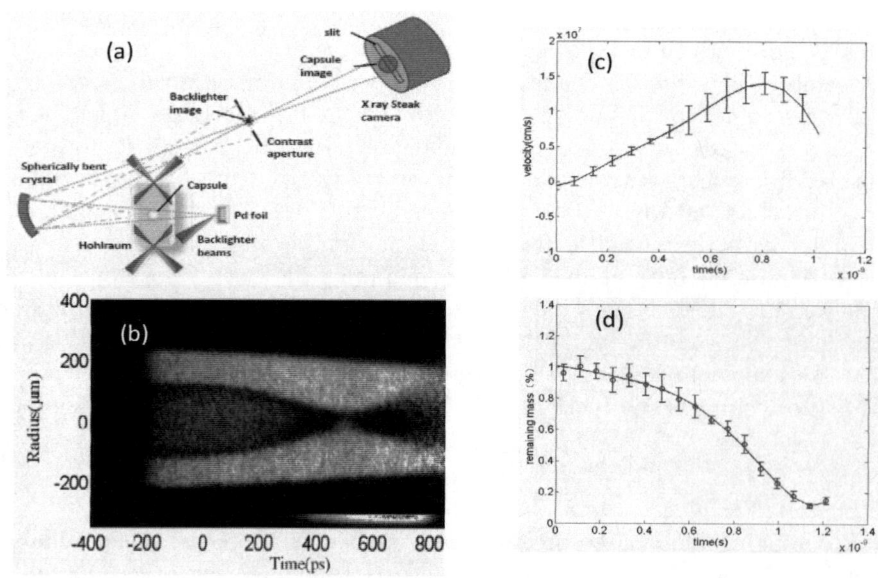


**Figure 7.** (a) A schematic of the experimental setup for the imploding process measurement; (b) 3.14keV monochromatic streaked radiograph; (d) Implosion velocity; and (d) Residual mass.

### *4.4. K-shell photoabsorption edge of strongly coupled matter*

Warm dense matter (WDM) is a non-ideal plasma with full and partial degenerate and exhibits the continuum lowing of efficient ionization potential, phase transition, pressure ionization, etc., it is very important to measure WDM properties for target simulations. Here K-edge shift of Potassium Chloride (KCl) of $Ek = 2.8keV$ is investigate experimentally. A sandwich sample (KCl) target (figure 8a) is compressed (figure 8c) by shocks driven by x rays from "dog bone" hohlraum (figure 8), then short pulse backlighting (130$ps$ laser from SG-II) is applied to get the time-resolved absorption spectroscopy from the compressed KCl sample (figure 8b). The red shift of K edge up to 11.7 eV and broadening of 15.2 eV from backlighting spectra in the maximal compression, resulting in K-shell ionization energy increase and continuum lowering, are clearly observed, and it shows that the WDM KCl is in partial degeneracy [4].

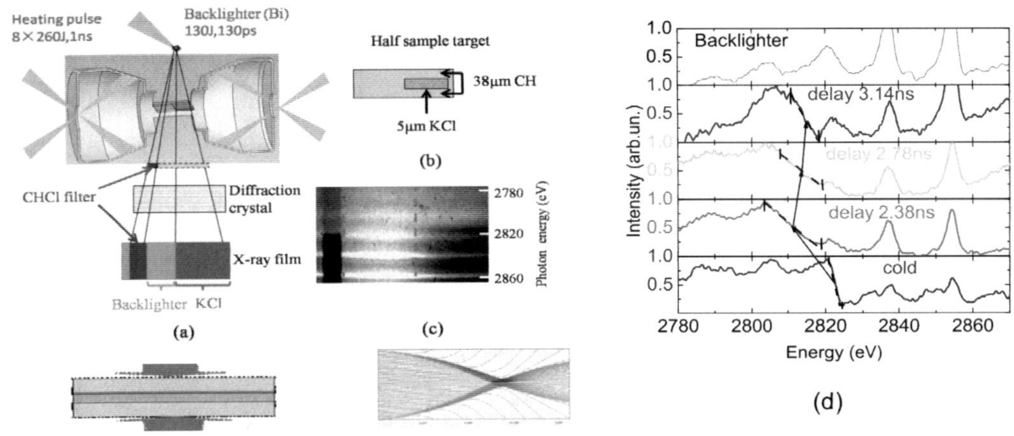


**Figure 8.** The "dog-bone" hohlraum schematic for partial degeneracy experiments: (a) the sandwich sample (KCl) target, (b) typical spectral image, (c) the shock compressed sample (KCl), (d) the time-resolved absorption spectroscopy from the compressed KCl sample.

## Final remarks

In summary, we are performing the ignition target design to improve the conventional indirect central ignition target and to explore new ignition schemes involving hybrid drive, and as well will provide physical data for the SG-IV ignition driver design. Target physics research on SG-III is a quite important step towards ignition, and would find and solve substantial issues in science and technology before transiting to ignition. The SG-IV ignition facility with laser energy of around 1.5MJ for third harmonics is being designed. China is going to be toward moderate/high gain ignition and burning goal around 2020.

## References

[1] J.D. Lindl *et al.*, Phys. Plasmas **11**, 339( 2004).
[2] M. Keskinen, Phys.Rev.Lett. **103**, 055001(2009).
[3] W. Theobald *et al.*, Phys. Plasmas **19**, 102706(2009).
[4] Y. Zhao *et al.*, Phys.Rev.Lett. **111**, 155003 (2013).

Journal of Physics: Conference Series **688** (2016) 012030 doi:10.1088/1742-6596/688/1/012030

# Orion: a high contrast user facility

**D I Hillier, C N Danson, S J Duffield, D A Egan, S P Elsmere, M T Girling, E J Harvey, N W Hopps, M J Norman, S J F Parker, P T Treadwell, D N Winter, T H Bett**

Plasma Physics Technology Centre, AWE, Aldermaston, Reading RG7 4PR, UK

David.hillier@awe.co.uk

**Abstract**. The Orion facility consists of two synchronized laser systems: two CPA (Chirped Pulse Amplification) beamlines each deliver 500J to target in a 0.5ps pulse (1PW) at 1054nm; and ten long pulse beamlines each deliver 500J in 0.1-5ns temporally shaped pulse at 351nm. One of the CPA beamlines has the option to be frequency doubled at sub-aperture to produce 100J laser pulses with a nanosecond contrast of ~ $10^{14}$. Further work is under way to enhance the contrast of both CPA beamlines in the first harmonic.

## 1. Introduction

The Orion laser facility [1] was commissioned in 2012, received its "authority to operate" in early 2013, and has been running as an operational facility ever since. The facility consists of two distinct laser systems coupled together into the same target chamber. Orion has two CPA (chirped pulse amplification) beamlines which produce sub-picosecond, petawatt pulses and ten, temporally shaped nanosecond beamlines operating at 351nm. In addition one of the CPA beamlines can be frequency doubled at reduced aperture for high contrast operation [2]. The combination of long and short pulse beamlines allows a wide range of experimental conditions to be realized.

## 2. System operation

The nanosecond beamlines are seeded by a distributed feedback fiber laser from which 100ps to 5ns temporally shaped pulses are sculpted by a pair of integrated optical modulators. These are initially amplified using a pre-amplifier module (PAM) [3] consisting of a regenerative amplifier, spatial shaping, 2D smoothing by spectral dispersion (SSD) system and a 4-passed rod amplifier. The output from the PAM is split using a series of polarizers and half-wave plates to seed the 10 main beamlines. The main beamlines consist of four 200mm diameter Nd:Glass (phosphate) disc amplifiers. The disc amplifiers are 4-passed using a reverser configuration with a deformed mirror to compensate for on-shot thermal aberrations. Each amplification stage is isolated using Pockels cells to prevent parasitic lasing. The beamlines are each capable of generating 500J at 351nm in a 1ns square pulse. The beams are focused on to target in two 50 degree cones using 1.2m focal length lenses. Focal spot conditioning is achieved using continuous kinoform phase plates. Orion has two PAMs each seeded by an independent pulse shaping system; these can be used to either seed five beamlines each or one PAM can seed all ten beamlines.

The short pulse beamlines consist of a 200fs mode-locked oscillator, the output from which is split into two then stretched to 6ns using Offner triplet stretchers. These pulses are amplified using a three stage optical parametric amplifier [4] to ~100mJ before being 4-passed through a pair of Nd:glass (phosphate and silicate) rod amplifiers then a series of Nd:glass (phosphate) disc amplifiers of increasing aperture to ~700J. The beam is compressed to ~500fs using a pair of gold diffraction gratings then focused onto target with a f/3 off-axis parabola to achieve an intensity of ~$10^{21}$ $Wcm^{-2}$.

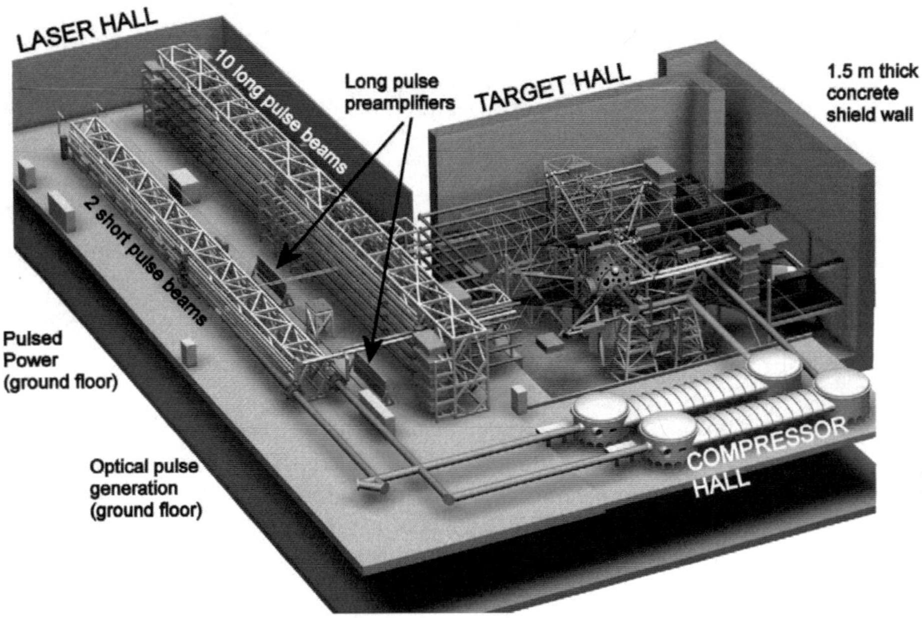


**Figure 1 The Orion laser facility**

Orion can be operated in a range of configurations depending on the experiment: The base line configuration has one PAM seeding all 10 long pulse beamlines and the two short pulse beamlines operating independently. If two different pulse shapes are required or large delays between beamlines then each set of 5 beamlines can be seeded by a separate PAM. Alternatively either one or both of the short pulse beamlines can be seeded with shaped long pulses from PAM 2 with the output from the beamline being picked off prior to the compressor and sent down a separate path via frequency tripling crystals to target.

## 3. Contrast enhancement

CPA lasers incorporating optical parametric amplifiers (OPA) suffer from a reduction in their temporal contrast due to parametric fluorescence. On Orion this manifests itself as a pedestal with ~$10^{-8}$ of the power of the compressed pulse and the same duration as the OPA pump pulse.

In order to enhance the contrast we frequency double the output of one of our short pulse beamlines (SP1) post compression. Conversion efficiency varies roughly linearly with intensity in the saturated/ back-converting regime and quadratically in the low conversion efficiency regime. Hence the pedestal converts far less efficiently than the main pulse and the contrast increases dramatically.

The apparatus for frequency converting the beam line is installed in a vacuum chamber inserted between the compressor and target chambers (Figure 2). The 600mm diameter beam from the compressor is first reduced in size to 300mm using a segmented apodiser consisting of four pieces of absorbing glass with a solid metal back. The beam is then reflected off a 1054nm mirror and through a 3mm thick, 320mm diameter KDP crystal cut for type I second harmonic generation. Post conversion the beam is reflected off three dichroic mirrors all of which are high reflectors (HR) at 527nm and highly transmissive (HT) at 1054nm to provide extinction of the 1054nm light. The leakage thought the first of these contains the bulk of the unconverted 1054nm light and is dumped into a slab of absorbing glass. The leakage through the second mirror is reduced using a Galilean telescope and used for diagnostics on the 527nm beam.

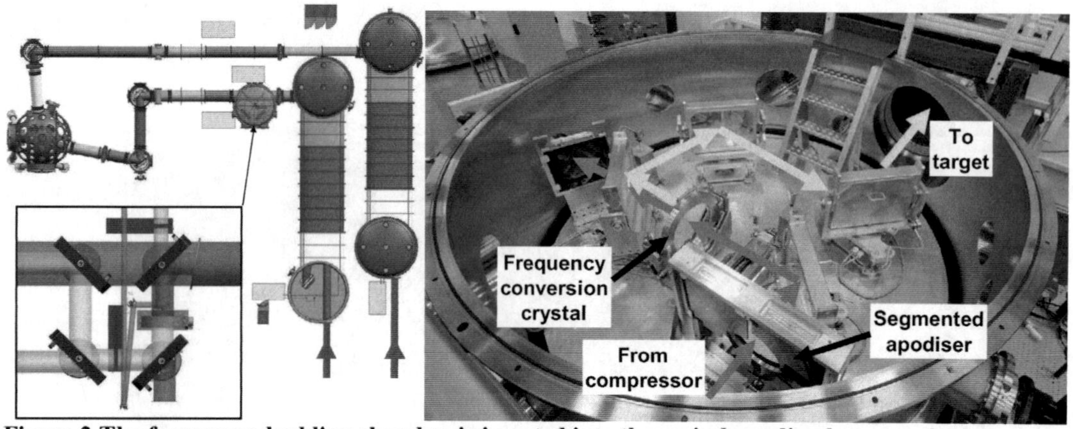


**Figure 2 The frequency doubling chamber is inserted into the main beamline between the compressor and target chambers.**

After leaving the conversion chamber the beam is reflected off two further mirrors, both designed to be HR at 1054 and 527nm, before being focused onto target using a dichroic f/6 off-axis parabola which is HR at 527nm and HT at 1054nm. The intensity of any first harmonic light at target is reduced by a factor of $\sim 10^2$ by each dichroic mirror and the parabola giving a total extinction of $\sim 10^8$.

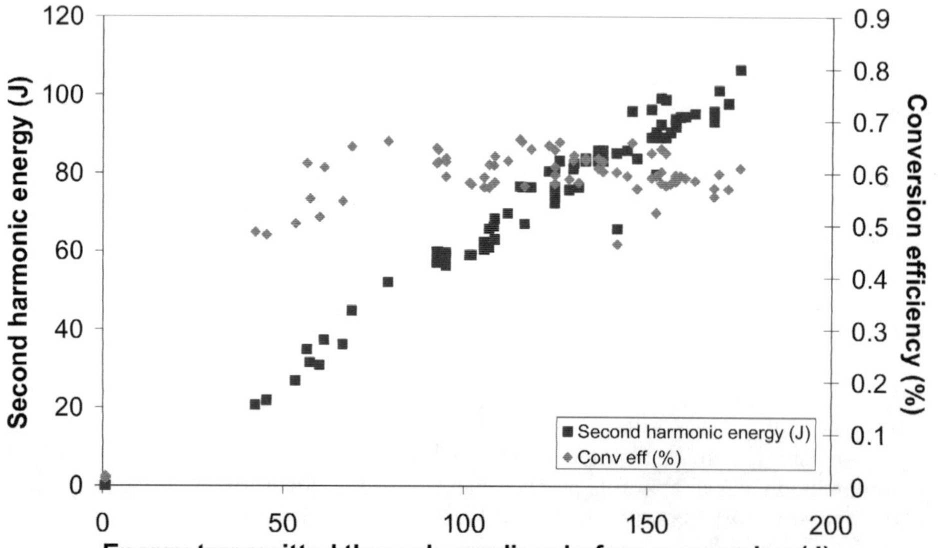


**Figure 3 Output energies and conversion efficiencies from the frequency doubled beamline**

The conversion crystal tuning angle was set by retro-reflecting light from the surface of the crystal then de-tuning by ~1mrad in the slow axis to minimize back reflections in the beamline. This detuning angle is small enough to have only a negligible effect on the conversion efficiency.

A series of full system shots of increasing energies were fired giving a maximum of 100J in the second harmonic and conversion efficiencies of up to 65% (Figure 3). The system was then used to carry out a range of target experiments with shots mostly fired at full energy.

Contrast measurements were performed in the first and second harmonic using the two photodiode technique [5]. This showed that the contrast was improved by a factor of $10^6$ to give a contrast of ~$10^{14}$. Making a measurement with this dynamic range required careful calibration of the filtering, extensive shielding of the photodiodes and running the diagnostics station beyond its damage threshold.

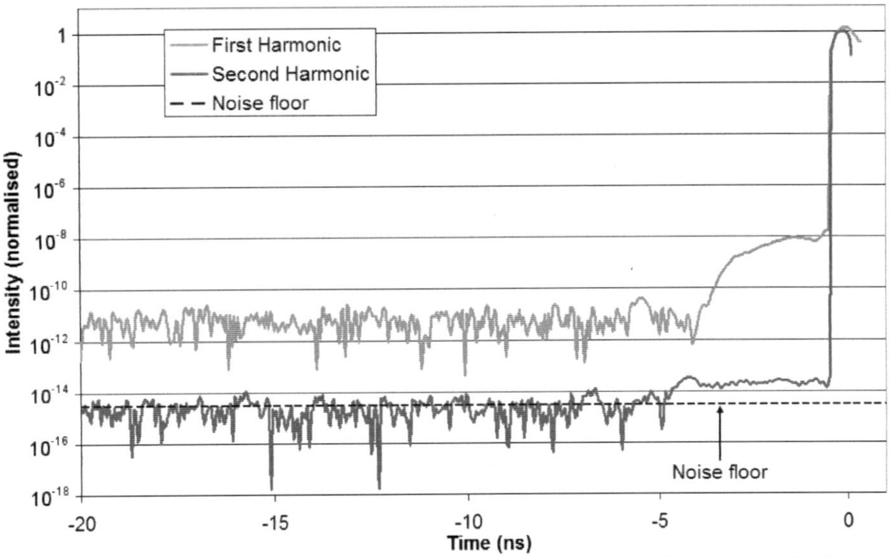


**Figure 4 Frequency doubling the beamline enhances the contrast by 6 orders of magnitude**

## 4. Future work

A program of work is currently underway to improve the contrast of the Orion beamlines in the first harmonic in order to increase the energy available to the target. We are in the process of implementing a short pulse optical parametric amplifier (SPOPA) similar to those described by Dorrer [6] and Musgrave [5]. Our SPOPA is a modular unit that will be installed into our beamline between the short pulse oscillator and main stretcher. It takes the output from the oscillator and splits it into two: one pulse is amplified using a regenerative amplifier then frequency doubled to give a 6ps, 500μJ pump beam. The other beam is stretched to match the duration of the pump pulse using a Martinez stretcher. The two beams are mixed in an LBO crystal to give clean 10uJ output to seed the main OPA. This will be used to enhance the nanosecond contrast by the gain of the OPA giving a predicted contrast of ~$10^{11}$. If used in conjunction with our frequency doubling system this should allow us to deliver a ~100J, 527nm pulse with a ~$10^{20}$ contrast to target.

## 5. Conclusions

The main beamlines of the Orion laser facility are now fully commissioned and the facility is performing experiments [7]. One of the short pulse beamlines is capable of being frequency doubled allowing it to provide 100J, 527nm laser pulses with a nanosecond contrast of ~$10^{14}$.

## References

[1] Hopps N W *et al* 2013 *Appl. Opt.* **52**, 3597–3607
[2] Hillier D I *et al* 2013 *Appl. Opt.* **52**, 4258-4263
[3] Hillier D I, Winter D N, and Hopps N W 2010 *Appl. Opt.* **49**, 3006–3013
[4] Girling M T *et al* 2010 *Proc. SPIE* 7721, 772122
[5] Musgrave I *et al* 2010 *Appl. Opt.* **49**, 6558–6562
[6] Dorrer C *et al* 2007 *Opt. Lett.* **32**, 2143–2145
[7] Hoarty D J *et al* 2013 *Phys. Rev. Letts.* **110**, 265003

# Laser-Plasma Interactions in Drive Campaign targets on the National Ignition Facility

**D E Hinkel[1], D A Callahan[1], J D Moody[1], P A Amendt[1], B F Lasinski[1], B J MacGowan[1], D Meeker[1], P A Michel[1], J Ralph[1], M D Rosen[1], J S Ross[1], M B Schneider[1], E Storm[1], D J Strozzi[1], and E A Williams[1]**

1) Lawrence Livermore National Laboratory, Livermore, CA 94550, USA
LLNL-CONF-646457

E-mail: hinkel1@llnl.gov

**Abstract.** The Drive campaign [D A Callahan *et al.*, this conference] on the National Ignition Facility (NIF) laser [E. I. Moses, R. N. Boyd, B. A. Remington, C. J. Keane, R. Al-Ayat, *Phys. Plasmas* **16**, 041006 (2009)] has the focused goal of understanding and optimizing the hohlraum for ignition. Both the temperature and symmetry of the radiation drive depend on laser and hohlraum characteristics. The drive temperature depends on the coupling of laser energy to the hohlraum, and the symmetry of the drive depends on beam-to-beam interactions that result in energy transfer [P. A. Michel, S. H. Glenzer, L. Divol, *et al., Phys. Plasmas* **17**, 056305 (2010).] within the hohlraum. To this end, hohlraums are being fielded where shape (rugby vs. cylindrical hohlraums), gas fill composition (neopentane at room temperature vs. cryogenic helium), and gas fill density (increase of ~ 150%) are independently changed. Cylindrical hohlraums with higher gas fill density show improved inner beam propagation, as should rugby hohlraums, because of the larger radius over the capsule (7 mm vs. 5.75 mm in a cylindrical hohlraum). Energy coupling improves in room temperature neopentane targets, as well as in hohlraums at higher gas fill density. In addition cross-beam energy transfer is being addressed directly by using targets that mock up one end of a hohlraum, but allow observation of the laser beam uniformity after energy transfer. Ideas such as splitting quads into "doublets" by re-pointing the right and left half of quads are also being pursued. LPI results of the Drive campaign will be summarized, and analyses of future directions presented.

## 1. Introduction

The radiation environment for an indirect drive capsule at the National Ignition Facility (NIF) [1] is the hohlraum (Figure 1). A hohlraum ("empty room") is typically a cylinder composed of high-Z material such as gold. A capsule is held in the center of the hohlraum by a tent, and is surrounded with low-Z gas, which acts to tamp the wall as it expands. Laser Entrance Holes (LEH), centered in the cylinder endcaps, are where laser beams enter the hohlraum. The beams propagate to the hohlraum wall where they are converted to x-rays, which drive the capsule implosion. The job of the hohlraum is to provide an appropriately "hot" radiation environment for the capsule implosion, and to provide sufficient radiation symmetry that the capsule remains nearly spherical as it implodes.

Laser-plasma interactions (LPI) impact both of these hohlraum aspects ("hot", and "round"). To reduce the impact of LPI we have modified three aspects of the hohlraum to investigate performance improvements. To improve hohlraum performance (both "hot" and "round"), we have modified the hohlraum fill composition (He replaced by $C_5H_{12}$, shot at room temperature), hohlraum fill gas density (increased by > 1.5 for the HiFoot design, and decreased to near-vacuum for 2-shock HDC capsules), and the hohlraum shape (cylinder to rugby) (c.f. Figure 1). In Section 2 we discuss recent progress on the impact of LPI on keeping the hohlraum "hot". In Section 3, we present a snapshot in time of the impact of LPI on capsule implosion sphericity. In Section 4 we summarize our results.

**Figure 1.** (a) An inertial confinement fusion (ICF) hohlraum provides a hot and symmetric radiation environment for the imploding capsule, suspended in its center. Laser-plasma interactions impact both of these aspects of the radiation environment. To keep the radiation temperature "hot", we have modified the hohlraum gas fill composition and the hohlraum gas fill density. (b) We have also modified the hohlraum shape from cylindrical to rugby.

## 2. Laser-plasma interactions in NIF hohlraums

As the laser beams propagate into the hohlraum toward the wall, they interact with the plasma environment. Stimulated Raman scatter (SRS) [2] occurs when a beam resonantly scatters off self-generated electron plasma waves. Stimulated Brillouin scatter (SBS) [2] occurs when resonant scatter off self-generated ion acoustic waves occurs. Both of these processes can lead to a direct loss of laser energy coupling to the hohlraum wall, and thus a drive reduction.

The Full Aperture Backscatter Station (FABS) [3] measures light scattered out of the hohlraum back down the beam line on one $30^0$ and one $50^0$ quad. The Near Backscatter Imager (NBI) [3] is a plate that surrounds these two quads, collecting near-backscattered light. On one $23^0$ quad there is a plate between the four beam lines of a quad that captures near-backscattered light. During 2011-2012, over the time when shock-timed implosions were fielded at NIF as part of the National Ignition Campaign (NIC), the measured SRS light from the $23^0$ and $30^0$ cones (16 quads in all) was nearly constant at ~ 150 kJ. Inner cone SBS adds approximately an additional 10-50 kJ of scattered light in the first three-quarters of 2011. In the fall of 2011, the NIC moved to a slightly larger hohlraum (5.75 mm in diameter vs 5.44 mm) which, when optimized, resulted in inner bean re-pointing. The inner cones were moved slightly outward, and thus intersect less capsule ablator material as they propagate toward the wall. The resulting inner cone SBS was subsequently reduced to 10 kJ or less.

The NIF hohlraum outer beams ($44.5^0$ and $50^0$ cones) create a gold bubble, in which SBS predominantly occurs. In 2011 and early 2012, outer cone SBS varied from 5 to ~ 35 kJ. When the laser pulse at peak power was lengthened in March, 2012, outer cone SBS increased to 40-50 kJ. On average, the diagnosed backscatter in 2011-2012 suggests backscatter energies of ~ 200 kJ.

The laser drive diagnostic (DrD) provides another measure of SBS. Here, a grating on the output of the main debris shield provides a diagnostic beam sample of the input beam (~ 0.1%). SBS from the hohlraum that propagates back up a beam line reflects off the second harmonic crystal at the 4% level, and is detected by the DrD. NIF DrDs have measured SBS on 57 of 192 beam lines since 2009. Data analysis (P. A. Michel and B. J. MacGowan, [4]) shows that the DrD data correlates with FABS measurements, at approximately a 1:1 ratio (+/- 10%). Further, SBS variations correlate with hohlraum perturbations such as when beams strike diagnostic windows (instead of the hohlraum wall). There are also SBS variations that correlate with the topology of cross-beam energy transfer [5]. For example, a dropped quad will result in modified power transfer to other quads, and thus different SBS levels. Further, DrD measurements indicate less SBS on the $44^0$ than on the $50^0$ beams.

Within the last year, we have modified the hohlraum to investigate performance improvements, with the goals of: (i) increasing the electron temperature within the hohlraum, and thereby reducing backscatter; (ii) reducing hohlraum wall motion, thereby improving inner beam propagation; (iii) fielding a hohlraum that provides sufficient radiation symmetry without cross-beam energy transfer.

With respect to (i) above, we have modified the hohlraum gas fill composition. To show efficacy, we fielded room temperature hohlraums with a neopentane gas fill. This fill, when ionized, has a larger effective Z, increasing inverse bremsstrahlung of the laser light, and thus increasing the electron temperature. Simulations show ~ 300 eV increase in the SRS region, with a larger change in the LEH region. In Figure 2, the nominal hohlraum has ~ 150 kJ of SRS and ~ 20 kJ of SBS (labelled He low foot). The SRS is reduced to ~ 60 kJ when the gas fill is changed to neopentane. However, the SBS on the outer beams increases to ~ 65 kJ. This SBS increase can be mitigated by borating the gold wall, thereby increasing ion Landau damping. This is a promising avenue for reducing backscatter losses.

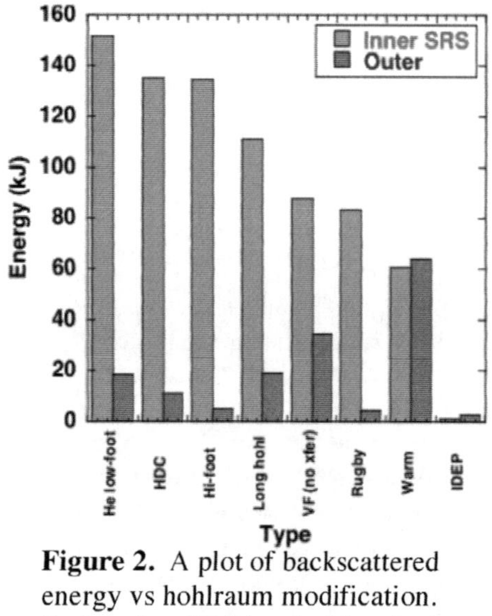


**Figure 2.** A plot of backscattered energy vs hohlraum modification. Red depicts SRS levels, blue depicts SBS levels.

To address (ii) above, we have fielded hohlraums where we have varied the gas fill density. In hohlraums where the capsule has an HDC ablator or a high foot drive, the gas fill density is increased. Here, both SRS and SBS are decreased from that in the nominal low foot hohlraum. Further we have also fielded the Indirect Drive Exploding Pusher (IDEP), a near vacuum target with very low gas fill density. This target demonstrated ~ 99% laser energy coupling, where both SRS and SBS were significantly decreased.

Rugby hohlraums have been successfully fielded with a very small wavelength separation between the inner and outer beams. The larger waist of the rugby (7 mm versus 5.75 mm for cylinder) better accommodates wall and ablator plasma blow-off, and electron densities remain low within the target, and thus the inner beams can propagate to the wall at the waist without the use of cross-beam energy transfer. In these targets, the inner beam SRS levels decreased by nearly a factor of two, and outer beam SBS levels were lower than in the nominal hohlraum. Inner beam SRS is quite similar to a cylindrical target without any cross-beam transfer [labelled as VF (no xfer)]. Future studies will continue to research items (i)-(iii).

## 3. Radiation symmetry in NIF hohlraums

NIF routinely uses cross-beam energy transfer (CBET) to achieve time-integrated P2 symmetry, i.e., to mitigate the tendency of the central hotspot to be oblate. CBET occurs where laser beams overlap, predominantly in the laser entrance holes. At NIF, the outer beams are incident on the hohlraum at a wavelength $\lambda_0$-$\Delta\lambda$, and the inner beams at a wavelength of $\lambda_0$. Forward Brillouin scatter transfers energy from the outer to the inner beams, which have a more difficult time reaching the wall because of their longer path length through cooler plasma. Since CBET scales as $I^*n_e/T_e$, where I is the laser intensity, $n_e$ is the electron plasma density, and $T_e$ is the electron temperature, it most readily occurs during the early part of the laser pulse (when $T_e$ is low) and at peak power, when the laser intensity is high. Simulations of CBET predict that the laser beams are spatially non-uniform, and thus this process is introducing a spatio-temporal intensity distribution. To assess the impact of these non-uniformities, we have changed the hohlraum shape from cylinder to rugby. The rugby shape allows for inner beam propagation without CBET. In May, 2013, we shot a rugby at full energy (1.3 MJ) and power (380 TW). This hohlraum showed good laser energy coupling (93%), but poor inner beam propagation.. Experimental diagnostics suggest significant wall motion, which impedes inner beam propagation. Our next rugby hohlraum will be shot with modified outer beam pointing (move inward by 500 μm), which is predicted to improve the radiation symmetry without using CBET.

The room temperature hohlraums with neopentane gas fill did not only show better energy coupling (as discussed in Section 2), but also achieved decent radiation symmetry at lower levels of CBET (smaller $\Delta\lambda$ difference between inner and outer beams). The challenge here is to field a cryogenic hohlraum with higher-Z fill. To this end, we plan to shoot a rugby hohlraum with 2.45 % atomic fraction neon in the helium gas fill, which will not freeze out at cryogenic temperatures.

## 4. Summary

We have investigated hohlraum performance improvements by modifying hohlraum shape as well as gas fill density and composition. The challenge here is to optimize across shape, fill composition, and fill density to find an optimized hohlraum with reduced backscatter losses, with minimal cross-beam energy transfer, and with good radiation symmetry. Rugby hohlraums hold promise for achieving radiation symmetry without large levels of cross-beam transfer. We plan to investigate other shapes, and to research different strategies for increasing the electron temperature within the hohlraum, as well as to perform a scan in gas fill density.

## Acknowledgments

We thank the Hohlraum Drive team for their valuable input. The authors would also like to acknowledge the valuable input of the NIF laser and target diagnostic scientific and operations teams as well as that of the LLNL and GA target fabrication and assembly teams. This work was performed under the auspices of the US Department of Energy by Lawrence Livermore National Laboratory under Contract DE-AC52-07NA27344.

## References

[1] Moses E I, Boyd R N, Remington B A, Keane C J, and Al-Ayat R 2009 *Phys. Plasmas* **16** 041006.
[2] Hinkel D E *et al.* 2011 *Phys. Plasmas* **18** 056312, and references therein.
[3] Moody J D *et al.* 2010 *Rev. Sci. Instrum.* **81** 10D921.
[4] Michel P A, *Anomalous Absorption Conference* presentation, July 2013, Stevenson, WA USA.
[5] Michel P *et al.*, 2010 *Phys. Plasmas* **17** 056305.

# The first data from the Orion laser: measurements of the spectrum of hot dense aluminium

**D J Hoarty[1], P Allan[1], S F James[1], C R D Brown[1], L M R Hobbs[1], M P Hill[1], J W O Harris[1], J Morton[1], M G Brookes[1], R Shepherd[2], J Dunn[2], H Chen[2], E Von Marley[2], P Beiersdorfer[2], H K Chung[3], R W Lee[4], G Brown[3], J Emig[3].**

1 Directorate of Research and Applied Science, AWE plc, Reading, RG7 4PR, UK
2 Lawrence Livermore National Laboratory, 7000 East Avenue, Livermore, CA94550, USA
3 Nuclear Data Nuclear Data Section, Division of Physical and Chemical Sciences, International Atomic Energy Agency, P. O. Box 100, A-1400, Vienna, Austria
4 Institute for Material Dynamics at Extreme Conditions, University of California, Berkeley, USA.

David.Hoarty@awe.co.uk

**Abstract.** The newly commissioned Orion laser system has been used to study dense plasmas created by a combination of short pulse laser heating and compression by laser driven shocks. Thus the plasma density was systematically varied between 1 and 10g/cc by using aluminium samples buried in plastic foils or diamond sheets. The aluminium was heated to electron temperatures between 500eV and 700eV allowing the plasma conditions to be diagnosed by K-shell emission spectroscopy. The K-shell spectra show the effect of the ionization potential depression as a function of density via the delocalization of n=3 levels and disappearance of n=3 transitions in He-like and H-like aluminium. The data are compared to simulated spectra, which account for the change in the ionization potential by the commonly used Stewart and Pyatt prescription; a simple ion sphere model and an alternative due to Ecker and Kröll suggested by recent X-ray free-electron laser experiments. The experimental data are in reasonable agreement with the model of Stewart and Pyatt, but are in better agreement with a simple ion sphere model. The data indicate that the Ecker and Kröll model overestimates substantially the ionization potential depression in this regime.

## 1. Introduction/Background

A new laser facility called Orion [1] has been commissioned at AWE in the UK which combines long pulse, nanosecond timescale, and short pulse, picosecond timescale, laser beam technologies. As part of the facility commissioning a series of experiments was carried out where samples were heated to electron temperatures over the range 500eV to 700eV and the plasma density was varied systematically over an order of magnitude between 1-10g/cc. The samples were aluminium layers buried in either plastic foils or diamond sheets. At such high densities the electrostatic field of neighbouring ions and electrons acts on the outer levels of the ions lowering the ionization energy of the dense plasma ions relative to the isolated ion case. The experiments described here show the effect of ionization potential depression (IPD) on the emitted spectra as a function of density by the disappearance of line emission from the n=3 levels in He-like and H-like aluminium ions. There has been a renewal of interest in IPD due the findings of recent X-ray free electron laser experiments (XFEL) at the Linac Coherent Light Source (LCLS) facility [2].

The Orion experiments extend work carried out on the HELEN laser between 2006 and 2009. Experiments on HELEN showed that a thin buried layer in plastic could be heated to high temperature in excess of 500eV and in some cases up to 1keV [3]. The conditions were diagnosed by modelling the details of the aluminium K-shell emission in the 1.5-3.0 keV energy range from aluminium ionized to He-like and H-like configurations by modelling the spectra with the atomic kinetics code FLY [4] and its successor FLYCHK [5]. Measurements were made using both a 1.5 ps temporal resolution streak camera coupled to a focussing crystal spectrometer [6] and time-integrating X-ray crystal spectrometers recording onto X-ray film or BAS-TR Fuji image plates [7]. Examples of the spectrum of aluminium recorded on HELEN are given Ref [3]. The first experiments on Orion were carried out to investigate the heating of buried layer targets using the infrared beam. These experiments confirmed the findings of the HELEN experiments that coupling of laser energy to heating the bulk solid target is poor in the presence of significant pre-pulse [3]. The effectiveness of the second harmonic conversion in improving solid target heating was demonstrated in a series of shots using targets with an overcoat of plastic between the laser and the buried layer, which was varied between 5μm up to 35μm on a shot by shot basis. The rear plastic coating was 2μm as in the previously reported HELEN experiments [3]. The data showed heating to temperatures above 500eV up to a depth of 25μm and a sharp fall off in peak electron temperature between 25 and 35μm [3].

## 2. Compressed plastic experiments on Orion

In the planar compression and short pulse heating of the buried layer targets the ablation pressure from the long-pulse laser drives a shock wave through the foil and into the buried layer [3]. The material between the buried layer and the laser is sufficiently thick that the laser does not ablate through to the buried layer before the end of the laser pulse. In order to heat the compressed foil the short-pulse beam is initiated when the transiting shock is between the buried layer position and the rear face of the foil. Combined long-pulse compression and short-pulse heating was applied to plastic targets where the ablative layer irradiated with the long-pulse beams was a 12μm thick coating of plastic and on the other side of the buried aluminium tracer the coating was 10μm of plastic. The shock driven through the 12μm ablator by long-pulse laser ablation compressed the layer and the plastic tampers. Before shock breakout at the rear of the target the short-pulse beam heated the target to high temperature. The experiment used Orion long-pulse beams each with a pulse length of 0.5ns FWHM at 0.35μm wavelength and energy of 180-200J. The short-pulse beam was converted to second harmonic operation as described above. The long pulses were shaped with an approximately 100ps rising edge. The emissivity of the aluminium sample was obtained in absolute units (J/keV/sr/cm$^2$/s) as described in Ref [3] along with further details of the experiment. Preheating of the aluminium layer caused it to expand into the surrounding plastic ahead of the arrival of the shock. However, experiments using two of the Orion beams were able to demonstrate compression above twice solid in aluminium and heating to a temperature in excess of 500eV. The spectral ranges of the instruments were set to record the K-shell emission from aluminium ionized to He-like and H-like configurations. The time-integrating spectrometers recorded the n=1-2 and n=1-3 transitions; the time-resolved data covered the n=1-3 only, due to restrictions on the spectral dispersion and streak camera slit length with the streak camera. This overlap in the spectrum allowed comparison of ratios from time-integrated and time-resolved data as in the HELEN experiments. The time-integrated data are shown in Fig. 1 along with a fit using the FLY code. In Fig 1 the prediction was scaled by 1.17 to achieve the agreement shown in the figure. This was within the experimental error of 30% for the measured emissivity.

The line-ratios of the n=1-3 transitions in the 1.8-2.2keV energy range are used to infer the temperature and the overall width of the line shape is used to establish the density. This is accurate despite the disagreement at line centre where the code predicts a dip in the profile that is much more pronounced than in the experimental data. The peak electron temperature inferred from the data is

550eV and the density 6g/cc. The data from the 1.5ps resolution streak camera are shown in the inset plot in Fig. 1 along with a curve generated using the FLY code [4]. The best fit to the time-resolved data is at a slightly higher temperature of 600eV and a density of 6g/cc compared to the time-integrated case. This slightly higher peak temperature in the time-resolved data is typical of the comparison between time-integrated and time-resolved data where the best fit to the time-integrated data tended to be 50-80eV lower than the time-resolved data on the same shot. Sensitivity studies have shown that the spectra are more sensitive to temperature than density [3]. The density fit at 7g/cc is a poorer fit to the $He_{\beta}$ line width and the 5g/cc is narrower than the experimental width of $Ly_{\beta}$. Fits using FLYCHK [5] and the ALICE [8] atomic physics and line-shape codes infer the same conditions. Similar experiments in the same campaign that used different samples of aluminium buried in diamond sheets have demonstrated even higher temperatures and densities. These experiments are described in the next section

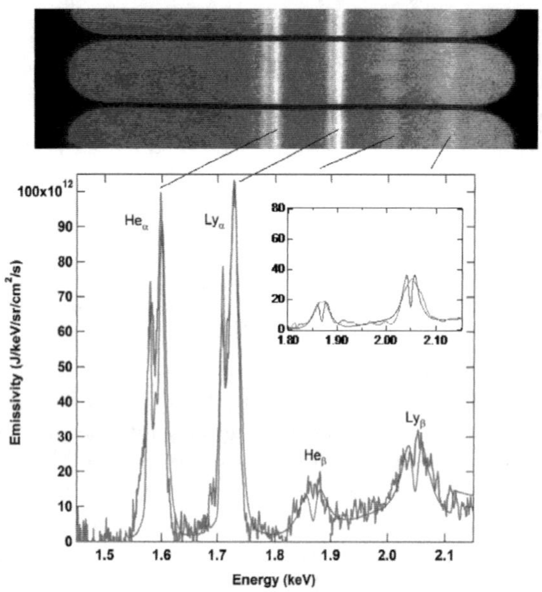


Fig 1: Time-integrated data from long-pulse compression and short-pulse heating of aluminium buried in plastic foil. The spectrometer data is shown and a lineout of the spectrum recorded. Also shown is the best fit to the data using the FLY atomic kinetics and line-shape code. The fit indicates a peak electron temperature of 550eV and a peak density of 6g/cc which meets the criteria for the milestone. The predicted emissivity has been scaled by a factor 1.17.

## 3. Experiments with aluminium samples in diamond

The maximum density achievable in the buried layer is determined by pressure equilibration with the compressed tamper material when the material in the layer expands into the tamper after heating and ionization [3]. The targets with high-density tampers used a combination of diamond, which was grown as a wafer, and diamond-like carbon, which was coated onto the target. To get the two diamond layers to bond a thin coating of silicon carbide was applied between them. The diamond targets were designed to prevent expansion of the aluminium buried sample into the surrounding tamper due to preheat by the long-pulse beams during the transit of the shock wave through the ablator layer. The 10μm diamond sheet was used as the target ablator for the long-pulse beam and the short-pulse beam irradiated the 4μm diamond-like carbon side to heat the compressed target through to the buried layer. First tests of the diamond targets were to examine the response to short-pulse heating alone. Comparison of this data to FLY calculations allowed the temperature and density conditions in the aluminium to be inferred. The peak electron temperature was 750±50eV and the material density was inferred to be 3.5±0.5g/cc. Experiments using a long-pulse beam of Orion to drive a shock through the diamond had the laser wavelength, pulse shape, focal spot and energy identical to that of the beams used in the aluminium in plastic foil experiments described above. The relative delay between the long-pulse and short-pulse beam was varied to step through the time history of the shock compression and short pulse heating of the target. The results are shown in Fig. 2 for the spectral range covered by the streak camera, which is the n=3 transitions in He-like and H-like aluminium ions. Figure 2 shows the effect of shock compression on the line emission from the aluminium dot buried in diamond. The long-pulse beam is initiated before the short-pulse beam and the drives a shock wave which takes 280ps to arrive at the aluminium layer. This shock transit time was confirmed by examination of the background silicon emission on selected targets from the batch of diamond targets as described below.

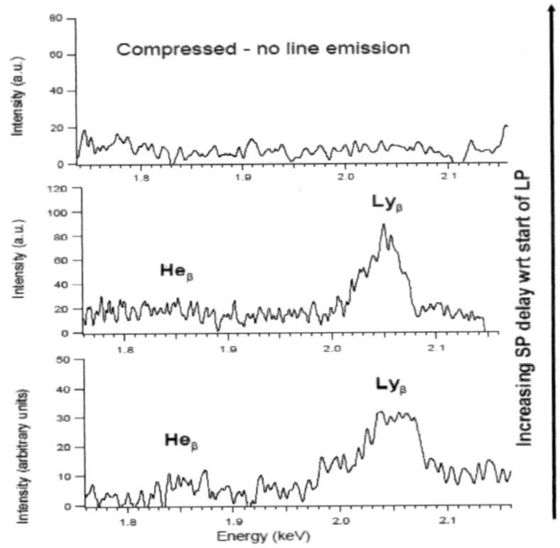


Fig 2: The graphs are streaked spectra taken at different long pulse-short pulse delay. Once compressed the n=1-3 transitions from the aluminium disappear.

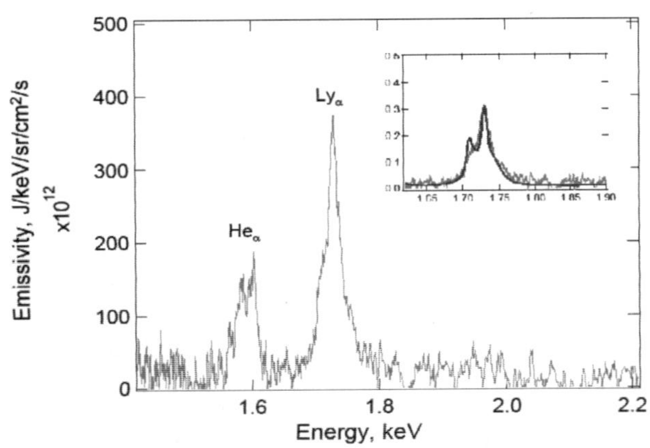


Fig 3: Time-integrated data from compressed aluminium in diamond. The curve is a lineout through the spectrum after image processing to remove background noise and allow for image plate and filter responses. The inset curve shows a fit to the experimentally measured $Ly_\alpha$ line (red curve) of the prediction from the atomic physics and line-shape model FLY (blue curve) indicates a density of 9±1g/cc. A peak electron temperature from the He/Ly line ratio was inferred to be 700±50eV.

The bottom curve of Fig. 2 was taken from streak data where the long-pulse shock had not yet arrived at the aluminium and shows clear emission from $He_\beta$ and $Ly_\beta$ lines. The middle curve was also taken prior to the shock arrival and shows the same lines clearly. However, in the top curve where the long pulse was initiated 300ps early the shock has compressed the sample but not yet broken out from the rear surface of the diamond sheet. The streak data at this time do not show the line emission seen at earlier times. This disappearance of the n=1-3 lines in the spectrum is due the delocalization of the n=3 level in the He-like and H-like aluminium ions so that the levels blend into the free-bound continuum. In the absence of the n=1-3 transitions in the streak-camera data the n=1-2 transitions from He-like and H-like ions in the time-integrated data from compressed aluminium in diamond, shown in Fig. 3, were used to infer the sample conditions. Note that, as in the time-resolved data, the time-integrated data show no signature of the n=1-3 transitions from the compressed aluminium. The inset curve in Fig. 3 shows the lineout through the time-integrated data for $Ly_\alpha$ compared to a predicted line-shape using the FLY model. The line-shape predicted by FLY is the best fit to the time-integrated data in the $Ly_\alpha$ line at 9±1g/cc and 700±50eV. The temperature was obtained from the ratio of He-like to H-like emission. However, it should be noted that the FLY code predicts that the n=1-3 transitions are still clearly in evidence at 9g/cc and 700eV and are not delocalized as indicated in the data. Also, the satellite emission on the low energy side of the line, which is due to doubly-excited states of He-like ions, is clearly over-estimated in the simulation. Further evidence of the conditions is available from the shock transit times and the radiation-hydrodynamics simulations.

It was possible to measure the shock transit time in the diamond targets because of a flaw in some of them where the silicon wafer that the diamond sheet was grown on was not completely etched away. This left a residue of silicon on the face irradiated by the long-pulse beam within the long-pulse spot. On these targets there was a silicon emission signal marking the leading edge of the long pulse beam. The accuracy of the Orion timing system was confirmed by changing the inter-beam delay in 10ps increments and measuring the change in the measured delay with the streak camera using the diamond targets with a silicon residue. By establishing the shock transit time accurately the predictions of the NYM radiation hydrodynamics calculations [9] could be benchmarked with respect to the measured shock transit time. It was found that the irradiance used in the NYM code had to be reduced by 20% from the measured value in order to match the measured shock transit time. The calculated temperatures and densities from the NYM simulations, adjusted to reproduce the observed shock transit time predict values close to the conditions of 9±1g/cc and 700±50eV inferred from the time-integrated spectral measurements [3]. This confirms the aluminium in compressed diamond conditions inferred from the spectral measurements.

## 4. Interpretation of the aluminium spectra

Traces of the aluminium spectra measured with the streak camera and crystal spectrometer in the range 1.8-2.2keV are shown in Fig. 4. The measurements at the various densities sampled experimentally show the conditions at which the n=3 levels delocalize. Shown alongside the experimental data are predictions of the FLYCHK code using both the Stewart and Pyatt (SP) [10] and Ecker and Kröll (EK) [11] models of ionization potential depression (IPD), with the same prescription for SP and EK models used here as in Ref. [2]. The curves of both experiment and simulation are in arbitrary units and have been scaled in intensity for clarity. The experimental curves from bottom to top were fit at the following conditions of the total aluminium material density and peak electron temperature: 1.2±0.4g/cc, 550eV; 2.5±0.3g/cc, 650eV; 4±0.5g/cc, 700eV; 5.5±0.5g/cc, 550eV; 9±1g/cc, 700eV. The simulations use the measured density, but the temperature in all cases is 700eV. The data show that n=3 transitions are clearly observed up to densities of at least 6g/cc, indicating that the IPD shift has not delocalized the n=3 levels at this density. The spectral data and radiation-hydrodynamics simulations suggest that at a density between 8 and 10g/cc the n=3 levels are delocalized and that the n=1-3 line transitions disappear from the spectrum [12]. The simulations using the SP prescription for IPD are in reasonable agreement with experiment but suggest that delocalization of the n=3 levels, from bound states to the continuum, would occur at a slightly higher density (11.6g/cc) than indicated by experiments. This curve is the top curve of the SP plots in Fig. 4. In contrast the simulations with the EK model for IPD predict delocalization of n=3 at much lower density due to the larger predicted shift in the bound-free edge due to ionization potential depression.

In summary, aluminium spectra have been measured at high temperature and density using samples buried in plastic foils and diamond sheet. The absolute emissivity is within 20% of predicted values and a systematic study of the aluminium spectrum varying the sample density, by a combination of short-pulse heating and long-pulse compression, has shown the conditions at which the n=3 levels in He-like and H-like lines delocalize and the transitions merge into the continuum. The data show disagreement with the predicted conditions at which delocalization of n=3 levels occurs in simulations of a time-dependent atomic kinetics model FLYCHK using a model of ionization potential depression due to Ecker and Kröll suggested by recent XFEL experiments [2]. The data indicate that at the range of the conditions sampled in these experiments the EK model overestimates the IPD shift significantly. The experimental data are in reasonable agreement with spectra predicted by FLYCHK using the SP model of IPD.

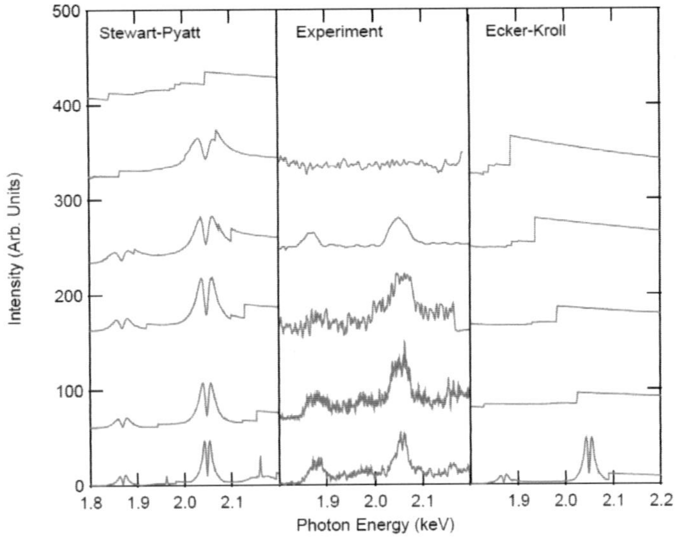


Fig 4: A compilation of the experimental data at different densities compared to FLYCHK predictions using Stewart and Pyatt and Ecker and Kröll treatments of ionization potential depression. The density of the curves from bottom to top is 1.2, 2.5, 4, 5.5 and 9g/cc. The SP plot has an additional curve at 11.6g/cc.

**References**

[1] N W Hopps et al, Proc. Of SPIE Vol 7916, 79160c DOI:10:1117/12.874252 (2011)
[2] O Ciricosta et al, Phys. Rev. Lett. **109,** 065002 (2012)
[3] D J Hoarty et al, High Energy Density Physics **9**, 661 (2013)
[4] R W Lee and J T Larsen, J. Quant. Spectrosc. Radiat. Trans. **56**, 535 (1996)
[5] H K Chung et al, High Energy Density Phys., **1**, 3 (2005)
[6] C R D Brown et al, Phys. Rev. Letts. **106,** 185003 (2011)
[7] N Izumi et al, Rev. Sci. Instrum. **77**, 10E325 (2006)
[8] E Hill, private communication
[9] P D Roberts et al., J. Phys. D **13**, 1957 (1980)
[10] J C Stewart and K d Pyatt Jr. Astrophys. J. **144**, 1203 (1966)
[11] G Ecker and W Kröll, Phys. Fluids **6**, 62 (1963)
[12] D J Hoarty et al, Phys. Rev. Letts. **110,** 2656003 (2013)

# Enhancement of fast electron energy deposition by external magnetic fields

**J J Honrubia[1], M Murakami[2], K Mima[3], T Johzaki[4], A Sunahara[5], H Nagatomo[2], S Fujioka[2], H Shiraga[2] and H Azechi[2]**

[1] School of Aerospace Engineering, Polytechnic University of Madrid, Spain
[2] Institute of Laser Engineering, Osaka University, Japan
[3] The Graduate School for Creation of New Photonics, Hamamatsu, Japan
[4] Graduate School of Engineering, Hiroshima University, Japan
[5] Institute for Laser Technology, Osaka, Japan

E-mail: javier.honrubia@upm.es

**Abstract.**
Recently, generation of external magnetic fields of a few kT has been reported [Fujioka et al. Scientific Reports 2013 **3** 1170]. These fields can be used in fast ignition to mitigate the large fast electron divergence. In this summary, two fast ignition applications are briefly outlined. The first one deals with electron guiding by external B-fields applied at the end of the shell implosion of a re-entrant cone target. Preliminary results show that the B-field strength at the time of peak ϱR may be sufficiently high for fast electron guiding. The second application deals with guiding of fast electrons in magnetized wires surrounded by plasma. Results show a significant enhancement of electron energy deposition at the end of the wire, which is particularly important for low-Z wires.

## 1. Introduction

Recently, magnetic fields have been proposed for guiding fast electrons towards the compressed core of fast ignition targets [1]. B-fields of a few kT are required to guide fast electrons towards the dense fuel. Two approaches have been proposed so far for generating so strong magnetic fields: i) magnetic flux amplification [2] and ii) field generation by laser-driven external coils [3]. In the first approach, it has been reported that, after the flux amplification, fast electrons are guided towards the dense core by an increasing magnetic field resulting in a significant 'backscattering' of those electrons due to the mirroring effect [1]. The second approach is based upon generating B-fields by a laser-driven external single-turn coil [3] at a time close to the end of the target implosion and offers the possibility to generate the optimal field distribution for fast electron guiding. The main issue is to ensure magnetic field penetration up to the dense core over times of the order of a few nanoseconds.

## 2. Model for B-field penetration in plasma.

A model for resistive diffusion and amplification of external magnetic fields has been developed. This model is based on post-processing the data obtained from radiation-hydrodynamic codes. It does not account for the fields generated in the plasma. Instead, the model deals with the penetration of external

[1] To whom any correspondence should be addressed.

B-fields in the plasma, which is important for fast electron transport and energy deposition. The following induction equation is solved for the B-field:

$$\frac{\partial \mathbf{B}}{\partial t} = \nabla \times (\mathbf{v} \times \mathbf{B}) + \frac{\eta}{\mu_0} \nabla^2 \mathbf{B} \tag{1}$$

where **v** is the fluid velocity and $\eta$ the plasma resistivity. The relative importance of each term is given by the magnetic Reynolds number $R_m = \mu_0 L v/\eta$, where $L$ is the typical flow scale-length. If $R_m >> 1$, the advection term is dominant, whereas if $R_m << 1$ the diffusion term prevails. Once **B** is obtained from Eq. (1), the Ampère law provides the plasma current density induced in the plasma, from which ohmic heating, plasma temperature and resistivity are obtained. The induction equation is solved semi-implicitly by using a preconditioned bi-conjugate gradient solver algorithm for matrix inversion [4]. It has been applied to two different situations found in electron-driven fast ignition targets: i) fast electron propagation in cone-targets with external B-fields and ii) fast electrons guiding in magnetized wires.

## 3. Fast electron propagation in cone-targets with external magnetic fields.

The B-field diffusion model described above provides the distribution of the external B-field in the target at the time of the peak ρR, just when fast electrons are generated at the cone tip. The model target used in our calculations is sketched in Fig. 1, where a re-entrant cone is attached to the direct-drive ignition scale capsule shown in Fig. 2a. It contains a DT shell of 0.3 mg, which is imploded with a peak velocity of 160 km/s. The r-t flow diagram of the 1-D implosion is depicted in Fig. 2b, where the 24 ns pulse duration is remarkable. Note that most of the implosion takes place without external magnetic fields, which are turned on just 1 ns before the end of the implosion by a coil sited in the position shown in Fig. 1. This timing has been chosen in order to avoid B-field configurations that 'backscatter' fast electrons due to the mirroring effect. Ideally, it is assumed that the coil current is 5 MA during 1 ns, generating a 3 kT B-field at the cone tip placed 450 mm away from the coil.

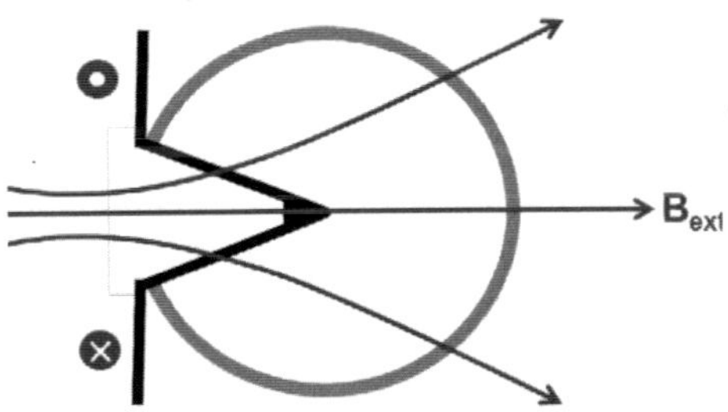


**Figure 1**. Sketch of the cone target used in the simulations. The coil is sited at the left of the shell and separated from the plasma by a metallic plate.

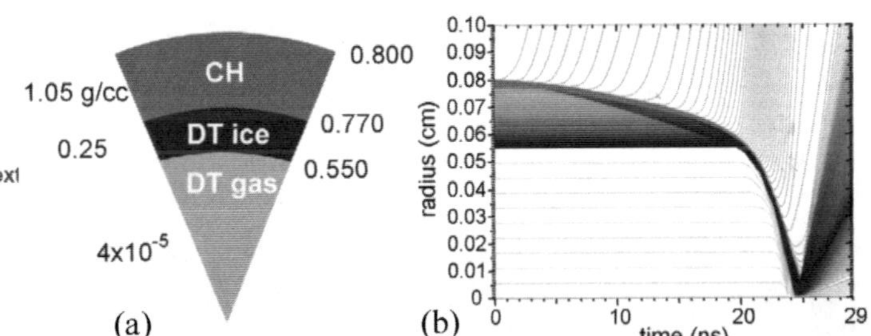


**Figure 2**. Target and r-t diagram of the target used in simulations. Shell radii are shown at the right of the target sketch in mm. The target is illuminated by a 0.2 TW pre-pulse followed by the main pulse with a peak power of 40 TW.

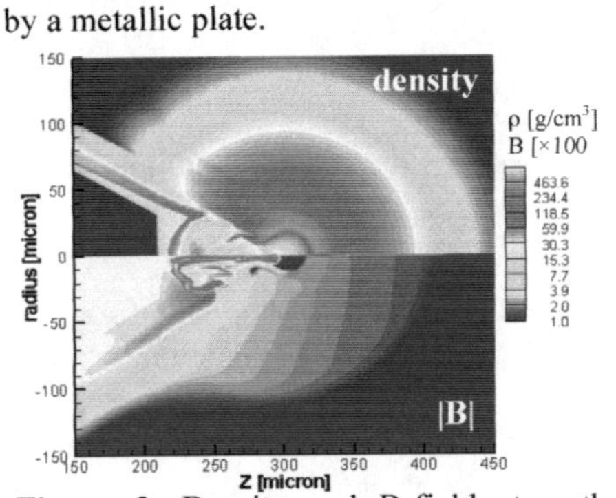


**Figure 3**. Density and B-field strength near the peak compression time of the capsule shown in Fig. 2.

The B-field in the target depends on the diffusion of the external B-field through the cone walls and the shell, and its amplification during the shell implosion. As shown in Eq. (1), high-resistivity materials should be used to ensure the diffusion of the B-field along the cone walls. Here, solid gold with an initial temperature of 10 eV has been chosen. The B-field distribution near the time of peak compression is shown in Fig. 3. Initially, the B-field diffuses along the cone and through the dense and cold part of the imploding shell, which has a relatively high resistivity. As the B-field is generated 1 ns before the time of peak ρR, its amplification is much lower than that reported in [2]. However, there is a small but non-negligible amplification factor of 2-3. It is worth noting that the B-field peaks where fast electrons are generated, and its value near the axis is about 1 kT, that may be sufficient to reduce divergence or even guide fast electrons, depending on their energy. Note also the oscillations of the B-field near the axis, which may affect fast electron transport. Our preliminary simulations show that imposing external magnetic fields near the end of the implosion may enhance fast electron energy deposition and coupling efficiency. However, more detailed studies including fast electron transport are required to fully assess this possibility.

**4. Fast electron guiding in wires.**

The use of resistive wires to guide fast electrons along relatively large distances has been proposed recently [5,6]. It was extended to specially engineered targets by Cai *et al.* [7]. In the context of fast ignition, we have performed a study of electron transport in wires of different materials – diamond-like carbon (DLC) at 3 g/cm$^3$, and aluminium, copper and gold at solid density– surrounded by an ideal imploded DT density distribution. The target used in our simulations is shown in Fig. 4, which is similar to that used by Solodov et al. [8] for fast electron divergence control. As beam collimation in wires is caused by radial resistivity gradients [5], high-Z materials should be more effective for fast electron transport between the cone tip and the compressed core. It is assumed that 1 ns before the time of peak ρR, a laser-driven coil placed as shown in Fig. 1 generates a magnetic field, which diffuses resistively along the cone tip and the wire. The initial temperature of the cone and the wire is 10 eV, while that of the DT is 100 eV. Thus, the B-field penetration in the DT fuel is quite low due to its lower resistivity. At the maximum compression time, a fast electron beam is generated at the cone tip and guided towards the compressed core by the diffused B-field. It is assumed that fast electrons

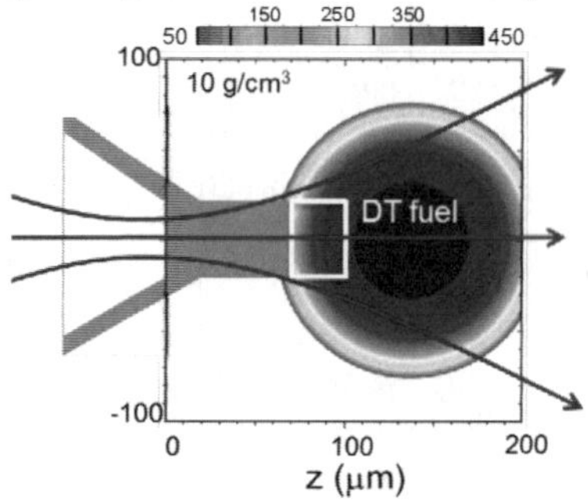


**Figure 4**. Wire target used to study the enhancement of the energy deposition by external magnetic fields. The density distribution of the DT is super-Gaussian with a peak value of 450 g/cm$^3$. The tip diameter is 40 μm, its thickness 20 μm and the length of the wire 50 μm. The cone and the wire are of the same material: DLC, Al, Cu or Au. Fast electrons with an exponential energy distribution, a mean energy of 3.5 MeV and a divergence half-angle of 55° impinge on the left target surface. The electron beam energy is 50 kJ. The *ignition region* for comparison of electron energy deposition is defined as the 30x40 μm white box shown within the DT.

are generated by a laser beam with a peak intensity of $5.5\times10^{20}$ W/cm$^2$ and a wavelength of 0.525 μm. The beam has a 20 μm (FWHM) width super-Gaussian density distribution in radius of and the fast electron temperature is obtained by the ponderomotive scaling. The pulse length is 14 ps and the laser

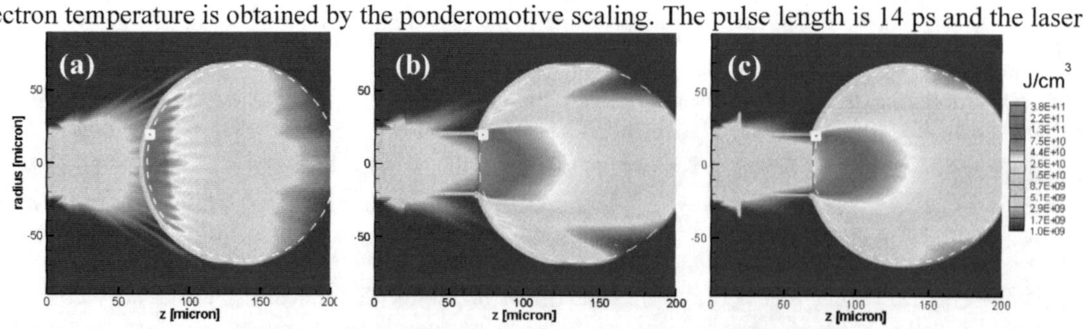


**Figure 5**. Fast electron energy deposition in the cases of: a) no wire nor B-field, b) with DLC wire without B-field, and c) with DLC and B-field generated in a coil of 1 mm diameter located 450 μm away from the cone tip sketched in Fig. 4. Electron transport calculations have been performed with the 2D hybrid code PETRA [9].

to fast electron conversion efficiency is 50%. The fast electron energy deposition is shown in Fig. 5. Without wire and external magnetic fields, it is dominated by the fast electron divergence and the coupling efficiency between electron beam and core is too low, as shown in Fig. 5a. Resistive wires enhance the energy deposition in the *ignition region* substantially due to the electron confinement within the wire [5], even for the low-Z DLC wire of Fig. 5b. However, the magnetic field induced at the DLC wire edge due to the resistivity gradient is not strong enough to trap all fast electrons in the wire, resulting in a significant loss of fast electrons, as shown in Fig. 5b. On the contrary, light materials are more effective for electron transport due to the lower electron energy deposition, scattering and ohmic heating. Assuming that a B-field is generated by a laser-driven coil similar to that described in [3], and that this field diffuses along the wire for 1 ns, Fig. 5c shows how the energy deposition in the ignition region and in the whole core is substantially enhanced due to the fast electron guiding by the imposed B-field.

The electron beam coupling efficiencies are shown in Fig. 6 for all the cases analysed. Note their enhancement due to the non-magnetized wire. When external magnetic fields are on, there is a substantial enhancement of the coupling efficiencies, especially for low-Z materials, for which the enhancement is similar to that produced by the wire itself. It is worth noting that coupling efficiencies higher than 20% can be achieved for DLC wires and B-fields of about 6 kT on target, which correspond to 10 MA current on the single-turn coil.

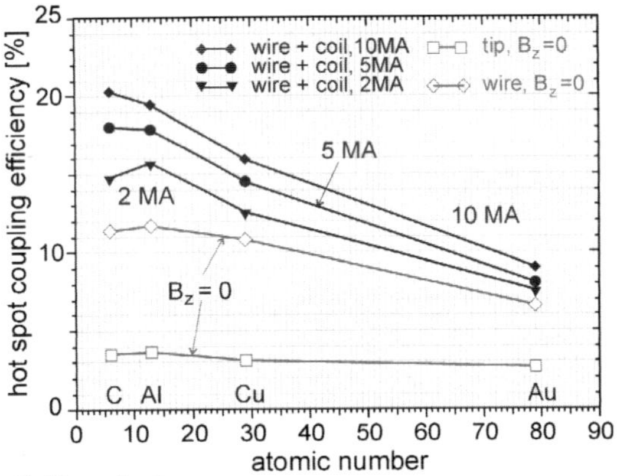


**Figure 6**. Fraction of the electron beam energy deposited in the "ignition region" shown in Fig. 4 for different cone tip and wire materials. The coupling efficiencies without wire and without B-field are shown in red, whereas those obtained with wire and without external B-field are drawn in light blue. The coupling efficiencies with wire and external B-field are plotted in dark blue with the current of the laser-driven coil as a parameter. The currents of 2, 5 and 10 MA correspond to a B-field on the target left surface of 1.2, 3 and 6 kT, respectively.

## 5. Conclusions

Regarding the fast electron transport in magnetized targets, the preliminary results shown motivate a more detailed study of the B-field penetration for a full assessment of the option proposed here. This study should include cone material selection, optimal time to apply the magnetic field and fast electron energy deposition in the magnetized imploded target.

Our results on fast electron guiding in magnetized wires show that external B-fields improve substantially the coupling efficiency, in particular for wires of low-Z materials. Note that this effect is more beneficial than it appears to be because lower electron beam energies mean, lower short pulse laser intensities, lower electron energies and shorter penetration range. The survival of the wire during the fuel compression is still an open question.

Future developments include the extension of the post-processor for B-field calculations and the implementation of a MHD model in radiation-hydrodynamic codes.

## References

[1] Strozzi D *et al.* 2012 *Phys. Plasmas* **19** 072711
[2] Chang P Y *et al.* 2011 *Phys. Rev. Lett.* **107** 035006
[3] Fujioka S *et al.* 2013 *Scientific Reports* **3** 1170
[4] Anderson D V, Koniges A E and Shumaker D E 1988 *Comput. Phys. Comm.* **51**, 391
[5] Robinson A P L and Sherlock M 2007 *Phys. Plasmas* **14** 083105
[6] Kar S *et al.* 2009 *Phys. Rev. Lett.* **102** 055001
[7] Cai H-B 2011 *Phys. Plasmas* **18** 023106
[8] Solodov A *et al.* 2010 *Controlling the Divergence of Laser-Generated Fast Electrons Through Resistivity Gradients in Fast-Ignition Targets* BAPS 2010 DPP CO6.9
[9] Honrubia J J *et al.*, 2006 *Laser Part. Beams* **24** 217

Journal of Physics: Conference Series **688** (2016) 012034 doi:10.1088/1742-6596/688/1/012034

# Supplemental heating of conventional Inertial Confinement Fusion

**B R Thomas, S J Hughes, W J Garbett, N J Sircombe**
AWE Plc, Reading, UK, RG7 4PR
E-mail: stephen.hughes@awe.co.uk

**Abstract.** We report a new ICF scheme whereby a capsule is imploded to near ignition conditions and subsequently flooded with hot electrons generated from a short-pulse laser-plasma interaction so as to heat the whole assembly by a few hundred eV. The cold dense shell pressure is increased by a larger factor than that of the hot spot at the capsule core, so that further heating and compression of the hot spot occurs. We suggest it may be possible to drive the capsule to ignition by the pressure augmentation supplied by this extra deposition of energy.

## 1. Introduction

In conventional ICF [1] a shell is accelerated by a rising pressure pulse or series of shocks along a low adiabat. At the end of the implosion, most of the kinetic energy of the inner, unablated part of the shell is converted into compression energy. Because of the nature of the acceleration process in a thin shell, which involves a density and temperature gradient through the layer, the innermost layer is on a higher adiabat than most of the 'fuel' and this, plus the initial gas fill of DT in the shell, forms a 'hot-spot' inside the cold compressed fuel.

Another approach [2] uses a lower implosion velocity to obtain lower compression in a larger fuel mass, and injects energy from an external source to generate the hot spot. There are more stringent requirements on $\rho r$ and $T_{HS}$ as the fuel is at constant density rather than constant pressure whilst the hot spot is being formed. As a consequence, hot spot formation is accompanied by a shock generated in the spot being sent out into the colder fuel. Extra energy is required in the hot spot to compensate for this work done against the cold fuel.

We propose combining elements of both approaches: an indirect drive, conventional implosion to stagnation at which point external energy is injected to raise the temperature of the whole target. This should have the effect of increasing the pressure in the cold fuel whilst only increasing the pressure of the hot spot by a few percent. The result should be a further compression of the hot spot to a higher $\rho r$ and temperature. The scheme does not offer a route to significantly higher gain and complicates a normal ICF implosion. It should be seen as a way to either remediate lack of pressure at minimum volume or to allow reduction in the implosion velocity to reduce deleterious mix.

## 2. Heating and energy budget

We consider a target based on a NIF ignition design as in figure 1. Hydrodynamic simulations of the capsule in 1D using the NYM code indicate that dudding the $4^{th}$ shock to reduce peak velocity

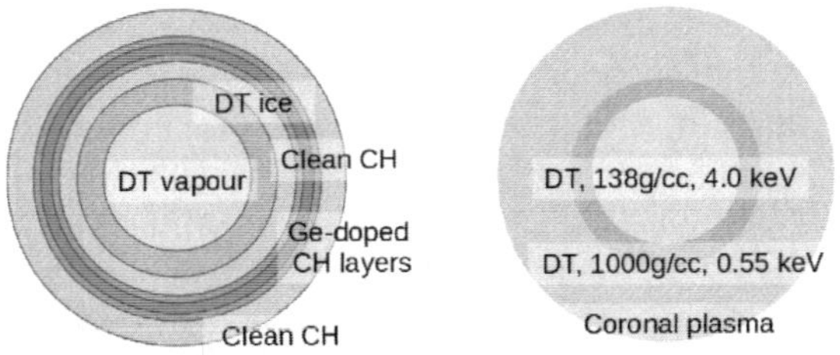


**Figure 1.** The left-hand panel shows a Rev5 Ignition capsule material configuration used for the NYM simulations. The right-hand panel shows the idealised state at stagnation.

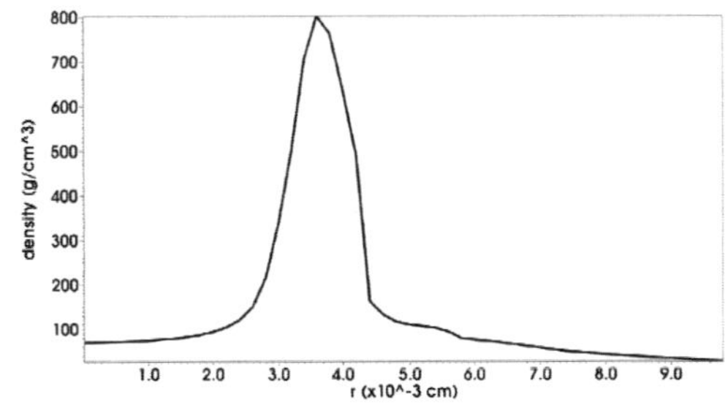


**Figure 2.** 1D density profile of the imploded capsule at around 20 ns extracted from NYM and used in electron heating simulations.

to 325 kms$^{-1}$, compared with current design mode requirements of 370 kms$^{-1}$. Assuming pressure $P_{\rm stag} \sim v^{2.8}$ this correlates to a 44% shortfall in required pressure.

The final hot-spot radius is about 35 $\mu$m in such an implosion. The main fuel shell has a density of about 1000 gcm$^{-3}$ and a thickness of about 8 $\mu$m so its $\rho r \simeq 0.8$ gcm$^{-2}$. The hot spot $\rho r = 0.3$ so the total fuel $\rho r = 1.1$. Here we consider the total fuel mass at the end of the implosion and assume the ablator (mostly CH) is expanded to low density and only the DT is left.

### *2.1. Heating the cold fuel*

The idealised situation at stagnation is an isobaric configuration of a dense shell surrounding hot fuel. Taking $\rho R_{\rm hs} = 0.3$ and $\rho R_{\rm tot} = 1.5$ gcm$^{-2}$ with $\rho_2 = 1000$ gcm$^{-3}$ and $T_1 = 4$ keV gives the configuration as shown in figure 1. We assume that both regions will be heated uniformly, leading to a pressure disparity. Hydrodynamic motion follows, with the main fuel further compressing the hotspot. Final conditions are again isobaric, with the new hotspot dimensions, pressure, densities and temperatures: $\{\rho'_1, \rho'_2, R'_1, R'_2, T'_1, T'_2\}$.

Assuming that mass is conserved seperately for hotspot and main fuel regions:

$$\begin{aligned} \rho_1 R_1^3 &= \rho'_1 R_1'^3 \,, \\ \rho_2 (R_2^3 - R_1^3) &= \rho'_2 (R_2'^3 - R_1'^3) \,. \end{aligned} \tag{1}$$

Pressure balance is $\rho'_1 T'_1 = \rho'_2 T'_2$ and energy balance is:

$$\begin{aligned} \left(\frac{3\Delta E}{4\pi}\right) + \rho_1 R_1^3 \epsilon_{\rm DT} T_1 + \rho_2 (R_2^3 - R_1^3) \epsilon_{\rm DT} T_2 = \\ \rho'_1 R_1'^3 \epsilon_{\rm DT} T'_1 + \rho'_2 (R_2'^3 - R_1'^3) \epsilon_{\rm DT} T'_2 \,. \end{aligned} \tag{2}$$

We make further simplifying assumptions: the hotspot temperature increase is negligible; and the the outer radius does not change significantly. Equation 2 then simplifies to:

$$\begin{aligned} (\rho'_1 - \rho_1) r_2^3 &= \left(\frac{3\Delta E}{4\pi \epsilon_{\rm DT} T_1}\right) \,, \\ \frac{\rho'_1}{\rho_1} &= 1 + \left(\frac{3\Delta E}{4\pi \epsilon_{\rm DT} T_1 \rho_1 r_2^3}\right) \,, \end{aligned} \tag{3}$$

yielding the pressure enhancement $\rho'_1/\rho_1$ for a given energy input $\Delta E$. To achieve a $\sim 40\%$ enhancement in $P$ requires $E \sim 5.5$ kJ and a temperature rise of 280 eV. NYM simulations indicate that artificial heating of the dense fuel layer by this amount generates significant yield, although the heating must be fast to prevent radiative losses cooling the shell.

### *2.2. Electron penetration and $T_{\rm hot}$*

Various formulae for electron range have been proposed. A possible spread of values [3] is 0.3 to 0.6 $T_{\rm hot}$ gcm$^{-2}$ where $T_{\rm hot}$ is in MeV. To penetrate the fuel from one side with $\rho r_{\rm tot} = 1.1$ gcm$^{-1}$ we need $T_{\rm hot} = 3.67$ MeV at the lower end of this range and $T_{\rm hot} = 1.83$ MeV at the upper. If $T_{\rm hot}$ is given by the common ponderomotive form [4] these values translate into irradiances of $1.61 \times 10^{20}$ Wcm$^{-2}$ and $4.01 \times 10^{19}$ Wcm$^{-2}$ respectively for $\lambda = 1.06$ $\mu$m.

If the laser energy to hot electron coupling efficiency is 0.3 and the hot electron coupling efficiency to the core is $f_c$ we need $8.17/f_c$ kJ per side of laser energy over a cross-sectional area of $5.8 \times 10^{-5}$ cm$^{-2}$. This means a pulse length of $0.875/f_c$ ps at the higher irradiance (shorter range) or $3.52/f_c$ ps at the lower irradiance (longer range).

Realistically, it appears we may be talking about a total short-pulse energy requirement of $\sim 50$ kJ but it should be noted that some allowances have been made for losses in this estimate.

## 3. Numerical simulation

To reduce the uncertainties in the analytic model we used numerical simulation to determine whether electron heating of the cold fuel by $\sim 300$ eV is viable at stagnation time. The density profile from the `NYM` 1D implosion calculation at stagnation time (figure 2) was used as an initial condition for investigating the electon source and transport physics associated with the short pulse laser.

The scale length of the coronal plasma is very much larger than is practical in a PIC code such as `EPOCH`, and the resulting electrons would almost certinly be at too high an energy and flood too large a volume. Instead we anticipate that renewed work on channelling (e.g. [5]) will provide sufficient control to allow the electron source to be located at a point much closer to the target and with a shorter plasma scale length. With this in mind, `EPOCH` simulations used a simplified plasma profile with a 20 $\mu$m scale length to represent the end of a channel. Figure 3 shows the electron density profile after 0.4 ps of illumination.

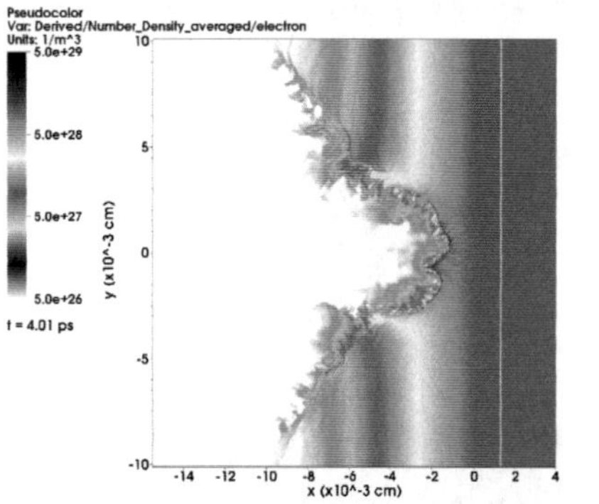


**Figure 3.** Results from `EPOCH` simulation of a 20 $\mu$m scale-length plasma to generate a source for heating calculations. The white line indicates the probe plane where particles were sampled.

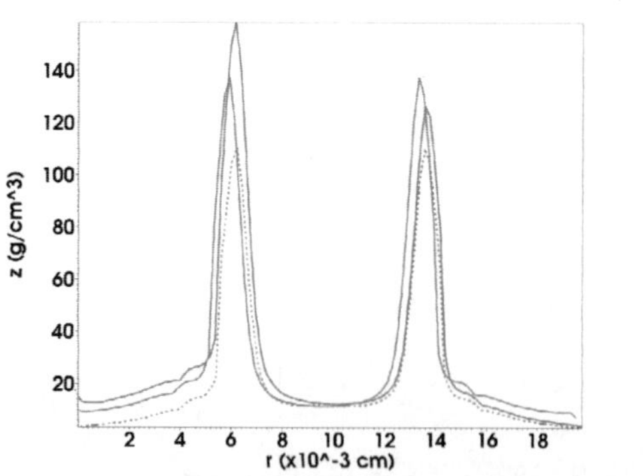


**Figure 4.** Pressure change profiles with an `EPOCH` source. Red lines are for a single heating beam, green for double heating. Continuous lines are in the $z$-axis, dashes are in $r$ to highlight the assymmetry.

To compute the energy deposition and heating, the PIC source was captured at a probe plane (shown in figure 3) and passed explicitly into the hybrid-electron transport code `THOR` using the approach described in [6] together with a radial density profile from the 1D `NYM` simulations.

Ideally, isotropic heating of the cold fuel by many short-pulse beams would provide a smooth heating profile, however this is probably impractical with indirect drive schemes on current

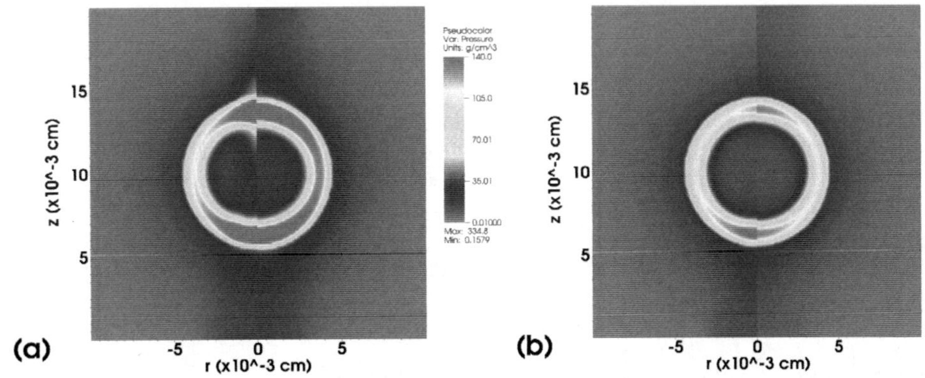


**Figure 5.** Comparison of one and two beam heating schemes with (a) an analytic ponderomtive source and (b) an EPOCH source. In each panel the left-hand reflection is with a single beam from $-z$ and the right-hand with twin beams from $\pm z$.

facilities. Instead, we consider here the use of one or two electron sources at the equator of the capsule in an indirect drive configuration.

A comparison of the pressure enhancements predicited in THOR calculations is shown in 5 comparing an analytic ponderomtive source (a) with an explicit PIC source (b) and the use of one or two heating beams. In case (a) the heating is significant, and with two beams is well above the 40% enhancement required. With a more realistic PIC source the heating is less pronounced beacuse the electron source diverges more and varies its pitch angle in time.

Pressure change profiles are shown in figure 4. The PIC source is less efficient than the ponderomtive source and there is a marked assymmetry that suggests more work is needed to understand the compatibility of the scheme with indirect drive configurations.

## 4. Conclusions

This initial study of the supplemental heating scheme suggests that it is practical, but that the major obstacle may be delivering the short pulse energy to the compressed fuel. Although channelling schemes under consideration for fast ignition show some potential this would be the next step in determining the viability of the scheme, and determining if practical electron sources can be generated that will heat with suitable efficiency. Further integrated modelling using THOR inline in a hydrodynamics code is also warrented to demonstrated whether the anisotropic electron heating will lead to significant capsule yield. Alternate capsule designs that are tailored to this late-time heating concept may also be interesting to pursue. It is also worth noing that the scheme could be used a near-ignition directly driven capsule with easier access for the short-pulse heating.

## References

[1] Lindl J D, Amendt P, Berger R L, Glendinning S G, Glenzer S H, Haan S W, Kauffman R L, Landen O L and Suter L J 2004 *Phys. Plasmas* **11** 339
[2] Tabak M, Hammer J, Glinsky M E, Kruer W L, Wilks S C, Woodworth J, Campbell E M, Perry M D and Mason R J 1994 *Phys. Plasmas* **1** 1626
[3] Deutsch C, Furukawa H, Mima K, Murakami M and Nishihara K 1996 *Phys. Rev. Lett.* **77** 2483
[4] Wilks S C, Kruer W L, Tabak M and Langdon A B 1992 *Phys. Rev. Lett.* **69** 1387
[5] Mironov V et al 2012 *Plas. Phys. Cont. Fus.* **54** 095008
[6] Sircombe N J, Hughes S J and Ramsay M G 2013 *New. J. Phys.* **15** 025025

Journal of Physics: Conference Series **688** (2016) 012035 doi:10.1088/1742-6596/688/1/012035

# Computational Study of Magnetic Field Amplification in Laser-Produced Shock Waves Relevant to Supernova Remnants

**Ayako Ishii**[1], **Shiori Kishita**[1], **Naofumi Ohnishi**[1], **Yasuhiro Kuramitsu**[2,3], **Taichi Morita**[3], **Youichi Sakawa**[3], **and Hideaki Takabe**[3]

[1]Department of Aerospace Engineering, Tohoku University, Sendai 980-8579, Japan
[2]Department of Physics, National Central University, Taoyuan 320, Taiwan
[3]Institute of Laser Engineering, Osaka University, Suita 565-0871, Japan

E-mail: ishii@rhd.mech.tohoku.ac.jp

**Abstract.** We have developed a two-dimensional magneto-radiation hydrodynamics code with an induction equation solver to analyze laboratory-astrophysics experiments relevant to a supernova remnant (SNR) environment. The computed magnetic field was amplified by a order of magnitude from the background level as expected for accelerating fields of cosmic rays in the SNR context. A part of the amplification of the magnetic field is caused by Richtmyer-Meshkov instability (RMI) that stretches a contact discontinuity. However, the maximum magnetic field is smaller than the theoretically predicted limit since the non-linear RMI cannot sufficiently grow and the field is amplified by the compression effect rather than the stretching in the present experimental condition.

## 1. Introduction

High energy cosmic rays are indispensable information to understand astrophysical phenomena and are emitted from various high-energy-density events in the universe. The supernova remnant (SNR) is a typical example of the acceleration field of such cosmic rays, in which strong shocks propagate in an inhomogeneous interstellar medium. Synchrotron X-ray emissions were observed in a shell of the SNR RX J1713.7-3946 decaying with a one-year timescale [1]. For justifying electron cooling of the synchrotron emissions in such a short timescale, strong magnetic fields are needed in a few-hundred-times greater level than the background one of $10^{-10}$ T. On the other hand, the background magnetic field may be amplified by accompanied turbulence [2]. Since there are density inhomogeneities of an interstellar medium in front of a supernova-driven shock wave, some eddies are formed by Richtmyer-Meshkov instability (RMI), extend a density interface, and then strengthen the magnetic field. In theoretical and numerical study, magnetic fields are amplified up to two orders of magnitude from those of the background by the RMI [3, 4]. Several laboratory-astrophysics experiments relevant to the SNR environment were performed by high-power lasers [5, 6]. However, recent experiments taking into account shock waves, magnetic fields, and density inhomogeneities could not obtain no clear indication of the field amplification. Numerical analysis is helpful to examine an evolution of the magnetic field in the laser-produced plasma flows and to design the feasible experiments. So, we newly developed a two-dimensional magneto-radiation hydrodynamics (MRHD) code including the induction equation, and simulated for the experiments of a counter flow from double plastic targets irradiated by high-power lasers.

## 2. Numerical method

We use a two-dimensional RHD code, RAICHO [7], whose basic equations are two-dimensional Euler equations including energy source terms and realistic EOS. The source terms are estimated for electron/ion thermal conduction, X-ray radiation transfer, and laser absorption in a laser-produced plasma. The radiative transfer equation is simultaneously solved by the MGFLD with non-LTE (CRE) opacity. Numerical fluxes are estimated by the AUSM-DV approximate Riemann solver with second-order spatial accuracy adopted by the MUSCL approach. Two temperatures for electrons and ions are taken into account by solving additional energy equations for a thermally non-equilibrium medium. The laser absorption of inverse-bremsstrahlung is calculated by the two-dimensional ray-tracing.

The evolution of the magnetic field is estimated by the standard magneto-hydrodynamics (MHD) fashion, so the MOC-CT method [8] was installed to the original code. Numerical simulations were conducted for the experiments in which a planar plastic target was irradiated by high-power laser beams and another target was ablated by the X-ray radiation from the laser-heated target (Fig. 1). Ambient gas is nitrogen whose initial density is $4.6 \times 10^{17}$ cm$^{-3}$. To numerically resolve these targets in a relatively wide computational region, we introduced unequally-spaced computational grids. In the experiments, several lasers were focused on a point with an offset from the target surface. With no offset, the lasers were focused on the target surface, so we modelled a single laser perpendicular to the surface in the corresponding simulation. On the other hand, since the lasers are split on the surface with any offsets, we input two lasers with a beam separation in the simulation, while the lasers are assumed to be parallel to each other. Note that the geometry of the numerical simulation is different from that of the experiment due to non-axisymmetric computation.

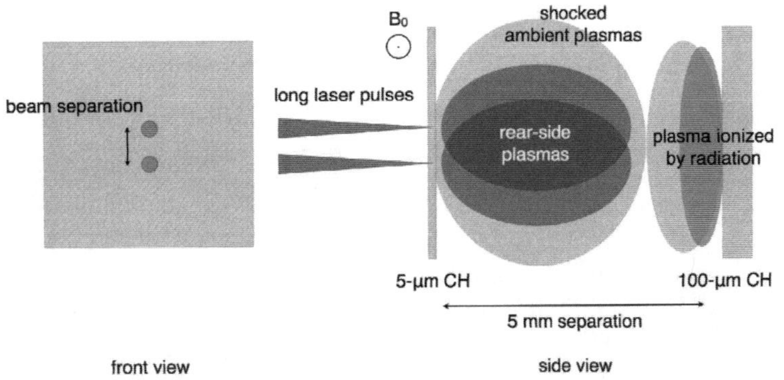


**Figure 1.** Simulation conditions.

## 3. Results

We have simulated the laser-produced plasma until 20 ns for the cases with 1-mm laser offset and without one, i.e. the corresponding beam separation. The laser intensities are $0.2 \times 10^{14}$ W/cm$^2$/beam in the case with laser offset and $0.3 \times 10^{14}$ W/cm$^2$/beam in the case without one. The spot radius is 100 $\mu$m and the pulse width is 1 ns in the both cases. One spot is separated from another by 500 $\mu$m in the case with offset. The experimentally-observed shadowgraphs at 20 ns after laser irradiation are shown in Fig. 2 (a) (with 1-mm offset) and in Fig. 3 (a) (without offset). Computed density and magnetic field are also shown in Fig. 2 (b)

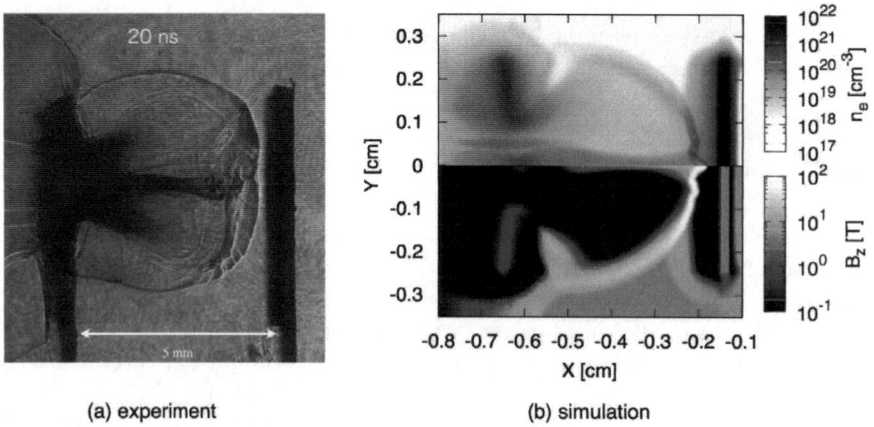


**Figure 2.** (a) Shadowgraph of the experiment and (b) contours of computed number density of electrons (top) and magnetic field (bottom) in the case with offset at 20 ns.

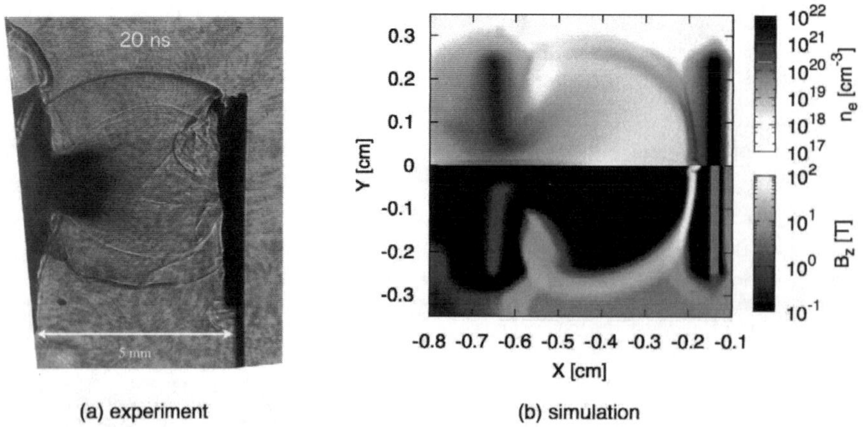


**Figure 3.** (a) Shadowgraph of the experiment and (b) contours of computed number density of electrons (top) and magnetic field (bottom) in the case without offset at 20 ns.

and Fig. 3 (b). Shock wave structures and contact discontinuities observed in the experiments are roughly reproduced in the simulations. Without offset, the laser-produced plasma simply expands from the rear side of the left target, and the growth of non-uniformity of the contact discontinuity is not so significant even after the interaction with the shock wave. On the other hand, with offset, a jet-like plasma is formed and collides with a counter flow generated by the radiation. The magnetic field becomes high at the area in which the laser-produced flow interferes with the radiation-produced one. The RMI seems to develop at the contact surface between the jet from the left target and the shock from the right target. The magnetic field profiles on the symmetry axis ($y = 0$) are shown in Fig. 4. The magnetic field is remarkably amplified near the contact surface. With offset, the magnetic field is globally large as compared with the case without offset because of the non-uniformity growing due to the jet propagation. The magnetic field is amplified to almost 40 times greater level than the initial condition of 3.5 T in our simulation. However, this value is one-order smaller than the theoretically predicted limit [4]. This may come from the fact that the magnetic field in this configuration is amplified by mainly the compression effect but the stretching of the contact surface because of the insufficient

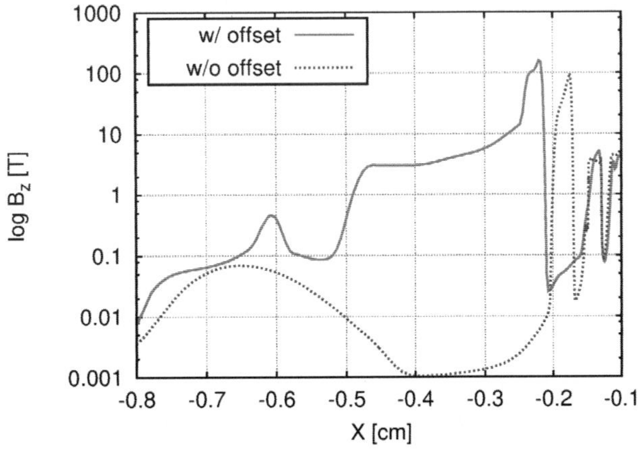


**Figure 4.** Magnetic field profiles in the symmetry axis at 20 ns.

evolution of the RMI. Here, the induction equation for the ideal MHD in a scalar form can be written as follows;

$$\frac{1}{2}\frac{\partial}{\partial t}|\boldsymbol{B}|^2 = -\boldsymbol{B}\cdot(\boldsymbol{v}\cdot\nabla)\,\boldsymbol{B} + \boldsymbol{B}\cdot(\boldsymbol{B}\cdot\nabla)\,\boldsymbol{v} - |\boldsymbol{B}|^2\nabla\cdot\boldsymbol{v}, \tag{1}$$

where the second and third terms of the right-hand side represent the stretching and the compression of the magnetic field, respectively [4]. The field amplification by the stretching term is expected in a highly nonlinear regime of the RMI. So, we should explore a suitable configuration of the experiment for observing the efficient field amplification using the developed code.

## 4. Conclusion

We have performed numerical analysis of the laser-plasma experiments relevant to the magnetic field amplification in the SNR. We developed two-dimensional MRHD code solving the induction equation by the MOC-CT method. The simulation results are in good agreement with the shadowgraphs of the experiments in terms of the jet-like plasma configuration and the shock wave propagation. Although the simulation results show that the magnetic field can be one-order amplified from the background level, the resultant amplification is lower than the theoretical limit yet. This is because that the growth of the non-linear RMI is not sufficient in the present condition. Thus, we should examine another condition resulting in more efficient field amplification and design a feasible experiment.

## References

[1] Uchiyama Y et al. 2007 *Nat. Lett.* **449** 576
[2] Giacalone J and Jokipii J R 2009 *Astrophys. J.* **701** 1865
[3] Inoue T, Yamazaki R and Inutsuka S 2009 *Astrophys. J.* **695** 825
[4] Sano T et al. 2012 *Astrophys. J.* **758** 126
[5] Kuramitsu Y et al. 2011 *Astrophys. Space Sci.* **336** 269
[6] Kuramitsu Y et al. 2012 *Phys. Rev. Lett.* **108** 195004
[7] Ohnishi N, Nishikino M and Sasaki A 2006 *J. Phys. IV France* **133** 1193
[8] Stone J M and Norman M L 1992 *Astrophys. J. Suppl. Ser.* **80** 791

# Target Monitoring and Plasma Diagnosis using 2$\omega$ probe beam for CANDY

**Katsuhiro Ishii[1], Yoshitaka Mori[1], Ryohei Hanayama[1], Suisei Nakayama[1], Shinichiro Okihara[1], Kazuhisa Fujita[1], Takashi Sekina[2], Nakahiro Sato[2], Takashi Kurita[2], Toshiyuki Kawashima[2], Hirofumi Kan[2], Osamu Komeda[3], Naoki Nakamura[3], Takuya Kondo[3], Manabu Fujine[3], Hirozumi Azuma[4], Tatsumi Hioki[4], Mitsutaka Kakeno[4], Tomoyoshi Motohiro[4], Yasuhiko Nishimura[1,5], Atsushi Sunahara[6], Yasuhiko Sentoku[7], Eisuke Miura[8], and Yoneyoshi Kitagawa[1]**

[1]The Graduate School for the Creation of New Photonics Industries, 1955-1 Kurematsu-cho, Nishi-ku, Hamamatsu, Shizuoka 431-1202 Japan
[2]Development Bureau, Hamamatsu Photonics K.K. 1820, Kurematsu-cho, Nishi-ku, Hamamatsu 431-1202 Japan
[3]Advanced Material Engineering Div., TOYOTA Motor Corporation, 1200, Mishuku, Susono, Shizuoka 410-1193 Japan
[4]TOYOTA Central Research and Development Laboratories, Inc., 41-1Yokomichi, Nagakute, Aichi 480-1192 Japan
[5]Toyota Technical Development Corp., 1-21 Imae, Hanamoto-cho, Toyota, Aichi 470-0334 Japan
[6]Institute for Laser Technology, 1-8-4 Utsubo-honmachi, Nishi-ku, Osaka 550-0004 Japan
[7]Department of Physics, University of Nevada, Reno 1664 N VIRGINIA ST Reno, NV 89557
[8]National Institute of Advanced Industrial Science and Technology, 1-1-1 Umezono, Tsukuba, Ibaraki 305-8568 Japan

E-mail: ishii@gpi.ac.jp

**Abstract.** We have developed the shadowgraph and interferometer with second-harmonic of heating pulses laser to observe target and plasma in highly-repetitive fusion reaction experiments. In the deuterated polystyrene ($(C_8D_8)_n$ double foil experiment, we confirm implosion plasma and plasma collision. In target injection experiment at a 1 Hz rate, we measure the position of the flying deuterated polystyrene beads at the moment of laser pulse illumination and observe the plasma generation by counter-illumination by 0.63 J, 800 nm, and 104 fs laser pulses.

## 1. Introduction

To realize repetitive-mode fast-ignition fusion mini-reactor which we named CANDY, we have conducted a series of fusion reaction experiments with highly-repetitive diode-pumped laser system. Our roadmap for realizing a nuclear fusion power has been shown in the literature [1]. First, we developed the DPSSL-pumped laser HAMA [2] to demonstrate repetitive fusion reaction [3] and fast heating inertial confinement fusion (ICF) in which both implosion and heating pulses are required [4]. In addition, we recently included a target injection system into HAMA and succeeded in the fusion reaction by irradiating the laser on the flying CD beads targets [5]. In highly-repetitive fusion reaction experiments require not only highly-repetitive laser system but also highly-repetitive measurement systems. To observe target and plasma, we have developed the shadowgraph and interferometer with second-harmonic of heating pulses laser, $2\omega$ probe system. In this paper, we show some experimental results of target monitoring and plasma diagnosis using $2\omega$ probe system. In the double foil experiment, implosion plasma and plasma collision are demonstrated. In target injection experiment at 1 Hz, the position of the flying target at the moment of implosion is measured and the generating plasma is observed.

## 2. $2\omega$ probe system

We have developed a fusion reaction experiments with highly-repetitive diode-pumped laser system [1-3]. An LD-pumped high-repetition-rate HAMA laser, where an LD-pumped Nd-solid state laser's [6] output is coupled to a Ti:sapphire laser, generate the amplified chirped 3.6 J-800 nm in the 0.4-ns pulse. It is divided into two beams, one for imploding and another for heating the imploded core. The imploding beam is divided into two beams (Long beam-1 and Long beam- 2) and is transported using different paths to the target chamber. The remaining beam is time-delayed and pulsecompressed to a 110-fs Gaussian beam by two pairs of gold-coated plane gratings. This beam is then divided into two heating beams (Short beam-1 and Short beam- 2). Short beam-1 is co-aligned to Long beam-1 and Short beam-2 to long beam-2, respectively. A cross-beam divider stage enables us to coalign the four beams to counter-illuminate both sides of a target.

A part of the pulse-compressed beam is converted to the second harmonics (400-nm wavelength) and used as probe beam. Figure 1 (a) and (b) show a schematic diagram of $2\omega$ probe system, $2\omega$ probe beam illuminating part and target imaging part. The collimated second harmonics with beam diameter of 5mm illuminate the target perpendicularly to the Short and Long beams. The path length of the second harmonics is adjusted so that the second harmonics and Short beam-1 -2 arrived simultaneously. Transmitted second harmonics is imaged on the ICCD camera via the interferometer [7]. Blocking one path of the interferometer, ICCD takes shadowgraphs.

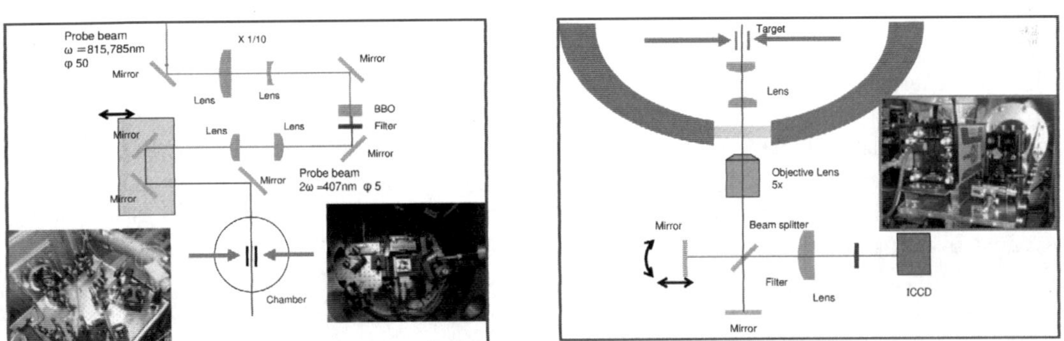


(a) (b)

**Figure 1.** A schematic diagram of $2\omega$ probe system.

## 3. Double foil experiment

The target consists of two parallel 2-mm square plain foils. The foil is made of deuterated polystyrene $((C_8D_8)_n$ and supported with stainless steel. The separation (gap) between the foils is adjustable by changing the thickness of stainless steel. Figure 2 show the shadowgraph of CD double foil before imploding and heating. We set the focal position of the beams at the center between the double foils. The focal spot size is 32 μm. The intensities are $5.9\pm0.5\times10^{13}$ and $1.9\pm0.3\times10^{17}$ [W/cm$^2$], respectively.

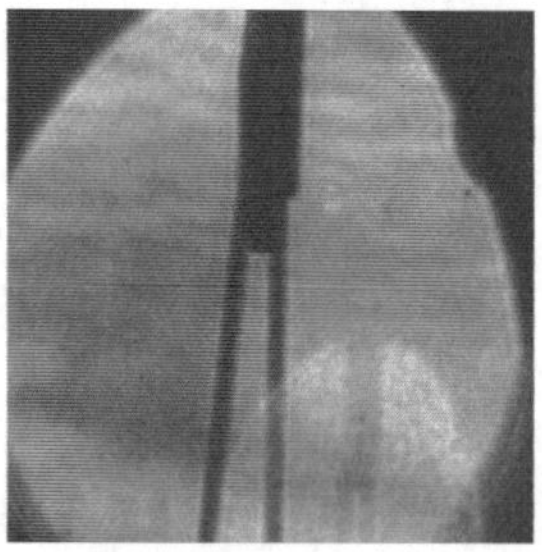

**Figure 2.** Shadowgraph of CD double foil before imploding and heating.

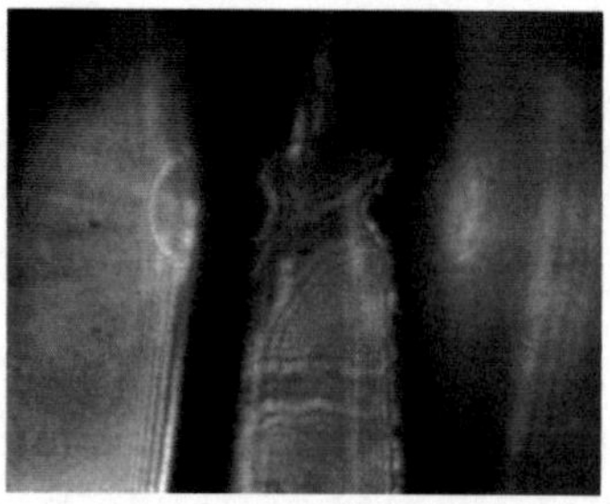

**Figure 3.** A snapshot 0.4 ns after the compression beam peak. The foil thickness is 50 μm and the gap separation is 300 μm.

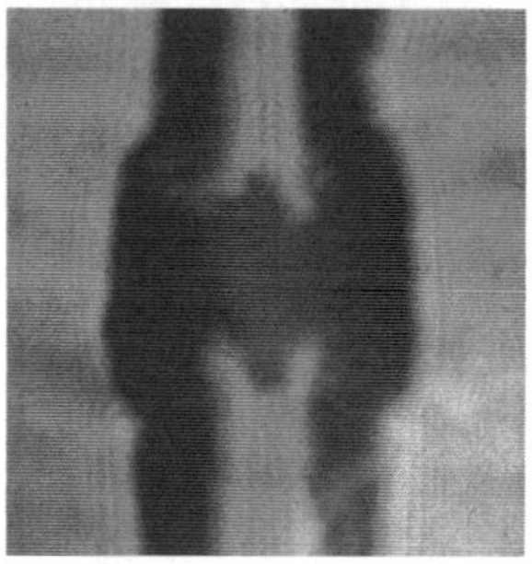

**Figure 4.** A snapshot 1.3 ns after the compression beam peak. The foil thickness is 11 μm and the gap separation is 100 μm.

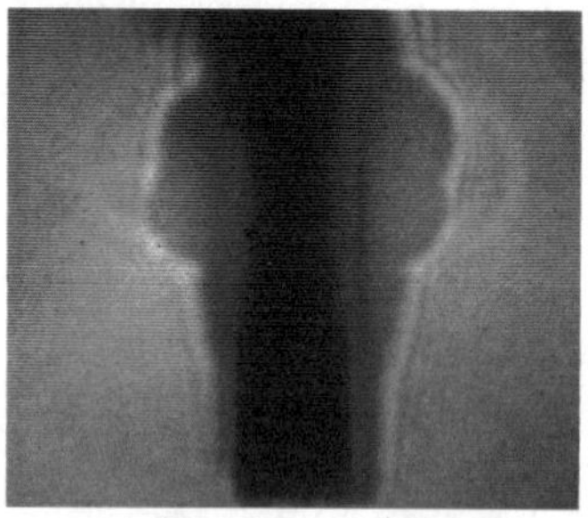

**Figure 5.** A snapshot of plasma generated when stainless steel with the thickness of 100 μm is illuminated by Long beam-1 and -2 with 0.1J, 800 nm, 0.4-ns pulse-width and Short beam-1 and -2 with 0.5J, 800 nm 110 fs pulse-width 1.3 ns after Long beams.

Figure 3 shows a snapshot 0.4 ns after the compression beam peak. The foil thickness is 50 μm and the gap separation is 300 μm. Long beam-1 and -2 with 0.1 J, and 800 nm, 0.4 ns pulse-width illuminate the target. Imploded plasma inside the foil is confirmed. The collision between two counter-injected plasmas is captured a 1.3 ns after the compression beam peak in Fig. 4. The foil thickness is 11 μm and the gap separation is 100 μm. It is demonstrated that the compressed core is formed. Figure 5 shows the plasma generated when stainless steel with the thickness of 100 μm is illuminated by Long beam-1 and -2 and Short beam-1 and -2 with 0.5J, 800 nm, and 110 fs pulse-width 1.3 ns after Long beams. It is taken at the moment of Short beams arrive at target. Small dips on the center of spreading plasma is observed. They result from the drilling by short pulse laser propagating in plasma.

## 4. Target injection experiment

In this section, a flying solid spherical deuterated polystyrene ($C_8D_8$ or CD) bead is illuminated by Short beam-1 and -2 with 0.63 J, 800 nm, and 104 fs pulse-width. CD beads have a diameter of 1.1 ± 0.1mm and a sphericity of 99%. Figure 6 shows the target injector, which is installed in the center of the vacuum chamber. A target loader stores more than 10,000 CD targets at a time. The targets stored inside the target loader freefall by gravity onto a 110-mm-diameter disk that rotates at 6 rpm. Each target on the disk is automatically conveyed to the exit hole and falls on a parabolic trajectory to the laser focal point of the HAMA laser 18 cm below the exit hole at 1 Hz. The hole is shaped as an ellipse with major and minor axes of 4.0 and 2.0 mm, respectively, and a depth of 8.0 mm. $2\omega$ probe system takes snapshot at the moment when Short beam-1 and -2 illuminate target.

Figure 7 shows snapshots of flying target. Dislocation of the target or bad timing of injection cause an image of the probe beam shown in Figure 7(a). When the target is inside the probe beam at the moment of Short beams illumination but is dislocated from Short beams, a shadow of the target is observed shown in Figure 7(b). When the Short beams hit the target, a shadow of the target and a irradiation from plasma occur. Shadows from small bumps on the surface of beads also observed, which comes from the scattering by plasma generating from illumination by Short beam-1 and -2. 2w probe system can measure the position of target at the moment of Short beams illumination and whether the Short beam-1 and -2 hit the target or not. When more than 600 targets is injected at a 1 Hz rate, approximately 43% of the targets were irradiated by the HAMA laser.

## 5. Summary

For highly-repetitive fusion reaction experiments, we have developed the shadowgraph and interferometer with second-harmonic of heating pulses laser, which can use target monitoring and plasma diagnosis. In the double foil experiment, implosion plasma and plasma collision are confirmed. In target injection experiment at a 1 Hz rate, we measure the position of the flying target at the moment of implosion and observe the plasma generation by counter-illumination by 0.63 J, 800 nm, and 104 fs laser pulses. The measurement system can be apply for flying shell target which is a candidate of laser fusion target, though flying beads target is used in this paper.

## References

[1] Kitagawa Y 2013 *et al. Plasma Fusion Res.* **8** 2404000
[2] Mori Y 2013 *et al. Nucl. Fusion* **53** 073011
[3] Kitagawa Y 2012 *et al. Phys. Rev. Lett.* **108** 155001
[4] Kitagawa Y 2011 *et al. Plasma Fusion Res.* **6** 1306006
[5] Komeda O 2013 *et al. Plasma Fusion Res.* **8** 1205020
[6] Sekine T 2013 *et al. Opt. Express* **21** 8393
[7] Mori Y 2012 *et al. Appl. Phys. Express* **5** 056401

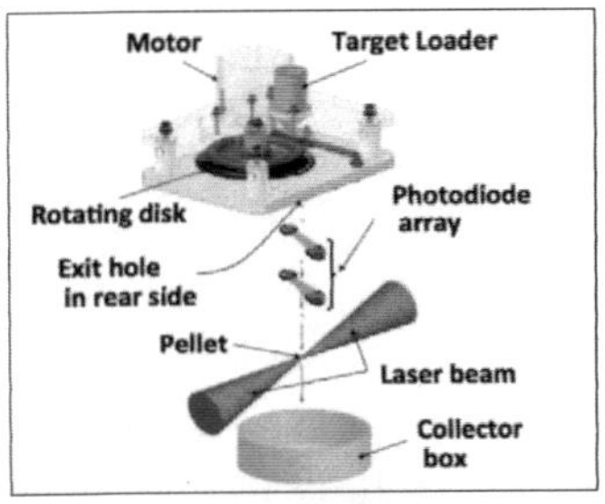


**Figure 6.** Target injection system.

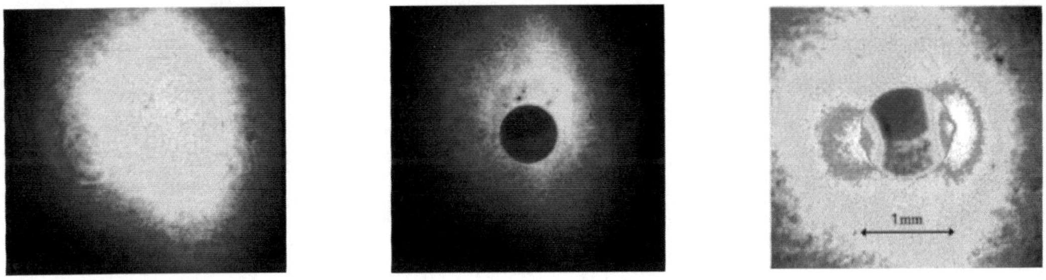


(a) (b) (c)

**Figure 7.** Snapshots of flying target.

# Cool-down performance of the new apparatus for fuel layering demonstrations of FIREX targets

**A Iwamoto[1], T Norimatsu[2], M Nakai[2], H Sakagami[1], H Shiraga[2] and H Azechi[2]**

[1] National Institute for Fusion Science, 322-6 Oroshi, Toki, Gifu 509-5292, Japan
[2] Institute of Laser Engineering, Osaka University, 2-6 Yamada-oka, Suita, Osaka 565-0871, Japan

E-mail: iwamoto.akifumi@LHD.nifs.ac.jp

**Abstract**. FIREX targets have been developed under two layering strategies: foam shell and cone guide laser heating methods. Basic studies have been conducted by the collaboration research between ILE and NIFS. Then the next stage requires the characterization of a layered solid fuel. The present system is at the disadvantage of optical observations. Therefore, a new apparatus is designed to solve it. Glass windows with a wide aperture are installed for an interferometer and a microscope. To isolate the vibration from a cryocooler, active vibration control units are equipped, and flexible thermal conductive links are utilized. Furthermore, a quick target exchange mechanism is applied to deal with different types of FIREX targets. A target holder is detachable from a main vacuum chamber. A metal gasket with not fixing bolts but a load of ~ thousand newtons on ensures GHe leak tightness for target cooling. Eventually, the design temperature of 10.00 K at a target container has been achieved. The cool-down performance indecates that the new apparatus provides a cryogenic environment for fuel layering demonstrations.

## 1. Introduction

Fast Ignition Realization EXperiment (FIREX) targets have been developed under the collaboration between the Institute of Laser Engineering (ILE), Osaka University and the National Institute for Fusion Science (NIFS). A typical target consists of a 500 μm plastic shell, a gold cone guide and a glass fill tube as shown in figure 1. To complete a FIREX target, ILE and NIFS have two fuel layering strategies: a foam shell method [1-4] and a cone guide heating technique for a polystyrene (PS) shell [5]. The final goal is to form a uniform solid DT or $D_2$ fuel layer with ~20 μm in thickness on the inner surface of the plastic shell. To date, preliminary studies have been conducted: the reduction of residual void spaces in a solid fuel layer within an aerogel material [4] and the layering demonstration in a 2 mm PS shell prototype target [5]. The formation of a 20 μm uniform solid fuel layer in the FIREX targets is a remaining issue.

The present apparatus provides a large multipurpose space for cryogenic experiments [6]. The lowest temperature is available at ~12 K, and vibrations from a refrigerator are successfully reduced to less than 10 μm at the target position. The mechanical insulation utilizing flexible thermal conductive links is valid to be compatible with cooling performance. Other ideas to control the vibrations are also employed. Eventually, it is possible to observe a target with several 10 μm resolution. Its limit mainly depends on dimensions of the apparatus. For the next step, higher resolution is demanded to characterize a layered solid fuel with 20 μm in thickness. Based on the cryogenic and the mechanical components developed for the vibration control, we built a new apparatus especially designed for the characterization of the solid fuel layer. Furthermore, it has another feature of a quick target exchange mechanism to deal with different types of FIREX targets. In this paper, its design and cool down performance are described.

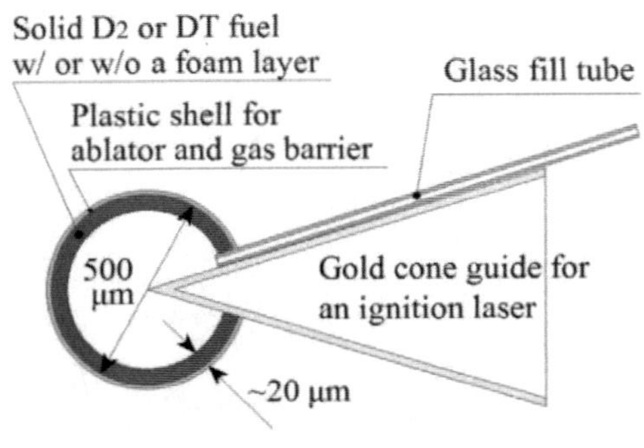


**Figure 1.** Typical FIREX target with a PS or a foam shell. All parts are glued by an epoxy adhesive with gas leak tightness.

## 2. Design of the new apparatus

The configuration of the new apparatus is shown in figure 2. The Gifford-McMahon (GM) cryocooler RDK-415D (Sumitomo Heavy Industries, Ltd.) is used to cool down the whole system. The refrigeration capacity is 1.5 W at 4.2 K. Its disadvantage is generating low frequency vibrations. Therefore, the cryocooler and the vacuum chamber with a target can are supported by individual structures with vibration control units. Flexible copper braids are also applied as thermal conductive links to mechanically insulate from each other. These should provide a low vibration environment. Four viewing windows are equipped for optical observations. To deal with different types of FIREX targets, the detachable target holder makes an easy and quick target exchange possible.

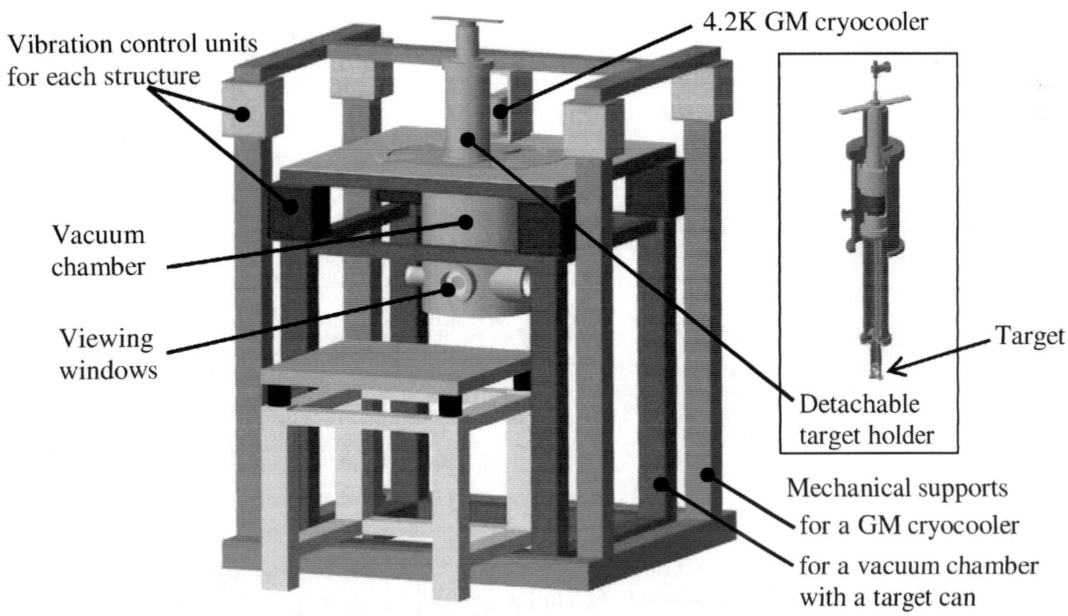


**Figure 2.** Schematic of the new apparatus.

Figures 3 and 4 show the detail of the target can and the cross section of the apparatus, respectively. A target is contained in a gold plated copper container with regulated temperature, which is connected to the second stage of the cryocooler via flexible thermal conductive links, and cooled by ambient gaseous helium (GHe). The target can is thermally insulated from room temperature with a thermal shield, which is connected to the first stage. For the characterization of a layered solid fuel, a target rotation mechanism on the z-axis is equipped, and the shortest distance between the target centre and room temperature is designed to be 80 mm to apply the digital microscope VHX-100 (Keyence Corporation) with the long-distance zoom lens VH-Z50L (Keyence Corporation). To realize an easy and quick target exchange, the metal gasket U-TIGHTSEAL® with an indium plated copper sheath (Usui Kokusai Sangyo Kaisha, Ltd.) is applied without fixing bolts. The target holder can apply a load of some thousand newtons (N) on the metal gasket to ensure GHe leak tightness in a cryogenic environment. A top handle can regulate the loading force. A gasket is valid for several time uses without its replacement.

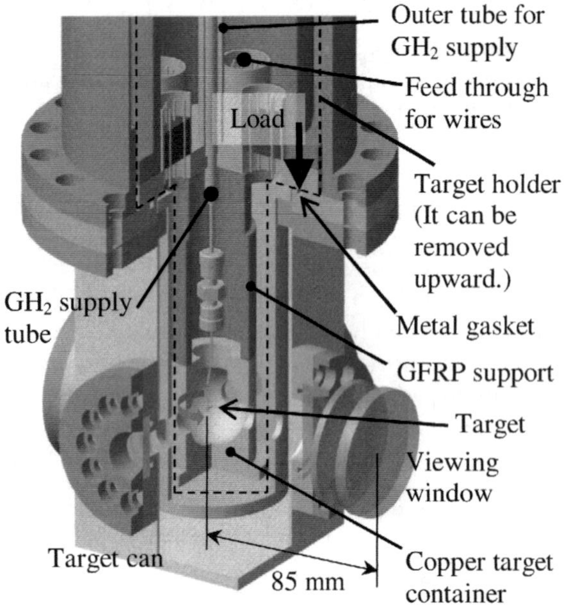


**Figure 3.** Detailed view in the target can. A heater and a temperature sensor are embedded in the copper container.

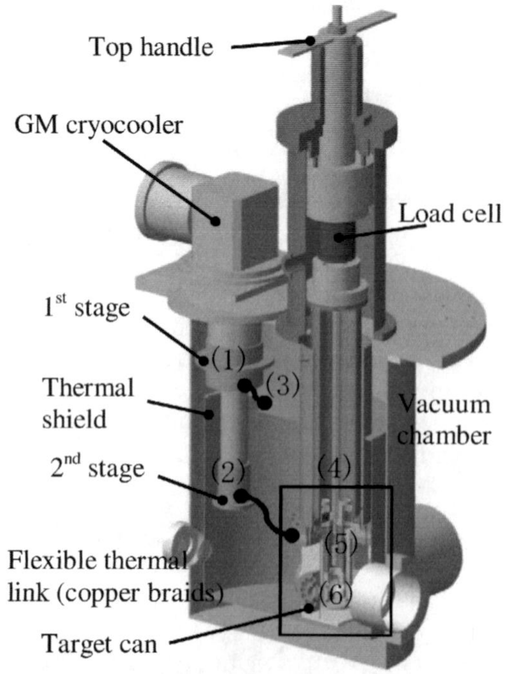


**Figure 4.** Cross section of the new apparatus. Temperature sensors are attached on the first stage (1), the second stage (2), the thermal shield (3), the outer tube for $GH_2$ supply (4), the GFRP support (5) and the copper target container (6). A load cell measures the pressure loaded to a metal gasket. The rectangular area corresponds to the detailed view of the target can shown in figure 3.

## 3. Cool-down performance

Cool-down performance was tested without a target. Temperature sensors are attached at the positions represented in figure 4. Cernox™ sensors (Lake shore Cryotronics, Inc.) are used except for a silicon diode on the second stage, which are calibrated within several 10 mK. After the evacuation of the vacuum chanber and the target can, the system was cooled down. Figure 5 represents the cool-down performamce. The target container is designed to be able to cool a target at less than 10.00 K. 10 hours were required to reach 10.00 K. Eventually, the lowest temperature was (1) 31.32 K, (2) 4.86 K, (3) 36.95 K, (4) 101.44 K or less, (5) 19.84 K and (6) 9.72 K at each point. The heat leak to the target can must be more than 1.5 W accoding to the specification of the cryocooler. Major heat leaks have been estimated to be via cryogenic supporting structures. The cool-down performance indecates that the new apparatus provides a cryogenic environment for fuel layering demonstrations.

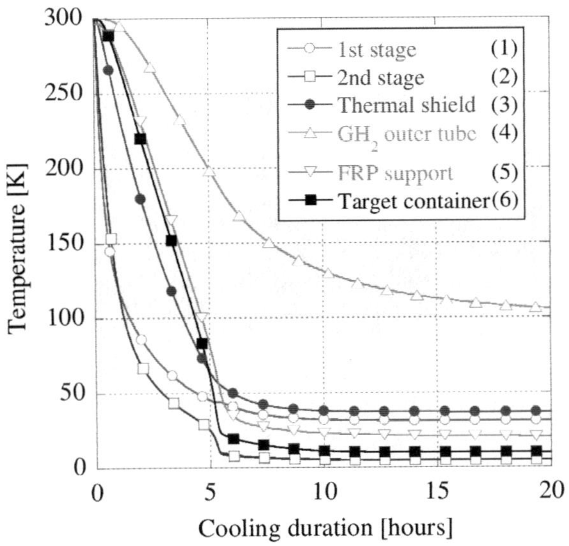


**Figure 5.** Cool-down curves.

## 4. Summary

The new apparatus, which has two features: the design for the optical observation by the digital microscope and the quick target exchange mechanism, was built for the characterization of a solid fuel formed in a FIREX target. Its cooling performance was tested without a target. 10 hours were required to cool it down to 10.00 K, where is the design temperature at the target container. The lowest temperature reaches 9.72 K. The cool-down performance indecates that the new apparatus provides a cryogenic environment for fuel layering demonstrations.

**Acknowledgements**
I would like to thank Mr. Y. Suzuki and Mr. H. Kobayashi at Taiyo Nippon Sanso Corporation for building the new apparatus.

**References**

[1] Sacks R A and Darling 1987 Direct drive cryogenic ICF capsules employing D-T wetted foam *Nucl. Fusion* **27** 447-52
[2] Norimatsu T, Kato Y, Nakai S and Kubo U 1989 Target fabrication for laser fusion research in Japan *J. Vac. Sci. Technol.* **A7** 3 1165-68
[3] Iwamoto A, *et al.* 2008 Study on a fuel layering sequence of the foam target for the FIREX project *J. Phys.: Conf. Ser.* **112** 032067
[4] Iwamoto A, Fujimura T, Nakai M, Norimatsu T, Sakagami H, Shiraga H and Azechi H 2013 *Nucl. Fusion* **53** 083009
[5] Iwamoto A, Fujimura T, Nakai M, Nagai K, Norimatsu T, Azechi H, Maekawa R and Sakagami H 2010 Study on possible fuel layering sequence for FIREX target *J. Phys.: Conf. Ser.* **244** 032039
[6] Iwamoto A, Maekawa R, Mito T, Okamoto M, Motojima O, Sugito S, Okada K, Nakai M, Norimatsu T and Nagai K 2006 Cool-down performance of the apparatus for the cryogenic target of the FIREX project *Fusion Eng. Des.* **81** 1647-52

Journal of Physics: Conference Series **688** (2016) 012038 doi:10.1088/1742-6596/688/1/012038

# Laser-matter interaction in cluster medium in the radiation dominated regime

**N. Iwata[1], Y. Kishimoto[1], F. Wu[1] and Y. Fukuda[2]**

[1] Graduate School of Energy Science, Kyoto University, Gokasho, Uji, Kyoto 611-0011, Japan
[2] Advanced Photon Research Center, JAEA, 8-1 Umemidai, Kizugawa, Kyoto 619-0215, Japan

**Abstract.** We study the interaction between laser and cluster medium in high intensity regime $10^{22-24}$ W/cm$^2$ using a particle based integral code (EPIC3D). By introducing four targets consisting of the same mass, i.e. same packing fraction, but having different internal structure, we investigate the effect of cluster on the acceleration dynamics comparing with that of thin film. In the radiation pressure dominated regime, the cluster medium exhibits a higher maximum energy than that achieved by the simple piston mechanism due to the additional accerelation by the Coulomb explosion. The optimum cluster radius for ion acceleration is found to exist depending on the laser power irradiated.

## 1. Introduction

The interaction between high power laser and matter has opened up various kinds of application such as high energy particle acceleration, generation of intense radiations from tera-heltz to EUV and x-ray, neutron production, etc [1]. Here, the state of material is a key ingredient which determines the characteristics of the interaction, and has to be chosen properly according to the purpose. For instance, besides solid and gas, cluster is interested, which exhibit prominent features in the interaction, i.e., the existence of the linear cluster mode, higher nonlinear absorption and particle acceleration due to the Coulomb explosion [2, 3, 4]. Recently, high energy ion acceleration has been realized in the interaction between a cluster medium and high intensity laser [5]. These phenomena have so far been investigated using laser intensities up to around $10^{21}$ W/cm$^2$.

Here, we investigate the interaction between laser and cluster medium extending the intensity higher than $10^{22}$ W/cm$^2$ up to $10^{24}$ W/cm$^2$. In this regime, ions become relativistic being accerelated by the radiation pressure incorpolating with the Coulomb explosion dynamics. Using a fully-relativistic electromagnetic particle-in-cell (PIC) code (EPIC3D), we study the interaction between laser and targets consisting of the same mass, i.e. same packing fraction, but having different cluster radius, by comparing them with thin film.

## 2. Simulation condition

In analyzing the laser-cluster interaction in the radiation dominated regime, we carry out a two-dimensional PIC simulation. We employ the periodic and outgoing boundary conditions in $x$ and $y$ directions with the size of $L_x = 128$ and $L_y = 2048$ in normalized unit, which correspond to $l_x = 1.28$ $\mu$m and $l_y = 20.48$ $\mu$m, respectively. The mesh number in $x$ and $y$ directions are $N_x = 128$ and $N_y = 2048$. A laser pulse with the wavelength $\lambda_L = 820$ nm excited by the antenna located at $y = 0.02$ $\mu$m propagates in the $y$-direction with linear polarization in the $x$ direction. The laser field is uniform in the transverse direction while the Gaussian profile in time with the dulation $\tau = 40$ fs (FWHM) is assumed. Here, we

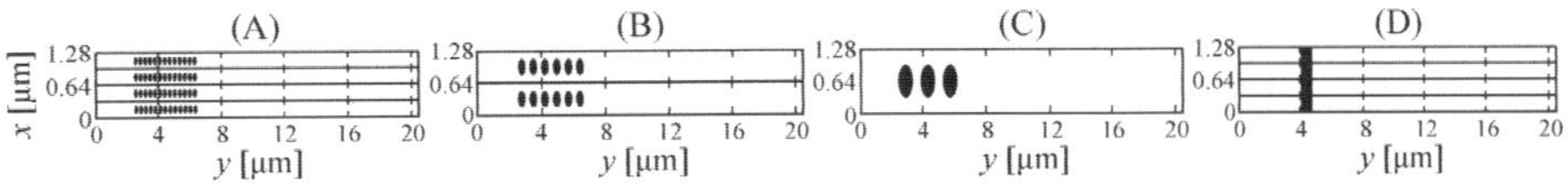


**Figure 1.** Initial plasma distributions in the EPIC simulation for cases (A) multi cluster medium with cluster radius $r_{cl}$ = 80nm, (B) $r_{cl}$ = 160nm, and (C) $r_{cl}$ = 320nm, and (D) solid thin film.

consider five cases of laser amplitude, i.e., $a_0$ =50, 200, 400, 600 and 800, where laser intensity ranges from $I = 5.1 \times 10^{21}\mathrm{W/cm^2}$ ($a_0$ = 50) to $1.3 \times 10^{24}\mathrm{W/cm^2}$ ($a_0$ = 800). The corresponding transverse excursion lengths of electron and carbon ion with the charge state $Z = 6$, $\xi_e$ and $\xi_i$, are found to range from $\xi_e = 6.5$ $\mu$m and $\xi_i = 0.3$ nm for $a_0 = 50$ to $\xi_e = 104$ $\mu$m and $\xi_i = 4.7$ nm for $a_0 = 800$.

In this study, we introduce three cluster targets consisting of the same mass, i.e. same packing fraction, but having different cluster radius as shown in Figs. 1 (A), (B) and (C). We also consider a solid thin film as shown in Fig. 1 (D) for comparison. Here, we model the cluster by a fully ionized uniform density plasma column occupying an area of radius $r_{cl}$ =80, 160 and 320 nm for cases (A), (B) and (C), respectively, in the $x$-$y$ plane. The species of the cluster is assumed to be solid carbon ($Z$ = 6) whose density is that of the diamond, i.e., $n_{cl}^{(i)} = 1.76 \times 10^{23}$ cm$^{-3}$. The electron density of the cluster satisfies $n_{cl}^{(e)} = Zn_{cl}^{(i)}$ and $n_{cl}^{(e)}/n_c = 637.4$, where $n_c$ is the cutoff density defined by $n_c \equiv m_e\omega_L^2/(4\pi e^2)$. The skin depth inside of the cluster for fully ionized state is $\delta_e = 5.17$ nm. Such clusters are regularly distributed in the region $2.56 \leq y \leq 6.40$ $\mu$m. The packing fraction defined by $f = N_{cl}\pi r_{cl}^2/S$ is given by $f = 0.21$ for all the cases (A)-(C), where $S = 3.20 \times 3.84$ $\mu$m$^2$ is the area occupied and $N_{cl}$ is the number of clusters in the area $S$. Then, the average density of electron $n_{av}^{(e)}$ in the area $S$ is given by $n_{av}^{(e)}/n_c = 130.4$ ($n_{av}^{(e)} = 2.16 \times 10^{23}$ cm$^{-3}$). Hence, the cluster medium is overdense in average if the relativistic effect is not taken into account. In case (D), i.e., the case of thin film, we also assume the same solid carbon plasma with $Z = 6$ that is uniformlly distributed in the region $4.09 \leq y \leq 4.87$ $\mu$m. Here, we set the film thickness $l_{film}$ = 785 nm so that the total mass included in the medium is same as that in the cluster medium. The relative relation among the electron and ion excursion lengths, incident laser wavelength, film thickness, cluster radius, and skin depth is given by $\xi_e > \lambda_L > l_{film} > r_{cl} \gg \delta_e \sim \xi_i$ in all the situations considered in this study.

## 3. Laser-matter interactions in multi cluster medium and solid thin film

First, we investigate the interactions for the cluster medium (B) and thin film (D) in Fig. 1 in the case of two laser intensities, $a_0$ = 200 and 800. The time histories of electron and ion energies, field energy and total energy in the system, and also spatial profiles of electromagnetic field $E_x$ and ion charge density normalized by $en_c$ at $t$ = 80 fsec are shown for the cluster medium (B) in the case of $a_0$ = 200 (Figs. 2 (B1) and (B2)) and of $a_0$ = 800 (Figs. 2 (B3) and (B4)), respectively. The corresponding figures for the thin film (D) are shown in Figs. 2 (D1) and (D2) for $a_0$ = 200 and in Figs. 2 (D3) and (D4) for $a_0$ = 800, respectively. The initial density distribution is also shown by dotted line in Figs. 2 (B2) and (B4) where six clusters along the $y$-axis can be seen while single thin film in Figs. 2 (D2) and (D4). For comparison, the time history of field energy in vacuum and the corresponding spatial profile of $E_x$ at $t$ = 80 fsec are shown for $a_0$ = 200 in Figs. 2 (I) and (II), where laser front emitted from antenna reaches to the right-hand side boundary and escape from the system at $t \sim 100$ fsec. At first, we study cases of the thin film (D) and cluster medium (B) for $a_0$ = 200. In the case of thin film (D), after the laser hits the target as seen in Fig. 2 (D1), the electron energy increases initially and ion energy does subsequently. Then, the electron energy saturates and decreases while the ion energy keeps increasing gradually, suggesting that ions are accelerated by the target normal sheath acceleration (TNSA) at the rear surface. This feature can be seen in the ion density distribution in Fig. 2 (D2) at $t$ = 80 fsec where the large amount of ions are pushed in the forward direction. As found from the decrease of the total

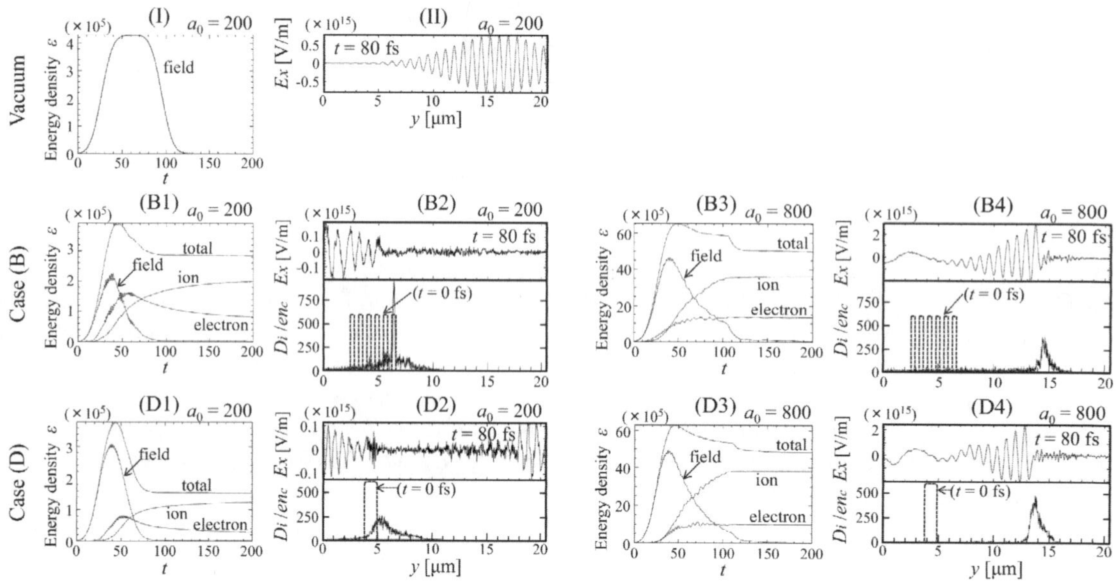


**Figure 2.** Time history of particles, field and total energies, and spatial profiles of electromagnetic field $E_x$ and normalized ion density at $t$ = 80 fsec for $a_0$ = 200 and 800 in the cases of vacuum ((I) and (II)), cluster medium (B) ((B1)-(B4)), and thin film (D) ((D1)-(D4)).

energy at $t \sim 45$ fsec from the maximum value, approximately half of the incident laser energy is found to be reflected while partially transmitted due to the relativistic effect as seen in Fig. 2 (D2). On the other hand, in the case of cluster medium (B), the dynamics is qualitatively similar whereas energy partition is found to be significantly different. Firstly, the energy absorption by electron and then the conversion to ion energy reach almost double, which is found from the fact that the decrease of the total energy from the maximum value at $t \sim 45$ fsec is small compared with that observed in Fig. 2 (D1). As seen in Fig. 2 (B2), the initial discrete cluster distributions are disintegrated except that of the most rear side. Here, more ions are found to be pushed not only in forward direction but also backward direction. This is due to the fact that the cluster Coulomb explosion takes place in both directions though forward direction is stronger due to the radiation pressure force.

Next, we consider the same cases (D) and (B) but for $a_0$ = 800. The case of the thin film is shown in Fig. 2 (D3) and (D4). In this case, after the laser hits the target, electron and ion energy simultaneously increases as seen in Fig. 2 (D3). Interestingly, even after the electron energy saturates, ion energy keeps increasing to a certain level, which is balanced with the decrease of the laser field energy. Namely, it is found that the laser field energy is directly transferred to that of ions. As found in Fig. 2 (D4), this corresponds to the situation that the laser pulse pushes the whole thin film consisting of a bunch of ions, which is referred to as laser piston by radiation pressure. The Doppler shifted long wavelength reflected laser light can also be seen. The ion energy distribution function at $t$ = 200 fsec is shown in (D) in Fig. 3 (II). A quasi-monoenergetic component exhibiting a energy hump around $\epsilon_i$ = 9 GeV can be seen. The maximum ion energy, which is approximately 12 GeV, is also indicated in (D) in Fig. 3 (I) together with other laser amplitudes including the case of $a_0$ = 200 which is around 3 GeV. On the other hand, in the case of the cluster medium (B), the dynamics is qualitatively similar to those in Figs. 2 (D3) and (D4). Namely, initial cluster distribution is disintegrated and pushed forward by the radiation pressure. The ion energy in Fig. 2 (B3) is slightly smaller than that in case (B) whereas the ion bunch is preceded compared with that observed in Fig. 2 (D4) at $t$ = 80 fsec. It is interesting to note in Fig. 3 (II) that the quasi-monoenergetic component locates around 10 GeV which is slightly larger than the case (D) while

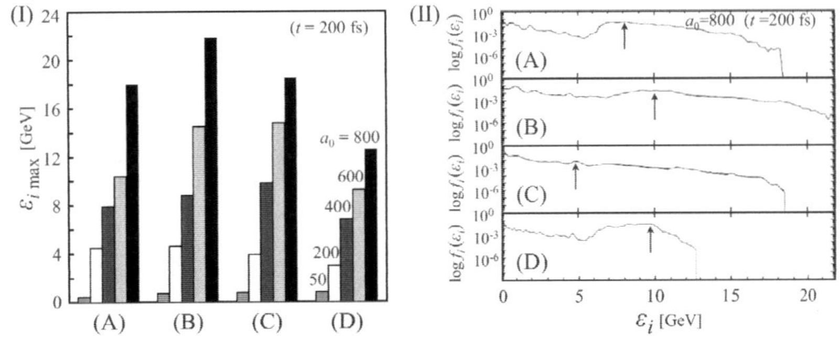


**Figure 3.** (I) Maximum ion energy obtained in each medium at $t$ = 200 fs for $a_0$ =50-800. (II) Energy distribution of ions in each medium for $a_0$ = 800 at $t$ = 200 fs.

the maximum ion energy reaches to approximately 22 GeV. This feature is also shown in (B) in Fig. 3 (I). This is found to result from the acceleration due to the Coulomb explosion of clusters which is added to that of the laser piston.

## 4. Dependence of ion energy on the internal structure of targets

Here, we summarize the ion energy distribution for $a_0$ = 800 and the maximum ion energy achieved in the interaction for different laser intensities including the case of (A) $r_{cl}$ = 80 and (C) 320 nm in Fig. 1 in addition to those of $r_{cl}$ = 160 nm and thin film. Note that the energy of the ion bunch due to the laser piston is estimated as (A) 8 GeV, (B) 10 GeV, (C) 5 GeV and (D) 9 GeV per ion. It is found that the maximum ion energy achieved in the cluster media (A), (B) and (C) leads to higher values than that in the solid thin film (D) for for $a_0 \geq 200$. This tendency can be explained from the Coulomb explosion of clusters contained in the medium. To estimate the effect of the Coulomb explosion, here we evaluate the maximum Coulomb potential that can be achieved by single cluster. We assume that all the electrons are excluded from the cluster to infinity and only ions remain with a spatially uniform distribution. In such a case, the potential energy outside of the cluster ($r \geq r_{cl}$) is given by $\phi(r) = 4\pi Zen_i r_{cl}^3/r$ [4, 6]. Then, ions at the cluster surface $r = r_{cl}$ expand to infinity due to the Coulomb repulsion force and gain kinetic energy of $\epsilon_i = Ze\phi(r_{cl})$ which is proportional to $r_{cl}^2$. Therefore, we can estimate the maximum ion energy as $\epsilon_{i\ \mathbf{max}} = 4\pi Z^2 e^2 n_i r_{cl}^2$ whose values are obtained for radii (A) 80 nm, (B) 160 nm and (C) 320 nm as 0.24, 0.98 and 3.92 GeV, respectively. In Fig.3 (I), the increases of $\epsilon_{i\ \mathbf{max}}$ in cluster medium compared with the thin film are 0.9-9.0 GeV in the case of $a_0$ = 800, which are in the same order as the maximum Coulomb potential of cluster calculated in the above. Thus, the large maximum ion energy achieved in the cluster media can be regarded as a result of the Coulomb explosion inside of the media. In other words, the higher internal free energy of the cluster media is used to accelerate the ions to higher energy. It is interesting to note that the relation between the maximum ion energy and cluster radius is different depending on $a_0$. Namely, in the case of $a_0$ = 200, medium with smaller cluster radius, e.g. (A) and (B), exhibit larger $\epsilon_{i\mathbf{max}}$ than case (C). In contrast, in the case of $a_0$ =400 and 600, larger $\epsilon_{i\ \mathbf{max}}$ is achieved in the medium with larger cluster radius. Finally, in the case of $a_0$ = 800, the case (B) shows the largest $\epsilon_{i\mathbf{max}}$ compared with the other cases. The details will be studied in future work.

## 5. Conclusion

Based on the PIC simulation, we studied the interaction between laser field in the regime of $a_0$ = 50-800 and targets consisting of same mass but having different internal structure, i.e., cluster media with different cluster radii and solid thin film. Comparison of ion energies achieved in the interactions with different targets shows the effects of the Coulomb explosion inside of the cluster media. The results indicate that the internal structure, which corresponds to the free energy of the target medium, is important in determining the interaction dynamics and the resulting ion accerelation. In this study, we have not included the radiation damping effect, which will be important in discussing intense radiation emissions from cluster medium. The ion maximum energy shown here is obtained in an ideal one-dimensional situation where transverse expansions of laser field and target are not taken into account. These issues are devoted to future work.

## References

[1] G. A. Mourou, T. Tajima and S. V. Bulanov, Rev. Mod. Phys. **78**, 309 (2006)
[2] T. Tajima, Y. Kishimoto, and M. C. Downer, Phys. Plasmas **6**, 3759 (1999)
[3] T. Ditmire et al., Nature **398**, 489 (1999)
[4] Y. Kishimoto, T. Masaki and T. Tajima, Phys. Plasmas **9**, 589 (2002)
[5] Y. Fukuda et al., Phys. Rev. Lett. **103**, 165002 (2009)
[6] K. Nishihara, H. Amitani, M. Murakami, S. V. Bulanov and T. Zh. Esirkepov, Nucl. Instrum. Methods Phys. Res. A **464**, 98 (2001)

Journal of Physics: Conference Series **688** (2016) 012039 doi:10.1088/1742-6596/688/1/012039

# X-ray emission spectroscopy of well-characterised non-LTE plasmas

**A C Bourgaux[1], S Bastiani-Ceccotti[1], P Audebert[1], J R Marquès[1], L Vassura[1], T Vinci[1], F Dorchies[2], P M Leguay[2], H K Chung[3], C Bowen[4], V Dervieux[4], P Renaudin[4], V Silvert[4] and S Jacquemot[1,4]**

[1] LULI, Ecole Polytechnique, F-91128 Palaiseau cedex, France
[2] CELIA, 351 cours de la libération, F-33405 Talence cedex, France
[3] IAEA NDS, Vienna International Center, A-1400 Vienna, Austria
[4] CEA DAM Ile de France, Bruyères le Châtel, F-91297 Arpajon, France

**Abstract.** This paper will present an experimental platform developed on LULI2000 to measure x-ray emission of non-LTE plasmas in well-defined hydrodynamic conditions thanks to implementation of a whole set of diagnostics, including time-resolved electronic and ionic Thomson scattering and self-optical pyrometry. K-, L- and M-shell spectra will be presented and the methodology, that has been developed to analyze them, discussed.

## 1. Introduction

Accurate calculations of multi-charged ion dynamics and radiative properties of non-LTE plasmas are of paramount importance as they impact simulations, and thus understanding, of a large variety of laboratory phenomena, as those encountered in indirectly-driven inertial confinement fusion or those involved in secondary radiation sources. Atomic kinetics modelling has of course made significant progress this last decade for mid to high atomic numbers (Z), with the development of hybrid computational methods which mix detailed and statistical approaches; however, relevant experiments are still required to benchmark them, as pointed out in conclusions of the $7^{th}$ NLTE code comparison recently published [1,2]. With this in prospect, and relying on the team's expertise, a dedicated experimental platform was recently set up and progressively improved on the LULI2000 laser facility [3-5]. It aims at recording x-ray spectral emission of laser-produced plasmas whose hydrodynamic parameters (namely the electronic density and temperature) are simultaneously and independently measured to provide valuable spectroscopic data.

## 2. Experimental set-up

For the most recent campaigns described here, the emitting plasmas were created through irradiation of solid targets by frequency-doubled neodymium laser "pump" beams ($\lambda$=0.53$\mu$m) focused with hybrid phase plates (HPP) to $\Phi$400$\mu$m focal spots. Laser intensity varied from 5 $10^{13}$ to 2 $10^{14}$ W/cm$^2$ by using one or two laser beams (then forming a 10° angle). Pulse duration was set to 1.5ns, with a roughly flat-top temporal profile. The targets were in fact small dots (figure 1) of aluminium, potassium bromide, niobium and tungsten deposited on a plastic substrate to allow time- and space-resolved self-optical pyrometry (SOP) observing visible signature of rear-side shock break-out and subsequent propagation. A very thin layer of gold (5$\mu$m) inserted into the plastic substrate protected this diagnostic from harder radiation. Lateral density gradients were expected to be kept low in the

produced plasmas, and thus the plasma expansion kept mono-dimensional, thanks to the small diameter of the dot. KeV x-ray spectra were recorded, with a spectral resolution of 2 eV, by a conical Potassium Acit Phthalate (KAP) crystal coupled with either an image plate (IP), for time-integrated images, or a streak camera, for time resolution. The spectral ranges covered allow observing the Al 2-1 K-shell and Br 3-2 L-shell transitions (1.6-1.8 keV) as well as the Nb 3d-2p L-shell and W 5f-3d M-shell transitions (2.55-2.9 keV).
If aluminium was chosen for the simplicity of its K-shell spectrum, allowing validation of the experimental set-up and diagnostics, the choice of bromine and tungsten was motivated by the use of these two elements in fusion research, the first one as ablator dopant, the second one as potential reactor's first wall.

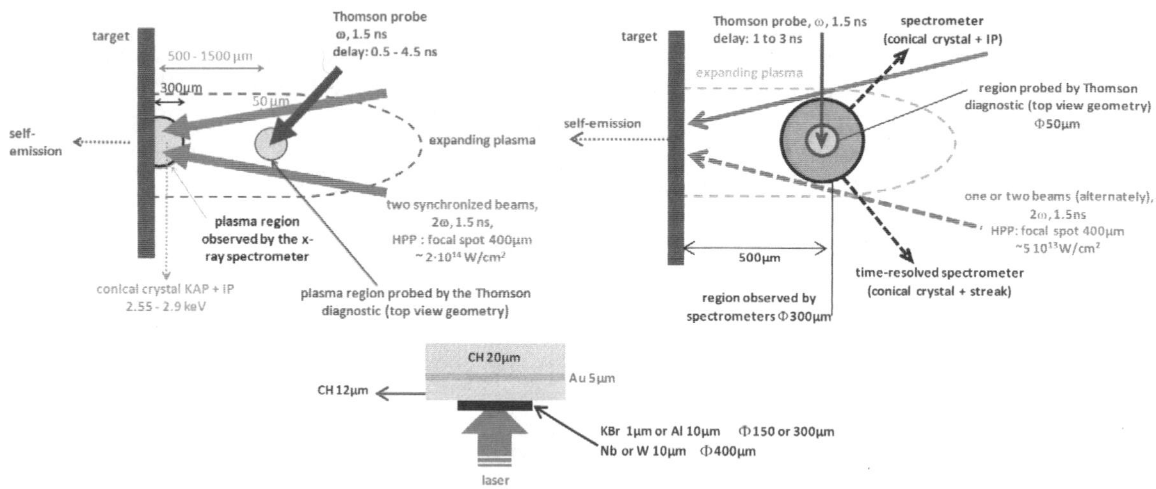


**Figure 1.** (top) Experimental set-ups used for Nb and W (left) and for Al and KBr (right) and (bottom) sketch of the dot targets.

An unconverted low-energy auxiliary laser beam, with the same temporal profile than the "pump" beams but delayed by a few ns, was used as a Thomson probe. The signal, backscattered on the ionic acoustic waves and the electronic plasma waves, was relayed to two spectrometers coupled with streak cameras, the first one at high spectral resolution to observe the ionic component and the second one to record the full electronic component. The probed volume was located at a distance from the target surface varying between +500 and +1500µm, with an extent of ~50µm. Note that the volume probed by Thomson scattering for the Nb-W experiment is disconnected from the emissive region, and probably colder as farther in the plasma corona, while, for the Al-KBr experiment, it is probably hotter, spatial integration of the spectrometers being done on a wider volume (see figure 1).

## 3. Data analysis

The first step in the analysis process is to determine the exact hydrodynamic conditions in the emissive region. As the laser energy absorbed by the target is not precisely known (for instance due to parametric instabilities), the laser intensity on target ($I_{las}$) is first determined thanks to a radiative-hydrodynamic mono-dimensional (1D) simulation by fitting the measured speed of the shock breaking through the rear surface of the target, as recorded by SOP (figure 2). Once $I_{las}$ fixed, the temporal evolution of the electronic density ($n_e$) and temperature ($T_e$) in the volume probed Thomson scattering are extracted from the simulation and compared to the values directly derived from the diagnostic. The Doppler shift of the ionic satellites allows evaluating the plasma speed (figure 3). The analysis for one Nb shot (#47) is shown on figure 4.
Despite the adjustment of the laser intensity made for MULTI, the electronic density is slightly

overestimated in one dimension. A better agreement is found with DUED 2D, pointing out the presence of some bi-dimensional effects despite the dot target design. These 2D effects are probably also responsible of the clear over-estimation of the product of the mean ionization charge (Z*) by the electronic temperature in MULTI, when compared with DUED. The steeper temporal profile of the experimental plasma speed evidences an incorrect treatment of the plasma expansion in the codes, with a clear underestimation of the plasma slowing-down, probably related to under-estimated radiative loss rates in the corona. If ionization is assumed to be frozen (the Thomson diagnostic probing here after the peak of the laser pulse), thus under-estimation is also seen on the temporal evolution of $Z^*T_e$, with an experimental temperature decreasing much more rapidly than the numerical one (-3eV/ns instead of -1eV/ns).

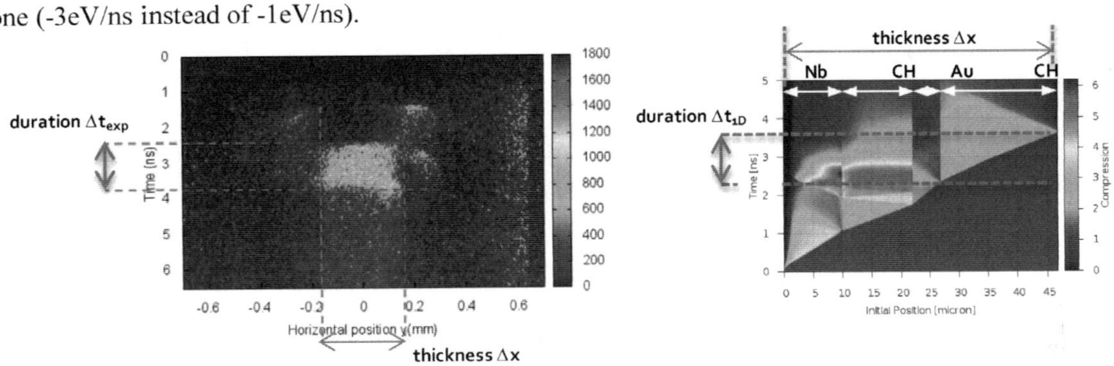


**Figure 2.** Rear-side self-emission of a Nb dot target (shot #47 - left) and (right) simulated compression ratios showing the propagation of the shock front from the front surface of the dot (x=+0) to the rear side; to restitute the shock speed (*i.e.* to have $\Delta t_{exp} \approx \Delta t_{1D}$), the laser intensity has to be decreased - for this specific shot - from 1.6 to 0.5 $10^{14}$ W/cm$^2$ in MULTI 1D [6].

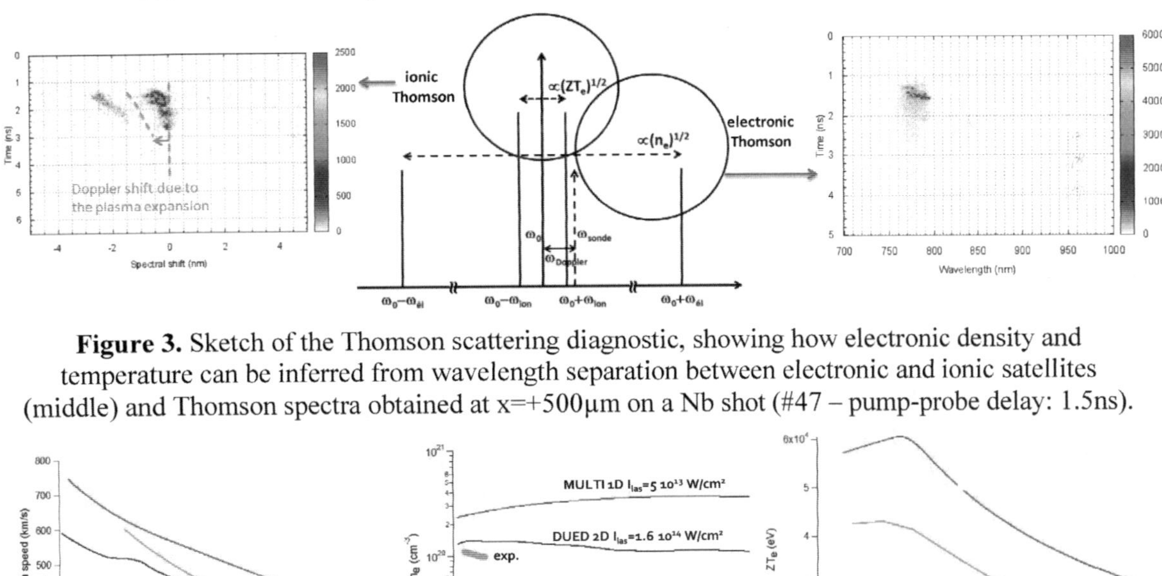


**Figure 3.** Sketch of the Thomson scattering diagnostic, showing how electronic density and temperature can be inferred from wavelength separation between electronic and ionic satellites (middle) and Thomson spectra obtained at x=+500μm on a Nb shot (#47 – pump-probe delay: 1.5ns).

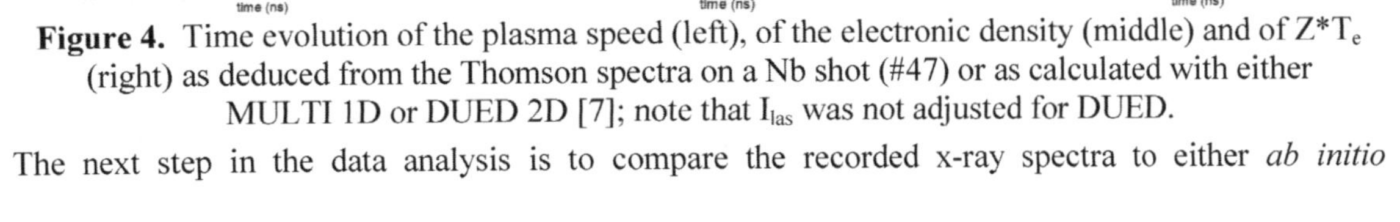


**Figure 4.** Time evolution of the plasma speed (left), of the electronic density (middle) and of $Z^*T_e$ (right) as deduced from the Thomson spectra on a Nb shot (#47) or as calculated with either MULTI 1D or DUED 2D [7]; note that $I_{las}$ was not adjusted for DUED.

The next step in the data analysis is to compare the recorded x-ray spectra to either *ab initio*

calculations performed with an atomic kinetics code (FLYCHK [8], AVERROES [9] or PRIMSPECT [10]) for a given set of ($T_e$, $n_e$) parameters, chosen to be representative of the experimental hydrodynamic conditions.
Work is in progress for Nb but a quite good agreement has already been found in the case of KBr, for which the Thomson data are correctly reproduced by MULTI simulations (figure 5), giving confidence in the whole methodology.

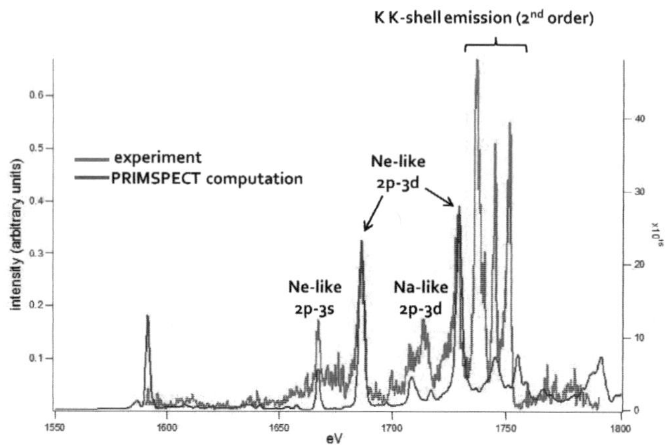


**Figure 5.** Comparison of an experimental KBr x-ray spectrum and of the PRIMSPECT Br one, for a temperature of 600 eV and a density of $10^{20}$ $cm^{-3}$; the main neon-like 3-2 resonant lines are clearly isolated and reasonably well-reproduced.

## 4. Conclusion

The LULI2000 platform has been developed to benchmark atomic physics packages. Thanks to its set of diagnostics, it is already giving high-quality data and providing valuable information, but some improvements are still required. Among them, frequency tripling of the Thomson beam will allow probing a denser region, farther from the corona, where the hydrodynamic simulations will be less sensitive to 2D or corona's effects. In addition, going beyond the single set of parameters as input of any of the atomic kinetics codes by post-processing the hydrodynamic simulations with a radiative transfer module (TRANSPEC [9] or SPECT3D [10]) will allow taking into account temporal or/and spatial integrative aspects of the spectrometers.

## 5. References

[1] Chung H-K, Bowen C, Fontes C J, Hansen S B and Ralchenko Yu 2013 *HEDP* **9** 645
[2] Hansen S, Armstrong G S J, Bastiani-Ceccotti S, Bowen C, Chung H-K, Colgan J P, de Dortan F, Fontes C J, Gilleron F, Marquès J-R, Piron R, Peyrusse O, Poirier M, Ralchenko Yu, Sasaki A, Stambulchik E and Thais F 2013 *HEDP* **9** 523
[3] Chenais-Popovics C, Malka V, Gauthier J-C, Gary S, Peyrusse O, Rabec Le Gloahec M, Matsushima I, Bauche-Arnoult C, Bachelier A and Bauche J 2002 *Phys. Rev. E* **65** 046418
[4] Bastiani-Ceccotti S, Kontogiannopoulos N, Marquès J-R, Tzortzakis S, Lecherbourg L, Thais F, Matsushima I, Peyrusse O and Chenais-Popovics C 2007 *HEDP* **3** 20
[5] Bastiani-Ceccotti S, Renaudin P, Dorchies F, Harmand M, Peyrusse O, Audebert P, Jacquemot S, Calisiti A and Benredjem D 2010 *HEDP* **6** 99
[6] Ramis R, Schmalz R and Meyer-ter-Vehn J 1988 *CPC* **49** 475
[7] Atzeni S 1986 *CPC* **43** 10; Atzeni S, Schiavi A, Califano F, Cattani F, Cornolti F, Del Sarto D, Liseykina T V, Macchi A and Pegoraro F 2005 *CPC* **169** 153
[8] Chung H-K, Chen M H, Morgan W L, Ralchenko Y and Lee R W 2005 *HEDP* **1** 3
[9] Peyrusse O 2000 *J. Phys. B* **33** 4303; Peyrusse O 2001 *JQSRT* **71** 571
[10] MacFarlane J J, Golovkin I E, Wang P, Woodruff P R and Pereira N A 2007 *HEDP* **3** 181

# A Single-shot Terahertz Time-domain Spectroscopy Instrument for Intense Laser System

**Z Jin[1], A Wada[2], J H Shin[2], N Yugami[3] and R Kodama[1, 2, 4]**
[1] Photon Pioneers Center, Osaka University, 2-1 Yamada-oka, Suita, Osaka, 565-0871 Japan
[2] Graduate School of Engineering, Osaka University, 2-1 Yamada-oka, Suita, Osaka, 565-0871 Japan
[3] Department of Advanced Interdisciplinary Sciences, Center for Optical Research & Education (CORE), and Optical Technology Innovation Center (OpTIC), Utsunomiya University, Yoto 7-1-2, Utsunomiya, Tochigi, 321-8585 Japan
[4] Institute of Laser Engineering, Osaka University, 2-6 Yamada-oka, Suita, Osaka, 565-0871 Japan

E-mail: jin-z@ppc.osaka-u.ac.jp

**Abstract.** A single-shot terahertz time-domain spectroscopy method is developed using an echelon-pair. Terahertz waveform is encoded to a beamlets array by electro-optic effect, with a total time window of 37.7 ps and a time step of 94.3 fs. Comparing with spectral encoding methods, this technique directly encodes terahertz waveform to spatial resolution, which is more suited as a terahertz diagnosis for high power ultra-short laser systems with inherent spectral fluctuations.

## 1. Introduction

Terahertz time-domain spectroscopy (TTDS) is a widely used diagnosis tool of various Terahertz (THz) sources. Traditional TTDS is using a multi-shot scanning method to obtain the temporal waveform of THz wave, which limits its applications in high power laser and THz source experiments as those sources are usually of low repetition rates or have shot-to-shot fluctuations. Recently, single-shot TTDS is of interests to many researches such as intense THz generations, ultrafast relativistic electron beam diagnostics [1-3], etc. Most of the schemes for single-shot detection of THz fields exploit the spectral encoding with a chirped probe pulse [4], which requires a rather stable output spectrum of the probe laser. For those widely used short pulse Chirped Pulse Amplification (CPA) laser systems, spectral shaping methods, like a birefringent filter [5] or an acousto-optic programmable gain control filter [6], are often used to obtain a broadband spectrum so that the pulses can be compressed to tens of femtosecond duration, as the limit of Fourier transform. These spectral shaping components meanwhile increase the instability of laser spectrum, making the spectral encoding TTDS more difficult. In this paper, we present a direct "spatial encoding" single-shot TTDS scheme by using an echelons pair proposed by Kim et.al. [7]

## 2. Experiments and results

The two reflective echelons used in our experiments are made of aluminum with a surface flatness better than $\lambda/5$. The reflective surfaces are designed with 20 steps each and step heights of 400 μm (echelon 1) and 20 um (echelon 2). As shown in figure 1, 400 (20×20) beamlets can be produced with a total size of 10 mm ×10 mm, a time step of ~94.3 fs and a total window of ~37.7 ps.

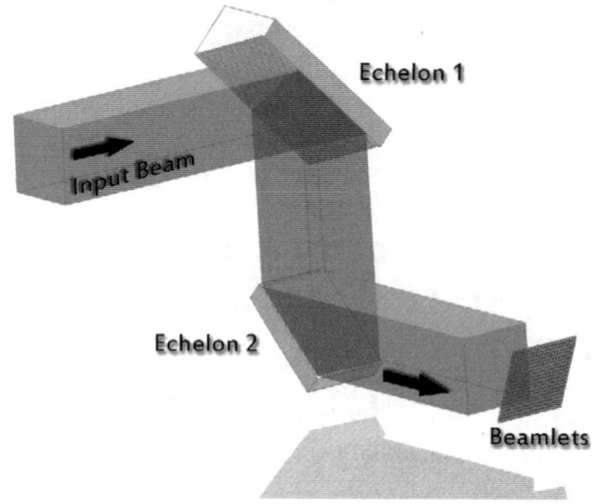


**Figure 1.** Schematic of reflective echelons pair.

A preliminary experiment was performed at the P3 laser system of the Graduate School of Engineering, Osaka University. The experimental configuration is shown as figure 2. A 10mJ, 800nm, 60fs laser beam is set up as a typical pump-probe electro-optic (EO) sampling system [8]. The pump beam is focused into ambient air by a lens with a focal length f=200 mm. A type I beta barium borate (BBO) crystal is placed after the focal lens to generate a second harmonic (2ω) laser pulse together with the fundamental one (ω). A polymer filter is used to transmit THz wave from the 2-color laser generated plasma channel [9] to be collected by an off-axis parabola (OAP), while blocking the residual ω and 2ω pulses. The probe arm after passing the reflective echelons pair is sent through a half-wave plate and a polarizer to change the beam polarization to be same with the THz wave. The beam is then collinearly focused onto a 1mm thick <110> ZnTe crystal with the synchronized THz pulse. The THz electric field is encoded to the different position of the beamlets, which is detected by a 16 bit CCD camera after passing another polarizer with a cross polarization. The two lenses (lens 1 and lens 2) are set as an image-relay to reproduce the image of the echelons pair to the camera while keeping the collinear focus with the OAP on the ZnTe.

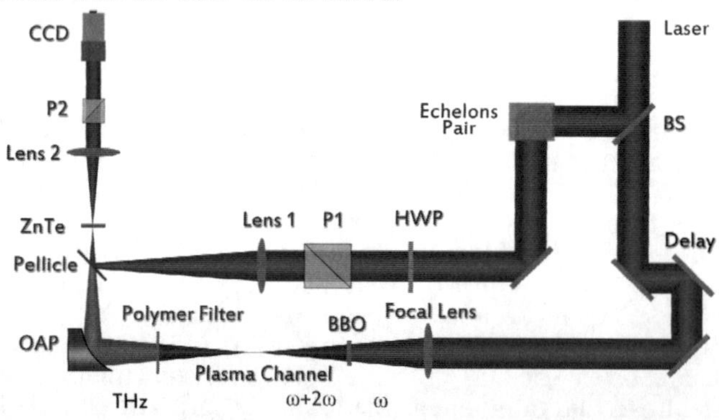


**Figure 2.** Experimental setup for single-shot TTDS of 2-color laser generated THz wave in the air.

A typical CCD image of the beamlets is shown as figure 3a. Note that the shape of each beamlet will be square when well imaged, which is difficult to be separated, as they are quite close to each other. When we shift the CCD off the image plane slightly, the beamlets' shapes change to circles with a much more clear boundary, while the relative intensity remaining similar.

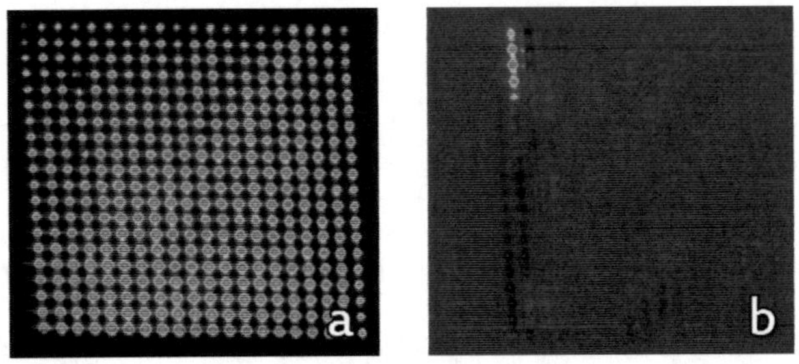


**Figure 3.** (a) Typical beamlets image without the cross polarizer P2. (b) Beamlets image obtained by comparing the difference of with and without the presence of THz field.

A sample image of the compared difference measured with and without the THz pulse clearly shows the intensity modulated by the THz field, as shown in figure 3b. The intensities of each beamlet are integrated and read out row-by-row following the time step. The single-shot THz waveform is then obtained with a signal-to-noise ratio better than 3 orders, as shown in figure 4a. The measured waveform shifts as we adjust the delay between the pump and probe beams, which agrees well with the designed time steps.

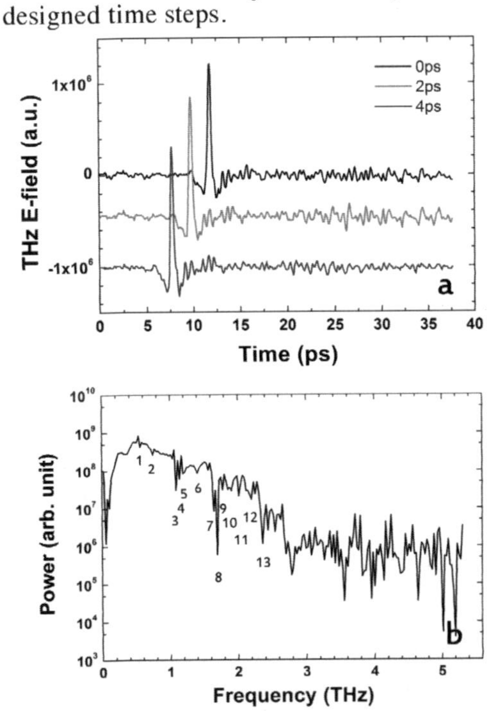


**Figure 4.** (a) Measured THz waveforms with different delay times between pump and probe beams. (b) The corresponded amplitude spectra. The numbered peaks give the absorption lines from the water vapor in the air.

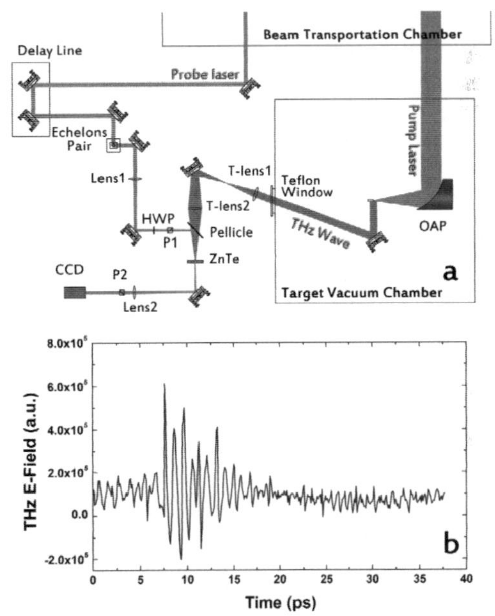


**Figure 5.** (a) Single-shot TTDS setup for THz diagnosis in laser-solid interaction experiment in vacuum chamber. (b) A typical measurement result of a multi-cycle THz waveform generated by laser-accelerated electron beam.

The corresponded THz spectrum is shown as figure 4b, by performing the Fourier transform of the temporal waveform. Multiple absorption peaks can be found which are mostly contributed from the water vapor in the ambient air. The numbered peaks between 0.2 THz to 2.4 THz show good consistence with the results from Ref. 10.

This single-shot TTDS method was also used to diagnose the THz wave generated in laser-solid interaction experiment. The experimental setup is shown as figure 5a. The P3 laser works at high-power single-shot mode, outputting 600 mJ, 27 fs pulses before target. The pump laser is focused by an OAP to a solid copper target with a focal spot diameter of 5 μm and a peak intensity of $\sim 1\times10^{20}$W/cm$^2$. The laser-accelerated electron beam coupling with a periodic metal structure excites a coherent radiation in the THz range by a scheme similar with Smith-Purcell radiation [11]. The THz wave, after passing a Teflon window, is then collected and focused onto the ZnTe crystal. The probe beam reflected out from the beam transportation chamber by a pellicle beam splitter with 3% of the total laser energy is then collimated with the same setup as the air-plasma experiment. Figure 5b gives a typical single-shot measurement result of the multi-cycle THz waveform generated when electron beam swapping the periodic target surface.

As a direct spatial encoding method, the measurement noise is mostly from the fluctuations of the beam pattern and the beam pointing stability. With careful controls of the laser parameters and system vibrations, the stability of our P3 laser's far field focal spot is better than few micrometers, and the RMS laser power stability is within 5%. However, with spectral encoding methods, an additional spectral instability will also be induced. In our case, to achieve an ultra-short laser pulse lesser than 30 fs, an acousto-optic programmable gain control filter (Mazzler) is installed inside the regenerative amplifier cavity, which controls the gain of the cavity to obtain an output laser spectrum broader than 70 nm. This spectral shaping optics also increases the instability of the spectrum to be worse than 20%, which is not easy to be reduced. We have tried to use spectral encoding measurement in our system. The THz waveform can't be clearly revealed, as the noise level is too high.

## 3. Conclusion

In conclusion, using a reflective echelons pair, we have demonstrated a single-shot TTDS method, which encodes the THz field directly to the spatial distribution, allowing a stable THz waveform measurement with high power ultra-short laser systems. With this scheme, we succeeded in the diagnosis of the THz wave generated by two-color laser ionized air-plasma with a 0.16 TW laser power and by laser-solid interaction with a high laser power of more than 20 TW inside vacuum system.

## Acknowledgments

This work is supported by the Core Research for Evolutional Science and Technology, Japan Science and Technology Agency.

## References

[1] Casalbuoni S, Schmidt B, Schmüser P, Arsov V and Wesch S 2009 *Phys, Rev. ST Accel. Beams* **12** 030705
[2] Berden G, Jamison S P, MacLeod A M, Gillespie W A, Redlich B and van der Meer A F G 2004 *Phys. Rev. Lett.* **93** 114802
[3] van Tilborg J, Schroeder C B, Tóth Cs, Geddes C G R, Esarey E and Leemans W P 2007 *Opt. Lett.* **32** 313
[4] Jiang Z P and Zhang X C 1998 *Appl. Phys. Lett.* **72** 16
[5] Barty C P J, Korn G, Raksi F, Rose-Petruck C, Squier J, Tien A C, Wilson K R, Yakovlev V V and Yamakawa K, 1996 *Opt. Lett.* **21** 219
[6] Verluise F, Laude V, Cheng Z, Spielmann C and Tournois P 2000 *Opt. Lett.* **25** 575
[7] Kim K Y, Yellampalle B, Taylor A J, Rodriguez G and Glownia J H 2007 *Opt. Lett.* **32** 1968
[8] Wu Q. and Zhang X C 1995 *Appl. Phys. Lett.* **67** 3523
[9] Kim K Y, Taylor A J, Glownia J H and Rodriguez G 2008 *Nature Photonics* **2** 605
[10] Xin X, Altan H, Saint A, Matten D and Alfano R R 2006 *J. Appl. Phys.* **100** 094905
[11] Jin Z, Chen Z L, Zhuo H B, Kon A, Nakatsutsumi M, Wang H B, Zhang B H, Gu Y Q, Wu Y C, Zhu B, Wang L, Yu M Y, Sheng Z M and Kodama R 2011 *Phys. Rev. Lett.* **107** 265003

# Electron beam guiding by strong longitudinal magnetic fields

**T. Johzaki[1], K. Mima[2], S. Fujioka[3], H. Sakagami[4], A. Sunahara[5], H. Nagatomo[3], H. Shiraga[3]**

[1]Graduate School of Engineering, Hiroshima University, Higashi-Hiroshima, Japan
[2]The Graduate School for the Creation of New Photonics Industries, Hamamatsu, Japan
[3]Institute of Laser Engineering, Osaka University, Suita, Osaka, Japan
[4]Institute for Laser Technology, Osaka, Japan
[5]National Institute for Fusion Science, Toki, Japan

E-mail: tjohzaki@hiroshima-u.ac.jp

**Abstract.** In electron-driven fast ignition, the guiding of fast electron beam having significantly large beam divergence is one of the most critical issues for efficient core heating. To guide the fast electron beam to the core, we consider to externally apply longitudinal magnetic fields. From the 2D PIC simulations applying uniform magnetic fields, it was shown that the field strength of 1~10kT is required for efficient guiding for the heating laser intensity of $10^{18}$~$10^{20}$W/cm$^2$.

## 1. Introduction

One of the most crucial issues of electron-driven fast ignition is efficient core heating by the laser produced fast electron beam. The main factors in preventing efficient heating are (1) too high fast electron energy and (2) too large beam divergence. The fast electron energy could be controlled by eliminating pre-plasma generation and by using heating laser with shorter wavelength. On the other hand, as for the beam divergence, it is difficult to control the angular spread of fast electrons since the laser-plasma interactions are the non-linear phenomena. So there are some ideas on guiding of fast electron beam with large divergence angle using self-generated [1-3] or externally-applied magnetic fields [4]. In the present paper, on the basis of 2D PIC simulations, we evaluate the effects of externally-applied longitudinal magnetic fields on fast electron generation and propagation.

## 2. Simulation set up

The target is initially fully ionized carbon foil, and the electron number density $n_e$ is $100n_{cr}$ ($n_{cr}$ is the laser critical density). The exponential-profile pre-plasma with scale length of 1μm is attached on the target front surface. The initial profile of $n_e$ is shown in **Fig.1**. The $p$-polarized laser pulse with $\lambda_L$ = 1.06 μm wavelength and $I_L$ = $10^{18}$ ~$10^{19}$ W/cm$^2$ intensity irradiates the target from the left boundary. The transverse intensity profile is the Gaussian with a spot size of 5 μm FWHM. The laser rises in $10T_0$, where $T_0$ is the laser period. After that the laser amplitude is kept constant during $30T_0$, and then is dropped to zero in

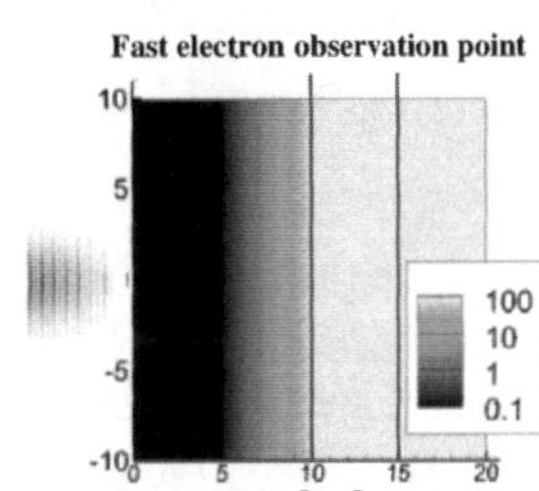


**Figure 1.** Initial profile of electron number density normalized by $n_{cr}$. The observation points of fast electrons are indicated by lines.

$10T_0$. A typical simulation time is $100T_0$, which corresponds to about 0.36 ps for $\lambda_L$ = 1.06 μm. The external magnetic field is uniformly applied in x-direction. The generated fast electrons are observed at the two different points indicated in Fig.1; one is the target surface, and the other is 5μm inside the target. The simulations were carried out by varying the applied magnetic-field strength $B_{x,ext}$ and $I_L$.

## 3. Results and discussion

The transverse momentum distributions ($p_y$, $p_z$) of fast electrons observed at the target surface for $I_L$=$10^{19}$W/cm$^2$ and $B_{x,ext}$ = 0, 1, 10 kT are plotted in **Fig.2**. In 2D simulations, the momentum spread in the laser B-field direction (z-direction) is negligibly small for the case without external field ($B_{x,ext}$=0). When the external magnetic field is applied, the fast electrons gyrate in the plane perpendicular to the applied field direction. Thus, the spread in *y*-direction (laser polarization direction) becomes smaller, and it in *z*-direction larger. **Figure 3** shows the momentum distribution in $p_x$-$p_t$ plane, where $p_t = (p_y^2+p_z^2)^{1/2}$, observed at the target surface. Compared with the case without the external field, for the case with the external fields, the angular spread in the transverse direction in the momentum space becomes larger and the number of fast electrons having the momentum along x-axis is reduced because of the addition of gyration motion. These results show that the external field enhances the angular spread of fast electron beam in the momentum space.

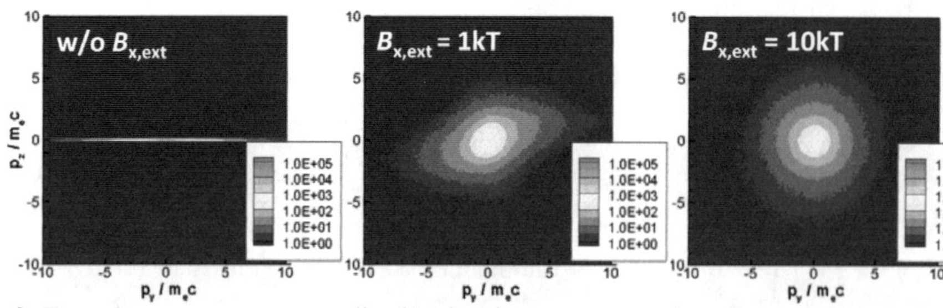


**Figure 2**. Fast electron momentum distribution in transverse plane ($p_y$,$p_z$) observed at the target surface for $I_L$ = $10^{19}$W/cm$^2$ and $B_{x,ext}$ = 0, 1, 10 kT.

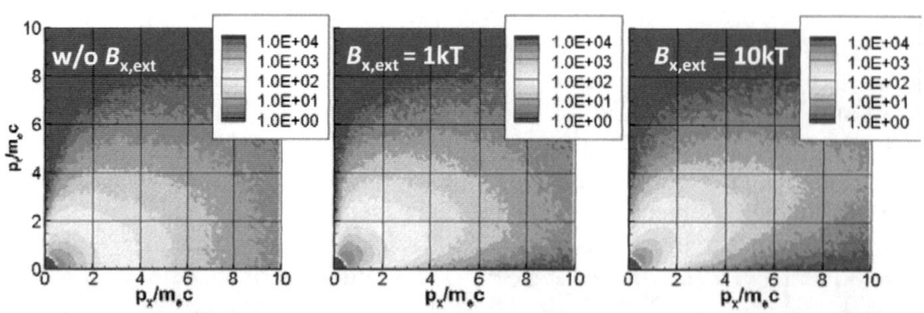


**Figure 3**. Fast electron momentum distribution in t in $p_x$-$p_t$ plane, where $p_t = (p_y^2+p_z^2)^{1/2}$, observed at the target surface for $I_L$ = $10^{19}$W/cm$^2$ and $B_{x,ext}$ = 0, 1, 10 kT.

As the role of external magnetic fields, we do not expect the reduction of angular spread but expect the suppression of spatial divergence. To evaluate the spatial divergence, we plot the spatial profiles of fast electron energy density observed at 176 fs in **Fig.4**. For the case without the external field, the fast electron beam spatially diverges with its propagation. Under the sufficiently strong external fields, however, the fast electrons are trapped by the external fields and flow along the magnetic field lines.

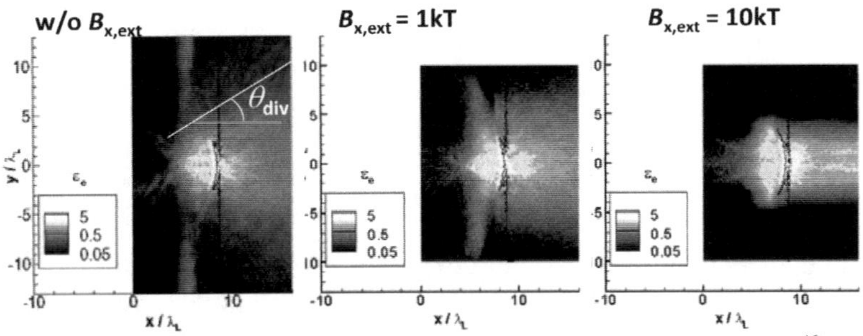


**Figure 4.** Spatial profiles of fast electron energy density at $t$ = 176fs for $I_L = 10^{19}$W/cm$^2$ and $B_{x,ext}$ = 0, 1, 10 kT.

For the case with the sufficiently strong external fields, the Larmor radius becomes smaller with increasing $B_{x,ext}$, which results in smaller spatial divergence.

From the parametric simulations by varying $I_L$ and $B_{x,ext}$, we evaluated the dependence of $\theta_{div}$ (spatial divergence angle) on $B_{x,ext}$ for $I_L = 10^{18}$, $10^{19}$, $10^{20}$ W/cm$^2$ (**Fig.5**). With increasing the laser intensity, the generated fast electron energy becomes higher, and then the required $B_{x,ext}$ for the beam guiding increases. It is found that the required magnetic field strength for beam collimation is $B_{ext}$ > a few kT for the laser intensity with $I_L = 10^{18} \sim 10^{20}$W/cm$^2$.

The external field strength required for $\theta_{div} \sim 0$, $B^0_{x,ext}$, is plotted as a function of $I_L$ in **Fig.6**. It is found that $B^0_{x,ext}$ is proportional to $I_L^{1/3}$. This dependency is understood from a following simple model explanation. For the beam guiding, the fast electron gyration radius $r_L$ should be smaller than a certain value, $r_0$ (for instance, the heating spot radius). The required field strength for this condition is

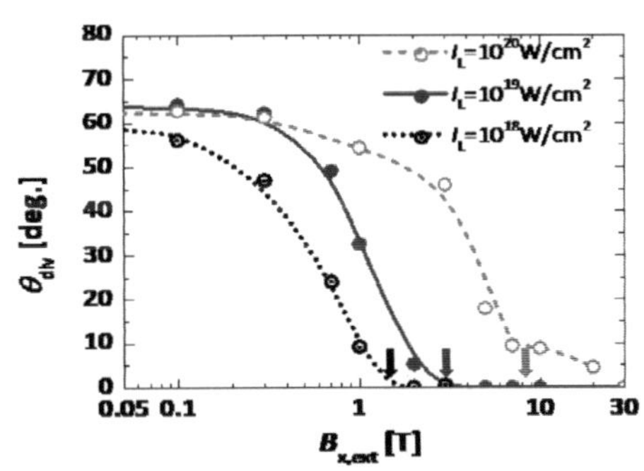


**Figure 5**. Dependence of spatial divergence angle $\theta_{div}$ on external field strength $B_{x,ext}$ for $I_L = 10^{18}$, $10^{19}$, $10^{20}$W/cm$^2$.

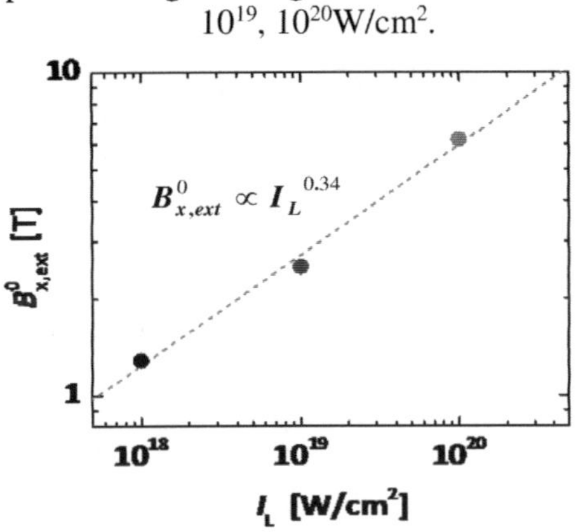


**Figure 6**. B-field strength required for $\theta_{div} = 0$, $B^0_{x,ext}$, as a function of $I_L$.

$$B_{x,ext} < \frac{p}{er_0}, \quad (1)$$

where $p$ is the fast electron momentum. In the relativistic region, $p$ is approximately proportional to the kinetic energy $E$;

$$p = mc\sqrt{\left(\frac{E}{mc^2}+1\right)^2 - 1} \approx \frac{E}{c}, \quad (2)$$

where $m$ and $c$ are the electron rest mass and the speed of light. On the other hand, the fast electron energy (*slope temperature*) evaluated here is proportional to $I_L^{1/3}$ (also, the experimental observation showed the same intensity dependency [5, 6]). The required field strength thus depends on $I_L^{1/3}$;

$$B_{x,ext}^0 \propto p \propto E \propto I_L^{1/3}. \quad (3)$$

## 4. Summary

For guiding of the fast electron beam with large angular divergence, we have evaluated the effects of externally-applied longitudinal magnetic fields on the fast electron generation and transport by 2D PIC simulations. It was found that angular spread in momentum distribution of fast electron beam is enhanced due to the external fields. However, the spatial divergence as the results of the fast electron transport is significantly reduced since the fast electrons with a large angular spread in the momentum space are trapped by the magnetic fields and propagate along the magnetic field lines. The strength of magnetic fields required for beam guiding depends on the irradiated laser intensity, $B_{x,ext}^0 \propto I_L^{1/3}$. The required field strength is 1~10kT for $I_L = 10^{18}$ ~ $10^{20}$W/cm$^2$.

Here, we assumed very simple situation (*e.g.*, uniform external fields, a plane foil target, a small laser spot, a short pulse duration and so on). Further investigation, where the more realistic situation is assumed, is required.

**Acknowledgments**

This work was partially supported by JSPS KAKENHI 25400534. The authors are grateful for the support of the computer room of ILE and the cyber-media center at Osaka University.

**References**

[1] Johzaki T, Nagatomo H, Sunahara A, Cai H-B, Sakagami H, Nakao Y and Mima K 2011 *Nucl. Fusion* **51**, 073022 .

[2] Norreys P A Green J S, Lancaster K L, Robinson A P R, Scott R H H, Perez F, Schlenvoight H-P, Baton S, Hulin S, Vauzour B, Santos J J, AdamsD J, Markey K, Ramakrishna B, Zepf M, Quinn M N, Yuan X H, McKenna P, Schreiber J, Davies J R, Higginson D P, Beg F N, Chen C, Ma T and Patel P 2010 *Plasma Phys. Control. Fusion* **52** 124046

[3] Johzaki T, Sunahara A, Fujioka S, Nagatomo H, Sakagami H and Mima K 2011 Fast Electron Beam Guiding for Effective Core Heating *IFSA 2011* September 2011, Bordeaux-Lac, France

[4] Strozzi D J, Tabak M, Larson D J, Divol L, Kemp A J, Bellei C, Marinak M M and Key M H *Phys. Plasmas* 2012 **19** 072711

[5] Beg F N, Bell A R, Dangor A E, Danson C N, Fews A P, Glinsky M E, Hammel B A, Lee P and Norryes P A 1997 *Phys. Plasmas* **4** 447

[6] Tanimoto T, Habara H, Kodama R, Nakatsutsumi M, Tanaka K A, Lancaster K L, Green J S, Scott R H H, Sherlock M, Norreys P A, Evans R G, Haines M G, Kar S, Zepf M, King J, Ma T, Wei M S, Yabuuchi T, Beg F N, Key M H, Nilson P, Stephens R B, Azechi H, Nagai K, Norimatsu T, Takeda K, Valente J and Davies J R 2009 *Phys Plasmas* **16** 062703

# A new symmetry model for hohlraum-driven capsule implosion experiments on the NIF

**O. Jones, R. Rygg, R. Tomasini, D. Eder, A. Kritcher, J. Milovich, L. Peterson, C. Thomas, M. Barrios, R. Benedetti, T. Doeppner, T. Ma, S. Nagel, A. Pak, J. Field, N. Izumi, S. Glenn, R. Town, D. Bradley**

Lawrence Livermore National Laboratory, P. O. Box 808, Livermore, CA 94551

Jones96@llnl.gov

**Abstract**. We have developed a new model for predicting the time-dependent radiation drive asymmetry in laser-heated hohlraums. The model consists of integrated Hydra capsule-hohlraum calculations coupled to a separate model for calculating the crossbeam energy transfer between the inner and outer cones of the National Ignition Facility (NIF) indirect drive configuration. The time-dependent crossbeam transfer model parameters were adjusted in order to best match the P2 component of the shape of the inflight shell inferred from backlit radiographs of the capsule taken when the shell was at a radius of 150-250 um. The adjusted model correctly predicts the observed inflight P2 and P4 components of the shape of the inflight shell, and also the P2 component of the shape of the hotspot inferred from x-ray self-emission images at the time of peak emission. It also correctly captures the scaling of the inflight P4 as the hohlraum length is varied. We then applied the newly benchmarked model to quantify the improved symmetry of the N130331 layered deuterium-tritium (DT) experiment in a re-optimized longer hohlraum.

## 1. Introduction

Minimizing low mode asymmetry in the radiation drive is required to efficiently implode ICF capsules at high convergence [1]. Low mode drive asymmetry of NIF implosions has primarily been diagnosed by measuring the shape of filtered x-ray emission images near the time of peak x-ray emission [2]. Other experimental techniques have provided information on the early-time symmetry, prior to the launch of the main laser pulse [3,4], but there has been no direct measurement of the shape of the compressed shell as it converges prior to stagnating and forming the hotspot. However, recently a technique was developed to obtain backlit pinhole images of imploding capsules in NIF [5]. With the addition of this capability, which is commonly referred to as the two-dimensional convergent ablation or "ConA-2D" technique, there is now data available during the full duration of the radiation drive that can be used to constrain our symmetry prediction models.

We calculate the symmetry of NIF implosions by modeling the integrated hohlraum-capsule system using the Hydra radiation hydrodynamics code [6]. Experimentally we control the lowest even mode Legendre asymmetry (P2) by varying the wavelength separation ($\Delta\lambda$) between the inner and outer beam cones to control the amount of cross-beam energy transfer (CBET) that occurs near the laser entrance hole where the inner and outer cone laser beams intersect. CBET is modeled in a separate calculation [7] using plasma parameters extracted from a Hydra simulation with no laser backscatter removed. We have found that the CBET model can accurately predict the radiation symmetry during the low power portion of the laser pulse as inferred from the experimental shock symmetry. However, the calculated CBET at high power during the peak of the laser pulse is generally larger than inferred from experiment and so is moderated by setting a threshold density perturbation value (δn/n) for the CBET [7]. When the resonant process that drives CBET causes the local δn/n to exceed that threshold, no further transfer is allowed in that portion of the laser beam crossing volume. A second Hydra calculation is then done with the measured input laser power modified by the measured laser backscatter and calculated crossbeam energy transfer. The input drive is further modified in order to match shock timing, trajectory, and neutron bang time data from other experiments [8].

In this work we introduce a laser-power-dependent δn/n value for the CBET that is a best fit to all the available symmetry data. We find that with this adjustment the integrated calculations correctly predict the low mode (P2, P4) inflight and stagnation shape of NIF implosion experiments. We used this model together with experimental data to improve the drive symmetry. The re-optimization resulted in a longer hohlraum and a smaller Δλ.

## 2. Power-dependent saturation model that matches P2

A series of ConA-2D experiments were done to determine the scaling of inflight shape with various control parameters. The baseline experiment was done in a 5.75 mm diameter, 9.43 mm length hohlraum and used a 360 TW peak power variant of the NIF N120802 436 TW laser pulse. The Δλ settings were 6.6 A separation between the 30° beams and the outer (44.5° and 50°) beams, and 8.1A separation between the 23° beams and the outer beams. To determine the dependency of shape on peak power, a second experiment was done at a lower peak power (250 TW). We were able to match the scaling with laser power of the inflight and stagnation P2 shape using a δn/n that depends on the laser power, as shown in Fig. 1.

| P (TW) | δn/n |
|---|---|
| 0 | 1.e-3 |
| 50 | 1.e-3 |
| 100 | 5.e-4 |
| 250 | 4.e-4 |
| 360 | 3.e-4 |
| 500 | 3.e-4 |

a)

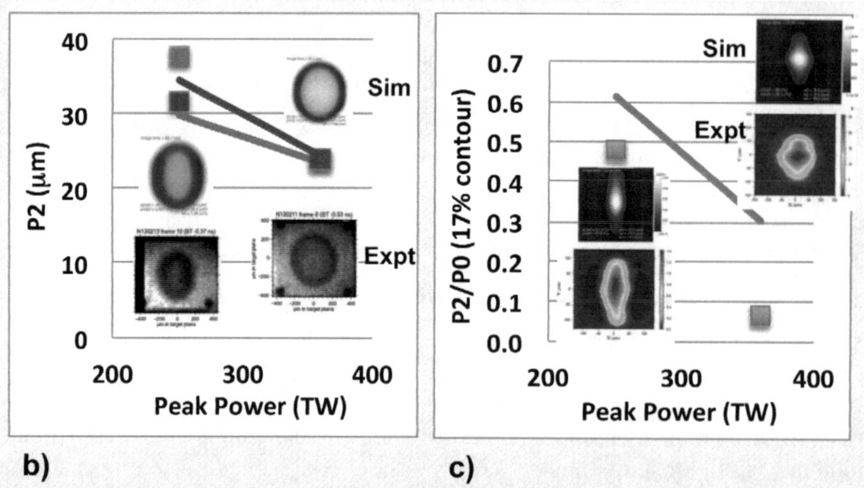


Figure 1 – a) CBET saturation vs. laser power, b) P2 distortion at 200 μm vs. peak power (red is limb minimum, blue is max slope, squares are experimental data), c) self-emission P2/P0 vs. peak power (squares are experimental data).

## 3. Model predictions of P4 compared to experiment

Backlit images of the baseline case described above also indicated there was an easily detectable inflight positive P4 distortion of ~10 μm when the average shell radius was ~200 μm. Modeling indicated that a longer hohlraum would be required to minimize the inflight P4. To determine the scaling with length, a series of ConA-2D experiments were done with hohlraum length varied from 9.13 mm (300 μm shorter than baseline) to 10.43 mm (1000 μm longer than baseline). As the hohlraum length was varied, the outer beam pointing (where beams cross the hohlraum axis) was changed so that the outer cone position stayed fixed relative to the LEH. In contrast, the inner cone pointing was not changed, which meant that the distance in z between the inner and outer spots varied with the hohlraum length. As a result of this pointing, for the longest hohlraums (10.13 and 10.43 mm) the LEH diameter was increased from 3.1 mm to 3.38 mm.

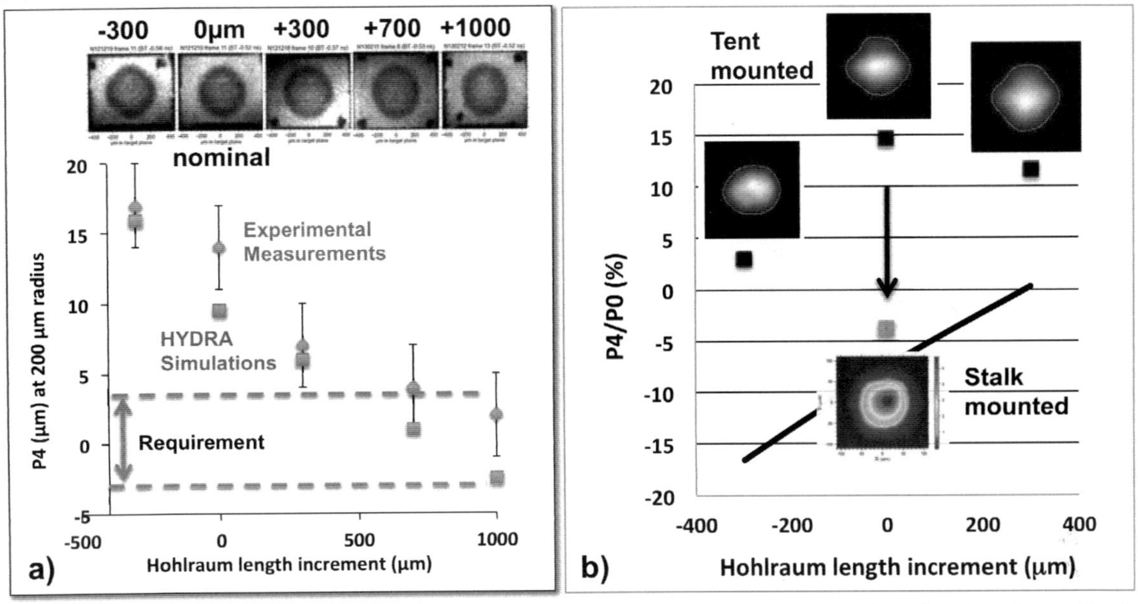


Figure 2 – a) P4 distortion of shell at 200 µm, b) P4/P0 of self-emission 17% contour at time of peak emission (line is simulation, squares are the data).

Figure 2a shows the P4 component in µm of the minimum transmission limb contour when the limb minimum average radius is 200 µm as a function of the hohlraum length. We see that as the hohlraum is lengthened, the magnitude of the P4 component is reduced and that the length scaling is well predicted by the integrated Hydra model. However, Figure 2b shows that the P4 component of the x-ray self-emission did not agree at all with the model for capsules that were mounted to the hohlraum in the usual way using a thin (~50 nm thick) tent. For example, the baseline 9.43 mm long hohlraum had a measured P4/P0 of +15%, whereas the expected value was -8%. The backlit images appeared to show a feature distorting the shell at an angle of ~45°, very close to where the support tent lifts off from the capsule. To test the theory of whether the tent imprint was causing the x-ray images to have an apparent positive P4, an experiment was done with the 9.43 mm hohlraum in which the capsule was held with a stalk instead of a tent. The result was the orange square in Figure 2b, which showed that the emission P4/P0 changed from +15% to -5%, which is much closer to the predicted -8%.

## 4. Predicted shape improvement of the re-optimized 10.13 mm hohlraum

The hohlraum length scan experiments showed that the hohlraum length and $\Delta\lambda$ that we were using for most of the 2012 ignition experiments was not optimal. The hohlraum was too short and had too much $\Delta\lambda$, resulting in measureable positive P2 and P4 inflight distortions of the compressed shell.

Using the newly benchmarked symmetry model as a guide, the hohlraum length was increased from 9.43 mm to 10.13 mm, and the color separation was changed from 6.6/8.1A to 4.5/6.0A. Because the hohlraum length had been changed, a shock timing experiment was done to retune the arrival times of the first three shocks. A subsequent ConA-2D experiment with this configuration (shot N130314) showed that the inflight shape had been changed as predicted by the bench-marked model and was very round in flight, so a cryogenically layered DT capsule was shot with the same configuration (shot N130331) to determine the effect of these improvements on overall performance (ion temperature, yield, etc.).

Since there was not a DT shot with the 9.43 mm hohlraum at the same peak laser power and energy to compare to, we used the new integrated symmetry model to quantify the improvement. To estimate the performance of the 9.43 mm hohlraum, we used the measured laser pulse shape and capsule dimensions for the N121210 ConA-2D experiment, which used a gas-filled symmetry capsule, but replaced the inner 14.2 μm of the ablator with an equivalent mass (68.8 μm) of frozen 50/50 DT. For the retuned 10.13 mm hohlraum we simply did a post-shot simulation of the actual DT shot (N130331).

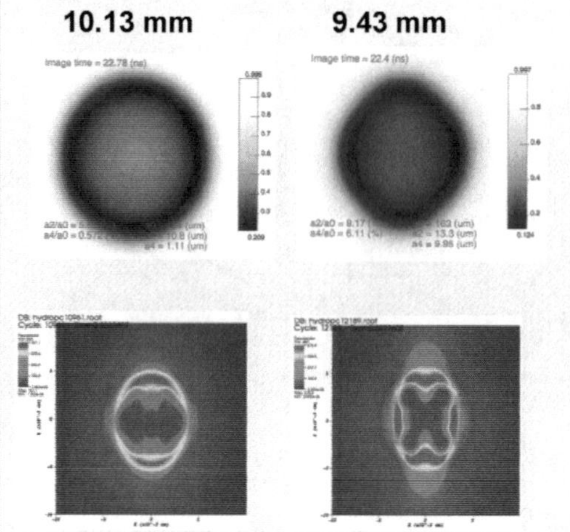


| | 10.13 mm Simulation | 9.43 mm Simulation |
|---|---|---|
| Yield | 1.64e15 | 1.15e15 |
| Y2D/Y1D | 0.44 | 0.26 |
| Tion (keV) | 3.12 | 2.91 |
| Vel (μm/ns) | 291 | 293 |
| Adiabat | 1.61 | 1.43 |
| Res KE (kJ) | 0.334 | 0.534 |
| ρR rms | 39.0 | 47.6 |
| Hotspot rms | 10.1 | 30.2 |

**Figure 3 – Comparison of calculated performance of DT implosions in 10.13 mm and 9.43 mm long hohlraums**

In Figure 3 we compare the calculated DT performance in the two hohlraums. On the left are simulated inflight radiographs and density slices at bang time for each hohlraum. The yield and ion temperature are higher for the re-optimized 10.13 mm hohlraum, even though the 1D performance is slightly worse (lower velocity, higher fuel adiabat). We also calculate the 1D yield by artificially symmetrizing the 2D calculation and then find the ratio of the 2D calculated yield to the 1D calculated yield, which is a measure of the portion of the yield degradation due to low mode drive asymmetry. This ratio is improved for the 10.13 mm hohlraum, although it could still be further improved. The 10.13 mm hohlraum also has less residual kinetic energy at the time of maximum compression, lower ρR rms, and lower hotspot rms, all of which indicate a rounder, more efficient implosion. Note that the experimental yield of N130331 was 3e14 or 18% of calculated, indicating that low mode drive asymmetry was not the primary cause of yield degradation for this experiment.

## 5. Conclusions

We used data from backlight radiography of NIF implosions to develop a new power-dependent CBET saturation model. Using this model coupled with integrated Hydra calculations we can accurately predict the low mode P2 and P4 components of the inflight and stagnation shape of the implosion due radiation drive asymmetry. This model predicts the scaling of P4 distortion with hohlraum length. We used to model to show that the N130331 DT implosion in a 10.13 mm long hohlraum had improved symmetry compared to a comparable implosion in a 9.43 mm long hohlraum.

## References

[1] J. D. Lindl, *Inertial confinement fusion* (Springer, New York, NY, USA, 1998).
[2] G. Kyrala, et al, Phys. Plasmas 18, 072703 (2011).
[3] E. Dewald, J. Milovich, C. Thomas, et al., Phys. Plasmas 18, 092703 (2011)
[4] J. D. Moody, H. F. Robey, P. M. Celliers, et al., Phys. Plasmas, submitted
[5] J. R. Rygg, O. S. Jones, J. E. Field, et al., PRL, submitted
[6] M. M. Marinak, G. D. Kerbel, N. A. Gentile, et al., Phys. Plasmas 8, 2275 (2001).
[7] P. Michel, S. H. Glenzer, L. Divol, et al., Phys. Plasmas 17, 056305 (2010).
[8] O. S. Jones, C. J. Cerjan, M. M. Marinak, et al., Phys. Plasmas 19, 056315 (2012).

# Remarks to SBS PCM based self-navigation of laser drivers

**M Kalal[1], L Matena[1], HJ Kong[2], M Martinkova[3], and S Cha[2]**

[1] FNSPE, Czech Technical University in Prague, Prague, Czech Republic
[2] Department of Physics, KAIST, Daejeon, Republic of Korea
[3] Extreme Light Infrastructure, Institute of Physics, AS CR, Prague, Czech Republic

E-mail: kalal@fjfi.cvut.cz

**Abstract.** A novel technology of self-navigation of laser drivers on injected inertial fusion energy pellets employing phase conjugating mirrors based on stimulating Brillouin scattering was recently proposed. Its feasibility as well as various implications were gradually studied and working solutions to potential problems were always suggested. As this technology could help to overcome several burning key issues of inertial fusion (e.g., a sufficiently precise navigation of laser drivers on injected pellets in the case of a direct drive scheme and decreased requirements on high-repetition high-power lasers) it gradually started to attract a carefully measured tentative interest among the major inertial fusion oriented laboratories and projects. In this paper the next step in this research path will be reported. It concerns the resulting phase and amplitude structures created by multiple low energy drivers (glints) illuminating the pellet during the first stage of the process after their reflection and a subsequent superposition on the collecting/focusing final optics. It was demonstrated that with a large number of such drivers acting simultaneously from many angles the situation gets somewhat complicated and requires more detailed studies/suggestions of suitable configurations.

## 1. Introduction

There are still many problems to be solved in inertial confinement fusion development, before a functional inertial fusion power plant can be built. An ignition has not been achieved, yet, not even on stationary pellets with indirect drive. Despite the fact that some encouraging progress has been recently reported by NIF. However, a stationary pellet is not suitable to be used in a power plant. One possible solution is to inject the pellet into the reaction chamber and irradiate it during its fly at the right moment by many laser beams. This might highly improve the repetition rate, but also brings the question of how to aim the lasers on the moving pellet with a necessary accuracy and timing. Tracking the pellet and aiming at it either by moving heavy final optics or by adjusting refractive index of suitably chosen optical elements is an option for consideration. But it might prove to be too difficult to handle in practice.

Therefore, an alternative approach was recently proposed for evaluation. This approach employs phase conjugating mirrors (PCM) generated by stimulated Brillouin scattering (SBS). If successfully mastered, it might require no active aiming at all. For more details regarding a graduate development of this technology kindly see the following papers [1, 2, 3, 4]. An existence of this technology was already acknowledged by major international specialists in their overview paper dedicated to inertial fusion [5]. It might be interesting to mention that recently a different phase conjugation based beam steering technology started to be studied by another group. In their approach the laser beam-steering mechanism is controlled by a four-wave mixing process [6].

Principles of SBS PCM self-navigation technology is schematically illustrated in Fig. 1.

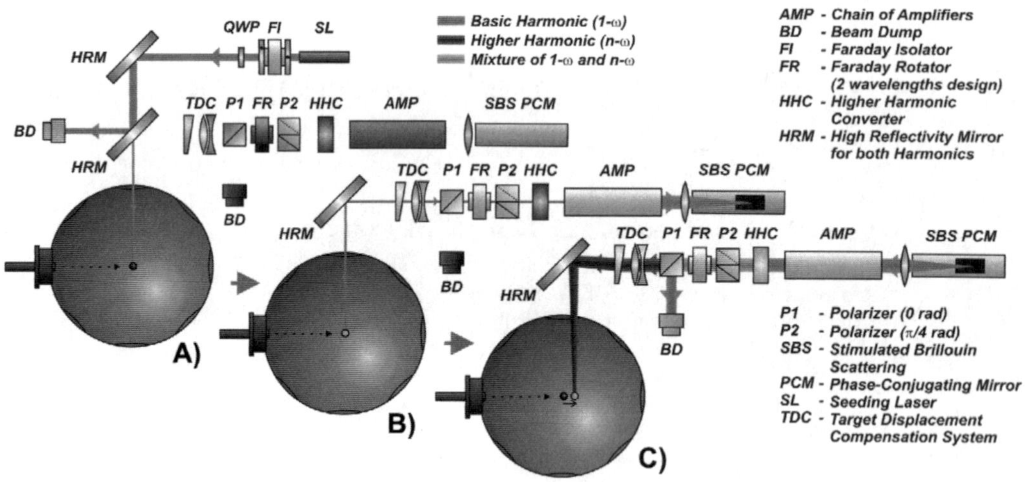


**Figure 1.** A scheme of one laser channel during three different stages of its operation [4]

**A)** When the injected pellet is approaching its best interaction position, a low energy seeding laser pulse (glint - red line) is sent to illuminate the pellet.

**B)** Reflected seeding laser pulse is collected by the focusing optics and amplified on its way to SBS PCM cell. Its energy is below threshold thus no higher harmonic conversion occurs.

**C)** Amplified pulse is reflected by SBS PCM cell, amplified again on its return, major part of its energy (now above the threshold) is converted to higher harmonic (blue line) and automatically aimed at the moving pellet by the target (pellet) displacement compensation system (TDC) for its final high power irradiation. The unconverted part is removed from the system by a specially designed Faraday insulator (red line).

In order to ensure a sufficient symmetry of the final high energy irradiation of the injected pellet, it seems unavoidable that the pellet would have to be, at first, illuminated by many low energy laser beams (glints) from many directions as well. In an ideal case using the same number of glints as it would be the number of the high energy beams. Therefore, the resulting light on any of these entrance windows (on its way out) would consist of contributions created by reflections of many glints from the pellet (approximately one half of their total number), which would superimpose. A crucial question thus arises: would the characteristics of the collected light be suitable for its amplification and PCM reflection? Providing some light to this question is the main purpose of this paper.

## 2. Model, calculations and results

Geometry in which the problem was studied is similar to (at least some) NIF parameters. The target chamber was assumed spherical with 10 m in diameter. The pellet was a 100 % reflective sphere 4 mm in diameter positioned exactly in the center of the target chamber. The light reflected from the pellet was superimposed on the plane outside the target chamber were the final optics would be placed. The selected distance from the center of the target chamber was in our calculations set to 10 m. Every collecting area was a square 400 x 400 mm (this particular shape was selected only to simplify the calculations). The laser beam (glint) cross-section was a square (not particularly important), aimed at the center of the chamber, wide enough to illuminate the whole pellet. In practice, for safety reasons, it would be wise (unless some specific care taken using other means) to distribute the windows in such a way that no laser light (not reflected by the pellet) could be fired directly into another window. A proper positioning of collecting areas (depending on their number) was dealt with by a special software developed for this purpose. Collecting areas were not considered flat for the calculation, but spherically curved due to the fact that behind every such area there has to be a lens to make the beam parallel.

As for the SBS PCM reflection the most suitable polarization of the incoming light is linear, a need for having polarizers in the beam line arises. A simple way to ensure the amplitude of the reflected glint would always stay the same, and yet the light would remain linearly polarized after it has been collected, is to use a circularly polarized light for the glints.

Due to the fact that analytical solutions for the phase and the amplitude (intensity) would be difficult to find even in the most simplified scenarios, and because it would be beneficial that the same approach could be used to analyze even more complicated scenarios later on (narrow beam, non-spherical target displaced from the center of the target chamber etc.), a numerical model based on Monte Carlo simulations was developed. Its principle is rather a simple one. A random ray from the beam is selected and followed to the target. If the target is hit, the ray is reflected according to the law of reflection and followed to the collecting area. Provided this area is hit as well, all necessary parameters are calculated and saved. This is repeated until enough rays hit the collecting area. Parameters that have to be calculated are the path traveled by the rays before hitting the collecting area (to calculate the phase) and the amplitude. The amplitude of the ray would normally stay unchanged (total reflection), but as it would eventually pass through a polarizer behind the collecting area (final optics), its orientation must be taken into account (not needed for the circularly polarized glints).

It is relatively easy to calculate the path traveled by every ray that hits the selected collecting area. However, working this way means we only know the traveled path in the points that were actually hit. And these are essentially randomly scattered. It is necessary to divide the collecting area into a regular grid, and calculate the traveled path (and any other quantity that might be needed) on this grid. Typically 10 million randomly selected rays were tested for every collecting area, and the rays that actually hit that area were counted. Subsequently, contributions from individual glints were superimposed over the whole grid one by one using the following formulae:

$$E = \sqrt{E_1^2 + E_2^2 + 2E_1E_2\cos(\alpha_2 - \alpha_1)} \quad (1)$$

$$\tan\alpha = \frac{E_1\sin\alpha_1 + E_2\sin\alpha_2}{E_1\cos\alpha_1 + E_2\cos\alpha_2} \quad (2)$$

Here $E$ and $\alpha$ are the amplitude and the phase of the resulting wave and $E_1, E_2, \alpha_1, \alpha_2$ are amplitudes and phases of the two interfering waves (in a given point).

Due to space limitation only a very brief description was provided. For details kindly see [7].

## 3. Conclusions

As it was found (see Fig. 2), the amplitude of the light after superposition changes rapidly across the collecting area (and its quick variations are present regardless of the number of beams used). In order to verify whether the proposed self-navigation technology can work, further research needs to be done in order to find out how would the collected light behave in amplifiers, whether the resulting beam would be able to reflect successfully on the phase conjugating mirror, and whether the phase and the amplitude structure of the light after its

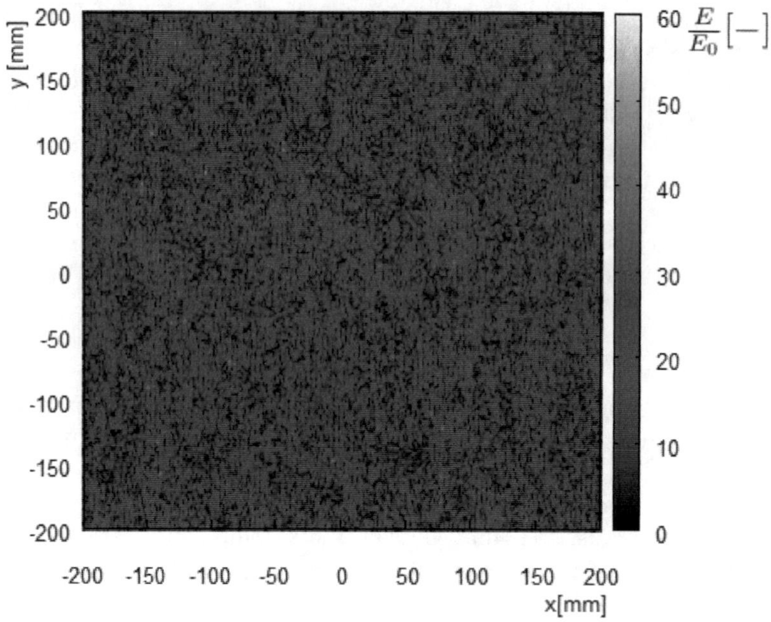


**Figure 2.** Amplitude structure in the collecting area after illuminating the pellet with 400 beams (expressed as a ratio to the amplitude of one beam only).

SBS PCM reflection would be good enough for the beam to compress the target. It seems likely that the amplitude structure (as well as the phase - not shown here) might pose an issue during these steps, and methods will have to be looked for to make the parameters of the laser driver more even. The most promising scenario seems to be when every driver beam line could work with just one (directly backwards) reflection (as illustrated in Fig. 1). However, such scenario might not be easy to realize in practice. It should be also noted that all the calculations were done in geometrical optics approximation. Thus reality might not be so unfavorable after all. But under any circumstances our findings might attract some attention to this particular issue.

**Acknowledgments**
On the Czech side funding from MSM 6840770039, RVO 68407700 is gratefully acknowledged. On the Korean side this work was supported by "Dual Use Technology Program" at Agency for Defense Development (ADD) of the Republic of Korea (UM12012RD1) and by the National R&D Program through the National Research Foundation of Korea (NRF) funded by the Ministry of Education, Science and Technology (2013-036083).

## References

[1] Kalal M *et al* 2010 *J. Kor. Phys. Soc.* **56** 184
[2] Kalal M *et al* 2010 *J. Fusion Energy* **29** 527
[3] Web page of IAEA - look for: *IFE/P6-15 Kalal* inside of the PDF document: http://www-naweb.iaea.org/napc/physics/FEC/FEC2012/html/proceedings.pdf
[4] Kalal M and Slezak O 2011 *Proceedings of SPIE* **8080** 808021-1
[5] Mima K *et al* 2011 *Nucl. Fusion* **51** 094004
[6] Kameyama N and Yoshida H 2013 *Fusion Sci. Technol.* **63** 120
[7] Matena L 2013 *Bachelor Thesis* Czech Technical University in Prague (in English - PDF version available upon request)

# Conceptual design of sub-exa-watt system by using optical parametric chirped pulse amplification

**J. Kawanaka, K. Tsubakimoto, H. Yoshida, K. Fujioka, Y. Fujimoto, S. Tokita, T. Jitsuno, N. Miyanaga, and Gekko-EXA Design Team**

Institute of Laser Engineering, Osaka University
2-6 Yamadaoka, Suita, Osaka 565-0871 Japan

E-mail: kawanaka@ile.osaka-u.ac.jp

**Abstract.** A 50 PW ultrahigh-peak-power laser has been conceptually designed, which is based on optical parametric chirped pulse amplification (OPCPA). A 250 J DPSSL and a flash-lamp-pumped kJ laser are adopted as new repeatable pump source. The existed LFEX-laser with more than ten kilo joules are used in the final amplifier stage and the OPCPA with the 2x2 tiled pump beams in random phase has been proposed with several ten centimeter aperture. A pulse duration of amplified pulses is set at less than 10 fs. A broadband OPCPA with ~500 nm of the gain spectral width near 1 μm is required. A partially deuterated KDP (p-DKDP) crystal is one of the most promising nonlinear crystals and our numerical calculation ensured such ultra-broad gain width. p-DKDP crystals with several deuteration ratio have been successfully grown.

## 1. Introduction

Ultrahigh peak power lasers have been actively developed over the world and opened new applications in science, engineering and medical fields. Recently, sub-exa-watt laser projects of ELI (Extream Light Infrastructure) in EU and XCELS (Exawatt Center for Extreme Light Studies) in Russia were planned. The laser developments are deeply discussed and the attractive applications are shown.

In our institute, high peak power laser of LFEX has been developed as a heating laser for fast ignition in fusion energy research, which is a single shot, huge laser system by using chirped pulse amplification with large glass slabs as a laser material.[1] On the other hand, a conceptual design of ultrahigh-peak-power laser "Gekko-EXA" has started in 2010 toward sub-exa-watt peak power and has been continued to make its reality high. The brief conceptual design and some principal techniques are shown in this paper.

## 2. System outline

Basic concept of ultrahigh peak power generation is shortening laser pulse duration below 10 fs rather than increasing pulse energy, and a chirped pulse amplification using an optical parametric process is chosen due to a high contrast ratio of signal pulse against pre-pulses and pedestals, lower heat loading than laser amplification, and perspective of new technology. A block diagram is shown in fig. 1. The final goal of this system is 10 fs, 50 PW, and 20 fs, 25 PW will be tested at the first step. The whole system consists of a front end, pump sources, and an OPCPA chain. The front end includes a fs-oscillator, a pulse amplifier, a white light generator, a phase dispersion compensator and a pulse stretcher. A part of fs pulses from the front end supplies to pump sources. The pulses are temporally stretched and are amplified again. The amplified pulses are used as a seed pulse of pump lasers of a diode-pumped solid-state laser (DPSSL, 250 J, 100 Hz), a flash-lamp pumped split-disk laser (5 kJ, 0.01 Hz) and the LFEX laser(12 kJ, single shot). An OPCPA chain has three stages, which corresponds to each pump laser. Each stage generates different output of 1 PW/10 fs/ 100 Hz, 20

PW/10 fs/0.01 Hz, and 50 PW/10 fs/0.0001 Hz, respectively. Also, sub ns and ns pulses are available in different pulse energy levels by using pump lasers independently. This variation of the output power of Gekko-EXA will meet more variable applications. The fs-oscillator is used as seed pulses of pump sources and of white light generation for OPCPA chain to reduce timing jitter in the OPA process. The 12 kJ laser technology has been already developed in the LFEX laser basically. Novel technologies of the DPSSL, the kJ split disk laser, ultra-broadband OPCPA and pDKDP crystals are discussed in the following.

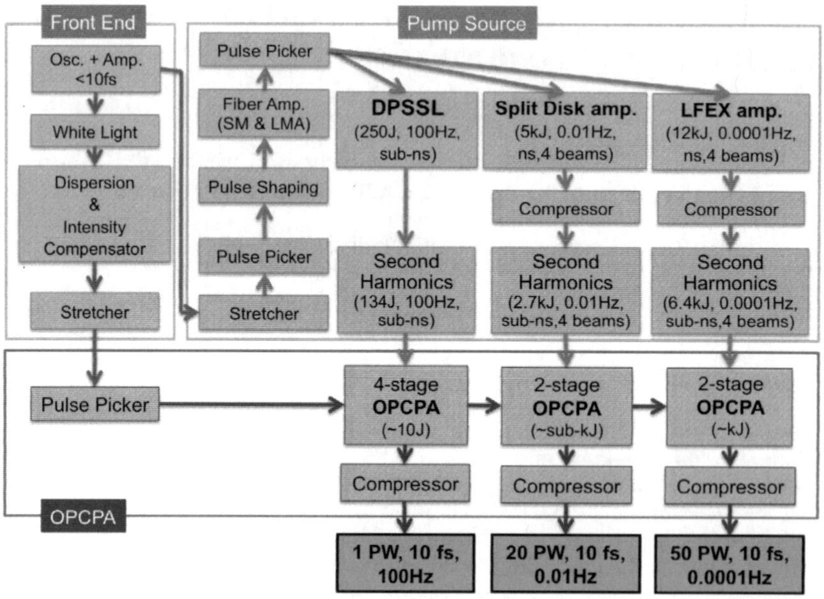


**Figure 1.** Block diagram of Gekko-EXA laser system.

## 3. Pump sources

### *3.1. 250 J/100 Hz Yb:YAG DPSSL*

Cryogenic Yb:YAG is used as a laser material due to the improved thermal strength, the controlled saturation fluence below damage threshold, and the efficient laser operation without re-absorption at low temperature. [2] An amplifier scheme is TRAM (Total-Reflection Active-Mirror).[3] TRAM is a composite ceramics, which consists of a non-doped YAG prism and a Yb:YAG thin layer, shown in fig. 2. The thin layer is effective in a reduction of temperature rise as well as "Thin Disk" concept [4]. The prism separates the input and output surfaces to improve damage threshold on the surface compared with "Thin Disk". Ten TRAMs are required for 250 J pulse energy. A size of Yb:YAG layers is 50 cm x 10 cm x 1 cm for single beam, and smaller size for muti-beams. The operation temperature of TRAM is liquid nitrogen temperature.

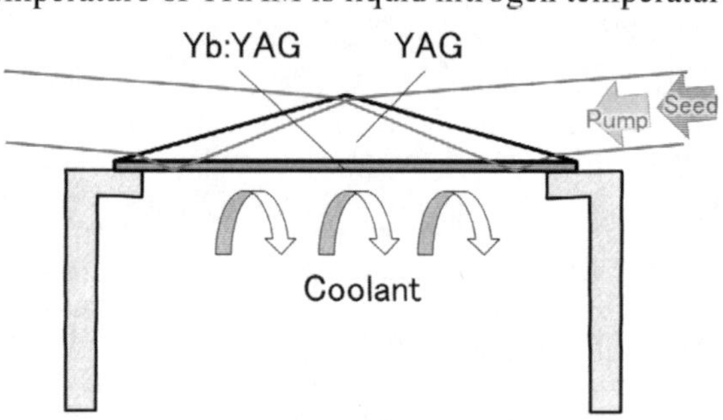

**Figure 2.** Schematic diagram of TRAM and fabrication in centimeter-size.

### *3.2. Flash-lamp-pumped kJ split disk laser*

Neodymium-doped phosphate glass is generally used in huge glass lasers such like NIF, GEKKO XII, and LFEX due to the obtainable excellent material quality and the high storage energy capability. The thermal strength is, however, considerably less than crystals. The Nd:glass slabs used in these system are generally cooled by nitrogen gas flow. To improve repetition rate up to 0.01 Hz, liquid flow cooling is adopted, called split disk amplifier. Figure 3(a)(b) and Table 1 show the brief amplifier design and the specifications, respectively, for the conventional disk amplifier and the split disk amplifier. Showing in fig. 3(b), three or four slabs in a split disk amplifier are required with a large size of 200 x 400 x 10~15 $mm^3$ for Nd:glass. A smaller size of 70 x 140 x 10 $mm^3$ is used for Nd:YAG to avoid parasitic oscillation loss, and 3 x 3 tiled beams are arranged for 1 kJ, which corresponds to 27 to 36 YAG slabs per one amplifier. The output pulse energy was calculated by using six amplifiers after 1- and 2-pass, as a function of seed pulse energy, in fig. 3(c). 2-pass amplification with 1 J seed energy results in more than 1 kJ at small signal gain (SSG) of 2.5 for Nd:YAG. About 1 kJ is expected after 2-pass amplification with 10 J seed at SSG=1.5 of Nd:glass. 2 x 2 tiled beams of these lasers will realize 4 to 5 kJ.

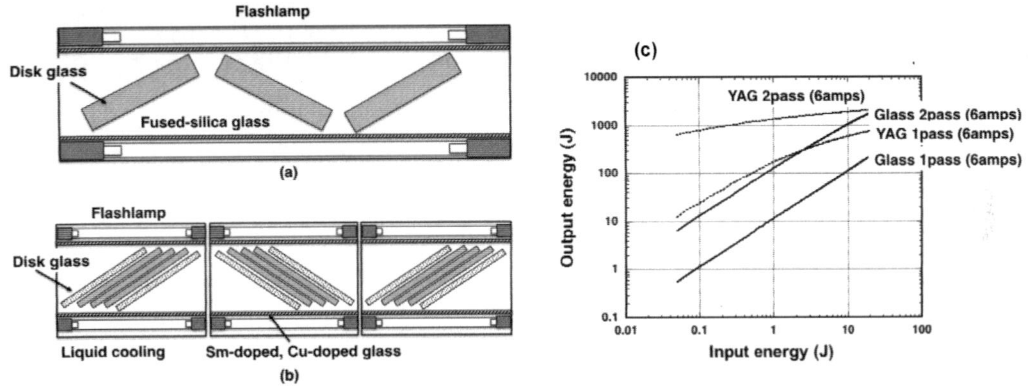


**Figure 3.** Schematic layout of (a) conventional disk amplifier and (b) split disk amplifier by liquid cooling. (c) Numerically calculated output energy with Nd:YAG and Nd:glass.

**Table 1.** Comparison table of (a) conventional disk amplifier and (b) new split disk amplifier.

| Repetition rate (Hz) | Conventional Single shot | New 0.1-0.01 Hz |
|---|---|---|
| Laser material | Laser glass | YAG or Laser glass |
| Cooling method | $N_2$ gas | Liquid (FC or Water) |
| Structure | Split Disk | Split Disk |
| Pumping circuit | Capacitor + Ig SW | Capacitor + IGBT |
| Charging time | 3-5 min. | 0.1 sec |
| Life time | 1 x 10 5 shots | 1 x 10 8 shots |
| Flashlamp | Bore 15mm x Arc 1300mm, 20kV | Bore 10-15mm x Arc 300mm, 3 kV |
| Improvement of efficiency | Ce-doped silica for flashlamp | (1) Nd, Cr: YAG<br>(2) Cu-doped silica tube, Sm-doped glass tube |

## 4. Ultra-broadband OPCPA with p-DKDP

An OPCPA chain has three stages, where different peak power of 1 PW, 20 PW and 50 PW operates at 100 Hz, 0.01 Hz, and 0.0001 Hz, respectively. More than 500 J is required to achieve 50 PW for 10 fs pulse duration. In the OPCPAs with high pulse energy, multi-beams are used as a pump beam. Our experimental results of OPCPA with randomly phased multi-pump-beams ensured that a signal was amplified in energy without distortion of its original phase.[5] KDP crystal in large size of several ten centimeter is obtainable to be suitable for such high pulse energy operation. An available spectral gain width is, however, about 200 nm, which is not enough. Because the pulse duration is 10 fs and ultrabroad gain width of 500 nm at 1 μm is required to obtain a high contrast ratio. A partially deuterated KDP (p-DKDP) crystal is focused on as a nonlinear crystal for ultra-broadband OPA. Figure 4 shows numerical results of gain with 65% p-DKDP by 526.5 nm pump. Non-collinear angle is 0.25° and the crystal thickness is 13 mm. The estimated gain bandwidth is 600 nm. Numerical results at each stages are shown in Table 3. The p-DKDP crystals are grown in centimeter size at different deutration ratio, shown in fig. 4.

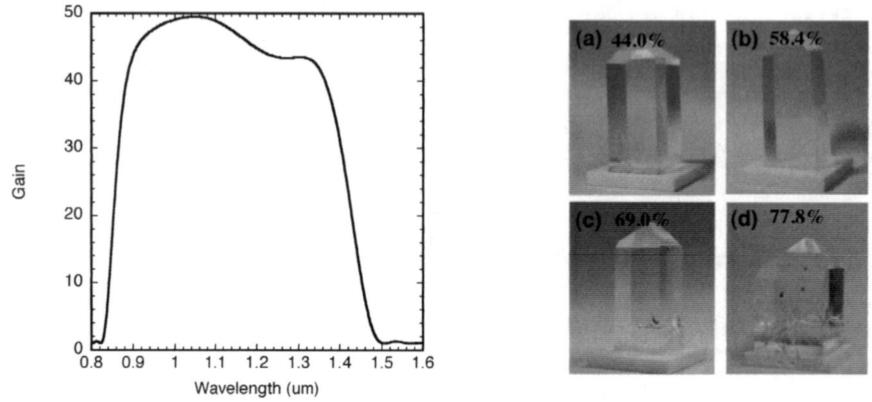


**Figure 4.** (a) Gain spectraum of OPCPA by 65% p-DKDP. (b) Grown p-DKDP.

**Table 3.** Numerical calculation results of OPCPA by three wave mixing equations.

| | Pump Wavelength (nm) | Pump energy (J) | Seed output energy (J) | Beam size (mm) | Deuteration ratio (%) | Length of crystal (mm) |
|---|---|---|---|---|---|---|
| 1PW,100Hz | 515 | 134 | 20 | 180x180 | 45 | 30,30,30, 30,15 |
| 10PW,0.01Hz | 526.5 | 2700 | 220 | 400x400 | 65 | 15 |
| 50PW,0.0001Hz | 526.5 | 6400 | 940 | 800x800 | 65 | 15 |

## 5. Conclusions

Ultrahigh peak power laser of Gekko-EXA has been conceptually designed toward 50 PW, 10 fs. The system is based on OPCPA with tiled multi-beams in high pulse energy operation and with p-DKDP crystals for ultrabroad gain spectrum. Three types of pump sources are proposed, 250 J/100 Hz, 5 kJ/0.01Hz and 12 kJ/0.0001Hz. The various laser output from the Gekko-EXA will be a useful tool for many applications in science, engineering and medical fields. The conceptual design will be brushed up more.

## References

[1] J. Kawanaka, N. Miyanaga, H. Azechi, T. Kanabe, T. Jitsuno, K. Kondo, Y. Fujimoto, N. Morio, S. Matsuo, Y. Kawakami, R. Mizoguchi, K. Tauchi, M. Yano, S. Kudo, and Y. Ogura, Journal of Physics : Conference Series, The fifth International Conference on Inertial Fusion Sciences and Applications (IFSA 2007), vol. 112, 032006 (2008).
[2] J. Kawanaka, Y. Takeuchi, A. Yoshida, S. J. Pearce, R. Yasuhara, T. Kawashima, and H. Kan, Laser Physics vol. **20**, No. 5, 1079-1084 (2010).
[3] H. Furuse, J. Kawanaka, K. Takeshita, N. Miyanaga, T. Saiki, K. Imasaki, M. Fujita, and S. Ishii, Opt. Lett. 34(21), 3439-3441 (2009).
[4] A. Giesen, H. Hugel, A. Voss, K. Wittig, U. Brauch, and H. Opower, Appl. Phys. B **58**, (1994), 365.
[5] T. Kurita, K. Sueda, K. Tsubakimoto, and N. Miyanaga, Opt. Exp. Vol. **18**, 14541 (2010).

# Ion beam control in laser plasma interaction

**S. Kawata[1], T. Izumiyama[1], D. Sato[1], T. Nagashima[1], M. Takano[1], D. Barada[1], Y. J. Gu[1,2], Y. Y. Ma[1,3], Q. Kong[2], P. X. Wang[2], and W. M. Wang[4]**

[1] Department of Advanced Interdisciplinary Sciences, Utsunomiya University, Utsunomiya, Japan
[2] Institute of Modern Physics, Fudan University, Shanghai, China
[3] Department of Physics, National University of Defence Technology, Changsha, China
[4] Institute of Physics, CAS (Chinese Academy of Sciences), Peking, China

E-mail: kwt@cc.utsunomiya-u.ac.jp

**Abstract**. By a two-stage successive acceleration in laser ion acceleration, our 2.5-dimensional particle-in-cell simulations demonstrate a remarkable increase in ion energy by a few hundreds of MeV; the maximum proton energy reaches about 250MeV. The ions are accelerated by the inductive continuous post-acceleration in a laser plasma interaction together with the target normal sheath acceleration and the breakout afterburner mechanism. An intense short-pulse laser generates a strong current by high-energy electrons accelerated, when an intense short-pulse laser illuminates a plasma target. The strong electric current creates a strong magnetic field along the high-energy electron current in the plasma. During the increase phase in the magnetic field strength, the moving longitudinal inductive electric field is induced by the Faraday law, and accelerates the forward-moving ions continously. The multi-stage acceleration provides a unique controllability in the ion energy and its quality.

## 1. Introduction

Ion beam has been used in cancer treatment, ion beam inertial fusion, material sciences and other purposes, and has a unique preferable feature to deposit its main energy, for example, inside a human body, so that cancer cells could be killed by the ion beam [1, 2]. However, conventional ion accelerators tend to be huge in its size and its cost. In this paper a control of laser produced ion beam is discussed for a future compact intense-laser ion accelerator.

The issues in the laser ion accelerator include the energy efficiency from the laser to the ions, the ion beam collimation, the ion energy spectrum control, the ion beam bunching, the ion particle energy control, the intense-laser repetitive operation, the laser and target alignments, etc. The energy efficiency from the laser to ions was improved by using a solid target with a fine sub-wavelength structure [3] or by a near-critical density gas plasma [1, 2]. The ion beam collimation was realized by holes behind the solid target, and the ion beam bunching [2] was successfully realized by the target normal sheath acceleration (TNSA).

When an intense laser illuminates a target, electrons in the target are accelerated and leave from the target; temporarily a strong electric field is formed between the high-energy electrons and the target ions, and then the target ions are accelerated gradually. In this paper we discuss on the control of the ion energy spectrum and the ion particle energy based on a two-stage ion acceleration. A staged laser acceleration would provide a flexibility to control the ion beam quality.

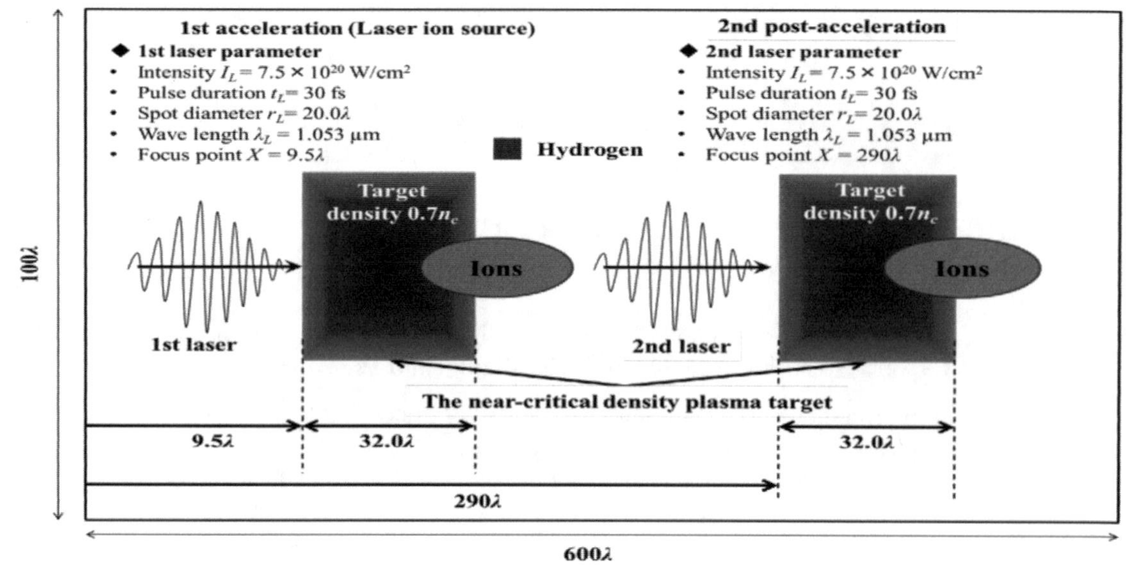


**Figure 1**. Simulation model for the two-stage proton acceleration.

## 2. Two-stage proton acceleration

Figure 1 presents our particle simulation model for the two-stage acceleration. In this specific case we employ the two identical plasma targets and the identical laser pulses. The Gaussian-laser pulse intensity is 7.5x10$^{20}$W/cm$^2$, the pulse length is 30fs and the target density is 0.7n$_c$. The other parameter values are shown in Fig. 1.

When the intense laser pulse propagates through the plasma, it accelerates a part of electrons. The electrons form a high current and generate the azimuthal magnetic field around the laser axis. In the increase phase in the azimuthal magnetic field strength, the strong inductive electric field was generated by the Faraday low. The ions were accelerated by the inductive electric field inside the plasma, as well as the TNSA mechanism, the dipole vortex one [4] and the breakout afterburner (BOA) one [5].

Figure 2 shows the simulation results for the transverse magnetic field, the inductive acceleration field, the TNSA field and the magnetic vortex acceleration field at $t$=140fs and 220fs for the first target illuminated by the first laser. At the later time, the BOA field or the charge-separation acceleration field appears as shown in Fig. 3. The phenomena in Figs. 2 and 3 are identical in the second target illuminated by the second laser. Figure 4 shows the history of the maximal proton energy in the two-stage acceleration. The proton energy reaches 253MeV at $t$=3000fs.

In the first stage the TNSA, magnetic vortex and BOA accelerations contribute mainly to the proton acceleration in the first stage. In the second stage the inductive electric field inside the plasma target also contributes to the continuous proton acceleration. The TNSA and magnetic vortex acceleration fields also appear at the right side of the second target. In addition, the continuous BOA acceleration contributes to the further proton acceleration.

Figure 5 presents the energy spectra of the proton beam generated at the first target at $t$=1400fs and 3000fs. At $t$=1400fs, the proton beam does not yet enter the second target, and then the beam protons are further accelerated in the second target. The energy spectrum at $t$=3000fs shows a preferable energy peak at around 250MeV.

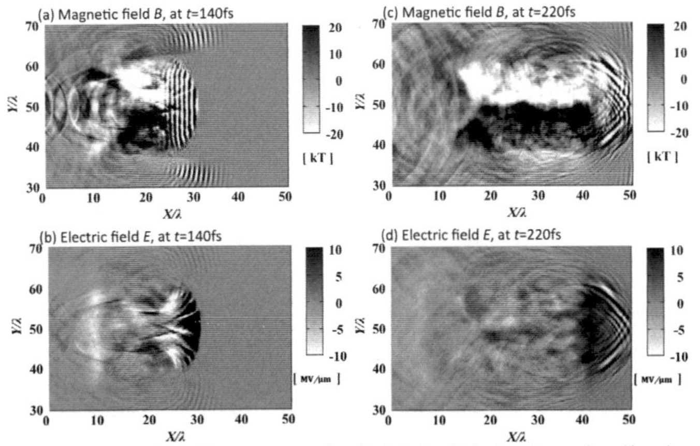


**Figure 2.** (a) The magnetic field $B$, (b) the longitudinal acceleration electric field $E$ at $t$=140fs, (c) $B$ and (d) $E$ at $t$=220fs for the first target illuminated by the first laser. The inductive field is observed inside the plasma. The TNSA and magnetic vortex acceleration fields appear at the right side of the target. The interaction phenomenon is identical for that in the 2nd target irradiated by the 2nd laser.

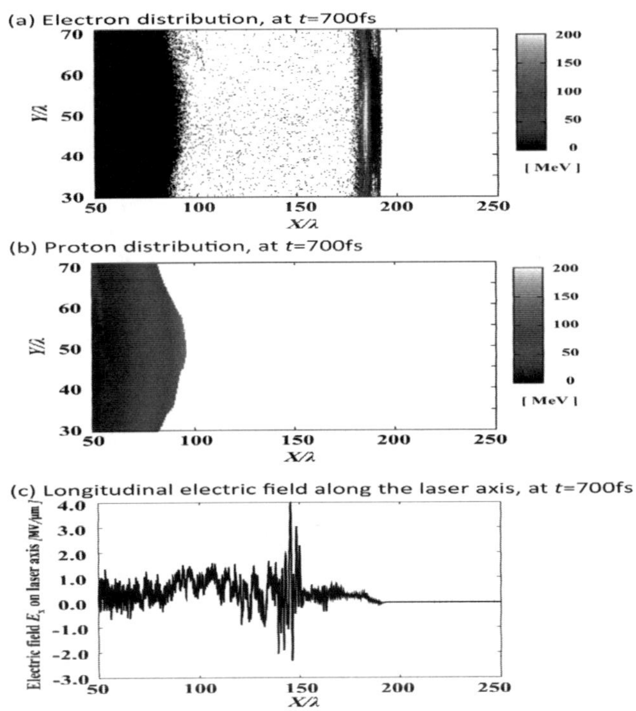


**Figure 3**. Simulation results for (a) the electron distrobution, (b) the proton distribution and (c) the longitudinal acceleration field along the laser axis at $t$=700fs. At this time the BOA field contributes to the proton further acceleration. The 1st laser interacts with the first target. The interaction phenomenon is identical for that in the 2nd target irradiated by the second laser.

## 3. Discussions and conclusions

In our work we employed the two targets and the two laser pulses for the staged laser ion acceleration. The staged acceleration provides a flexibility to control the ion beam quality, though the laser illumination timing and the second laser alignment with the proton beam generated by the first target should be taken care of.

In addition, the first laser should not give a serious influence to the second target. The target length was selected to optimize the proton acceleration efficiency and also to minimize the first laser ember effect to the second target. Actually in our simulations the first laser has successfully almost no effect to the second target. At the same time the first target debris should not affect seriously the second laser propagation. In this paper the distance between the two targets was selected in order to keep this effect small. During the second laser propagation, the first target debris was well removed. We have also checked these effects to obtain the effective post acceleration at the second target.

In order to realize a real laser ion accelerator for a daily use [2], the ion beam should be repetitively and stably generated, should be collimated [2, 6, 7], and also must be controlled well. The ion beam bunching [2] would be further required depending on the usage purpose. To this end, further studies are required [2, 8].

In this paper we have presented the two-stage laser ion acceleration and the remarkable ion energy increase as well as the mono-energetic spectrum. The present results shown in this paper would demonstrate the flexibility of the multi-stage acceleration.

**Acknowledgments**
This work was partially supported by MEXT, JSPS, CORE (Center for Optical research and Education, Utsunomiya Univ.), ASHULA project (JSPS Asia Core to Core Program), and ILE/Osaka Univ. The authors also would like to extend their acknowledgements to Prof. J. Limpouch, Prof. O. Klimo, Prof. A. Andreev, Prof. K. Tanaka, Prof. Y. K. Ho, Prof. Z.M. Sheng, Ptof. W. Zhang, Prof. J. Xie, Prof. H. Zhang and Prof. D. Dai.

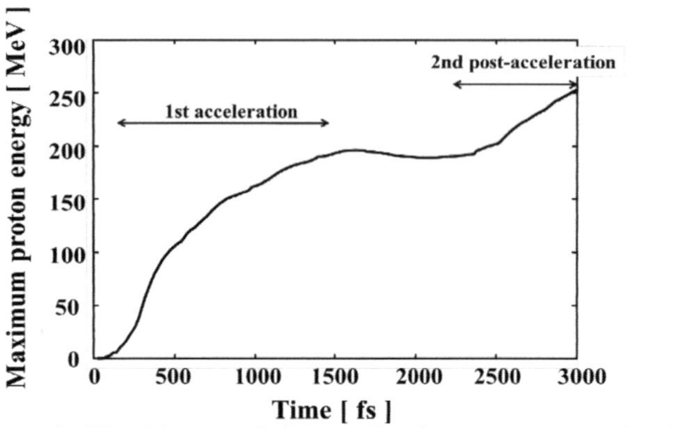


**Figure 4**. The history of the maximal proton energy in the two-stage laser ion acceleration.

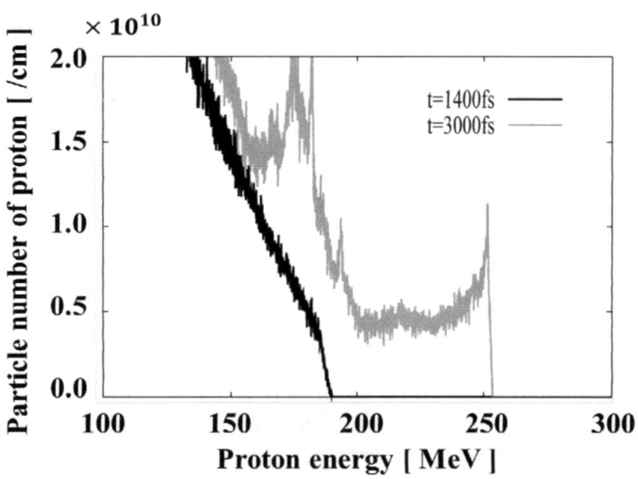


**Figure 5**. The proton beam energy spectra at *t*=1400fs and 3000fs in the two-stage laser ion acceleration. After *t*=1400fs, the proton beam generated at the first target enters the second target. After the post acceleration, the high energy part of the proton beam forms a preferable energy peak at about 250MeV.

## References

[1] Bulanov S V, Esirkepov T Zh, Khoroshkov V S, Kuznetsov A V and Pegoraro F 2002 *Phys. Lett.* A **299** 240

[2] Kawata S, Izumiyama T, Nagashima T, Takanao M, Barada D, Kong Q, Gu Y J, Wang P X, Ma Y Y and Wang W M 2013 *Laser Therapy* **22** 103

[3] Nodera Y, Kawata S, Onuma N, Limpouch J, Klimo O, Kikuchi T 2008 *Phys. Rev.* E **78** 046401

[4] Nakamura T, Bulanov S V, Esirkepov T Z and Kando M 2010 *Phys. Rev. Lett.* **105** 135002

[5] Yin L, Albright B J, Hegelich B M and Fernández J C 2006 *Laser Part. Beams* **24** 291

[6] Bulanov S V and Esirkepov T Zh, 2007 *Phys. Rev. Lett.* **98** 049503

[7] Sonobe R, Kawata, S Miyazaki S, Nakamura M and Kikuchi T 2005 *Phys. of Plasmas* **12** 073104

[8] Macchi A, Borghesi M and Passoni M 2013 *Rev. Mod. Phys.* **85** 751

# Beyond Extreme Ultra Violet (BEUV) Radiation from Spherically symmetrical High-Z plasmas

°**Kensuke Yoshida[1], Shinsuke Fujioka[1], Takeshi Higashiguchi [2], Teruyuki Ugomori[1], Nozomi Tanaka[1], Masato Kawasaki[2], Yuhei Suzuki[2], Chihiro Suzuki[3], Kentaro Tomita[4], Ryouichi Hirose[5], Takeo Eshima[5], Hayato Ohashi[2], Masaharu Nishikino[6], Enda Scally[7], Hiroaki Nshimura[1], Hiroshi Azechi [1], Gerard O'Sullivan[7]**

E-mail: yoshida-k@ile.osaka-u.ac.jp

Abstract. Photo-lithography is a key technology for volume manufacture of high performance and compact semiconductor devices. Smaller and more complex structures can be fabricated by using shorter wavelength light in the photo-lithography. One of the most critical issues in development of the next generation photo-lithography is to increase energy conversion efficiency (CE) from laser to shorter wavelength light. Experimental database of beyond extreme ultraviolet (BEUV) radiation was obtained by using spherically symmetrical high-Z plasmas generated with spherically allocated laser beams. Absolute energy and spectra of BEUV light emitted from Tb, Gd, and Mo plasmas were measured with a absolutely calibrated BEUV calorimeter and a transmission grating spectrometer. $1.0 \times 10^{12}$ W/cm$^2$ is the optimal laser intensity to produced efficient BEUV light source plasmas with Tb and Gd targets. Maximum CE is achieved at 0.8% that is two times higher than the published CEs obtained with planar targets.

## 1. Introduction

Continuous development of semiconductor device technologies gives large impacts on our society. Photo-lithography is a key technology for volume manufacture of high performance and compact semiconductor devices. Smaller and more complex structures in devices can be fabricated by using shorter wavelength light in the photo-lithography. Current photo-lithography uses 193 nm-wavelength ArF laser as a light source. 13.5 nm-wavelength extreme ultraviolet (EUV) is a light of the next generation photo-lithography [1]. The transmittance of EUV light in a material is significantly low, thus EUV lithography (EUVL) system is composed of reflective optics. Mo/Si multilayer mirrors are used in the EUVL system because reflectivity of a Mo/Si multilayer mirror is relatively high (70%) at 13.5 nm. It is proposed to switch wavelength of a EUVL light source from 13.5 nm to a shorter one for enhancement of resolution of the EUVL. The next generation EUV light is called as "beyond EUV (BEUV)" in the

[1] Institute of Laser Engineering, Osaka University (2-6 Yamadaoka, Suita, Osaka 565-0871)
[2] Utsunomiya University (350 Minemachi, Utsunomiya, Tochigi 321-8505)
[3] National Institute for Fusion Science (322-6 Oroshicyo, Tokishi, Gihu 509-5292)
[4] Kyusyu University (6-10-1 Hakozaki, Higashiku, Fukuokashi 812-8581)
[5] Tohoku University (2-1-1 Katahira, Aobaku, Sendaishi, Miyagi 980-8577)
[6] Japan Atomic Energy Agency (4-49, Muramatsu, Toukaimura, Nakagun, Ibaraki 319-1184)
[7] University College Dublin (UCD), Ireland (Belfield, Dublin 4)

Journal of Physics: Conference Series **688** (2016) 012046 doi:10.1088/1742-6596/688/1/012046

EUVL community. Development of BEUV light source has been started to extend the capability of the EUVL [2]. As is shown in Fig. 1, reflectivity of a La/$B_4C$ multilayer mirror is relatively high in the BEUV range, it is more than 50% at 6.7 nm, and bandwidth (BW) of its reflectivity is about 0.6% of the central wavelength, therefore 6.7 nm is a candidate wavelength for the BEUV lithography.

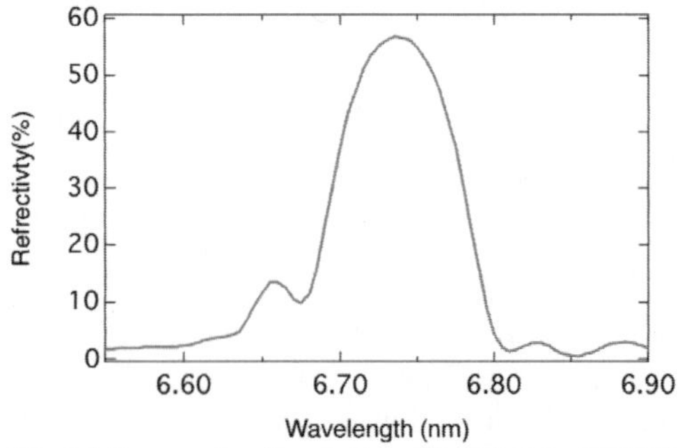

Fig.1 Measured reflectivity of La/$B_4C$ in BEUV range

Optical system of the lithography is composed of, at least, eight mirrors to reduce image aberration, this means only $0.5^8 = 0.4\%$ of the source power is utilized in the BEUVL processes. The BW of BEUV light, which is transmitted through the BEUVL system, becomes narrower (0.3%) than the BW of a single multilayer mirror owing to multiple reflections. One of the most important issues in BEUVL development is to maximize the energy conversion efficiency (CE) from laser to BEUV light. In this study, the CE is defined as energy ratio between 6.7 nm ± 0.3% BW light and the laser.

## 2. Generation of spherically symmetrical plasmas

Laser-produced plasma (LPP) is one of intense EUV and BEUV radiation sources. A target material becomes a high temperature plasma by irradiation of intense laser pulse. Nd:glass laser system, GEKKO-XII at Institute of Laser Engineering, Osaka University, was used in this study. GEKKO-XII laser facility consists of twelve laser beams, whose wavelength, pulse shape, duration and energy are 1.053 $\mu$m, Gaussian, 1.3 ns of FWHM, and 12 J in total, respectively. The twelve laser beams are allocated at twelve faces of the regular dodecahedron to irradiate spherical targets uniformly. The twelve laser beams were focused on spherical Tb, Gd, Mo targets and the laser focal point was displaced based on the target diameter such that the target was illuminated uniformly. Mean squared error of the non-uniformity was 1.8%, which was uniform enough for this research [3]. Laser intensity on the target surface was varied from $3 \times 10^{11}$ to $3 \times 10^{13}$ W/cm$^2$ by adjusting diameters of the spherical targets from 100 $\mu$m 1000 $\mu$m. 2 $\mu$m-thick metallic layer is coated on spherical polystyrene balls.

There are three advantage of generation of q spherically symmetric plasma. Kinetic energy loss caused by lateral plasma expansion along the target surface is suppressed, high CE can be obtained due to reduction of the additional kinetic energy losses. Spherical symmetry of the plasma makes it easy to compare the experimental results with theoretical model and simulation results. The plasma can be observed simultaneously from multiple directions with various diagnostics.

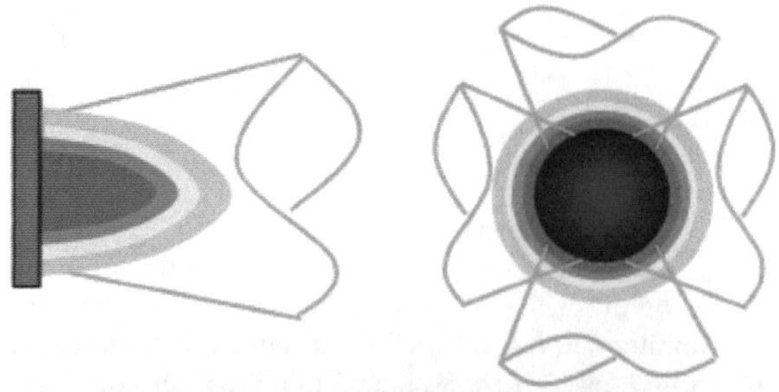
Fig.3 Schematics of laser produced plasmas
(Left) Multi-dimensional plasma expansion causes additional energy loss in the case of a plane target irradiated by a one-dimensional laser beam.
(Right) One-dimensional spherical plasma is generated with a spherical target irradiated by multiple laser beams.

## 3. Energy conversion efficiencies from laser to 6.7 nm light.

Emission spectra from the plasmas depend on the atomic number. According to previous studies, laser-produced Gd, Tb and Mo plasma emits BEUV light near 6.7 nm [4]. Absolute radiant energy of BEUV radiation was measured with a BEUV calorimeter that consists of a Mo/$B_4C$ multilayer mirror and a Zr filter, and a x-ray photodiode as shown in Fig. 4. Dependence of the CE on laser intensity is summarized in Fig.5 (a). The maximum CEs of 0.8% was obtained with Tb and Gd plasmas generated with the laser intensity of 1 x $10^{12}$ W/cm$^2$. The maximum CE (0.45%) obtained with Mo plasmas is much lower than those with Tb and Gd plasmas.

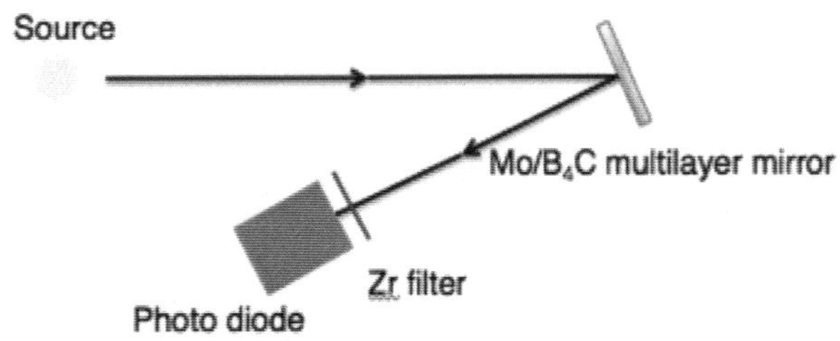


Fig.4 Schematic of energymeter

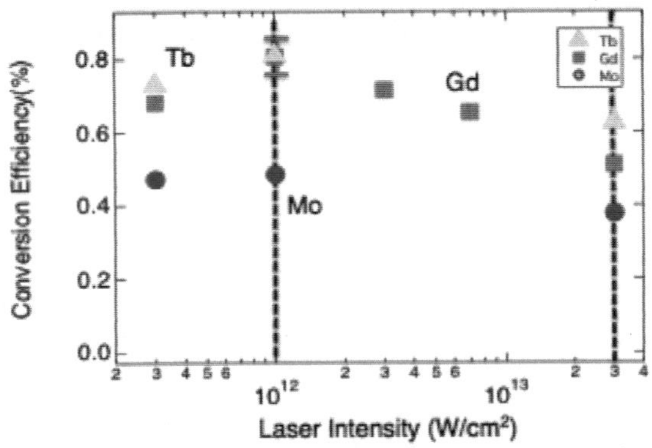


Fig.5 (a) Dependence of CEs on laser intensities and target material those obtained with spherically symmetry plasmas.

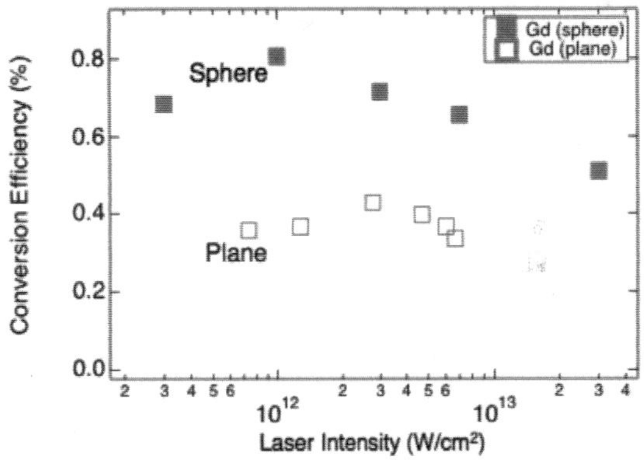


Fig.5 (b) Comparison of CEs from spherical targets (closed square) and those from planar targets (open square). The CEs obtained with spherical plasmas is two times higher than those obtained with planar targets.

Fig.5 (b) shows CEs obtained with spherical (this study) and planar (previous study [5] Gd target. The planar target experiment was conducted by using the same calorimeter and a 10 ns Nd:YAG with a maximum energy of 420 mJ at the wavelength of 1064 nm. This laser beam was injected normally onto solid planar Gd targets and after each laser shots the target surface was replenished. The CEs obtained with spherical plasmas is two times higher than those obtained with planar targets. This result approves that kinetic energy loss caused by multi-dimensional plasma expansion is suppressed in one-dimensional spherically symmetry plasmas.

## 4. Emission spectra of Tb, Gd and Mo plasmas

The BEUV emission was observed with time-integrated transmission grating spectrometer (TGS) coupled with an absolutely calibrated charge coupled device camera. Figure 6 shows BEUV spectra those are normalized with laser energy. The spectra were obtained with Gd, Tb, and Mo plasmas for 1 x $10^{12}$ and 3 x $10^{13}$ W/cm$^2$ of laser intensities. Emission peak of Gd, Tb spectra are located at 6.7 nm. A Mo plasma also emits 6.7 nm, however, the peak at 6.7 nm is relatively weak. BEUV emission from Gd and Tb plasmas decrease with increasing laser intensity, while shorter wavelength emission (< 2 nm) increases. Spectral shape of the Mo plasma becomes flat vaguely in 3-11 nm range with increasing laser intensity.

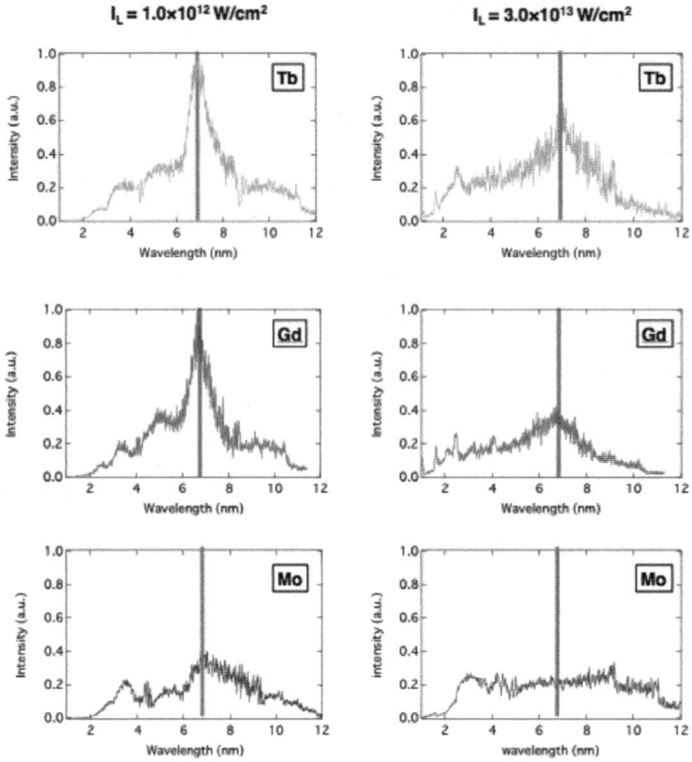


Fig.6 Comparison of BEUV spectra emitted from Tb, Gd, and Mo plasmas. Left and right spectra are obtained with $1.0\times10^{12}$ and $3.0\times10^{13}$ W/cm$^2$ of laser intensities, respectively.

According to the experimental results, optimal laser intensity is 1 x $10^{12}$ W/cm$^2$, electron temperature of laser-produced plasmas is too high to emits BEUV light for more than the optimal laser intensity. [6] Emission lines from highly charged (> 30) ions were observed in a wavelength range of 1 - 3 nm. [7] It shows that the laser energy is consumed for over-ionization, resulting in a degradation of BEUV-CEs.

## 5. Conclusion and future plans

$1.0 \times 10^{12}$ W/cm$^2$ is the optimal laser intensity to generate efficient BEUV light source. Maximum CE is achieved at 0.8% that is two times higher than the published CEs obtained with planar targets. Shorter wavelength emission increases due to higher plasma temperature in the case of the $3.0 \times 10^{13}$ W/cm$^2$ irradiation compared to the optimal intensity case. These data are expected to be of use in future plasma modeling for BEUV source development. It is necessary to clarify the parameters of BEUV source plasmas for further improvement of BEUV-CEs.

## Reference

[1] G. Tallents *et al.*, Nature Photonics **4**, 809 (2010).
[2] J. Benschop, in Proceedings of the 2010 International Workshop on EUVL (EUV Litho Inc., Maui, Hawaii, 2010).
[3] Y. Shimada, H. Nishimura , M. Nakai , K. Hashimoto , M. Yamaura *et al.*, Appl. Phys. Lett. **86**, 051501 (2005).
[4] G. M. Zeng *et al.*, J. Appl. Phys., **72**, 3355 (1992).
[5] T. Cummins *et al.*, Appl. Phys. Lett. , **100**, 061118 (2012).
[6] Bowen Li, T. Otsuka, T. Higashiguchi, N. Yugami, W. Jiang *et al.*, Appl. Phys. Lett. **101**, 013112 (2012).
[7] T. Higashiguchi *et al.*, Appl. Phys. Lett. **100**, 014103 (2012).

# Numerical Simulation on Measurement of Optical and Thermal Properties for Warm Dense Matter Generated by Isochoric Heating with Pulsed Power Discharge Device

**Takashi KIKUCHI, Ryota HAYASHI, Takuya TAKAHASHI, Fumihiro TAMURA, Kazumasa TAKAHASHI, Toru SASAKI, Tsukasa ASO, and Nob. HARADA**

Nagaoka University of Technology, Japan

E-mail: tkikuchi@vos.nagaokaut.ac.jp

**Abstract.** Property data in warm dense matter (WDM) are important to optimize implosion dynamics in a fuel pellet of inertial confinement fusion (ICF). A table-top pulsed power discharge device with isochoric heating using a sapphire hollow capillary was proposed, and was used to generate the extreme state of matter with a well-defined condition. We investigated numerically to generate the WDM by using the pulsed power discharge device. The numerical model was developed by time-dependent one-dimensional thermal diffusion with radiative transfer of multi-group approximation, and the numerical simulation was carried out according with the experimental condition. The achieved temperature of the numerical simulation result was confirmed by the previous experimental result. Also, the radiation energy density was shown at each group of the wavelength of emission.

## 1. Introduction

Warm dense matter (WDM) is in a transition regime from a solid to plasma, and the property data are important to control implosion dynamics in a fuel pellet of inertial confinement fusion (ICF) [1]. Because, to irradiate an energy driver such as intense lasers, high power X-ray, and high-current heavy ion beams, the material phase of the fuel pellet changes from solid to plasma conditions.

Since the WDM regime is in an extreme high pressure situation, it is difficult to create stationary the condition with a measurable setup. Consequently, the properties in the WDM regime are unclear.

Pulsed power discharge devices were used to generate the extreme state of matter from a solid to plasma. For the generation of WDM with a well-defined condition, the apparatus with isochoric heating using a sapphire hollow capillary as a rigid body wall was proposed by using a table-top pulsed-power device [2, 3]. In the experimental apparatus, the emission from the heated sample was observable due to the transparent sapphire capillary.

In our previous studies [4, 5, 6, 7], we numerically investigated to generate the WDM by using pulsed power discharge devices to obtain the properties of the WDM. In this study, the

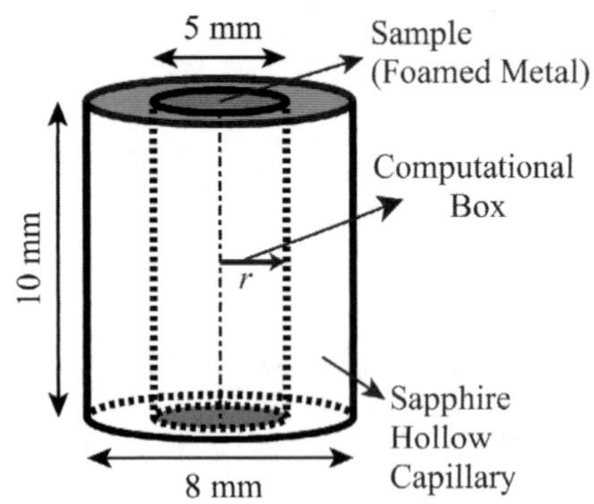


**Figure 1.** Experimental setup in WDM generation for isochoric heating with pulsed power discharge and calculation region.

numerical model with the thermodynamics and the radiation transfer is developed, and the calculation is demonstrated to obtain the thermal and optical properties of the WDM.

## 2. Numerical model

Time-dependent one-dimensional thermal diffusion equation with cylindrical symmetry configuration is numerically solved to simulate the WDM generation in the compact pulsed power discharge experiment [2, 3]. The details of the numerical model used in this paper were explained in Ref. [5].

In the experimental setup [2, 3], we can assume to calculate the phenomena in the foam/plasma ignoring the hydrodynamics, because the fluid dynamics of the sample plasma is limited by the capillary. For this reason, we only calculate the thermodynamics in the foam/plasma without the fluid dynamics of plasma. Figure 1 shows the experimental setup in the WDM generation for the isochoric heating with the pulsed power discharge and the calculation region in this study. In the apparatus, a foamed copper is used as a sample, and is surrounded with a hollow sapphire capillary. The computational box is adjusted in the experimental setup [4, 5, 6, 7]. The inner region ($0 < r < 2.5$ mm) is the foamed copper as a sample, and the outer region (2.5 mm $< r <$ 4 mm) is the sapphire as a rigid capillary. The capillary length is 10 mm.

The density of the foamed copper is 0.1 times the solid density (8920 kg/m$^3$). The mass density of the sapphire is 3970 kg/m$^3$ as the solid. Since the sample is a foamed material, we assumed that the skin effect can be ignored. As a result, the discharge current distribution is assumed as uniform in the copper region.

The initial temperature is set as 300 K by a room temperature in the whole computational region. The conventional thermal property data of copper in solid, liquid, and gas phases are given by Refs. [8, 9, 10]. In the sapphire region, the material parameters for numerical simulation are 42 W/m-K for the thermal conductivity and 750 J/kg-K for the specific heat, as room temperature values.

The radiative transfer equation is solved by the diffusion approximation [11]. The diffusion equation for the radiation energy density $E^g$ is given by

$$\frac{\partial E^g}{\partial t} - \frac{1}{r}\frac{\partial}{\partial r}\left( r \frac{c}{3\,\kappa_{\mathrm{R}}^g}\frac{\partial E^g}{\partial r}\right) = \left(4\pi B^g - cE^g\right)\kappa_{\mathrm{P}}^g, \tag{1}$$

along the radius $r$ in the cylindrical coordinate with a multi-group approximation in the frequency domain group $g$ [1]. Here, ${\kappa_{\mathrm{R}}}^g$ and ${\kappa_{\mathrm{P}}}^g$ are the Rosseland and the Planck absorption coefficients at $g$ th group [12], and $B^g$ is the blackbody intensity.

| Group $g$ | Wavelength $\lambda$ |
|---|---|
| 1 | $> 800$ nm |
| 2 | $800 \sim 680$ nm |
| 3 | $680 \sim 520$ nm |
| 4 | $520 \sim 370$ nm |
| 5 | $370 \sim 340$ nm |
| 6 | $< 340$ nm |

**Table 1.** Relation between group number and frequency domain for multi-group approximation. Here, the wavelength $\lambda$ of radiation is defined by the speed of light $c$ / the frequency $\nu$.

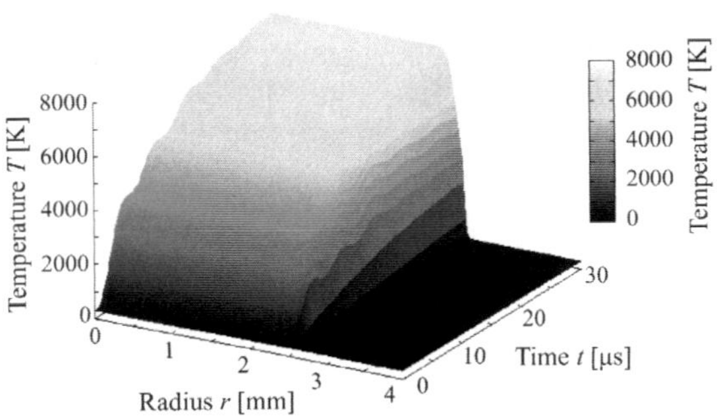


**Figure 2.** Temperature of sample during pulsed power discharge.

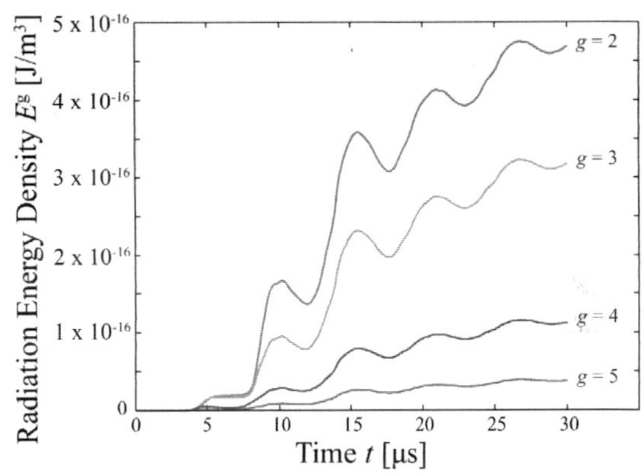


**Figure 3.** Radiation energy density at outer edge at each group.

According to the measuring instrument in the experimental setup [13], the frequency domain is divided and grouped as Table 1.

The $\kappa_\nu(\rho, T, Z^*, \nu)$ is the frequency $\nu$ dependent absorption coefficient [14], where $\rho$, $T$, and $Z^*$ are the density, the temperature, and the mean ionization degree (0.5 is assumed from the previous experimental result [2, 3]), respectively.

The radiative transfer with the flux limited diffusion approximation [15] is calculated by [5]

$$\frac{\partial E^g}{\partial t} = -\frac{c}{r}\frac{\partial E^g}{\partial r}, \tag{2}$$

in the sapphire region. Eqs. (1) and (2) are solved numerically by the implicit method.

The input power history is given by the corresponding experimental data [2, 13].

## 3. Simulation result

Figure 2 shows the temperature of the sample during the pulsed power discharge. The previous experimental result indicated that the sample temperature increased up to 5000 K at around 10 $\mu$s [2]. The calculation result shows that the temperature of the interior of the sample achieved to 5000 K at 10 $\mu$s. As a result, the calculation result corresponds to the previous experimental result until 10 $\mu$s.

However, the temperature for the calculation result indicated to be achieved over 7000 K after 10 $\mu$s. It is implied that the disagreement between the experimental and numerical results was caused due to lack of accurate thermodynamic property data, because of the boiling point for Cu is about 6000 K.

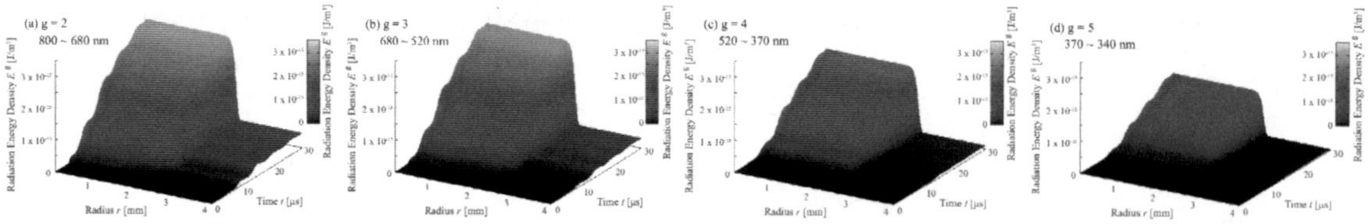


**Figure 4.** Radiation energy density distribution histories at each group during pulsed power discharge, (a) for $g = 2$, (b) for $g = 3$, (c) for $g = 4$, (d) for $g = 5$, respectively.

Figures 3 and 4 show the radiation energy density distributions at outer edge and in the entire of the calculation region at each group. Due to the measurement system [13], it is expected that the emission intensity from the WDM is observed experimentally such as shown in Fig. 3. Since the WDM is in the optically thick, the radiation energy density is uniform distribution at each group in the sample region except around the interface between the sample and the sapphire capillary.

In comparison with the experimental and numerical results in one-group approximation, the observed emission intensity corresponded qualitatively to the calculated radiation energy density [13]. When the fast multi-group analyzing spectroscope by ellipsometry observes experimentally the emission intensity from the generated WDM, the numerical results as shown in Fig. 3 are compared to the experimental results.

## 4. Conclusion

To obtain the property data of WDM, both the experimental and numerical approaches were proposed. In this study, the numerical modeling and the numerical simulation were carried out according with the experimental setup by the pulsed power discharge device with isochoric heating. The achieved temperature of the numerical simulation result was confirmed by the previous experimental result. Consequently, the numerical simulation was useful to understand the thermodynamic properties during the discharge in comparison with the experimental results. To improve the opacity model and so on, the radiative transfer calculation will be quantitatively confirmed with the experimental results.

## Acknowledgments

This work was supported from MEXT Grant-in-Aid for Scientific Research, and by Program for High Reliable Materials Design and Manufacturing in Nagaoka University of Technology, and by JSPS Grant-in-Aid for Challenging Exploratory Research No.25630418.

## References

[1] Atzeni S and Meyer-ter-vehn J 2004 *The Physics of Inertial Fusion: Beam Plasma Interaction, Hydrodynamics, Hot Dense Matter* (New York: Oxford University Press)
[2] Amano Y, *et al.* 2012 *Rev. Sci. Instrum.* **83** 085107
[3] Amano Y, *et al.* 2013 *EPJ Web Conf.* **59** 16005
[4] Kikuchi T, *et al.* 2011 *Proc. Plasma 2011* (Kanazawa) 23P162-R
[5] Kikuchi T, *et al.* 2013 *to be published in NIFS Proc.* ***94***
[6] Kikuchi T, *et al.* 2012 *Proc. EAPPC2012 / BEAMS2012* (Karlsruhe, Germany) P2C-3, Paper_320
[7] Kikuchi T, *et al.* 2013 *NIFS Proc.* **93** 1
[8] Ho C Y, *et al.* 1972 *JPCRD* **1(2)** 279; Ho C Y, *et al.* 1974 *JPCRD* **3(Suppl.1)** 1
[9] NIST Standard Reference Database Number 69, http://webbook.nist.gov/chemistry/
[10] Chase M W 1998 *J. Phys. Chem. Ref. Data* **9** 1
[11] Castor J I 2007 *Radiation Hydrodynamics* (New York: Cambridge University Press)

[12] Drska L and Sinor M 1996 *Computer Phys. Comm.* **997** 1
[13] Saito H, *et al.* 2012 *Proc. 11th APPCPST and 25th SPSM* (Kyoto, Japan) 1-P67; Kikuchi T, *et al. submitted to Transactions on GIGAKU 2*
[14] Mancic A, *et al.* 2010 *High Energy Density Phys.* **6** 21
[15] Levermore C D and Pomraning G C 1981 *Astrophysical J.* **248** 321

# Understanding the stagnation and burn of implosions on NIF

J D Kilkenny[5], J A Caggiano[1], R Hatarik[1], J P Knauer[2], D B Sayre[1], B K Spears[1], S V Weber[1], C B Yeamans[1], C J Cerjan[1], L Divol[1], M J Eckart[1], V Yu Glebov[2], H W Herrmann[3], S Le Pape[1], D H Munro[1], G P Grim[3], O S Jones[1], L Berzak-Hopkins[1], M Gatu-Johnson[4], A J Mackinnon[1], N B Meezan[1], D T Casey[1], J A Frenje[4], J M Mcnaney[1], R Petrasso[4], H Rinderknecht[4], W Stoeffl[1] and A B Zylstra[4].
[1]Lawrence Livermore National Laboratory, Livermore, CA 94550, USA
[2]Laboratory for Laser Energetics, Univ. of Rochester, Rochester, NY 14623, USA
[3]Los Alamos National Laboratory, Los Alamos, NM 87545, USA
[4] Plasma Science and Fusion Center, Massachusetts Institute of Technology, Cambridge, MA 02139, USA
[5]General Atomics, San Diego, CA, 29186, USA

**E-mail:** kilkenny@fusion.gat.com

**Abstract.** : An improved the set of nuclear diagnostics on NIF measures the properties of the stagnation plasma of implosions, including the drift velocity, areal density (ρr) anisotropy and carbon ρr of the compressed core. Two types of deuterium-tritium (DT) gas filled targets are imploded by shaped x-ray pulses, producing stagnated and burning DT cores of radial convergence ($C_r$) ~ 5 or ~20. Comparison with two-dimensional modeling with inner and outer surface mix shows good agreement with nuclear measurements.

## 1. Introduction

The x-ray drive, inertial confinement fusion approach at the National Ignition Facility (NIF) uses x-rays to implode a multi-layered, spherical shell at the center of the hohlraum. For the ignition goal the radiation drive is shaped to launch 4 staged shocks to control the adiabat and to achieve high compression of the cryogenically layered fuel [1]. However for low foot ignition designs the measured capsule neutron yield is substantially lower than simulations [2]. It has been known for decades, based on theory [3] and experiments with high power lasers [4, 5] that implosions are unstable to the Rayleigh-Taylor (RT) instability which can lead to ablator material mixing into the hot spot and reducing the yield. In an effort to understand the yield discrepancy DT gas filled plastic capsules are imploded as simulants of the cryogenically layered implosions that are fielded in an "ignition" hohlraum [1]. The gas fill results in a lower convergence, less stressing implosion than a layered implosion. Improved nuclear diagnostics measure the drift velocities and carbon (from plastic ablators) areal densities of the imploded cores as well as the normal attributes, yield, ion temperature, and down scattered fraction. With improved modelling good agreement with simulations is seen, for example the neutron yield agrees with simulations to ~ 80%.

## 2. Gas Filled Implosions to Emulate Layered Implosions

Two types of gas filled implosions are discussed here. The first type is a set of x-ray drive "exploding pushers". These easier implosions achieve low areal density with a high yield and a low drift velocity.

Described elsewhere [6] these are single shock, $C_r$ ~5 x-ray driven implosion producing a well-understood stagnation plasma. Here they are used to test the upgraded diagnostics of section 3.

The second type is a set of higher convergence, $C_r$ ~ 20, and more unstable implosions. The design consists of four shocks and a target like the layered implosions [1], but with a thicker plastic shell to emulate the DT ice, and a DT gas fill. Simulations of this second type represent the conditions of the actual experiment, including the as-shot capsule metrology, capsule surface roughness, and the tent (the thin foils which hold the capsule in the hohlraum) as seeds for the growth of hydrodynamic instabilities. The 2D zoning resolves up to a spherical harmonic mode of l=100. As in reference [2] the drive is reduced to match the measured implosion timing, kinematics, and low-mode asymmetry.

## 3. The improved set of neutron diagnostics

Recent upgrades to NIF's nuclear diagnostics allow better measurements of the properties of implosions. The upgrades were validated on the $C_r$ ~ 5 single-shock implosions, and then used on the $C_r$ ~ 20 four-shock implosions.

*3.1. Neutron Time of Flight (nToF) detectors:* The initial set of NIF diagnostic had two spectrally resolving nToFs [8] with 20 m neutron flight paths. A recently added nToF with an 18 m flight path and near the south pole of the chamber gives three nearly orthogonal views of the implosion with spectrally resolving detectors. Modifications to the original design on all three include faster scintillators [9] and faster recording, allowing measurements of the drift velocity of the neutron source and the carbon back-scattering edges [10]. As before these nToFs also measure yield, "ion temperature $T_{ion}$ " and down scattered neutrons as described in reference [11]. Note that the temporal and spatial averaging of the nToFs gives a perceived "ion temperature" which includes an inhomogeneous velocity contribution as well as the thermal motion. However "$T_{ion}$" from the simulations in this paper also includes this factor-at least to the fidelity of the simulation.

Additionally these nToFs measure the average drift velocity for the first time. For example if the neutron emitting imploded core has an average drift velocity towards a detector it will slightly increase the neutron velocity in the laboratory frame resulting in an earlier arrival. To demonstrate this ability to measure drift velocities three shots were performed with a symmetric drive, a 4% upwards, and 4% downwards drive asymmetry as shown in figure 1.

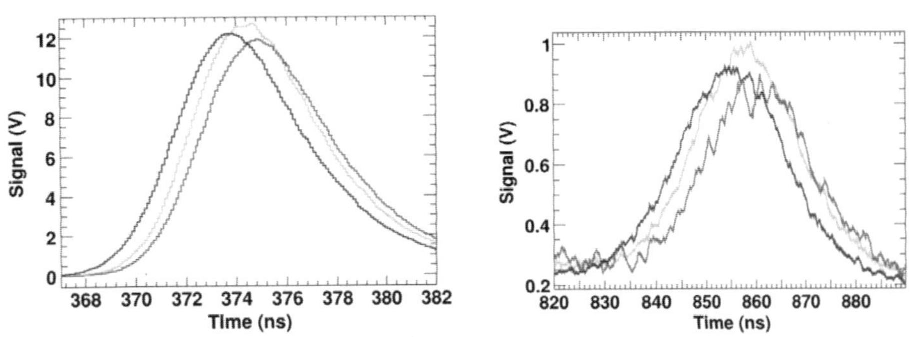


Figure 1: nToF signals near the south pole for DT neutrons (left) and DD neutrons (right). The blue trace is for a deliberate downwards $P_1$ drive asymmetry of 4%( N130625), the green trace is for a symmetric drive (N130507) and the red trace is for a deliberate upwards $P_1$ drive asymmetry of 4% (N130814).

To calculate the drift velocities the bang time, when neutron production peaks is measured by time resolved x-ray emission and DT-fusion gamma-ray emission to an accuracy of about 100 psec. The velocity of the neutrons in the lab frame uses the measured $T_{ion}$ [12]. The drift velocities are obtained by fitting the arrival time for both the DT and the DD neutrons and in km/s are 76, 1, and -76 and

62, -12 and -78 respectively for the down, symmetric and up implosions. The drift velocities increase, are close to zero, and decrease as expected with the imposed asymmetry. Additionally the DD and DT velocities are close to agreeing within the fitting error of ~ 10 km/s.

*3.2. Flange Neutron Activation Diagnostic System (fNADS):* The fNADS system on NIF [13] uses $^{90}Zr(n,2n)^{89}Zr$ activation measured by zirconium (Zr) slugs mounted on 19 target chamber flange covers. Measured activation anisotropy can be due to anisotropy in the flux of un-scattered neutrons or a drift velocity as this activation cross-section increases with neutron energy. The relative variation in the activation, $A/A_{ave}$ for the two asymmetric experiments of section 3.1 are shown in Figure 2 showing an increase in activation in the direction of motion. If the measured velocities are used to correct the anisotropy in yield the remaining anisotropy due to areal density variations is obtained. Some level of consistency with the down-scattered ratio is obtained but is under further investigation.

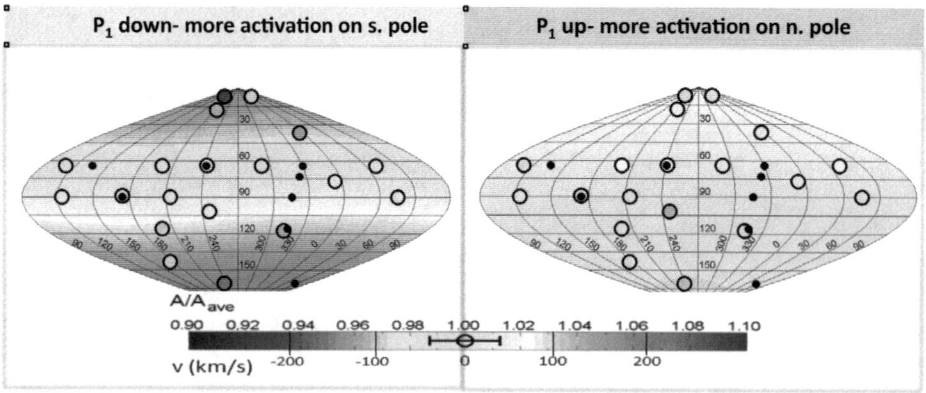


Figure 2: The variation in relative activation $A/A_{ave}$ for the asymmetric experiments of section 3.1.

*3.3. The Gamma Reaction History (GRH) detector:* In addition to the DT-fusion gamma rays the GRH diagnostic measures the 4.4 MeV gamma line from the $^{12}$**C(*n,n'*γ)** reaction [14]. The GRH was absolutely calibrated by measuring the 4.4 MeV γs emitted from a puck of carbon held 5.5 cm from a calibrated exploding pusher target. On the implosions described here, carbon nuclei remaining from the plastic ablator are excited by 14 MeV DT neutrons and emit 4.4 MeV γ s, thus measuring the $^{12}C$ areal density. Figure 3 shows the good agreement between GRH measured and calculated (Hydra) 4.4 MeV γ yields for the single shock (bottom left) and four shock implosions (top right).

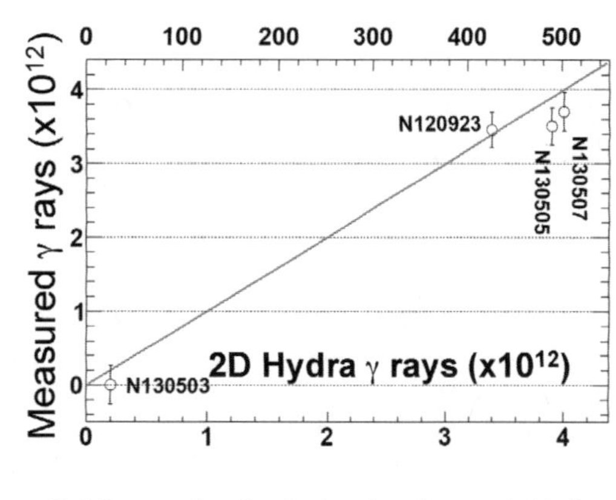


Figure 3: Measured and calculated carbon γ yields from GRH. Upper horizontal axis is the $^{12}C$ ρr (mg/cm$^2$).

## 4. Comparison of experiments with simulations and conclusions

The DD and DT yields are compared to simulations in figure 4. These simulations follow those in reference [2] but with a mix prescription for the inner surface as it is decelerated by the gas, representing 0.05 times the distance to the fall line. The fall line fraction is a simple parameterized prescription to introduce time dependent mix based on the Read and Youngs [7] mix measurements. The fall line mix parameter of 0.05 was chosen to match more sophisticated modeling. The yields are calculated in 1D, 2D and in 2D with this mix prescription showing excellent agreement for the single shock implosions and good agreement (80%) for the four-shock system. The nToF-derived $T_{ion}$ from the DD and DT burn is also compared to simulations post processed for a simulated $T_{ion}$ in figure 4 showing excellent agreement with the simulations. The stronger dependence on $T_{ion}$ for the DT reaction over time and spatial averaging of the nToFs gives a larger measured "$T_{ion}$ " from the DT reaction than the DD reaction. The differences are 400 eV and 200 eV for the single and four shock implosions respectively.

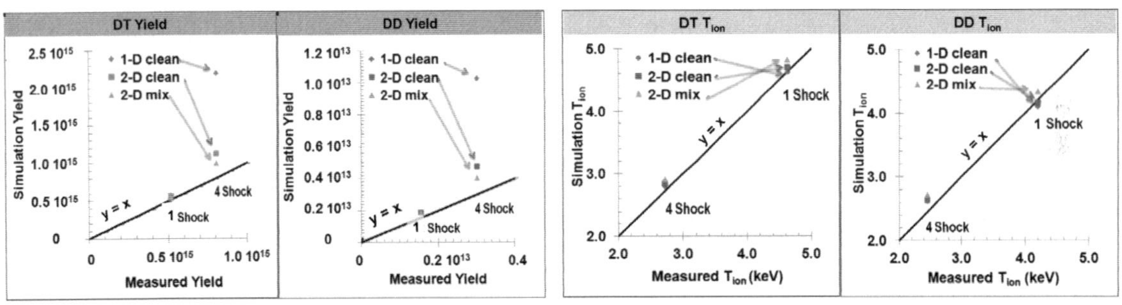


Figure 4: Calculated and measured DT and DD neutron yields (left) and corresponding instrumentally measured $T_{ions}$ for the one shock implosion ( N130503) and a four shock implosion (N130505).

This work was performed under the auspices of the U.S. Department of Energy by Lawrence Livermore National Laboratory under Contract DE-AC52-07NA27344.

**References**

[1] M. J. Edwards et. al., Phys. of Plasmas **20**, 070501 ( 2012).
[2] D. S. Clark et. al., Phys. of Plasmas **20**, 056318 (2013).
[3] J.D. Lindl and W.C. Mead, Phys. Rev. Lett. **34**, 1273 (1975).
[4] J.S. Wark, J.D. Kilkenny, et. al., Appl. Phys. Lett. **48**, 969 (1987).
[5] J. D. Kilkenny, Phys. Fluids B**2**, 1400 (1990).
[6] S. LePape, submitted to Phys. Rev. Lett.
[7] K. I. Read, 1984 *Physica D* **12**, 45 and D. L.Youngs, 1984 *Physica D* **12**, 32.
[8] V. Yu. Glebov et. al., Rev. Sci. Instrumen. **81**,10D325 (2010).
[9] R. Hatarik et. al., Rev. Sci. Instrumen. **83**, 10D911 (2012).
[10] C. J. Forrest et. al., Rev. Sci. Instrumen. **83**, 10D919 (2012).
[11] R. A. Lerche et. al., Rev. Sci. Instrumen. **81**, 10D319 (2010).
[12] L. Ballabio, J. Kallne and G. Gorini, Nuclear Fusion, **38**, 1723 (1999).
[13] C. B. Yeamans et. al., Rev. Sci. Instrumen. **83**, 10D313 (2012).
[14] D. B. Sayre et. al., Rev. Sci. Instrumen. **83**, 10D905 (2012).

# Progress toward a unified kJ-machine CANDY

**Yoneyoshi Kitagawa[1], Yoshitaka Mori[1], Osamu Komeda[1,3], Ryohei Hanayama[1], Katsuhiro Ishii[1], Shinichiro Okihara[1], Kazuhisa Fujita[1], Suisei Nakayama[1], Takashi Sekine[2], Nakahiro Sato[2], Takashi Kurita[2], Toshiyuki Kawashima[2], Takeshi Watari[2], Hirofumi Kan[2], Naoki Nakamura[3],Takuya Kondo[3], Manabu Fujine[3], Hirozumi Azuma[4], Tomoyoshi Motohiro[4], Tatsumi Hioki[4], Mitsutaka Kakeno[4], Yasuhiko Nishimura[5], Atsushi Sunahara[6], Yasuhiko Sentoku[7], Eisuke Miura[8], Yasunobu Arikawa[9], Takahiro Nagai[9], Yuki Abe[9], Satoshi Ozaki[10] and Akira Noda[11]**

[1]The Graduate School for the Creation of New Photonics Industries, 1955-1 Kurematsu-cho, Nishi-ku, Hamamatsu, Shizuoka 431-1202 Japan
[2]Development Bureau, Hamamatsu Photonics K.K. 1820, Kurematsu-cho, Nishi-ku, Hamamatsu 431-1202 Japan
[3]Advanced Material Engineering Div., TOYOTA Motor Corporation, 1200, Mishuku, Susono, Shizuoka 410-1193 Japan
[4]TOYOTA Central Research and Development Laboratories, Inc., 41-1Yokomichi, Nagakute, Aichi 480-1192 Japan
[5]Toyota Technical Development Corp., 1-21 Imae, Hanamoto-cho, Toyota, Aichi 470-0334 Japan
[6]Institute for Laser Technology, 1-8-4 Utsubo-honmachi, Nishi-ku, Osaka 550-0004 Japan
[7]Department of Physics, University of Nevada, Reno 1664 N VIRGINIA ST Reno, NV 89557
[8]National Institute of Advanced Industrial Science and Technology, 1-1-1 Umezono, Tsukuba, Ibaraki 305-8568 Japan
[9]Institute of laser Engineering, Osaka University, 2-6 Yamadaoka, Suita,Osaka 565 Japan
[10]National Institute for Fusion Science, 322-6 Oroshi Toki, Gifu 509-5292 Japan
[11]Advanced Research Center for Beam Science, Institute for Chemical Research, Kyoto University, Gokasho, Uji, Kyoto 611-0011 Japan

E-mail: `kitagawa@gpi.ac.jp`

**Abstract.** To construct a unified experimental machine CANDY using a kJ DPSSL driver in the fast-ignition scheme, the Laser for Fast Ignition Experiment (LFEX) at Osaka is used, showing that the laser-driven ions heat the preimploded core of a deuterated polystyrene (CD) shell target from 0.8 keV to 2 keV, resulting in $5 \times 10^8$ DD neutrons best ever obtained in the scheme. 4-J/10-Hz DPSSL laser HAMA is for the first time applied to the CD shell implosion-core heating experiments in the fast ignition scheme to yield neutrons and also to a continuous target injection, which yields neutrons of $3 \times 10^5$ n/4$\pi$sr n/shot.

## 1. Introduction

To realize the Inertial Confinement Fusion (ICF) power plant, one key issue is a high-repetition-rate laser of kilojoules or greater. Others are the fuel fabrication and high-repetition injection, and ignition-high gain physics. Power plant technology, such as an innovative wall materials, will need to be developed. We divide the roadmap for achieving a fusion plant into three phases. The zeroth phase involves developing kJ drivers. The first phase is to develop a breakeven machine that uses a 100-kJ driver. Here the fuel is compressed to 100× the solid density or more. The second phase is to demonstrate a commercial reactor. The fuel may be compressed to around 500×.

The zeroth goal is to construct an unified experimental machine CANDY. We have developed a laser-diode-(LD-)pumped laser system HAMA with a repetition rate of 10 Hz[1, 2]. We have done the fast ignition experiment at OSAKA to test this scheme. The Laser for Fast Ignition Experiment (LFEX), the most intense laser in the world, heated the preimploded core of a deuterated polystyrene (CD) shell target and increased the core temperature from 0.8 keV to 2 keV. We also present the target injection and engagement results for the first time[3, 4].

## 2. Direct Ion Heating of Laser-Imploded Core in a Fast-Ignition Scheme

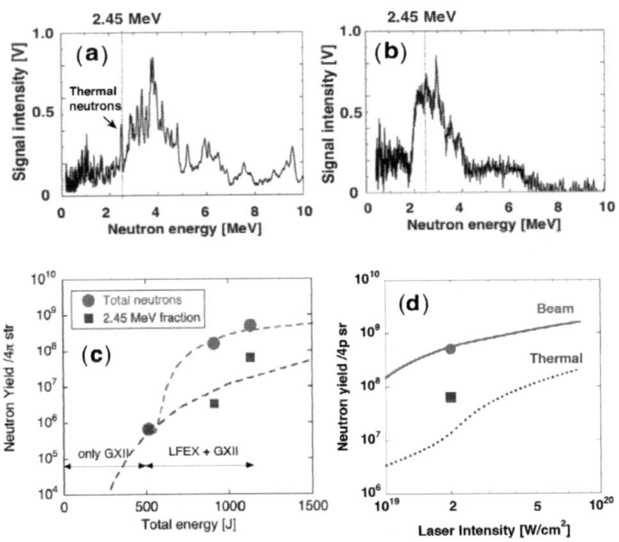


**Figure 1.** (A) Neutron TOF signal from Scintillator 1. The sensitivity is 1 count/5.6 neutrons. The solid angle of the detector is $1.4 \times 10^{-4}$ sr. The small neutron peak at 2.45 MeV is thermal one. (b) Spectrum from Scintillator 2. (c) Neutron yields from Scintillator 1 as a function of LFEX + GXII lasers energy. Red dots are the total yields and blue squares are 2.45-MeV fraction of the yield. Vertical error is 16%. Blue line is the calculated implosion due to the thermal fusion effect. Red line is heated by LFEX using a beam fusion model. (d) STAR 1D: Beam fusion (Dotted line) and thermal fusion (solid line) yields versus LFEX intensity on target. Red circle and blue solid squares are the experiments from (b).

LFEX, faced to a core as close as possible, directly heated a preimploded core, enhancing $D(d,n)^3He$-reacted neutron (DD neutron) yields by a factor of 1000 ($5 \times 10^8$ n/4$\pi$ sr), the best ever obtained in fast-ignition scheme. A CD shell was polarily imploded by two counter beams (1.3 ns-515 J) of the GEKKO XII (GXII) green laser. Then, we focused the 600-J/1.5-ps LFEX laser onto the naked core from the side, vertical to the laser axis. The core temperature increased by a factor of two, i.e., from 0.8 keV to 2 keV. The laser-driven ions (deuteron and carbon$^{+6}$, too) heated the core. Beam-fusion neutrons and thermal neutrons are simultaneously observed. Figure 1 (a) shows neutron TOF spectrum from the gated Liquid Scintillator (Scintillator 1) at 13.35 m apart from the target in the direction of -69.13° (right-forward) to the LFEX incidence. The 4$\pi$ sr yield is $5.1 \pm 0.8 \times 10^8$ neutrons against 613-J LFEX + 515-J GXII. The small neutron peak at 2.45 MeV is thermal one with a fraction of 13% or $6.4 \times 10^7$ n/4$\pi$ sr. The spectral width gives us the core temperature of 2 keV. (b) is the spectrum from Scintillator 2 at 2.5 m in 90° (bottom) to the LFEX axis: the same Shot. (c) is neutron yield from Scintillator 1 as a function of LFEX + GXII lasers energy. GXII 515-501 J. Red dots are the total yields and blue squares are 2.45-MeV fraction of the yield. Vertical error is 16%. Blue line is the calculated

Journal of Physics: Conference Series **688** (2016) 012049 doi:10.1088/1742-6596/688/1/012049

implosion due to the thermal fusion effect. Red line is heated by LFEX using a beam fusion model.

STAR ID hydrocode predicts that deuterons are related to beam fusion and carbons to thermal fusion. (d) is STAR 1D resutls. Dot lines and solid lines are beam-fusion originated and thermal fusion originated yields, respectively. Red circle and blue solid squares are the experiments from (b). This scheme is a potential path to fast-ignite the core at high gain fusion.

## 3. High-rep. rate fusion using fast heating of a compactly imploded CD core

A 4J/0.4-ns output of a laser-diode-pumped high-repetition laser HAMA is divided into four beams, two of which counter-illuminate double-deuterated polystyrene foils separated by 100 $\mu$m for implosion[5]. The remaining two beams, compressed to 110 fs for fast heating, illuminate the same paths. As shown in Figs. 2(c) and (d), we clearly found that the heating pulses heat the imploded core, emitting x-ray radiations >20 eV, and yielding some $10^3$ thermal neutrons. [PRL 108, 155001 (2012)][2]

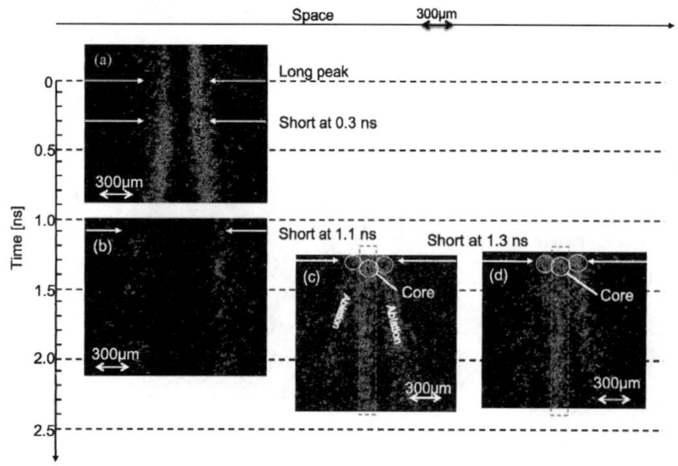


**Figure 2.** X-ray streak image of the imploding-heating foils (a) at 0.3 ns delay. No core is formed. (b) At 1.1-ns delay, no core emission is seen. (c) and (d) At 1.3-ns delay, the core is heated. Three bright emissions are at the central core (circled) and at the peripheral (side) areas (circled). Ablation is the ablated plasma. Au cathode detects emissions not less than 20 eV.

## 4. Laser engagement of 1-Hz-injected pellets, channel borings, and neutrons[4]

Deuterated polystyrene ($C_8D_8$ or CD) bead pellets, after free-falling for a distance of 18 cm at 1 Hz, are successfully engaged in flight by two counter laser beams from a diode-pumped, ultra-intense laser HAMA[1]. The snapshot of a flying pellet at the instance of engagement by using a $2\omega$ harmonic laser probe is in Fig. 3. The laser energy on the pellet, pulse duration, and wavelength are 0.63 J per beam, 104 fs, and 811 nm, respectively, and the intensity is $4.7 \times 10^{18}$ W/cm$^2$. The irradiated pellets produce D(d,n)$^3$He-reacted neutrons (2.45 MeV DD neutrons) with a yield of $9.5 \times 10^4/4\pi$ sr/shot at maximum. The laser is found out to bore a straight channel through the 1-mm-diameter beads. A 10$\mu$m-diameter clear open channel is discovered inside the engaged pellet. It is not observed when the beads are attached to any metal support. The results of pellet injection, repetitive laser illumination, and hole boring indicate potentially useful technologies and findings for the next step in realizing inertial fusion energy.

Currently, the observed pellet placement accuracy is such that only 20% of the pellets are within 0.49 mm of the focal point on the laser perpendicular direction, leading to an average hit rate of less than 20%. To produce DD neutrons routinely, the laser intensity is required to be more than $10^{18}$ W/cm$^2$. Neutrons more than $1.0 \times 10^4/4\pi$ sr were observed for 2.5 % of these pellets. Improvements in the mechanical function are required to increase the placement accuracy. This result represents a step toward CANDY and the laser fusion.

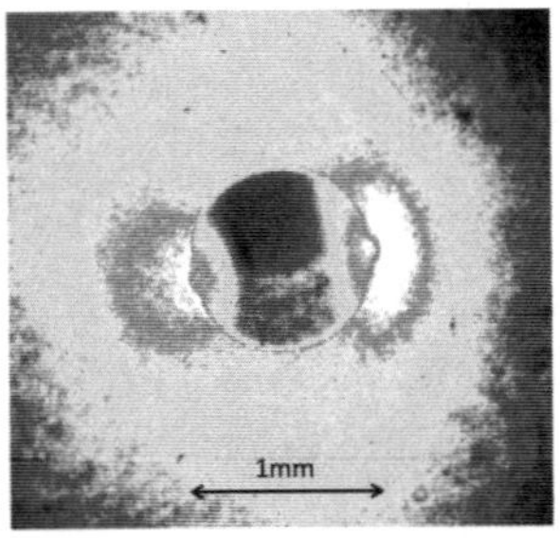


**Figure 3.** Snapshot of a flying pellet engagement by laser probe. The probe is perpendicular to the counter beam axis. The two counter beams from HAMA simultaneously irradiate the surface of a falling pellet. CCD camera with an IF filter (394 nm), opened for 20 ns, captures both the probe shadow and self-emissions. The probe is synchronized with the two counter beams.

## 5. CANDY Concept

The zeroth goal is to construct an unified experimental machine CANDY. DT or DD cryogenic fuel pellet is injected at 10 Hz, to which the counter implosion beams are engaged, followed coaxially by the fast heating beams. Neutron Yield is $5 \times 10^{12}$/shot and energy gain is around 0.7%. A liquid Pb-Li blanket is to catch some mount of neutrons and radiations.

| | | |
|---|---|---|
| Neutron yield | DT | $5 \times 10^{12}$/shot |
| | DD | $5 \times 10^{10}$/shot |
| Implosion laser | Energy | 2 kJ/shot |
| | Wavelength | 500 nm |
| | Repetition rate | 10 Hz |
| Heating laser | Energy | 2 kJ/shot |
| | Wavelength | 1000 nm |
| Gain | DT | 0.007 [190 W] |
| | DD | $1.5 \times 10^{-5}$ [0.3 W] |

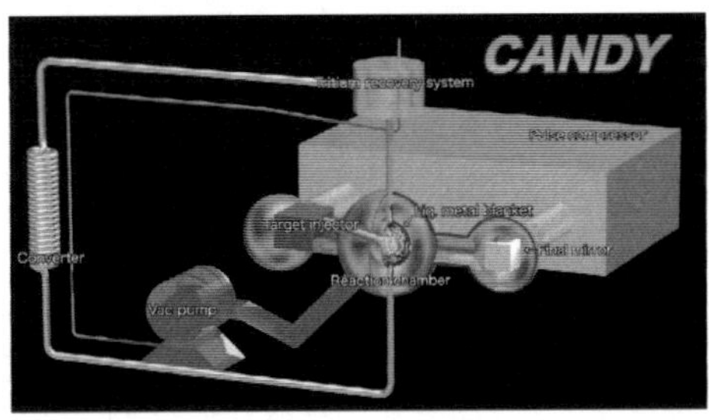


**Figure 4.** Conceptual Design of CANDY.

## 6. Summary

To construct CANDY using a kJ DPSSL driver in the fast-ignition scheme, we pursue the fast ignition and the high repetition laser implosion. We succeed in the continuous target injection and engagement. LFEX results show us the ion contribution to the fast ignition.

This study was partially conducted under the collaboration program of the Institute of Laser Engineering (ILE), Osaka University. We would like to thank the LFEX construction and operation, GXII operation, measurement, and target fabrication teams at the ILE. Y.S is supported by NNSA/US DOE under Contracts DE-FC02-04ER54789 and DE-SC0008827.

*References*

[1] Kitagawa Y *et al.* 2011 *Plasma and Fusion Research: Letters* **6** 1306006
[2] Kitagawa Y *et al.* 2012 *Phys. Rev. Lett.* **108** 155001
[3] Komeda O *et al.* 2013 *Plasma and Fusion Research: Rapid Communications* **8** 1205020
[4] Komeda O *et al.* 2013 *Scientific Reports* 6 September
[5] Mori Y *et al.* 2013 *Nucl. Fusion* **53** 073011

Journal of Physics: Conference Series **688** (2016) 012050 doi:10.1088/1742-6596/688/1/012050

# Precision Measurement of Delbrück Scattering via Laser Compton Scattered $\gamma$-rays

**J K Koga[1] and T Hayakawa[2]**

[1]Quantum Beam Science Directorate, Japan Atomic Energy Agency, Kizugawa, Kyoto, Japan
[2]Quantum Beam Science Directorate, Japan Atomic Energy Agency, Tokai, Ibaraki, Japan

E-mail: koga.james@jaea.go.jp

**Abstract.** Precision measurements such as the muon anomalous magnetic moment have indicated deviations from the standard model and have in turn prompted higher precision theoretical calculations. Delbrück scattering is the scattering of photons off the Coulomb field of nuclei via virtual electron-positron pairs and has been measured using $\gamma$-rays from radioactivities and following neutron capture reactions. However, because low flux $\gamma$-rays from nuclear transitions have been used in the low photon energy regime fairly large uncertainty exists in the data. In addition, due to the complexity and time consuming nature of the theoretical calculation the scattering cross sections are obtained from tables with interpolation between the tabular values. In recent years high flux $\gamma$-ray sources via laser Compton scattering (LCS) using energy-recovery linacs have been proposed. These sources allow measuring the Delbrück scattering with high precision. We will present our own independent calculations for the scattering cross section and show what precision can be obtained using the new LCS $\gamma$-ray sources in the low photon energy regime.

## 1. Introduction

In recent years deviations between quantum electrodynamics (QED) calculations and experimental measurements such as the muon anomalous magnetic moment have hinted at new physics beyond the standard model [1]. The verification of which has required having high precision theoretical calculations [2]. Delbrück scattering, which is the scattering of $\gamma$-rays off the Coulomb field of nuclei via virtual electron-positron pairs [3], has been experimentally measured and also shows deviations from the lowest order QED calculations for large Z nuclei [4]. In the low energy regime ($<$ 20 MeV) unpolarized sources were used and there was a fairly large uncertainty in the data [4]. Because of the difficulty of the theoretical calculation, the scattering cross sections can currently only be obtained from tables [5].

Soon high flux highly monoenergtic highly polarized $\gamma$-ray sources via laser Compton scattering (LCS) will become available [6, 7]. As a result, high precision measurements of the Delbrück scattering cross section will be possible and will also require high precision theoretical calculations. In this paper our own calculations for the scattering cross section are presented.

These calculations and comparison with experimental results could determine whether or not deviations from QED occur. In addition these calculations could be used to determine the cause of the expectedly large index of refraction of $\gamma$-rays measured for Silicon attributed to Delbrück scattering [8]. If confirmed by our calculations, this could lead to the development of $\gamma$-ray lenses which would have applications in nuclear physics [9] and in astrophysics [10].

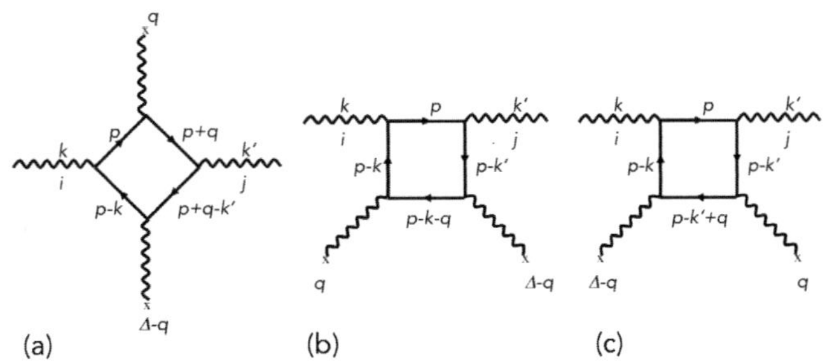


**Figure 1.** Lowest order Feynman diagrams for Delbrück scattering. The incoming and outgoing photons are represented by their 4-vector momenta, $k$ and $k'$, and polarization directions, $i$ and $j$, respectively. $\Delta = k' - k$ the momentum transfer. $x$'s are the Coulomb field where $q$ is the momentum of the field.

## 2. Delbrück scattering

We use natural units where $c = \hbar = 1$. Figure 1 shows the lowest order Feynman diagrams for Delbrück scattering [3]. The differential scattering cross section calculated from these diagrams is given by [3]:

$$\frac{d\sigma}{d\Omega} = \frac{1}{(4\pi)^2}|\mathcal{M}_0^{(D)}|^2 \tag{1}$$

where

$$\mathcal{M}_0^{(D)} = ie^6 Z^2 \int \frac{d^3q}{(2\pi)^3} \frac{M_0(k,k',q)}{(\vec{q}^2)(\vec{\Delta}-\vec{q})^2}, \tag{2}$$

$e$ is the electron charge, $Z$ is the atomic number of the scattering nucleus, $k = (\omega, \vec{k})$ and $k' = (\omega, \vec{k}')$ are the incoming and outgoing photon 4-vectors , respectively, $\vec{\Delta}$ is the momentum transfer taken as $|\vec{\Delta}| = 2\omega \sin(\theta/2)$ and $\theta$ being the scattering angle, and $q$ is the momentum of the Coulomb field, with

$$\begin{aligned} M_0(k,k',q) = \frac{i\delta_{ij}}{24\pi^2} + \int \frac{d^4p}{(2\pi)^4}\{ & \frac{\mathrm{Tr}[\gamma_i(\not{p}+m)\gamma_0(\not{p}+\not{q}+m)\gamma_j(\not{p}+\not{q}-\not{k}'+m)\gamma_0(\not{p}-\not{k}+m)]}{(p^2-m^2)[(p+q)^2-m^2][(p+q-k')^2-m^2][(p-k)^2-m^2]} \\ & + \frac{\mathrm{Tr}[\gamma_i(\not{p}+m)\gamma_j(\not{p}-\not{k}'+m)\gamma_0(\not{p}-\not{k}-\not{q}+m)\gamma_0(\not{p}-\not{k}+m)]}{(p^2-m^2)[(p-k')^2-m^2][(p-k-q)^2-m^2][(p-k)^2-m^2]} \\ & + \frac{\mathrm{Tr}[\gamma_i(\not{p}+m)\gamma_j(\not{p}-\not{k}'+m)\gamma_0(\not{p}-\not{k}-\not{q}'+m)\gamma_0(\not{p}-\not{k}+m)]}{(p^2-m^2)[(p-k')^2-m^2][(p-k-q')^2-m^2][(p-k)^2-m^2]}\} \end{aligned} \tag{3}$$

where the $\gamma$'s are the Dirac matrices, $\not{p} \equiv \gamma_\mu p^\mu$, $q' = \Delta - q$, $m$ is the electron mass, $(i, j)$ refer to the photon polarization directions, and "Tr" refers to the trace of the matrix expressions.

The explicit expressions including cancellations of similar terms are given elsewhere [3].

The integrals over $p$ in Eqs. 3 can be calculated using the specialized package of routines, LoopTools [11]. The multi-dimensional integrals over $q$ in Eq. 2 can be calculated via Monte Carlo methods in the package of routines, CUBA [12].

## 3. Preliminary Results

Here we present preliminary numerical results for scattering element $M_0(k, k', q)$ for the first trace (trace 1) of Eq. 3. Figure 2 shows the geometry we choose for our calculations. For simplicity we calculate the particular component where the initial photon is polarized

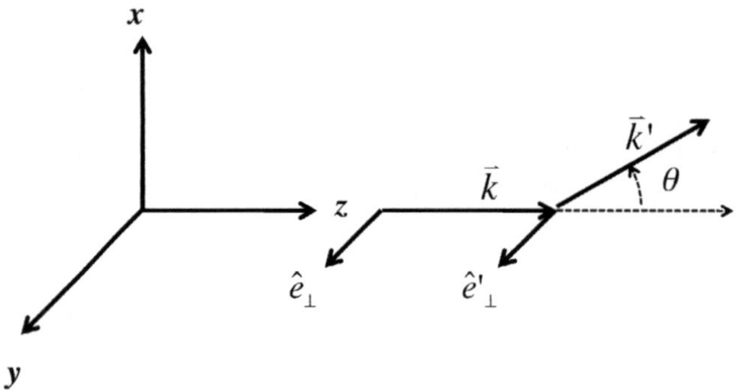


**Figure 2.** Geometry for Delbrück scattering. $\vec{k}$ and $\vec{k}'$ are incoming and outgoing photon wavenumber vectors where the incoming photon is assumed to be propagating along the $z$-axis and both photon energies are equal to $\omega$. $\theta$ is the scattering angle. $\hat{e}_\perp$ and $\hat{e}'_\perp$ are the initial and final photon polarization vectors, respectively, where $\perp$ means that the polarizations are perpendicular to the scattering plane.

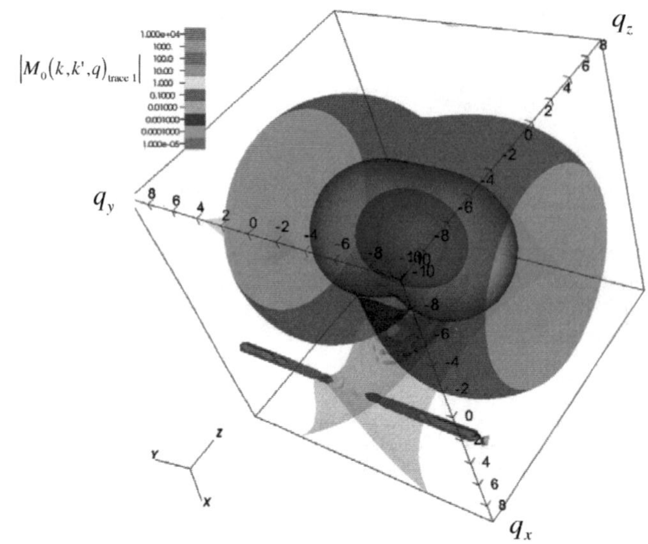


**Figure 3.** Preliminary values of $|M_0(k, k', q)_{\text{trace 1}}|$ for a photon energy of 2.09 MeV with $\theta = 24.6°$ where the axes are in units of MeV.

perpendicular to the scattering plane with polarization vector, $\hat{e}_\perp$, and the final photon is also polarized perpendicular to the scattering plane with polarization vector, $\hat{e}'_\perp$. Figure 3 shows preliminary values of $|M_0(k, k', q)|_{\text{trace 1}}$ for a photon energy of 2.09 MeV with $\theta = 24.6°$ where $(q_x, q_y, q_z)$ are in units of MeV. We have calculated values for $q$ up to $\pm 10$ MeV using a $50 \times 50 \times 50$ size grid in the $q_x$, $q_y$, and $q_z$ directions, respectively. As apparent from the figure near $q_x \approx 4$, $q_z \approx -8$, and for all $q_y$, there are regions where the values are large at single grid points. For different values of scattering angle and photon energies these regions differ and can have very large values. These are most likely numerical instabilities, which typically occur in loop calculations, and are currently under investigation. One possible source of numerical instability is the occurrence of small Gram determinants, which could occur when obtaining the coefficients used to represent the tensor integrals in Eq. 3 in terms of simpler scalar integrals [13]. We are investigating packages such as CutTools [14], which handle these and other types of numerical instabilities. In the next phase of the calculation we will work on eliminating these instabilities, calculate the values for $|M_0(k,k',q)|$ of all the traces in Eq. 3, calculate the integral $\mathcal{M}_0^{(D)}$ using the CUBA library[12], and then compare with published values [5].

## 4. Discussion and Conclusions

Proposed future LCS $\gamma$-ray sources will have high total flux ($10^{13}$ ph/s), almost 100% polarization, small energy spread (FWHM $10^{-3}$), and be tunable in energy up to 13.2 MeV [6, 7]. These sources may allow measuring only the Delbrück scattering component among the various scattering mechanisms occurring at low photon energies with previously unavailable high precision scanning specific energies and scattering angles. Since accuracy goes as $1/\sqrt{N}$ where $N$ is the scattered photon numbers, obtaining high accuracy measurements of the cross sections should be possible with these new sources.

In this paper we have presented the calculation method for obtaining the Delbrück scattering cross section using packaged routines. We have presented some preliminary calculations of the scattering matrix for the first trace in Eq. 3, which show some numerical instability, for a simple configuration. In our next calculations we will examine the possibilities to suppress these instabilities, calculate the scattering matrix for all three traces, obtain the integrated cross section and examine the regimes where Delbrück scattering can be measured accurately for specific energies and scattering angles.

## References

[1] Bennett G W *et al.* (Muon (g-2) Collaboration) 2006 *Phys. Rev. D* **73**(7) 072003 URL `http://link.aps.org/doi/10.1103/PhysRevD.73.072003`
[2] Aoyama T, Hayakawa M, Kinoshita T and Nio M 2012 *Phys. Rev. Lett.* **109**(11) 111808 URL `http://link.aps.org/doi/10.1103/PhysRevLett.109.111808`
[3] Cheng H, Tsai E C and Zhu X 1982 *Phys. Rev. D* **26**(4) 908–921 URL `http://link.aps.org/doi/10.1103/PhysRevD.26.908`
[4] Schumacher M 1999 *Radiation Physics and Chemistry* **56** 101 – 111 ISSN 0969-806X URL `http://www.sciencedirect.com/science/article/pii/S0969806X99002893`
[5] Falkenberg H, Hünger A, Rullhusen P, Schumacher M, Milstein A and Mork K 1992 *Atomic Data and Nuclear Data Tables* **50** 1 – 27 ISSN 0092-640X URL `http://www.sciencedirect.com/science/article/pii/0092640X9290023B`
[6] Hajima R, Hayakawa T, Kikuzawa N and Minehara E 2008 *Journal of Nuclear Science and Technology* **45** 441–451 (*Preprint* `http://www.tandfonline.com/doi/pdf/10.1080/18811248.2008.9711453`) URL `http://www.tandfonline.com/doi/abs/10.1080/18811248.2008.9711453`
[7] Habs D, Tajima T and Zamfir V 2011 *Nuclear Physics News* **21** 23–29
[8] Habs D, Günther M M, Jentschel M and Urban W 2012 *Phys. Rev. Lett.* **108**(18) 184802 URL `http://link.aps.org/doi/10.1103/PhysRevLett.108.184802`
[9] Thirolf P and Habs D 2014 *The European Physical Journal Special Topics* **223** 1213–1219 ISSN 1951-6355 URL `http://dx.doi.org/10.1140/epjst/e2014-02175-1`
[10] Lacki B C, Horiuchi S and Beacom J F 2014 *The Astrophysical Journal* **786** 40 URL `http://stacks.iop.org/0004-637X/786/i=1/a=40`
[11] Hahn T and Pérez-Victoria M 1999 *Computer Physics Communications* **118** 153 – 165 ISSN 0010-4655 URL `http://www.sciencedirect.com/science/article/pii/S0010465598001738`
[12] Hahn T 2005 *Computer Physics Communications* **168** 78 – 95 ISSN 0010-4655 URL `http://www.sciencedirect.com/science/article/pii/S0010465505000792`
[13] Yundin V 2012 *Massive loop corrections for collider physics* Ph.D. thesis Humboldt-Universität zu Berlin, Berlin, Germany
[14] Ossola G, Papadopoulos C G and Pittau R 2008 *Journal of High Energy Physics* **2008** 042 URL `http://stacks.iop.org/1126-6708/2008/i=03/a=042`

# Control of imploded core plasma by changing beam arrangement of Gekko XII

**M Koga[1], Y Ishii[2], T Sogo[2], H Shiraga[2], H Nagatomo[2], T Ozaki[3], S Fujioka[2] and H Azechi[2]**

[1] Graduate School of Engineering, University of Hyogo, Japan
[2] Institute of Laser Engineering, Osaka University, Japan
[3] National Institute for Fusion Science, Japan

koga@eng.u-hyogo.ac.jp

**Abstract.** We proposed the new laser arrangement to implode a target, a 12-beam implosion, and tried it in a fast ignition integration experiments. The concept of the 12-beam implosion is as follows. The three laser beams near a cone are shifted not to irradiate a cone and a long gold cone is used to prevent the irradiation of those three laser beams with fundamental frequency. In experiments, time-resolved two-dimensional X-ray images of an imploded core plasma were obtained by an X-ray framing camera. It is found that the spherical symmetry of the imploded core plasma is improved and the motion of the imploded core plasma towards the cone is suppressed in half in the 12-beam implosion. We believe that the 12-beam implosion will become a useful concept in controlling the imploded core plasma.

## 1. Introduction

Fast ignition is one proposed technique for achieving a high fusion energy gain factor in inertial fusion research [1-3]. The Fast Ignition Realization Experiment (FIREX-I) project is going on at Osaka University by using a Gekko XII laser system and an LFEX laser system. Usually, three laser beams of the Gekko XII are not used for implosion because they irradiate a cone. However, it was found that this laser unbalance caused an ununiform implosion and made the imploded core plasma move toward the cone [4]. This moving core plasma may collide with the cone tip and break it. This may disturb the generation of fast electrons from the cone tip and become a serious problem for effective heating. Therefore, we proposed the new laser arrangement to implode a target, a 12-beam implosion, and tried it in a fast ignition integration experiments.

## 2. The concept of a 12-beam implosion

Figure 1 shows a schematic diagram of a target and configurations of the Gekko XII laser beams. The figure has rotational symmetries through 120 degrees. For example, three beams (Beam 3, 5, 9) have the same positional relationship to the target and each beam disposed deviated by 120 degree in the cone axis. It is clearly seen that the three laser beams near the cone (Beam 3, 5, 9) irradiate the edge of the cone.

Therefore, we changed beam configuration as shown in Fig. 2. The three laser beams (Beam 3, 5, 9) are shifted not to irradiate a cone. We cannot focus a laser beam tightly after a random phase shifter because of its optical property. Tight focus of this additional three beams is needed for cone clearance. Therefore we did not use the random phase shifters for additional three beams which are different from conventional implosion beams. In addition, we extended the length of the cone by 1.7 mm in order to prevent irradiation by those three beams with fundamental frequency. (The conventional cone size is shown as a broken line in Fig. 2.) This change does not make target alignment particularly difficult.

The energy of three beams (Beam 3, 5, 9) was determined as follows. First of all, we estimated the area of irradiation from the focal length of the beams. From the calculation, it was found that the area of irradiation by the beams near the cone became 0.72 times smaller than that of the other beams. Secondly, the energies of the other beams are decreased by about 20 % because of random phase shifters. Therefore, in order to conform the energy density of the beam near the cone to that of the other beams, the energy of the beam near the cone set to 60 % of that of the other beams.

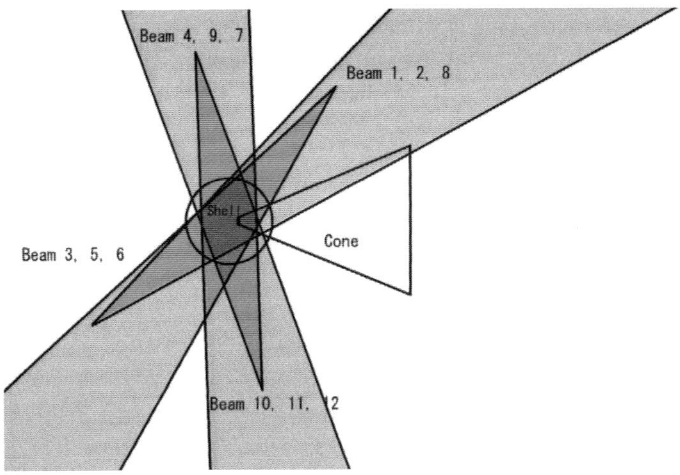


**Figure 1.** Schematic diagram of a target and an original configuration of the Gekko XII laser beams in the 9-beam implosion.

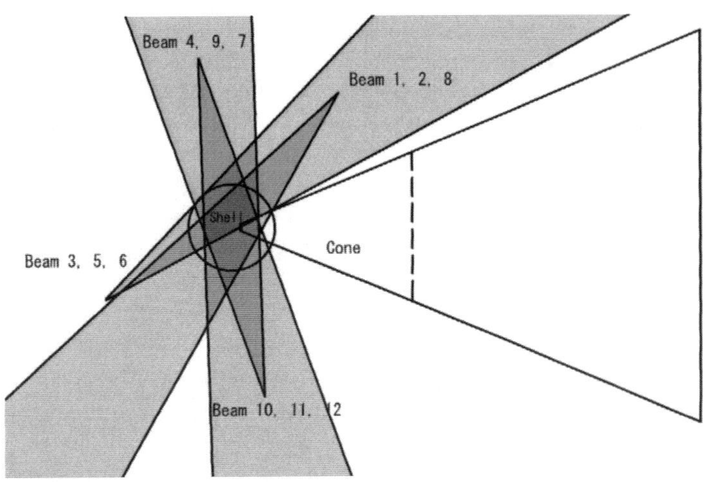


**Figure 2.** Schematic diagram of a target and a configuration of the Gekko XII laser beams in the 12-beam implosion. The conventional cone size is shown as a broken line.

## 3. Experimental setup

The experiments were performed using the Gekko XII and the LFEX laser system at the Institute of Laser Engineering, Osaka University. The wavelength of the Gekko XII laser was 0.53 μm and the pulse duration was 1.5 ns. The total energy of the GXII implosion laser was about 2 kJ. The wavelength of the LFEX laser is 1.05 μm and the pulse duration was set to 1~5 ps.

An X-ray framing camera (XFC) [5-7] was used to obtain the time-resolved two-dimensional X-ray images of the imploded core plasma. Details of our XFC system are described in ref. [7]. An array of 10 μm diameter pinholes was used to create separate two-dimensional X-ray images. The blast shield made of 25-μm-thick beryllium was attached in front of the pinhole disk in order to shield it from debris and soft X-rays. Platinum X-ray mirrors were introduced to prevent noise caused by high energy X-rays. Only thermal X-rays (< 5keV) were reflected by these mirrors. The direct incidence of high energy X-rays through pinhole disk was shielded by a lead block. Two-dimensional images were projected onto a multi-channel-plate (MCP) with a size of 40 mm × 50 mm. The MCP had two gold striplines across its face. An electric pulse was sent down a stripline to obtain the image under each pinhole only when the pulse was present at that location on the MCP. The intensified images were recorded with a CCD camera. The time interval between frames was about 80 ps and the spatial resolution was 17.3 μm. XFC was set perpendicular to the cone axis.

Targets were deuterated plastic shells with gold cones. The diameter of the shells was 500 μm. Long cones, whose length were 2.7 mm, were used in 12-beam implosion experiment. Conventional cone, whose length were 1 mm, were used in the other experiments. The distance between the cone tip and the center of its shell was set to 50 μm.

## 4. Results and Discussion

Figure 3 (a) and 3 (b) show XFC images of imploded core plasmas obtained in the usual 9-beam implosion and 12-beam implosion, respectively. The position of the chamber center is indicated by a white cross. The position of the cone tip is marked with a white dotted line. In every experiment, the tip position was checked before it was set and we checked that the shell center was located at the chamber center before it was shoot. Therefore, we can guarantee their position in XFC images. In Fig. 3, it is clearly seen that the imploded core plasma shows more spherical form in the 12-beam implosion than that in the 9-beam implosion. We defined the edge of the imploded core plasma as the point where the X-ray intensity exceeded the noise level. The radius of the imploded core plasma was calculated by using this edge position and the center of mass position. The spherical symmetry of the imploded core plasma was quantified by the ratio between the standard deviation and the mean value of the imploded core plasma radius. The calculated value was 0.09 in the 12-beam implosion, while it was 0.14 in the 9-beam implosion. This means that the spherical symmetry of the imploded core plasma is certainly improved in the 12-beam implosion.

As shown in Fig. 3 (a), in the 9-beam implosion, the imploded core plasma is generated near the cone tip and moves toward the cone. On the other hand, as shown in Fig. 3 (b), in the 12-beam implosion, the imploded core plasma is generated away from the cone tip and stays there. In ideal case, three beams near the cone balance with opposite three beams and the imploded core plasma will be formed at the center. In this case, additional three beams might be too intense to balance and generate the imploded core plasma away from the cone tip. The velocity of the imploded core plasma was estimated from the time variation of the center of mass position. In the 12-beam implosion, it is found to be suppressed to about a half ($0.45 \times 10^7$ [cm/s] ) of that in the 9-beam implosion ($1.20 \times 10^7$ [cm/s] ). Two dimensional simulation [8, 9] also confirmed this idea. We are going to publish a paper including this simulation result soon.

Figure 4 shows high energy X-ray intensity observed by an electron spectrometer. As shown in Fig. 4, the photon number of high energy X-ray in the 12-beam implosion may be slightly high relative to it in the 9-beam implosion. If preformed plasma is generated inside of the cone prior to an irradiation of LFEX laser, it is heated overly by irradiation of LFEX laser and generates high energy X-ray by bremsstrahlung radiation. Above mentioned result may suggest the fear of the preformed plasma generated inside of the cone by irradiation of three beams near the cone or shock waves generated by those beams. Further study is needed to discuss this result.

## 5. Summary

The new beam layout, a 12-beam implosion, was tested with Gekko XII and LFEX lasers. Time-resolved two-dimensional X-ray images of the imploded core shows that the spherical symmetry of the imploded core plasma is improved and the motion of the imploded core plasma towards the cone is suppressed in the 12-beam implosion. Although further study is needed to clear the high X-ray generation problem, we think that the 12-beam implosion must become a useful concept to control the imploded core plasma.

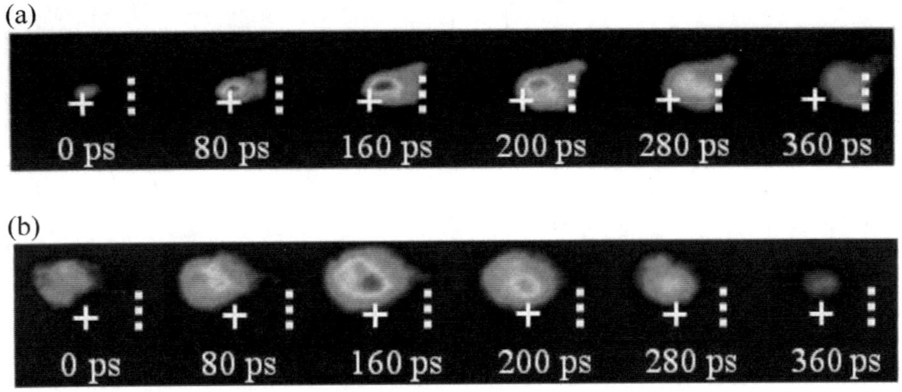


**Figure 3.** XFC images of imploded core plasmas obtained in (a) the conventional 9-beam implosion and (b) the 12-beam implosion, respectively. The position of the chamber center is indicated by a white cross. The position of the cone tip is marked by a white dotted line.

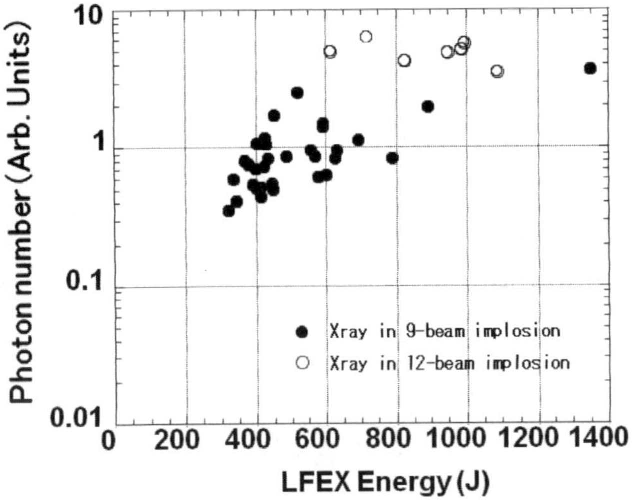


**Figure 4.** The high energy X-ray intensity observed by an electron spectrometer.

## References

[1] N. G. Basov, S. Y. Guskov, and L. P. Feoktistov: J. Soviet Laser Research 13, (1992) 396.
[2] M. Tabak, J. Hammer, M. E. Glinsky, W. L. Kruer, S. C. Wilks, J. Woodworth, E. M. Campbell,

and M. D. Perry: Phys. Plasmas 1, (1994) 1626.
[3] R. Kodama et al.: NATURE 412, (2001) 798.
[4] J. Zhong, H. Shiraga, and H. Azechi: Rev. Sci. Instrum. 79, (2008) 10E907.
[5] D. K. Bradley, P. M. Bell, J. D. Kilkenny, R. Hanks, O. Landen, P. A. Jaanimagi, P. W. McKenty, and C. P. Verdon: Rev. Sci. Instrum. 63, (1992) 4813.
[6] J. D. Kilkenny: Rev. Sci. Instrum. 63, (1992) 4688.
[7] M. Koga, T. Fujiwara, T. Sakaiya, M. Lee, K. Shigemori, H. Shiraga and H. Azechi: Rev. Sci. Instrum. 79, (2008) 10E909.
[8] H. Nagatomo, T. Johzaki, A. Sunahara, H. Shiraga, and K. Mima: J. Phys. France IV 133, (2006) 397.
[9] H. Nagatomo, T. Johzaki, A. Sunahara, and K. Mima: J. Plasma Phys., 72, (2006) 791.

# Velocity evolution of electro-magnetically driven shock wave for beam-dissociated hydrogen interaction experiment

**Kotaro Kondo, and Yoshiyuki Oguri**

Research Laboratory for Nuclear Reactors, Tokyo Institute of Technology,
2-12-1 Ookayama, Meguro-ku, Tokyo 152-8550 JAPAN

E-mail: `kotaro@nr.titech.ac.jp`

**Abstract.** We present the velocity measurements in electro-magnetic shock tube for beam interaction experiment by three methods; laser refraction, photodiode for self-emission, and high speed framing camera. The laser refraction showed that the average shock velocity was 6.7 km/s when the initial pressure was 1000 Pa and the initial charging voltage was 16 kV. The self-emissions from piston discharge plasma were measured by photodiodes and by high speed framing camera. The measurements showed that the duration between shock and piston was up to 8 microseconds with a 400-mm propagation in the shock tube, which is enough time as dissociation target for beam interaction experiment.The complementary velocity measurement is significant for understanding the electro-magnetically driven shock physics.

## 1. Introduction

Heavy ion fusion, ion-driven fast ignition [1, 2], heavy-ion-driven high energy density physics experiment and astrophysics [3] require the knowledge of stopping power. Grisham [4] proposed $\sim$ MeV/u ion beams driven high energy density physics experiment using a subrange target and showed a possibility of the high homogeneous energy deposition to the target. In the proposal, the interaction between the projectile and the target becomes strong and the stopping power depends strongly on the electronic state of the target because the projectile velocity is close to the electron velocity of the target. The change of the stopping power due to the target condition is of interest to the moderate energy ion-driven high energy density physics experiment. However, the stopping power with the effects of dissociation of the target molecules have not been experimentally investigated, whereas that in the plasma target were studied [5, 6]. The main purpose of our research is measurement of the stopping power with the dissociation. We selected the hydrogen as the dissociated target because the molecule has only valence electron.

In order to measure the stopping power in matter, the target condition should be well-defined. We proposed electro-magnetically driven shock wave as a tool to produce dissociated targets because an electro-magnetic shock tube generated one-dimensional steady shock wave [7]. In this method, Rankine-Hugoniot relations with the shock wave velocity can give the physical condition behind the shock. Previous studies [8, 9] showed that the duration of the shock compression region between the shock front and the discharge plasma working as a piston is required to be the order of microseconds for the stopping power measurement. A long shock tube can generate the shocked matter with long duration time since the shock-compressed area

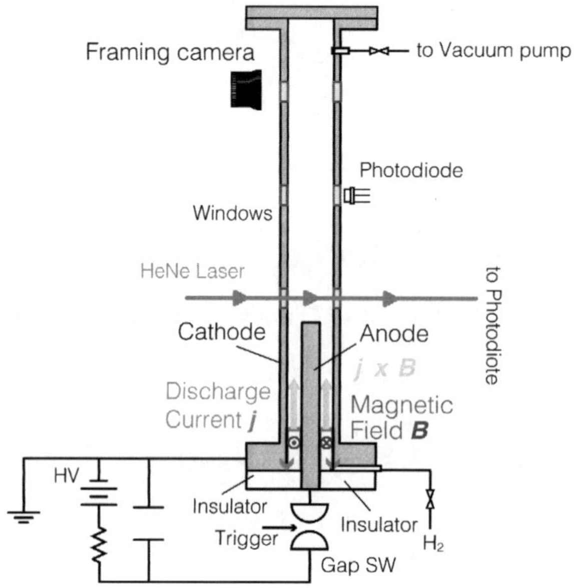


**Figure 1.** This experimental setup is shown including electro-magnetically driven shock tube, equivalent circuit, and velocity measurement system with laser refraction, photodiode, and framing camera.

between the shock and the piston increases with the propagation length of the shock wave. An electro-magnetic shock tube with long propagation section were developed. Precise velocity measurement in electro-magnetic shock tube is required to identify the physical state and the duration between shock and piston. In this paper, velocity measurements by the laser refraction, photodiode for self-emission, and framing camera are presented.

## 2. Experimental setup

### *2.1. Electro-magnetically driven shock tube*

We shows the experimental setup in Fig. 1. The hydrogen gas pressure in the electro-magnetic shock tube was from 300 to 1000 Pa. The electrodes were connected with capacitors with a total capacitance of 3.5 $\mu$F and a gap switch. The charging voltage was applied from 16 to 18 kV. The central electrode, which worked as the anode with a 12-mm outer diameter, was 250 mm length. The electrode length was designed to generate long duration between shock and piston for beam interaction experiment considering the previous experiments [9]. The inner diameter of the cathode was 30 mm. After the trigger signal was sent to the gap switch, the discharge plasma was formed between the electrodes. The discharge current sheet accelerated by the magnetic pressure produced a shock wave. When the shock wave has a proper velocity, we have the dissociated hydrogen behind the shock wave.

### *2.2. Velocity measurement system*

We have three methods; laser refraction, photodiode for self-emission, and framing camera as velocity measurement. The laser refraction, which needs only simple and compact devices, gives the shock wave transit time. He-Ne lasers as probe were led to some photodiodes (S5972, Hamamatsu, Inc.) via the optical windows on the outer hollow cylinder. The density gradient at the shock front caused a laser refraction when the shock reached to the window position. At this moment, the photodiode detected the change of the laser signal. From the two photodiode

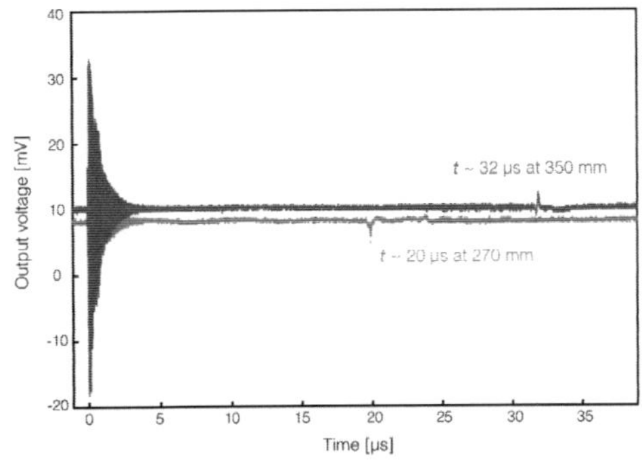


**Figure 2.** Photodiode output signals by laser refraction.

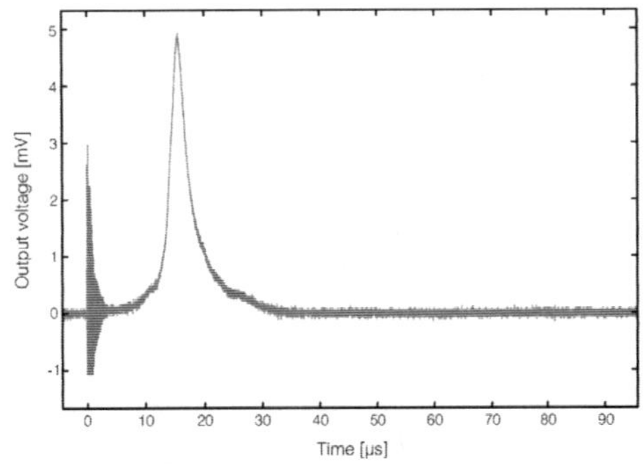


**Figure 3.** Photodiode output signal from self-emission by the piston discharge plasma. The photodiode is placed at 270 mm from the bottom of the electrodes.

signals with two different positions, we obtained the time of flight between the two positions of the shock wave. In this measurement, the averaged shock velocities were estimated.

Photodiodes and framing camera behind the optical windows on the outer hollow cylinder detected the self-emission from discharge plasma which worked as piston to drive the shock wave. The high speed framing camera system had three CCDs (C6588, Hamamatsu, Inc.) with the high-speed gated image intensifier unit (C2925-01, Hamamatsu, Inc.) which realize to take three pictures simultaneously. A delay pulse generator (DG535, Stanford Research System, Inc.) controlled the framing camera and supplied the trigger signal to the gap switch for the main discharge.

## 3. Experimental results

### *3.1. Laser refraction*

Output signals from the two photodiodes which were placed at 270 mm and at 350 mm from the bottom of the electrodes are shown in Fig. 2. The horizontal axis stands for the time after the main discharge starts. The initial hydrogen gas pressure was 1000 Pa and the initial charging voltage was 16 kV. The signal baselines from two photodiodes were 8 mV and 10 mV, respectively because the CW lasers were used. We observed discharge noise signals from the both photodiodes until 5 $\mu$s. The changes of the photodiode signals at 20 $\mu$s and at 32 $\mu$s by laser refraction were observed. The averaged shock velocity is estimated to be 6.7 km/s.

### *3.2. Photodiode for self-emission*

In order to estimate the duration between shock front and piston, we directly measure the self-emission from the piston discharge plasma by photodiodes. The photodiode output signal from baseline is shown in Fig. 3. The initial hydrogen gas pressure was 300 Pa and the initial charging voltage was 16 kV. Noise signal was observed until 5 $\mu$s due to the main discharge. The transit time of the piston discharge plasma is expected to be corresponding to the peak time of the output.

### *3.3. Framing camera*

The framing camera gives the three pictures from the self-emission by the piston discharge plasma in Fig. 4. The white dashed circles show the effective windows. Emission except the effective window was reflection of the discharge plasma. Fig. 4 shows that the discharge plasma

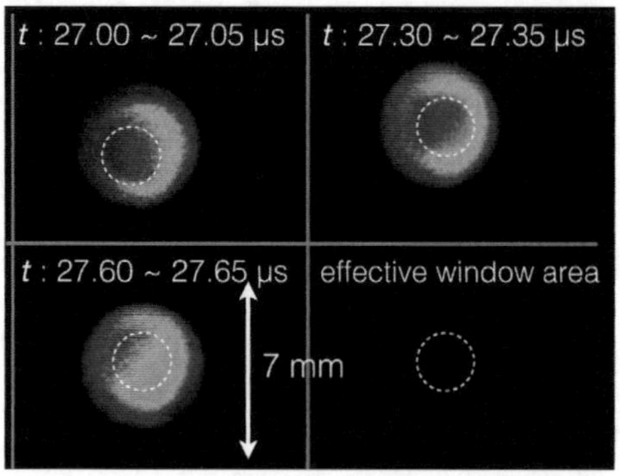


**Figure 4.** $t$ represents the time after the main discharge starts. Each exposure time is 50 ns. The window is placed at 350 mm from the bottom of the electrodes.

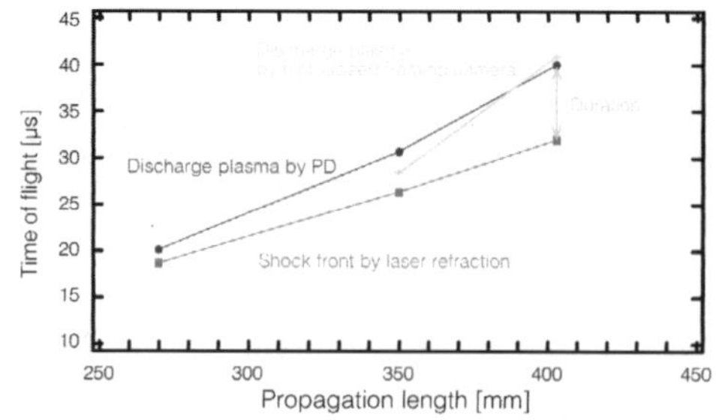


**Figure 5.** The relationship between time of flight and propagation length between the bottom of electrodes and window is shown.

arrived at 350 mm from the bottom of the electrode on $\sim$ 27 $\mu$s. The initial hydrogen gas pressure was 400 Pa and the initial charging voltage was 17 kV in this measurement.

### *3.4. Duration between shock and piston*

The velocity measurements by three methods is shown in Fig. 5. The initial pressure was 400 Pa and the initial charging voltage was 17 kV in the measurement as shown in Fig. 5. The duration was up to 8 $\mu$s with a 400 mm propagation in the shock tube, which was enough for synchronization with projectile. The transit time of the piston discharge plasma by photodiode is agreement with that by the framing camera. On the other hand, the emission from shock compression region was not observed. In this condition, the calculation from the EOS code[8, 9] shows that temperature at the shock compression region is not high due to low Mach shock, which is supposed to be the too weak emission from this region.

## 4. summary

The shock wave velocity and the piston discharge plasma velocity were measured by the three methods. The experiment showed the duration between the shock and the piston increased with the propagation length. No emission from shocked matter was observed because the temperature was not high. In this experimental condition, the laser refraction method is useful to detect the shock front. The complementary velocity measurement is helpful to understand the electromagnetically driven shock physics for the stopping power measurement.

### *4.1. Acknowledgments*

We would like to thank S. Matsui for preparation of the experimental devices.

## References

[1] M. Tabak *et.al.*, Phys. Plasmas 1, 1626 (1994).
[2] M. Roth *et.al.*, Phys. Rev. Lett. 86, 436 (2001).
[3] V. Tatischeff *et.al.*, Astrophys. J. 504, 874 (1998).
[4] L. R. Grisham, Phys. Plasmas 11, 5727 (2004).
[5] A. Frank *et.al.*, Phys. Rev. Lett. 110, 115001 (2013).
[6] A. Frank *et.al.*, Phys. Rev. E 81, 026401 (2010).
[7] K. Kondo *et.al.*, Rev. Sci. Instrum. 77, 036104 (2006).
[8] J. Hasegawa *et.al.*, Nucl. Instr. and Meth. A 606, 205 (2009).
[9] K. Kondo *et.al.*, Nucl. Instr. and Meth. A 733, 1 (2014).

# Numerical Modeling of Complex Targets for High-Energy-Density Experiments with Ion Beams and other Drivers

Alice Koniges[1,a], Wangyi Liu[1], Steven Lidia[1], Thomas Schenkel[1], John Barnard[2], Alex Friedman[2], David Eder[2], Aaron Fisher[2], and Nathan Masters[2]

[1] Lawrence Berkeley National Laboratory
[2] Lawrence Livermore National Laboratory

**Abstract.** We explore the simulation challenges and requirements for experiments planned on facilities such as the NDCX-II ion accelerator at LBNL, currently undergoing commissioning. Hydrodynamic modeling of NDCX-II experiments include certain lower temperature effects, e.g., surface tension and target fragmentation, that are not generally present in extreme high-energy laser facility experiments, where targets are completely vaporized in an extremely short period of time. Target designs proposed for NDCX-II range from metal foils of order one micron thick (thin targets) to metallic foam targets several tens of microns thick (thick targets). These high-energy-density experiments allow for the study of fracture as well as the process of bubble and droplet formation. We incorporate these physics effects into a code called ALE-AMR that uses a combination of Arbitrary Lagrangian Eulerian hydrodynamics and Adaptive Mesh Refinement. Inclusion of certain effects becomes tricky as we must deal with non-orthogonal meshes of various levels of refinement in three dimensions. A surface tension model used for droplet dynamics is implemented in ALE-AMR using curvature calculated from volume fractions. Thick foam target experiments provide information on how ion beam induced shock waves couple into kinetic energy of fluid flow. Although NDCX-II is not fully commissioned, experiments are being conducted that explore material defect production and dynamics.

## 1 Introduction

The Neutralized Drift Compression Experiment II (NDCX-II) is an induction accelerator. A key feature of NDCX-II is time-of-flight longitudinal "drift compression" of a section of the beam to achieve a high intensity pulse; this is a process analogous to chirped-pulse compression of a short-pulse laser beam. A head-to-tail velocity gradient is imparted to the ion beam by a set of induction cells (accelerating elements), and the pulse then compresses as it drifts down a beam line, in a neutralizing plasma environment which provides space-charge compensation. NDCX-II will accelerate a beam containing 20-70 nC of Li+ or He+ ions to 1.2-3 MeV and compress it into a sub-nanosecond pulse at the target with an 1-mm radius spot size when fully commissioned. The machine is an induction linac with custom voltage waveforms to control the longitudinal space charge forces and compress the pulse. Overall longitudinal compression factors of ~500X are required to achieve a 1-ns pulse, but most of that compression occurs in the accelerator [1]. In order to minimize the accelerator length and to optimize the usage of pulsed power, an initial stage of non-neutral beam compression will shorten the pulse length from 500 ns to less than 70 ns by utilizing the first few induction cells after the injector. A diagram of NDCX-II is given in Fig. 1 with the details of the end station and induction cell given.

[a] e-mail: aekoniges@lbl.gov

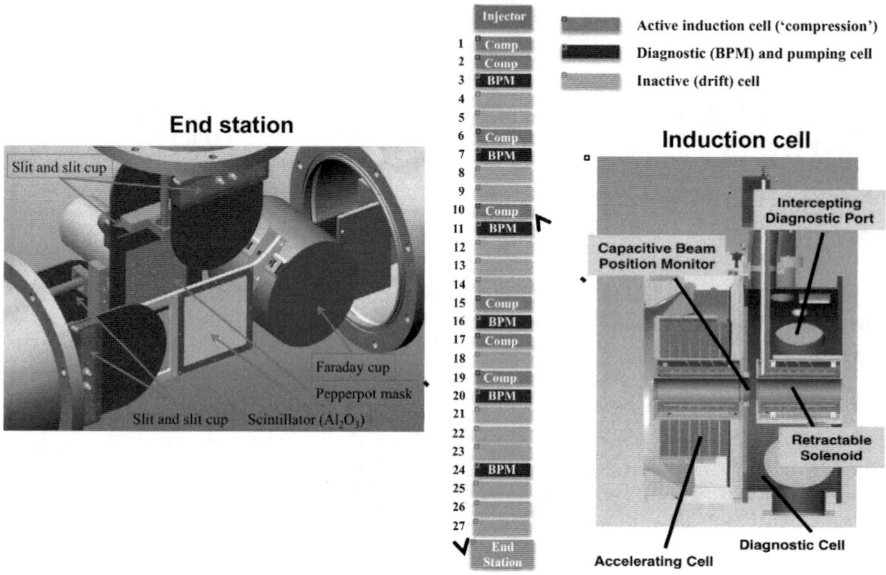


**Fig. 1.** NDCX-II consists of 27 induction, diagnostic, and drift cells for acceleration and pulse shaping.

## 2 Axisymmetric target models

When the range of the ion beam exceeds the thickness of the material being heated (thin targets), the material can be heated quite uniformly. If the timescale is sufficiently short, and the ion energy is slightly above the peak energy in dE/dX (the Bragg peak) when it enters the foil and slightly below the Bragg peak when it exits the foil the resulting temperature will be quite uniform, as the local temperature will be proportional to dE/dX. Over time a rarefaction wave propagates towards the center of the foil, and the heated material propagates rapidly outward at speeds characteristic of the sound speed. Characteristics of these rarefaction waves in ion heated metallic foils and foams can be measured via pyrometry, imaging, VISAR, and/or X-ray imaging diagnostics. By comparing the temperature, velocity, and density information obtained from these diagnostics and comparing to hydrodynamic simulations with different candidate equations of state we may distinguish between the candidate equations of state. Although the initial energy of NDCX-II will be below the Bragg peak for most materials and the hydrodynamic timescale will be somewhat shorter than the pulse duration, simulations show that approximately uniform conditions will still be created. Several experiments for NDCX-II can be modeled in 2D exploiting the axial symmetry present in the ion beam and target. Figure 2 shows 2D axisymmetric ALE AMR simulation of an expanding foil for NDCX-II parameters[2]. The simulation results of density and temperature compare well with results from a standard Inertial Fusion Energy modeling code, Hydra.

In the warm dense matter regime, the material does not expand significantly during heating, thus achieving a mostly uniform density/temperature phase. We model this state using the ALE-AMR code to solve the plasma/fluid equations given here in indicial notation ($i, j, k = 1, 2, 3$) as:

$$\frac{D\rho}{Dt} = -\rho\nabla\cdot\mathbf{U} = -\rho U_{i,i} \tag{1}$$

$$\rho\frac{D\mathbf{U}}{Dt} = \frac{1}{\rho}\nabla\cdot\boldsymbol{\sigma} = \frac{1}{\rho}\sigma_{ij,j} \tag{2}$$

$$\frac{De}{Dt} = \frac{1}{\rho}Vs:\dot{\varepsilon} - P\dot{V} = \frac{1}{\rho}V\left(s_{ij}\dot{\varepsilon}_{ij}\right) - P\dot{V} \tag{3}$$

where $\frac{D}{Dt} = \frac{\partial}{\partial t} + \mathbf{U}\cdot\nabla$ is the substantial derivative, $\rho$ is the density, $\mathbf{U} = (u, v, w)$ is the material velocity, $t$ is time, $\boldsymbol{\sigma}$ is the total stress tensor, $P$ is the pressure, $e$ is the internal energy, $V$ is the relative volume

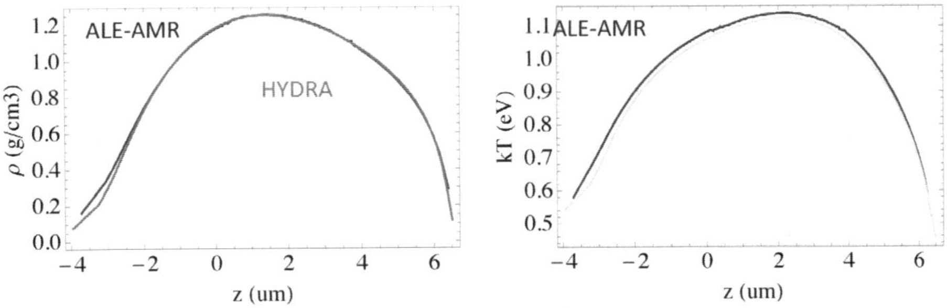


**Fig. 2.** Density and temperature profiles at the completion of the 1 ns, 2.8 MeV, Li ion heating pulse along the radial center of an Al foil. Fluence of 20 J/cm2.

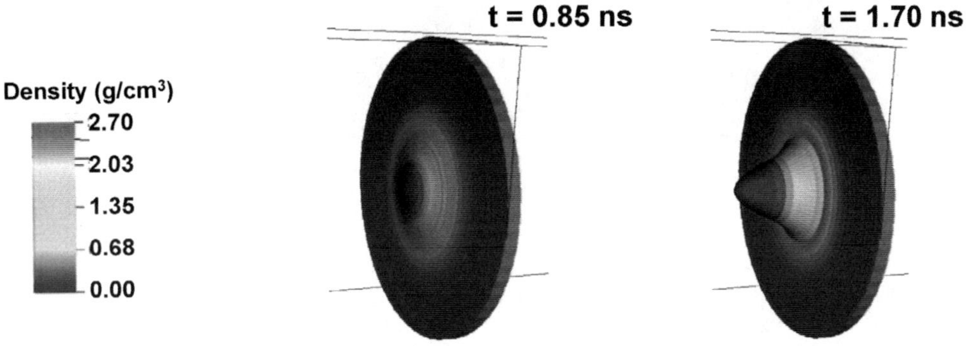


**Fig. 3.** ALE-AMR 2D simulation of thin foil heating experiment after the completion of the heating pulse at 0.85 ns (left) and after at 1.70 ns (right). Density contours are shown in color. The longitudinal scale is exaggerated relative to the transverse. The radius of the simulated target is 1 mm, while the thickness of the original target is 1.0 $\mu m$.

($\rho V = \rho_0$ where $\rho_0$ is the reference density), $s$ is the deviatoric stress, and $\dot{\epsilon}$ is the strain rate tensor defined as $\dot{\epsilon}_{ij} = \frac{1}{2}\left(\frac{\partial U_i}{\partial x_j} + \frac{\partial U_j}{\partial x_i}\right)$. For details see Koniges, et al.[3]. Thermal conduction is modeled using a finite element diffusion solver[4]. In Fig. 3 results of an ALE-AMR simulation confirm that there is little hydrodynamic motion during the heating phase. Note: longitudinal scale is exaggerated.

## 3 Droplet formation models

Some physical processes such as droplet formation and material failure are intrinsically 3D in nature. Droplets were observed in earlier NDCX-I experiments. When the target bunches up into droplets, much of the incident beam current passes through the target to be detected by the current transformer downstream of the target. Another indication of droplet formation comes from the intensity measurements where after a few microseconds, the light intensity from the target begins to decrease. One explanation for this effect is the reduction in projected surface area, as the approximately 200-nm thick foil breaks into of order 1000 nm-sized radius droplets. It is important to model this process of droplet formation and subsequent dynamics. The 3D ALE-AMR code has ion, x-ray and laser deposition, radiation hydrodynamics, thermal diffusion, anisotropic material strength with material time history, and advanced models for fragmentation including void insertion. When solid material fails in the context of a material failure condition in the code, a small volume of void is inserted in the zone. If the zone is stretched due to tensile forces or a divergent velocity field, the majority of the volume increase in the zone is due to void growth. The interface reconstruction scheme allows for void regions in neighboring zones to merge and can produce fragments of material surrounded by void[5]. We have

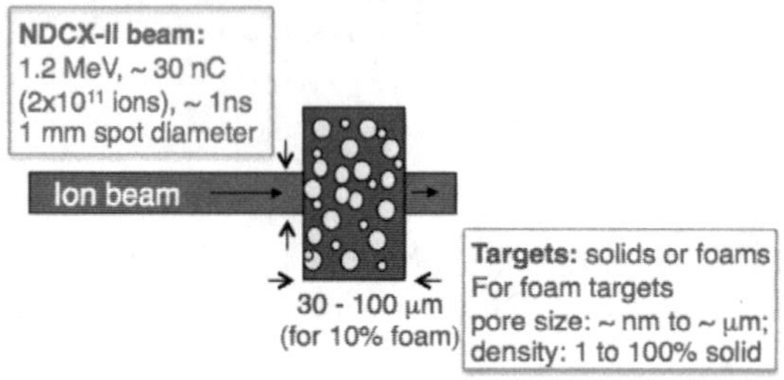


**Fig. 4.** Potential NDCX-II foam experimental configuration.

implemented a similar approach for modeling breakup of droplets. The full discussion of this model is being presented in a longer paper.

## 4 Current and future experiments

In its current state, NDCX-II can be used for lower beam power density experiments. The short excitation pulse with tunable intensity and time profile enables pump/probe type studies of defect dynamics in a broad range of materials[6]. Some capabilities of ALE-AMR including material histories and multiple levels of refinement to support different physics models at the finest scale may enable modeling of such experiments in the future. We are exploring the use of a hierarchical material model format introduced within ALE-AMR's flexible material modeling framework to study defect formation and dynamics. Additional experiments proposed for NDCX-II require full commissioning, namely full beam power and focusing capability. A potential experimental configuration is diagramed in Fig. 4. ALE-AMR is also a candidate for modeling such experiments. The experiments would be used to study warm dense matter over a range of densities by using a combination of solid and foam targets.

## 5 Acknowledgements

We acknowledge the National Energy Research Scientific Computing Center, supported by the Office of Science, U.S. Department of Energy under Contract No. DE-AC02-05CH11231. Work by LBNL under DE-AC02-05CH11231 was supported by the Director, Office of Science of the U.S. Department of Energy and the Petascale Initiative in Computational Science and Engineering. Work by LLNL was performed under the auspices of the U.S. Department of Energy by Lawrence Livermore National Security, LLC, Lawrence Livermore National Laboratory under Contract DE-AC52-07NA27344.

## References

1. A. Friedman, J. Barnard, R. Cohen, D. Grote, S. Lund, W. Sharp, A. Faltens, E. Henestroza, J. Jung, J. Kwan et al., Phys. Plasmas **17**, 056704 (2010)
2. A.E. Koniges, N.D. Masters, A.C. Fisher, R.W. Anderson, D.C. Eder, T.B. Kaiser, D.S. Bailey, B. Gunney, P. Wang, B. Brown et al., Journal of Physics: Conference Series **244**, 032019 (2010)
3. A. Koniges, N. Masters, A. Fisher, D. Eder, W. Liu, R. Anderson, D. Benson, A. Bertozzi, Plasma Science and Technology **17**(2), 117 (2015)
4. A. Fisher, D. Bailey, T. Kaiser, D. Eder, B. Gunney, N. Masters, A. Koniges, R. Anderson, Plasma Science and Technology **17**(2), 109 (2015)
5. D. Eder, A. Fisher, A. Koniges, N. Masters, Nuclear Fusion **53**(11), 113037 (2013)
6. T. Schenkel, S. Lidia, C. Weis, W. Waldron, J. Schwartz, A.M. Minor, P. Hosemann, J. Kwan, Nuclear Instruments and Methods in Physics Research Section B: Beam Interactions with Materials and Atoms **315**, 350 (2013)

# Limitation of the plasma channel due to the frequency blueshift

**Hideyuki Kotaki[1], Yukio Hayashi[1], Michiaki Mori[1], Masaki Kando[1], James K. Koga[1], and Sergei V. Bulanov[1,2]**

[1]Japan Atomic Energy Agency, Kizugawa, Kyoto 619-0215, Japan
[2]A. M. Prokhorov Institute of General Physics RAS, Moscow, Russia

E-mail: kotaki.hideyuki@jaea.go.jp

**Abstract.** In laser-plasma interaction, there are physical mechanisms, such as quantum beam generation, relativistic self-focusing of laser pulses, frequency blueshift of laser pulses, etc. Long plasma channel formation via self-focusing is an important mechanism for generating a stable electron beam. The frequency blueshift changes the critical power of the self-focusing. The plasma channel is limited by the change of the power. The limit of the channel causes a fixed blueshift of the laser pulse.

## 1. Introduction

Laser-plasma interaction includes a lot of physics, such as quantum beam generation, laser self-focusing, frequency shift, etc. Laser WakeField Acceleration (LWFA) [1], which is based on the effect of plasma wave excitation in the wake of an intense laser pulse, is now regarded as a basis for the next-generation of charged particle accelerators. Electron bunches have been accelerated up to 1 GeV by LWFA [2, 3]. In experiments, it has been demonstrated that LWFA is capable of generating electron bunches with high quality [4, 5, 6, 7, 8].

In order to generate a bunch with high quality, required for applications, the electrons should be duly injected into the wakefield and this injection should be controllable. The injection can happen spontaneously, due to a longitudinal or transverse break of the wake wave, caused by its strong nonlinearity [9, 10, 11] and with cluster-gas targets [12]. This regime leads to the acceleration of fast particles, although in an uncontrolled way. Recently we generated a stable electron beam by using an Argon and Nitrogen gas target in the self-injection scheme [15, 16]. In the scheme of self-injection, the long channel improves the stability of the electron beam energy, pointing, and divergence.

On the other hand, in laser-plasma interactions, there is a phenomenon of blueshifting of the laser pulse. The frequency blueshift has two mechanisms, which are ionization blueshift [17, 18, 19] and photon acceleration [20, 21]. The ionization blueshift occurs due to a frequency shift at the boundary between neutral gas and plasma. It occurs in an ionization front. The ionization blueshift is independent of the interaction length and the laser intensity. Photon acceleration is a frequency blueshift in a plasma wake wave. The shift depends on the interaction length, the plasma density, and the laser intensity. From experimental results, a fixed blueshift has been observed [17]. When the target gas is changed, the parameters of ionization should change. Nevertheless, the spectrum of the transmitted laser light was the same, even though the target gases were different.

In addition, relativistic self-focusing is one of the important physics for laser-plasma interaction [22]. When the laser power is high enough, the laser pulse self-focuses. The critical power, $P_{cr}$, is $P_{cr} = 16.2(\omega_0/\omega_p)^2$ [GW], where $\omega_p = (4\pi n_e e^2/m_e)^{1/2}$ is the plasma frequency, $\omega_0$ is the laser frequency, and $n_e$ is the plasma density. The self-focusing is related to $n_e$ and $\omega_0$. The plasma channel is produced in a balance between focusing and de-focusing.

In plasma, the laser pulse is self-focused with frequency blueshift, which includes both the ionization blueshift and the photon acceleration. The plasma channel ends when the laser power becomes smaller than the critical power for self-focusing. In this letter, the effect of the frequency blueshift of the laser pulse in the plasma channel is studied in order to elucidate the fixed blueshift and the plasma channel length limitation.

## 2. Experimental setup and condition

The experiments have been performed with a Ti:sapphire laser system at the Japan Atomic Energy Agency (JAEA) - Advanced Photon Research Center (APRC) named JLITE-X [23]. The laser contrast ratio within picosecond timescales is $10^6$. The contrast ratio within nanosecond timescales is significantly suppressed to the order of $10^8$. The laser pulses, which are linearly polarized, with 200 and 133 mJ energy are focused onto a 3-mm-diameter helium gas-jet by an off-axis parabolic mirror (OAP) with the focal length of 646 mm (f/22). The $1/e^2$ diameter of the focal spot is 32 $\mu$m. The energy concentration within this region is 60%. The pulse width of the laser pulse, $\tau$, is 40 fs. The estimated peak irradiances from the measurement data, $I_0$, are $7.5 \times 10^{17}$ and $5.0 \times 10^{17}$ W/cm$^2$ in vacuum corresponding to a dimensionless amplitude of the driver laser fields $a_0 = 8.5 \times 10^{-10}\lambda_0[\mu\text{m}]\sqrt{I_0[\text{W/cm}^2]} = 0.6$ and 0.5, where $\lambda_0$ is the laser light wavelength of 800 nm. The spectrum of the transmitted laser pulse is measured with a spectrometer. Figure 1 shows the spectrum of the transmitted laser light. The spectrum shifts with the interaction length, $L$. The transmitted laser power is measured with a power meter on a linear stage.

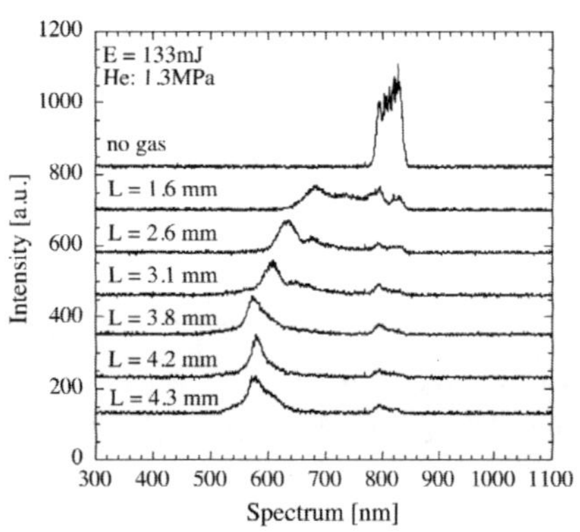


**Figure 1.** Spectrum of the transmitted laser light. The spectrum shift increased with increasing interaction length, L.

## 3. Experimental results and discussion

### *3.1. Gas-length vs shift*

The frequency blueshift has two mechanisms, ionization blueshift and photon acceleration. The ionization blueshift is independent of the interaction length and the laser intensity. The photon acceleration depends on the interaction length, the plasma density, and the laser intensity. Figure 2 shows the peak value of the spectrum of the transmitted laser light. The first shift of 50 nm at $L = 0$ is the ionization blueshift. It is independent of the interaction length. The spectrum shift is increasing with $L$. The shifts are proportional to $L$ and depend on the laser intensity. These should be caused by the photon acceleration. The shifts are saturated when $L > 3.8$ mm. The shifts may finish due to the ending of the plasma channel.

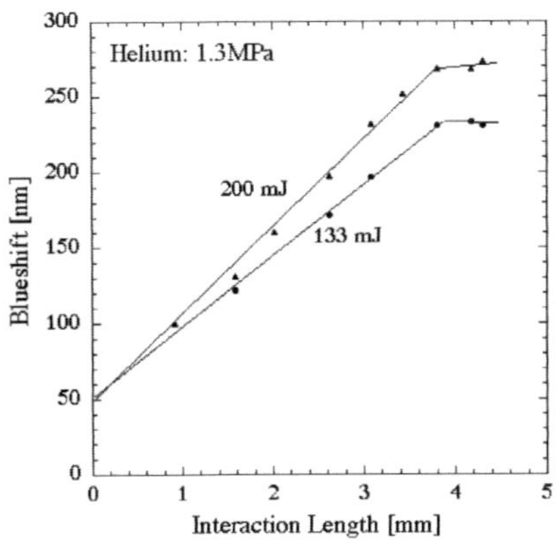


**Figure 2.** The peak value of the transmitted laser spectrum. The first shift of 70 nm at $L = 0$ is the ionization blueshift. The spectrum shift, which is increasing with $L$, is the photon acceleration.

## 4. Comparison of $P$ vs $P_{cr}$

In the interaction, the laser power decreases due to the plasma generation by the laser pulse. The laser spectrum blueshifts are due to (i) ionization blueshift and (ii) photon acceleration. These effects change $P_{cr}$ and the transmitted laser power, $P$. The plasma channel ends when $P < P_{cr}$ due to the end of the relativistic self-focusing. Figure 3 shows the relationship between the transmitted laser power, P, and the critical power, $P_{cr}$. $P_{cr}$ is calculated from the plasma density and the measured transmitted laser spectrum. When $n_e$ is less than $1.5 \times 10^{19}$ cm$^{-3}$, the laser pulse has no self-focusing due to low plasma density. When $n_e$ is between $2.0 \times 10^{19}$ cm$^{-3}$ and $4.0 \times 10^{19}$ cm$^{-3}$, the difference of $P$ and $P_{cr}$ is caused by the scattering of the laser pulse by gas after the plasma channel ends. When the plasma density is high enough, the plasma channel continues to the end of the gas-jet. When $n_e$ is higher than $4.0 \times 10^{19}$ cm$^{-3}$, $P$ is in good agreement with $P_{cr}$, because the plasma channel continues to the end of the gas-jet. These results show that the laser pulse in plasma is blueshifted and the plasma channel length is limited due to the spectrum shift. In addition, the blueshift stops when $P < P_{cr}$. The blueshift is also limited due to the stop of self-focusing, and the spectrum of the transmitted laser light is fixed.

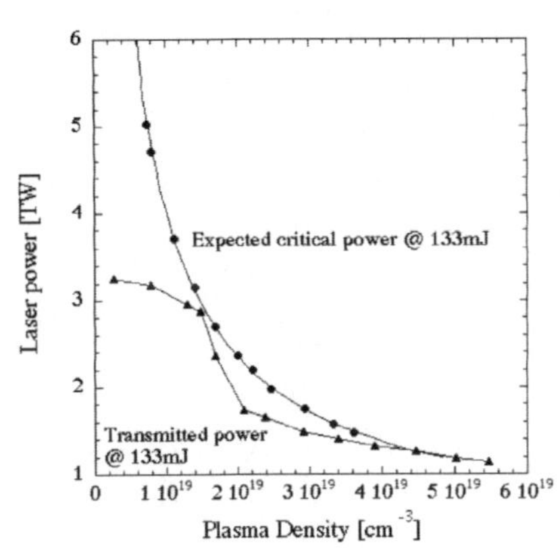


**Figure 3.** Relationship between the transmitted laser power, P, and the critical power, $P_{cr}$. $P_{cr}$ is calculated from the plasma density and the measured transmitted laser spectrum.

## 5. Conclusions

In conclusion, we observe two types of the blueshift, (a) ionization blueshift and (b) photon acceleration. The intense laser pulse produces a plasma channel due to the self-focusing. The plasma channel corresponds to a balance between self-focusing and defocusing. In the channel, the laser power decreases and the frequency is blueshifting. When $P < P_{cr}$, the balance is broken and the plasma channel disappears. The end of self-focusing limits the length of the plasma channel and fixes the spectrum of the transmitted laser light.

## 6. Acknowledgments

This work was supported by the Japan Society for the Promotion of Science under a Grants-in-Aid for Scientic Research (Contract No. 23600010)

## References

[1] Tajima T and Dawson J M 1979 *Phys. Rev. Lett.* **43** 267
[2] Leemans W P, Nagler B, Gonsalves A J, Tth Cs, Nakamura K, Geddes C G R, Esarey E, Schroeder C B and Hooker S M 2006 *Nat. Phys.* **2** 696
[3] Hafz N A M, Jeong T M, Choi I W, Lee S K, Pae K H, Kulagin V V, Sung J H, Yu T J, Hong K-H, Hosokai T, Cary J R, Ko D-K and Lee J-M 2008 *Nat. Photonics* **2** 571
[4] Miura E, Koyama K, Kato S, Saito N, Adachi M, Kawada Y, Nakamura T and Tanimoto M 2005 *Appl. Phys. Lett.* **86** 251501
[5] Mangles S P D, Murphy C D, Najmudin Z, Thomas A G R, Collier J L, Dangor A E, Divall E J, Foster P S, Gallacher J G, Hooker C J, Jaroszynski D A, Langley A J, Mori W B, Norreys P A, Tsung F S, Viskup R, Walton B R and Krushelnick K 2004 *Nature* **431** 535
[6] Geddes C G R, Toth Cs, Tilborg J van, Esarey E, Schroeder C B, Bruhwiler D, Nieter C, Cary J and Leemans W P 2004 *Nature* **431** 538
[7] Faure J, Glinec Y, Pukhov A, Kiselev S, Gordienko S, Lefebvre E, Rousseau J-P, Burgy F and Malka V 2004 *Nature* **431** 541
[8] Yamazaki A, Kotaki H, Daito I, Kando M, Bulanov S V, Esirkepov T Zh, Kondo S, Kanazawa S, Homma T, Nakajima K, Oishi Y, Nayuki T, Fujii T and Nemoto K 2005 *Phys. Plasmas* **12** 093101
[9] Bulanov S V, Pegoraro F, Pukhov A M and Sakharov A S 1997 *Phys. Rev. Lett.* **78** 4205
[10] Pukhov A and Meyer-ter-Vehn J 2002 *Appl. Phys.* B **74** 355
[11] Zhidkov A, Koga J, Hosokai T, Kinoshita K and Uesaka M 2004 *Phys. Plasmas* **11** 5379
[12] Fukuda Y, Akahane Y, Aoyama M, Hayashi Y, Homma T, Inoue N, Kando M, Kanazawa S, Kiriyama H, Kondo S, Kotaki H, Masuda S, Mori M, Yamazaki A, Yamakawa K, Echkina E Yu, Inovenkov I N, Koga J and Bulanov S V 2007 *Phys. Lett.* A **363** 130
[13] Bulanov S V, Naumova N, Pegoraro F and Sakai J 1998 *Phys. Rev.* E **58** R5257
[14] Geddes C G R, Nakamura K, Plateau G R, Toth Cs, Cormier-Michel E, Esarey E, Schroeder C B, Cary J R and Leemans W P 2008 *Phys. Rev. Lett.* **100** 215004
[15] Mori M, Kondo K, Mizuta Y, Kando M, Kotaki H, Nishiuchi M, Kado M, Pirozhkov A S, Ogura K, Sugiyama H, Bulanov S V, Tanaka K A, Nishimura H and Daido H 2009 *Phys. Rev. ST Accel. Beams* **12** 082801
[16] Mori M, Kando M, Kotaki H, Hayashi Y, Bulanov S V, Koga J K, Kondo K, Pirozhkov A S, Nishimura H and Nagashima K 2011 *J. Phys. Soc. Jpn.* **80** 105001
[17] Koga J K, Naumova N, Kando M, Tsintsadze L N, Nakajima K, Bulanov S V, Dewa H, Kotaki H and Tajima T 2000 *Phys. Plasmas* **7** 5223
[18] Koga J K 2009 *J. Opt. Soc. Am.* **26** 930
[19] Savage R L Jr, Joshi C and Mori W B 1992 *Phys. Rev. Lett.* **68** 946
[20] Wilks S C, Dawson J M, Mori W B, Katsouleas T and Jones M E 1989 *Phys. Rev. Lett.* **62** 2600
[21] Murphy C D, Trines R, Vieira J, Reitsma A J W, Bingham R, Collier J L, Divall E J, Foster P S, Hooker C J, Langley A J, Norreys P A, Fonseca R A, Fiuza F, Silva L O, Mendona J T, Mori W B, Gallacher J G, Viskup R, Jaroszynski D A, Mangles S P D, Thomas A G R, Krushelnick K and Najmudin Z 2006 *Phys. Plasmas* **13** 033108
[22] Sun G Z, Ott E, Lee Y C and Guzdar P 1987 *Phys. Fluids* **30** 526
[23] Mori M, Pirozhkov A, Nishiuchi M, Ogura K, Sagisaka A, Hayashi Y, Orimo S, Fukumi A, Li Z, Kado M and Daido H 2006 *Laser Phys.* **16** 1092

# Shock Hugoniot measurements of CH at Gbar pressures at the NIF

**A. L. Kritcher[1], T. Doeppner[1], D. Swift[1], J. Hawreliak[1], J. Nilsen[1], J. Hammer[1], B. Bachmann[1], G. Collins[1], O. Landen[1], C. Keane[1], S. Glenzer[2], S. Rothman[3], D. Chapman[3], D. Kraus[4], R.W.Falcone[4]**

[1]Lawrence Livermore National Laboratory, Livermore, CA, [2]SLAC Accelerator National Laboratory, Menlo Park, CA, USA,[3]Atomic Weapons Establishment, Reading, United Kingdom, [4]Department of Physics, University of California Berkeley, Berkeley, CA, USA

E-mail: `kritcher2@llnl.gov`

**Abstract.** Laboratory measurements of the shock Hugoniot at high pressure, exceeding several hundred Mbar, are of great importance in the understanding and accurate modeling of matter at extreme conditions. In this work we present a platform to measure the material properties, specifically the single shock Hugoniot and electron temperature, at extreme pressures of ∼Gbar at the National Ignition Facility (NIF). In these experiments we launch spherically convergent shocks into solid CH, using a Hohlraum radiation drive. X-ray radiography is applied to measure the shock speed and infer the mass density profile, enabling determining of the material pressure and Hugoniot equation of state. X-ray scattering is applied to measure the electron temperature through measurement of the electron velocity distribution.

## 1. Introduction

Shock Hugoniot measurements of matter at high energy density are important for understanding the structure and evolution of gas-giant planets [1, 2, 3] or highly evolved stars [4]. At Gbar pressures atomic shell effects [5] may come into play which can change the predicted compressibility at given pressure due to pressure and temperature ionization. In addition, benchmarking dense matter models in the regime that is relevant for fusion energy experiments, which assemble material to the highest densities and pressures ever achieved in the laboratory [6, 7] (hundreds of g/cc and >100 Gbar) is important. Previous experiments have reached Gbar conditions via underground nuclear explosions [8]. Near Gbar conditions have also been observed in x-ray driven colliding Au foils [9]. In these experiments, the shock Hugoniot was not directly measured and potential preheat of the target foil could significantly impact interpretation of the results. More recent experiments have reached Gbar pressures in laser driven cone-in-shell targets, but had issues with shock uniformity and stability, resulting in a non ideal platform for a detailed Equation of State (EOS) measurement [10].

We create Gbar conditions in the laboratory using high-energy lasers at the National Ignition Facility (NIF) [11] and present a stable platform well suited for shock Hugoniot measurements. In these experiments we drive a strong convergent shock wave through solid spherical targets in an indirectly driven laser geometry and characterize the Hugoniot and electron temperature using x-ray radiography [12, 13, 14] and x-ray Thomson scattering [15, 16, 17, 18, 19, 20, 21, 22]. We measure the mass density profile and shock speed from the radiographic signal, and infer

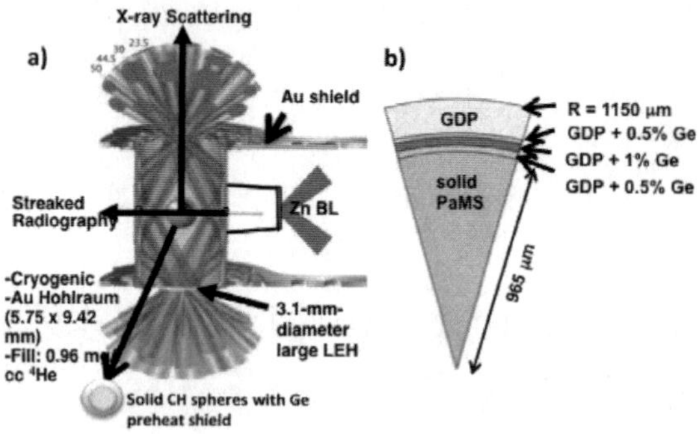


**Figure 1.** a) Schematic of the experimental setup. Here solid targets of CH are shock compressed using a indirectly driven hohlraum radiation source. The plasma is probed with x-ray radiography in the equatorial direction and x-ray scattering in the polar direction. b) Wedged cross-section of the solid PaMS target with a Ge doped GDP (Glow Discharge Polymer) ablator.

the pressure in the shocked region using the Hugoniot relations. In these experiments preheat of the interior of the target does not have a strong impact on interpretation of the results. The calculated preheat due to absorption of Au Mband radiation from the Hohlraum is less than 1eV, which is a small perturbation to the total pressure. In the following sections, we will present experiments at the NIF to employ these techniques, under an accepted NIF basic science shot-time allocation.

## 2. Experimental Setup and Method

In these experiments we launch a strong shock wave into a solid ball of polystyrene using a Hohlraum radiation drive. This radiation drive is created using 184 laser beams that heat gold Hohlraum walls and create a black body cavity at radiation temperatures of up to 280 eV. The Au wall motion due to heating and expansion is mitigated by filling the Hohlraum with 0.96 mg/cc of $He^4$. We track the shock wave traveling through the target via streaked x-ray radiography as a function of radius and time. The x-ray radiography backlighter is created via laser irradiation of a Zn foil mounted outside of the Hohlraum. Two slits at the waist of the hohlraum enable transmission of the backlighter photons to the target and to the x-ray streak camera in the equatorial direction. The same backlighter photons are also used for x-ray Thomson scattering in the polar direction. An Au shield is used to block the direct line of sight of the backlighter photons to the scattering spectrometer in the polar direction. See Fig.1 a) for a schematic of the experimental configuration. The solid targets consist of a 965$\mu$m radius solid sphere of PAMS (Poly(Alpha-MethylStyrene)), covered by a graded doped Ge preheat shield, Fig.1 b). The preheat shield is used to mitigate pre-heating of the PAMS by Hohlraum Au M-band emission.

The time-resolved x-ray radiography will be used to measure the shock speed $u_s$ and mass density $\rho$ profiles. Rankine-Hugoniot relations relate the pressure $P$ and density $\rho = 1/\nu$ in the shocked state to the unshocked material (subscript 0), given $u_s$, $P = u_s^2\rho_0(1 - \rho_0/\rho)$. Thus, by measuring $\rho$ and $u_s$ we infer the pressure and constrain the shock Hugoniot. The calculated error bars on this measurement are about 4.5% in both density and pressure. In addition we measure x-ray Thomson scattering from free electrons, or Compton scattering. The Compton feature, downshifted in energy, provides information on the electron velocity distribution through Doppler broadening. If the plasma is non-degenerate, probing the velocity distribution through

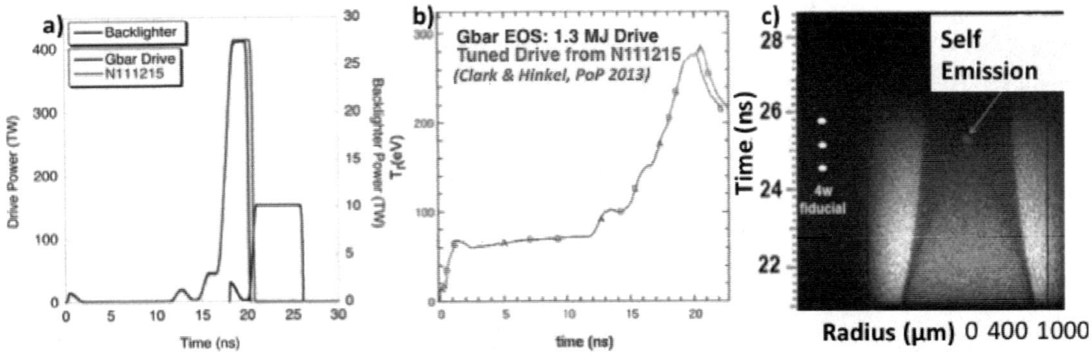


**Figure 2.** a) Power profiles vs time of the laser drive (red/blue) and backlighter beams (black). b) Radiation temperature as a function of time, derived from integrated Hohlraum simulations for post shot N111215 (red) and the modified Gbar EOS drive (blue). c) Simulated radiograph using the Gbar 1.3 MJ radiation drive.

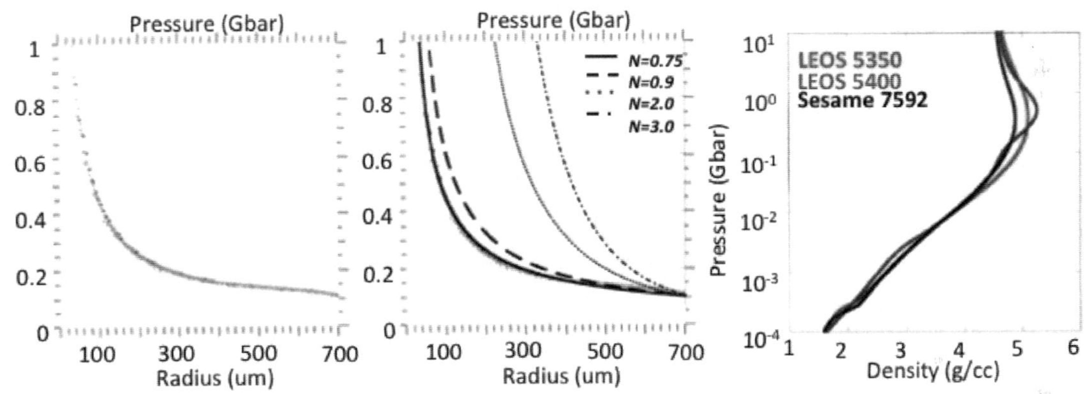


**Figure 3.** (Left hand side) Pressure at the shock front as a function of shock front radius, using LEOS 5350 (red/magenta), LEOS 5400 (green) and Sesame 7592(black). (Right hand side) The same plot as on the left hand side, with analytical expressions of the pressure increase as a function of radius, where $P = P_0(r_0/r)^N$.

measurement of the Compton feature width, provides a measure of the electron temperature.

## 3. Radiation Hydrodynamic Simulations

Radiation hydrodynamic simulations have been preformed using the code HYDRA [23] in capsule-only simulations that use a frequency dependent radiation drive surrounding capsule. The radiation drive is derived from integrated Hohlraum simulations which includes the non-Planckian behavior of the hohlraum. We start from a radiation drive, derived from experiment N111215, that has been tuned in a post-shot manner [24] to match experimental observables in ICF implosions for a similar laser pulse shape. This radiation drive was modified to account for using eight of the drive beams to irradiate the backlighter foil. The radiation drive is also adjusted for using a laser drive energy of 1.3 MJ in the first experiments, compared to 1.45 MJ for shot N111215, see Fig.2 b).The corresponding laser power profiles and radiation temperatures as a function of time are shown in Fig.2 a) and b). Also shown is the backlighter power profile as a function of time, which is delayed from the drive beams to capture the shock convergence at the target center. A simulated radiograph, as a function of radius and time, including realistic noise and backlighter functions shows that we can resolve the shock front down to small radii, of <100 $\mu$m, see Fig. 3 a).

The simulations have been performed using three choices of Equations of State (EOS) models LEOS 5350 (GDP), LEOS 5400 (GDP), and Sesame 7592 (polystyrene) and tabular OPAL opacities [25]. Figure 3 (left hand side) shows the predicted pressure as a function of radius using LEOS 5350 and 5400, Sesame 7592 and the radiation drive shown in see Fig.2 b). Reaching Gbar pressures in a planar shock wave geometry is currently not feasible at the NIF for the mm scale samples required to employ diagnostics such as x-ray radiography. Therefore, we employ a sphericaly convergent shock wave geometry where convergent flow increases the shock pressure. These simulations show that we can reach near Gbar pressures in this platform using a 1.3 MJ Hohlraum radiation drive.

The pressure in a converging spherical shock wave increases as the radius decreases, and can be approximated by [26],

$$P \propto r^{-2\beta} \tag{1}$$

where $\beta^{-1} = 1/2 + 1/\gamma + (\gamma/2(\gamma-1))^{1/2}$ is an approximation for strong shock waves and $\gamma$ is the adiabatic index. For $\gamma = 5/3$, $\beta = 0.45$ and for $\gamma = 4/3$, $\beta = 0.375$ . The increase in pressure due to convergence is smaller than for a sound wave because the shock wave is non-isentropic with a fraction of the energy being dissipated into heat. This analytical relationship is shown in the center plot of Fig.2 b), over-plotted on the simulated curves. Here the pressure, $P$, is a function of the initial pressure, $P_0$, radius, $r$, and initial radius, $r + 0$: $P = P_0(r_0/r)^N$. For the solid curve $N = 0.75$, the dashed curve $N = 0.9$, the dotted curve $N = 2$, and the dash dot curve $N = 3$. Here, we see that the curve with $N = 0.75$, and $\beta = 4/3$, best fits the simulated pressure increase, which may be influenced by radiative preheat of the shock front at high convergence ratios. Measurement of the shape of the mass density profile will provide indication on the onset of radiative preheat at the shock front. The pressure as a function of radius is weakly dependent on choice of equation of state. However, the compressibility of the material at a given pressure is strongly dependent on choice of EOS this can be seen on the right hand side of Fig.3 for the three choices of EOS models considered. Since we will be measuring both the density and pressure simultaneously, we can benchmark these EOS models.

In summary, we have developed a platform to create and probe matter at Gbar pressures at the NIF. In these experiments we measure the shock speed, mass density profile, and infer the pressure, thereby constraining the single shock Hugoniot of samples at pressures of 100Mbar-Gbar. In addition, experimental measurement of the pressure increase as a function of radius can provide information on the adiabatic index. This work was performed under the auspices of the U.S. Department of Energy by Lawrence Livermore National Laboratory under Contract No. DE-AC52-07NA27344 and supported by Laboratory Directed Research and Development Grant No. 13-ERD-073.

## References

[1] L. Stixrude, Phys. Rev. Lett. **108**, 055505 (2012).
[2] S. Seager, M. Kuchner , C. A. Hier-Majumder, and B. Militzer, Astrophys. J. **669**, 1279 (2007).
[3] D.C. Swift, J. Eggert, D.G. Hicks, S. Hamel, K. Caspersen,*et al.* Astrophys. J. 744, 59 (2012)
[4] V. Fortov, Extreme States of Matter: On Earth and in the Cosmos, 224 (2011).
[5] J. C. Pain, High En. Den. Phys., **3** 204 (2007).
[6] J. Nuckolls *et al.*, Nature, **239** (1972).
[7] J. D. Lindl *et al.*, Phys. Plasmas **11**, 339. (2004).
[8] Vladimirov AS, Voloshin NP, Nogin VN, Petrovtsev AV, Simonenko VA., Sov Phys - JETP Lett **39** (1984).
[9] R. Cauble, Phys. Rev. Lett. **70**, 14 (1993).

[10] K. Shigemori *et al.*, Appl. Phys. Lett. **102**, 183501. (2013).
[11] E. I. Moses, R. N. Boyd, B. A. Remington, C. J. Keane, and R. Al-Ayat, Phys. Plasmas **16**, 041006 (2009).
[12] D. G. Hicks, B. K. Spears, D. G. Braun, R. E. Olson, *et al.*, Rev. Sci. Instrum. **81**, 10E304 (2010).
[13] D. G. Hicks, N. B. Meezan, E. L. Dewald, A. J. Mackinnon, *et al.*,Phys. Plasmas 19, 122702 (2012)
[14] D. C. Swift, J. A. Hawreliak, *et al.*, Proc. American Physical Society, CP1426 pp.477-480 (2011).
[15] S. H. Glenzer and Ronald Redmer, Rev. Mod. Phys. **81**, 1625 (2009).
[16] A. L. Kritcher, T. Doeppner, C. Fortmann, T. Ma,*et al.* , Phys. Rev. Lett. **107**, 015002 (2011).
[17] A. L. Kritcher, Paul Neumayer,John Castor,Tilo Dppner,*et al.*, Science, **322**, 69 (2008).
[18] L. B. Fletcher, A. Kritcher, A. Pak, T. Ma, T. Dppner,*et al.*,Phys. Plasmas 20, 056316 (2013)
[19] G. Gregori, S. H. Glenzer, K. B. Fournier, K. M. Campbell,*et al.*, Phys. Rev. Lett. **101**, 045003 (2008).
[20] H. J. Lee, P. Neumayer, J. Castor, T. Dppner, R. W. Falcone, *et al.*,, Phys. Rev. Lett. **102**, 115001 (2009).
[21] B. Barbrel, M. Koenig, A. Benuzzi-Mounaix, E. Brambrink, *et al.*,, Phys. Rev. Lett. **102**, 165004 (2009).
[22] D. Riley, N. C. Woolsey, D. McSherry, I. Weaver, A. Djaoui, and E. Nardi , Phys. Rev. Lett. **84**, 1704 (2000).
[23] M. Marinak, S. W. Haan, T. R. Dittrich, R. E. Tipton, and G. B. Zimmerman, Phys. Plasmas **5**, 1125 (1998).
[24] D. S. Clark,D. E. Hinkel, D. C. Ederl, O. S. Jones, S. W. Haan, *et al.*,, Phys. Plasmas 20, 056318 (2013).
[25] C. A. Iglesias and F. J. Rogers, Astrophys. J. 464, 943 (1996).
[26] F. Winterberg, *The Rel. of Thermonuclear En. by Inertial Confinement*, World Scientific Publishing (2010).

# Spherical shock in the presence of an external magnetic field

**Y Kuramitsu[1], S Matsukiyo[2], S Isayama[2], D Harada[2], T Oyama[2], R Fujino[2], Y Sakawa[3], T Morita[3], Y Yamaura[3], T Ishikawa[3], T Moritaka[3], T Sano[3], K Tomita[4], R Shimoda[4], Y Sato[4], K Uchino[4], A Pelka[5], R Crowston[6], N Woolsey[6], G Gregori[7], M Koenig[8] D W Yuan[9], C L Yin[9], Y T Li[9], K Zhang[10], J Y Zhong[10], F L Wang[10], N Ohnishi[11], K Nagamine[12], H Yoneda[12], and H Takabe[3]**

[1] Department of Physics, National Central University, Jhongli, Taiwan
[2] Department of Earth System Science and Technology, Kyushu University, Kasuga Japan
[3] Institute of Laser Engineering, Osaka University, Suita, Japan
[4] Department of Applied Science for Electronics and Materials, Kyushu University, Kasuga, Japan
[5] Helmholtz-Zentrum Dresden-Rossendorf, Dresden, Germany
[6]Department of Physics, University of York, York, UK
[7]Department of Physics, University of Oxford, Oxford, UK
[8] Laboratoire pour l'Utilisation des Lasers Intenses, Ecole Polytechnique, Palaiseau, France
[9] Institute of Physics, Chinese Academy of Sciences, China
[10] Institution National Astrophysical Observatories, CAS, China
[11] Department of Aerospace Engineering, Tohoku University, Sendai, Japan [12] University of Electro-Communications, Tokyo, Japan

E-mail: yasu@ncu.edu.tw

**Abstract.** We investigate spherical collisionless shocks in the presence of an external magnetic field. Spherical collisionless shocks are common resultant of interactions between a expanding plasma and a surrounding plasma, such as the solar wind, stellar winds, and supernova remnants. Anisotropies often observed in shock propagations and their emissions, and it is widely believed a magnetic field plays a major role. Since the local observations of magnetic fields in astrophysical plasmas are not accessible, laboratory experiments provide unique capability to investigate such phenomena. We model the spherical shocks in the universe by irradiating a solid spherical target surrounded by a plasma in the presence of a magnetic field. We present preliminary results obtained by shadowgraphy.

## 1. Introduction

Collisionless shocks are universal phenomena in space and astrophysical plasmas, where supersonic or super Alfvénic plasma flows encounter sounding media, such as shocks in supernova remnants. A remarkable feature of collisionless shocks is the nonthermal acceleration of energetic particles or cosmic rays. It is widely believed that the cosmic rays are accelerated in the shock environments, where one of the most important factors is the field orientation against the shock normal. However, one of the most uncertain values in the astrophysical observations is the magnetic field. For instance, the relation between the field orientation and the acceleration

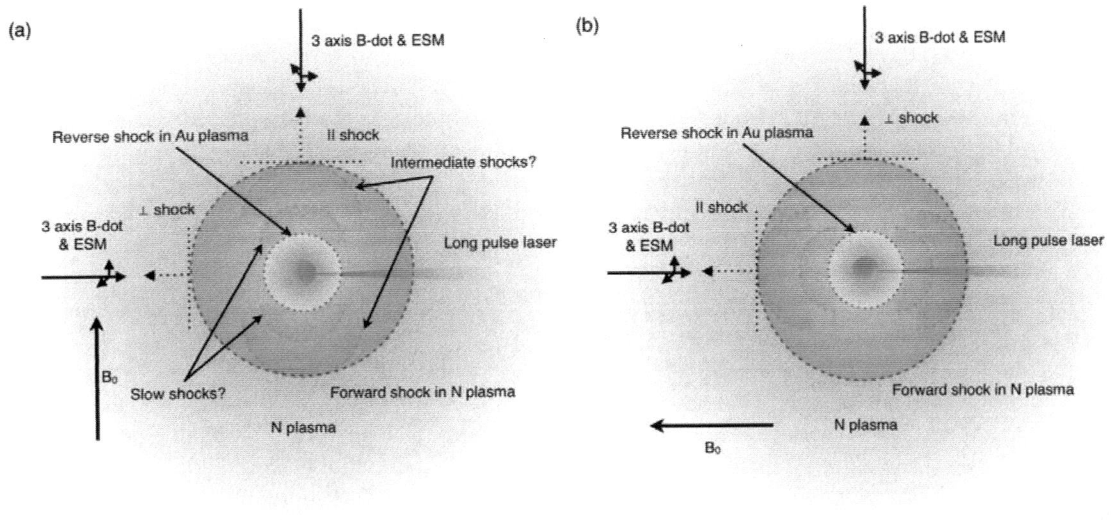


**Figure 1.** Schematics of the experimental conditions. In order to avoid the anisotropy caused by the laser, two magnetic field geometry will be applied.

efficiency of the cosmic rays is a long-standing and fundamental unsolved problem. So far, numerical simulations had been only ways to investigate such the problem [1, 2, 3].

Recently laboratory simulations of space and astrophysical phenomena have been actively investigated with high-power and ultra intense lasers. Collsionless shocks have been successfully created in laboratory plasmas. Shock density and emission jumps and the time evolution were obtained with optical imaging [4, 5]. A turbulent shock electric field was observed with proton radiography [6, 7]. Together with the global imaging, local observation of the magnetic field has demonstrated the field origin in the universe [8]. Local quantities of plasmas such as the density, velocity and temperature for ions and electrons can be obtained by collective Thomson scattering measurements [9, 10] Moreover, it is also possible to measure the energy distribution functions of accelerated electrons with electron spectrometer [11, 12]. These are unique and great advantages of laboratory investigation of space and astrophysical phenomena, that is, the simultaneous observations of the global and local quantities and the energy distribution functions of particles. With these capabilities we investigate the shock propagations and the particle acceleration in the presence of a magnetic field, and their dependence on the field orientation. Furthermore, we will investigate the possibility to prove the existence of an intermediate shock, which is believed to exist in theoretical and numerical studies [13, 14, 15], however, no one has been experimentally observed it.

## 2. Experiment

The experiment is carried out with Gekko XII laser facility at Osaka University. Three beams of long pulse laser (energy of 100 - 120 J at $3\omega$, temporal and spatial Gaussian shape with 500 ps pulse duration and 300 $\mu$m spot, F/15 for each beam) irradiate a spherical solid target in order to create relatively isotropic plasma expansion. The experimental setup is shown schematically in Fig. 1. The target environment is gas filled with nitrogen (N), which is ionized with the radiations from the laser-solid interaction. In order to efficiently ionized the ambient gas a high-Z materials is used to enhance the radiations. We use a gold (Au) coated plastic balloon target. The ionized N ($3 \times 10^{17}$ cm-3) will be magnetized by an external magnetic field. The magnetic field is applied with two permanent magnet ($\sim 0.6$ T on surface with 20 mm diameter) to provide a relatively uniform and straight magnetic field $\sim 0.04$ T for the region of interest. The shock formations and propagations in a magnetized plasma strongly depend on the orientation of the magnetic field with respect to the shock normal direction. We measure these difference with optical imaging, collective Thomson scattering (CTS), and magnetic induction probes (also know as B-dot probes). The electron density and distribution images are obtained with interferometer and shadowgraphy, respectively. Images of emissions from plasmas also obtained with spatial and temporal resolutions. The time evolution is obtained with self-emission streaked optical pyrometer (SOP) and streaked shadowgraphy. We measure the ion feature with CTS, providing us the local information of the plasma bulk velocity, the sound velocity and the electron density. The two channels of B-dots provide three vectors, local magnetic field in two different positions, corresponding to two different angles between the magnetic field and the shock normal.

We expect to observe the different shock conditions at the different angles, however, the direction of laser incidence also provides an effective anisotropy. In order to distinguish the effects of the magnetic field from the laser, we rotate the magnets to have two different magnetic field orientation keeping all the other experimental conditions. If the magnetic field is dominant over the laser incidence, anisotropy in shock propagations and the shock magnetic field are also rotate so as the magnetic field rotation. If the laser effect is dominant, even though the external magnetic field is rotated, we will observe nominally identical images and local magnetic field.

## 3. Results

Figure 2 shows images obtained from shadowgraphy with different conditions as (a) from 23 ns from main laser arrival and the nominal laser incidence has 30 degrees from the magnetic field, (b) the same as (a) except 28 ns from the main laser arrival and with different field of view, and (c) the same as (b) except the magnetic field is perpendicular to the laser incidence. The solid arrows and the solid circles in the images represent the directions of the magnetic field and the positions of the target (500 $\mu$m diameter) with the stalk before the shots. As marked by numbers two or three sharp lines can be recognized. These correspond to large density jump and hence possibly multiple shock waves. In Fig. 2 (a) three shock like structures degenerated as pointed by the dashed arrow. The degeneration of the shock like structures are seen in Fig. 2 (b) and 2 (c). The dashed lines in Figs. 2 (b) and 2 (c) show the tangential lines on the degeneration point and the normal lines to them. Since in Fig. 2 (c) the magnetic field is rotated 60 degrees from Fig. 2 (b), if the degeneration results from the shock propagation in the different field geometry, the dashed lines have to rotate 60 degrees. Our results show that there are some difference but not as large as expected. The degeneration is not purely determined by the magnetic field but the laser mostly affects the shock geometries. The number of shocks are in most cases two.

## 4. Discussions and summary

The simplest interpretation of these structures are the forward and reverse shocks. When there are three structures, one might consider the contact discontinuity as one of them. However, it is not necessary to degenerate the shock waves if they are a pair of forward-reverse shock. One might think that the multiple shocks come from the multiple laser beams. The shock propagation velocities are about 500 km/s from the time of the images taken and the positions of the shock fronts. The spatial separation of the two or three shock fronts are $\sim 1$ mm, corresponding to the time needed for the shock waves propagate is $\sim 2$ ns. This is much larger than the main laser jitter, and thus, the multiple shocks do not correspond to the number of laser beams. In magnetohydrodynamics there can be 7 discontinuities when a magnetized plasma explodes in a magnetized plasma. In our experiment the Au plasma is unlikely magnetized due to its large mass and high velocity. Moreover, the Larmor radius of the N plasma is also very large, at least of the order or larger than the field of view of the shadowgraphy. The electrons of N plasma are well magnetized, and therefore, we have to consider the electron dynamics to understand

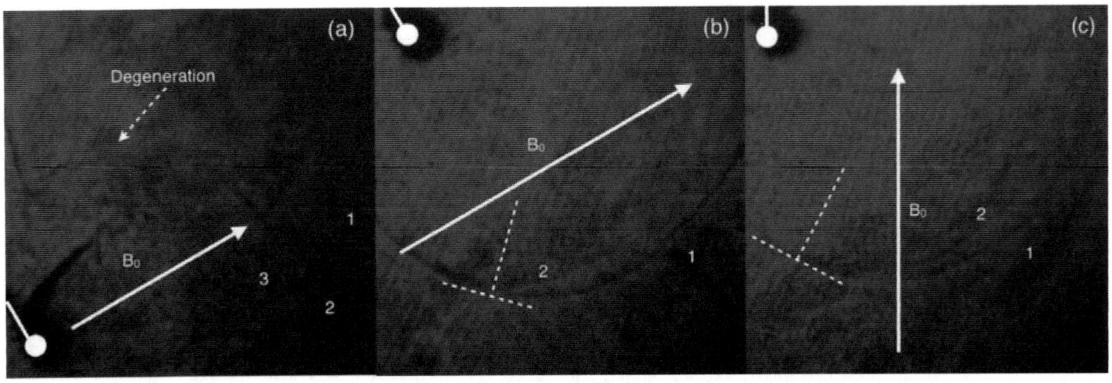


**Figure 2.** Images taken with shadowgraphy.

the observed structures. We need to characterize the N plasma with radiative hydrodynamic simulations. The radiation is stronger on the ablation side than the opposite side to the laser irradiation. Furthermore, the laser beams can directly ionize the N gas only on their paths. Anisotropies naturally arise from the laser configuration. In order to determine what kind of discontinuities they are and how the anisotropies arise, further analyses and experiments are required together with the aid of theoretical and numerical study.

We report the first experimental investigation on anisotropy of spherical shock in the presence of an external magnetic field. By irradiating a spherical solid target in an ambient media with laser, spherical shocks are excited in the presence of magnetic field. There are two or three shock like structures observed with optical diagnostics. These shock waves anisotropically expand and show degeneration on some directions. When magnetic field is rotated, the degeneration occurs in different place, but not as much as the change of magnetic field. This indicates that the laser configuration also affects the shock geometry. We have shown the first preliminary data of spherical shock wave, providing us rich and challenging phenomena to investigate. It should be noted that by changing the conditions of laser, target and magnetic field one can investigate variety phenomena, such as MHD shocks including the intermediate shock, micro-shocks governed by electron dynamics, and the shocks in SNRs.

**Acknowledgments**
We would like to appreciate the technical supports from Institute of Laser Engineering.

**References**

[1] Matsukiyo S, Ohira Y, Yamazaki R and Umeda T 2011 ApJ **742** 47
[2] Matsukiyo S and Scholer M 2011 *Journal of Geophysical Research (Space Physics)* **116** A08106
[3] Matsukiyo S and Hada T 2009 ApJ **692** 1004–1012
[4] Morita T et al. 2010 *Physics of Plasmas* **17** 122702–+
[5] Kuramitsu Y et al. 2011 Phys. Rev. Lett. **106** 175002
[6] Kuramitsu Y et al. 2012 *Phys. Rev. Lett.* **108**(19) 195004
[7] Kugland N L et al. 2012 *Nature Physics* **8** 809–812
[8] Gregori G et al. 2012 Nature **481** 480–483
[9] Ross J S et al. 2012 *Physics of Plasmas* **19** 056501 (pages 8)
[10] Morita T et al. 2013 *Physics of Plasmas* **20** 092115
[11] Kuramitsu Y et al. 2011 *Physics of Plasmas* **18** 010701–+
[12] Kuramitsu Y et al. 2011 *Phys. Rev. E* **83** 026401
[13] Wu C C 1990 J. Geophys. Res. **95** 8149–8175
[14] Hada T 1994 Geophys. Res. Lett. **21** 2275–2278
[15] Karimabadi H 1995 *Advances in Space Research* **15** 507–520

# Science and code validation program to secure ignition on LMJ

**E Lefebvre[1], C. Boniface, M. Bonnefille, A. Casner, C. Esnault, D. Galmiche, P. Gauthier, F. Girard, R. Gisbert, J.-P. Leidinger, P. Loiseau, L. Masse, P.-E. Masson-Laborde, P. Mignon, M.-C. Monteil, P. Seytor, V. Tassin**
CEA/DAM, DIF, 91297 Arpajon, France

E-mail: erik.lefebvre@cea.fr

**Abstract.** The CEA/DAM ICF experimental program is currently conducted on LIL and Omega with the goal of improving our simulation tool, the FCI2 code. In this effort, we focus on typical ICF observables: hohlraum radiation drive history, capsule core shape and neutron emission history, hydrodynamic instability growth. In addition to integrated experiment, specific designs are also helpful to pinpoint a particular phenomenon. In this article, we review our current efforts and status, and our future projects on Omega and LMJ.

## 1. Introduction

The current baseline target for ignition on LMJ [1] features a Ge-doped CH capsule placed in a rugby shaped hohlraum [2], the latter being illuminated by 160 laser beams placed along two cones per hohlraum half, with 40 beams (10 quads) per cone. The results obtained during the National Ignition Campaign on the NIF pose several challenges to our understanding of hohlraum and capsule physics at the MJ laser energy level. Prominent among these are the reported drive deficit resulting in unexpectedly low peak capsule velocity [3], and exacerbated mix amounts in the DT hot spot [4]. Despite subtle differences in laser and target design for the LMJ and the French ignition program, it is likely that the failure modes reported on NIF could also stand in the way of ignition on LMJ. These results therefore prompt us to evolve our modelling capabilities in order to be able to better understand the NIC campaign finding, and provide a safer basis for design optimization for LMJ.

Integrated ICF target simulations show various sensitivities to a number of physical or numerical options, ranging from the choice of specific equations of state or opacities for the target material, to electron conduction or non-LTE atomic physics models, or target meshing choices. To prevent fallacious agreement between experimental results and post-shot simulations by *ad hoc* modelling choices, we try to identify a *single* optimal set of code and physical parameters that

- maximizes post-shot agreement
- gives robust simulation results in a reasonable computation time
- is grounded on the best physical model available

The final choice, and its abilities to operate predictively, depends of course on the diversity of the experimental database on which the modelling is tested. While past settings were solely based on Omega and LIL results, our current effort is to expand our database to NIF results, effectively asking for a model working point that is as relevant at 30 kJ as at 1.8 MJ laser energy.

[1] To whom any correspondence should be addressed.

Integrated ICF experiments such as those carried out during the NIC campaign are efficient at identifying the existence of shortcomings in the modelling, but are not best suited to solve these puzzles, since so many physical phenomena are entangled in these experiments. Using LIL and Omega, we therefore strive to carry out experiments that help isolate specific physics problems in order to progressively validate our models piece by piece.

## 2. Radiation drive

Our workhorse code for integrated radiation hydrodynamics simulations is FCI2, a 2D lagrangian code including B-field generation and nonlocal electron conduction, and modeling NLTE effects in high-Z materials with the GonRad effective model. With those physics models and other numerical options, this code has been benchmarked against ~ 30 experiments in which the x-ray flux emitted from a laser-irradiated target was measured using broadband spectrometers such as Dante and DMX. These experiments range from open gas-filled or empty hohlraums on LIL, possibly featuring shields, gas-filled or empty hohlraums on Omega and NIF. The peak thermal radiation flux (in the 0-2 keV range) is always well modelled, within or very close to the error bars of the measurement. Agreement on the M-band flux, on the other hand, shows more variability.

In addition to the x-ray fluxes, LPI measurements (SRS and SBS spectra) are progressively included among the observables we consider, since they reflect the plasma dynamics inside the hohlraum and bring about additional criteria to judge the quality of the modeling. As an example, figure 1 illustrates a Stimulated Raman Scattering spectrum measured on Omega in a methane-filled rugby-shaped hohlraum, for which the typical "μ-shaped" structure is well accounted for by the linear gain calculation with the Piranah LPI post-processor. This structure can be linked to backscattering from the denser gas region compressed between the expanding ablator and wall materials, close to the target equator (region (2) in figure 1(c)), whereas the fainter, broader and earlier structure originates from regions closer to the laser entrance hole (region (1) in figure 1(c)). However, this modelling can also sometimes be proved wrong, even for relatively small variations with respect to the present experiment, such as a slightly lower fill gas pressure.

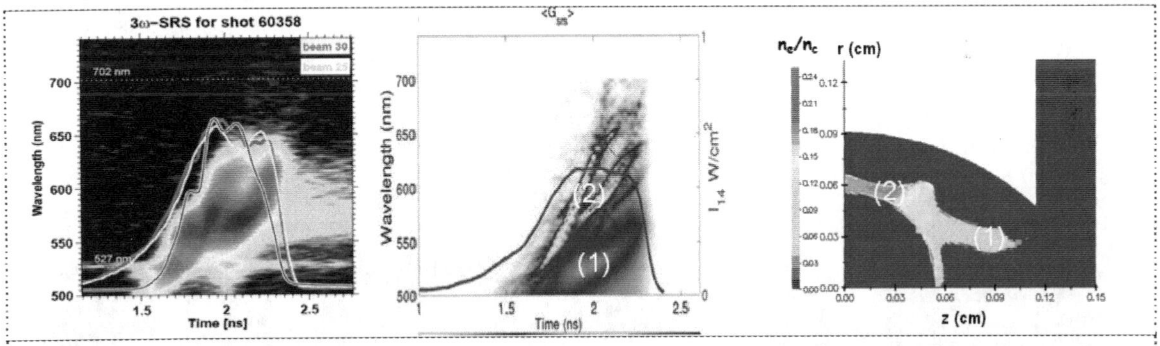


**Figure 1.** Typical Stimulated Raman Scattering spectrum measured on the Omega laser in a gas-filled, rugby-shaped hohlraum (a), linear SRS gains calculated by our LPI postprocessor, Piranah (b), and localization of the origin of the main spectrum components in the hohlraum plasma (c).

Measurements obtained on integrated ICF experiments – with their complex hydrodynamics and limited view of the gold emission regions – can be usefully complemented by specific, simplified experiments such as the one we carried out on Omega in June 2013 using gold-coated spheres. This campaign was a continuation of 2006 experiments for which the Dante and DMX broadband spectrometers would give conflicting measurements at high irradiance. Our latest measurements now support the highest conversion efficiency from the 2006 campaign. Additional data using halfraums and planar foils were also gathered. This choice of target should enable challenging our simulation capabilities on a wide range of physical situations, some giving rise to B-field generation and some not, some in which radiation reabsorption is important and others in which it is not, and some amenable to lagrangian hydrodynamic modelling, while others require Arbitrary Lagrange Euler (ALE) methods. From 2016 onwards, this type of experiment is bound to be continued on LMJ, with interesting additional measurements of B-field generation in open or hohlraum geometry made possible by proton-radiography with the PETAL short pulse beam [petal]. The LIL LPI campaign is another example of an experiment specifically designed to improve our hohlraum modeling capability by addressing a particular physical phenomenon, and is described in more details elsewhere [6].

## 3. Capsule core shape

FCI2 with its standard settings has also been benchmarked against a series of capsule implosion experiments, most of them on Omega (in empty or gas-filled, rugby or cylinder hohlraums), as well as some NIC experiments (and most notably the rugby hohlraum shots of March and May 2013 [7]. Core shape is measured at maximum X-ray emission by decomposing the 2D capsule emission contour at 17% of its maximum on Legendre polynomials. Mode-2 asymmetry (so-called P2/P0) is our main observable here. In this respect, experiments and calculations are found to be in good agreement in empty hohlraums on Omega, while agreement is more variable on gas-filled experiments. Closer analysis reveals that the offset between measured and computed mode-2 asymmetry seems to be correlated to the fraction of inner-cone energy that gets absorbed in the outer-cone gold bubble or around the LEH, with higher amounts of such ill-deposited energy giving rise to poorer agreement on symmetry. The analysis of this correlation is still ongoing.

On NIF experiments, it comes as little surprise that capsule symmetry is a very strong function of cross-beam energy transfer [8]. For the same reason, symmetry is also a strong function of backscattered energy, in particular from the inner cones. In the analysis of rugby shot N130318, for which 40 kJ of SRS were measured on the inner cones [7], we observed that asymmetry could be changed (numerically) from P2/P0=-30% to P2/P0=-54% by subtracting an extra 20 kJ from the inner cone energy (figure 2). To justify this variation on backscattered energy, one should keep in mind that the usual way of handling backscatter in integrated simulations, i.e. by subtracting the backscattered power from the incident one, is an approximation, as backscattered photons actually propagate and are absorbed in part of the hohlraum plasma. An order of magnitude for this effect is assessed by computing, in a simulation for shot N130318, the collisional absorption of photons at the peak SRS wavelength propagating backward out of the hohlraum from their point of backscattering. Around the peak of the laser pulse, we find that ~20% of the backscattered energy is actually reabsorbed in the plasma, prompting us to revise the methodology for dealing with backscattered energy in our radiation hydrodynamics simulations.

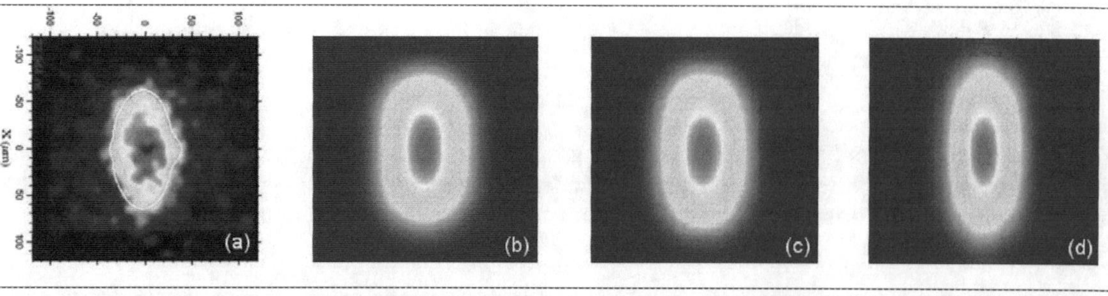


**Figure 2.** Experimental hot-spot shape measured on NIF for rugby shot N130318, P2/P0=-41% [7] (a), FCI2 simulated shape using the net laser power (incident minus backscattered), P2/P0=-30% (b), and with additional energy subtracted from the inner beams: -13 kJ, P2/P0=-39% (c), -20 kJ, P2/P0=-54% (d).

## 4. Capsule bang time

On a dozen of experiments, on Omega and on NIF, in cylindrical as well as rugby-shaped hohlraums, the time of peak neutron emission has been measured and compared to FCI2 simulations. The simulated bang time is consistently early compared to the measurement, on average by 200 ps on Omega and by 500 ps on NIF. On Omega, making up for this offset would require a systematic overestimate of the x-ray drive by 4%, or a systematic underestimate of the ablator thickness by 4 µm – two hypotheses that are deemed improbable. On NIF, this mismatch is very similar to that identified by LLNL and which led to the use of "multipliers" to relate the flux seen by the capsule to that computed in the hohlraum calculation [3]. This shortcoming casts doubts on our understanding and modelling of both the hohlraum and the capsule, and should therefore be solved with high priority. To address this problem, specific experiments have been conducted on the Omega laser, in which 1D (streaked) and 2D radiographs of a converging plastic capsule are measured and compared to simulations. The analysis of these data is ongoing, and initial results are presented elsewhere [9].

## 5. Hydrodynamic instabilities

The unexpected mix levels measured on NIF [4] highlight the need for improved understanding of hydrodynamic instability growth. Strong sensitivity to details of the radiation drive have been observed in numerical simulations of hydrodynamic instabilities of a typical NIF capsule, resulting in perturbations that do or do not invert during compression for radiation temperature changes on the order of 10 eVs in the picket or in the trough or the radiation drive pulse. We will carry out converging hydrodynamic instability experiments on Omega in 2014, leveraging on our implosion dynamics platform [9], to extend our database to higher convergence ratio than could be achieved on Nova. Obviously, the NIF "hydro growth" experiments [4] will also provide a wealth of information on this subject. However, simpler, planar geometry experiments can also be relevant here. An experimental concept derived from the Ablative Rayleigh-Taylor Science Use of NIF proposal [10] has been worked out to produce a surrogate flow to that of a NIF capsule, and we believe such experiments would be helpful in understanding the early phase of hydrodynamic instabilities in a fusion capsule. Such planar geometry will also be more amenable to early experiments on LMJ in the near future.

## 6. Conclusions

The CEA/DAM experimental ICF program is currently executed on Omega and LIL. Through the post-shot modelling of these experiments, and of NIF results, we aim at identifying the set of code parameters and physics options that leads to best agreement with the experiments over a large database of results. Integrated ICF experiments provide a wealth of data but their analysis can be complicated by the entanglement of various physical processes. In complement to these experiments, we therefore strive to design simpler experiments that focus on a specific physics issue. Examples of such efforts that are being carried out on Omega and will soon be transferred from LIL to LMJ have been given in this article.

## References

[1] Lion C 2010 *J. Phys. Conf. Series* **244** 012003; Miquel J-L, *et al.*, these proceedings
[2] Lafitte S and Loiseau P 2010 *Phys. Plasmas* **17** 102704
[3] Callahan D, *et al.* 2012 *Phys. Plasmas* **19** 056305; Callahan D, *et al.*, these proceedings
[4] Regan S P, *et al.* 2012 *Phys. Plasmas* **19** 056307; Remington B, *et al.*, these proceedings
[5] Batani D, *et al.* 2013 *Acta Polytechnica* **53** 103-109
[6] Loiseau P, *et al.*, these proceedings
[7] Ross S, *et al.*, these proceedings; Amendt P, *et al.*, these proceedings
[8] Michel P, *et al.* 2010 *Phys. Plasmas* **17** 056305
[9] Casner A, *et al.*, these proceedings
[10] Casner A, *et al.* 2013 *Phys. Plasmas* **19** 082708

# Modeling of radiative and quantum electrodynamics effects in PIC simulations of ultra-relativistic laser-plasma interaction

**M Lobet[1,2], E d'Humières[2], M Grech[3], C Ruyer[1], X Davoine[1], L Gremillet[1]**

[1]CEA, DAM, DIF, F-91297, Arpajon, France
[2]Univ. Bordeaux-CNRS-CEA, CELIA, UMR 5107, 33405, Talence, France
[3]LULI, École Polytechnique-CNRS-CEA-Université Paris VI, UMR 7605, École Polytechnique, 91128 Palaiseau, France

E-mail: mathieu.lobet@cea.fr

**Abstract.** Next generation of ultra-intense laser facilities will lead to novel physical conditions ruled by collective and quantum electrodynamics effects, such as synchrotron-like emission of high-energy photons and $e^+e^-$ pair generation. In view of the future experiments performed in this regime, the latter processes have been implemented into the particle-in-cell code CALDER.

## 1. Introduction

The forthcoming ten petawatt laser systems, VULCAN-10PW in the United Kingdom (300 J, 30 fs) and APOLLON in France (150 J, 15 fs), will pave the way to the exawatt ELI project aiming to reach intensities in excess of $10^{25}$ Wcm$^{-2}$. This will lead to unexplored laser-matter scenarii, characterized by coupled plasma and quantum electrodynamics (QED) processes. In order to enhance these novel effects, a favourable configuration relies upon the interaction of the laser pulse with counter-propagating, multi-GeV electrons. At laser intensities up to $10^{23}$ Wcm$^{-2}$, the electron dynamics enters the so-called radiation-dominated regime characterized by strong production of high-energy photons via nonlinear Compton scattering. Interacting back with the laser field, these photons can lead to efficient production of $e^-e^+$ pairs (via the Breit-Wheeler mechanism). The resulting dense pair plasma, beyond its fundamental interest, can potentially modify the overall laser-plasma interaction [2].

## 2. Modelling

The importance of QED effects on a relativistic electron of mass $m_e$, charge $e$ and energy $\varepsilon_e = \gamma_e m_e c^2$, and on a photon of energy $\varepsilon_\gamma = \gamma_\gamma m_e c^2$, both interacting with a monochromatic plane wave $(\mathbf{E}, \mathbf{B})$, depends on the respective Lorentz-invariant quantum parameters:

$$\chi_e = \frac{\gamma_e}{E_s}\left[{\mathbf{E}_\parallel}^2/\gamma_e^2 + (\mathbf{E}_\perp + \mathbf{v_e}\times\mathbf{B})^2\right]^{1/2}, \qquad \chi_\gamma = \frac{\gamma_\gamma}{E_s}\left[(\mathbf{E}_\perp + \mathbf{c}\times\mathbf{B})^2\right]^{1/2}, \tag{1}$$

where $E_s = m_e^2 c^3/e\hbar \simeq 1.3\times 10^{18}$ V/m is the Schwinger field beyond which the electromagnetic field can no longer be treated classically. The energy distribution of the production rate of photons is given by [4]

$$\frac{d^2N_\gamma}{dtd\gamma_\gamma} = \frac{1}{\pi\sqrt{3}}\frac{\alpha_f}{\tau_c\gamma_e^2}\left[\int_{2y}^{+\infty} K_{1/3}(s)ds - (2+3\chi_\gamma y)\,K_{2/3}(2y)\right], \tag{2}$$

where $\alpha_f \simeq 1/137$ is the fine structure constant, $\tau_c = \hbar/m_ec^2$ the Compton length and $y = \chi_\gamma/[3\chi_e(\chi_e - \chi_\gamma)]$. For $\chi_e \ll 1$ ($I < 10^{22}$ Wcm$^{-2}$, $\gamma_e \sim 2000$), large numbers of photons are emitted, each taking away a small fraction of the electron energy. The emission process can be therefore treated classically and modelled as a continuous damping reaction. By constrast, in the quantum regime, for $\chi_e \sim 1$ ($I \sim 10^{23}$ Wcm$^{-2}$), fewer photons are emitted, but their energies can be comparable to that of the electron, leading to strong recoils and discontinuous trajectories. Finally, for high-energy photons ($\varepsilon_\gamma > 2m_ec^2$) interacting with the laser field, the energy distribution of the probability to decay into $e^+e^-$ pair is given by [4]

$$\frac{d^2N_\pm}{dtd\gamma_e} = \frac{1}{\pi\sqrt{3}}\frac{\alpha_f}{\tau_c\gamma_\gamma^2}\left[\int_{2y}^{+\infty} K_{1/3}(s)ds - (2-3\chi_\gamma y)\,K_{2/3}(2y)\right]. \tag{3}$$

Here, $\gamma_e$ refers to the energy of the created electron.

## 3. Numerical implementation

Particle-in-cell (PIC) codes are efficient tools for simulating the kinetic evolution of plasmas subjected to intense electromagnetic waves [3]. The PIC code CALDER has been enriched with physical models describing photon and pair generation. In the PIC description, particles are represented by super-particles with given density, shape and momenta. Radiative and QED models are applied on the super-particles and we similarly define the notion of super-photons generated during the emission process. In the classical radiative regime, the model considered is based on Sokolov's scheme for radiation damping [5]:

$$\frac{d\mathbf{p_e}}{dt} = \mathbf{f}_l - e(\delta\mathbf{v_e}\times\mathbf{B}) - P_{rad}\frac{\mathbf{p_e}}{\gamma m_ec^2},\quad \frac{d\mathbf{x_e}}{dt} = \mathbf{v_e}+\delta\mathbf{v_e},\quad \delta\mathbf{v_e} = \frac{\tau_0}{m_e}\frac{\mathbf{f}_l - \mathbf{v_e}(\mathbf{v_e}\cdot\mathbf{E})}{1+\tau_0(\mathbf{v_e}\cdot\mathbf{E})/m_ec^2} \tag{4}$$

where $\mathbf{f}_l$ is the Lorentz force vector and $\tau_0 = 2e^2/(12\pi\varepsilon_0 m_ec^3)$. The radiated power is calculated using the quantum equation (2):

$$P_{rad} = \int_1^{\gamma_e}\frac{d^2N_\gamma}{dtd\gamma}\gamma m_ec^2 d\gamma \tag{5}$$

In the quantum radiative regime, the Monte Carlo (MC) description is well-suited to treating the stochastic character of the photon generation [6]. This is valid provided that the field is uniform and quasi-static during the emission process. Propagating electrons are assigned an optical depth $\tau_e$, evolving with time according to the field and the electron energy,

$$\frac{d\tau_e}{dt} = \int_0^{\chi_e}\frac{d^2N}{d\chi dt}d\chi. \tag{6}$$

In the MC algorithm, a discrete photon is emitted when $\tau_e = \tau_e^f$, where the total optical depth $\tau_e^f$ is sampled from $\tau_e^f = -\log\xi$ with $\xi \in ]0,1]$ a uniform random number. The photon quantum paramater $\chi_\gamma$ is calculated by inverting $P_\gamma(0\to\chi_\gamma) = \xi'$ where

$$P_\gamma(0\to\chi) = \int_0^{\chi}\frac{d^2N_\gamma}{dtd\chi}d\chi \Big/ \int_0^{\chi_e}\frac{d^2N_\gamma}{dtd\chi}d\chi \tag{7}$$

is the the photon cumulative distribution function and $\xi' \in [0,1]$ is a new random parameter. The photon is emitted to the direction parallel to the electron velocity with wave vector $\mathbf{k}$ (since the angular spread is $\sim 1/\gamma_\gamma \ll 1$). The electron momentum $p_e^f$ therefore becomes

$$\mathbf{p}_e^f = \mathbf{p}_e - \bar{h}\mathbf{k} = \left(p_e - \frac{\varepsilon_\gamma}{c}\right)\frac{\mathbf{p}_e}{|\mathbf{p}_e|}. \tag{8}$$

A similar MC scheme is applied to the super-photons to model pair creation using the probability function (3). Each photon is assigned an optical depth $\tau_\gamma$ and a randomly sampled final one $\tau_\gamma^f$.When $\tau_\gamma = \tau_\gamma^f$, the super-photon is deleted and a $e^-e^+$ pair is created at the current position. Solving $P_\pm(0 \to \chi_-) = \xi''$ gives the newly-created electron quantum parameter $\chi_-$, $\xi''$ being randomly calculated in $]0,1]$ and $P_\pm$ the cumulative distribution function associated to the pair generation. The positron quantum parameter is $\chi_+ = \chi_\gamma - \chi_-$. As the MC radiative model leads to the generation of many low-energy photons unable to decay into pairs and prejudicial for the numerical efficiency (most notably in configurations prone to pair cascading), we usually restrict the creation of super-photons to energies above $2m_ec^2$. Moreover, we have worked out a coalescence scheme ensuring energy and momentum conservation on average.

## 4. Validation test

As a test simulation, we consider the photon and pair generation by electrons of initial energy $\varepsilon_e/m_ec^2 = 3000$ in a constant magnetic field $B = 7.48$ MT. The initial quantum parameter is equal to $\chi_e \simeq 5$. The initial synchrotron frequency and radius are respectively equal to $\omega_B^{-1} = 2.3$ fs and $R \simeq 0.7$ $\mu$m. We define $f_+, f_-, f_\gamma$ the energy distributions of the positrons, electrons and photons, respectively. The time evolution of the latter can be described by the following set of equations [7]

$$\frac{df_\gamma}{dt}(\gamma_\gamma, t) = \int_{\gamma_\gamma}^{+\infty} \frac{d^2N_\gamma}{dtd\gamma_\gamma}(\gamma_\gamma,\gamma)f_\pm(\gamma)d\gamma - 2\int_2^{\gamma_\gamma}\frac{d^2N_\pm}{dtd\gamma_\pm}(\gamma,\gamma_\gamma)f_\gamma(\gamma_\gamma)d\gamma \tag{9}$$

$$\begin{aligned}\frac{df_\pm}{dt}(\gamma_\pm, t) &= -\int_0^{\gamma_\pm}\frac{d^2N_\gamma}{dtd\gamma_\gamma}(\gamma_\pm,\gamma)f_\pm(\gamma_\pm)d\gamma + \int_{\gamma_\pm}^{+\infty}\frac{d^2N_\gamma}{dtd\gamma_\gamma}(\gamma,\gamma-\gamma_\pm)f_\pm(\gamma)d\gamma \\ &+ \int_{\gamma_\pm+1}^{+\infty}\frac{d^2N_\pm}{dtd\gamma_\pm}(\gamma_\pm,\gamma)f_\gamma(\gamma)d\gamma.\end{aligned} \tag{10}$$

The time evolution of the total energy, density and averaged quantum parameter is plotted in Fig. 1 for each species. Figure 2 displays the corresponding energy distributions. Very good agreement is obtained between the numerical solutions of Eqs. (9) and (10) and the MC-PIC simulations. The interaction starts in the quantum regime. The firstly created photons have an energy close to initial electron energy, $\gamma_\gamma \sim 10^3$. Consequently, the radiated energy significantly grows during the first rotation at the expense of the electron energy (Fig. 1a). The average photon quantum parameter is close to $\chi_\gamma \sim 1$ (Fig 1c), which is favourable for pair production. In turn, newly generated particles radiate their energy away, hence leading to pair cascading (Fig. 1b). When entering the semi-classical regime, the production rate of positrons tends to stabilize since the average photon energy drops below the pair creation threshold ($\chi_\gamma < 10^{-1}$). Radiation losses, originally responsible for pair creation, now only act as a cooling mechanism. Note that in the quantum regime, the stochasticity, intrinsic to the QED processes stretches the electron energy spectrum, whereas in the semi-classical regime, radiation losses tend to shrink it back.

## 5. Conclusion

We have presented the implementation in a PIC code of the photon and pair generation via the Compton scattering and Breit Wheeler processes. These physical models will be used in forthcoming papers to study a number of laser-plasma scenarii at ultra-relativistic intensities.

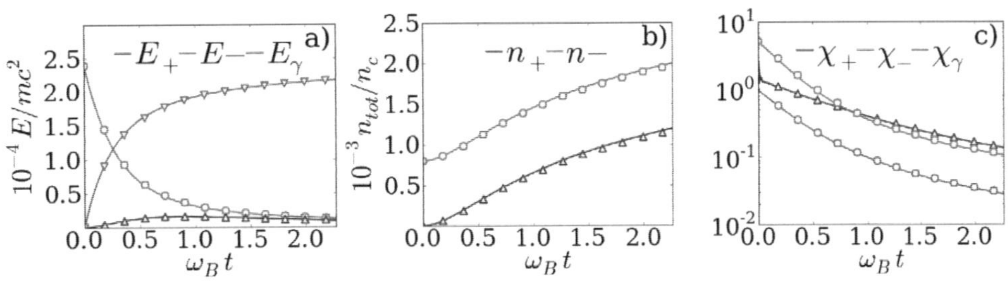

**Figure 1.** Time evolution of the total energy (a), density (b) and quantum parameter (c) from the PIC simulation (lines) and the numerical solution of Eqs. (9) and (10) (markers).

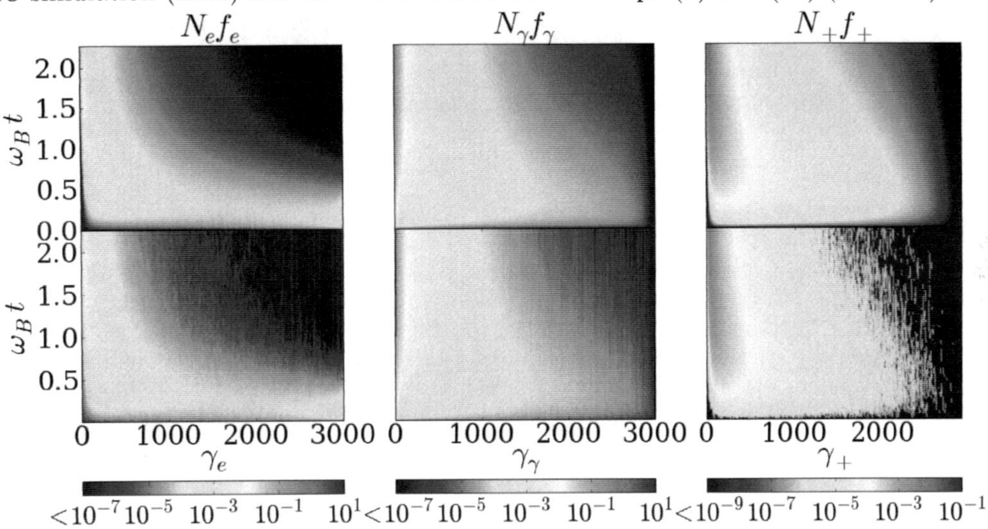

**Figure 2.** Time evolution of the energy distributions for electrons (right), photons (center) and positrons (left) from the PIC simulation (bottom) and the numerical solution of Eqs. (9) and (10) (top).

**Acknowledgments**
The authors acknowledge support by the French Agence Nationale de la Recherche (LABEX PALM-ANR-10-LABX-39) and thanks R Duclous, A Debayle for interesting discussions. The PIC simulations were performed using HPC resources at TGCC/CCRT (Grant No. 2013-052707).

## References

[1] G. A. Mourou *et al.* 2006 *Rev. Mod. Phys.* **78** 309 ; A. Macchi *et al.* 2013 *Rev. Mod. Phys.* **85** 751
[2] A. R. Bell and J. G. Kirk 2008 *Phys. Rev. Lett.* **101** 200403 ; S. V. Bulanov *et al.* 2011 *NIMPRA* **660** 31-34 ; A. Di Piazza *et al.* 2012 *Rev. Mod. Phys.* **84** 1177
[3] C. K. Birdsall and A. B. Langdon 1985 *Plasma Physics via Computer Simulation* (New York: mcgraw-Hill) ; E. Lefebvre *et al.* 2003 *Nucl. Fusion* **43** 629-633
[4] A. I. Nikishov and V. I. Ritus 1964 *JETP* **19** 529 ; T. Erber 1966 *Rev. Mod. Phys.* **38** 626
[5] I. V. Sokolov *et al.* 2009 *Phys. Plasmas* **16** 093115
[6] R. Duclous *et al.* 2011 *Plas. Phys. Control Fusion* **53** 015009 ; N.V. Elkina *et al.* 2011 *Phys. Rev. ST Accel. Beams* **14** 054401
[7] I. V. Sokolov 2010 *Phys. Rev. Lett.* **105** 195005 ; J. G. Kirk *et al.* 2009 *Plas. Phys. Control Fusion* **51** 085008 ; N. Neitz *et al.* 2013 *Phys. Rev. Lett.* **111** 054802

Journal of Physics: Conference Series **688** (2016) 012059 doi:10.1088/1742-6596/688/1/012059

# Simulation of laser-plasma interaction experiments with gas-filled hohlraums on the LIL facility

**P. Loiseau[1], P.-E. Masson-Laborde[1], D. Teychenné[1], M.-C. Monteil[1], M. Casanova[1], D. Marion[1], G. Tran[1,2], G. Huser[1], C. Rousseaux[1], S. Hüller[2], A. Héron[2] and D. Pesme[2]**

[1] CEA, DAM, DIF, 91297 Arpajon, France

[2] Centre de Physique Théorique, CNRS, École Polytechnique, 91128 Palaiseau, France

E-mail: pascal.loiseau@cea.fr

**Abstract.** Laser-plasma interaction is a major issue for achieving ignition in inertial confinement fusion schemes, and still a major concern for the upcoming french laser mégajoule (LMJ) program. In order to mitigate the deleterious effects due to laser-plasma instabilities (LPI), clearly evidenced during the recent US National Ignition Campaign conducted on the National Ignition Facility, we use the LIL facility as a demonstrator for LPI studies. In this article, we focus on preliminary results regarding the propagation of a typical LMJ quadruplet through gas-filled hohlraums. Results on hohlraum energetics will then be discussed.

## 1. Introduction

One major goal of the Laser MegaJoule facility (LMJ) [1], currently under completion in France, is to achieve fusion ignition and thermonuclear gain from a target driven with a laser. First attempts to achieve ignition will be undertaken as soon as the facility is 2/3 completed. In this configuration the laser is able to deliver up to 1.2 MJ and 390 TW at peak power, at 0.351 $\mu$m. An ignition indirect drive target has been designed for the 2/3 LMJ step [2]. It requires 0.9 MJ and 260 TW of laser energy and power, to achieve a 300 eV radiative temperature in a rugby-shaped hohlraum and to give a yield of about 20 MJ. This achievement relies firstly on the mitigation of laser plasma instabilities (LPI). First experiments conducted on the National Ignition Facility [3] and recent ones [4] illustrate the fact that LPI is not the sole reason for implosion degradation, but it does obscure the understanding of the target performance. Such issues can be addressed by conducting laser-plasma interaction campaigns on the LIL facility, using various lengths of gas-filled hohlraums.

The LIL facility is a prototype of one quadruplet of the LMJ. Three different gas-filled hohlraums have been designed in order to mimic plasma conditions that are expected along two particular beam paths in ignition hohlraums [5,6]. We focus on the open cylinder targets, 1.5- or 4-millimeter long and 1 atm neo-pentane gas-filled. The target material is mainly gold. The LIL quadruplet is aligned along the hohlraum axis and delivers a 6-ns long pulse with 15 kJ at $3\omega$. Many optical diagnostics were activated during the campaign, see Fig. 1, giving a full temporal and angular description of parametric instabilities such as stimulated Brillouin (SBS) or Raman (SRS) scattering. X-ray diagnostics were used for measuring the radiative temperature, and transmission beam diagnostics were activated. Finally, this whole set of diagnostics allows a full measurement of hohlraum energetics.

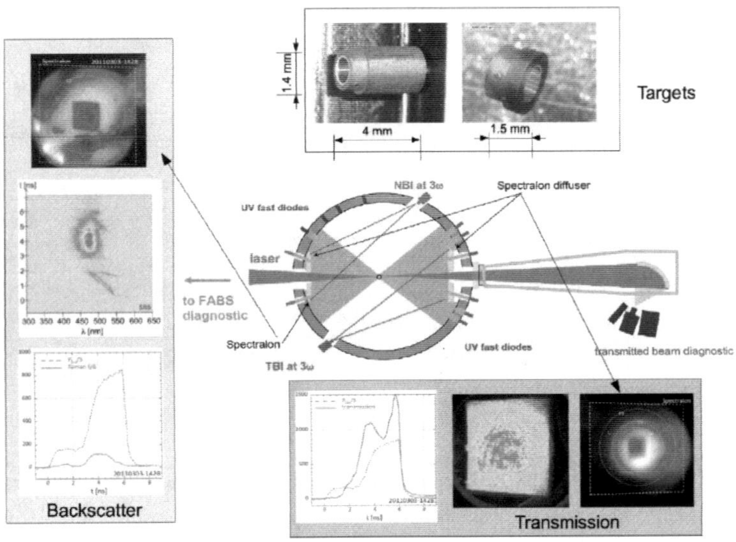


**Figure 1.** LIL diagnostics and targets.

In the following, we will present and discuss hydrodynamic calculations together with preliminary results of the LPI campaign.

## 2. Simulation of LIL experiments

The modelling of laser-plasma interaction experiments is usually made by using radiative-hydrodynamic codes, such as FCI2. In this case, the laser-plasma interaction is often reduced to a collisional absorption of the electromagnetic wave - via the inverse Bremsstrahlung mechanism -, and the laser propagation is described by ray-tracing packages. This modelling is adapted to the long hydrodynamical time and spatial scales that are favored in nanosecond experiments. However, laser-plasma interactions that arise during the experiment, are not coherently taken into account, nor the rich laser-plasma interaction in general. In order to simulate experimental results, such as the radiative temperature characterizing the drive on an ignition capsule, we need to remove the energy losses due to LPI, essentially the backscattered energies, from the incident laser pulses delivered by the facility. Using LIL's results, we address the question to what extent this usual recipe is realistic or not.

Backscattered measured energies on the LIL facility, essentially from SRS, were respectively $\sim 30\%$ and $\sim 5\%$ of the laser incident energy for the 4-mm (long) and 1.5-mm (short) hohlraums. We measured a moderate effect of the nominal optical smoothing scheme, based on a spectral dispersion and a focusing grating. Simulations were done using a modified laser pulse, taking into account the SRS temporal behaviour. As expected, the calculated maximum radiative temperatures are correct.

Once the simulation is done, it is now interesting to compare the SRS spectrum to the simulated one. SRS spectra both reveal the electron density and temperature evolutions. Those quantities are seldom measured on large laser facilities. Fig. 2 shows the calculated Raman spectra for the long (a) and short (b) hohlraums, and the SRS experimental wavelength time evolution. The early part of the signal, up to 3 ns, is related to a density channel created by the

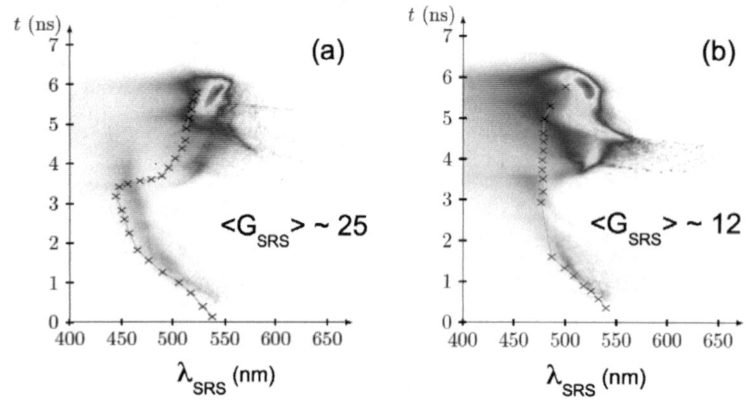


**Figure 2.** Simulated SRS spectra for the long (a) and short (b) hohlraums. The connected points correspond to the time evolution of the measured backscattered wavelength.

laser. The later part of the SRS signal is caused by hydrodynamic motion inside the hohlraum, where shock waves rebound off the walls. A frequency shift between calculated and measured wavelength is clearly seen for both hohlraums. This could be the signature of nonlinear kinetic effects, those effects being beyond the validity range of a linear gain calculation.

Transmission measurements indicate that only $\sim$ 1.5 kJ and $\sim$ 4.5 kJ laser energies exit respectively long and short hohlraums. In radiative-hydrodynamic codes, the transmission is defined as the unabsorbed laser energy. We find that simulations predict $\sim$ 4.0 kJ and $\sim$ 8.0 kJ respectively for long and short hohlraums. This discrepancy is surprising and highlights the inacurracy of laser beam propagation models in hot plasmas. A slight benefit from optical smoothing is only observed in the shorter hohlraum. Some insights can be found in the time integrated beam transmission diagnostic where strong spatial incoherence is observed, especially for short hohlraums.

## 3. Discussion

The previous simulation results exhibit conflicting conclusions: the amount of energy injected in the simulation is right regarding radiative temperature measurements, but SRS calculated spectra are shifted, and too much laser energy is unabsorbed (transmitted) in simulation. Regarding SRS spectra, several reasons can be raised. First, because hydrodynamic conditions were not measured, plasma conditions can be miscalculated. Besides, due to laser self-focusing or filamentation, laser intensities may also be incorrect, affecting plasma conditions. During the second stage of the pulse, the laser power is four time higher and could lead to self-focusing of the beam. Actually, Raman spectra are quite similar whatever the optical smoothing scheme is, indicating that self-focusing is not solely responsible for the discrepancy. Varying atomic physics packages and electron thermal conduction models did not improve the spectrum. Also, removing the measured SRS power is a poor model of laser-plasma coupling. In particular, such a model underestimates the real amount of SRS energy by not considering a potential SRS light absorption inside the plasma. Accordingly, Raman light absorption can modify local electron temperature. Because linear gain calculations do not take into account non-linear frequency shifts arising from a strong modification of the electron distribution function, SRS calculated spectra miss the right resonance.

This confirms the need of a kinetic Raman model. We are developing a three-wave coupling

Raman solver in multi-dimensions with nonlinear kinetic capabilities [7, 8]. We show in Fig. 3 first results, in one-dimension (1D) for a short hohlraum. The plasma parameters are taken at 4 ns corresponding to the observed Raman peak power. The matching conditions for the three waves are imposed at $10\%n_c$, corresponding to the maximum density, and $5\%n_c$ corresponding to the expected resonant density deduced from the experimental spectrum. SRS wavelengths ($k\lambda_D$ parameter) are repectively 540 nm (0.27) and 480 nm (0.4) for the $10\%n_c$ and $5\%n_c$ cases. Both simulations produce a high 1D reflectivity, around 20%. This was expected for the $10\%n_c$ case but not for the $5\%n_c$ case. Moreover, the resonance at low density disappears when kinetic effects are not taken into account. This interesting result proves that our modeling is able to catch those new features and could explain the experimental spectrum. Finally, we use the 3D paraxial code HERA [9, 10] to investigate the propagation of the LIL quad, by means of massivelly parallel simulations. The Raman model is intended to be introduced shortly in HERA.

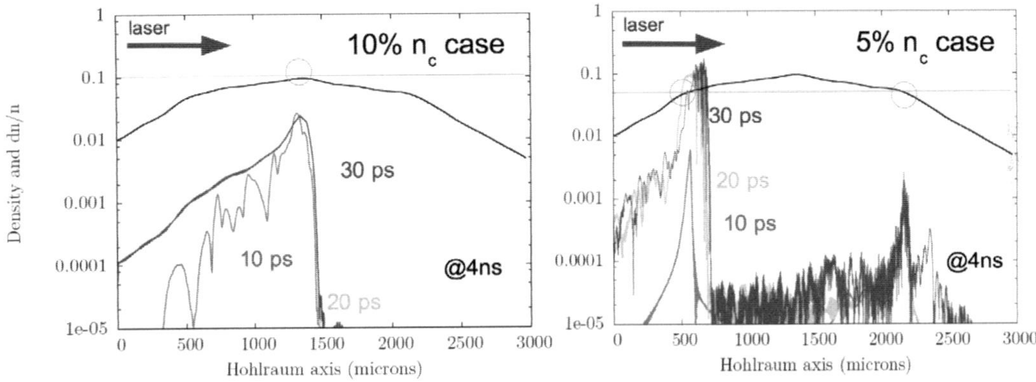


**Figure 3.** Spatial localization of Langmuir waves at 10, 20 and 30 ps for two matching electron densities: $5\%n_c$ and $10\%n_c$. The wave amplitude is given in relative density variation $\delta n/n$. The laser propagates from the left to the right.

## 4. Conclusion

The mitigation of LPI is a major issue for ignition achievement. It relies on our capabilities to model LPI but also on our capability to predict the plasma parameters such as the electron density and temperature. Radiative-hydrodynamic simulations of LIL experiments emphasize the difficulty of predicting all the measured quantities. As a matter of fact, single beam propagation experiments on LIL illustrate the difficulty of securing a safe inner beam propagation in ignition hohlraums.

## References

[1] C. Lion, J. Phys.: Conf. Ser. 244, 012003 (2010).
[2] S. Laffite and P. Loiseau, Phys. Plasmas 17, 102704 (2010).
[3] S. H. Glenzer *et al.*, Phys. Rev. Lett. 106, 085004 (2011).
[4] D. Callahan and D. Hinkel talks, IFSA Conference (2013).
[5] O. Morice *et al.*, J. Phys.: Conf. Ser. 112, 022037 (2008).
[6] A. Casner *et al.*, J. Phys.: Conf. Ser. 244, 032042 (2010).
[7] H. X. Vu *et al.*, Phys. Plasmas 9, 1745 (2002).
[8] P-E. Masson-Laborde *et al.*, Phys. Plasmas 17, 092704 (2010).
[9] P. Loiseau *et al.*, Phys. Rev. Lett. 97, 205001 (2006).
[10] P. Ballereau *et al.*, J. Scient. Comput. 33, 1 (2007).

# HiLASE: a scalable option for Laser Inertial Fusion Energy

**A Lucianetti[1], V Jambunathan[1], M Divoky[1], O Slezak[1], M Sawicka[1], J Pilar[1], S Bonora[1,2] , P D Mason[3], P J Phillips[3], K Ertel[3], S Banerjee[3], C Hernandez-Gomez[3], J L Collier[3], C Edwards[3], M Tyldesley[3], and T Mocek[1]**
[1]HiLASE Project, Institute of Physics AS CR, 18221 Prague, Czech Republic
[2]CNR-IFN Laboratory for Ultraviolet and X-ray Optical Research, 35131 Padova, Italy
[3]STFC Rutherford Appleton Laboratory, Didcot, OX11 7 FZ, United Kingdom

E-mail: lucianetti@fzu.cz

**Abstract.** The goal of the HiLASE project is to design and optimize parameters for 100 J/ 10 Hz Yb:YAG laser amplifiers that are scalable to the kJ regime. The HiLASE power amplifier design is based on a cryogenic, gas-cooled multi-slab concept. Simulation results of the 10 J pre-amplifier agree very well with experimental measurements. In order to fulfil the very demanding requirements, which include wall-plug efficiency > 12% and repetition rates up to 10 Hz, HiLASE and RAL teams are closely working together and developing the approach described here.

## 1. Introduction

The energy challenge is one of the most urgent – and difficult – tests facing our society today. High energy DPSSL programs like HiPER (High Power laser Energy Research) [1] in Europe, LIFE (Laser Inertial Fusion Engine) [2] in the United States of America and HALNA (High Average-power Laser for Nuclear Fusion Application) [3] in Japan are scientific programs dedicated to demonstrate the feasibility of laser driven fusion as a future energy source. Such laser drivers should be able to deliver 1 to 10 kJ at a repetition rate about 10 Hz and a wall-plug efficient of more than 10 %.

A high repetition rate, single-beam, diode-pumped 100 J class laser system is now under development at the Central Laser Facility (CLF) in collaboration with the Institute of Physics (IoP), Prague. This high energy pulsed laser amplification system for the HiLASE project (High average power pulsed LASErs) is based on a gas-cooled cryogenic, diode-pumped Yb:YAG multi-slab architecture [4]. The pre-amplifier and the power amplifier consist of two stacks of quadratic ceramic Yb:YAG slabs. Cold helium gas flowing at initial temperature of about 170 K is forced through the gaps between the slabs for cooling. The amplifiers are end or face-pumped from both sides. Employing slabs with increasing doping level towards the centre of the amplifier reduces the required overall thickness for a given maximum gain coefficient and also equalises the heat load for all slabs [5,6].

[1] To whom any correspondence should be addressed.

## 2. Optical layout of the HiLASE laser system and preliminary experimental results

The HiLASE multi-slab laser system consists of temporally-shaped fiber front end, regenerative amplifier, 10 J/10 Hz cryogenically-cooled pre-amplifier and 100 J/10 Hz cryogenically-cooled power amplifier. Model predictions developed at HiLASE [5] indicated maximum output energy of 10 J at pump energy of 48 J. This is in very good agreement with the experimental results recently obtained at CLF, as shown in Figure 1.

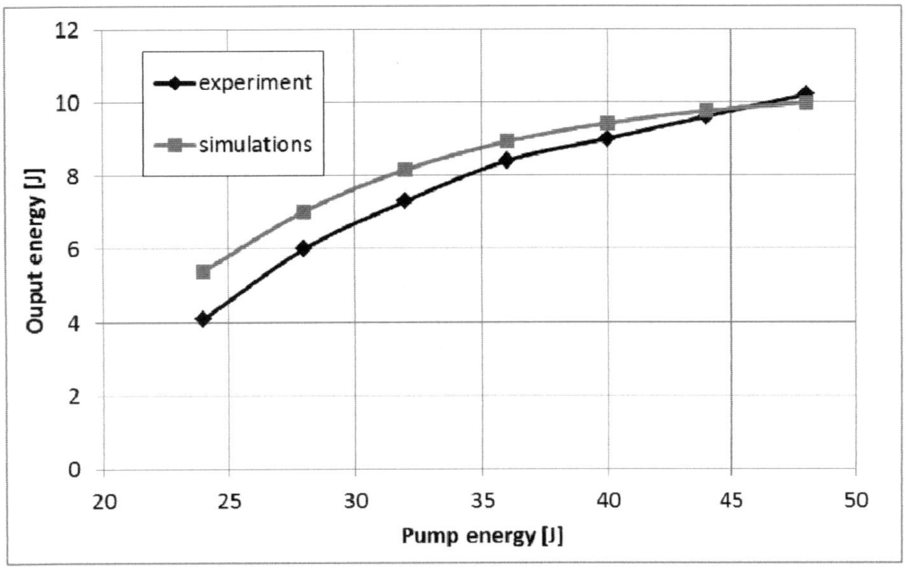


**Figure 1.** Output pulse energy as function of pump energy at 10 Hz repetition rate.

An isometric view of the 10 J pre-amplifier and 100 J power amplifier systems is shown in Figure 2.

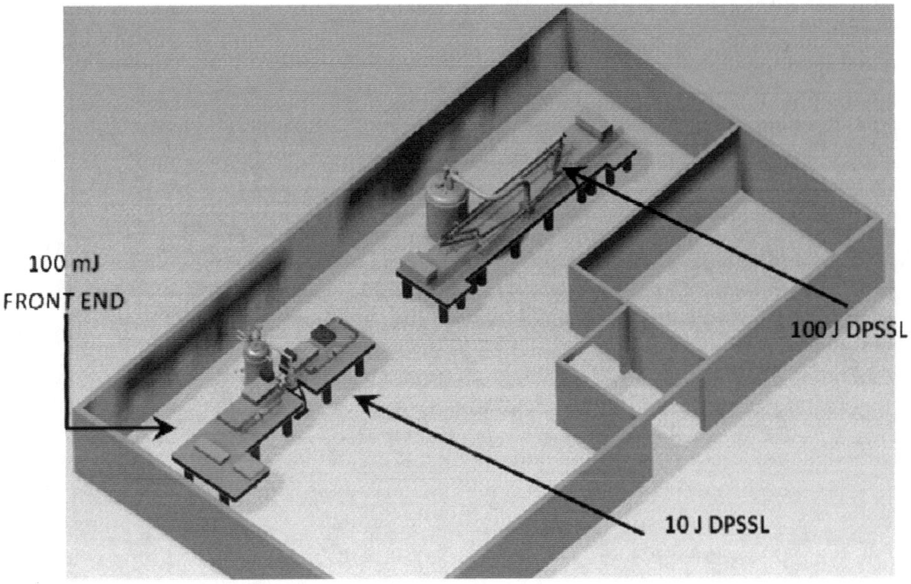


**Figure 2.** Isometric view of 10 J and 100 J/ 10 Hz HiLASE laser systems.

The details for a cryogenic gas-cooled 100 J / 10 Hz laser are summarized in Table 1.

**Table 1.** Output parameters for 100 J / 10 Hz HiLASE laser.

| Parameter | Specification |
|---|---|
| Pulse energy | > 100 J |
| Average output power | > 1 kW |
| Pulse length (FWHM) | 2-10 ns |
| Pulse shape | Programmable (150 ps steps) |
| Repetition rate | 1-10 Hz |
| Output beam size | 51 mm x 51 mm |
| RMS modulation | < 1 % |
| Wavefront quality | lambda/10 |
| E-o efficiency | > 12 % |

The baseline power amplifier design consists of 6 slabs with three different doping levels. In order to ensure that the impact of ASE loss in the amplifier is kept to a minimum, the doping levels have been calculated to ensure that the product of the small signal gain coefficient ($g_0$) and the diagonal transverse dimension of the square slabs (D) remains below 3.0 at all positions through the stack of gain slabs.

## 3. Thermal analysis of HiLASE slab amplifiers

We have undertaken extensive energetics, thermal and fluid-mechanical modelling in order to optimize the parameters of various amplifier configurations. Figures 3 and 4 show the calculated gas flow into the amplifier head and the predicted temperature distribution in the 100 J-class amplifier slab, respectively. The inlet flow velocity is 10 m/s and maximum of 33 m/s is achieved as the gas flows through the amplifier.

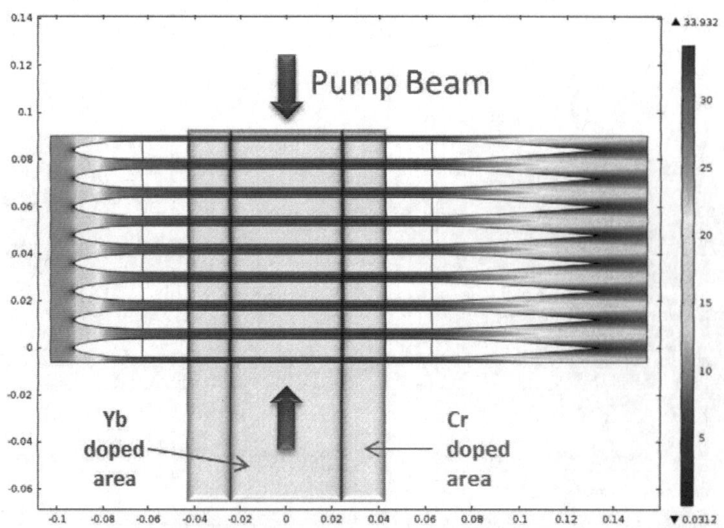


**Figure 3.** Calculated gas flow into the amplifier head.

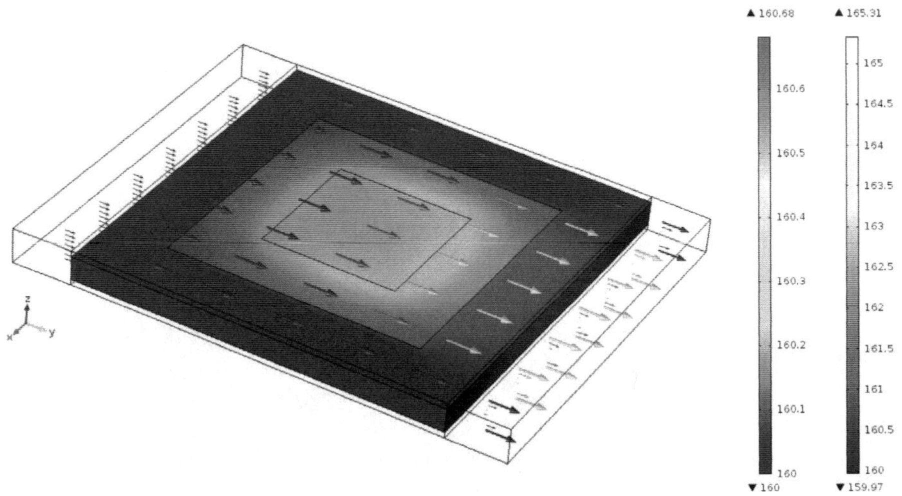


**Figure 4.** Predicted temperature distribution in a 100J-class HiLASE amplifier slab.

The variation of temperature across the pumped area is less than 1.5 Kelvin. The asymmetric temperature profile in the gain medium is due to the fact that the majority of the absorbed pump power is deposited as heat in the cladding and, in addition, the heat exchange coefficient decreases along the helium flow direction.

## 4. Summary

A conceptual design for a fully diode pumped, 100J-class DPSSL amplifier system capable of efficient generation of nanosecond pulses at 10 Hz repetition rate has been developed. Preliminary experimental results on the 10 J pre-amplifier agree very well with model predictions. This offers the potential for high electrical-to-optical efficiency of more than 12 %, which should enable the target overall efficiency necessary for cost-effective IFE production to be achieved.

## 5. Acknowledgments

This work benefitted from the support of the Academy of Sciences of the Czech Republic and of the Czech Republic's Ministry of Education, Youth and Sports to the HiLASE (CZ.1.05/2.1.00/01.0027), DPSSLasers (CZ.1.07/2.3.00/20.0143), and Postdok (CZ.1.07/2.3.00/30.0057) projects co-financed from the European Regional Development Fund.

## 6. References

[1] Chanteloup J C, Albach D, Lucianetti A, Ertel K, Banerjee S, Mason P D, Hernandez-Gomez C, Collier J L, Hein J, Wolf M, Koerner J and Le Garrec B J 2010 *J. Phys.: Conf. Ser.* **244** 012010.

[2] Bayramian A et al. 2011 *Fusion. Sci. Technol.* **60** 28-48

[3] Yasuhara R, Kawashima T, Sekine T, Kurita T, Ikegawa T, Matsumoto N, Miyamoto, Kan H, Yoshida H, Kawanaka J, Nakatsuka, Miyanaga N, Izawa Y and Kanake T 2008 *Opt. Lett.* **33** 1711-1713

[4] Banerjee S, Ertel K, Mason P D, Phillips PJ, Siebold M , Loeser M, Hernandez-Gomez C and Collier J L 2012 *Opt. Lett.* **12** 2175-2177

[5] Sawicka M, Divoky M, Novak J, Lucianetti A, Rus B and Mocek T 2012 *J. Opt. Soc. Am. B* **29** 1270-1276

[6] Slezak O, Lucianetti A, Divoky M, Sawicka M, and Mocek T 2013 *IEEE. J. Quantum Electron.* **49** 960-966

Journal of Physics: Conference Series **688** (2016) 012061 doi:10.1088/1742-6596/688/1/012061

# Collimation of laser-produced proton beam

**M. Takano[1], T. Nagashima[1], T. Izumiyama[1], Y. J. Gu[1, 2], D. Barada[1], Q. Kong[2], P. X. Wang[2], Y. Y. Ma[3], W. M. Wang[4], and S. Kawata[1]**
[1] *Utsunomiya University, Utsunomiya 321-8585, Japan*
[2] *Fudan University, Shanghai 200433, China*
[3] *National University of Defense Technology, Changsha 410073, China*
[4] *Inst. of Physics, CAS(Chinese Academy of Sciences), Beijing 100190, China*

E-mail: mt136626@cc.utsunomiya-u.ac.jp, kwt@cc.utsunomiya-u.ac.jp

**Abstract.** In intense laser plasma interaction for particle acceleration several issues remain to be solved. In this paper we focus on a collimation of ion beam, which is produced by a laser plasma interaction. In this study, the ion beam is collimated by a thin film target. When an intense short pulse laser illuminates a target, target electrons are accelerated, and create an electron cloud that generates a sheath electric field at the target surface. Such the ion acceleration mechanism is called the target normal sheath acceleration (TNSA). The TNSA field would be used for the ion beam collimation by the electric field. We have successfully obtained a collimated beam in our particle-in-cell simulations.

## 1. Introduction

Accelerator is used in a various fields, such as industrial development, medical treatment, basic researches, etc: for example, a cancer treatment by an ion beam produced by the accelerator. The ion beam has the Bragg peak characteristic, and deposits the ion beam energy inside material. It would be possible to kill only the cancer cells without severe damages of normal cells [1, 2]. However, the conventional ion accelerator tends to be huge in its cost and its size. Therefore, laser accelerator has been proposed as an alternative one [3]. Laser driven particle accelerator is expected to contribute to the development of compact cost-effective accelerators. However, there are various issues in laser particle acceleration at present. The issues in the laser ion acceleration include the ion beam divergence, the energy spectrum control, the energy efficiencies from the laser to the ions [4], the ion beam collimation [2, 5-9] and the ion beam bunching [2]. The energy efficiency from the laser to ions was improved by a solid target with a fine sub-wavelength structure [8, 9] or by a near-critical density gas plasma [2].

In this paper, we study on an improvement of the quality of the ion beam in the laser-driven accelerator, and especially on the ion beam collimation. The ion beam collimation is realized by the holes behind the solid target in this paper [5-9].

Figure 1 presents the target model employed in this paper. A laser-produced proton beam is introduced to the solid structured aluminium target. At the target rear surface, larger holes are implemented to collimate the ion beam. The hole size is the order of the ion beam or larger than that for the effective collimation. At the laser side the target in Fig. 1 has many sub-wavelength holes, which contribute to the efficient laser energy absorption [2, 8, 9] and so to contribute to create the collimation

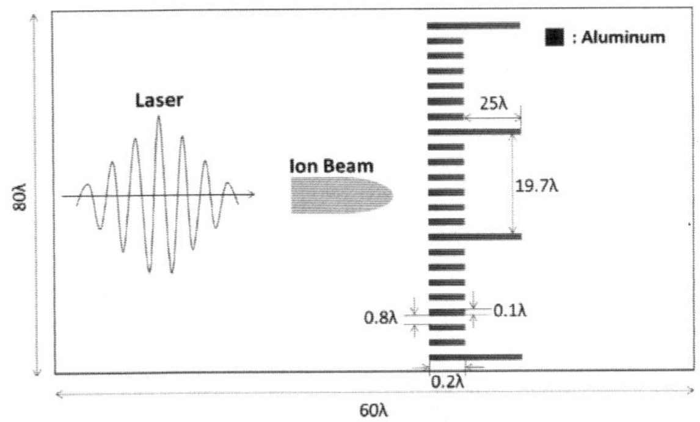


**Figure 1.** Simulation model for the collimation.

field effectively. In this paper, we performed 2.5-dimensional particle-in-cell simulations [10] to study the ion beam collimation.

In addition to the collimation device in Fig. 1, S. Lund and his colleagues have proposed another collimation device using many thin foils [11]. The thin foils prevent the electron return current and also shields the ion beam self charge by the mirror charge so that the ion beam self magnetic field contributes to the ion beam collimation.

## 2. Simulation model for ion beam collimation

Figure 1 presents our particle simulation model for the ion beam collimation. The simulation box is 60 $\lambda$ in the longitudinal direction and 80 $\lambda$ in the transverse direction. The laser wave length $\lambda$ is 1.053 $\mu$ m, the laser intensity is $1.0 \times 10^{19} \sim 1.0 \times 10^{20}$ W/cm$^2$, the laser pulse duration 100fs and the laser spot diameter 50.0 $\lambda$. The laser transverse and temporal profiles are in the Gaussian distribution.

The aluminium target density has a linear gradient scale from $0n_c$ to the maximum density of $462n_c$ with the linear density gradient over the target thickness in $x$. Here $n_c$ is the critical density. The laser illuminates the target from the left. The other parameter values are shown in Fig. 1. At the target rear the transverse TNSA (target normal sheath acceleration) electric field is generated as shown in Fig. 2 [5, 6].

The incoming proton beam is produced by a laser plasma interaction beforehand. This beam diameter is 10 $\lambda$. The hole diameter at the thin film target is larger than that of the ion beam. The distributions of the original proton beam divergence angle is about -10~10 degrees.

## 3. Ion beam collimation

Figure 2 shows the simulation results of the transverse direction electric field generated by the laser irradiation. The electric field required for the collimation of the ion beam is generated between the walls very clearly.

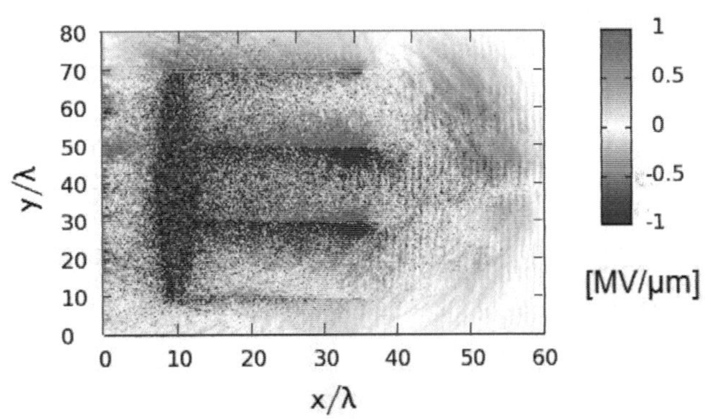


**Figure 2.** The transverse electric field is successfully generated, and reduces the proton divergence.

Figure 3 shows the results for the proton beam collimation. Initially the proton beam had the wide angle divergence (solid line). If the collimation device was not used, the angle divergence was increased as shown in Fig. 3(b) (short-dotted line). However, when the collimation device was employed, the angle divergence was reduced successfully as shown in Fig.3(c) (long-dotted line) and the proton beam width was kept short as shown in Fig. 4. Full width at half maximum is 1.01 degree as shown in Fig.3(c) (long-dotted line).

Figure 4 shows the results for the spatial distribution of the protons. If the collimation device was not used, the proton beam width was increased as shown in Fig. 4(b). However, when the collimation device was employed, the proton beam width was kept thin as shown in Fig.4(c).

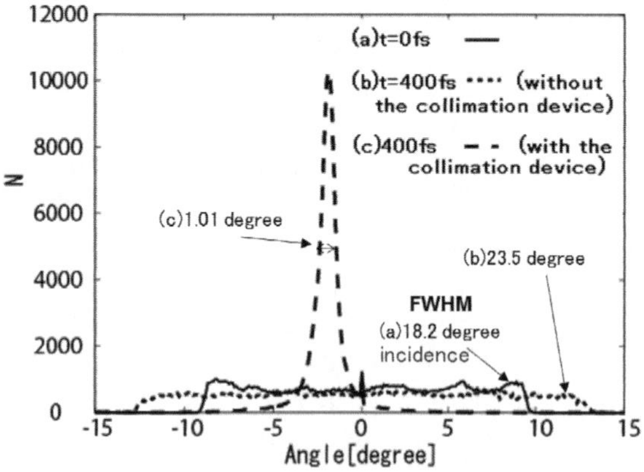


**Figure 3.** Divergence angle distributions for (a) the original proton beam (solid line), for (b) the proton beam without the collimation device (short-dotted line) and for (c) the proton beam with the collimation device (long-dotted line). The collimation device reduces the proton transverse divergence successfully.

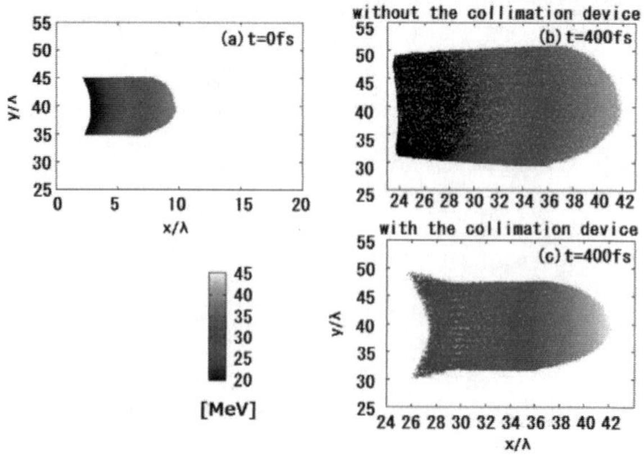


**Figure 4.** Spatial distributions of protons for (a) the original proton beam at *t*=0fs, for (b) the proton beam without the collimation device and for (c) the proton beam with the collimation device at *t*=400fs. The collimation device reduces the proton beam divergence.

We also examine the effect of laser intensity on the collimation. Figure 5 shows the results for the divergence angle distributions for the proton beam at 400fs at the laser intensity of (a) $1.0\times10^{19}$ W/cm$^2$, (b) $5.0\times10^{19}$ W/cm$^2$ and (c) $1.0\times10^{20}$ W/cm$^2$. At the laser intensity of $1.0\times10^{19}$ W/cm$^2$, the electric field required for the collimation of the ion beam is not enough to generated between the wall. The collimation device does not reduces well the proton transverse divergence as shown in Fig. 5(a). At the laser intensity of $1.0\times10^{20}$ W/cm$^2$, the excessive electric field is created between the walls; the ion beam is over focused, and the focused beam protons diverge again after the focusing as shown in Fig. 5(c). At the laser intensity of $5.0\times10^{19}$ W/cm$^2$, the collimation device reduces the proton transverse divergence successfully as shown in Fig. 5(b) (solid line).

Figure 6 shows the results for the spatial distributions of the protons at 400fs at the laser intensities of (a) $1.0\times10^{19}$ W/cm$^2$, (b) $5.0\times10^{19}$ W/cm$^2$ and (c) $1.0\times10^{20}$ W/cm$^2$. The arrows in Fig. 6 indicate the velocity direction of typical protons.

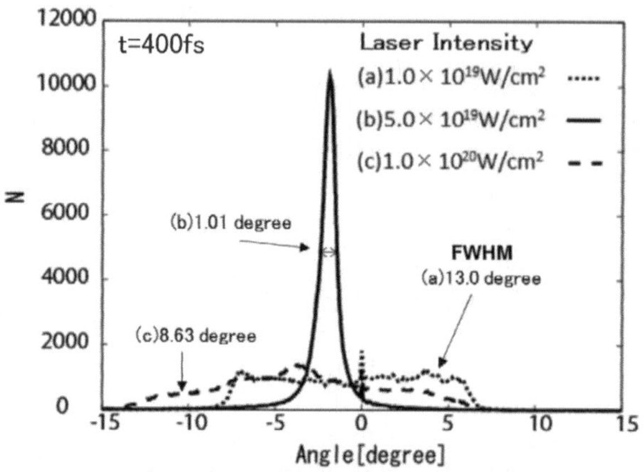


**Figure 5.** Divergence angle distributions at (a) the laser intensity of $1.0\times10^{19}$ W/cm$^2$ (short-dotted line), at (b) $5.0\times10^{19}$ W/cm$^2$ (solid line) and at (c) $1.0\times10^{20}$ W/cm$^2$ (long-dotted line). The laser intensity of $5.0\times10^{19}$ W/cm$^2$ reduces the proton transverse divergence successfully.

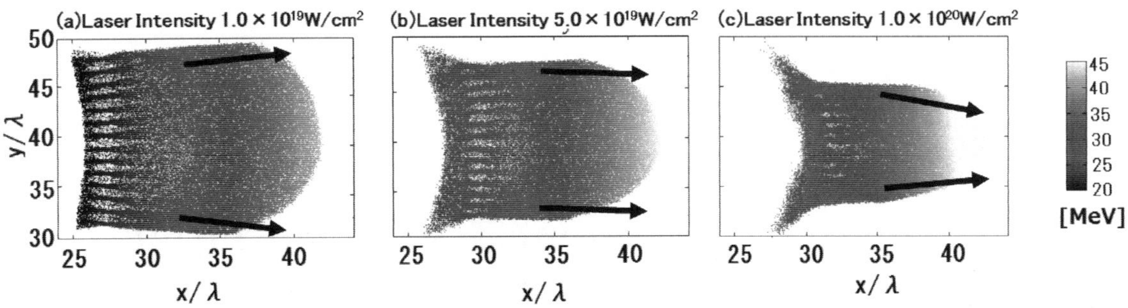


**Figure 6.** Spatial distributions of protons at (a) the laser intensity $1.0\times10^{19}$ W/cm$^2$, at (b) the laser intensity $5.0\times10^{19}$ W/cm$^2$ and for (c) the laser intensity $1.0\times10^{20}$ W/cm$^2$. The laser intensity $5.0\times10^{19}$ W/cm$^2$ reduces the proton transverse divergence successfully. The arrows indicate the velocity direction of typical protons.

## 4. Conclusions

We proposed the collimation device in intense laser plasma interaction. The simulation results demonstrated that the collimation device reduced the proton divergence successfully. We found that there is an optimum value for the laser intensity in the thin film target collimation device.

**Acknowledgements**

This study is party supported by MEXT, JSPS, CORE (Center for Optical Research and Education, Utsunomiya Univ.), ASHULA project (JSPS Asia Core to Core Program: Asian Core Program for High Energy Density Science Using Intense Laser Photons), and ILE / Osaka University.

## References

[1] Bulanov S V and Khoroshkov V S 2002 *Plasma Phys. Rep.* **28** 453
[2] Kawata S, Izumiyama T, Nagashima T, Takanao M, Barada D, Kong Q, Gu Y J, Wang P X, Ma Y Y and Wang W M 2013 *Laser Therapy* **22** 103
[3] Tajima T and Dawson J M 1979 *Phys. Rev. Lett.* **43** 267
[4] Robinson A P, Bell A R and Kingham R J 2006 *Phys. Rev. Lett.* **96** 035005
[5] Sonobe R, Kawata S, Miyazaki S, Nakamura N and Kikuchi T 2005 *Phys. Plasmas* **12** 073104
[6] Miyazaki S, Kawata S, Sonobe R and Kikuchi T 2005 *Phys. Rev. E* **71** 056403
[7] Nakamura M, Kawata S, Sonobe R, Kong Q, Miyazaki S, Onuma N and Kikuchi T 2007 *J. Appl Phys.* **101** 113305
[8] Nodera Y, Kawata S, Onuma N, Limpouchm J, Klimo O and Kikuchi T 2008 *Phys. Rev. E* **78** 046401
[9] Takahashi K, Kawata K, Satoh D, Ma Y Y, Barada D, Kong Q and Wang P X 2010 *Phys. Plasmas* **17** 093102
[10] Lasinski B F, Langdon A B, Hatchael S P, Key M H and Tabak M 1999 *Phys. Plasmas* **6** 2041
[11] Lund S M, Cohen R H and Ni P A 2013 *Phys. Rev. ST Accel. Beams* **16** 044202

Journal of Physics: Conference Series **688** (2016) 012062 doi:10.1088/1742-6596/688/1/012062

# Collective scattering of an incident monochromatic circularly polarized wave in an unmagnetized non-equilibrium plasma

**S Matsukiyo[1], Kuramitsu[2], K Tomita[3]**

[1] Department of Earth System Science and Technology, Kyushu University, 6-1 Kasuga Koen, Kasuga, Fukuoka, 816-8580, Japan

[2] Department of Physics, ational Central University, o. 300, Jhongda Rd., Jhongli City, Taoyuan 320, Taiwan

[3] Department of Applied Science for Electronics and Materials, Kyushu University, 6-1 Kasuga Koen, Kasuga, Fukuoka, 816-8580, Japan

E-mail: `matsukiy@esst.kyushu-u.ac.jp`

**Abstract.** Collective scattering of an incident monochromatic circularly polarized electromagnetic wave in an unmagnetized non-e uilibrium plasma is investigated. The wave e uation of the scattered waves are numerically solved by giving wave spectra of the incident electromagnetic wave and a variety of longitudinal fluctuations of a plasma. In addition to the well known symmetric ion and electron features of fre uency spectrum in an e uilibrium plasma, some asymmetries are found in a fre uency spectrum when anisotropic longitudinal wave propagation or the effect of the presence of a beam mode are taken into account.

## 1. Introduction

Thomson scattering has been widely used as a diagnostic technique to measure local parameters of a laboratory plasma [1]. A spectral density function of scattered waves provides collective features as well as non-collective ones of a plasma. The collective effects become remarkable if the wavelengths of incident and scattered waves are longer than the Debye length. An observed spectrum of scattered waves is fitted conventionally with the modeled spectral density function based on the assumption that the plasma is more or less quiet, i.e., in equilibrium [2].

Recent laboratory astrophysics, on the other hand, covers highly nonlinear and non-equilibrium phenomena like collisionless shocks, magnetic reconnection, etc. In such a plasma the scattering process of incident probe laser should be very complicated and scattered wave spectrum may be quite different from that in an equilibrium plasma. For instance, a beam component of a plasma is often seen upstream of a high Mach number collisionless shock [3]. Some incident ions are reflected at a high Mach number shock due to strong electrostatic potential so that the reflected ions behave as a beam. Such a beam may drive a variety of instabilities. The resultant local plasma should be highly turbulent and anisotropic. However, the theoretical framework of Thomson scattering in nonlinear and non-equilibrium plasmas has not been well established. In this study collective scattering of an incident monochromatic circularly polarized electromagnetic wave in some non-equilibrium plasmas is investigated by using numerical simulations.


Published under licence by IOP Publishing Ltd

## 2. Formulations and Basic Assumptions

We consider an unmagnetized electron-proton plasma, governed by the following equations.

$$n_j m_j \left(\frac{\partial}{\partial t} + \mathbf{V}_j \cdot \nabla\right) \mathbf{V}_j = n_j q_j \left(\mathbf{E} + \frac{\mathbf{V}_j \times \mathbf{B}}{c}\right) - T_j \nabla n_j \tag{1}$$

$$\frac{\partial n_j}{\partial t} = -\nabla \cdot (n_j \mathbf{V}_j), \quad \nabla \cdot \mathbf{E} = 4\pi\rho. \tag{2}$$

$$\nabla \cdot \mathbf{B} = 0, \quad \frac{\partial \mathbf{B}}{\partial \mathbf{t}} = -c\nabla \times \mathbf{E} \tag{3}$$

$$\nabla \times (\nabla \times \mathbf{E}) + \frac{1}{c^2}\frac{\partial^2 \mathbf{E}}{\partial t^2} = -\frac{4\pi}{c^2}\frac{\partial}{\partial \mathbf{t}} \tag{4}$$

Here, $\mathbf{V}_j = (u_j, \mathbf{v}_j)$ is the velocity of $j$-th partice species in which $u$ and $\mathbf{v}$ denote longitudinal and transverse components of the velocity, $n_j$ the density, $q_j$ the particle charge, $m_j$ the particle mass, $T_j$ the isotropic temperature which is assumed to be constant, $c$ the speed of light, and $\mathbf{E}$ and $\mathbf{B}$ are the electric and the magnetic fields, respectively. The current and the charge densities are given by $\quad = \sum_j q_j n_j \mathbf{v}_j$ and $\rho = \sum_j q_j n_j$.

It is assumed that a circularly polarized electromagnetic incident wave propagating parallel to the $x-$direction is written as $\mathbf{E}_I = (\hat{E}_I/\sqrt{2})e^{\ \phi_I}\hat{\mathbf{e}} + c.c.$ $(\phi_I = k_I x - \omega_I t,\ \hat{\mathbf{e}} = (\hat{\mathbf{y}} - i\hat{\mathbf{z}})/\sqrt{\mathbf{2}})$. One can easily confirm that the corresponding amplitudes in magnetic field and velocity are given by $\hat{B}_I = i(k_I c/\omega_I)\hat{E}_I$ and $\hat{v}_{Ie} = -ie\hat{E}_I/m_e\omega_I$, respectively. This incident wave may be scattered by the plasma containing longitudinal fluctuations given by $\delta X = \sum_{\mathbf{k}}(\delta\hat{X}(\mathbf{k})/2)e^{\ \phi(\mathbf{k})} + c.c.$ $(\phi(\mathbf{k}) = \mathbf{k}\cdot\mathbf{r} - \omega t)$, where $X$ can be either density, $n_j$, longitudinal velocity, $u_j$, or electrostatic field, $E_x$. The scattered waves are expressed as $\delta\mathbf{E}_S = \sum_{\mathbf{k}_S}(\delta\hat{\mathbf{E}}_S(\mathbf{k}_S)/\sqrt{2})e^{\ \phi_S(\mathbf{k}_S)}\hat{\mathbf{e}} + c.c.$ $(\phi_S(\mathbf{k}_S) = \mathbf{k}_S\cdot\mathbf{r} - \omega_S t)$.

Let us assume that the amplitude of the incident wave dominates those of the scattered waves. Then, we obtain an wave equation for the incident wave as $c^2\nabla\times(\nabla\times\mathbf{E}_I) + \partial^2\mathbf{E}_I/\partial t^2 = -\omega_p^2\mathbf{E}_I$, where $\omega_p^2 = \omega_{pe}^2 + \omega_p^2$ and $\partial\mathbf{v}_{Ij}/\partial t = (q_j/m_j)\mathbf{E}_I$ have been used. The scattered waves, on the other hand, satisfy

$$\nabla \times (\nabla \times \delta\mathbf{E}_S) + \frac{1}{c^2}\frac{\partial^2 \delta\mathbf{E}_S}{\partial t^2} = -\frac{4\pi}{c^2}\frac{\partial \delta}{\partial t}, \tag{5}$$

where $\delta \quad = en_0(\delta\mathbf{v} \ - \delta\mathbf{v}_e) + e(\delta n \ \mathbf{v}_I - \delta n_e\mathbf{v}_{Ie})$. Note that $\delta\mathbf{v} \ \approx \mathbf{v}_I \ \approx 0$ is a good approximation, since the transverse fluctuations are quite high frequency electromagnetic modes. Hence, if we assume $\nabla = \hat{\mathbf{x}}\partial/\partial x$, the wave equation for the scattered waves is written as

$$\left(-\frac{\partial^2}{\partial x^2} + \frac{1}{c^2}\frac{\partial^2}{\partial t^2} + \frac{\omega_{pe}^2}{c^2}\right)\delta\mathbf{E}_S = \frac{4\pi e}{c^2}\frac{\partial}{\partial t}\left(\mathbf{v}_{Ie}\delta n_e\right). \tag{6}$$

If the density fluctuation is based on relatively high frequency Langmuir waves, the corresponding $\delta E_x$ satisfies $(\partial^2/\partial t^2 - v_{te}^2\partial^2/\partial x^2 + \omega_{pe}^2)\delta E_x = -(\omega_{pe}^2/c)(\mathbf{v}_{Ie}\times\delta\mathbf{B}_S + \delta\mathbf{v}_e\times\mathbf{B}_I)_x$, where $v_{te}^2 = T_e/m_e$. When the right hand side (RHS) is non-negligible, this represents the equation for stimulated Raman scattering. But if the incident wave is too weak to disturbe the plasma so that the RHS is negligible, $\delta E_x$ may be merely expressed as the superposition of longitudinal waves satisfying the dispersion relation $\omega^2(k) = \omega_{pe}^2 + k^2 v_{te}^2$. Hence, $\phi(k) = kx \pm t\sqrt{\omega_{pe}^2 + k^2 v_{te}^2}$. Eq.(6), then, represents the evolution of the waves scattered by collective electron motions.

If the density fluctuation is based on relatively low frequency ion acoustic (IA) waves, the corresponding $\delta u$ satisfies $(\partial^2/\partial t^2 - C_s^2\partial^2/\partial x^2)\delta u = [\partial(\delta \ _S \times \mathbf{B}_I + \ _I \times \delta\mathbf{B}_S)_x/\partial t]/(n_0 m \ c)$, where $C_s^2 = (T_e + T \ )/m \ $. When the RHS is non-negligible, this represents the equation for stimulated Brillouin scattering. But if the incident wave is too weak to disturbe the plasma so that the RHS is negligible, $\delta u$ may be merely expressed as the superposition of longitudinal waves satisfying the dispersion relation $\omega^2(k) = k^2 C_s^2$. Note that in such a low frequency regime $\delta n \approx \delta n_e$ ($\delta u \approx \delta u_e$), since the electron inertial is negligible. Eq.(6), then, represents the evolution of the waves scattered by collective ion motions.

Without losing generality, one can write as $\mathbf{E}_I = \hat{E}_I(\cos\phi_I, \sin\phi_I)$, $\mathbf{v}_I = (e\hat{E}_I/m_e\omega_I)(\sin\phi_I, -\cos\phi_I)$, and $\delta n_e = \sum_k (kn_0/\omega)\delta\hat{u}\cos(\phi+\phi_{0k})$, where $\hat{E}_I$ and $\delta\hat{u}$ are real. Here, $\phi_{0k}$ denotes the initial phase of each longitudinal mode. Substituting the above into eq.(6) leads to

$$\left(-\frac{\partial^2}{\partial x^2} + \frac{\partial^2}{\partial t^2} + 1\right)\delta\mathbf{E}_S = -\frac{1}{\omega_I}\sum_k k\delta\hat{u}(\cos(\phi+\phi_{0k}+\phi_I), \sin(\phi+\phi_{0k}-\phi_I)). \tag{7}$$

Here, space and time have been normalized to $c/\omega_{pe}$ and $\omega_{pe}^{-1}$, the scattered electric field to $\hat{E}_I$, respectively. The above equation is numerically solved for a number of different plasma conditions in the following.

## 3. Numerical results

We first assume that the incident wave frequency is $\omega_I = 8.9$ and IA and electron thermal velocities in the homogeneous thermal plasma are $C_s = 0.01$ and $v_{te} = 0.1$, respectively. The amplitudes of the IA waves are assumed to be one order larger than those of the Langmuir waves, while the different wave modes on the same dispersion branch have equal amplitudes. The initial phase, $\phi_{0k}$, is random. In Fig.1(a) the frequency spectrum of the waves propagating in negative $x$-direction (backward scattered waves) is shown. The dominant two peaks near the incident wave frequency, $\omega_I$, are due to the scattering by collective ion motions of the plasma. In other words, they are the result of the coupling between the incident wave and the IA waves. The subdominant two peaks are due to the scattering by collective electron motions of the plasma, which are the result of the coupling between the incident wave and the Langmuir waves. In a collective Thomson scattering the former peaks are often called as ion feature and the latter as electron feature. In the following only the ion feature is focused.

A plasma is assumed to be turbulent and the amplitude spectrum of the velocity fluctuations is given by a power law, $\delta\hat{u}(k) = Ak^{-2}$, where $A$ denotes a constant. The integrated total amplitude $W = \int_{|k_{min}|}^{|k_{max}|}\delta\hat{u}dk$ is a parameter. Here, $k_{\;\;n}$ and $k_{\;\;x}$ are the minimum and the maximum wavenumbers of the longitudinal fluctuations. They are fixed throughout the following analysis as $k_{\;\;n} = 2\pi/51.2$ and $k_{\;\;x} = 20\pi$. The incident wave frequency is $\omega_I = 11.8$.

Fig.1(b) represents the case that the IA waves are isotropic, i.e., the same spectra are assumed for all the four quadrant in $\omega - k$ space with $W = 0.1$. In this case the symmetric two peaks of ion feature are clearly seen. Fig.1(c) shows the case that the IA waves propagating in negative $x$-direction, i.e., the second and the fourth quadrant in $\omega - k$ space, are weakened to have smaller amplitudes by one order of magnitude. Then, the higher frequency peak ($\omega_S > \omega_I$) is reduced and the ion feature becomes asymmetric.

The influence of the presence of a drifting beam is shown in Fig.1(d) and (e). Here, the longitudinal fluctuations whose dispersion relation is $\omega = ku_d$ ($u_d = 0.2$) is added. Fig.1(d) shows the case that a beam instability is well developed and the amplitude spectrum of the beam mode is the same type of power law as the IA waves. The beam mode amplitudes are one order of magnitude larger than those of the isotropic IA waves. Apparently, in addition to the usual two ion feature peaks, one strong peak at $\omega_S \sim 8$ is seen. When a beam instability has not been well developed, it is often the case that only the limited region in $k$ has wave power. Such a case is shown in Fig.1(e), in which the amplitude spectrum of the longitudinal

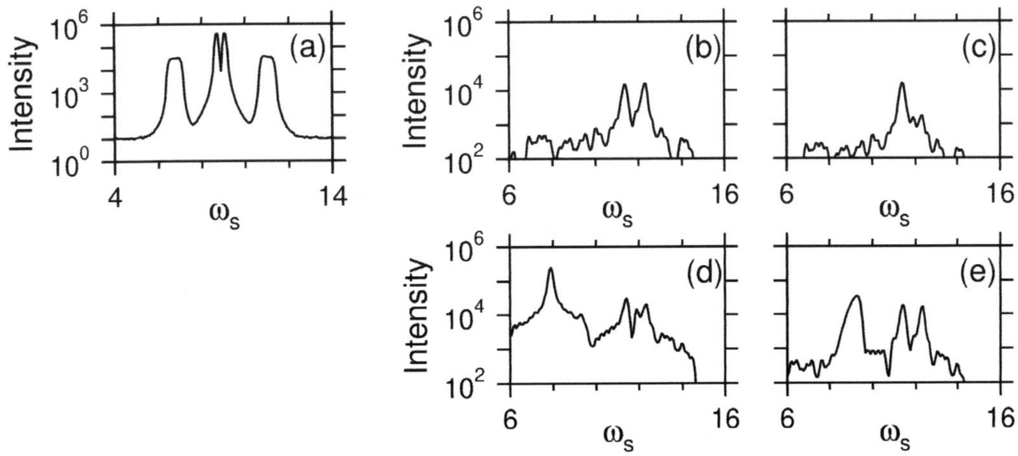


**Figure 1.** Frequency spectra of scattered waves in a variety of plasma conditions obtained by solving eq.(7). The longitudinal fluctuations in a plasma are given by (a) random noise on Langmuir and IA modes, (b) power law fluctuations of isotropic IA waves, (c) same as (b) but the waves propagating in negative $x$-direction have one order of magnitude smaller amplitudes, (d) power law fluctuations of a beam mode and isotropic IA waves, (e) same as (d) but the beam mode has a Gaussian spectrum centered at $k = k_I/2$.

fluctuations is given by a Gaussian centered at $k = k_I/2$. In contrast to Fig.1(d), the third peak in the frequency spectrum is shifted toward higher frequency.

## 4. Summary

The scattering of a monochromatic circularly polarized electromagnetic wave in an unmagnetized plasma is discussed. The wave equation of the scattered waves is solved numerically for a number of cases with different plasma conditions. The process considered here is essentially same as that of the collective Thomson scattering of an incident laser light in a laboratory plasma. Unlike an equilibrium plasma, the frequency spectrum of the scattered waves show asymmetric nature in non-equilibrium plasmas. As well known, when the IA waves have anisotropy in their propagation direction, the so-called ion feature of the scattered wave spectrum becomes asymmetric. If a plasma contains a beam, an additional peak appears in the frequency spectrum of the scattered waves.

*Acknowledgments*

We should like to thank Y. Sakawa and T. Morita for useful discussions. This work was supported by the joint research project of the Institute of Laser Engineering, Osaka University. S.M. is supported by a Grant-in-Aid for Scientific Research (C) 25400479.

## References

[1] Froula D H, Glenzer S H, Luhmann Jr. C, and Sheffield J 2011 *Plasma Scattering of Electromagnetic Radiation* Elsevier
[2] Evans D E, and Katzenstein J 1969 *Rep. Prog. Phys.* **32** 207
[3] Sorasio G, Marti M, Fonseca R, and Silva L O 2006 *Phys. Rev. Lett.* **96** 045005

Journal of Physics: Conference Series **688** (2016) 012063 doi:10.1088/1742-6596/688/1/012063

# Nonlinear motion of a current-vortex sheet in MHD Richtmyer-Meshkov instability

**Chihiro Matsuoka[1], Katsunobu Nishihara[2], and Takayoshi Sano[2]**

[1]Department of Physics, Graduate School of Science and Technology, Ehime University, 2-5 Bunkyocho, Matsuyama, Ehime, Japan
[2]Institute of Laser Engineering, Osaka University, 2-6, Yamada-oka Suita, Osaka, 565-0871, Japan

E-mail: matsuoka.chihiro.mm@ehime-u.ac.jp

**Abstract.** onlinear motion of a vortex sheet applied an initially homogeneous magnetic field is investigated using the vortex method. We show that the existence of a vortex sheet induces a strong magnetic field and non-uniform current at the fluid interface. This system can be a model of MHD Richtmyer-Meshkov instability (MHD RMI). We present some interfacial profiles in MHD RMI and discuss the difference between the pure vortex sheet in ordinary RMI and the current-vortex sheet in MHD RMI, in which it is shown that the magnetic field suppresses the roll-up of the vortex sheet. We also find that the strong magnetic field amplification is caused by the stretch of the interface.

## 1. Introduction

We consider a fluid interface with density and tangential velocity jumps such that a homogeneous (constant) magnetic field is imposed in the whole region including the interface. Then the sharp gradient of the magnetic field between the interface induces a non-uniform current-vortex sheet on the fluid interface. This can be a model to explain the extraordinary strong magnetic field amplification in Super Nova Remnants (SNR) [1, 2]. In spite of its importance in astrophysics and plasma physics, few models to explain this extraordinary magnetic amplification mechanism in SNR exist so far. In this paper, we extend the vortex sheet motion in the Richtmyer-Meshkov instability (RMI) [3, 4, 5] (refer to figure 1) into the magnetohydrodynamic flows (MHD RMI) and show that a non-uniform current-vortex sheet causes the strong magnetic field amplification and its growth rate is determined by the stretch; i.e., the geometry, of the interface.

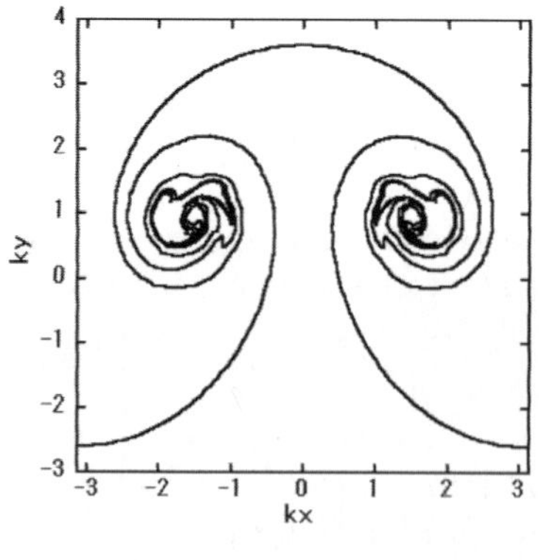


**Figure 1.** Interfacial profile of pure RMI, where the Atwood number $A = -0.2$ and $k$ denotes the wave number ($kx$ and $ky$ are dimensionless coordinates).

## 2. Governing equations

We consider a fluid interface with density and tangential velocity jumps in two-dimensional inviscid and incompressible magnetohydrodynamic flows. This sharp density jump induces two different magnetic fields $\boldsymbol{B}_1$ and $\boldsymbol{B}_2$ between the interface, where the upper fluid (fluid 2) is assumed to be lighter than the lower one (fluid 1). The governing equations in each fluid region $i$ $(i = 1, 2)$ are given by

$$\frac{\partial \boldsymbol{u}}{\partial t} + (\boldsymbol{u} \cdot \nabla)\boldsymbol{u} - \frac{1}{\rho\ \mu}(\boldsymbol{B} \cdot \nabla)\boldsymbol{B} = -\frac{1}{\rho}\nabla P. \tag{1}$$

$$\frac{\partial \boldsymbol{B}}{\partial t} = \nabla \times (\boldsymbol{u} \times \boldsymbol{B}\ ), \tag{2}$$

$$\nabla \cdot \boldsymbol{u} = 0, \tag{3}$$

$$\nabla \cdot \boldsymbol{B} = 0, \tag{4}$$

$$\mu\ \boldsymbol{J} = \nabla \times \boldsymbol{B}\ , \tag{5}$$

where the constants $\rho$ and $\mu$ $(i = 1, 2)$ are the density and permeability in fluid $i$ (we set $\mu_1 = \mu_2 = \mu$ throughout this paper), $\boldsymbol{u}$ is the velocity related to the velocity potential $\phi$ as $\boldsymbol{u} = \nabla\phi$ , $\boldsymbol{B}$ is the magnetic field, $\boldsymbol{J}$ is the current density, $P$ is the total pressure described by $P_1 = p_1 + \boldsymbol{B}_1 \cdot \boldsymbol{B}_1/(2\mu)$ and $P_2 = p_2 + \boldsymbol{B}_2 \cdot \boldsymbol{B}_2/(2\mu)$, in which $p$ is the fluid pressure.

Imposing the total pressure continuous condition $P_1 = P_2$ at the interface and assuming that the current localizes only on the interface (i.e., the current does not flow in the bulk; this condition is true if and only if the initial magnetic field is imposed to be parallel to the initial interface), equation (1) yields the evolution equation for the circulation $' \equiv \phi_1 - \phi_2$:

$$\frac{\mathrm{d}\,'}{\mathrm{d}t} = 2A\frac{\mathrm{d}\Phi}{\mathrm{d}t} - A\boldsymbol{q} \cdot \boldsymbol{q} + \frac{A + 2\tilde{\alpha}}{4}\boldsymbol{\gamma} \cdot \boldsymbol{\gamma} - \tilde{\alpha}A\boldsymbol{\gamma} \cdot \boldsymbol{q} + R\bar{\boldsymbol{B}} \cdot \boldsymbol{j}_s, \tag{6}$$

where $\boldsymbol{\gamma} = (\gamma_x, \gamma_y) = \boldsymbol{u}_1 - \boldsymbol{u}_2$ $(\gamma = \boldsymbol{\gamma} \cdot \boldsymbol{t} = \partial\,'/\partial s$, where $s$ and $\boldsymbol{t}$ are length and the unit tangent of the interface, respectively) is the vortex sheet strength, $A = (\rho_2 - \rho_1)/(\rho_1 + \rho_2)$ is the Atwood number, $\tilde{\alpha} = \tilde{\alpha}(A)$ $(|\tilde{\alpha}| \leq 1)$ is an artificial parameter of a function of $A$, and

$$\frac{\mathrm{d}}{\mathrm{d}t} = \frac{\partial}{\partial t} + \boldsymbol{u}^+ \cdot \nabla, \quad \boldsymbol{u}^+ = \boldsymbol{q} + \frac{\tilde{\alpha}}{2}\boldsymbol{\gamma},$$

in which $\boldsymbol{q} = (U, V)$ is the average fluid velocity $\boldsymbol{q} = (\boldsymbol{u}_1 + \boldsymbol{u}_2)/2$ and $\boldsymbol{u}^+$ is the interfacial velocity. Here, $\bar{\boldsymbol{B}} = (\boldsymbol{B}_1 + \boldsymbol{B}_2)/2$, $\boldsymbol{j}_s = \boldsymbol{B}_1 - \boldsymbol{B}_2$, and $R = 2B_0^2/[\mu(\rho_1 + \rho_2)v_{l\ n}^2] = (v\ /v_{l\ n})^2$, where $v_{l\ n}$ is the linear growth velocity of the pure RMI [4, 5] and $v$ is the Alfven velocity for the initial homogeneous magnetic field $B_0$.

Now we assume that the initial magnetic field is applied parallel to the interface; i.e., $\boldsymbol{B}_0 = B_0\boldsymbol{t}$. Then we can prove that the magnetic field $\boldsymbol{B}$ is always parallel to the interface (details of the proof are omitted here) and the induction equation (2) can be rewritten as

$$\begin{aligned}
\frac{\mathrm{d}B_1^t}{\mathrm{d}t} &= \frac{\tilde{\alpha} - 1}{2s_\theta}\gamma B_{1,\theta}^t + \frac{B_1^t}{s_\theta}\left(q_\theta^t + \frac{\gamma_\theta}{2}\right), \\
\frac{\mathrm{d}B_2^t}{\mathrm{d}t} &= \frac{\tilde{\alpha} + 1}{2s_\theta}\gamma B_{2,\theta}^t + \frac{B_2^t}{s_\theta}\left(q_\theta^t - \frac{\gamma_\theta}{2}\right),
\end{aligned} \tag{7}$$

where $\theta$ is the Lagrange parameter that parameterizes the interface and

$$B^t{}_{,\theta} = \frac{\partial B^t}{\partial \theta} \quad (i = 1, 2), \quad q_\theta^t = \frac{\partial}{\partial \theta}(\boldsymbol{q} \cdot \boldsymbol{t}).$$

Differentiating equation (6) with respect to $\theta$, we obtain the following Fredholm integral equation of the second kind:

$$\frac{\mathrm{d}\gamma}{\mathrm{d}t} = \frac{2A}{s_\theta}\left(x_\theta\frac{\mathrm{d}U}{\mathrm{d}t} + y_\theta\frac{\mathrm{d}V}{\mathrm{d}t}\right) - \frac{(1+\tilde{\alpha}A)\gamma}{s_\theta^2}(x_\theta U_\theta + y_\theta V_\theta) + \frac{A+\tilde{\alpha}}{4s_\theta}(\gamma^2)_\theta + R(\bar{\boldsymbol{B}}\cdot\boldsymbol{j}_s)_\theta, \tag{8}$$

where the subscript $\theta$ denotes the differentiation with respect to $\theta$. The last term of this equation denotes the force that originates from the magnetic tension term $\boldsymbol{B}\cdot\nabla\boldsymbol{B}$ in (1), which acts to straighten the interface. The interfacial velocity; i.e., the velocity of the current-vortex sheet $\mathrm{d}\boldsymbol{x}/\mathrm{d}t = (\mathrm{d}x/\mathrm{d}t, \mathrm{d}y/\mathrm{d}t)$ is given by

$$\frac{\mathrm{d}x}{\mathrm{d}t} = U + \frac{\tilde{\alpha}x_\theta}{2s_\theta}\gamma, \quad \frac{\mathrm{d}y}{\mathrm{d}t} = V + \frac{\tilde{\alpha}y_\theta}{2s_\theta}\gamma \tag{9}$$

Now we introduce the complex form of the average velocity $\boldsymbol{q}$ as $q^* = U - iV$. Then we obtain the vortex induced velocity due to the Biot-Savart law (Birkhoff-Rott equation) [4, 5]:

$$q^* = U - iV = \frac{1}{2\pi i}\mathrm{PV}\int_{-\infty}^{\infty}\frac{\gamma(\theta')s_\theta(\theta')d\theta'}{z(\theta)-z(\theta')}. \tag{10}$$

Solving the evolution equations (7), (8), (9), and (10) simultaneously, we can determine the motion of a current-vortex sheet.

## 3. Numerical results

In this section, we present some numerical results of equations (7), (8), (9), and (10) solved by using the vortex method [4]. The initial condition is set to

$$x(\theta,0) = \theta, \quad y(\theta,0) = 0; \quad \gamma(\theta,0) = -2\sin\theta,$$
$$\boldsymbol{B}_j(\theta,0) = B_0\hat{\boldsymbol{e}_x} \quad (j=1,2), \tag{11}$$

where $\hat{\boldsymbol{e}_x}$ is the unit vector in the $x$-direction and we set $B_0 = 1$ throughout this paper. Here, we select $\tilde{\alpha} = -A^2$ and $\tilde{\alpha} = -0.05$ for the calculations of the Atwood numbers $A = -0.2$ and $A = -0.7$, respectively. The grid points $N$ is set to $N = 512$ for all calculations.

Figure 2 shows the distribution of the surface current $\boldsymbol{j}_s = \boldsymbol{B_1} - \boldsymbol{B_2}$ for the Atwood numbers $A = -0.2$ and $A = -0.7$. We see that a non-uniform current flows on the interface. In figure 2 (a), the strong current concentrates in the neighborhood of the vortex core at which the vortex sheet has the maximum vorticity. In figure 2 (b), the sign of the current changes at the core, between which the magnitude of the magnetic field $\boldsymbol{B_1}$ and $\boldsymbol{B_2}$ is reversed. The roll-up of the interface in (a) is weaker than that of the pure RMI in figure 1, in spite of the same Atwood number. This fact suggests that the existence of the magnetic field suppresses the growth of a vortex sheet.

We show the maximum values of $R|\boldsymbol{B}/\boldsymbol{B_0}|^2$ for various $R$ in figure 3 (a). The mamimum value of the magnetic field is realized at the spike and $\boldsymbol{B} = \boldsymbol{B}_2$; the magnetic field on the side of fluid 2 (the lighter fluid). The values $R|\boldsymbol{B}/\boldsymbol{B_0}|^2$ attain almost the same value $\beta$ $(1 < \beta < 10)$ irrelevant to $R$, which indicates that the magnetic field is amplified up to $|\boldsymbol{B}| > R^{-1/2}$. Figure 3 (b) shows the value of $RG_s^2$, where $G_s = G_s(t)$ is defined as the value at a spike of the temporal integration of the $s$-derivative of the tangential velocity $\boldsymbol{u}_2\cdot\boldsymbol{t}$:

$$G_s(t) = \left[\int\frac{1}{s_\theta}\left(q_\theta^t - \frac{\gamma_\theta}{2}\right)\mathrm{d}t\right]_{\theta=0} = \left[\int\frac{\partial(\boldsymbol{u}_2\cdot\boldsymbol{t})}{\partial s}\mathrm{d}t\right]_{\theta=0}. \tag{12}$$

This value corresponds to the stretch rate at a spike in the neighborhood of the interface of fluid 2. From the comparison of (a) and (b), we see that the maximum magnetic field is determined by the value of $G_s$.

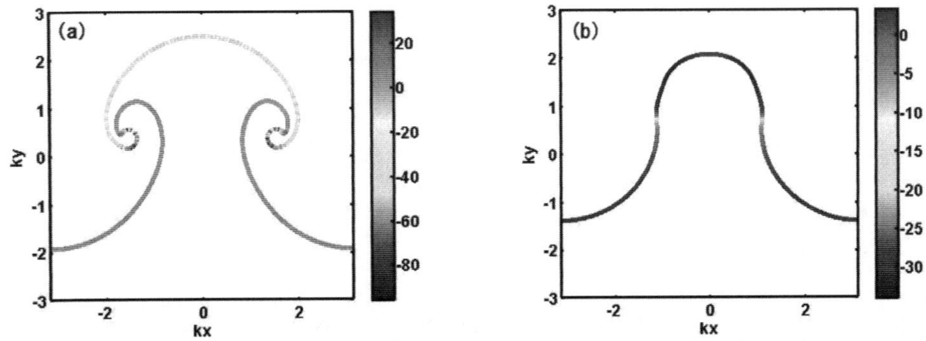


**Figure 2.** Surface current $\boldsymbol{j}_s = \boldsymbol{B}_1 - \boldsymbol{B}_2$; (a) $A = -0.2$ and (b) $A = -0.7$, where both figures are taken at time $t = 5.0$ and the color bar denotes the strength of $\boldsymbol{j}_s$: $j_s = B_1^t - B_2^t$.

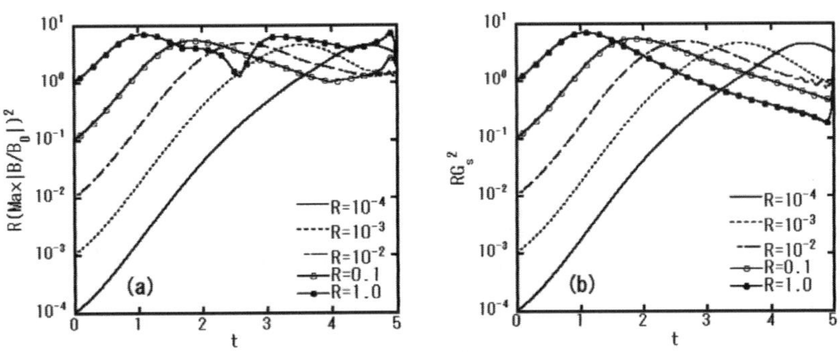


**Figure 3.** Comparison of the (a) maximum value of the magnetic field $R|\boldsymbol{B}/\boldsymbol{B}_0|^2$ and (b) value of $RG_s^2$, where $A = -0.7$.

## 4. Conclusion and discussions

We have investigated the motion of non-uniform current-vortex sheets in MHD-RMI. We have found that the strong magnetic field amplification is realized at the fluid interface and this amplification is caused by the stretch of a vortex sheet. We also found that the magnetic field suppresses the roll-up of the vortex sheet. We have presented the numerical results for $A = -0.2$ and $A = -0.7$ in this paper; however, various interesting phenomena are found in other Atwood numbers; especially for $A = 0$ and $A = -1.0$. The details of the calculations for these Atwood numbers will be published elsewhere.

## References

[1] Inoue T, Yamazaki R and Inutsuka S 2009 *Astrophys. J.* **695** 825.
[2] Sano T, ishihara K, Matsuoka C and Inoue T 2012 *Astrophys. J.* **758** 126.
[3] Richtmyer R D 1960 *Commun. Pure Appl. Math.* **13** 297; Meshkov E E 1969 *Sov. Fluid Dynamics* **4** 101.
[4] Matsuoka C and ishihara K 2006 *Phys. Rev. E* **73** 026304.
[5] ishihara K, Wouchuk, J G, Matsuoka C, Ishizaki R and Zhakhovskii V V 2010 *Phil. Trans. Roy. Soc. A* **368** 1769 (references therein).

Journal of Physics: Conference Series **688** (2016) 012064 doi:10.1088/1742-6596/688/1/012064

# Results from neutron imaging of ICF experiments at NIF

**F E Merrill[1,3], C R Danly[1], D N Fittinghoff[2], G P Grim[1], N Guler[1], P L Volegov[1] and C H Wilde[1]**
[1]Los Alamos National Laboratory, PO Box 1663, Los Alamos, NM, USA
[2]Livermore National Laboratory, PO Box 808, Livermore, CA, USA

E-mail: fmerrill@lanl.gov

**Abstract.** In 2011 a neutron imaging diagnostic was commissioned at the National Ignition Facility (NIF). This new system has been used to collect neutron images to measure the size and shape of the burning DT plasma and the surrounding fuel assembly. The imaging technique uses a pinhole neutron aperture placed between the neutron source and a neutron detector. The detection system measures the two-dimensional distribution of neutrons passing through the pinhole. This diagnostic collects two images at two times. The long flight path for this diagnostic, 28 m, results in a chromatic separation of the neutrons, allowing the independently timed images to measure the source distribution for two neutron energies. Typically one image measures the distribution of the 14 MeV neutrons, and the other image measures the distribution of the 6-12 MeV neutrons. The combination of these two images has provided data on the size and shape of the burning plasma within the compressed capsule, as well as a measure of the quantity and spatial distribution of the cold fuel surrounding this core. Images have been collected for the majority of the experiments performed as part of the ignition campaign. Results from this data have been used to estimate a burn-averaged fuel assembly as well as providing performance metrics to gauge progress towards ignition. This data set and our interpretation are presented.

## 1. Introduction

The National Ignition Facility (NIF) is currently investigating the possibility of achieving thermonuclear ignition of a deuterium-tritium (DT) plasma [1]. The present strategy is to generate the required densities and pressures through indirect ablative compression of plastic (CH) capsules containing DT ice layers, which surround DT gas. In this design, the DT ice compresses a pocket of DT gas to temperatures of ~10 keV with total fuel densities, ρR, of greater than ~1 g/cm$^2$. With the appropriate fuel assembly conditions, this system is predicted to result in fusion processes starting in the compressed DT gas, generating a hotspot, which then initiates the propagation of an ignition front traveling through the surrounding "cold" high density DT fuel. This reaction would release over a megajoule of fusion energy and more than $10^{17}$ neutrons.

Tuning the implosion to achieve ignition requires the measurement of plasma conditions at very small length and time scales. The hot spot is predicted to have a 40-50 μm diameter and to maintain fusion conditions for ~100 ps, while the surrounding cold fuel is expected to be ~100 μm in diameter. Establishing the appropriate conditions for ignition, such as symmetry, hot spot and cold fuel volume,

[3]To whom any correspondence should be addressed.

has required an iterative tuning campaign. These measurements must be made at non-ignition conditions, allowing the tuning effort to move from initial non-igniting implosions towards conditions required to achieve ignition.

The neutron imaging system is composed of an aperture array, used to form the neutron images, and a detector system that is used to measure the neutron flux passing through the pinhole array. The detector system [2,3] is capable of collecting two images, which are independently timed. With the detector positioned 28 m from the neutron source, the neutron arrival time is correlated to the neutron energies, allowing measurement of the neutron source distributions from two energy ranges by gating the detectors. Typically, one detector is gated to view the 14-MeV neutrons (13-17 MeV), which are generated from DT fusion processes, while the second detector is gated to measure the source distribution of lower energy neutrons typically in the range from 6-12 MeV. The ratio of the lower energy neutrons to 14 MeV neutrons is referred to as the down scatter ratio. These lower energy neutrons are predominantly DT fusion neutrons that have scattered in the surrounding cold fuel and, therefore, provide information on the distribution of cold fuel surrounding the hotspot. These two measurements provide the size and shape of the hotspot source as well as the surrounding cold fuel.

The scattering of 14 MeV neutrons, which have been generated by fusion reactions in the hot spot, within the surrounding cold DT fuel, results in a correlation between scattering angle and neutron energy. The kinematics of these scattering processes results in the detection of neutrons that have scattered in the cold fuel located on the same side of the fuel assembly as the neutron imaging line of site. Therefore, neutron images collected from scattered neutrons at lower energies provide a measure of the cold fuel distribution on the imaging side of the fuel assembly.

## 2. Results of Point Design Tuning Campaign

At NIF, dozens of experiments have been performed to tune the point design capsule performance toward thermo-nuclear ignition [4,5]. Neutron images were collected from each cryogenically layered implosion experiment. The neutron source distribution was calculated from each 14-MeV and 6-12 MeV neutron image, providing size and shape information of the hot-spot and cold fuel assemblies, respectively. Contours at 17% of the peak intensity were fit with Legendre polynomials to characterize the size and shape of each neutron source distribution [2,3]. The results of this fit provides a measure of the lowest mode of hot-spot asymmetry, P2/P0, which is the ratio of the second-order to the zeroth-order Legendre polynomials that are required to fit the 17% contour. Several experiments were performed to reduce this asymmetry [6], as implosion symmetry is believed to be one of the important parameters in achieving ignition. Figure 1 shows a plot of neutron yield as a function of hot-spot source asymmetry P2/P0 for NIF point design experiments. There is no clear correlation observed between the neutron yield and the hot-spot asymmetry, in disagreement with previous experiments and code predictions for capsule performance [4]. It is important to note, however, that implosion symmetry is not the only change in each of these experiments and other performance characteristics could be dominating performance, which would hide the underlying correlation between shape and performance.

It is instructive to look at the fuel assembly for a subset of the experiments that have been summarized in Figure 1. The images that have been chosen here are of one of the most symmetric fuel assemblies (NIF shot number N120321) and one of the most oblate fuel assemblies (NIF shot number N130331). The symmetric shot resulted in a neutron yield of $4.2\text{x}10^{14}$ neutrons, an ion temperature of 3 keV and a down scatter ratio of 6%, while the oblate shot yielded $3.0\text{x}10^{14}$ neutrons, an ion temperature of 3 keV and a down scatter ratio of 4%,. There are many variables beyond shape that can impact performance. One example is mix; where hydrodynamic instabilities move the CH capsule and dopant material into the hotspot and reduces the yield [7]. In the experiments shown in Figure 2, the inferred mix values [8] were 370±150 ng and 50±50 ng for the symmetric and asymmetric experiments, respectively. Although not directly observed, it is likely that processes such as mix of the ablator material into the hotspot has led to a larger impact on performance than the shape of the hotspot, explaining the lack of correlation between hotspot symmetry and the resulting yield.

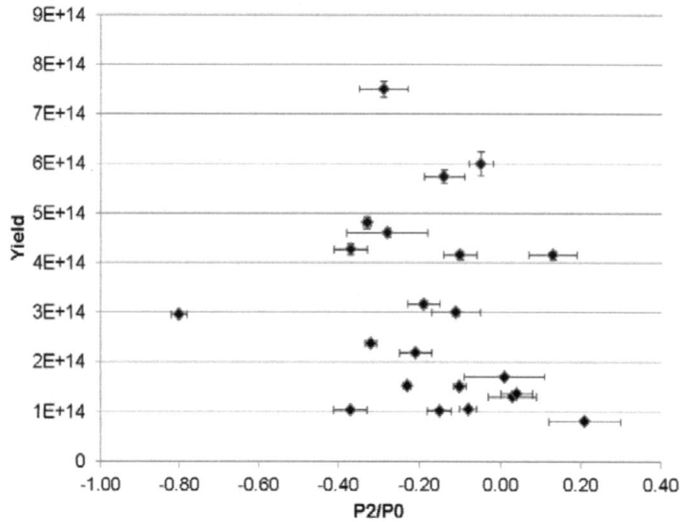


**Figure 1:** Neutron yield versus asymmetry, P2/P0. There is no clear correlation between yield and symmetry in disagreement with previous experimental results and simulation predictions. This lack of correlation may be explained by other performance parameters dominating the yield performance of the ignition point design.

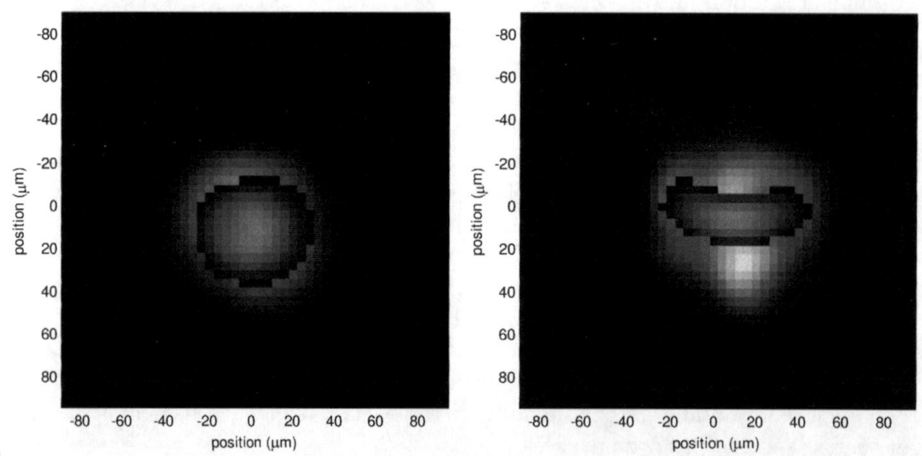


**Figure 2:** Left: fuel assembly measured from N120321 with a P2/P0=-0.10 and Right: fuel assembly from N130331 with P2/P0=-0.80. In these images the hotspot sources are shown in the red channel and are overlaid with an image of the cold fuel, which is displayed in the cyan channel. The shot on the left resulted in a yield of $4.2 \times 10^{14}$ neutrons, an ion temperature of 3 keV and a down scatter ratio of 6%, while the shot on the right resulted in $3.0 \times 10^{14}$ neutrons, an ion temperature of 3 keV and a down scatter ratio of 4%. Red intensity is proportion to yield while cyan intensity is proportion to the product of yield and down scatter ratio.

## 3. Results of High-Foot Tuning Experiments

More recently a series of "high foot" experiments has been executed at NIF to study the performance of implosions that intentionally drive the fuel along a higher adiabat than the point design. It has been predicted that these experiments should result in reduced mix of the ablator into the hot spot. Three images from the initial tuning experiments in this series (N130802, N130710 and N130812) are shown in Figure 3. The yields from these experiments were $4.8 \times 10^{14}$, $1.0 \times 10^{15}$ and $2 \times 10^{15}$ neutrons, respectively, while the inferred mix levels were 0±150 ng, 0±150 ng and 84±35 ng respectively. Ion temperature and down scatter ratio are reported in the figure caption below. In these experiments there is a clear correlation between the symmetry of the fuel assembly and performance as well as an observation of reduced ablator mix. The emergence of this correlation is consistent with mix, rather than the shape of the fuel assembly, dominating the performance of the point design.

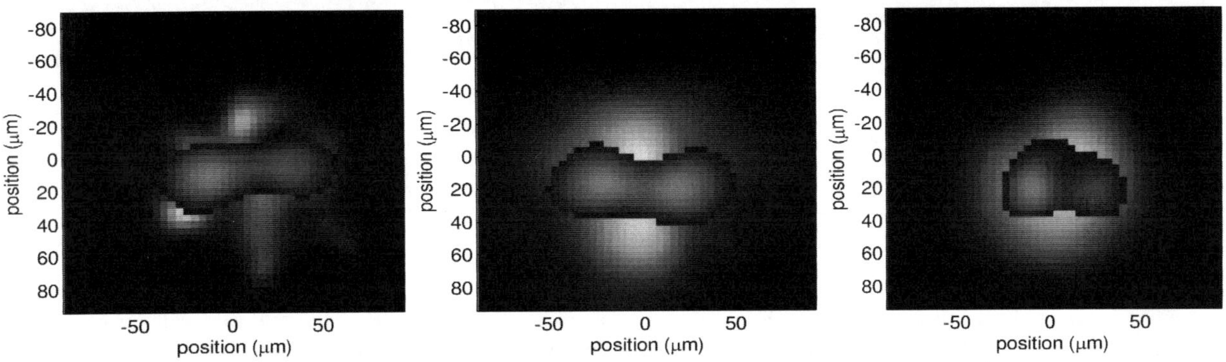


**Figure 3:** Hotspot sources are shown in the red overlaid on images of the cold fuel shown in cyan. Left (N130802) resulted in a P2/P0=-0.67, $4.8 \times 10^{14}$ neutrons, ion temperature of 3 keV and down scatter ratio of 3%, middle (N130710): a P2/P0=-0.25, $1 \times 10^{15}$ neutrons, ion temperature of 3.5 keV and down scatter ratio of 3% and on the right (N130812): P2/P0=-0.25, $2.4 \times 10^{15}$ neutrons, ion temperature of 4 keV and down scatter ratio of 4%. Red intensity is proportion to yield while cyan intensity is proportion to the product of yield and down scatter ratio.

## 4. Conclusions

Throughout the tuning efforts at NIF, the fuel assembly has been measured through the collection of neutron images for a large number of experiments. For the point design experiments, there is no observed correlation between performance, as measured by yield, and the shape of the hot spot. This lack of a correlation suggests that other processes are dominating the performance. Although not observable through neutron images, one possible process is mixing of the ablator material into the hot spot. This hypothesis is strengthened through the recent measurements of the high-foot series of experiments, where less mix has been observed along with an emerging correlation between symmetry of the fuel assembly and performance as measured by yield.

[1] S. H. Glenzer, et al., Physics of Plasmas **19** (5), 056318 (056315) (2012)
[2] F. E. Merrill, et al., Review of Scientific Instruments **83** (10), 10D317 (316) (2012)
[3] P. Volegov, et al., Review of Scientific Instruments **85** (2), 023508 (2014)
[4] S. W. Haan, et al., Phys. Plasmas 18, 051001 (2011); doi: 10.1063/1.3592169
[5] M. J. Edwards, et al., Phys. Plasmas **18**, 051003 (2011); doi: 10.1063/1.3592173
[6] P. Michel, et al., Phys. Rev. Lett. 102, 025004 (2009)
[7] S. P. Regan, Phys. Plasmas **19**, 056307 (2012); doi: 10.1063/1.3694057
[8] S. P. Regan, et al., Phys. Rev. Lett. 111, 045001 (2013)

Journal of Physics: Conference Series **688** (2016) 012065 doi:10.1088/1742-6596/688/1/012065

# Theoretical analysis for temperature dependence of laser-induced damage threshold of optical thin films

**K Mikami[1,2], S Motokoshi[3], T Somekawa[3], T Jitsuno[1], M Fujita[3], KA Tanaka[2], and H Azechi[1]**
[1] Institute of Laser Engineering, Osaka University, 2-6 Yamada-oka, Suita, Osaka 565-0871 Japan
[2] Graduate School of Engineering, Osaka University, 2-1 Yamada-oka, Suita, Osaka 565-0871 Japan
[3] Institute for Laser Technology, 1-8-4 Utsubo-honmachi, Nishi-ku, Osaka 550-0004 Japan
E-mail: kmikami@ile.osaka-u.ac.jp

**Abstract.** The temperature dependence of the laser-induced damage threshold on optical coatings was studied in detail for laser pulses from 123 K to 473 K at different temperatures. The laser-induced damage threshold increased with decreasing temperatures when we tested long pulses (200 ps and 4 ns). The temperature dependence, however, was reversed for pulses shorter than a few picoseconds (100 fs testing). We propose a scaling model with a flowchart that includes three separate processes: free-electron generation, electron multiplication, and electron heating. Furthermore, we calculated the temperature dependence of laser-induced damage thresholds at different temperatures. Our calculation results agreed well with the experimental results.

## 1. Introduction

Next generation high-power laser systems are expected to develop as one of the essential tools in several scientific and industrial fields. The Yb:YAG laser medium for cryogenically cooled laser systems is a typical material [1,2] and has been studied in ILE Osaka University [3-5]. In particular, the laser system using Yb:YAG ceramic is planned as a laser fusion driver.

The output power of a high-power laser system is limited by the laser-induced damage threshold (LIDT) because damaged optics cannot provide the required performance. Improvement of the LIDT is important to enhance the output power of a high-power laser system. A laser-induced damage mechanism (LIDM) has been studied for over 40 years [6]. A typical study of pulse width dependence of the LIDT was reported by Stuart *et al.* in 1995 [7]. That study revealed a new scaling law, $\tau^{1/2}$ *scaling* and the role of pulse width in the LIDM. In particular, a balance between linear and nonlinear phenomena was mentioned as an essential parameter to provide the scaling, which is an important consideration for high-intensity laser system design.

The role of temperature in the LIDM is extremely important for cryogenically cooled laser systems, and several studies for temperature dependence of LIDT have been reported [8-13]. However, the theoretical approach for the effect in the LIDM is insufficient to discuss to estimate the LIDT measured by different pulse widths at different temperatures.

[1] To whom any correspondence should be addressed. E-mail: kmikami@ile.osaka-u.ac.jp

In this study, we demonstrated the measurement of the LIDT at different temperatures with different pulse widths; furthermore, we suggested a model to explain the data.

## 2. Experiment

### *2.1. Experimental setup*

We evaluated the temperature dependence of the LIDT using a single-mode Nd:YAG laser (wavelength 1064 nm, pulse width 4 ns, single shot) and a Ti:Sapphire laser (wavelength 800 nm, pulse width 100 fs, 2 ps, and 200 ps, repetition frequency 10 Hz). Figure 1 shows the experimental setup. The irradiation energy of the laser pulse was adjusted with a half-wave plate and a polarizer. The mechanical shatter was used to pick up a pulse from the Ti:Sapphire laser. The laser pulse was focused by a lens (f = 200 mm), and the spot sizes were defined by the $1/e^2$ peak intensity and were 80 µm ($\lambda = 800$ nm) and 160 µm ($\lambda = 1064$ nm) in diameter. The spot size was measured by image transferring on a sample surface to a CCD camera using a lens (f = 600) with 3× magnification. The sample was held by a Cu holder and set in a vacuum chamber (~5 Pa) to prevent water condensation at low temperature. The incident angle was 0°. The sample's temperature was adjusted with liquid $N_2$ and a heater, and monitored with a Pt resistance temperature detector attached to the sample holder.

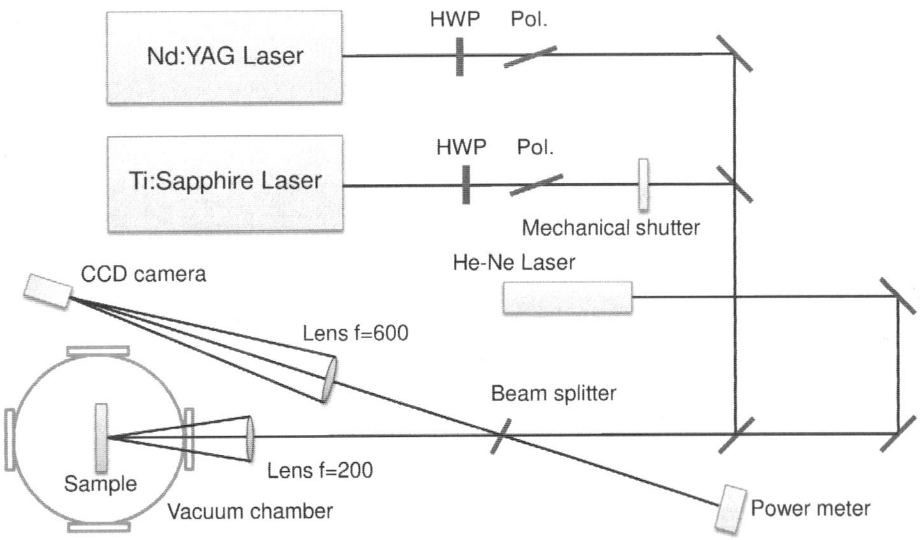


**Figure 1.** Experimental layout

### *2.2. Determination method for LIDT*

In this study, we employed the N-on-1 test method. A fixed site was irradiated by laser pulses. The fluence was gradually increased until damage occurred. The total shot number of laser pulses was approximately 15 shots at each test. The laser-induced damage was detected by plasma emission and transfer of the surface image with a co-aligned He–Ne laser. In the method for damage detection, we could detect a damage site with a minimum diameter of 2 µm. The measured LIDT was tested with three average values at the same experimental conditions.

### *2.3. Experimental sample*

Six types of single-layer coatings ($SiO_2$, $Al_2O_3$, $HfO_2$, $ZrO_2$, $Ta_2O_5$, and $MgF_2$) were prepared on optical polished silica glass substrates, with a root mean square roughness <0.8 nm, by electron beam evaporation. The optical thickness of the single layers had a quarter wavelength ($nd = \lambda/4$) at 800 nm.

### *2.4. Experimental result*

Figure 2 2 shows the temperature dependence of the LIDT for (a) $SiO_2$, (b) $Al_2O_3$, (c) $HfO_2$, (d) $ZrO_2$, (e) $Ta_2O_5$, and (f) $MgF_2$ single-layer coatings. The vertical axis indicates the normalized LIDT with the LIDT at room temperature. Table 1 shows the LIDT at room temperature. The LIDT for all samples decreased with decreasing temperature in 100 fs testing. With the 2 ps laser pulse, however, the dependence of some materials showed an increase in the LIDT with decreasing temperature. Additionally, under 200 ps and 4 ns laser pulses, the LIDT exhibited the opposite trend. The slope of the LIDT for each coating material was changed for the pulse width around a few picoseconds.

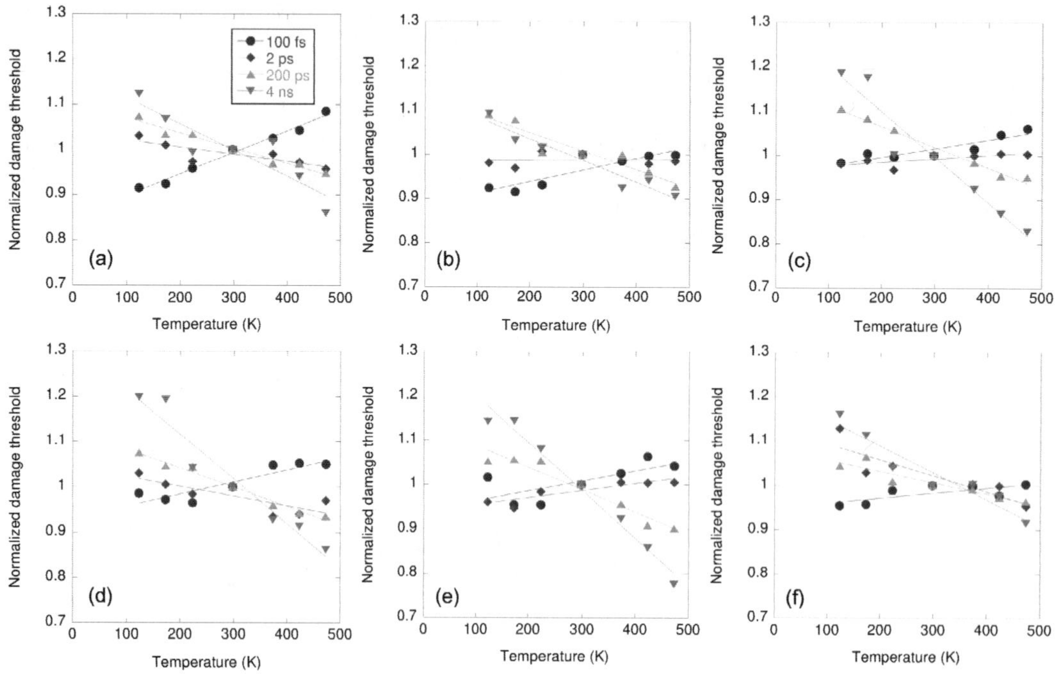


**Figure 2**. Temperature dependence of LIDT for (a) $SiO_2$, (b) $Al_2O_3$, (c) $HfO_2$, (d) $ZrO_2$, (e) $Ta_2O_5$, and (f) $MgF_2$ single-layer coatings

**Table 1.** LIDT at room temperature

| Pulse width | LIDT at room temperature (J/cm$^2$) | | | | | |
|---|---|---|---|---|---|---|
| | $SiO_2$ | $Al_2O_5$ | $HfO_2$ | $ZrO_2$ | $Ta_2O_5$ | $MgF_2$ |
| 100 fs | 0.55 | 0.39 | 0.36 | 0.35 | 0.30 | 0.65 |
| 2 ps | 2.0 | 1.43 | 1.4 | 1.44 | 1.4 | 1.5 |
| 200 ps | 5.6 | 437 | 4.5 | 4.5 | 4.3 | 5.3 |
| 100 fs | 22 | 22 | 9.1 | 10 | 7.1 | 22 |

## 3. Quantitative analysis

### *3.1. Theoretical model*

A flowchart was fabricated using the LIDM to calculate the LIDT, as shown in Fig. 3. The damage formation was constructed with three processes: free-electron generation, electron multiplication, and electron heating. The route in the flowchart was branched because of the irradiated pulse width. In the short-pulse region, with no more than a few picoseconds, the high intensity causes multiphoton ionization as a nonlinear optical phenomenon. By contrast, the lower intensity of the short pulse leads to linear optical phenomena, *i.e.*, photoionization and phonon–electron interaction, in the long-pulse region, which lasts more than a few picoseconds.

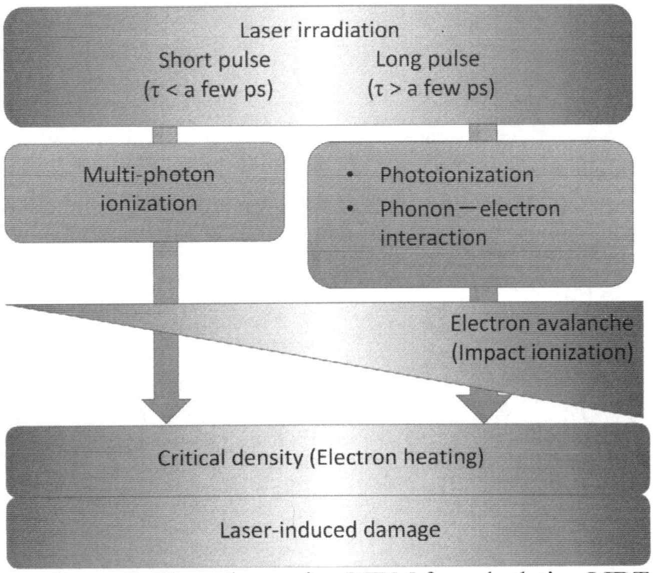


**Figure 3.** Flow chart using LIDM for calculation LIDT

Multiphoton ionization [7] is the transfer of electrons from the valence band to the conduction band with two or more photon absorptions (*multiphoton* absorption). Photoionization [14] is the transfer of electrons from the defect and/or impurity state to the conduction band with single photon absorption. Phonon–electron interactions, an important process at long pulse widths, were studied using the energy gain rate $G$ and average energy loss rate $L$ of an electron and the average electron model [11]. The number densities of generated electrons by multiphoton ionization, $n_m$, and phonon–electron interaction, $n_i$, can be calculated using Eqs. (1) and (2) [7,11].

$$n_i = \frac{G-L}{E_g V} = \frac{\left(e^2/m\right)\tau_{s(\varepsilon)}E^2 - \mathrm{h}\omega}{E_g V} \quad (1),$$

$$n_m = \sigma_4 N_S \left(\frac{c\varepsilon E^2}{2\mathrm{h}\nu}\right)^4 \left(\frac{\pi}{\ln 2}\right)^{0.5} \frac{\tau}{4} \quad (2),$$

where $E_g$ is the band gap energy, $V$ is the volume, $e$ is the charge of electron, $m$ is the electron mass, $\tau_{s(\varepsilon)}$ = 30 fs is the collisional time [7], $E$ is the electric field, $\omega$ is the optical phonon frequency [15], $\sigma_4$ = 2 × $10^{-144}$ $cm^8s^3$ is the cross section of four-photon absorption [7], $N_s$ is the atomic density, $c$ is the light speed, $\varepsilon$ is the dielectric constant, and $\upsilon$ is the laser frequency. In this study, we assumed $V$ to be a product of the irradiated spot size and sample thickness. The Rayleigh length is better than the assumption in the calculation when the sample thickness is longer than the Rayleigh length. Additionally, the number density of a free electron from photoionization, $n_p$, was estimated to be $10^{-14}$ $cm^{-3}$, assuming an impurity concentration on the order of 0.01 ppm to 1 ppm.

The generated free electrons are multiplied by impact ionization and electron avalanche. The process has a strong tendency to be dominant under irradiation with a long pulse laser. The multiplied number density is calculated using the number density of the generated free electron, volume $V$, pulse width $\tau$, and multiple rate $\beta$. An equation to calculate the multiple rates has been reported:

$$\beta = A\frac{E^2}{E_g \rho} \quad (3),$$

where $A$ = 17 is an arbitrary constant and $\rho$ is the electrical resistivity [16]. Eq. (3) can be used under the assumption that the multiplication of the free electrons is exponential because of adequate acceleration time (pulse width). When the acceleration time is not adequate, that is, no more than a few picoseconds, Eq. (3) is replaced by Eq. (4) [17],

$$\beta' \tau V = 2^{\tau/\tau_{s(\varepsilon)}} \quad (4).$$

Finally, the multiplied free electrons are heated with a laser light. The number density under electron heating is discussed with Bose–Einstein statistical mechanics. The number density is defined as critical density $n_{cr}$ and described as Eqs. (5) and (6) [18],

$$n_{cr} = \frac{p}{4\pi e^{*4}} \frac{\hbar\omega}{\tau_{s(\varepsilon)}\left(2N_W + 1\right)} \quad (5),$$

$$N_W = \left\{ \exp\left( \frac{\hbar\omega}{k_0 T} \right) \right\}^{-1} \quad (6),$$

where $p = (2m\varepsilon_e)^{1/2}$ is the momentum of free electrons, $e^* \sim e(1/\varepsilon)^{1/2}$ is the effective electron charge, $k_0$ is the Boltzmann constant, and $T$ is the temperature.

The conditional expressions were described as Eqs. (7) and (8), and calculated using Eqs. (1)–(6)

$$n_{cr} = \beta \tau V \left(n_p + n_i\right) \quad (7),$$

$$n_{cr} = \beta' \tau V n_m \quad (8).$$

Eqs. (7) and (8) can give the electric field that triggers the laser-induced damage. The given electric field is converted to intensity and fluence, that is, to the LIDT.

### *3.2. Calculation results*

The previous discussion enables us to model the temperature dependence of the LIDT. Figure 4 shows the measured and calculated temperature dependence of the LIDT for $SiO_2$ coatings at 100 fs and 4 ns. The solid line in Fig. 4 shows the results of the calculation with Eq. (7). The dashed line in Fig. 4 shows the results of the calculation with Eq. (8). The calculated results are in good agreement with the experimental results.

The model can further reveal the key parameter for the temperature dependence of the LIDT. Table 2 shows the calculated results using the proposed model for the variation when the temperature is changed from 473 K to 123 K.

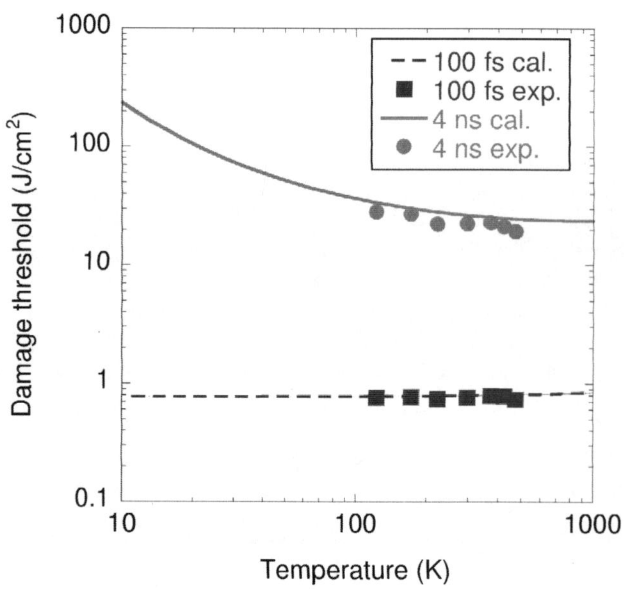


**Figure 4.** Measured and calculated temperature dependence of LIDT for $SiO_2$ coating.

When the LIDT in a short-pulse region is calculated by Eq. (8), the LIDT decreases with decreasing temperature. This is because the critical density at 123 K is 1.2 times lower than that at 473 K. That is, the free electrons at a higher temperature have a lower density than those at a lower temperature, and therefore, the LIDT decreases. When the LIDT in a long-pulse region is calculated by Eq. (7), however, the LIDT increases with decreasing temperature. The number density of free electrons by phonon–electron interaction at 473 K increases by 1.2 times because of the inhibited optical phonon frequency. However, the variation in phonon–electron interaction is smaller than the variation in multiple rates. Figure 5(a) shows the temperature dependence of electrical resistivity for $SiO_2$ [17], and Fig. 5(b) shows the calculated multiple rate. The temperature dependence of the LIDT in long-pulse testing is decided by that of the multiple rate, *i.e.*, the temperature dependence of the electrical resistivity. The electron avalanche in the long-pulse region can become more dominant in the LIDM than in the short-pulse region. The variation in the trend of the temperature dependence of the LIDT will be decided because of the electron avalanche process.

Additionally, the model for the temperature dependence of the LIDT may suggest a method of producing extremely high LIDT optics, at least several times higher than conventional optics, because the calculated LIDT (solid line in Fig. 4) exponentially increases with decreasing temperature. The LIDT at a low temperature can be improved mainly by controlling the electrical resistivity. If the temperature dependence of each parameter, particularly electrical resistivity, is fixed at a cryogenic condition, we might obtain extremely high LIDT optics.

**Table 2.** Calculation result for variation (temperature from 473 K to 123 K)

| Process | | Variation |
|---|---|---|
| Free-electron generation | Number density of free electron by photoionization | No change |
| | Number density of free electron by phonon-electron interaction | 1.2 times increase |
| | Number density of free electron by multi photon ionization | No change |
| Electron multiplication | Impact ionization | No change |
| | Multiple rate in electron avalanche | 2 times decreased |
| Electron heating | Critical density | 1.2 times decrease |

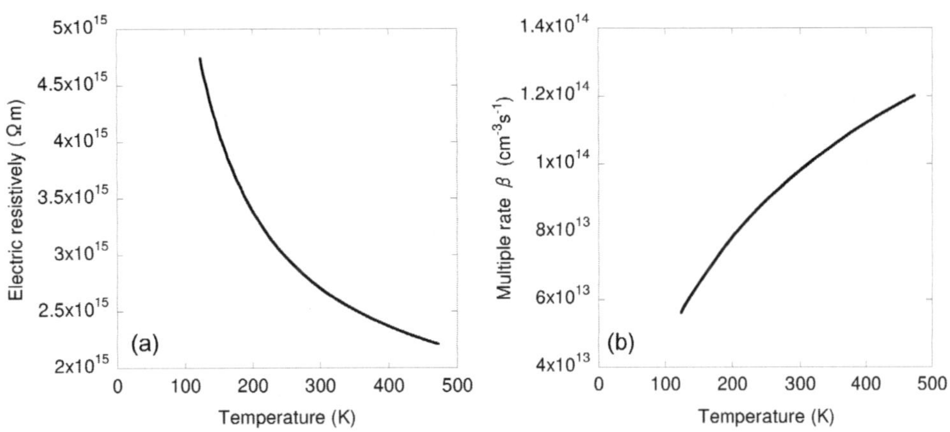


**Figure 5**. (a) Temperature dependence of the electric resistivity of $SiO_2$ and (b) temperature dependence of electron multiple rate.

## 4. Conclusion

In conclusion, we measured the temperature dependence of the LIDT of optical coatings at different pulse widths. The trend of the LIDT temperature dependence varied because of the irradiated pulse width. The calculated results obtained using the proposed model agreed well with the measured results. The electrical resistivity, which is related to the electron avalanche, is an essential parameter affecting the temperature dependence. To improve the LIDT of optical materials, it is particularly important to evaluate the temperature dependence of the electrical resistivity.

We may also propose a method of producing extremely high LIDT optics, at least several times higher than conventional optics, because the calculated LIDT (solid line in Fig. 4) exponentially increases with decreasing temperature. LIDT at a low temperature can be improved mainly by controlling the electrical resistivity. We can realize this new type of optics by combining existing optics and a cryostat, where the mirror substrate can be cooled down to either 77 K or 2 K with liquid $N_2$ or He. The new optics will have the potential to demonstrate high power that has not been achieved to date.

## 5. References

[1] Broun B C, Bruno T M and Singley M 2010 *Opt. Express* **18** 16573
[2] Broun B C, Singley J M, Yager E, Kowalewski, Guelzow J and Kuper J W 2008 *Proc. SPIE* **6952** 69520K
[3] Kawanaka J, Tokita S, Nishioka H, Fujita M, Yamakawa K, Ueda K and Y Izawa 2005 *Phys. Laser* **15** 1306
[4] Furuse H, Kawanaka J, Takeshita K, Miyanaga N, Saiki T, Imasaki K, Fujita M and S Ishii 2009 *Opt. Lett.* **34** 3439
[5] Kawanaka J, Takeushi Y, Yoshida A, Parce S J, Yasuhara R, Kawashima T and Kan H 2010 *Phys. Laser* **20** 1079
[6] Manekov A A 2008 *Proc. SPIE* **7132** 713202
[7] Stuart B C, Feit M D, Rubenchik A M, Shore B W Perry M D 1995 *Phys. Rev. Lett.* **74** 2248
[8] Mikami K, Motokoshi S, Fujita M, Jitsuno T, Kawanaka J and Yasuhara R 2009 *Proc. SPIE* **7504** 75041R
[9] Mikami K, Motokoshi S, Fujita M, Jitsuno T, Kawanaka J and Yasuhara R 2010 *J. Phys.: Conf. Ser.* **244** 032023
[10] Mikami K, Motokoshi S, Fujita M, Jitsuno T and Tanaka K A 2012 *Proc. SPIE* **8190** 81900A
[11] Manenkov A A 1978 *Natl. Bur. Stand. (U.S.) Spee. Publ.* **541** 455
[12] Markle L D and Kitriouns D 1988 *Phys. Rev. B* **38** 1473
[13] Plantenak J R, Walser R M and Becker M F 1987 *Natl. Bur. Stand. (U.S.) Spee. Publ.* **752** 216
[14] Braunlich P and Schmid A 1975 *Appl. Phys. Lett.* **26** 150
[15] Worlock J M and Fleury P A 1967 *Phys. Rev. Lett.* **19** 1176
[16] Roy S and Chakravorty 1995 *J. Mater. Res.* **9** 2314
[17] Seitz F 1949 *Phys. Rev.* **76** 1376
[18] Frohlich H and Paranjape B V 1956 *Proc. Phys. Soc. B* **69** 21

**Acknowledgments**

This study was funded by a Grant-in-Aid for Research Fellow from the Japan Society for the Promotion of Science (to K Mikami, Grant No. 25-275).

# Magnetized Fast ignition (MFI) and Laser Plasma Interactions in Strong Magnetic Field

Kunioki Mima[1], T.Johzaki[2], J.Honrubia[3], H.Nagatomo[4], T.Taguchi[5], A.Sunahara[6], and H.Sakagami[7], S.Fujioka[4], and G.Logan[4]

[1]The Graduate School for the Creation of New Photonics Industries, Hamamatsu, Japan

[2]Graduate School of Engineering, Hiroshima University, Higashi-Hiroshima, Japan

[3] School of Aeronautics, Technical University of Madrid, Madrid, Spain

[4]Institute of Laser Engineering, Osaka University, Suita, Osaka, Japan

[5]Faculty of Engineering, Setsunan University, Neyagawa, Osaka, japan

[6]Institute for Laser Technology, Osaka, Japan

[7]National Institute for Fusion Science, Toki, Japan

**Abstract**

In this paper, magnetized fast ignition (MFI) is proposed for improving the coupling efficiency of a heating laser to a core plasma. In the MFI, the external magnetic field is applied to reduce the hot electron energy and focus the dense hot electron flux to the core. The external magnetic field higher than 100T is generated by the laser driven coil and it is amplified by the implosion. The magnetic field at the tip of the cone is expected to reach higher than 10kT and the laser plasma interaction and the hot electron transport are modified. As the results of applying the external magnetic field, hot electron energy is reduced to less than 5MeV for the laser intensity of $10^{20}$W/cm$^2$ and the Weibel instability is suppressed to collimate the hot electron beam to the core.

## 1. Introduction

The critical issue of the fast ignition has been recognized to be the coupling efficiency of a short pulse laser to a hot spark. Namely, the ultra-intense laser plasma interaction (U-LPI) and the relativistic electron transport (RET) are the critical processes for the fast ignition. In the U-LPI, the generated relativistic electron energy is required to be no more than 5MeV, and in the RET, the diameter of the heating area should be less than 100$\mu$m. In this paper, the magnetized fast ignition (MFI) is proposed and the effects of a high external magnetic filed on the U-LPI are presented.

## 2. Concept of MFI

The schematic over-view of the MFI is shown in the Fig. 1. Here, the magnetic field higher than 10kT could be generated by using laser driven coil [1][2] and B-field compression with implosion [3]. For an example, the magnetic field at the center of a laser driven coil is assumed 1kT, then the magnetic field at the tip of the cone will be a few hundreds Tesla. When the magnetic field is compressed by 10 times in linear scale by the implosion, the magnetic field at the tip of the cone reaches a few tens of kilo Tesla, as shown in the figure 1. Here, magnetic fields are amplified by imploding plasmas, when the magnetic Raynolds number is high enough in the gas plasma contained in the capsule. Note also that magnetic field diffuses out because of the high resistivity of the shell and is carried out by the ablative flow.

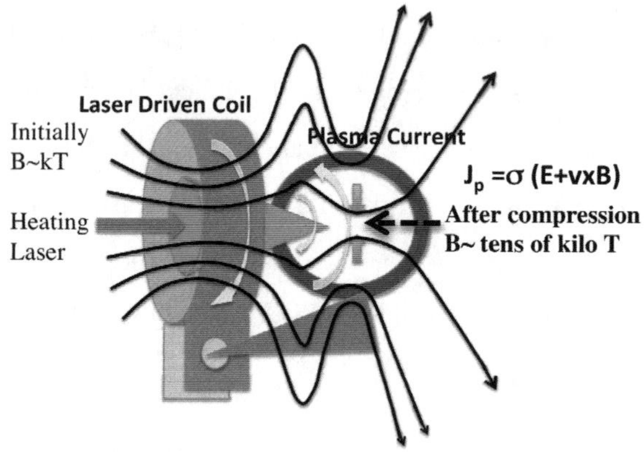


Fig.1 Geometry of Laser driven one turn coil and a cone shell target. Magnetic field is illustrated by solid lines and the circular arrows indicate induced currents in the coil and in the imploding plasma. After the compression of the magnetic field, the heating laser is injected.

In Fig. 1, the electron cyclotron frequency could be higher than the frequency of 1$\mu$m wavelength laser and a Right handed Circularly Polarized (R-CP) laser penetrates into over-dense plasmas as a whistler wave. This suggests that the laser energy is directly deposited in the over dense plasma and the average hot electron energy could be lowered. The transmittance and the reflectivity of R-CP laser for a plasma slab are estimated as follows. The refractive index of CP-em wave is given by,

$$N^2 = \left(\frac{ck}{\omega}\right)^2 = 1 - \frac{\omega_p^2}{\omega(\omega \mp \omega_c)} ,$$

where $\omega_p$ is the electron plasma frequency, $\omega_p$ is the electron cyclotron frequency, and – sign (+ sign) is for R-CP (Left handed CP).

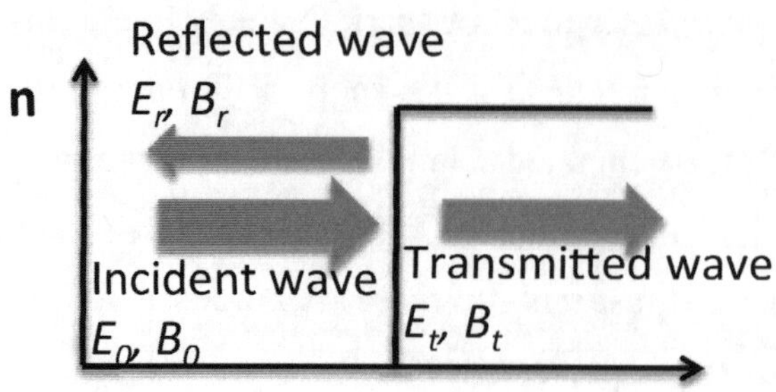


Fig.2 Geometry of laser interaction with a slab plasma. The external magnetic field is applied along X-axis.

At the boundary between vacuum and the plasma of electron density: $n>n_c$ (cut-off density), the electric fields, magnetic fields, reflectivity, and transmittance are estimated as follows.

The incident, reflected and transmitted wave electromagnetic fields are shown in Fig.2. The boundary conditions for the waves yield the relations of the amplitude of transmitted and reflected waves to the incident wave amplitude:

Transmitted wave: $\dfrac{E_t}{E_0} = \dfrac{2.0}{1.0+N}$, $\dfrac{B_t}{B_0} = \dfrac{N}{(1.0+N)}$

Reflected wave : $\dfrac{E_r}{E_0} = \dfrac{1.0-N}{1.0+N}$, $\dfrac{B_r}{B_0} = \dfrac{1.0-N}{1.0+N}$

Therefore, the reflectivity: R and the transmittance are obtained as,

$$R = \frac{(1.0-N)^2}{(1.0+N)^2} \text{ and } T = \frac{4N}{(1.0+N)^2} \text{ , respectively.}$$

For an example, these relations indicate that R=0.56 and T=0.44, when $\omega_c/\omega = 2$ (20kT) and n = $50n_c$. In this case, the electric field in the plasma is 1/4 of the incident laser electric field. Namely, the amplitude of the vector potential, a is 1/4. Therefore, the average energy of high energy electrons is also 1/4, which means the average electron energy could be 1~2MeV even for the laser intensity higher than $10^{20}$ W/cm$^2$. This will make the heat deposition efficiency higher in the hot spark. The effects of the magnetic field on the transport of the hot electrons are simulated by the hybrid simulation code[4]. It was demonstrated that the coupling efficiency could be higher than 20% for a moderate angular divergence of the hot electrons, when 1kT magnetic field is applied. The heat deposition density distributions are compared with and with out magnetic field as shown in Fig.3.

Fig.3 Heat deposition profiles of the 2D hybrid simulations for without (a) and with external magnetic field.

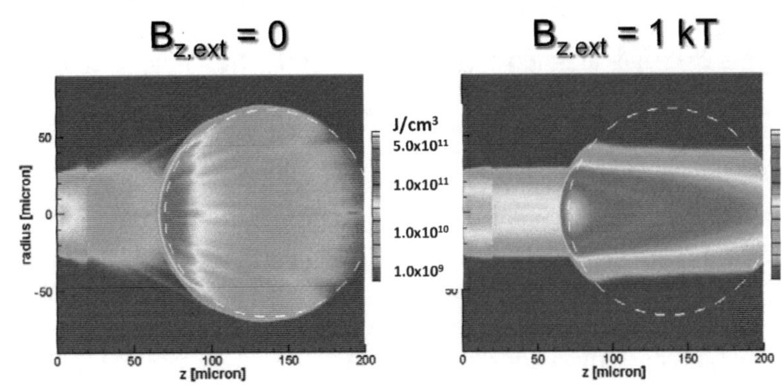


(a) (b)

In this hybrid simulations, we assumed that the cone is made by diamond like carbon (DLC), the total electron energy is 50kJ, the electron energy and angular spread are 3.5 MeV and 55°, respectively, and the core plasma density, the cone tip to the core edge distance is 50 $\mu$m, and hot spark density and area are 450 g/cm$^3$ and 40x30 $\mu$m, respectively.

The angular divergence of laser produced hot electrons may also be reduced by the sufficiently strong external magnetic field. One of the mechanisms of the enhancement of the angular divergence is the Weibel instability near the interaction region. The effects of the external magnetic field on the Weibel instability is under investigation[5]. The simulation result indicates that the growth rate of the Weibel instability is significantly reduced when the magnetic field exceeds 10kT. However, the secondary instability is not stabilized and the angular spread is reduced only for very high magnetic field higher than 20kT.

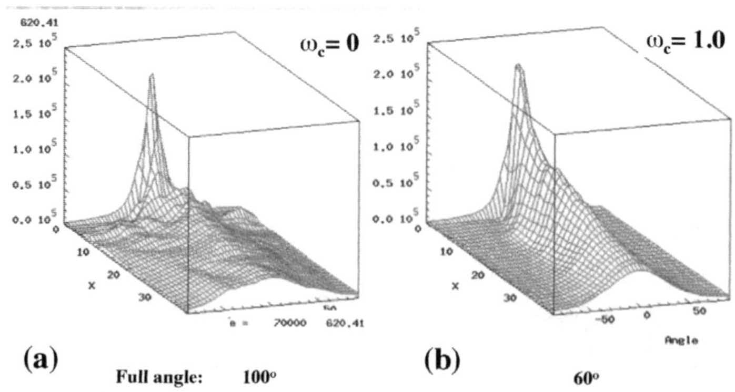


Fig.4 Angular diffusion of high energy electron in the magnetic field turbulence by the Webel instability in 40 $\mu$m propagation.
(a) Without B-field, (b) With B-field (30kT), [5]

In summary, the recent development on the laser driven coil and the amplification of magnetic field leads to a new idea: "Magnetized Fast Ignition" (MFI). In the MFI, it is expected that the heating laser penetrate into over dense plasma as a whistler wave and deposits the energy, the angular divergence of hot electrons is reduced by suppressing the Weibel instability with B-field, and the hot electrons are guided efficiently by the magnetic field. At sufficiently strong external magnetic fields, Those effects will significantly improve performance of the fast ignition.

Reference

1) H.Daido, K.Mima, etal., P.R.L 56(1986)846, and J.J. Applied Physics (1987) NO.8, 1290
2) S.Fujioka, etal., Scientific Rep.3(2013)1170
3) Chang, etal., PRL 107(2011)035006
4) T.Johzaki, etal., and J.Honrubia, etal., IFSA 2013 proceedings, to be published
5) T.Taguchi, etal., IFSA 2013 proceedings, to be published

# The Laser Mega-Joule : LMJ & PETAL status and Program Overview

**J-L Miquel, C Lion, P Vivini**

CEA/DAM Île de France – Bruyères le Châtel – 91297 Arpajon Cedex (France)

E-mail : jean-luc.miquel@cea.fr

**Abstract**. The laser Megajoule (LMJ), developed by the French Commissariat à l'Energie Atomique et aux Energies Alternatives (CEA), will be a cornerstone of the French Simulation Program, which combines improvement of physics models, high performance numerical simulation, and experimental validation. The LMJ facility is under construction at CEA CESTA near Bordeaux and will provide the experimental capabilities to study High-Energy Density Physics (HEDP). One of its goals is to obtain ignition and burn of DT-filled capsules imploded, through indirect drive scheme, inside rugby-shape hohlraum. The PETAL project consists in the addition of one short-pulse (ps) ultra-high-power, high-energy beam (kJ) to the LMJ facility. PETAL will offer a combination of a very high intensity multi-petawatt beam, synchronized with the nanosecond beams of the LMJ. This combination will expand the LMJ experimental field on HEDP. This paper presents an update of LMJ & PETAL status, together with the development of the overall program including targets, plasma diagnostics and simulation tools.

## 1. Introduction

The Laser Mega-Joule (LMJ) is part of the French Simulation Program developed by the Commissariat à l'Energie Atomique et aux Energies Alternatives (CEA). The Simulation program aims to improve the theoretical models and data's used in various domains of physics, by means of high performance numerical simulations and experimental validations.

LMJ offers unique capabilities for the Simulation Program, providing an extraordinary instrument to study High Energy Density Physics (HEDP) and Basic Science (equation of state, atomic physics, nuclear physics …). A large panel of experiments will be done on LMJ to study physical processes at temperatures from 100 eV to 100 keV, and pressures from 1 Mbars to 100 Gbars. Among these experiments, Inertial Confinement Fusion (ICF) is the most exciting challenge, since ICF experiments fix the most stringent specifications on LMJ's performances.

The PETAL project, part of the CEA opening policy, consists in the addition of one high-energy multi-Petawatt beam to LMJ. PETAL will provide a combination of a very high intensity beam, synchronized with the very high energy beams of LMJ. LMJ/PETAL will be an exceptional tool for academic research, offering the opportunity to study matter in extreme conditions.

## 2. LMJ's main characteristics and status

Designed to deliver 1.8 MJ of UV light on target with 240 beams, LMJ is under construction at CEA/CESTA at a primary stage of 176 beams. LMJ is a flashlamp-pumped neodymium-doped glass laser (1.053 μm wavelength) configured in a multi-pass power amplifier system. The 1.053 μm light is frequency converted to the third harmonic (0.351 μm) and focused, by means of gratings, on a target at the center of target chamber. LMJ will deliver shaped pulses from 0.3 ns to 25 ns with a maximum energy of 1.5 MJ and a maximum power of 400 TW of UV light on target.

The main building, commissioned in 2008, includes four similar laser bays, 128-meter long, situated in pairs on each side of the central target bay of 60-meter diameter and 38-meter height.

The four laser bays are completed since the end of 2013, and the laser slabs are now being installed. The 176 square 370x370 $mm^2$ beams are grouped into 22 bundles of 8 beams. In the switchyards, each individual bundle is divided into two quads, the basic independent unit for experiments, which are directed to the upper and lower hemispheres of the chamber.

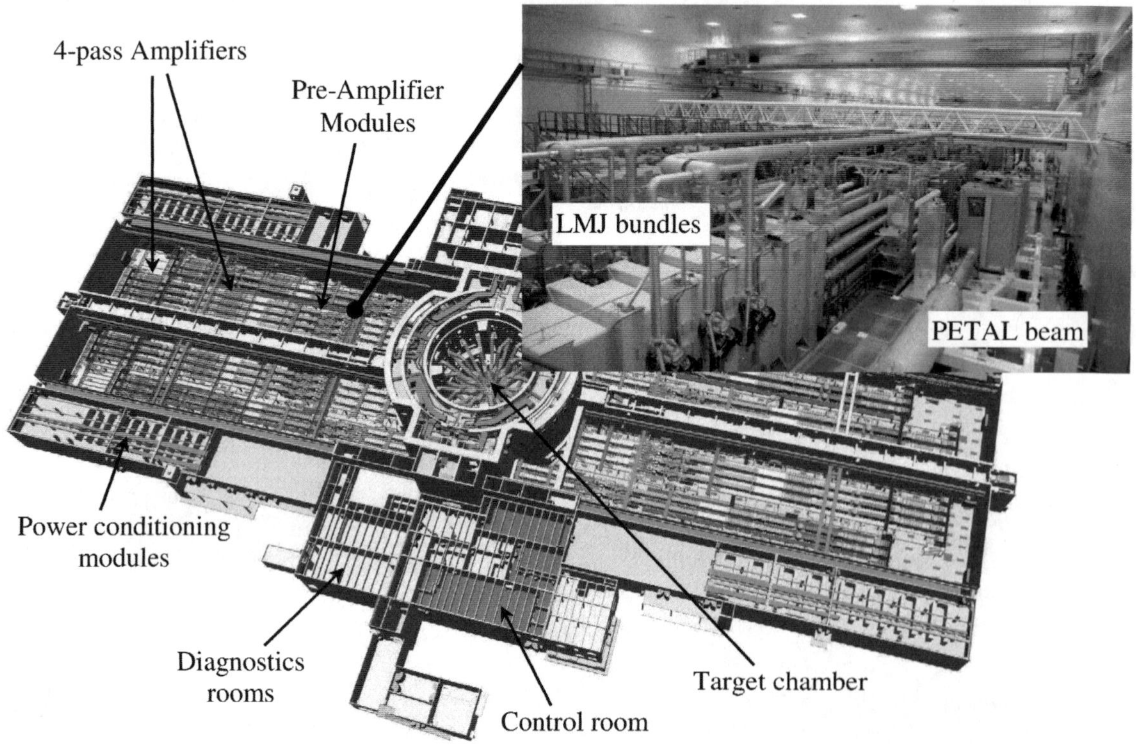


Figure 1 : Schematic view of the LMJ, and picture of one of the LMJ laser bays with 7 LMJ bundles (56 beams) and PETAL beam

At the center of the target bay, the target chamber, introduced in the building in November 2006, consists of a 10-meter diameter aluminum sphere, fitted with two hundred ports for the injection of the laser beams and the location of diagnostics. It is a 10 cm-thick aluminum sphere covered with a neutron shielding made of 40 cm thick borated concrete. LMJ is configured to operate in the “indirect drive” scheme for fusion, which directs the laser beams into cones in the upper and lower hemispheres of the target chamber. Forty quads enter the target chamber through ports that are located on two cones at 33.2° and 49° polar angles. Four other quads enter the target chamber at 59.5° polar angle, and will be dedicated to radiographic purpose.

Inside the chamber, 249 protection panels for X-ray and debris are installed (and will be removed for maintenance) by a robot. Half of the laser beam ports are today installed ; they include the final optics assembly : vacuum windows, debris shield and device to check the damages on optics.
A lot of equipments is required in the target area, among them two target positioning systems (TPS) are considered : a cryogenic TPS for fusion target and a non-cryogenic TPS for other experiments, the latter is already connected to the chamber. A set of ten diagnostic positioning systems will be installed,

they will position 150 kg diagnostic with a 50 μm precision. A Reference Holder will be used for the alignment of all beams, diagnostics and target.

## 3. PETAL : a multi-Petawatt beam coupled to LMJ

The PETAL project consists in the addition of one short-pulse (500 fs to 10 ps) ultra-high-power, high-energy beam (a few kJ compressed energy) to the LMJ facility. The PETAL/LMJ facility will be an exceptional tool for HEDP, basic science as well as for the physics of ignition, in relation with the HiPER project.

The PETAL design is based on the chirped pulse amplification (CPA) technique combined with optical parametric amplification (OPA). Further, it takes the benefits of the laser developments made for the high-energy LMJ facility allowing it to reach the kilojoules level.[1]

The front end consists in a standard Ti:sapphire oscillator delivering 3nJ /100 fs / 16 nm pulse at 1053 nm wavelength. The pulse is stretched to 9 ns in an Öffner stretcher in two passes, and then amplified in the pre-amplifier module (PAM) including two OPA stages and pump laser. A 150 mJ amplified signal pulse with a shot-to-shot stability of less than 2% has been demonstrated.

The PETAL amplifier section has the same architecture as the LIL/LMJ amplifier section using a single 370 × 370 $mm^2$ beam and will deliver up to 6 kJ. The main difference is the longitudinal chromatism corrector installed in the U-turn which allows to correct 1,5 ps in two passes.

The compressor is a two-stage system. A first compressor, in air atmosphere, is used to reduce the pulse duration from 1.8 ns to 350 ps in an equivalent double pass configuration. The output mirror is segmented in order to divide the initial beam into 4 subapertures which are independently compressed and synchronized into the second compressor in a single pass configuration under vacuum. These sub-apertures are coherently added using the segmented mirror with three interferometric displacements. The pulse duration is adjustable from 0.5 to 10 ps.

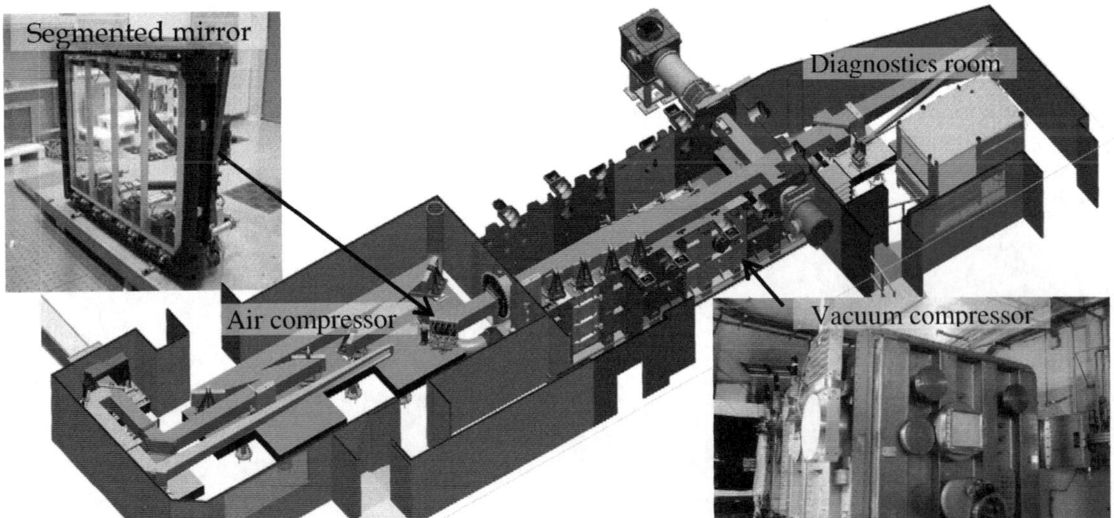


Figure 2 : Schematic view of the two stages compressor with pictures of the segmented mirror and vacuum box for the last compressor stage.

The focusing system consists in an off-axis parabola with a 90° deviation angle, followed by a pointing mirror. The focal length is 7.8 meters, and the focal spot has a 50 μm diameter, this will result in intensities above $10^{20}$ $W/cm^2$ on target.

The PETAL performances depend on the damage threshold of optics. Great efforts have been made on gratings in order to improve their strength. The effect of electric field on damages has been demonstrated [2], and the groove profile of PETAL multilayer dielectric gratings has been optimized in order to obtain a damage threshold above 4J/cm².But the current transport mirrors cannot sustain more than 2.4 J/cm² and limit the available energy on target at a 1-2 kJ level. New technologies are required to increase this value. Several ways of improvement are identified and under investigations.

## 4. Robust path toward ignition

LMJ is designed to provide the experimental capabilities to study HEDP. A large number of experiments will be done on LMJ, covering diverse aspects of Plasma Physics (radiation transport, opacities measurements …), Materials Science (Equation of state (EOS) measurements, mechanical properties …), Hydrodynamics (Rayleigh-Taylor instabilities, turbulence …), Atomic and Nuclear Physics. Using a wide variety of pulse shapes, it will be possible to bring matter to extreme conditions, with temperatures from $10^6$ to $10^9$ K, pressures from $10^9$ to $10^{13}$ atm, and densities up to $10^3$ g/cm$^3$.

Achieving fusion with LMJ is the most constraining for the facility and requires a coordinated program associating the facility itself, as well as optimized fusion targets, plasma diagnostics, and simulation tools.

### 4.1. *Ignition "point design"*

One of the LMJ aims is to obtain ignition and burning of a DT fuel included in a capsule which is imploded, inside a high-Z rugby-shape hohlraum, through indirect drive scheme. The capsule is made of an ablator, primarily constituted of plastic (CH) that contains the fuel composed of a cryogenically cooled deuterium-tritium (DT) shell surrounding a central DT gas region.

As laser plasma interaction, coupled to energetics performances, and hydrodynamic instabilities are critical issues in ignition target design, a particular attention has been devoted to (i) the mitigation of laser plasma instabilities to reduce backscattered energy, (ii) the design of accurate shape of the hohlraum to improve energetics performance, (iii) the structure of the ablator to reduce hydrodynamic instabilities during implosion.

Taking into account these issues, an ignition indirect drive target has been designed [3], it requires 0.9 MJ and 260 TW of absorbed laser energy and power, to achieve a temperature of 300 eV in a rugby-shaped hohlraum and give a yield of about 20 MJ.

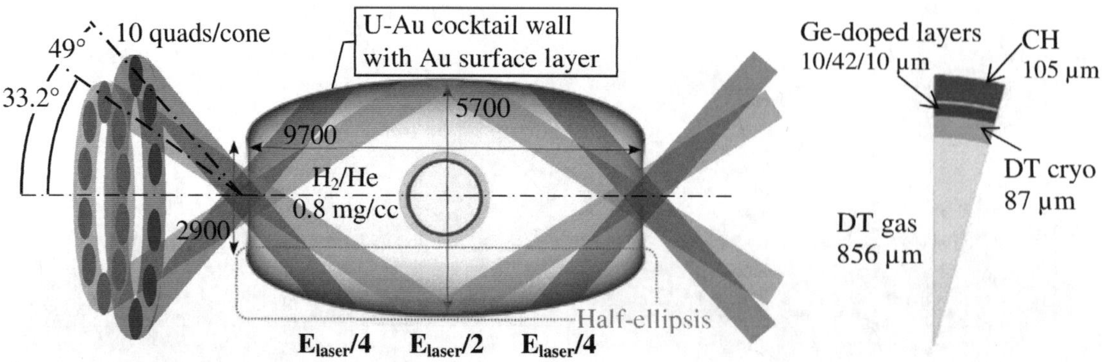


Figure 3 : Schematic view of the ignition target. The hohlraum is heated with 40 quads beams in two cones at angle 33.2° and 49° to the hohlraum axis. The ignition capsule uses several plastic layers of various thickness, some of them are doped with Germanium with varying concentration.

The hohlraum profile is made of two half ellipsis ; it has a 9.7 mm length, a 5.7 mm diameter and a 2.9 mm laser entrance hole (LEH) diameter. The hohlraum is made of U-Au cocktail wall with Au surface layer, and filled with a $H_2$/He gas mixture at a 0.8 mg/cc density to hold back the plasma from the hohlraum wall. The 40 quads enter the hohlraum through two LEH and are arranged in two cones at 33.2° and 49° providing a high ten-fold symmetry in the azimuthal plane. As the two inner cones on each side meet at the equatorial plane, the laser energy distribution is one half at the equator, and one quarter on both polar sides, which is the right energy allocation for rugby shaped hohlraum.

The capsule, with a 2.22 mm outside diameter, is made of several layers of plastic of various thickness, some of them are doped with Germanium, to absorb high-energy X-rays from the hohlraum, and with varying concentration to adapt the density gradient at the ablator-fuel interface. The inside fuel ice layer is 87μm thick and represents 0.228 mg of DT. The central part of the capsule is occupied by a 0.3 mg/cc density DT gas at a temperature of 18.3 K.

Alternative designs are under investigation with Si-doped ablators, laminated ablators [4] (stack of ~1μm thick doped and non-doped layers), or highly doped ablators (with a uniform Ge or Si distribution), and also larger target which require a bit more laser energy but less laser power and then could produce less constraints on optics.

During the LMJ power increase, starting at the end 2014, CEA will perform focused experiments dedicated to fundamental physics (EOS, opacity, transport), laser-plasma interaction, hohlraum physics, and capsule physics in order to better apprehend main aspects of fusion target behavior and improve predictive capabilities of the simulation tools. Then we will test new understanding, designs, and models in integrated implosion. This path forward will help us to secure our point design.

### *4.2. Plasma diagnostics development*

Over 30 photon and particle diagnostics are considered for LMJ, with high spatial, temporal and spectral resolution in the optical, X-ray, and nuclear domains. The early diagnostics, designed using the feedback of LIL's diagnostics, consist of :

• four hard and soft X-ray imaging systems (30 eV to 15 keV range) with a 15 to 150 μm spatial resolution and a 30 to 100 ps time resolution, providing 30 imaging channels ;

• a diagnostic set for hohlraum temperature measurements including an absolutely calibrated broadband x-ray spectrometer (30 eV to 20 keV range, 20 channels), a grating spectrometer (1 to 5 keV range), and a streaked pinhole imaging system of the emitting area ;

• an optical diagnostic set dedicated to EOS measurements including two VISAR, two SOP (Streaked Optical Pyrometer), and a reflectivity measurement ;

• a Full Aperture Backscatter System, and a Near Backscatter Imager to measure the power, spectrum, and angular distribution of backscatter light to determine the energy balance.

Other diagnostics are considered with enhanced spatial and spectral resolutions, as well as specific diagnostics for PETAL including high energy (100 keV-200 MeV) charged particles and hard X-ray (15 – 100 keV) spectrometers.

A new generation of framing and streak cameras is developed to take into account the harsh environment that will be encountered on LMJ and the electromagnetic perturbations due to PETAL.

A specific Metrology Line has been developed on the Soleil synchrotron facilities for the metrology of plasma diagnostics components using an XUV branch (30 eV to 2000 eV) and a hard X-rays branch (100 eV to 28 keV) for reference measurements. The global commissioning of diagnostics is made on the EQUINOX laser facility at CEA/Ile-de-France, and will start at the end of 2013.

### *4.3. Research program for LMJ targets*

The targets development requires numerous technologies : materials shaping, surface processing by Physical or Chemical Vapor Deposition (PVD, CVD) and electrochemistry, super critical extraction, micro-machining, laser machining, electro-machining, cryogenic studies, thermal studies, micro assembly, and characterizations. A lot of developments were made for LIL and OMEGA experiments which will be beneficial for LMJ targets. Now we are improving the assembling process using ultra-precision machining, and PVD molding technique.

We have developed a permeation filling scheme in which ignition targets are filled by permeation at room temperature under 500 bar of DT, then cooled to 20 K to withstand pressure and avoid leaks by permeation. The complete process have been tested and validated ; it guarantees the best target quality. A prototype of LMJ cryogenic equipment, the DEMOCRYTE demonstrator, has been developed for the validation of cryogenic performances, namely keep the ~18 K temperature of the DT with an inhomogeneity less than 70 μK.

However this permeation process cannot be applied for targets with impermeable layers, therefore the filling by capillaries is under study in order to expand our filling capabilities for cryogenic targets.

## 5. Conclusions

Since the building has been commissioned the LMJ is on schedule for the first experiments by end of 2014. The laser bundle assemblies of the amplifier section are already installed in the four laser bays, the mechanical frameworks of the target area is mounted around the target chamber, and most of the equipments are installed or under test. All specifications have been met on qualification sub-systems, and on the LIL facility.

The front end and compression stages of PETAL have been qualified on the LIL facility, and amplifier section, transport and focusing are now completed.

Starting in 2014 with one bundle and 4 diagnostics, LMJ will increase its capacities in the following years with the completion of other bundles and a full set of diagnostics. The improvement of physical and numerical models will accompany the experimental developments during the growth of LMJ capacities, paving a progressive way for robust ignition designs on LMJ.

LMJ/PETAL laser facility will be open to Academic research. The Institute of Laser Plasma (ILP) is working on the research program, which will be centred on 4 pillars : (i) High Energy Density Physics, (ii) Inertial Confinement Fusion, (iii) Laboratory astrophysics and nuclear physics and (iv) Acceleration and High Energy Physics.

First shots with PETAL are scheduled in 2015 and experiments combining LMJ and PETAL will be performed from 2016, giving us the possibility to address new physics.

## Acknowledgements

The PETAL project is being performed under the auspices of the Conseil Régional d'Aquitaine, of the French Ministry of Research and of the European Union and with the scientific supports of the Institut Lasers et Plasmas.

## References

[1] Blanchot N, Behar G, Berthier T, Bignon E, Boubault F, Chappuis C, Coüc H, Damiens-Dupont C, Ebrardt J, Gautheron Y et al. 2008 *Plasma Phys. Control. Fusion* **50** 124045.
Rouyer C, Blanchot N, Neauport J and Sauteret C 2007 *Opt. Express* **15** 2019-32.
Blanchot N, Bar E, Behar G, Bellet C, Bigourd D, Boubault F, Chappuis C, Coïc H, Damiens-Dupont C, Flour O, et al. 2010 *Opt. Express* **18** (10) 10088.
[2] Neauport J, Lavastre E, Razé G, Dupuy G, Bonod N, Balas M, de Villele G, Flamand J, Kaladgew S and Desserouer F 2007 *Opt. Express* **15** 12508.
[3] Laffite S and Loiseau P 2010 *Phys. Plasmas* **17** 102704.
[4] Masse L, Casner A, Galmiche D, Huser G, Liberatore S and Theobald M 2011 *Phys. Rev.* E **83** 055401.

# Characteristics of X-rays produced via laser Compton scattering using quasi-monoenergetic electron beam driven by laser-plasma acceleration

**Eisuke Miura, Ryunosuke Kuroda, and Hiroyuki Toyokawa**
ational Institute of Advanced Industrial Science and Technology (AIST)
AIST Tsukuba Central 2, 1-1-1 Umezono, Tsukuba, Ibaraki 305-8568, JAPA
E-mail: e-miura@aist.go.jp

**Abstract.** X-ray generation via laser Compton scattering has been demonstrated using a uasi-monoenergetic electron beam with a narrow energy spread driven by laser-plasma acceleration. A well-collimated X-ray beam with a divergence angle of approximately 5 mrad is produced. The X-ray photon number is estimated to be $2 \times 10^7$ per pulse. The peak energy of the X-rays is estimated to be 60 keV by a numerical simulation.

## 1. Introduction

In laser-plasma acceleration, electrons are accelerated by the electric field of a plasma wave driven by an intense laser pulse [1]. The electron pulse duration is extremely short, on the order of 10 fs, because the wavelength of the accelerating field is on the order of 10 $\mu$m. Furthermore, a compact electron accelerator can be realized using a high accelerating field of more than 100 GV/m. Such unique characteristics of laser-plasma acceleration enable us to produce all-optical, compact, femtosecond -ray sources. Ultrashort -ray sources have attracted much attention, because of their potential applications in investigating the ultrafast structural dynamics of materials through time-resolved -ray diffraction and spectroscopy. Various types of ultrashort -ray sources have been so far developed using a laser-accelerated electron beam [2, 3, 4, 5].

One type of ultrashort -ray sources is a laser Compton scattering (LCS) -ray source, which is produced by scattering a femtosecond laser pulse off a high-energy electron beam. LCS -ray sources can produce a well-collimated, quasi-monochromatic -ray beam. Several groups have so far reported LCS -ray generation using a laser-accelerated electron beam [3, 5]. Recently, the generation of high-energy -rays with a considerably large photon number has been reported. However, because the energy spreads of the electron beams were large, the -ray energy spreads were also large. For various applications, a quasi-monochromatic -ray source is desirable. To produce a quasi-monochromatic -ray source using LCS, a quasi-monoenergetic electron (QME) beam with a narrow energy spread is necessary [6, 7, 8, 9, 10]. In this paper, we report -ray generation via LCS using a laser-accelerated QME beam.

## 2. Experimental conditions

Figure 1 shows the experimental setup. The laser pulses for electron acceleration and LCS are referred to as "main pulse" and "colliding pulse". Both the pulses were horizontally polarized.

Journal of Physics: Conference Series **688** (2016) 012068 doi:10.1088/1742-6596/688/1/012068

The main pulse (800 nm, 700 mJ, 40 fs) was focused onto the edge of a 2-mm-long He gas jet using an $f/14$ off-axis parabolic mirror, and the intensity was $4.7 \times 10^{18}$ W/cm$^2$. The colliding pulse (800 nm, 140 mJ, 100 fs) was focused onto the exit of the main pulse from the gas jet using an $f/6$ off-axis parabolic mirror, and the intensity was $8.8 \times 10^{17}$ W/cm$^2$. The incident angle of the colliding pulse was 20° with respect to the propagation axis of the main pulse in vacuum referred to as "main laser axis". The plasma electron density was $1.5 \times 10^{19}$ cm$^{-3}$.

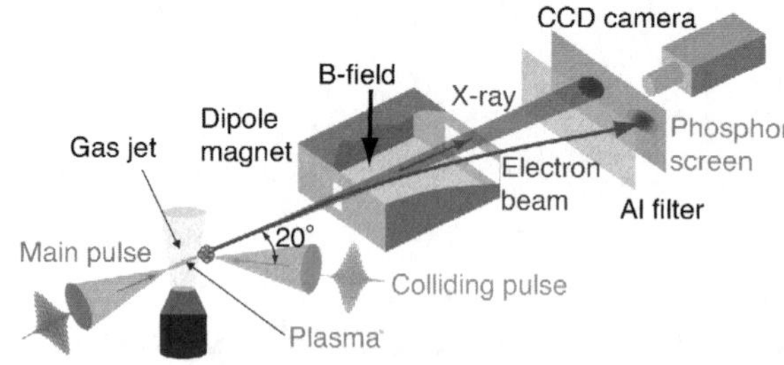


**Figure 1.** Experimental setup.

LCS -rays were scattered in the forward direction of electron beam emission. The electron beam was bent by a dipole magnet with a magnetic field of 0.27 T, and was spatially separated from the -rays. Both the -rays and electrons were incident on a phosphor screen (DR -HGH, Mitsubishi Chemical) through a 115-$\mu$m-thick Al filter. The phosphor images of the -rays and energy-resolved electrons were simultaneously captured with a single shot using a CCD camera. Electrons with energies higher than 30 MeV were observed. The charge of the electron beam was estimated from the calibrated sensitivity of the detection system [11]. The detection system was also sensitive to -rays with energies from 15 to 150 keV.

## 3. Experimental results

Figure 2(a) shows the phosphor image, when LCS -rays were produced. The image on the left side in Fig. 2(a) was the energy-resolved electron image of a QME beam, and the small spot near the main laser axis indicated by a cross was an LCS -ray image. Figure 2(b) shows the horizontal line profile of the phosphor image at the position, in which the LCS -ray image was observed. The dashed line indicates the position of the main laser axis. By extrapolating from the saturated intensity of the energy-resolved electron image, the peak energy and charge of the QME beam were estimated to be approximately 60 MeV and 70 pC, respectively. The divergence angles of the -ray beam estimated from the $e^{-2}$ intensity radius of the image were 6.2 and 4.2 mrad in the horizontal and vertical directions, respectively. In LCS, the beam divergence angle is given by $\sim 1/\gamma$, where $\gamma$ is the Lorentz factor of the electron energy. The divergence angle estimated from the electron energy of 60 MeV ($\gamma \sim 120$) is 8.3 mrad, which is close to the observed divergence angles. This supports that the small spot in Fig. 2(a) is the image of LCS -rays. The photon energy of the -rays within the scattering angle of 5 mrad was estimated to range from 30 to 120 keV taking the energy spread of the QME beam between 40 and 70 MeV into account.

The sensitivity to electrons of the detection system using a phosphor screen coupled with a CCD camera has been known [11]. On the other hand, the efficiency of the conversion from the energy deposited in the phosphor layer to the fluorescence for electrons is nearly equal to that for -rays for a similar type phosphor screen [12]. This indicates that the energies deposited in the phosphor layer determine the sensitivity of the phosphor screen. The sensitivity to -rays can be deduced from the energies deposited in the phosphor layer on the basis of the sensitivity to electrons. The deposited energies were calculated by the simulation code EGS5 [13]. The sensitivity to -rays with energies from 30 to 120 keV was approximately one-eighth that to electrons with energies of around 60 MeV. Integrating the counts of the image in Fig. 2(a) yielded the -ray photon number of $2 \times 10^7$ per pulse. The -ray photon number per pulse was comparable to those achieved in LCS -ray sources using rf accelerators [14].

-ray radiation can be produced by the betatron oscillation of electrons trapped in a plasma wave [2]. The critical energy of betatron radiation -rays was estimated to be approximately

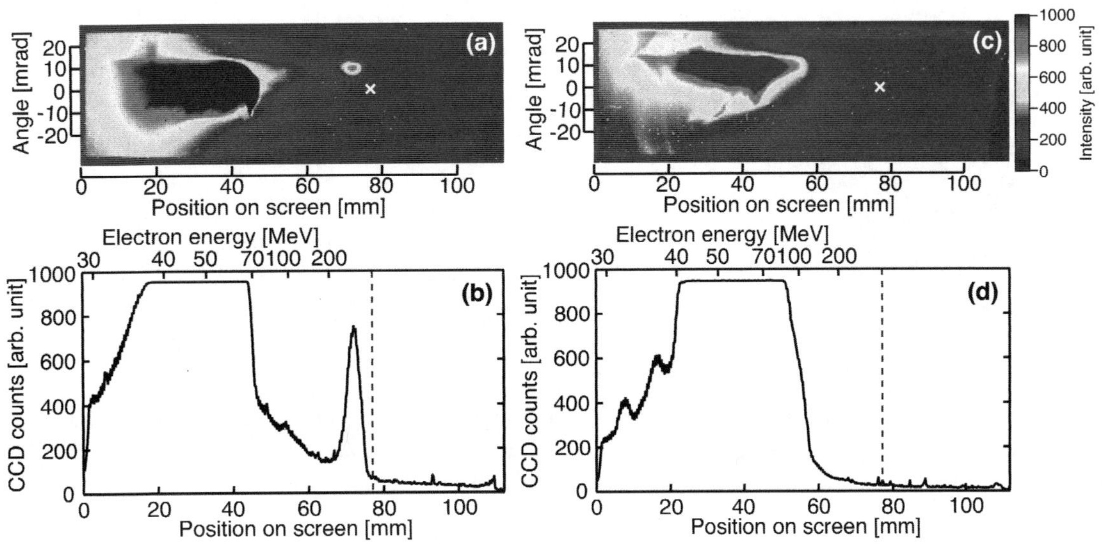


**Figure 2.** (a) Phosphor image and (b) the horizontal line profile of the image, when LCS  -rays were produced. (c) Phosphor image and (d) the horizontal line profile of the image observed without irradiating the colliding pulse.

7 keV for 60 MeV electrons under our experimental conditions. The sensitivity of the detection system to  -rays with energies lower than 7 keV was quite low. As shown in Figs. 2(c)(d), no spot images around the main laser axis were observed without irradiating the colliding pulse. Thus, the spot image in Fig. 2(a) was not the image of betatron radiation  -rays.

## 4. Numerical simulation results

Characteristics of the LCS  -rays were investigated using the simulation code CAIN [15]. Figure 3 shows (a) the  -ray beam pattern and (b) the spectrum of  -ray photons scattered within an angle of 5 mrad. The peak energy, relative full-width at half-maximum (FWHM) energy spread, charge, normalized emittance, and FWHM bunch length of the QME beam were assumed to be 60 MeV, 35%, 70 pC, 0.8 mm mrad, and 3 $\mu$m, respectively. The peak energy, relative FWHM energy spread, and charge were determined by the mean values of QME beam parameters obtained in the another experiment conducted under the same conditions. Other parameters were determined by the particle-in-cell simulation result.

As shown in Fig. 3(a), the  -ray beam pattern was vertically elliptical. In contrast, the observed beam pattern was horizontally elliptical. The reason may be the horizontally elliptical pattern of the QME beam caused by the interaction of the horizontally polarized main pulse with the accelerated electrons [16]. The divergence angles estimated from the $e^{-2}$ intensity radius were 9.3 and 14.1 mrad in the horizontal and vertical directions, respectively. The observed divergence angles were smaller than those of the simulation result. The reason may be the uncertainty in the estimation of the electron energy in the experiment. Errors in the determination of the electron energy were caused by the fluctuation of the incident angle of the electron beam to the magnet. When the emission direction of the QME beam, as well as that of the  -ray beam, was shifted to the left side from the main laser axis as shown in Fig. 2(a), the estimated electron energy was lower than the actual energy. A higher-energy electron beam can produce an  -ray beam with a smaller divergence angle.

The peak energy and photon number of the  -rays scattered within an angle of 5 mrad were 60 keV and $1.8 \times 10^7$, respectively. The photon number was in good agreement with the

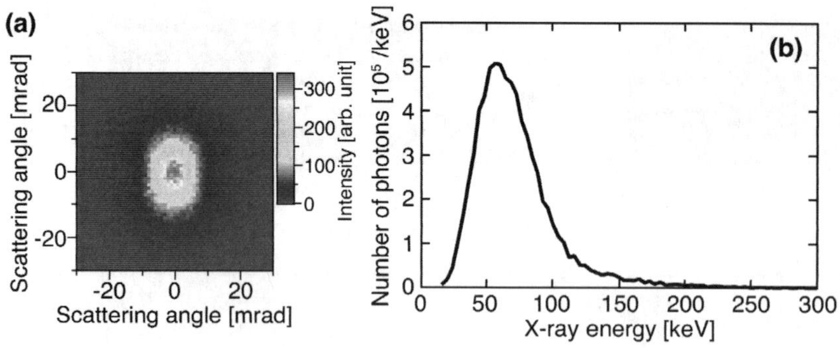


**Figure 3.** (a) -ray beam pattern and (b) spectrum of -ray photons scattered within an angle of 5 mrad calculated by the simulation code.

experimental result. The energy spread of the -rays was large, because the energy spread of the QME beam was large. The peak energy of the -rays was estimated to be 86 keV for 60 MeV electrons. The peak energy of the simulation result was lower than the theoretically predicted value. The -ray energies were shifted to lower energies because of the increase in the effective mass of electrons caused by the nonlinear interaction with the colliding pulse, the normalized vector potential of which was $\sim 0.6$, close to the relativistic intensity. In the -ray spectrum, a long tail was also observed in the high energy range. The tail was formed by the higher order harmonics component generated by the nonlinear interaction. The simulation result shows the effect of nonlinear scattering.

## 5. Summary

We have demonstrated -ray generation via LCS using a laser-accelerated QME beam. A well-collimated -ray beam with a divergence angle of approximately 5 mrad was produced. The -ray photon number was estimated to be $2 \times 10^7$ per pulse. Numerical simulation shows that the peak energy and photon number of the -rays scattered within an angle of 5 mrad were 60 keV and $1.8 \times 10^7$, respectively. The photon number was in good agreement with the experimental result. The numerical simulation result supports the experimental result.

## Acknowledgments

This work was partially supported by the Matsuo foundation and JSPS KAKENHI.

## References

[1] Tajima T and Dawson J M 1979 *Phys. Rev. Lett.* **43** 267
[2] Rousse A et al 2004 *Phys. Rev. Lett.* **93** 135005
[3] Schwoerer H et al 2006 *Phys. Rev. Lett.* **96** 014802, Phuoc K T et al 2012 *Nature Photon.* **6** 308, Mori Y et al 2012 *Appl. Phys. Express* **5** 056401, Chen S et al 2013 *Phys. Rev. Lett.* **110** 155003
[4] Schlenvoigt H -P et al 2008 *Nature Phys.* **4** 130
[5] Miura E et al 2012 *AIP Conf. Proc.* **1507**, ed Zgadzaj R Gaul E and Downer M C ( ew York: AIP) p 304
[6] Mangles S P D et al 2004 *Nature* **431** 535, Geddes C G R et al 2004 *Nature* **431** 538, Faure J et al 2004 *Nature* **431** 541
[7] Miura E et al 2005 *Appl. Phys. Lett.* **86** 251501
[8] Masuda S and Miura E 2008 *Appl. Phys. Express* **1** 086002
[9] Masuda S and Miura E 2009 *Phys. Plasmas* **16** 093105
[10] Miura E and Masuda S 2009 *Appl. Phys. Express* **2** 126003
[11] Masuda S et al 2008 *Rev. Sci. Instrum.* **79** 083301
[12] Glinec Y et al 2006 *Rev. Sci. Instrum.* **77** 103301
[13] Hirayama H et al 2007 The EGS5 code System *SLAC Report: SLAC-R-730* pp 1 - 424
[14] Ikeura-Sekiguchi H et al 2008 *Appl. Phys. Lett.* **92** 131107
[15] Chen P et al 1995 *Nucl. Instrum. Methods* A **355** 107
[16] Mangles S P D et al 2006 *Phys. Rev. Lett.* **96** 215001

# Wave guided laser wake-field acceleration in splash plasma channels

**Yoshio MIZUTA[1], Tomonao HOSOKAI[2,3], Shinichi MASUDA[2,3], Alexei ZHIDKOV[2,3], Nobuhiko NAKANII[2,3], Zhan JIN[2], Hiroki NAKAHARA[1], Tomohiro KOHARA[1], Kenta IWASA[1], Masaki KANDO[4], Sergei BULANOV[4], and Ryosuke KODAMA[1,2,5]**
1) Graduate School of Engineering, Osaka University, Japan
2) Photon Pioneers Center, Osaka University, Japan
3) Japan Science and Technology Agency (JST), CREST
4) Kansai Photon Science Institute, Japan Atomic Energy Agency, Japan
5) Institute of Laser Engineering, Osaka University, Japan
E-mail;mizuta-y@ile.osaka-u.ac.jp

**Abstract.** A transient plasma micro optics (plasma channel and focusing plasma optics-TPMO) in the LWFA provides controllable electron self-injections that result in production of higher quality bunches. In recent study of the TPMO, the deep, straight and short-lived plasma channels [splash plasma channel] were produced by picosecond and sub-picosecond laser pulses in the ponderomotive force dominant regime. Various techniques were used to characterize those channels in argon gas jets irradiated by moderate intensity, ~$10^{15\text{-}16}$ W/cm$^2$, laser pulses with their durations from sub-picoseconds.

## 1. Introduction

We study laser wake-field acceleration (LWFA) for generation of stable ultra-short electron bunches with a low beam divergence and a high charge. Such acceleration is a basis for high-temporal and high-spatial resolution imaging systems as well as the first stage of the ultra-high energy multistage acceleration technique. Using of plasma channels in the wake field acceleration has been shown to provide a controllable self-injection that result in the production of higher quality stable-electron beams. Moreover, the longer guiding of the laser pulses in a plasma channel leads to a higher energy of accelerated electrons. We have demonstrated the process of the formation of short-lived, deep, straight, and controllable plasma channels (splash plasma channels) produced by picosecond and sub picosecond laser pre-pulses in argon gas jets [1, 2]. To understand the guiding ability of splash channels, particle-in-cell (PIC) simulations which include the ionization were performed. Stabile quasi mono-energetic electron beams with their parameters determined by a splash channel structure were generated. Splash channels may be useful for the wave guided LWFA. We are developing a laser system which has double laser pulses for the channel-guided LWFA in Photon Pioneers Center, Osaka University. The channel-guided LWFA consists of two stages: (i) a splash plasma channel produced by a sub-picosecond laser pulse and (ii) the acceleration of electrons by a second, femtosecond high intensity (> $10^{19}$ W/cm$^2$) laser pulse. In this paper, we present our experimental plans and evaluation of the channel-guided LWFA.

## 2. Experimental Setup

The experiment was performed at JLITE-X 800nm Ti: Sappier CPA laser system at the JAEA-KPSI (Kansai Photon Science Institute, Japan Atomic Energy Agency) [3, 4]. An outline of the experimental setup is given in Fig. 1. A laser pulse with a diameter 30 mm is focused to the position of the ~100 μm front edge of the slit gas jet height of 1.5 mm form nozzle exit with f/5.9 off-axis parabolic mirror. In order to study the effects of the TPMO formation by the laser pulse, we controlled the nanosecond order laser pre-pulse by tuning the Pockels cell behind the regenerative amplifier of the laser system [4]. We also controlled the pulse durations from 300 fs to ~6 ps by turning the distance of the grating pair in the pulse compressor, to observe the effect of pulse duration. The focal spot size was 16 μm in full width at $1/e^2$ of maximum with 186 μm Rayleigh length. The maximum laser intensity on the target was estimated to be $3.5\times10^{15}$ W/cm$^2$ .We used an argon gas jet target, which was provided through a supersonic nozzle by a pulse valve (Smartshell Co.Ltd). The stagnation pressure of the pulse valve was 2.00 MPa. The neutral gas density is estimated to be $1.5\times10^{19}$ cm$^{-3}$. To observe the formation process of a TPMO, we used time-resolved interferometer. The wavelength of the probe laser was 400 nm, which was produced with a 2 mm-thickness BBO (barium metaborate) crystal. The interferograms were taken by a CCD camera.

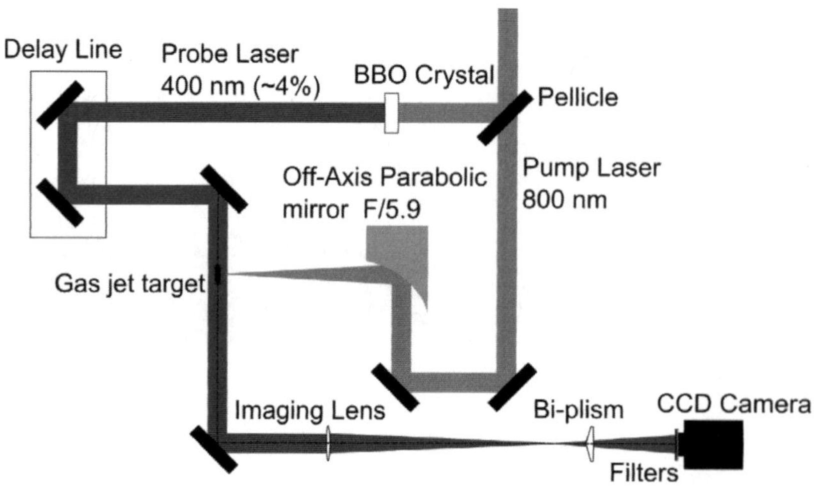


Fig. 1. Experimental setup

## 3. Experimental results

Fig.2 (a), (b) (left) show interferograms taken during the propagation of a CPA pulse of 300 fs through the gas jets. Laser pulse comes from the left side. The corresponding electron density distributions are also shown in Fig.2 (a), (b) (right). Time delay between main pulse and probing pulse, which is about 9 ps, was chosen to illustrate the most important stages of channel formation and relaxation. The electron density distribution was calculated from the phase shift. An IDEA code was used to obtain a phase shift map. A fast Fourier transform technique was used to retrieve the plasma-imposed phase shift. The phase-shift map is processed then with an Abel inversion algorism [5].

The maximum channel length was 600 μm. It was three times the Raleigh length. The channel diameter is about 15 μm. The maximal electron density was $4.9\times10^{19}$ cm$^{-3}$ and the minimum electron density was $1.0\times10^{18}$ cm$^{-3}$. The minimum electron density in the channel is almost 50 times as lower as the maximal.

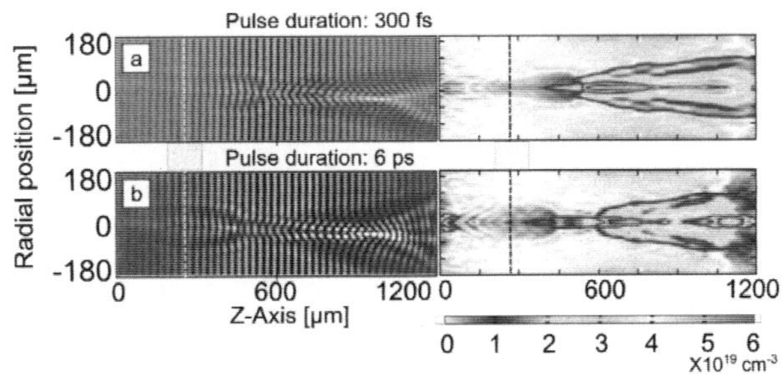


Fig. 2. Plasma channel
((a)The pulse energy 90 mJ and pulse duration 300 fs , (b)The pulse energy 90 mJ and pulse duration 6 ps)

## 4. PIC simulation of LWFA in a splash channel

To understand usability of the channels for the laser wake-field acceleration, we perform 2D particle-in-cell (PIC) simulations which included the optical field ionization of plasma. We use the variable particle weight approach [7-9] to calculate the plasma ionization.

PIC simulations are performed for a linearly polarized laser pulse propagating in Ar under 1/2 of the normal pressure or $N_0$=1.5x10$^{19}$ cm$^{-3}$, and for the laser, intensity $I$=3x10$^{19}$ W/cm$^2$, $\lambda$=0.8 µm, and 40 fs FHWM duration. The $s$-polarized laser pulse is numerically focused in diameter $D$=16 µm with the focal length, $f$~150 µm. The initial density profile is chosen to emulate the density profile from the experiment with 6 ps laser pulse (see chapter 3). While the channel diameter is kept constant $D$=37 µm, the channel depth is a parameter. In the PIC simulation we use the code FPLaser2D [10] exploiting the moving window technique. We use a 320 µm x150µm window and the spatial resolution $\lambda$/10; the kinetic cell is 2 times as large as the PIC one.

Electron density distribution for two channels with different depth is shown in Fig. 3 a, b. One can see that the ionization of higher density plasma periphery results in an essential narrowing of the splash channel. Therefore, the initial plasma structure cannot completely determine the electron self-injection and to provide the controllable self-injection at the long pulse guiding ionization effect have to be taken into account. Nevertheless finally, the electron acceleration can be controlled by the splash channel structure.

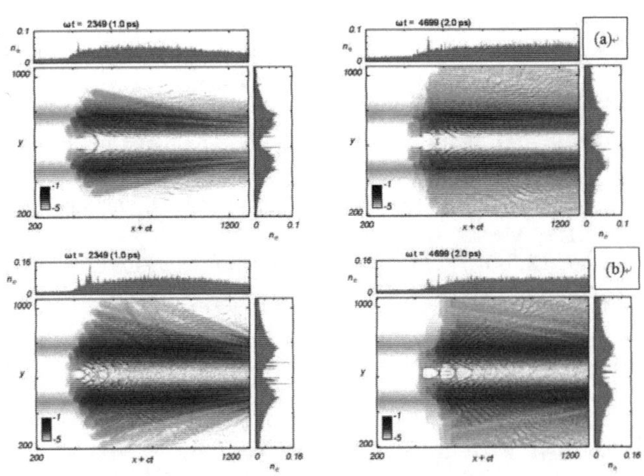


Fig. 3 Spatial distribution of electron density distribution in plasma channels with the different depths:
(a) Maximum electron density is 3 x10$^{19}$ cm$^{-3}$ and minimum density is 0.5 x10$^{18}$ cm$^{-3}$.
(b) Maximum electron density is 3 x10$^{19}$ cm$^{-3}$ and minimum density is 2.0 x10$^{18}$ cm$^{-3}$.

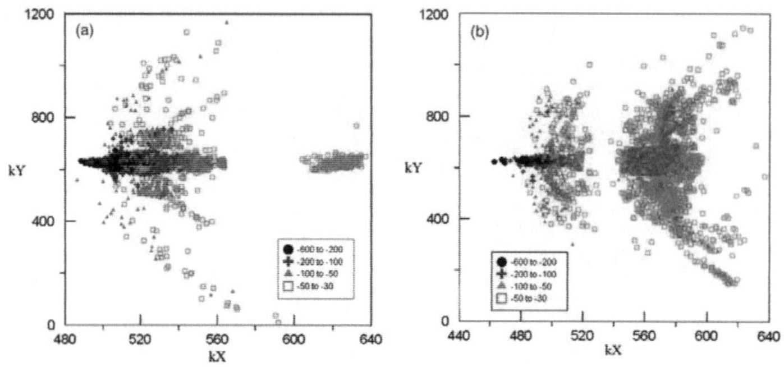


Fig. 4 Spatial distribution of accelerated electron momenta in two plasma channels: (a) minimal electron density is $0.5 \times 10^{18}$ $cm^{-3}$; (b) minimal electron density is $2.0 \times 10^{18}$ $cm^{-3}$.

In Fig.4 spatial distribution of momenta of accelerated electrons for two different plasma channels are given. One can see that in the shallower channel the maximal energy of electrons is expectedly lower than that in the deepest channel. The geometrical emittance and energy spread are also poor in the shallow channel. That means the formation of splash channel may result in the quality of the electron beams and their total charge. The energy spread also depends on the channel depth: it decreases from 30% for the shallow splash channel down to 8% for the deep channel.

## 5. Conclusion

We have analysed experimentally and numerically the process of formation of short-lived plasma channels (splash channels), produced by picosecond order laser pulses in argon gas jets and its result in the laser wake field acceleration. The quantitative measurements of splash-plasma channel parameters have been performed with the use of the picosecond time-resolved interferometry. In the present experiments we could produce a splash channel with its length ~ 600 μm, which is 3 times longer than the Rayleigh length. We anticipate that far longer channels with their lengths of the order of several mm can be created by this technique. The results of the calculation have confirmed the advantage of the channel-guided LWFA in the generation of the high quality, repeatable electron beams.

## References

[1] T. Hosokai, A. Zhidkov, A. Yamazaki, Y. Mizuta, M. Uesaka, and R. Kodama, Appl. Phys. Lett. **96**, 121501 (2010)

[2] Y. Mizuta, T. Hosokai, S. Masuda, A. Zhidkov, K. Makito, N. Nakanii, S. Kajino, A. Nishida, M. Kando, M. Mori, H. Kotaki, Y. Hayashi, S. V. Bulanov, and R. Kodama, Phys. Rev. ST Accel. Beams **15**, 121301 (2012)

[3] M. Mori, A. Pirozhkov, M. Nishiuchi, K. Ogura, A. Sagisaka, Y. Hayashi,S. Orimo, A. Fukumi, Z. Li, M. Kado, and H. Daido. LASER Phys. **16**, 1092(2006)

[4] M.Mori, *et al* IEEE TRANSACTIONS ON PLASMA SCIENCE, VOL. **36**, NO.4 AUGUST 2008

[5] H. Kotaki, M. Kando, T. Oketa, S. Masuda, James K. Koga, S. Kondo, S. Kanazawa,T. Yokoyama, and T. Matoba. Phys. Plasmas **9**, 1392 (2002)

[7] A. Zhidkov, T. Esirkepov, T. Fujii, K. Nemoto, J. Koga, S.V. Bulanov, Phys. Rev. Lett. **103**, 215003 (2009)

[8] A. Zhidkov, J. Koga, T. Esirkepov, T. Hosokai, M. Uesaka, T. Tajima, Phys. Rev. E **69**, 066408 (2004)

[9] A. Zhidkov and A. Sasaki, Phys. Plasmas **7**, 1341 (2000)

[10] R. R. Freeman, P. H. Bucksbaum, H. Milchberg, S. Darack, D. Schumacher, and M. E. Geusic, Phys. Rev. Lett. **59**, 1092 (1987)

Journal of Physics: Conference Series **688** (2016) 012070 doi:10.1088/1742-6596/688/1/012070

# Upgrade of repetitive fast-heating fusion driver HAMA to implode a shell target by using diode pumped solid state laser

**Yoshitaka MORI[1], Takashi SEKINE[2], Osamu KOMEDA[3], Yasuhiko NISHIMURA[1,4], Atsushi SUNAHARA[5], Eisuke MIURA[6], Suisei Nakayama[1], Ryohei HANAYAMA[1], Katsuhiro ISHII[1], Nakahiro SATO[2], Takashi KURITA[2], Toshiyuki KAWASHIMA[2], Hirofumi KAN[2], Naoki NAKAMURA[3], Takuya KONDO[3], Manabu FUJINE[3], Hirozumi AZUMA[7], Tatsumi HIOKI[7], Mitsutaka KAKENO[7], Tsutomu KAJINO[7], Tomoyoshi MOTOHIRO[7], Yasuhiko SENTOKU[8], and Yoneyoshi KITAGAWA[1]**

[1]The Graduate School for the Creation of New Photonics Industries, Hamamatsu, Shizuoka, 431-1202, Japan
[2]Hamamatsu Photonics, K. K., Hamamatsu, Shizuoka, 431-1202, Japan
[3]Advanced Material Engineering Div., TOYOTA Motor Corporation, Susono, Shizuoka, 410-1193, Japan
[4]Toyota Technical Development Corp., Toyota, Aichi, 470-0334, Japan
[5]Institute for Laser Technology, Nishi-ku, Osaka, 550-0004, Japan
[6]National Institute of Advanced Industrial Science and Technology, Tsukuba, Ibaraki, 305-8568, Japan
[7]TOYOTA Central Research and Development Laboratories, Inc., Nagakute, Aichi, 480-1192, Japan
[8]Department of Physics, University of Nevada, Reno, Nevada, 89557, USA

E-mail: ymori@gpi.ac.jp

**Abstract.** The HAMA is 1-Hz fast heating fusion driver pumped by a 10 J second-harmonic of diode-pumped Nd:glass laser: KURE-1. We have upgraded HAMA to realize an implosion of spherical shell target by using a remaining fundamental beam from KURE-1. This beam of 6 J/1 Hz is transported to the current counter irradiation system. The resulting beam includes three pulses in sequence: 2.2 J/15 ns and 0.7 J/300 ps for implosion, and 0.5 J/ 190 fs for heating. We estimate the implosion dynamics from 1-D radiation hydrodynamic code (START-1D). It indicates a possibility of tailored-pulse implosion by optimizing the beam spot sizes of imploding beams on the target surface. This upgrade leads to a demonstration of repetitive implosion and additional heating of a spherical shell target in accordance with a repetition of laser operation and that of a target injection system.

## 1. Introduction

Progress toward fusion ignition is kept going on the National Ignition Facility (NIF) by using 1.8 MJ, 500 TW single shot laser[1]. Following an effort of this scientific proof of ignition, engineering development needs to start on inertial fusion energy (IFE) by using a repetitive fusion driver.

A diode-pumped solid-state laser (DPSSL) is a promising candidate as a reactor driver for IFE because it can be operated at a high-repetition rate (>10 Hz) with high efficiency(>10 %)[2].

We have developed a repetitive fusion driver HAMA[3] by using a 10 J green DPSSL based on a road map[4]. This fusion driver has been realizing repetition-mode fusion experiments such as DD fusion reaction[5, 6], a compact fast core heating[7], and laser engagement of injected flying pellets[8, 9]. The HAMA is a Ti:sapphire laser pumped by a second-harmonic of diode-pumped Nd:glass laser: KURE-I[10]. A 3.8-J, 0.3-ns amplified chirped pulse is divided into four beams: two counter-irradiate a target, and the remaining two are pulse-compressed to 190 fs for heating the imploded target. HAMA succeeded in a compact fast core heating experiment by using double-foil deuterated polystyrene target with 11 $\mu$m thick and gap separation of 100 $\mu$m[7]. A beam energy of 1 J for implosion limits this target gap. By using a remaining fundamental beam from KURE-I of 6 J, we can approach to an imploding a spherical shell target of $\phi$ 500 $\mu$m with 7 $\mu$m thick, which is commonly used for ICF experiments[11] of kJ class single shot laser. By using 1-D radiation hydrodynamics code: STAR-1D[12], we have simulated a performance of an imploded core formation with HAMA laser including the remaining fundamental beam from KURE-1. This upgrade enables a repetitive implosion and additional heating of a spherical shell target in accordance with a repetition of laser operation and that of a target injection system. This paper describes an improvement of HAMA to accomplish the spherical shell target irradiation.

## 2. Laser configuration of HAMA upgrade

Figure 1 shows a diagram of HAMA including DPSSL irradiation. The HAMA is a Ti:sapphire laser pumped by a second-harmonic of diode-pumped Nd:glass laser: KURE-I[10]. It consists of a seed laser called BEAT[13] and a pump laser called KURE-I. The BEAT is a 10 Hz Ti : sapphire optical parametric chirped pulse amplification (OPCPA) system. KURE-I is a green 10 J/10 Hz DPSSL system based on a water cooled Nd : glass zigzag path slab amplification scheme [8].

In KURE-I, second-harmonic generation through $CsLiB_6O_{10}$ nonlinear crystal provides two beams; a second-harmonic:$2\omega$ of 8 J, and fundamental:$\omega$ of 6 J. Both beams are transported to a HAMA breadboard and split by a beam splitter. The $2\omega$ is then transported to the HAMA amplifier as a pumping beam to amplify the seed pulse from BEAT at a repetition of 1 Hz. In relation to the seed pulse, the amplified seed pulse of 3.8 J, 0.3 ns output is divided into two beams by the L-S beam splitter. One is "L-beam" used for implosion of shell target, the remaining beam is pulse-compressed to 190 fs results in "S-beam" used for heating the ablated target.

The $2\omega$ beam from KURE-1 arrives on the Ti:sapphire at a timing 12.4 ns in advance to the seed pulse from BEAT[13, 14]. The $\omega$, "K-beam", is transported toword the KURE-1 bread board to make a delay, resulting in an irradiation timing of 6 ns in advance to the S-beam. This K-beam is introduced into the L beam line by using the K-L beam combiner. The divergence of K-beam is controlled by a convex lens mounted on a K beam image relay system. Three beams, K, L, and S are split and combined by a 50:50 beam splitter/combiner in the compression chamber, then transported to the irradiation chamber as BEAM-1 and BEAM-2. The resulting beam of BEAM-1 or BEAM-2 includes three pulses in sequence : 2.4 J/15 ns, 0.55 J/300 ps, and 0.40 J/190 fs, respectively.

Figure 2 shows pulse shape in sequence and beam spot profiles of BEAM-1 and BEAM-2. Timing control between K and S is electrical synchronization, L and S is optical synchronization. All of six beams are focused on the chamber center to implode and heat a spherical shell target of $\phi$ 500 $\mu$m with 7 $\mu$m thick. The beam focusing alignments are accomplished simultaneously for OAP1 and OAP2 by using a counter beam focusing monitor to minimize the spot size of short beams, respectively. The resulting beam spot size ($1/e^2$) is 22 $\mu$m, 90 $\mu$m and 11 $\mu$m

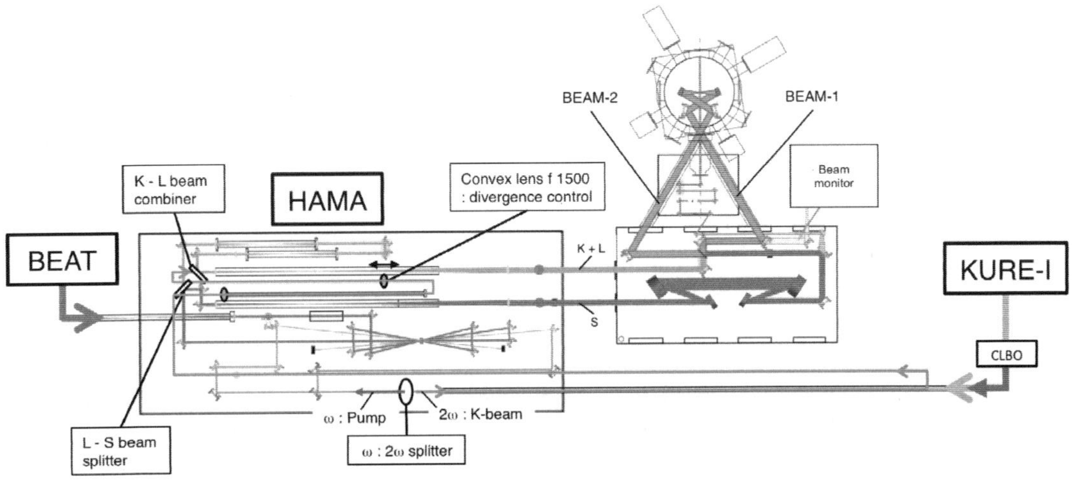


**Figure 1.** Diagram of HAMA including DPSSL irradiation.

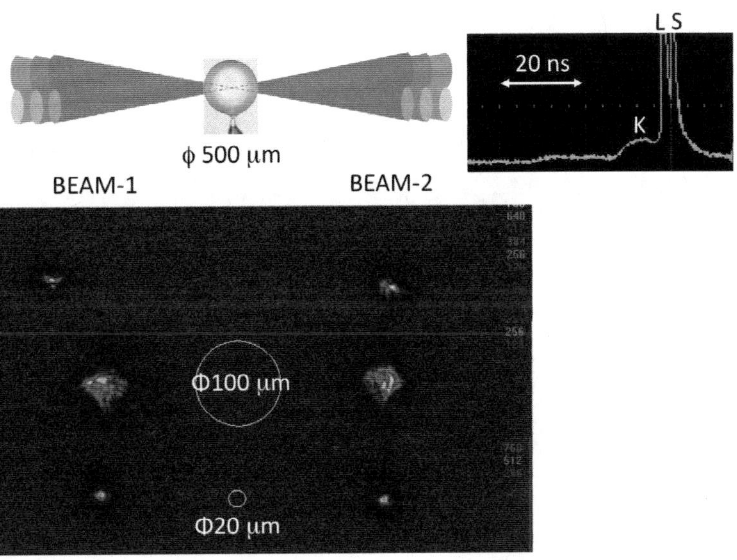


**Figure 2.** Pulse shape in sequence (upper right) and beam spot profiles (bottom).

for K, L, and S, respectively. For K-beam, spot size on the shell target surface is 100 $\mu$m in square. The spot size of L is 8 times larger than that of S, which indicates further improvement of an alinement technique or beam quality is requried in the future. Beam characteristics of there beams K, L and S includes pulse duration, energy, and intensity are given in Table. 1. These intensities are values on the focusing center. K and L-beams are used for implosion, and S-beams is for heating, respectively.

**Table 1.** Four beams energy and intensity on target

| | Imploding beam | | Heating beam |
|---|---|---|---|
| **Beam** | **K** | **L** | **S** |
| Pulse width (FWHM) | 15.2 ns | 300 ps | 192 fs |
| Energy (J) | 2.2 | 0.70 | 0.50 |
| Intensity (W/cm$^2$) | 1.9 $\times 10^{13}$ | 6.0 $\times$ $10^{13}$ | 4.5 $\times$ $10^{18}$ |

## 3. Implosion dynamics from STAR-1D

Implosion dynamics for K and L irradiation is evaluated from 1-D radiation hydrodynamics code: STAR-1D[12] by using the laser parameters given in Table 1. Figure 3 represents a hydrodynamic flowchart (radius-time diagram) for K-beam focusing (a) on center and (b) 540 $\mu$m in focus. In Fig. 3 (a), a foot pulse of K beam makes shell implosion and core formation before the main pulse of K beam arrives. This main pulse gives re-compression of expanding core and L beam in sequence provides another core formation. By contrast, as shown in Fig. 3 (b) where the beam spot size on the target surface is 2.5 times larger than that of the case (a), the main pulse of K-beam implodes the shell and L pulse in sequence compress the core resulting in a tailored pulse compression in similar. In Fig. 3 (b), the beginning of pulse " K " ablatively implode the shell. In sequence, the " L" at 38 ns drives a shock to implode the target inward, resulting in the implosion at 40.5 ns. The " heating " pulse, " S " , would be irradiated at this timing. We will scan an irradiation timing of S-beam with scanning range of 2 ns in experiments.

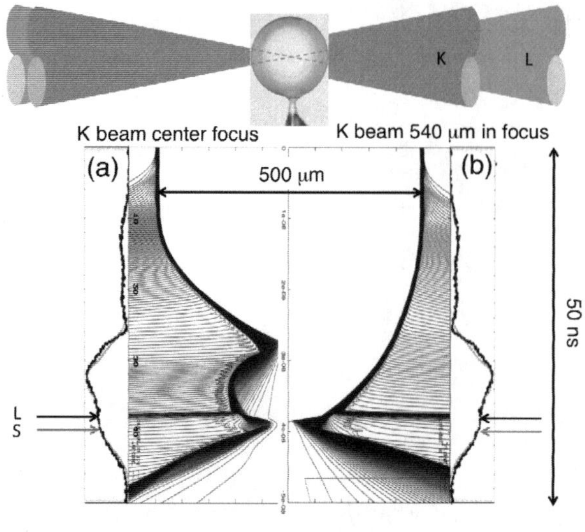


**Figure 3.** Hydrodynamic flowchart by START-1D.

## 4. Conclusions

We improved the HAMA performance to realize spherical shell target implosion and fast heating. The resulting focus intensity of three pulses in sequence is 1.9 $\times$ $10^{13}$ W/cm$^2$, 6 $\times$ $10^{13}$ W/cm$^2$, and 4.5 $\times$ $10^{18}$ W/cm$^2$, respectively. In these intensities, the hydrodynamics simulation indicates implosion of spherical shell target of $\phi$ 500 $\mu$m with 7 $\mu$m thick and formation of the imploded core at the center.

## Acknowledgments
Y. M. appreciate for Y. Takeuchi and Y. Hatano for their laser operation and S. Suita for target setting.

## References
[1] Moses E I *et al* 2009 *Phys. Plasmas* **16** 041006
[2] Orth C D, Payne S A and Krupke W F 1996 *Nucl. Fusion* **36** 75
[3] Mori Y *et al* 2013 *Nucl. Fusion* **53** 073011
[4] Kitagawa Y *et al* 2013 *Plasma Fusion Res.* **8** 2404000
[5] Kitagawa Y *et al* 2011 *Plasma Fusion Res.* **6** 1306006
[6] Komdeda O *et al* 2013 *Fusion Sci. Technol.* **63** 296
[7] Kitagawa Y *et al* 2012 *Phys. Rev. Lett.* **108** 155001
[8] Komeda O *et al* 2013 *Plasma Fusion Res.* **8** 1205020
[9] Komeda O *et al* 2013 *Sci. Reports* **3** 2561
[10] Sekine T *et al* 2013 *Opt. Express* **21** 8393
[11] Kitagawa Y *et al* 2005 *Phys. Rev. E* **71** 016403
[12] Sunahara A *et al* 2008 *Plasma Fusion Res.* **3** 043
[13] Mori Y and Kitagawa Y 2013 *Appl. Phys. B* **110** 57
[14] Mori Y *et al* 2007 *Int. J. Mod. Phys. B* **21** 572

# Proton imaging of an electrostatic field structure formed in laser–produced counter–streaming plasmas

**T. Morita[1], N. L. Kugland[2], W. Wan[3], R. Crowston[4], R. P. Drake[3], F. Fiuza[2], G. Gregori[5], C. Huntington[2], T. Ishikawa[6], M. Koenig[7], C. Kuranz[3], M. C. Levy[2,8], D. Martinez[2], . Meinecke[5], F. Miniati[9], C. D. Murphy[10], A. Pelka[7], C. Plechaty[2], R. Presura[11], N. Quirós[11], B. A. Remington[2], B. Reville[12], . S. Ross[2], D. D. Ryutov[2], . Sakawa[13], L. Steele[14], H. Takabe[13], . amaura[6], N. Woolsey[4], H.-S. Park[2]**

1. Department of Energy Engineering Science, Faculty of Engineering Sciences, Kyushu University, 6-1 Kasuga-koen, Kasuga, Fukuoka 816-8580, Japan
2. Lawrence Livermore ational Laboratory, P.O. Box 808, Livermore, California 94550, USA
3. Department of Atmospheric, Oceanic, and Space Sciences, University of Michigan, Ann Arbor, MI 48109, USA
4. University of York, York, orth Yorkshire YO10 5DD, United Kingdom
5. Department of Physics, University of Oxford, Parks Road, Oxford OX1 3PU, UK
6. Graduate School of Science, Osaka University, 1-1 Machikane-yama, Toyonaka, Osaka 560-0043, Japan
7. Laboratoire pour l'Utilisation des Lasers Intenses (LULI), École Polytechni ue-Univ, Paris VI, 91128 Palaiseau, France
8. Rice University, Houston, Texas 77251, USA
9. Physics Department, Wolfgang-Pauli-Strasse 27, ETH-Zürich, CH-8093 Zürich, Switzerland
10. University of Edinburgh, Edinburgh, EH8 9YL, United Kingdom
11. University of evada, Reno, evada 89557, USA
12. Centre for Plasma Physics, Queen's University Belfast, University Rd, Belfast BT7 1 , UK
13. Institute of Laser Engineering, Osaka University, 2-6 Yamada-oka, Suita, Osaka 565-0871, Japan
14. University of California, Davis, California 95616, USA

E-mail: morita@aees.kyushu-u.ac.jp

**Abstract.** We report the measurements of electrostatic field structures associated with an electrostatic shock formed in laser produced counter streaming plasmas with proton imaging. The thickness of the electrostatic structure is estimated from proton images with different proton kinetic energies from 4.7 MeV to 10.7 MeV. The width of the transition region is characterized by electron scale length in the laser produced plasma, suggesting that the field structure is formed due to a collisionless electrostatic shock.

## 1. Introduction

Shock waves are commonly observed in the universe, space plasmas, and laboratory plasmas. In particular, collisionless shocks play significant roles in particle acceleration, for example, in Earth's bow shock and supernova remnant shocks. In collisionless shocks, collisions between particles do not account for the formation mechanism and particle–field interactions are essential.

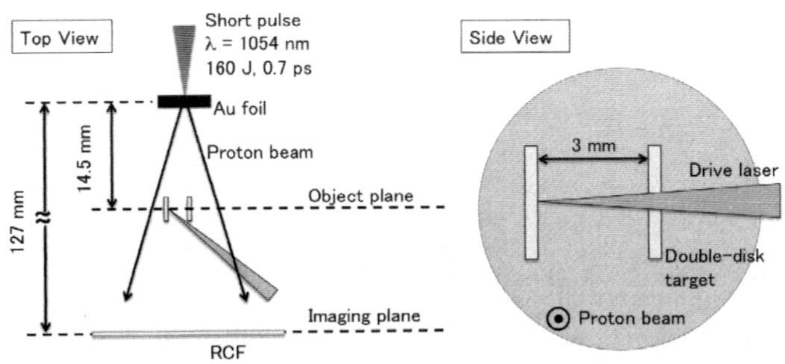


**Figure 1.** The target design and geometry of diagnostics for proton imaging. Two Mylar (CHO) disks were located in parallel separated by 3 mm, and one side of the target was irradiated by the drive laser. The laser-produced plasmas were diagnosed by a proton beam generated by focusing a short-pulse laser on an Au disk placed at 14.5 mm away from the plasma. The protons were detected by the stack of radiochromic films (RCFs) located at 127 mm from the proton source.

Therefore, the shock thickness is much smaller than ion-ion mean-free-path and a large electromagnetic field exists at the shock front. These shocks have been studied by observing emissions from astrophysical objects[1], and by in-situ measurements using satellites in space plasmas[2].

Laboratory experiments with high-power laser systems can be an alternative to observations or in-situ measurements by satellites[3]. Collisionless shocks have been produced and investigated in counter-streaming laser-produced plasmas[4]. They have been measured by optical diagnostics such as interferometry, shadowgraphy, optical pyrometry, and Thomson scattering to obtain the fundamental plasma parameters: density, temperature, charge state, and flow velocity[5]. However, the measurement of an electromagnetic field is indispensable for collisionless shock experiments.

Proton imaging is one of the methods to measure the field structures[6]; it is widely used for measuring the electric and magnetic fields in inertial confinement fusion experiments[7] and the field in laser-produced plasmas[8].

In this paper, we present the measurement of electric field structures formed at a shock with a proton beam generated by irradiating a thin solid foil by a short-pulse laser beam.

## 2. Experiment

The experiment was performed with the Jupiter laser facility at Lawrence Livermore National Laboratory. Double-disk Mylar (CHO) targets are used to produce counter-streaming plasmas as shown in Fig. 1. Each disk measures 2 × 2 mm and 0.5 mm in thickness, and two disks are separated by 3 mm. One of the disks (drive disk) was illuminated by a laser beam (drive laser), delivering 120 J on average in a 1.5 ns square pulse at a wavelength of 527 nm, with a focal spot of 500 $\mu$m in diameter. The other disk (second disk) was ablated by the radiation from the laser-produced plasma early in time, and by the plasma arriving later in time as described in the Refs. [4].

A laser pulse (160 J in 0.7 ps, wavelength of 1057 nm) was focused on an Au disk with a thickness of 25 $\mu$m to produce high-energy protons for radiography. The distance between the disk and the object plane, where the target was located, was $l = 14.5$ mm. The proton beam images electric and magnetic fields formed in the laser-produced plasmas on radiochromic films located at an imaging plane with a distance $L + l = 127$ mm from the Au disk, where $L$ is the distance between the object and the radiochromic films. The radiochromic films were stacked to detect protons in different energy ranges: 4.7, 7.0, 8.8, and 10.7 MeV with 10% energy width for each film.

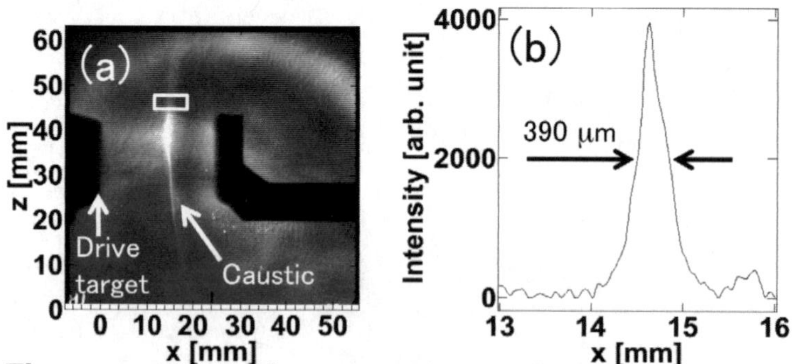


**Figure 2.** (a) The proton image taken at $t = 20$ ns at the proton energy of 4.7 MeV. (b) The line–out at $x \sim 15$ mm [square in (a)].

## 3. Result

Figure 2(a) shows the proton image at $t = 20$ ns at the proton energy of 4.7 MeV. A caustic structure is observed at $x \sim 15$ mm in front of the second disk. In general, this sudden change in proton intensity indicates a large electromagnetic field[6]. The line–out of the caustic for Fig. 2(a) is shown in Fig. 2(b). The widths of the caustics depend on the proton energy, i.e., high–energy protons make a thinner structure as shown in Fig. 3.

## 4. Discussion

The observed caustic structure corresponds to a shock front as discussed in previous works[4], and the left– and right–hand sides of this structure are upstream and downstream of a shock, respectively. Here, we assume that the protons are deflected by an electric field not by a magnetic field. If the shock is formed by a self-generated magnetic field due to electromagnetic Weibel instability, the shock thickness should be much larger ($c/\omega_{\rm pi} > 10$ mm [9]) than experimental observation. Also, a spherically symmetric flat–top potential is assumed in the downstream region at the vicinity of a shock.

The position $x$ on the imaging plane is expressed using the position $x_0$ on the object plane and the deflection angle $\alpha(x_0)$ as $x = L\,(x_0/l + \alpha(x_0))$[6]. The deflection angle $\alpha$ of the proton beam by a spherically symmetric flat–top potential is expressed as $\alpha(x_0) = (e\phi_0 x_0/2W\sqrt{a\delta})F(\tilde{\xi}) = (e\phi_0(\tilde{\xi}\delta + a)2W\sqrt{a\delta})F(\tilde{\xi})$[6], where $\tilde{\xi} = (x_0 - a)/\delta$, $\delta$ is the transition thickness, $a$ is the curvature radius of the potential structure, $W$ is the proton energy, $e\phi$ is the potential energy, and $F(\tilde{\xi}) = (2/\pi)\int_{-\infty}^{\infty} d\eta/((\eta^2 + \tilde{\xi})^2 + 1)$. In case of caustic formation on the imaging plane, $\partial x/\partial x_0 = 0$ is satisfied. Therefore, $1/l + (e\phi_0/2W\delta\sqrt{a\delta})[\delta F(\tilde{\xi}) + (\tilde{\xi}\delta + a)F'(\tilde{\xi})] = 0$. Generally, this equation has two solutions ($\tilde{\xi}_1$ and $\tilde{\xi}_2$), however, the caustic positions $x(\tilde{\xi}_1)$ and $x(\tilde{\xi}_2)$ can not be resolved in this experiment because of small separation between them, and these two caustics are observed as a single peak in the proton intensity. Therefore, the thickness of an electric field at the shock front ($\delta$) is comparable to

$$\delta \approx |x(\tilde{\xi_2}) - x(\tilde{\xi_1})|. \tag{1}$$

The thicknesses of the caustics in different energy bins are shown in Fig. 3 obtained as full widths at half maximums of the caustics for the energies of 4.7, 7.0, 8.8, and 10.7 MeV. The experimental data are fitted with Eq. (1) for various thicknesses. If this caustic is formed due to an electrostatic shock, the thickness is characterized by electron inertial length $c/\omega_{pe} \sim 2$ $\mu$m [$n_e \sim 5 \times 10^{18}$ cm$^{-3}$ evaluated from the interferometry (not shown)][9]. The solid and dotted lines show the fitted results with $\delta = 0.01$ $\mu$m $\sim \lambda_D$, where $\lambda_D$ is the Debye length, and 0.2 $\mu$m $\sim c/\omega_{pe}$, respectively. The fitted result with larger thickness of $\delta = 10$ $\mu$m (dotted line) shows

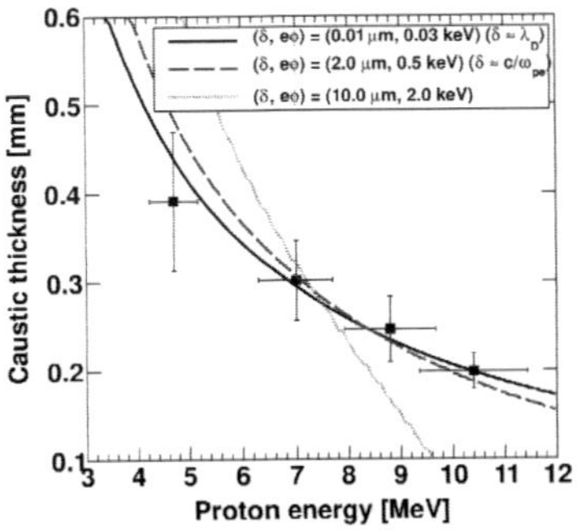


**Figure 3.** Caustic thicknesses as a function of proton energy. The thickness is evaluated as a full width at half maximum of each line–out as shown in Fig. 2. The best–fit results using various thickness $\delta$ are shown with solid ($\delta = 10$ nm $\sim \lambda_D$), dashed ($\delta = 2.0$ $\mu$m $\sim c/\omega_{pe}$), and dotted ($\delta = 10$ $\mu$m) lines.

large deviation from the experimental data, while the results from small thickness $\delta < c/\omega_{pe}$ shows good agreement with the experimental data, indicating that the structure is defined in an electron scale. Note that the ion inertial length is over hundreds of micron and is much larger than the thickness estimated here. Moreover, the ion–ion mean–free–path of counter–streaming plasmas is $\sim$200 $\mu$m[4] using the velocity of $v = 3$ mm/20 ns $= 150$ km/s. This small thickness of the electric field structure suggests that the caustic is formed by an electrostatic shock in collisionless counter–streaming plasmas.

## 5. Summary

We measured an electrostatic field structure formed in counter–streaming laser–produced plasmas with proton imaging technique. The thickness of the electrostatic structure is estimated from proton images with different proton kinetic energies from 4.7 MeV to 10.7 MeV. The width of the transition region is comparable to or less than the electron inertial length of laser–produced plasmas, suggesting that the field structure is formed due to a collisionless electrostatic shock.

## Acknowledgments

The authors acknowledge the dedicated technical support by the staff at Jupiter laser facility for the laser operation and diagnostics. This work was performed for the U.S. Department of Energy by the Lawrence Livermore National Laboratory, under Contract No. DE-AC52-07NA27344. Support was provided by LLNL LDRD Grant No. 11-ERD-054 and Japan Society for the Promotion of Science (JSPS).

## References

[1] Bamba A, Yamazaki R, Ueno M and Koyama K 2003 *The Astrophysical Journal* **589** 827 837; Aharonian F A, *et al* 2004 *Nature* **432** 75 7; Uchiyama Y, Aharonian F A, Tanaka T, Takahashi T and Maeda Y 2007 *Nature* **449** 576 8;

[2] Wenzel K, Reinhard R, Sanderson T and Sarris E 1985 *Journal of Geophysical Research* **90** 12 18; Sarris E T, Anagnostopoulos G C and Krimigis S M 1987 *Journal of Geophysical Research* **92** 12083 Anagnostopoulos G C 1994 *Physica Scripta* **T52** 142 151;

[3] Remington B, Drake R and Ryutov D 2006 *Reviews of Modern Physics* **78** 755 807

[4] Morita T *et al Physics of Plasmas* **17** 122702; Kuramitsu Y *et al* 2011 *Physical Review Letters* **106** 175002

[5] Ross J S *et al* 2006 *Review of Scientific Instruments* **77** 10E520; Ross J S *et al* 2012 *Physics of Plasmas* **19** 056501

[6] Kugland L *et al* 2012 *The Review of scientific instruments* **83** 101301

[7] Mackinnon A *et al* 2006 *Physical Review Letters* **97** 045001; Rygg J R *et al* 2008 *Science (New York, N.Y.)* **319** 1223 5; Li C K *et al* 2012 *Physical Review Letters* **108** 025001

[8] Borghesi M *et al* 2002 *Physics of Plasmas* **9** 2214; ilson P *et al* 2006 *Physical Review Letters* **97** 255001; Romagnani L *et al* 2008 *Physical Review Letters* **101** 025004; Kugland L *et al* 2012 *Nature Physics* **8** 809 812

[9] Kato T and Takabe H 2010 *Physics of Plasmas* **17** 032114; Kato T *et al* , 2008 *The Astrophysical Journal Letters* **681** 93-96

# Density filament and helical field line structures in three dimensional Weibel-mediated collisionless shocks

**Toseo Moritaka[1], Youichi Sakawa[2], Yasuhiro Kuramitsu[1], Taichi Morita[3], Yuta Yamaura[2], Taishi Ishikawa[2] and Hideaki Takabe[2]**

[1]Department of Physics, National Central University, Taoyuan City, Taiwan
[2]Institute of Laser Engineering, Osaka University, Osaka, Japan
[3]Department of Energy Engineering Science, Kyushu University, Fukuoka, Japan

E-mail: toseo@ncu.edu.tw

**Abstract.** Collisionless shocks mediated by Weibel instability are attracting attention for their relevance to experimental demonstrations of astrophysical shocks in high-intensity laser facilities. The three dimensional structure of Weibel-mediated shocks is investigated through a fully kinetic particle-in-cell simulation. The structures obtained are characterized by the following features: (i) helical magnetic field lines elongated in the direction upstream of the shock region, (ii) high and low density filaments inside the helical field lines. These structures originate from the interaction between counter-streaming plasma flow and magnetic vortexes caused by Weibel instability, and potentially affect the shock formation mechanism.

## 1. Introduction

Collisionless shocks are fundamental process in space and astrophysical plasmas and are present in various phenomena, such as supernova remnants and magnetospheric bow shocks[1]. Observations of supernova remnants suggest that collisionless shocks could be a major source of the energetic components of cosmic rays[2]. Experimental demonstrations of collisionless shocks have been also performed by using high-intensity lasers[3,4,5]. In the primary design of experiments, collisionless shocks are generated in unmagnetized plasmas counter-streaming at a high Mach number ejected from an opposing pair of thin films.

Weibel instability[6] is considered to play an important role in collisionless shock formation in unmagnetized plasmas. The unstable modes have wavenumber perpendicular to the streaming direction and resulting density filament structures extend along the streaming direction. Previous two dimensional particle-in-cell simulations[7,8] showed that the filament coalescence causes ion-scale turbulence and dominant shock dissipation. Fluctuations of electron current density initiate this coalescence process and the scale length of shock transition is much larger than the ion inertial length. High-resolution kinetic simulation is required to capture this multi-scale process. Three dimensional feature of Weibel-mediated shock formation has not been fully investigated because of the vast computational costs.

In this paper, we investigate the three dimensional structure of unmagnetized collisionless shocks through a fully kinetic particle-in-cell code highly optimized for massive parallel computers. The simulation demonstrates high and low density cylindrical structures enclosed

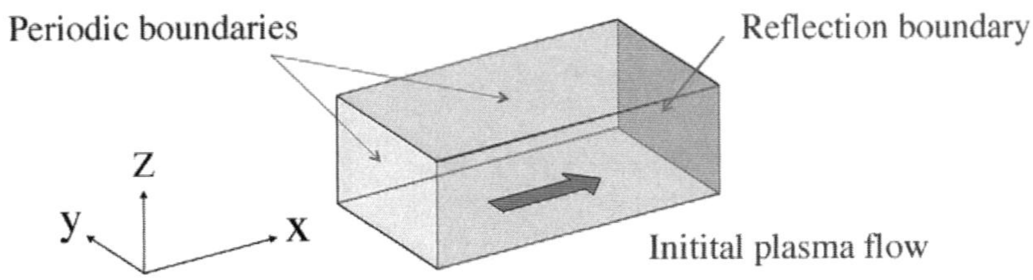


**Figure 1.** Schematic diagram of simulation domain.

with helical magnetic field lines. This paper is organized as follows. The simulation model is presented in section 2. The simulation results are shown in section 3. Finally, the present study is summarized in section 4.

## 2. Simulation model

A fully kinetic plasma simulation is performed to investigate the three dimensional structure of unmagnetized collisionless shocks. The simulation scheme is the standard relativistic particle-in-cell method. Newton-Lorentz and Maxwell equations are solved by Buneman-Boris and leap-flog schemes, respectively. Second order interpolation and charge conservation scheme[9] are employed to estimate Lorentz force acting on the computational particle and current density profile on the spatial grid.

Figure 1 shows the schematic diagram of the simulation domain. The simulation domain is set in a three dimensional Cartesian coordinate system $(x, y, z)$ and implemented on a $8000 \times 512 \times 512$ point grid. A reflection boundary condition is employed for the $+x$-boundary while a periodic condition is employed for the other boundaries. The origin of the x-axis is set at the reflection boundary. We consider a uniform plasma flowing in the $+x$-direction as the initial state of the system. The counter-streaming plasma flow is generated by the initial plasma flow and the reflected plasma flow.

The physical conditions are set as follows: The ion-to-electron mass ratio is 20. The initial electron and ion temperatures are the same. The Mach number of the initial flow is 10.7. The physical domain size is $(L_x, L_y, L_z) = (2000S_e, 128S_e, 128S_e)$, where $S_e$ denotes electron inertia length. The domain size, $L_x$, corresponds to hundreds of ion inertial length, $S_i$, i.e., several times as much as the shock transition length expected from the previous simulation[8]. For the recent laser experiment on Weibel instability in counter-streaming plasmas $(n \sim 10^{25}[1/\mathrm{m}^3])$[5], this scale $(O(100)S_i \sim O(1)[\mathrm{cm}])$ is comparable to or slightly larger than the measurement scale. Initially, 45 computational particles are employed for each particle species per spatial grid.

## 3. Simulation results

Figure 2 shows a birds-eye view of the electron density profiles at (A) $t\omega_{pe} = 824.4$ and (B) $t\omega_{pe} = 1331.8$, where $\omega_{pe}$ is the electron plasma frequency. The high density region expands from the reflection boundary as the reflected plasma flows over the incoming plasma. Density profiles in the x-y and x-z planes, namely, the in-plane density profiles, are predominantly formed of filament structures elongated in the x-direction. The density filaments become thick near the reflection boundary. These features are the same as those seen in the results of previous two dimensional simulations[6,7], and originate from Weibel instability.

Figure 3 shows electron density profiles in the y-z plane, namely, the out-of-plane density profiles, at $t\omega_{pe} = 1331.8$. Their positions on the x-axis are (A)$-400S_e$, (B)$-330S_e$ and (C)$-165S_e$ (see Fig.2). The profiles obtained correspond to the sectional structures of the density filament. The clustered regions of low density in panel (A) indicate the presence of low

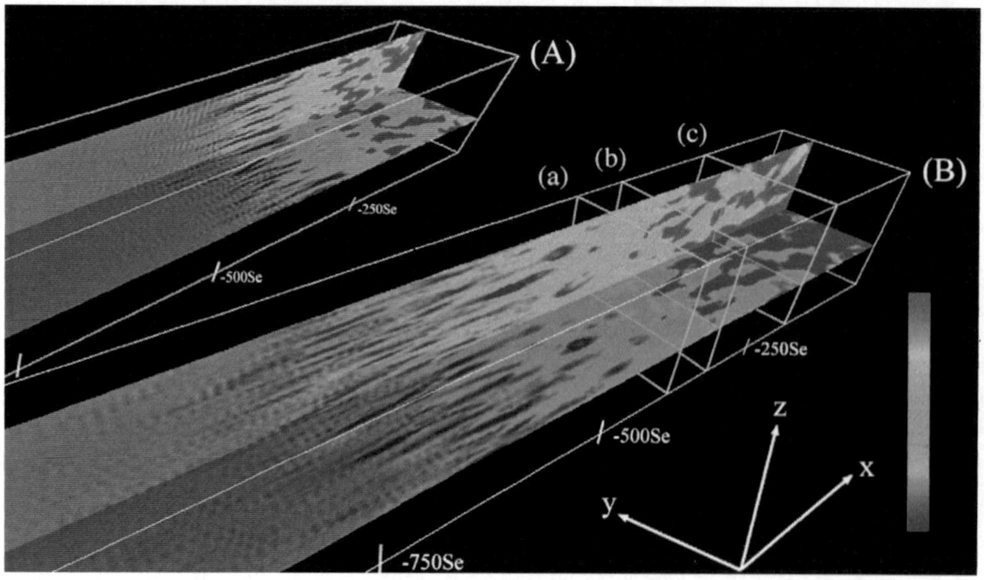


**Figure 2.** Birds-eye view of the electron density profile described by color contour plots for the x-y and x-z planes at (A) $t\omega_{pe} = 824.4$ and (B) $t\omega_{pe} = 1331.8$.

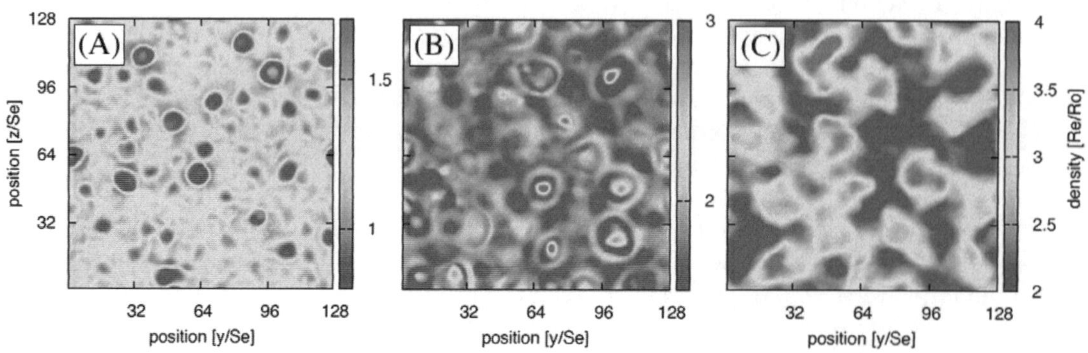


**Figure 3.** Color contour plots of electron density profiles in the y-z planes at (A)$x = -400S_e$, (B)$x = -330S_e$ and (C)$x = -165S_e$.

density cylindrical structures away from the reflection boundary (hereinafter referred to as the upstream side). On the other hand, the clustered regions of high density in panel (C) suggest high density filaments near the reflection boundary (hereinafter referred to as the downstream side). As shown in panel (B), intermediate structures with two-layered filaments are formed around the boundary between the upstream and downstream sides.

Figure 4 shows the extracted magnetic field lines (yellow lines) superposed on the electron density profiles (color contour) on the (A) upstream ($x = -550S_e \sim -300S_e$) and (B) downstream ($x = -300S_e \sim -50S_e$) sides. The magnetic field lines form helical structures in both regions. These structures derive from magnetic vortices due to Weibel instability elongated in the streaming plasmas. The difference in these structures is the relative density level inside the helical field lines. On the upstream side, the helical magnetic field lines is twisting around

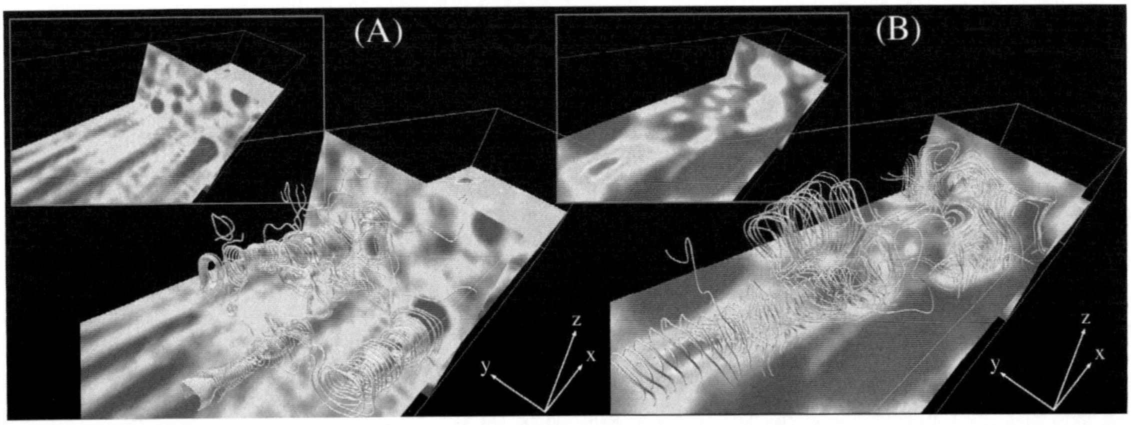


**Figure 4.** Birds-eye views of extracted magnetic field lines (yellow lines) and electron density profiles (color contour plots) for (A) the upstream and (B) the downstream side. Boxed thumbnails show the electron density profiles without the magnetic field lines.

the low density regions. These field lines would be generated from the high energy component of plasma flow from the downstream side. In contrast, the magnetic field lines on the downstream side twist around the high density filaments, such that the helical field lines are filled with plasma. This difference likely originates from two contrastive interaction processes: the plasma diversion away from the thin helical field lines on the upstream side and plasma convergence into the thick helical field lines on the downstream side.

## 4. Conclusion

The three dimensional structure of Weibel-mediated shocks was investigated through a large-scale particle-in-cell simulation. The simulation demonstrated two types of cylindrical density structures enclosed by helical magnetic field lines. The first type is thin cylindrical low density regions on the upstream side and the other is high density thick filaments on the downstream side. These structures could originate from the interaction between in-plane plasma flow and out-of-plane magnetic vortices due to Weibel instability. In the future, we will consider the impact of these structures on the efficiency of shock dissipation and particle acceleration.

## Acknowledgments

The calculations were performed using the Plasma Simulator at the National Institute for Fusion Science (NIFS13KNSS039) and the K-computer at the RIKEN Advanced Institute for Computational Science (hp120267).

## References

[1] Treuman R A 2009 *Astron. Astrophys. Rev* **17** 409
[2] Nikolic S, van de Ven G, Heng K *et al* 2013 *Science* **340** 45
[3] Kuramitsu Y, Sakawa Y, Morita T *et al* 2011 *Phys. Rev. Lett* **106** 175002
[4] Morita T, Sakawa Y, Tomita K *et al* 2013 *Phys. Plasmas* **20** 092115
[5] Fox F, Fiksel G, Bhattacharjee A, *et al* 2013 *Phys. Rev. Lett* **111** 225002
[6] Weibel E S 1959 *Phys. Rev. Lett* **2** 83
[7] Spitokovsky A 2008 *Astrophys. J* **673** L39
[8] Kato T N and Takabe H 2008 *Astrophys. J* **681** L93
[9] Esirkepov T Zh 2001 *Comp. Phys. Comm* **135** 144

Journal of Physics: Conference Series **688** (2016) 012073 doi:10.1088/1742-6596/688/1/012073

# The National Ignition Facility: Transition to a User Facility[1]

**E. I. Moses, J. Atherton, L. Lagin, D. Larson, C. Keane, B. MacGowan, R. Patterson, M. Spaeth, B. Van Wonterghem, P. Wegner, R. Kauffman**

Lawrence Livermore National Laboratory, Livermore, CA 94450

moses1@llnl.gov

**Abstract**. The National Ignition Facility (NIF) at Lawrence Livermore National Laboratory (LLNL) has been operational since March 2009 and has been transitioning to a user facility supporting ignition science, high energy density science (HEDS), national security applications, and fundamental science. The facility has achieved its design goal of 1.8 MJ and 500 TW of 3ω light on target, and has performed target experiments with 1.9 MJ at peak powers of 410 TW. The facility is on track to perform over 200 target shots this year in support of all of its user communities. The facility has nearly 60 diagnostic systems operational and has shown flexibility in laser pulse shape and performance to meet the requirements of its multiple users. Progress continues on its goal of demonstrating thermonuclear burn in the laboratory. It has performed over 40 indirect-drive experiments with cryogenic-layered capsules. New platforms are being developed for HEDS and fundamental science. Equation-of-state and material strength experiments have been done on a number of materials with pressures of over 50 MBars obtained in diamond, conditions never previously encountered in the laboratory and similar to those found in planetary interiors. Experiments are also in progress investigating radiation transport, hydrodynamic instabilities, and direct drive implosions. NIF continues to develop as an experimental facility. Advanced Radiographic Capability (ARC) is now being installed on NIF for producing high-energy radiographs of the imploded cores of ignition targets and for short pulse laser-plasma interaction experiments. One NIF beam is planned for conversion to two picosecond beams in 2014. Other new diagnostics such as x-ray Thomson scattering, low energy neutron spectrometer, and multi-layer reflecting x-ray optics are also planned. Incremental improvements in laser performance such as improved optics damage performance, beam balance, and back reflection control are being pursued.

## 1. Introduction

The National Ignition Facility (NIF), located at Lawrence Livermore National Laboratory (LLNL) in Livermore, California, is the world's largest and most energetic laser facility for research in inertial confinement fusion (ICF) and high energy density science (HEDS) in support of its national security

[1] This work performed under the auspices of the U.S. Department of Energy by Lawrence Livermore National Laboratory under Contract DE-AC52-07NA27344

mission [1]. NIF also performs experiments for fundamental science and other national security missions as well as ICF as a potential source of renewable energy. NIF is the first laser system designed to achieve ignition and thermonuclear burn of deuterium-tritium–filled ICF capsules. NIF has developed an unprecedented experimental capability, much of it developed as part of the National Ignition Campaign (NIC) [2]. NIC, started in 2005 and completed in 2012, was a comprehensive program for developing NIF as an ignition platform and transitioning NIF to a user facility. NIF is now a fully operational facility with users performing experiments for multiple missions

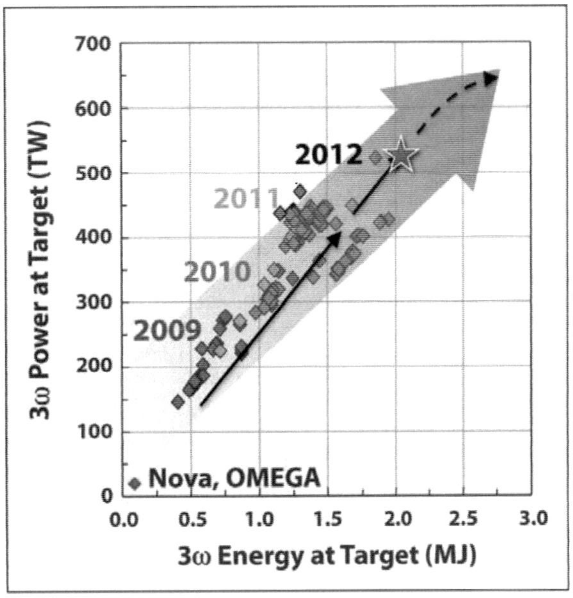


Figure 1. Progress in power and energy on NIF continues increasing to the NIF design point

Significant progress has been made since the last conference in developing NIF capabilities and in the science being performed at NIF. NIF laser performance has been demonstrated above its design goal of 1.8 MJ and 500 TW. NIF operability has improved increasing its shot rate each year and broadening the number of experimental platforms. NIF continues to expand its capabilities with an increasing number of diagnostics and facility enhancements. NIF has transitioned from NIC into a user facility and has broadened its HEDS program both for national security and fundamental science. In addition, progress continues on reaching its ICF goal of ignition in the laboratory. Recent experiments have measured neutron yields greater than $10^{15}$ neutrons and are greater than 50% of predictions from simulations. Analysis indicates that a significant fraction of the yield is from self-heating by alpha deposition.

This paper summarizes some of these advances since the last conference and provides an overview of capabilities that have been developed. Progress made in HEDS and toward ignition is reviewed. In depth analysis can be found in other papers in this conference and publications elsewhere.

## 2. NIF Facility

NIF is the most recent laser facility constructed at LLNL for ICF research. NIF consists of a 192 beam Nd-glass laser system installed in two laser bays, two switchyards to redirect the beams for indirect drive target irradiation, a target area designed for ignition experiments producing 20 MJ of fusion yield, and a control room for operations. NIF is designed to produce 1.8 MJ and 500 TW of 3ω light on target in a highly shaped pulse for indirect-drive fusion ignition experiments [3]. The NIF facility includes diagnostics systems, cryogenic target fielding capabilities, integrated control systems, and contamination controls systems required for performing integrated ignition experiments.

Since the last conference NIF has demonstrated that it meets all of its performance requirements and has transitioned to an operational facility. Since commissioning, NIF has steadily increased its power and energy as shown in Figure 1. At the time of Project completion in 2009, NIF was operating at 500 kJ and 200 TW of 3ω light. At the time of the last conference, NIF had attained ~1.3 MJ and 400 TW of 3ω light. In 2012, NIF demonstrated that it could achieve its design goals of 1.8 MJ and 500 TW of 3ω light on target in an ignition shaped pulse. Laser performance at these energy and power levels was demonstrated in the course of normal operation [4]. This improvement in performance and operations at the design point shown in Figure 1 is attributable to several advances in technology. Much of the increase has been enabled by improving optics quality and their resistance to damage. NIF is designed to operate at fluences of 8 J/cm$^2$ for 3ω light. When NIF was designed, damage levels for optics were ~2 J/cm$^2$. Through years of research, new processing techniques have been developed for improving NIF optics quality allowing NIF to operate at its design energy [5]. A system has been developed to block damage sites when damage does occur to prevent the damage site from increasing in size [6]. Techniques have also been developed to repair damage sites allowing optics to be recycled reducing overall operating costs [7]. Along with an active damage inspection program, other improvements in performance include controlling amplitude modulation in the laser pulse to reduce peak fluence and increase of the clear aperture of the beams to propagate more energy for a given fluence [8]. Additional improvements have been identified that could potentially allow NIF to operate at energy above 2 MJ.

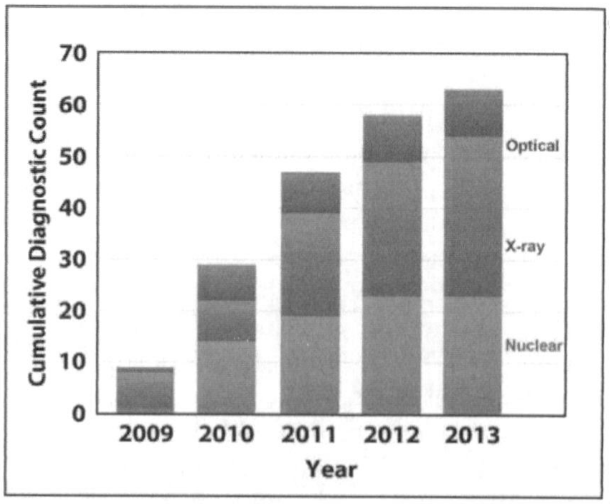


Figure 2. Target diagnostics on NIF continue to increase each year

The NIF facility has also greatly improved its experimental capabilities from Project completion and even since the last conference two years ago. For example, Figure 2 shows the number of target diagnostics available for experiments as a function of years. The number of diagnostics has increased by around a factor of five since Project completion and approximately 20 diagnostics have been added since the last conference. All of the NIC partners: LLNL, Los Alamos National Laboratory (LANL), Sandia National Laboratories (SNL), University of Rochester Laboratory for Laser Energetics (UR-LLE), General Atomics as well as a number of other collaborators including Lawrence Berkeley National Laboratory, the Massachusetts Institute of Technology (MIT), the U.K. Atomic Weapons Establishment (AWE) in England, and the French Atomic Energy Commission (CEA) in France have contributed to this effort. Some of the new capability recently added include the South Pole Bang Time (SPBT) diagnostic to measure the x-ray bang time of the capsule [9], additional neutron time-of-

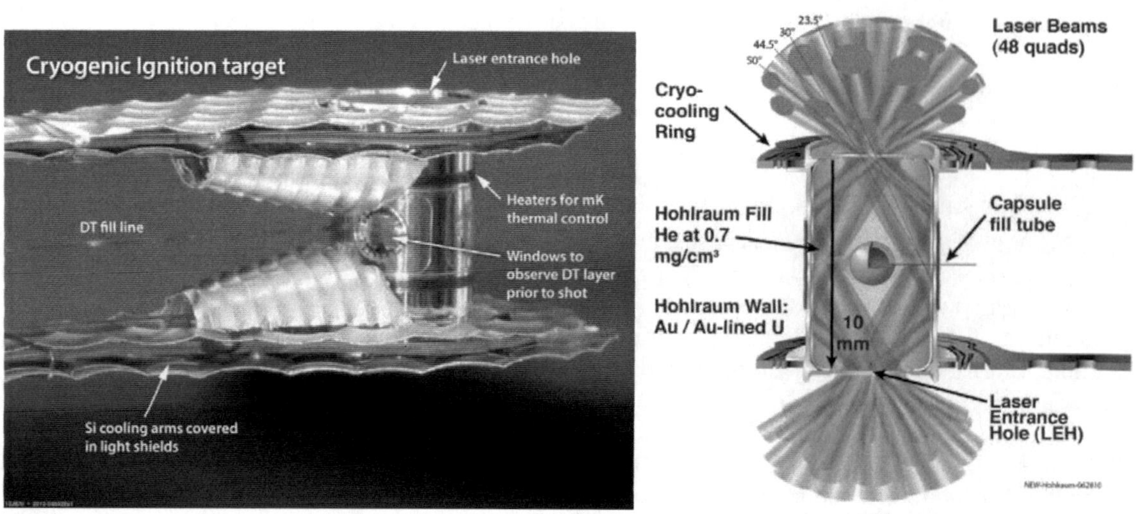


(a) (b)

Figure 3. A cryogenic hohlraum for ignition experiments (a) and a schematic of the target showing the irradiation geometry (b).

Figure 4. The large compressor vessel that will contain the full aperture gratings for compressing the ARC pulse to 30 ps.

flight (nToF) detectors to measure drift velocity of the implosion [10], an array of flanged-mounted neutron activation detectors (fNADS) to measure isotropy of the neutron emission [11], and Radioactive Gas Spectrometer (RAGS) fielded by SNL and LLNL to measure gaseous activation products [12]. More than 20 diagnostics may be fielded on a single experiment.

Advances also continue in target fabrication. Cryogenic targets are one of the most complex targets fielded at any facility. An example is shown in Figure 3 along with a schematic of an ignition target. The target contains apparatus for cooling, temperature monitoring, and thermal control needed for producing cryogenic deuterium-tritium (DT) layers inside of the capsule in the hohlraum. The target contains viewing windows for measuring the layer quality. The layer is characterized by x-ray radiography along three orthogonal axes in the hohlraum. Ignition quality layers can now be grown routinely. New capsule characterization methods have recently been developed that measure the surface of each capsule before being mounted in the target. Surface finish, manufacturing defects, and contamination particles can be detected and tract to correlate with target experiments data.

Laser capabilities are also expanding. Recently, the capability to delay beams up to 100 ns for producing x-ray backlighters was installed for radiography of hydrodynamic instability experiments. NIF is now installing the capability for producing short-pulse beams for high-energy x-ray radiography. ARC, advanced radiographic capability is being installed on four NIF beams, or a quad [13]. Each NIF beam is divided into two ARC beams that propagate pulses independently to target. ARC uses chirped pulse amplification to amplify and temporally compress the beam to produce a 1.5 kJ pulse of 1 energy in a 30 ps pulse in each ARC beam. Beam path hardware is presently being installed in the large compressor chamber that holds the large gratings for pulse compression shown in Figure 4. The chirped pulse oscillator has been installed and beams are being propagated through the main laser chain. The first four of the eight beams are planned to be operational in 2015.

## 3. NIF User facility

Since it was established as a construction project in 1995, NIF has been envisioned as a national user facility. Built in support of the Stockpile Stewardship Program (SSP) most of the experiments support SSP either for Inertial Confinement Fusion (ICF) research or for High Energy Density Stewardship Science (HEDSS). NIF also performs experiments supporting other national security applications and fundamental science. As NIF transitions to a user facility, management structures are being developed to serve the broad user community. Shot proposals from the different user groups are being peer reviewed for technical excellence and reviewed by the facility for execution. A NIF User group has been formed and has held two meetings for developing proposals for fundamental science. The first round of fundamental science proposals has been reviewed and ten proposals with sixteen principal investigators have been chosen. Experiments for some of the proposals have begun.

One of the fundamental science experiments is to measure the equation of state of carbon up to a Gigabar along its Hugoniot [14]. The experiment is a collaboration of scientists from University of California Berkeley, SLAC Accelerator National Laboratory, AWE, and LLNL. The experiment uses a hohlraum platform developed for ICF ignition experiments with the ICF capsule being replaced with a solid CH sphere. X-rays from the hohlraum ablate the outside of the sphere producing a spherically convergent shock in the CH sphere. The velocity of the shock front is measured using x-ray radiography as shown in Figure 5. The shock front is tracked by the increased absorption of the x-ray backlighter as the density of the material increases due to the shock front. When the shock converges at the center an x-ray flash is observed. By fitting the trajectory, the equation of state (EOS) of CH can be extracted. For this shot the pressure is derived to be 720 Mbars at the center of convergence.

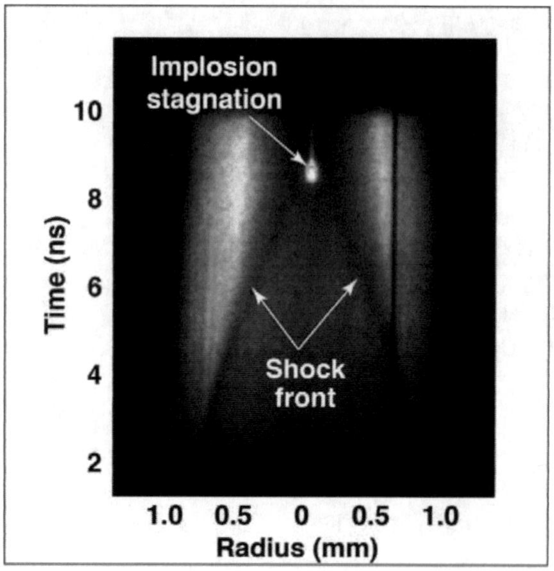


Figure 5. A streaked radiograph of a CH sphere show the radiation-driven shock front converging to produce a stagnation flash. Analysis shows that pressures of 720 Mbars were attained.

There is much synergy among the different user communities for developing experimental capabilities. For example, many of the platforms developed for ICF are used for HEDSS and fundamental science experiments. The same platform is used for studying EOS as part of the HEDSS Program and for fundamental science experiments. Of course diagnostics are used across all of the different users and much of the target fabrication development is applicable to many different research areas. An example of the cross connection of different research areas is illustrated in Figure 6. Solid Radiochemistry Collectors were being developed by ICF to measure activation products produced in

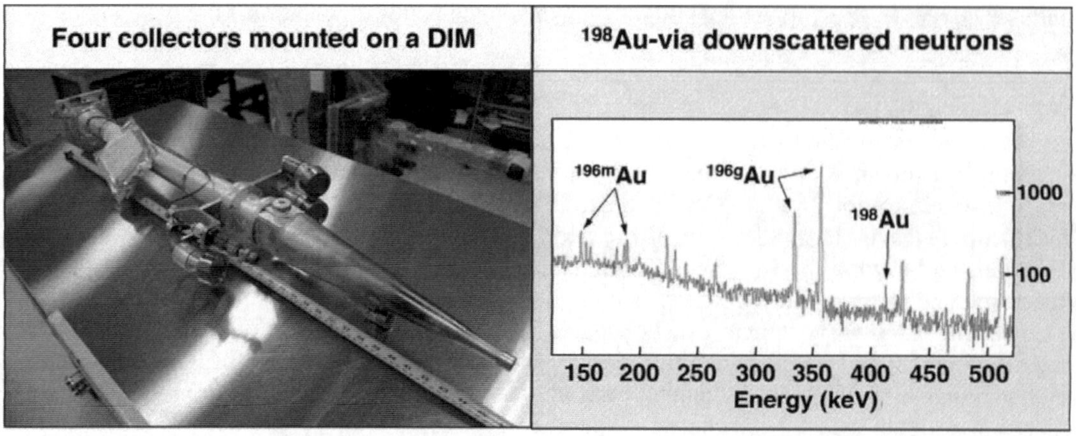


(a) (b)

Figure 6. Activated Au from the hohlraum has been collected on NIF. (a) Collectors are mounted on the x-ray imaging diagnostic. (b) An activated Au spectrum is shown. The relative abundance of activated products depends on the flux of low energy neutrons.

ICF ignition experiments. The collectors are discs of solid material placed near the target as shown in Figure 6a. After the shot the collectors are taken to a nuclear counting facility to measure the activated products. In testing the collectors, activated Au from the hohlraum debris is found as shown in Figure 6b. The ratio of Au activation products depends on the flux of low energy neutrons produced by down scattering the 14 MeV DT neutrons in the high-density shell material indicating that excitation of excited nuclear states of Au are involved. This data is presently being analyzed to extract cross sections of excited state nuclei. NIF is the only facility in the world that can produce neutron flux densities allowing studying excited nuclear states.

## 4. NIF Progress on Ignition

Achieving ignition and burn continues to be a major effort on NIF. Achieving ignition requires assembling the fuel to a hot spot areal density of ~0.3 g/cm$^2$ at temperatures of ~5-10 keV. Implosion velocities of 3-4x10$^7$ cm/s and fuel pressures of ~200 Gigabars are required in a low mix implosion [15,16]. The implosion is controlled by a combination of laser pulse shaping and target geometry that optimizes the fuel adiabat, implosion velocity, implosion shape, and hydrodynamic mix.

Initial experiments were done as part of the NIC. NIC experiments used indirect drive and a CH capsule with a laser pulse shape that produced an implosion having a low adiabat, close to Fermi degeneracy [17]. During NIC, there were 38 cryogenic layered ignition experiments in addition to a large number of experiments for measuring shock timing, shape, shell velocity, and hohlraum symmetry. A summary of progress in NIC is shown as the blue points in Figure 7 [18]. Initial experiments, begun in September 2010, used Ge-doped capsules and performed orders of magnitude below predictions from integrated simulations. After a number of experiments for optimizing shock timing and implosion symmetry and changing to Si-doped capsules that were found to be more efficient, capsule yields approaching 10$^{15}$ neutrons were obtained. These yields were still nearly an order of magnitude below the simulation predictions. Experiments then tried to improve performance by going to higher compression and higher velocity. Instead of improving performance, these experiments did not perform as well. In analyzing these results, it appeared that low-mode implosion asymmetry and hydrodynamic mix were degrading performance more than predicted by the simulations.

Experiments since the end of NIC have focused on understanding the implosion physics for improving target performance. New experimental platforms have been developed to investigate detailed physics of the implosion. One new platform is two-dimensional radiography of the imploding shell. These radiographs have shown that the implosions on NIC had a significant P4 inflight asymmetry. This asymmetry is shown in Figure 8 measured when the implosion is about 20% of its initial radius. More symmetric implosion can be produced if the hohlraum length is increased by 0.7-1.0 mm from the standard NIC length of 9.4 mm [19]. Symmetry on NIC was adjusted by

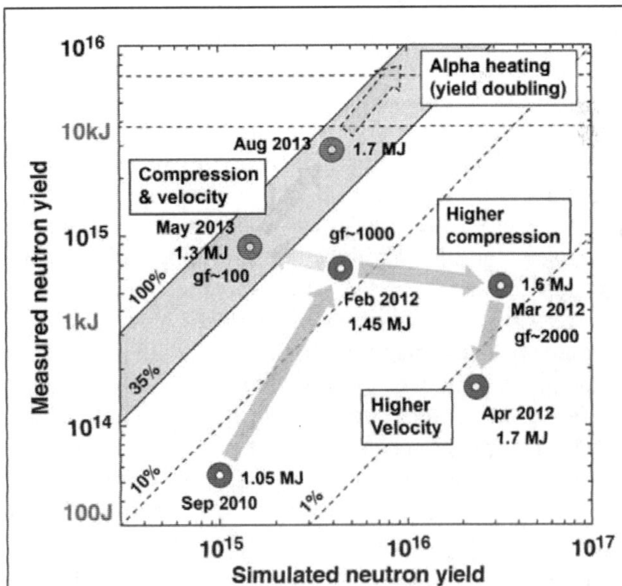


Figure 7. Progress toward ignition is shown as measured neutron yield versus yield predicted by simulations. The blue points show the progress during NIC with low adiabat implosions. The green points show the progress made recently with high adiabat implosions.

measuring the asymmetry of the hot spot at peak compression. These images only measured the shape of the hot central fuel. The surrounding cold fuel can have significant asymmetries reducing confinement and fuel areal density. These inflight radiography experiments also showed that the support structure, or tents, that hold the capsule in place provided a seed for hydrodynamic instabilities. Effort is under way to reduce the mass of the tents to reduce the seed for hydrodynamic instabilities.

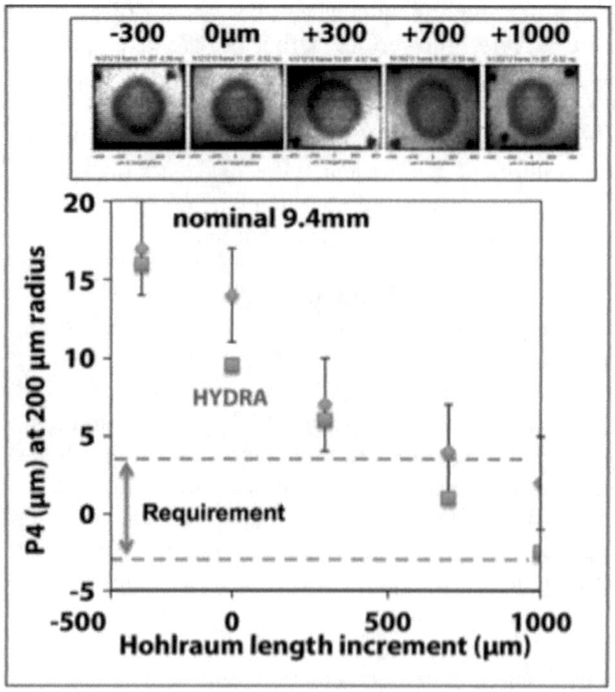


Figure 8. X-ray radiography measures the shape of the shell in flight.

Another example of experiments studying the underlying physics of ignition implosions is a campaign to measure the growth rate of hydrodynamic instabilities. In these experiments, single mode perturbations are imprinted onto an ignition capsule as shown in Figure 9 [20]. The capsule is mounted onto a reentrant cone and placed in the center of a hohlraum. The reentrant cone allows access for x–ray radiography of the imploding shell. The hohlraum is driven with a typical ignition pulse. Eight beams are used to produce a 5 keV backlighter source. Examples of x-ray radiographs taken when the shell has imploded to about half of its initial radius is shown in Figure 9. Single mode periods are clearly visible in the radiographs. Intensity modulations are analyzed to extract growth factors for comparison with simulations. Preliminary results indicate that the growth factors in CH capsules using the NIC ignition pulse are large as predicted by simulations. Also, these experiments show that implosions using pulses that produce a higher adiabat implosion are more stable with respect to mix as predicted by simulations.

Recent experiments have begun exploring the performance of higher adiabat implosions. These experiments use Si-doped CH capsules similar to those used for NIC experiments and are driven with laser pulses that have a "high foot", a higher initial pulse, compared to NIC experiments [17]. The initial pulse produces a shock that sets the initial adiabat of the cryogenic fuel. The higher adiabat implosion calculates to be more stable to ablation front hydrodynamic instabilities and have been confirmed in experiments as discussed above. The higher adiabat implosions are less efficient and calculate not to obtain as high of gain as NIC targets. Initial experiments show good performance and are summarized as the green data in Figure 7. Implosions have produced yields of greater than $10^{15}$ neutrons and are greater than 50% of predictions from simulations. These experiments indicate that a significant fraction of the yield is from self-heating by alpha deposition, a first for ICF implosions. Experiments are continuing extending the performance to higher intensity and higher velocity to understand the limits due to hydrodynamic instabilities and mix.

Experiments are also planned to explore alternative ablators. High-density carbon (HDC) and Be ablators calculate to produce more efficient implosions than CH ablators. HDC, or diamond capsules have been fabricated [21] and initial experiments have begun and have obtained high implosion velocities and good performance in gas-filled capsules [22]. Plans for cryogenic capsule implosions are being developed. Designs for Be ignition experiments are also being developed [23] and experiments are planned to begin next year.

Figure 9. Capsules with imprinted periodic perturbations are used to study hydrodynamic instabilities.

## 5. Summary

NIF is a fully operational facility supporting experiments for ICF and HEDS as well as fundamental science and other national security missions. It has demonstrated that it meets design goals of 1.8 MJ and 500 TW of 3ω light on target in a shaped ignition pulse. It can perform at design levels as part of routine operations. This high level of performance is a result of a number of advances for maintaining optics performance operating at design fluences. NIF has developed a vast array of experimental capabilities. It has commissioned over 60 different diagnostic systems. It routinely fields cryogenic layered targets for ignition experiments and has developed a number of other experimental platforms for use by its diverse user community. It continues to develop capability and plans to have four ARC beams operational in 2015 for short pulse high-energy x-ray backlighting capability.

NIF is transitioning into a national user facility for supporting its users in SSP, fundamental science, and other national security missions. NIF is developing a peer review process for ensuring that it produces world-class science. NIF has completed its first set of reviews of fundamental science proposals; ten experiments (with sixteen principal investigators) have been chosen. Experiments have begun on some of the proposals and they are already producing unique results.

Progress continues toward attaining ignition. At the completion of NIC, neutron yield was nearly an order of magnitude below predictions from integrated simulations. Low-mode implosion symmetry and hydrodynamic mix appeared to be degrading performance more than predicted by the simulations. Experiments since the end of NIC have focused on understanding the implosion physics for improving target performance. These include measurement of the inflight symmetry of the imploding capsule and quantitative measure of the mix growth factors. Implosion platforms have been developed that are more stable to mix although they do not calculate to perform as well as the NIC design. These have performed close to predictions from simulations in experiments to date. In recent experiments, a significant fraction of the yield appears to be from self-heating by alpha deposition, a first for ICF implosions. Experiments are continuing extending the performance to higher intensity and higher velocity to understand the limits due to hydrodynamic instabilities and mix. Target designs are being developed for alternative ablators such as HDC and Be that are potentially more efficient and experiments are planned to test these in the future.

## References

[1] G.H. Miller, E.I. Moses, C.R. Wuest, "The National Ignition Facility: enabling fusion ignition for the 21st century," Nucl. Fusion **44**, S228 (2004).

[2] E.I. Moses and the NIC Collaborators, "The National Ignition Campaign: Status and Progress," Nucl. Fusion **53**, 104020 (2013).

[3] J. R. Murray, "Overview of the National Ignition Facility," Supplement to Proceedings of SPIE Vol. 3492, 1 (1998).

[4] J. M. Di Nicola, S. N. Dixit, J. Heebner, et al., "The National Ignition Facility Laser Performance Status," IFSA 2013 proceedings.

Journal of Physics: Conference Series **688** (2016) 012073 doi:10.1088/1742-6596/688/1/012073

[5] T. I. Suratwala, P. E. Miller, J. D. Bude, et al., "HF-based etching processes for improving laser damage resistance of fused silica optical surfaces," Journal of the American Ceramic Society **94**, 416-428 (2011).
[6] J. Heebner, M. Borden, P. Miller, et al., "Programmable beam spatial shaping system for the National Ignition Facility," Proceedings of SPIE, Vol. 7916 (2011).
[7] N. Shen, P. E. Miller, J. D. Bude, et al., "Thermal annealing of laser damage precursors on fused silica surfaces," Optical Engineering **51**, 121817 (2012).
[8] S. C. Burkhart, A. A. S. Awwal, M. R. Borden, et al., "The National Ignition Facility: beam area increase," Proceedings of SPIE, Vol. 8602 (2013).
[9] D. H. Edgell, D. K. Bradley, E. J. Bond, et al., "South pole bang-time diagnostic on the National Ignition Facility," Rev. Sci. Instrum. **83**, 10E119 (2012).
[10] V. Yu. Glebov, C. Forrest, J. P. Knauer, et al., "Testing a new NIF neutron time-of-flight detector with a bibenzyl scintillator on OMEGA," Rev. Sci. Instrum. **83**, 10D309 (2012).
[11] D. L. Bleuel, C. B. Yeamans, L. A. Bernstein, et al., "Neutron activation diagnostics at the National Ignition Facility," Rev. Sci. Instrum. 83, 10D313 (2012); C. B. Yeamans, D. L. Bleuel and L. A. Bernstein, "Enhanced NIF neutron activation diagnostics," Rev. Sci. Instrum. **83**, 10D315 (2012).
[12] D. A. Shaughnessy, C. A. Velsko, D. R. Jedlovec, C. B. Yeamans, K. J. Moody, E. Tereshatov, W. Stoeffl and A. Riddle, "The Radiochemical Analysis of Gaseous Samples (RAGS) apparatus for nuclear diagnostics at the National Ignition Facility," Rev. Sci. Instrum. **83**, 10D917 (2012).
[13] J K Crane, G Tietbohl, P Arnold, et al., "Progress on converting a NIF quad to eight, petawatt beams for advanced radiography," J. Phys.: Conf. Ser. **244**, 032003 (2010).
[14] A. Kritcher, D. Swift, T. Doeppner, et al., "Probing matterat Gbar pressures at NIF," this conference.
[15] J. Nuckols, L. Wood, A. Thiessen, and G. Zimmerman, "Laser Compressiom of Matter to Super-High Densities: Thermonuclear (CTR) Applications," Nature **239**, 139 (1972).
[16] J. D. Lindl, "Development of the indirect-drive approach to inertial confinement fusion and the target physics basis for ignition and gain," Phys. Plasmas **2**, 3933 (1995).
[17] M. J. Edwards, J. D. Lindl, B. K. Spears, et al., "The experimental plan for cryogenic layered target implosions on the National Ignition Facility–The inertial confinement approach to fusion," Phys. Plasmas **18**, 051003 (2011).
[18] M. J. Edwards, P. K. Patel, J. D. Lindl, et al., "Progress towards ignition on the National Ignition Facility," Phys. Plasmas 20, 070501 (2013); M. J. Edwards, "The Ignition Physics Campaign on NIF: Status and Progress," this conference.
[19] O. Jones, R. Rygg, R. Tomasini, et al., " A new symmetry model for hohlraum-driven capsule implosion experiments on the National Ignition Facility," this conference.
[20] V. A. Smalyuk, J. Caggiano, D. Casey, et al., "Hydrodynamic Growth and Mix Experiments at National Ignition Facility," this conference.
[21] J. Biener, D.D. Ho, C. Wild, et al., "Diamond spheres for inertial confinement fusion," Nucl. Fusion **49**, 112001 (2009).
[22] S. J. Ross, "High-Density Carbon Ablator Experiments on the National Ignition Facility," this conference.
[23] A. Simakov, D. Wilson, A. Yi, et al., Beryllium Ignition Target Design for Indirect Drive NIF Experiments," this conference.

# High efficient ultrahigh acceleration of plasma blocks by PW-ps laser pulses for producing fusion flames in DT and HB11 of solid state density

**S Moustaizis[1], P Lalousis[2#], H Hora[3] and G H Miley[4]**

[1]Technical University of Crete, Science Department, Chania, Greece
[2]Institute of Electron Structure and Lasers FORTH, Heraklion, Greece
[3]Department of Theoretical Physics, University of New South Wales (Australia)
[4]Dept. Nuclear, Plasma & Radiol.Engineering, Univ. Illinois, Urbana IL, USA

E-mail: Lalousis Paraskevas [lalousis@iesl.forth.gr]

**Abstract.** Ultrahigh acceleration of plasma blocks in the range of $10^{20}$ cm/s$^2$ has been confirmed experimentally after this was long predicted as a non-thermal direct conversion of optical energy into plasma motion due to dominating nonlinear (ponderomotive) forces [1]. The use of laser pulses of more than PW power and ps or shorter duration can ignite a nuclear fusion flame in solid density deuterium tritium because the necessary energy flux of $>10^8$J/cm$^2$ according to the theory of Chu [2] is available [3]. For the studies of the necessary velocities of the generated fusion flames above 1000 km/s the detailed processes can be analyzed by using the advanced genuine two-fluid hydrodynamic model [4] where it was surprising that the ignition of the fusion flame by the picosecond interaction needs a comparably long development in the nanosecond range before the thermal processes result in shock fronts similar to the Rakine-Hugoniot theory. For the evaluation of power generation the problem of lateral energy losses was studied by using very high pulsed magnetic fields. The recently produced 10 Kilotesla magnetic fields [5] are very promising for solutions.

## 1. Introduction

For the aim to produce controlled nuclear fusion reactions for generation of energy, the use of lasers for initiating and/or driving the reaction is followed up by a number of different schemes where the interaction process of the laser with the irradiated fusion fuel can depend on the duration of hte laser pulse. A basic difference for fusion energy generation was opened with the use of Petawatt-picosecond (PW-ps) laser-plasma interaction [6] initiated since 1986 [7] in contrast to laser driven spherical compression and thermal ignition needing more than 1000 times solid state densities deuterium-tritium (DT) for nanosecond laser pulses and where a breakthrough is expected [8]. In contrast to these thermal processes with unavoidable losses, delays, inefficiencies and instabilities, the new option with PW-ps pulses uses the efficient direct conversion of laser energy into motion of plasma blocks where the thermal processes are nearly fully avoided. In contrast to this scheme, The direct conversion of laser energy by the nonlinear (ponderomotive) force into motion of plasma blocks was evident after dielectric properties of plasmas with optical constants into the Maxwellian stress tensor was introduced permitting the generalizing of the ponderomotive force up to the nonlinear force for explaining measured MeV ion energies at laser-plasma interaction including self-focusing [9]. Computations of plane geometry interaction of ps laser pulses with $10^{18}$W/cm$^2$ laser pulses arrived at ultra-high accelerations of plasma blocks above $10^{20}$cm/s$^2$ within picosecond interaction, clearly demonstrating that the thermal processes were of very minor influence [10].

These ultra-high accelerations were first measured by Sauerbrey [11] where the values of $2\times10^{20}$cm/s$^2$ were in perfect agreement with the theory by inclusion of dielectric swelling [1]. These accelerations were reproduced [12] and were up to fife orders of magnitudes higher than the measured acceleration by thermal pressures using nanosecond laser pulses. The essential necessity was providing contrast ratios above $10^8$ for suppression of relativistic self-focusing when using <ps laser pulses. This was supported by other independent measurements where plane wave front interaction was verified for confirming interaction with dielectrically increased skin layers [1]. When applying these results with ps laser pulses of powers above TW, PW (petawatt) and higher for igniting a

fusion flame in solid density uncompressed fusion fuel [13], the interesting results from plane geometry computations needed a solution how a section from the plane interaction with a cylindrical plasma can overcome lateral losses. For these purpose, a cylindrical magnetic confinemet was studied [14] by using magnetic fields up to 300 Tesla as reported in the following. This is a first evaluation with respect to the ignition of the fusion flame by plasma blocks produced by nonlinear force diven ps laser pulses. This is on the way to the recent generated magnetic fields above 10 kTesla [5].

## 2. Shock generation and penetration speeds of fusion flames

The system of equation describing the dynamics of the plasma particles interaction with electric and magnetic fields is based on the two fluid approximation. Fluid conservation equations for each plasma species plus Maxwell's equations are numerically computed using finite Volume Method (FVM). The computations are based on the plasma block acceleration from the dielectric increased interaction thickness of the laser irradiated skin depth given by optical properties. This differs from laser driven skind depth acceleration which is defined by plasma properties where the generation of separated electron and ion beams is essential [15]. The first computations on the ignition of a fusion flame in solid deuterium-tritium (DT) by a picoseconds energy deposition with an energy flux E* were performed using one-fluid hydrodynamics [2], improved later by two-fluid hydrodynamics [16]. The following resported results are based on a genuine two-fluid hydrodynamic code which was used [17][18] in order to study spatio-temporal details [19][20]. The ignition thresholds E* arrived in the range of $10^8$J/cm$^2$ for DT gives ultrahigh acceleration of plasma blocks which are in the range above $10^{19}$cm/s$^2$ [10][11][12] for incident ps laser laser pulses of intensities above $10^{18}$ W/cm$^2$.

The genuine two-fluid computations permitted to follow up the development of the electric field within the plasma for long times after the ps laser ignition had initiated the fusion flame but also the time dependence of the ion density showed the generation of the shock compression in agreement with the Rankine-Hugoniot theory [19], however with many more details than the analytical theory could cover. It was remarkable that the shock front did not build up immediately after the ps laser ignition of the fusion flame and it needed times of many hundred ps before the shock compression was generated. The very general code including thermal mechanisms with damping and equipartition between electrons and ions etc. led also to understand how the shock front depending on time was increasing its depth. The velocities of the shock front at ignition above the threshold were above 1000 km/s in DT at times of few ns after intiating the block ignition, decaying from initial velocites above 4000 km/s [20].

One example of the shock generation is shown in Figure 1 with the DT densities at a series of times as parameter. In this case a laser intenisty up to $10^{20}$W/cm$^2$ from a KrF laser was used.

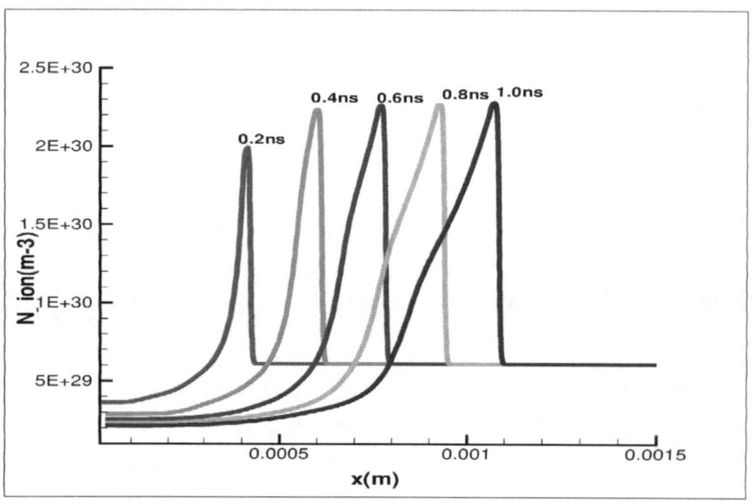


**Figure 1.** Ion density $N_i$ depending on the depth X in the initially cold fuel at different times after the ps laser irradiation

The shock generation of four times the initial density as known form the Rankine-Hugoniot theory has not been reached at 0.2 nanoseconds after picosecond laser ignition of the fusion flame was irradiated. The generation of the increasing thickness of the shock front is the result of the complete thermal computation with separate electron heating, heating of the ions by equipartition and the dominating pressure by th ion fluid.

## 3. Elimnation of lateral energy losses

One critial point of the igniton by a fusion flame with nonlinear force driven ultrahigh acceleration of plasma blocks is the conversion of the results with plane geometry presumptions to real power station conditions. One has to select a cylindrical section of the nonlinear force ignition instead the infinite plane geometry. Taking out a cylindrical section for the plane wave geometry has the problem of the lateral losses of confinement for the cylindrical fusion fuel.

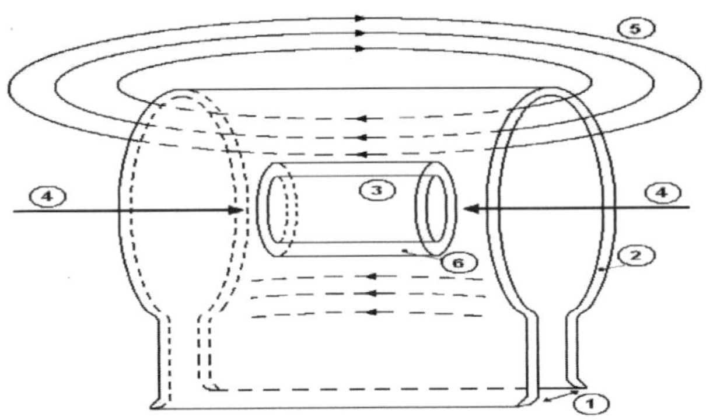


**Figure 2.** Production of a pulsed magnetic field 5 of 100Tesla if few nanosecond duration by a pulsed discharge 1 with the coil 2 for confining the cylindrical plasma 3 generated by ps laser pulse generated plasma blocks 4 from both sides in order to confine the plasma surface 6, according to [14].

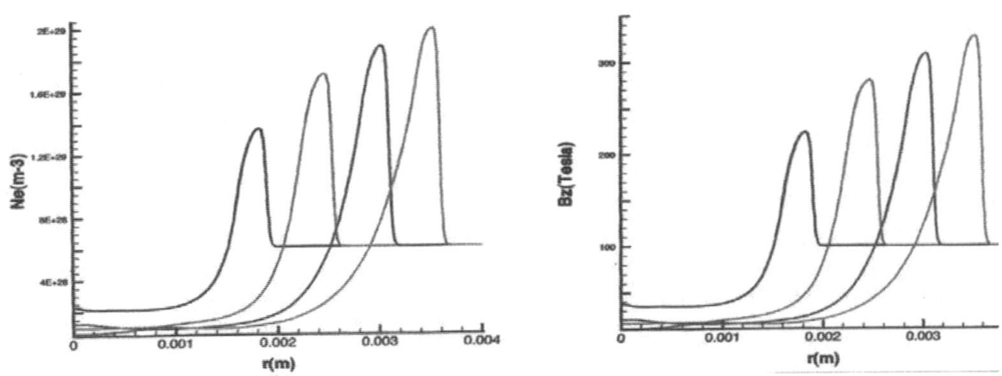


**Figure 3.** Computation of electron density $N_e$ and axial cylindric magentic field $B_z$ depending on radial cylindric coordinate r at times (from the left of the plots) 0.4, 0.8, 1.2 and 1.6 nanoseconds of a cylindrical solid density DT plasma of 1 mm radius at time zero located in an axially parallel magnetic field of 100 Tesla. At time zero a $10^{20}$ W/cm$^2$ KrF laser pulse of one ps duration produced a plama block for igniting a fusion flame.

Figure 2 describes the device for producing initially a magentic field of 100Tesla when a $10^{20}$ W/cm$^2$ KrF ps laser pulse initiates the nulear fusion flame in DT at the ends of the plasma cylinder. The ps interaction occurs in a fuel depth of 5 micrometer where the nonlinear force driven plasma block contains directed ions from the ultrahigh plasma block acceleration. Before the interaction, an external 100Tesla magnetic field is applied and results are plotted in Fig. 3 in the sequence of 0.4, 0.8, 1.2 and 1.6 ns. The genuine two-fluid hydrodynamic equations [16, 18] in contrast to the earlier computations [3] describe the motion [19] of the plasma by the electron density $N_e$ which is nearly identical with the ion density showing how the plasma pressure is driving the magentic field out from the center while simultaneously the magnetic field is steepening up in the periphery to 300 Tesla. The whole dynamic is described including the time dependents of the electric field and of the axial magentic field and rotational plasma motion due to **E**×**B** term, showing a trapping of the radial alpha fluid expansion and the resulting fusion gain.

The computations show the limitation of radial-cylindric energy losses and conditions for generating fusion energy of this scheme. Alternative to this cylindrical magnetic scheme to control losses of the fusion reactions one can use spherical geometry. This will be improved when using initial magentic fields of more than 10 kTesla which are now being developed [5].

It is possibile to completly suppress lateral losses by the spherical geometry instead of the plane irradiation of the solid density DT fusion fuel, by producing spherical plasma blocks by ultrafast acceleration with picosecond laser pulses by lossless conversion of optical energy into mechanical compression producing a space charge neutralized DT ion block directed to the spherical center with 80keV ions. Working with the genuine two-fluid hydrodynamics [17][18][19][20] and the fluid of the generated alpha particle fluid, the spherical irradiation of 125PW laser power compresses the plasma to 20 times the solid state of DT at the center of the sphere at 14.4 ps after the ps laser pulse ignited the fusion flame. Then a rarefaction wave moves in radial direction.

[1] Hora H et al 2007 *Physics of Plasmas* **14**, 072701
[2] Chu M S 1972 Physics of Fluids **15**, 412
[3] Hora H 2009 *Laser and Particle Beams* **27**, 207
[4] Lalousis P, Hora H, Eliezer S, Martinez-Val J-M, Moustaizis S, Miley G H, and Mourou G 2013 *Phys. Lett.* A **377**, 885
[5] Fujioka S, et al. *Scientific Reports*, **3**, 1177, DOI: 10.1038/srep01170.
[6] Hora H 2012 *Laser and Particle Beams* **30**, 325
[7] Strickland D and Mourou G 1986 *Opt. Commun.***56**, 212
[8] Hora H 2013 *Laser and Particle Beams* **31** 361
[9] Hora H 2013 Collective electron interaction at ultrafast acceleration of plasma blocks, in *High Power, high energy and high-intensity laser technology and research using extreme light:entering new frontiers with petawatt-class lasers* J. Hein, G. Korn and L.O. Silva eds. Proceedings of SPIE Vol. **8780**, paper 878024 /1-17.
[10] Hora H 1981 *Physics of Laser Driven Plasmas*. New York: John Wiley, p. 179
[11] Sauerbrey R 1996 *Physics of Plasmas* **3**, 4712
[12] Földes I B, J. Bakos J S, Gal K, Juhasz Z, Kedves M A, Koczis G, Szatmari S and Veres G 2000 et al. *Laser Physics* **10,** 264
[13] Hora H, Lalousis P, Moustaizis S 2014 *Laser and Particle Beams* **32,** 63
[14] Hora H, Lalousis P and Moustaizis S 2013 Nuclear fusion reactor with lateral confinement *German Patent Disclsure* DE 102012001734 A1; 2013.08.01
[15] Labaune C, Depierraux S, Goyon C, Loisel G, Yahia V and Rafelski J 2013 *Nature Communic.* **4**, 2506
[16] Lalousis P and Hora H 1983 *Laser and Particle Beams* **1**, 283
[17] Hora H, Malekynia B, Ghoranneviss M and Miley G H 2008 *Appl. Phys. Lett.* **93**, 011101
[18] Hora H, Lalousis P and Eliezer S 1984 *Phys. Rev. Letters* **53**, 1650
[19] Lalousis P, Hora H, Eliezer S, Martinez-Val J-M, Moustaizis S, Miley G M and Mourou G 2013, *Physics Letters* **339**, 885
[20] Lalousis P, Földes I and Hora H 2012 *Laser and Particle Beams* **30**, 233

# Development of a polar direct drive platform for mix and burn experiments on the National Ignition Facility

**T J Murphy[1], G A Kyrala[1], N S Krasheninnikova[1], P A Bradley[1], J A Cobble[1], I L Tregillis[1], K A D Obrey[1], J A Baumgaertel[1], S C Hsu[1], R C Shah[1], P Hakel[1], J L Kline[1], M J Schmitt[1], R J Kanzleiter[1], S H Batha[1], R J Wallace[2], S Bhandarkar[2], P Fitzsimmons[3], M Hoppe[3], A Nikroo[3] and P McKenty[4]**

[1]Los Alamos National Laboratory, P. O. Box 1663, Los Alamos, NM 87544 USA
[2]Lawrence Livemore National Laboratory, 7000 East Ave, Livermore, CA 94550 USA
[3]General Atomics, P. O. Box 85608, San Diego, CA 92186 USA
[4]University of Rochester Laboratory for Laser Energetics, 250 E. River Rd, Rochester, NY 14623 USA

E-mail: tjmurphy@lanl.gov

**Abstract.** Capsules driven with polar drive [1, 2] on the National Ignition Facility [3] are being used [4] to study mix in convergent geometry. In preparation for experiments that will utilize deuterated plastic shells with a pure tritium fill, hydrogen-filled capsules with copper-doped deuterated layers have been imploded on NIF to provide spectroscopic and nuclear measurements of capsule performance. Experiments have shown that the mix region, when composed of shell material doped with about 1% copper (by atom), reaches temperatures of about 2 keV, while undoped mixed regions reach about 3 keV. Based on the yield from these implosions, we estimate the thickness of CD that mixed into the gas as between about 0.25 and 0.43 $\mu$m of the inner capsule surface, corresponding to about 5 to 9 $\mu$g of material. Using 5 atm of tritium as the fill gas should result in over $10^{13}$ DT neutrons being produced, which is sufficient for neutron imaging [5].

## 1. Effects of mix on burn

Mix of plastic into the DT fuel of an ICF capsule adds non-burning material, diluting the fuel and reducing burn. At a sub-grid scale, the mix can be described as a distribution in concentrations of the DT within a region of mixed DT and plastic.

Turbulent mixing in ICF capsules must be described using models that attempt to capture the important physics in a computationally efficient manner, allowing large-scale simulations of imploding capsules. One such model is the BHR model [6] and its revisions [7, 8]. Results from this model have been used in burn models that use a probability distribution function [9] (PDF) to describe concentration variations at the sub-grid level to determine the effect on burn [10].

In order to measure the mix of materials at an interface and determine its effect, we are performing experiments leading to implosions of capsules with doped-deuterated inner layers and a tritium gas fill [4]. As this layer mixes into the gas, the x ray self-emission will be imaged using a multi-monochromatic imager (MMI) currently under development [11] to determine the amount and extent of mix [12] of the deuterated layer into the tritium gas. This mix will be compared to the source region for DT neutrons [5] which can only come from the region where the deuterium from the deuterated layer mixes with the tritium gas [13]. X-ray and neutron images of sufficient quality require 1 J/sr in the Ly-$\beta$ and He-$\beta$ x-ray lines of the dopant material and a DT neutron yield exceeding $10^{13}$ [5], respectively.

## 2. Defining the measure of mix

To illustrate the PDF model in a simplified way, assume a mixture of deuterium and tritium, initially separated. DT neutron yield increases with increasing mix of the two species. This can be calculated, noting that the yield from a volume element is dependent on the average of the product of the densities of deuterium and tritium, rather than the product of their averages.

$$Y_{DT} = \langle n_D n_T \rangle \langle \sigma v \rangle_{DT} V \tau \tag{1}$$

where, in general,

$$\langle n_D n_T \rangle \leq \langle n_D \rangle \langle n_T \rangle . \tag{2}$$

Defining the concentrations of D and T at any point as $c = n_D/n_i$ and $1 - c = n_T/n_i$, where $n_i$ is the ion density, and letting $c = \bar{c} + \tilde{c}$, the spatially averaged concentration and the spatially varying concentration within a computational volume, one can show:

$$\langle n_D n_T \rangle = \langle n_D \rangle \langle n_T \rangle \theta \tag{3}$$

where

$$\theta = \left[ 1 - \frac{\langle \tilde{c}^2 \rangle}{\bar{c}(1 - \bar{c})} \right] \tag{4}$$

so that

$$Y_{DT} = \theta \langle n_D \rangle \langle n_T \rangle \langle \sigma v \rangle_{DT} V \tau . \tag{5}$$

$\theta$ is referred to as "mixedness" and ranges from 0 to 1; $1 - \theta$ is the "degree of segregation" [14]. This parameter can be directly related to quantities produced by the BHR turbulence model [15].

The goal of these experiments is to image the x ray emission due to mix [11, 12] of doped deuterated plastic shell material into the tritium fill gas in an ICF capsule and image the DT neutrons [5] produced. Results will be compared to determine the mixedness ($\theta$) and how it compares to simulations utilizing the BHR2 turbulent mix model and the PDF burn model.

## 3. Copper-doped deuterated shell implosions

Two implosions were performed using capsules (figure 1) in which the inner 3 $\mu$m were deuterated and which contained a 1-$\mu$m copper-doped layer either at the inner surface or buried under 1 $\mu$m of undoped deuterated plastic. The capsules were filled with 5 atm of hydrogen. These implosions were performed using polar direct drive [1, 2]. Laser pointing that produces the necessary symmetry has been simulated [16] and symmetric implosions have been obtained. Total laser energy of 355 kJ on the capsules in a 2.15 ns square pulse was used. Power in the polar beams was reduced to 80% of the equatorial beams to improve the drive symmetry.

X ray spectroscopy (figure 2) using the SuperSnout [17] spectrometer shows a number of x-ray lines from the surface doped layer, while none are apparent from the buried layer. This indicates that less than 1 $\mu$m of shell material mixed into the gas. The $\beta$ lines of copper were too weak for imaging with the planned MMI but the line ratios indicate that the temperature in the mix region is about 2 keV. Detailed modeling suggests that changing from copper to a lower

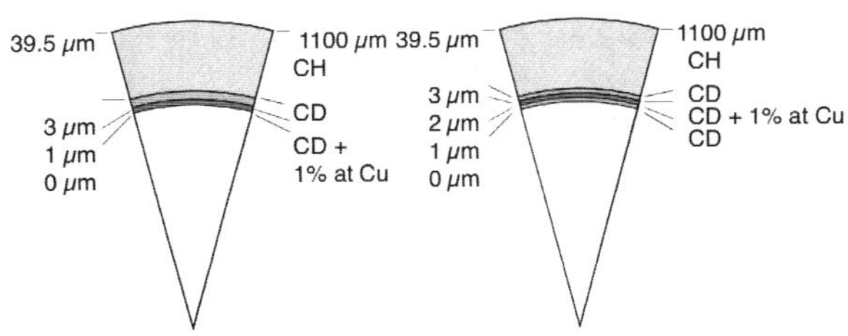


**Figure 1.** Deuterated shell capsules with (left) surface copper-doped layer and (right) buried copper-doped layer.

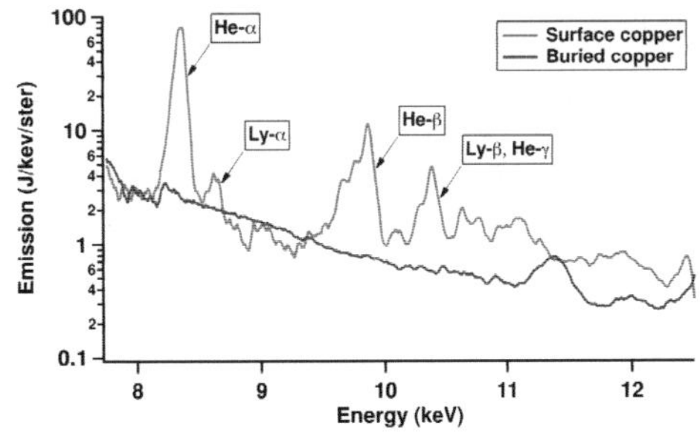


**Figure 2.** Time-integrated emission spectra for capsules with a 1 $\mu$m copper-doped layer and with a similar layer buried under 1 $\mu$m of unroped plastic.

atomic number dopant such as titanium or iron should produce sufficient x rays for imaging. Reduction of the concentration may be necessary to minimize the effect of the dopant on the implosion.

Neutron yield measurements from these implosions allow an estimate of the DT neutron yield that would have been produced had the capsules been filled with 5 atm of tritium gas instead of the hydrogen that was used for these experiments. If we assume that the yield comes from a layer of deuterated plastic of thickness $\Delta_o$ that uniformly mixes into the shell, then then the number of deuterium atoms that mix into the fuel is given by

$$N_d = 4\pi R_{io}^2 \Delta_0 n_{CD} \tag{6}$$

where $n_{CD}$ is the number density of CD in the unimploded capsule. From this, the DD neutron yield can be estimated:

$$\begin{aligned} Y_{DD} &= \frac{1}{2} n_D^2 \langle \sigma v \rangle_{DD} V_f \tau \\ &= \frac{1}{2} \left( \frac{N_D}{V_f} \right)^2 \langle \sigma v \rangle_{DD} V_f \tau \\ &= 6\pi C^3 R_{io} \Delta_o^2 n_{CD}^2 \langle \sigma v \rangle_{DD} \tau \end{aligned} \tag{7}$$

where the convergence ratio is defined as $C = R_{io}/R_{if}$.

The mix depth needed to explain a given neutron yield in this model is given by

$$\Delta_0 = \sqrt{\frac{Y_{DD}}{6\pi C^3 R_{io} n_{CD}^2 \langle \sigma v \rangle_{DD} \tau}}. \tag{8}$$

We can then estimate the DT yield resulting if the hydrogen fill is replaced by tritium at initial number density $n_{to}$ using equation 5:

$$\begin{aligned} Y_{DT} &= \theta \langle n_D \rangle \langle n_T \rangle \langle \sigma v \rangle_{DT} V_f \tau \\ &= \theta N_D \langle n_T \rangle \langle \sigma v \rangle_{DT} \tau \\ &= \theta 4\pi R_{io}^2 \Delta_o n_{CD} n_{To} C^3 \langle \sigma v \rangle_{DT} \tau \end{aligned} \tag{9}$$

Results from these equations are shown in table 1 assuming a convergence of 7, complete mixing of the deuterium in to the tritium, and that the burn time $\tau$ can be approximated by the measured x-ray emission width.

**Table 1.** Measurements used in determining the amount of deuterated plastic which contributes to the neutron yield in the deuterated shell implosions, and the estimated DT yield if the capsuled had been filled with tritium gas.

| Capsule | Surface doped layer | Buried doped layer | |
|---|---|---|---|
| $T_i$ | 2.00 | 3.03 | keV |
| $\langle\sigma v\rangle_{DD}$ | $3.00 \times 10^{-21}$ | $1.59 \times 10^{-20}$ | $cm^3\ s^{-1}$ |
| $\tau$ | 343 | 398 | ps |
| $Y_{DD}$ | $1.28 \times 10^9$ | $7.38 \times 10^9$ | neutrons |
| $\Delta_{max}$ | 0.26 | 0.25 | $\mu m$ |
| Estimated $Y_{DT}$ | | | |
| $\langle\sigma v\rangle_{DT}$ | $2.71 \times 10^{-19}$ | $1.81 \times 10^{-18}$ | $cm^3\ s^{-1}$ |
| $Y_{DT}$ | $1.85 \times 10^{12}$ | $1.39 \times 10^{13}$ | neutrons |

## 4. Summary

Data from these implosion experiments contribute to the design for a mix and burn platform on the National Ignition Facility. X-ray spectroscopic data has been used to determine the range of elements that would be suitable as a tracer for imaging the mix of a deuterated plastic layer into the fuel gas. Neutron data suggests that with a tritium fill, a sufficient neutron yield would be produced to allow neutron imaging.

## Acknowledgments

G. P. Grim provided additional experimental support. S. P. Regan and M. A. Barrios provided the x-ray spectra and S. Khan provided x-ray temporal emission data. D. L. Bleuel provided neutron yields. This work was performed under the auspices of the U.S. DOE by LANL under contract DE-AC52-06NA25396.

## References

[1] Skupsky S *et al.*, *Phys. Plasmas* **11**, 2763 (2004).
[2] Kyrala G A *et al*, *Laser Part. Beams* **23**, 187 (2005).
[3] Moses E I, *Fusion Sci. Tech.* **54**, 361 (2008).
[4] Schmitt M J *et al.*, *Phys. Plasmas* **20**, 056310 (2013).
[5] Grim G P *et al.*, *Phys. Plasmas* **20**, 056320 (2013).
[6] Besnard D, Harlow F H, Rauenzahn R M and Zemach C, "Turbulence Transport Equations for Variable-Density Turbulence and Their Relationship to Two-Field Models," Los Alamos National Laboratory Report No. LA-UR-12303, 1992 (unpublished).
[7] Banerjee A, Gore R A and Andrews M J, *Phys. Rev.* E **82**, 046309 (2010).
[8] Stalsberg-Zarling K and Gore R, "The BHR-2 Turbulence Model: Incompressible Isotropic Decay, Rayleigh-Taylor, Kelvin-Helmholtz, and Homogeneous Variable Density Turbulence," LA-UR-11-04773 (unpublished).
[9] Delichatsios M A and Mathews M K, *Fire Safety Science — Proc. of 2nd Intl. Symp.*, p 149, 1989.
[10] Fincke J R, unpublished.
[11] Kyrala G A *et al.*, *Proc. SPIE* **8850**, 88500P (2013).
[12] Welser-Sherrill L *et al*, *High Energy Density Physics* **5**, 249 (2009).
[13] Wilson D C *el al.*, *Phys. Plasmas* **18**, 112707 (2011).
[14] Danckwerts P V, *Appl. Sci. Research* **A3**, 279 (1952).
[15] Haines B M, Ristorcelli R and Grinstein F, submitted to *Phys. Rev.* E, LA-UR-12-27011.
[16] Krasheninnikova N S, Finnegan S M and Schmitt M J, *Phys. Plasmas* **19**, 012702 (2012).
[17] Regan S P *et al*, "Supersnout II diagnostic for the National Ignition Facility," *Rev. Sci. Instrum.*, in prep.

# Laser ion acceleration and neutron source in short-pulse solid-nanoparticle interaction

**K Nishihara[1,2], T Watari[1], K Matsukado[1], T Sekine[1], Y Takeuchi[1], M Takagi[1], N Satoh[1], T Kawashima[1] and H Kan[1]**

[1] Hamamatsu Photonics, K.K., 1820 Kurematsu, Hamamatsu 431-1202 Japan
[2] Institute of Laser Engineering, Osaka University, Suita, Osaka 565-1801 Japan

E-mail; nishihara@ile.osaka-u.ac.jp

**Abstract**. We propose both an efficient neutron source and an extremely high energy proton source using solid CD and CH nano-particles, respectively, irradiated by an intense laser light. With a use of 3-d PIC simulations, we obtain an optimum CD radius for a neutron source, 250 nm and required laser field of $a=eE/m\omega c \approx 2$, which results in D-D reaction rate of $<\sigma v> = 2\mathrm{x}10^{-16}$ $cm^3/s$, corresponding to an effective deuteron temperature of 500 keV to 1MeV. Reduction of neutron yield by pre-expansion is discussed. In a range of $a \approx 100$, laser radiation pressure surrounding the particles accelerates electrons in the forward direction. Protons following the electrons become directional high energy, for example, proton energy of 450 MeV is obtained within 130 fs in CH particle interaction with 700 nm in radius. More than 10% of total protons in CH-particles are accelerated forward. Proton energy continuously increases with time and with the increase of particle size and the direction is also collimated with time.

## 1. Introduction

After the observation of nuclear fusion from the explosion of deuterium clusters heated with a compact, high-repetition-rate table-top laser [1], a number of experiments were performed to obtain neutrons and high energy protons using clusters and small particles (many references in Ref. [2]). If the laser intensity is high enough to expel all of electrons from the small particles, Coulomb explosion occurs and ion energy distribution function (EDF) becomes proportional to square root of the energy $dE / dN \propto \sqrt{E}$ up to its maximum value proportional to square of the particle radius $E_{\max} \propto R_0^2$ [3]. Since D-D fusion cross section has its maximum value at the energy of a few MeV, high neutron yield can be obtained by choosing proper particle radius so that $E_{\max}$ becomes a few MeV. There exist a critical laser intensity to expel electrons for a given particle radius. We performed 3d PIC simulations, using REMP[4], to obtain optimum conditions for a neutron source for solid CD nano-particles in a range of the normalized laser intensity of $a = eE/m\omega c = O(1)$.

The radiation pressure plays an essential role for proton acceleration at high intensity regime of $a$ = 100 [5]. For CH particles comparable to a laser wavelength, laser light can penetrate into the particle due to both relativistic effect and its pressure. The radiation pressure accelerates efficiently electrons in the forward direction and high energy protons follows electrons. We show that protons following electrons become directional high energy and its energy continues to increase with time. It should be mentioned that we have also observed experimentally the generation of stable neutrons and high energy protons using solid CD and CH nano-particles [6].

Simulation box size was (120λ, 10λ, 10λ), where λ (=*c/f*) is incident laser wavelength. A particle located at 20λ from x-boundary and the center in y-z plane was irradiated by a linear polarized Gaussian pulse with FWHM of 40 $f^{1}$ and duration of +/- 30 $f^{1}$. Initial electron density is given as $\omega_{pe} = 6\omega$, namely electron plasma frequency is six times greater than the laser frequency.

## 2. Efficient neutron source and its conditions

### *2.1. Optimum conditions*

We performed 3d-PIC simulations for particle sizes of 200 – 500 nm in diameter and laser intensities of $a$ = 0.01 to 10. The optimum conditions found among those parameters are CD particle with 500 nm in diameter and the normalized laser electric field of $a = eE/m\omega c = 2$. Deuteron energy spectrum obtained is presented in Figure 1 with D-D fusion cross section, which shows that the deuteron energy spectrum has its peak near the energy corresponding to the maximum fusion cross section. The D-D reaction rates $<\sigma v>$ are estimated by averaging the cross section over the deuteron energy spectrum

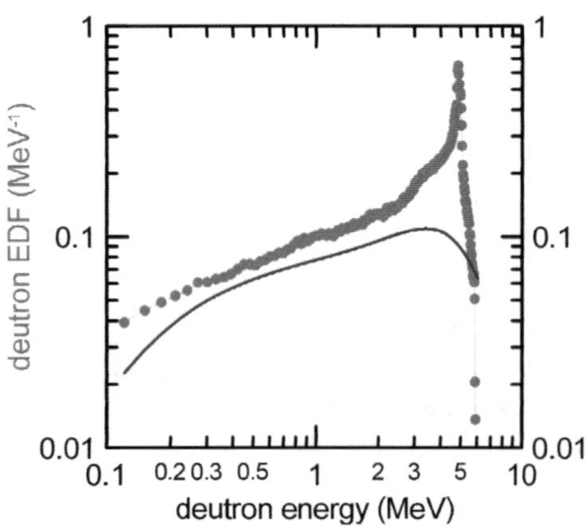


**Figure 1.** Deuteron energy spectrum for 500 nm CD particle irradiated by a laser of $a = eE/mc\omega = 2$ (dot) and cross section of D-D reaction (line).

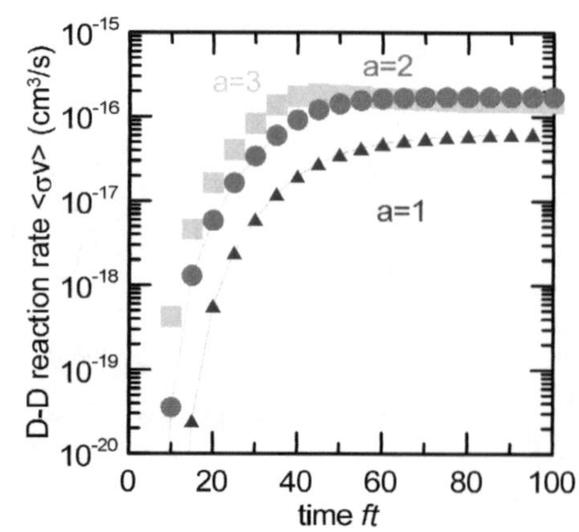


**Figure 2.** D-D reaction rate averaged over deuteron EDF of 500 nm CD irradiated by intensities of $a$ = 1, $a$ = 2, and $a$ = 3.

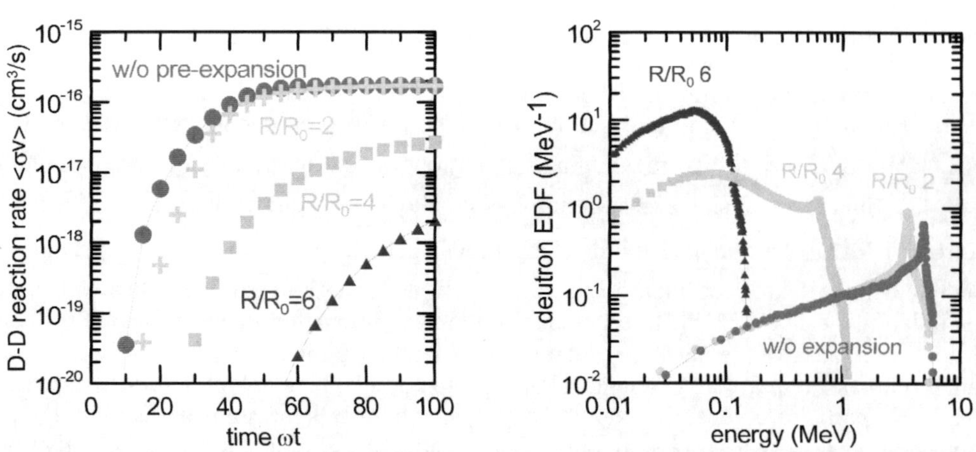


**Figure 3.** D-D reaction rate (left) and energy distribution function (right) for CD particles with pre-expansion, cross (cyan); R/R0 = 2, green (square); R/R0 = 4, triangle (blue); R/R0 = 6, and red (dot); without pre-expansion.

observed and are shown as function of time in Figure 2, for different *a* of 1, 2 and 3. For the case of *a* = 1, the cross section is small, while the cross section decreases with time for *a* = 3. The decrease is due to the deuteron energy too large for D-D fusion reaction.

The estimated D-D reaction rate of 2x10$^{-16}$ cm$^3$/s in neutron branch in D-D reaction corresponds to that of effective deuteron temperature of 500 keV to 1 MeV for a Maxwellian velocity distribution function.

### 2.2. *Suppression of D-D reaction rate by pre-expansion due to pedestal*

Pre-expansion due to pedestal or prepulse reduces ion energy. Reduction of D-D reaction rate is investigated by assuming initial density profile of a spherical isothermal expansion [7], $n(r) = n_0 (R_0/R)^2 \exp\left(-(r/R)^2\right)$, where $R/R_0$ is the ratio of the expanded radius to its initial value. As shown in Figure 3, the reduction is small for the case of $R/R_0$ = 2. This expansion may correspond roughly to the contrast ratio of $10^{-7\ \mathrm{to}\ -8}$ for 100 ps pedestal estimated from electron temperature scaling of $I = 0.6nT_e v_{Te}$, where $I$ is the pedestal pulse intensity, and $T_e$ and $v_{Te}$ are electron temperature and its thermal speed, respectively and assuming the expansion speed is given by ion sound speed [7]. However for $R/R_0$ = 4 and 6, the maximum energy and thus the reaction rate decreases very much as shown in Figure 3.

## 3. High energy directed proton source

Proton energy spectra and propagation direction are investigated for a laser with the normalized electric field of *a* = 100 interacting with CH particles with different sizes. Figure 4 shows a typical energy spectrum for a CH particle with 1.4 μm in diameter at time of *ft* = 95. It consists of three energy groups. Here we pay attention on the 1st group (hatched region), 386 MeV < *E* and the peak energy at $E_{peak}$ = 415 MeV indicated by arrow, since they are directed forward as shown by their solid angle distribution in Figure 5. The solid angle distribution at time *ft* = 95 becomes narrower than that at time *ft* = 75, it indicates that the acceleration direction is collimated with time. The radiation pressure accelerates efficiently electrons in the forward direction and high energy protons follows electrons and become directional high energy.

As shown blue squares in Figure 6, the ratio of the high energy protons to the total protons increases with the increase of the radius. For 1.4 μm CH, the high energy proton becomes greater than 10 % at *ft* = 95. It should be also mentioned that the 1st peak energy indicated by an arrow in Figure 4

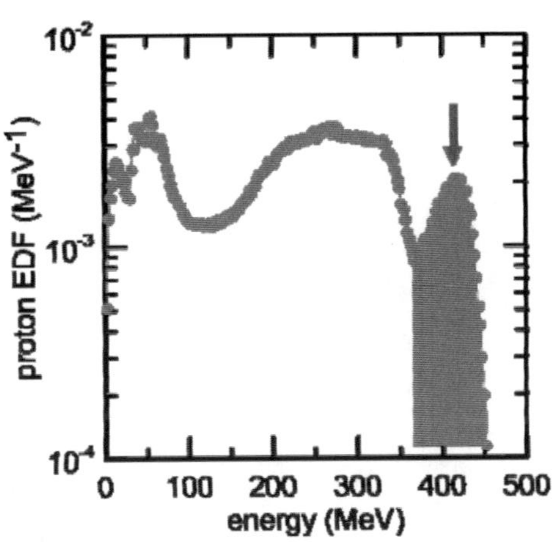


**Figure 4.** Proton energy distribution function at time *ft* = 95. There is a high energy group of 386 MeV < E (shaded).

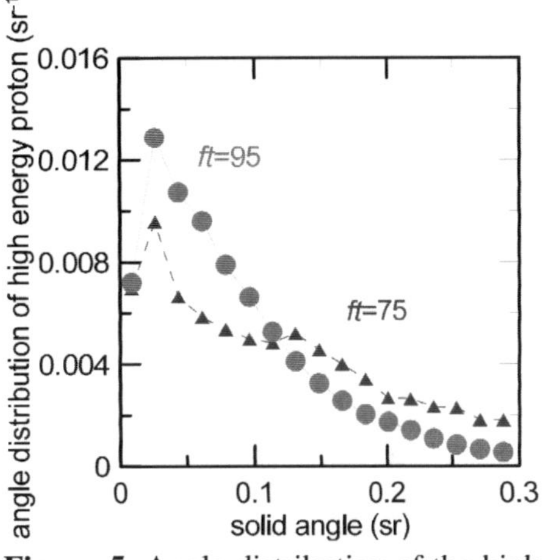


**Figure 5.** Angle distribution of the high energy protons in Figure 4 (E > 386 MeV) at *ft* = 75 (blue) and 95 (red).

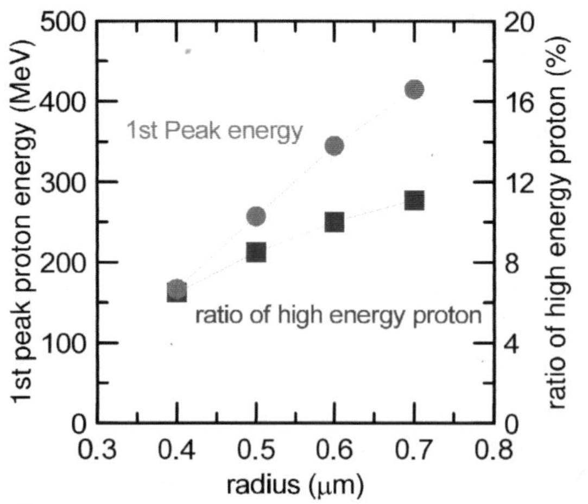


**Figure 6.** The 1st peak energy and ratio of the high energy protons to total protons for different CH radii at *ft* = 95.

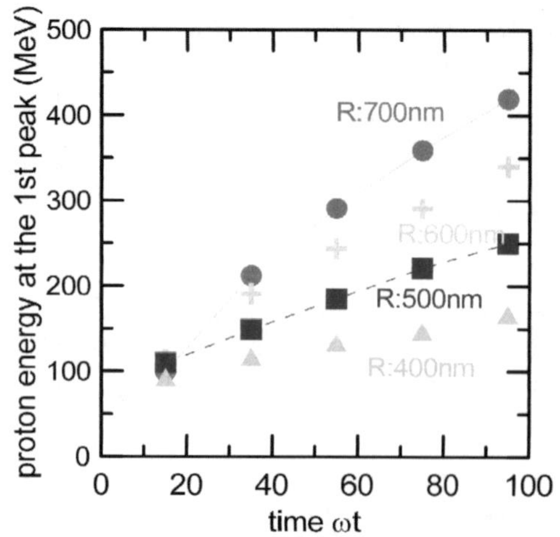


**Figure 7.** The 1st peak energy for different CH particle radii of 0.4, 0.5 0.6 and 0.7 μm as function of time.

increases with the increase of the radius.

Figure 7 shows the 1st peak energy as functions of time for various CH particle sizes. It should be noted that the proton energy increases with time even after the laser pulse duration of $60 f^{1}$ with $40 f^{1}$ FWHM. It indicates that the protons follow the electrons continuously accelerated by the laser pulse, which may be similar to the unlimited acceleration observed for mass limited target [5]. The peak energy increases as the radius is larger as shown in Figure 7. These results indicate that directed relativistic protons of GeV can be obtained using larger particles even at this laser intensity.

## 4. Conclusion

Optimum CD nano-particle size and laser intensity for maximizing D-D reaction rate were obtained at 500 nm in diameter and laser of $a = 2$. D-D reaction rate of $<\sigma v> = 2\times10^{-16}$ cm$^3$/s, which corresponding to effective deuteron temperature of 500 keV to 1 MeV, can be obtained. We also investigated the suppression of D-D reaction by pre-expansion of the particles due to pedestal. The contrast ratio should be less than $10^{-7 \text{ to } -8}$.

In radiation pressure dominant regime ($a$ = 100), protons following electrons become directional high energy of at least 450 MeV within 130 fs for CH particle with 1.4 μm in diameter. More than 10 % of total protons in the CH particles are accelerated forward. Proton energy increases with time and with the increase of the particle size, and the propagation direction is also collimated with time. These results indicate that relativistic (GeV) and collimated protons can be obtained at this laser intensity.

**Acknowledgments**

Simulations were performed on NEC SX-systems at the Cybermedia Centre of Osaka University.

**References**

[1] Ditmire T et al. 1999 *Nature* **398** 489
[2] Daido H et al. 2012 *Rept. Prog. Phys.* **75** 056401
[3] Nishihara K et al. 2001 *Nucl. Instrum. Methods Phys. Res. A* **464** 98
[4] Esirkepov T 2001 *Comput. Phys. Commum.* **135** 144
[5] Bulanov S 2010 *Phys. Rev. Lett.* **104** 135003
[6] Watari T et al. 2013 *This Proceeding* (to be published)
[7] Murakami M and Basko M M 2006 *Phys Plasmas* 13 012105

# Atomic process modeling in dense hydrogenic plasmas by nearest neighbor approximation

**T Nishikawa**
Graduate School of Natural Science and Technology, Okayama University, Okayama 700-8530, Japan

E-mail: nishikawa.takeshi@okayama-u.ac.jp

**Abstract.** Atomic process modelling for obtaining the electronic state densities of the bound and free electrons and calculating the resulting average ionization degrees of dense hydrogenic plasmas is studied. In the analysis of atomic processes in plasmas, the Saha equation is usually solved to obtain population of each ionic species of various electronic configurations and resulting average ionization degrees as a function of electron temperature and plasma density under the assumption that the plasmas are in a certain equilibrium, for example, the thermodynamic equilibrium. I had proposed an atomic model based on the nearest neighbor approximation to determine a finite state density of the bound state and a state density of the free electron in hydrogenic plasmas. Using the models, we had been able to calculate the population for ions of various electronic configurations in hydrogenic plasmas without any ad hoc parameters to determine the electronic state densities. Unfortunately, the models had not been able to apply for dense plasmas near solid density. In this paper, examining the effect of the Coulomb interaction between the two ions and the Saha equation, we discuss the applicability of the models for dense hydrogenic plasmas.

## 1. Introduction

I have already developed an atomic process modeling for obtaining the electronic state densities for the bound and free electrons, and calculating the resulting average ionization degrees of hydrogenic plasmas using nearest neighbor approximation [1]. Using the modeling, we have solved a long-running problem that we always encounter when we want to calculate the average ionization degrees of plasmas; "How many highly excited states we have to take into account in the atomic process model". However, when we apply the model for plasmas of near solid-density that are typically found in the high-energy density physics, the resulting average ionization degrees abruptly changed into the fully ionized values before the total number density of atom and ion is far less than that of the solid density of $5\times10^{22}$ cm$^{-3}$. These results are not physically acceptable. To fix the problem, I reexamine the state densities of the bound and free electrons of plasmas near solid density. By introducing the Coulomb correlation between the two ions and modifying the Saha equation, I could successfully obtain physically acceptable results of the average ionization degrees of hydrogenic plasmas.

## 2. Nearest neighbor approximation with Coulomb correlation

In the nearest neighbor approximation, I used the fact that the probability finding the nearest neighbor ion at $R$ in the spherical coordinates is expressed by the simple algebraic equation

$$P(R) = 4\pi R^2 N_{\rm p} \exp\left(-\frac{4}{3}\pi R^3 N_{\rm p}\right) \tag{1}$$

where $N_{\rm p}$ is the number density of perturbing particles [2]. Basically, $N_{\rm p}$ is equal to the number density of the ions, $N_{\rm i}$. For the special case of the fully ionized that we will often used in this paper, $N_{\rm p}$ is equal to the total number density of the ions and neutral atoms, $N_0$. To calculate the state density for the bound electron, the two hydrogenic ions are set next to each other. The potential profile around the ions is given by the superposition of the two ion's Coulomb potential. The hydrogenic ion has a

Journal of Physics: Conference Series **688** (2016) 012077 doi:10.1088/1742-6596/688/1/012077

classical bound state, and the energy is assumed to be unchanging even in the different potential profile from the Coulomb potential. In this case, the potential profile around the ions shows a saddle point. The electronic state is assumed that of a free electron for energies above this saddle point and that the energy values below the saddle point indicate bound states. The probability of the bound state of the hydrogenic ion, of which principal quantum number is $n$, is given by

$$w_n = \int_{8n^2/Z}^{\infty} P(R)dR \tag{2}$$

where $Z$ denotes the atomic number of the ion. For the free state, the state density $f(E)$ as a function of the electron's total energy of the potential and kinetic energy is calculated by

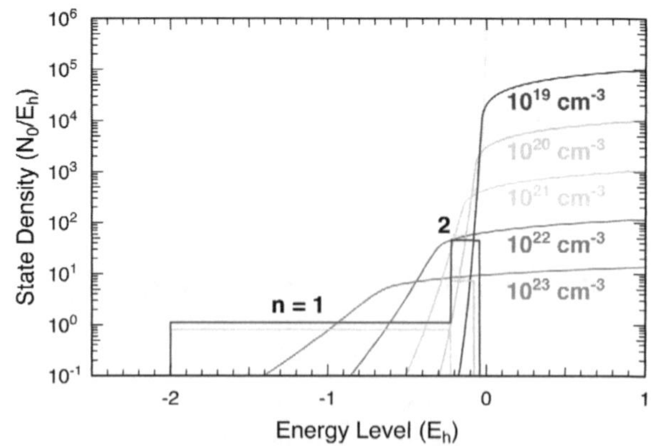


**Figure 1.** State Densities of free and bound electrons for various ion number densities of ion for fully ionized hydrogenic plasmas.

$$f(E) = \frac{8\pi\sqrt{2m_e^3}}{h^3}\int_0^{\infty} P(R)dR\int_0^{R_0} 2\pi r^2\,dr\int_0^{\pi}\sin\theta\,d\theta\sqrt{E + \frac{Ze^2}{4\pi\varepsilon_0}\left(\frac{1}{r} + \frac{1}{\sqrt{R^2 - 2Rr\cos\theta + r^2}}\right)} \tag{3}$$

where $m_e$, $h$, $e$, and $\varepsilon_0$ denote the mass of the electron, Planck's constant, the elementary charge, and the permittivity of free space, respectively. I used the relation between the electron's energy and momentum at $(r, \theta)$ in the spherical coordinates inside the ion sphere radius $R_0$ as $4\pi R_0^3 N_i / 3 = 1$. To accommodate the effect, we rewrite the relation between the electron's energy and momentum as

$$\frac{p^2}{2m_e} = E + \frac{Ze^2}{4\pi\varepsilon_0}\left(\frac{1}{r} + \frac{1}{\sqrt{R^2 - 2rR\cos\theta + r^2}}\right) \tag{4}$$

instead of its conventional form as $p^2 / 2m_e = E$. The first term inside the bracket in the right-hand side of equation (4) corresponds to the Coulomb potential of the ion that we are calculating its state densities, and the second term are due to the nearest neighbor ion's Coulomb potential.

Figure 1 shows the energy dependence of the state densities of the free and bound states of the hydrogenic ions for near its energy equal to zero using the nearest neighbor approximation. The energy is normalized by twice the hydrogenic ionization potential, i.e., Hartree units. By the atomic model using the nearest neighbor approximation, the free state density further exists below $E = 0$ because the model assumes that the electronic state is that of free electron for energies above the saddle point that is formed by the nearest neighbor ions. Expressing the continuity of the state densities from free to bound in the figure, I show the bound states as a reduced continuous bound state with a statistical weight $2n^2$ spread over the range between the neighboring states that leads to an area equal to its original statistical weight $2n^2 / (E_{n+1/2} - E_{n-1/2})$, i.e., the area below each step in figure 1 is equal to $2n^2$. It may be noted that the vertical axis represents the logarithm scale. The reduced continuous bound state profile converges to the asymptotic behavior of the number of bound state below $E = 0$ given by $2^{-3/2}(-E)^{-5/2}$. Using the model, the total amount of the electronic state density of bound and free is continuously changing as a function of the electron's energy. However, the free state density exists below its energy below $E = -0.5E_h$ for the case of $N_p = 10^{22}$ cm$^{-3}$. This means that the average ionization degrees calculated by the model does not become zero even in the extreme case of temperature equal to zero. We know the average ionization degrees of low temperature near 0 K is zero even in the case of solid density. This is partly responsible for neglecting the Coulomb interaction between two ions. In the original atomic model using the nearest neighbor approximation [1], the Coulomb correlation between ions is neglected simply because we would like to compare the results with the results obtained by the atomic model based on the plasma microfield [3]. In the atomic model based on the microfield, the Holtsmark field is used for the microfield that is derived from the same assumption used in obtaining the probability of equation (1). To examine the effect, I introduced the effect of the Coulomb interaction into the equation (1) as

$$P'(R) = k \times 4\pi R^2 N_p \exp\left(-\frac{4}{3}\pi R^3 N_p - \frac{1}{4\pi\varepsilon_0}\frac{Ze^2}{Rk_B T_i}\right) \tag{5}$$

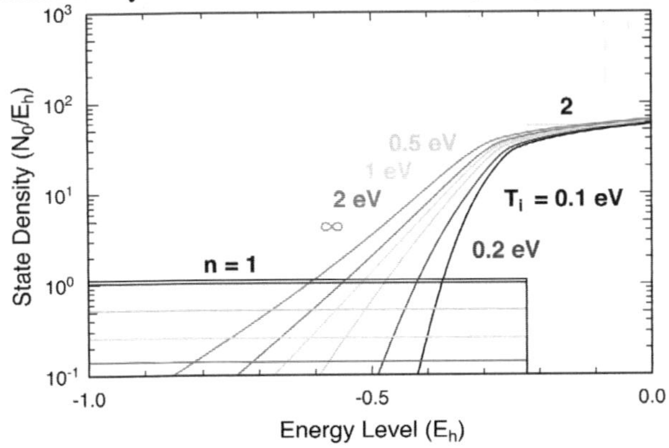


**Figure 2.** State Densities of free and bound for various ionic temperature for fully ionized hydrogenic plasmas. Number density of ion is $10^{22}$ $cm^{-3}$.

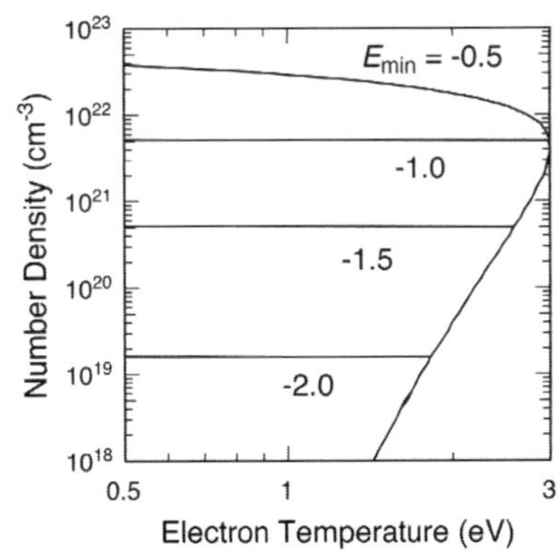


**Figure 3.** Conturs of average ionization degrees are 0.5 of hydrogenic plasmas for various cut-off minimum energy of free electron in equation (7).

where $T_i$ is the temperature of the ion. In equation (5), $k$ is introduced to satisfy

$$\int_0^\infty P'(R)dR = 1 \tag{6}$$

Figure 2 shows the state densities for the bound and free electrons for various ion temperatures that determine the Coulomb correlation between two ions. From figure 2, we can see that the state density of the free electron of lower energy increases as the temperature of the plasma becomes higher. When the temperature is low enough, the state density of free electron of its energy of $E$ = -0.5 $E_h$ is small enough to be calculated as its average ionization degrees to be zero.

Using the above-mentioned state densities for the bound and free electrons, we can calculate the average ionization degrees of hydrogenic plasmas. Though the Saha equation that has been already integrated by the energy of free electrons, usually used in the calculation of the average ionization degrees, we use the equation before the integration

$$\frac{N_{Z+1}N_e}{N_Z} = \frac{U_{Z+1}}{U_Z}\exp\left(-\frac{I_p}{k_BT_e}\right)\frac{8\pi\sqrt{2m_e^3}}{h^3}\int_{-\infty}^{\infty} f(E)\exp\left(-\frac{E}{k_BT_e}\right)dE \tag{7}$$

In equation (7), the free electron state density $f(E)$ defined by equation (3) is used instead of conventional $\sqrt{E}$ . Even if we use the equations, we cannot remove the difficulty that the average ionization degrees abruptly become zero. To clarify the origin of the difficulty, we calculate the contours of the average ionization degree of 0.5 for various lower limit of the free electron energy instead of its original lower energy of $-\infty$ defined in equation (7). In figure 3, the dependence of various lower limit of the free electron energy is shown. When we calculate the average ionization degrees with the lower energy limit of -0.5$E_h$, the resulting values are physically appropriate. As the lower energy limit becomes lower, the abrupt ionization shows in lower densities. This indicates that the difficulty is originated from the free state density below $E$ < -0.5$E_h$. Detailed investigation shows that the integration in equation (7) becomes unphysically overestimated if the electron temperature becomes substantially small. This means that the Boltzman factor in equation (7) becomes larger than unity when the electron energy level of the free sate are below the energy of the ground state. To remove the difficulty, we modify the Saha equation to be

$$\frac{N_{Z+1}N_e}{N_Z} = \frac{U_{Z+1}}{U_Z}\exp\left(-\frac{I_p}{k_BT_e}\right)\frac{8\pi\sqrt{2m_e^3}}{h^3}\left(\int_{-\infty}^{-I_p} f(E)dE + \int_{-I_p}^{\infty} f(E)\exp\left(-\frac{E}{k_BT_e}\right)dE\right) \tag{8}$$

In equation (8), the free electron is counted without the Boltzmann factor in the integration if its energy becomes smaller than the energy level of the ground state of hydrogenic ion. Using equation (8), two contours of the average ionization of hydrogenic plasmas are obtained, figure 4 and 5. The difference between two results is the treatment of the neutral atom. In calculating the average ionization degrees in figure 4, the number density of the neutral atom is omitted in calculating the equation (5) as a perturber, while in the calculation in figure 5, the number density of the neutral atom is added in $N_p$ in equation (5). For both cases, the free state densities are calculated inside the ion sphere radius $R_0$ as $4\pi R_0^3 N_p$ / 3 = 1. If we omit the neutral atom's contribution to the state density calculation, the number density of the perturber becomes smaller so that the resulting free-state density

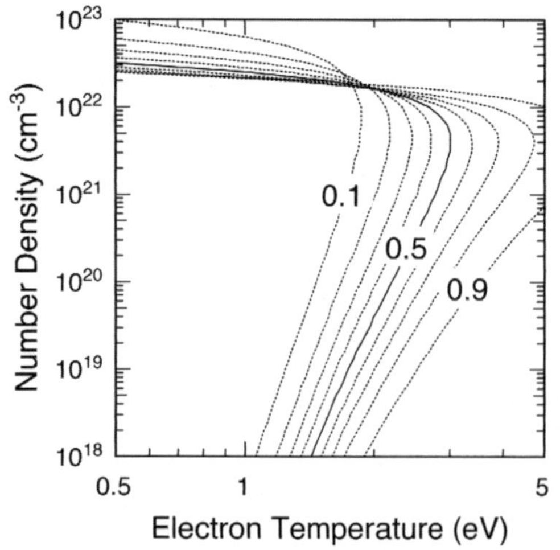


**Figure 4.** Conturs of average ionization degrees of hydrogenic plasmas. Number density of perturbing ion is that of ion.

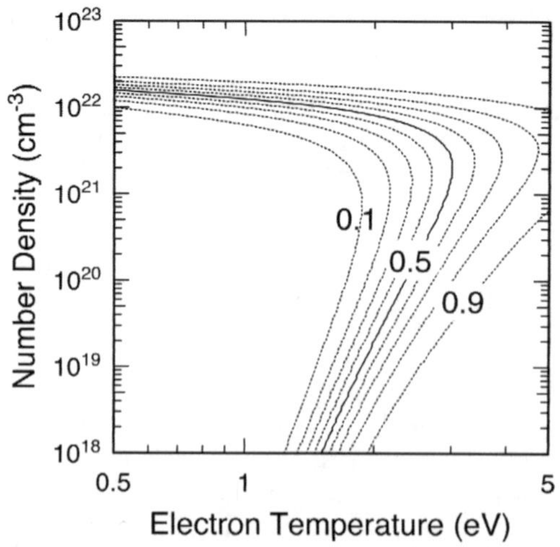


**Figure 5.** Conturs of average ionization degrees of hydrogenic plasmas. Number density of perturbing ion is that of ion plus atom.

becomes larger. For the case of the average ionization degrees of 0.1, the values of the average ionization degrees are equal to those obtained from its number density of ten times smaller values. Therefore the contours of the lower ionization are shifted to higher density region. When the average ionization degrees become large like 0.9, the difference would be small. For the lower density region, the neutral atom's volume in space becomes relatively small so that the resulting values of average ionization degrees for lower density region becomes smaller differences. In the high density near solid, the difference between the neutral atom and ion on the contribution to the state density calculation is seemed to be small. For the high density and low ionized region, more sophisticated treatment on the neutral atoms contribution to the state density calculation should be required. The classical treatment on the energy level of the bound electron should be also improved in next.
Thomas-Fermi model [4] is famous for calculating the average ionization degrees in dense plasmas, and many equations of state data used its results. Basically in the Thomas-Fermi model, they do not include the effect of the neighboring ion's potential in calculating the electronic state. In real plasmas, due to the potential nearby ion, the free state density becomes larger and the resulting values of the average ionization degrees are expected to be larger values than those obtained by the Thomas-Fermi model.

## Conclusions

We have reexamined the state density of the free and bound electron near its ionization threshold energy to reveal the difficulty that occurs in calculating the average ionization degrees near solid density by the nearest neighbor approximation. By introducing the Coulomb correlation between two ions and modifying the Saha equation, I could successfully obtain physically acceptable results of the average ionization degrees of hydrogenic plasmas. There still exists a detailed examination on the treatment of the neutral atom contribution to the state density and the energy level calculations of bound states.

## Acknowledgement

A part of this work was supported by JSPS KAKENHI Grant Number 23340185.

## References

[1] Nishikawa T in preparation
[2] Lochte-Holtgreven W 1968 *Plasma Diagnostics* (Amsterdam: North-Holland) p 81
[3] Nishikawa T 2012 Plasma and Fusion Research **7** 1401142
[4] For example, Feynman R P, Metropolis N, and Teller E 1949 Phys. Rev. **75** 1561

# Direct-driven target implosion in heavy ion fusion

**K. Noguchi**[1], **T. Suzuki**[1], **T. Kurosaki**[1], **D. Barada**[1], **S. Kawata**[1], **Y. Y. Ma**[1], **A. I. Ogoyski**[2]
[1] *Utsunomiya University, Utsunomiya 321-8585, Japan*
[2] *Varna Technical University, Varna 9010, Bulgaria*

E-mail: mt126649@cc.utsunomiya-u.ac.jp, kwt@cc.utsunomiya-u.ac.jp

**Abstract.** In inertial confinement fusion, the driver beam illumination non-uniformity leads a degradation of fusion energy output. A fuel target alignment error would happen in a fusion reactor; the target alignment error induces heavy ion beam illumination non-uniformity on a target. On the other hand, heavy ion beam accelerator provides a capability to oscillate a beam axis with a high frequency. The wobbling beams may provide a new method to reduce or smooth the beam illumination non-uniformity. First we study the effect of driver irradiation non-uniformity induced by the target alignment error (*dz*) on the target implosion. We found that *dz* should be less than about 130 μm for a sufficient fusion energy output. We also optimize the wobbling scheme. The spiral wobbling heavy ion beams would provide a promissing scheme to the uniform beam illumination.

## 1. Introduction

In direct drive inertial fusion [1], lasers or ion beams are used as an energy driver. The ion beam generation and absorption efficiencies are high; heavy ion beam (HIB) deposits its energy inside of the fuel target. HIB has preferable characteristics of the stabile and repetitive operation of HIB accelerator and a precise beam axis control as well as its high efficiency (~30-40%) of HIB generation. Therefore, it is considered that HIB would be a promising candidate of an energy driver in inertial fusion [2-4]. In this study, we employ Pb HIBs as the energy driver beams in order to study a beam illumination non-uniformity onto a spherical target.

In an actual inertial confinement fusion, a fuel target is irradiated by HIBs, when a fuel target is injected and aligned at the center of the fusion reactor [5]. A fuel target alignment error (*dz*) would appear; the target alignment error induces the HIBs illumination non-uniformity on the target. The beam illumination non-uniformity leads a degradation of fusion energy output.

In this paper, a direct-drive scheme is employed, and we present the HIBs illumination non-uniformity effect on the fuel target implosion. As a result, we found that *dz* should be less than about 130 μm for a sufficient fusion energy output.

## 2. Simulation model

Figure 1(a) shows a fuel target in heavy ion inertial fusion (HIF). The target consists of Al and C as a tamper and a HIB-energy absorber, respectively [6-9]. Figure 1(b) shows the input pulse employed. We call the lower power part in the input pulse as the foot pulse and the higher power one as the main pulse. The input Pb beam energy is 1.8MJ. We employ a 32-HIBs illumination system [10]. The HIBs illumination non-uniformity is evaluated by the global *rms*, including also the Bragg peak effect in the

Journal of Physics: Conference Series **688** (2016) 012078 doi:10.1088/1742-6596/688/1/012078

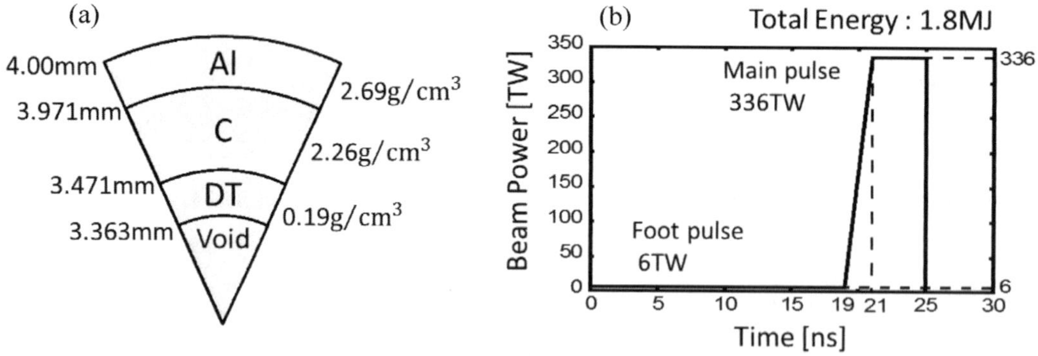


**Figure 1.** (a) Target structure and (b) The HIB input pulse.

energy deposition profile in the target radial direction. In HIF, the Bragg peak deposition area plays the most important role for the target implosion.

### 3. HIBs illumination non-uniformity Effect on fuel target implosion

First we show the implosion dynamics in order to verify the effect of the HIBs illumination non-uniformity on the target implosion, when the illumination non-uniformity is 2.87%. The target alignment error *dz* shows the displacement from the fusion reactor center. The HIBs used in this section are the Gaussian beams, whose radii are 3.3mm. Figures 2 show the mass-density distributions at (a) *t*=10ns, (b) *t*=28ns and (c) *t*=35ns. The HIBs deposit their energy mainly in the C layer by the Bragg Peak characteristic. The density valley appears in the C layer. In Fig. 2(a), the DT layer is accelrated further toward the center. Then, the high-density core is formed at the center of the target immediately after the void close.

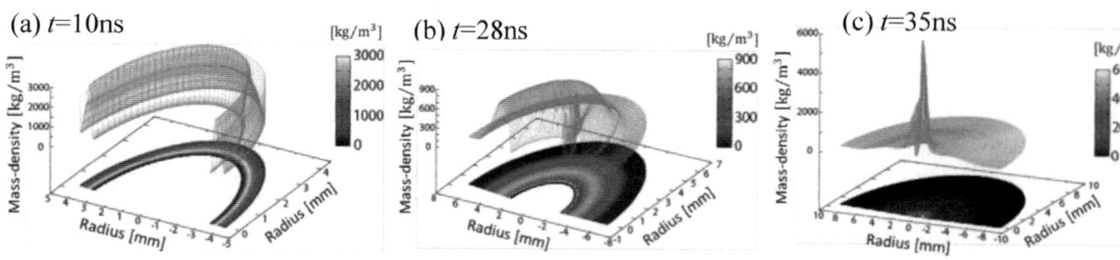


**Figure 2.** Mass-density distributions at (a) *t*=10ns, (b) *t*=28ns and (c) *t*=35ns.

In this section, we studied the effect of the HIBs illumination non-uniformity on the implosion. Figure 3 shows the fusion energy gain curve versus the target alignment error *dz*, and also the illumination non-uniformity versus *dz*. We obtained the relationship between *dz* and illumination non-uniformity by the precise HIBs illumination code in Ref. [11]. Figure 3 shows that the fuel ignition was not achieved beyond about *dz*=140μm. The results in Fig. 3 present that the target algnment error *dz* should be less than about 130 μm for a sufficient fusion energy output.

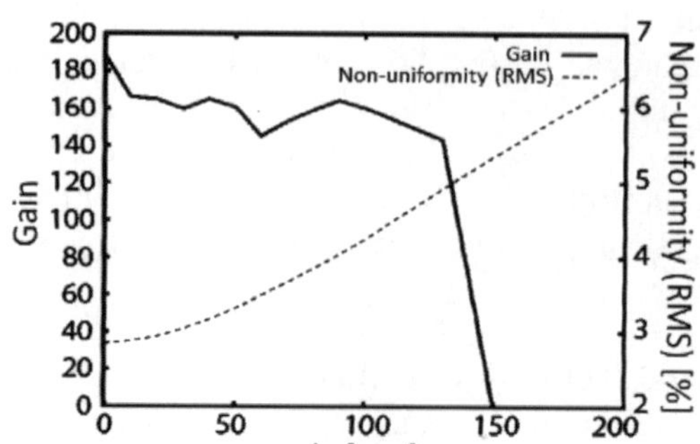


**Figure 3.** Fusion gain and illumination non-uniformity vs. target alignment error *dz*.

In order to verify the effect of the illumination non-uniformity, we have analyzed the non-uniformity of the ion temperature in the three cases of *dz*=0μm, 60μm and *dz*=150μm. Figure 4 shows the non-uniformities of ion temperature at each time. At the time of 5ns (Fig. 4(a)), the non-uniformity at the Al and C layers increases with the increase in the the target alignment error. As times passes, the effect of the HIBs illumination non-uniformity appears at the target inner part as shown in Figs. 4(b) and (c).

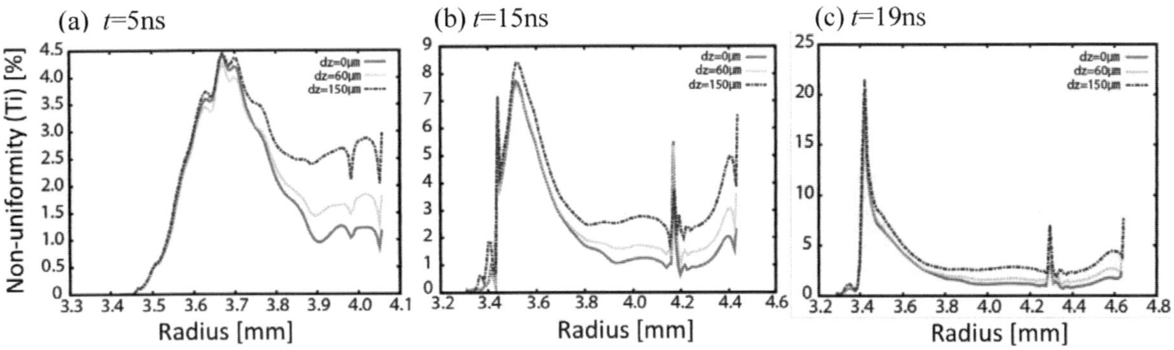


**Figure 4.** Non-uniformity of the ion temperature in the $\theta$ direction at (a) $t$=5ns, (b) $t$=15ns and (c) $t$=19ns.

Figure 5 shows the non-uniformities of the density at 21.8ns for the three cases of *dz*=0μm, 60μm and *dz*=150μm. In Fig. 5 the maximum non-uniformity becomes 24.5% for the target alignment error *dz*=150μm, though the maximum non-uniformity is less than about 20% for the cases of *dz*=0μm and 60μm. As a result, the fusion gain was not obtained for the target alignment error *dz*=150μm. However, for the cases of *dz*=0μm and 60μm a sufficient fusion energy output was obtained as shown in Fig. 3.

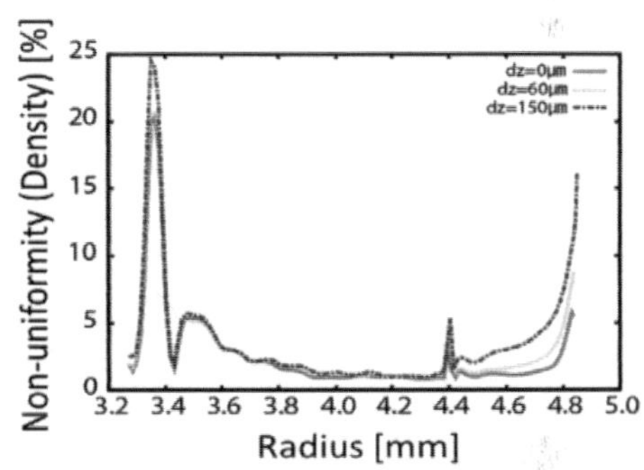


**Figure 5.** Non-uniformity of the mass-density in the $\theta$ direction at $t$=21.8ns.

## 4. Spiral wobbling HIBs illumination uniformity on fuel target implosion

In the Section 3 we found that the fuel target gain is quite sensitive to the driver beam illumination uniformity. In order to reduce the HIBs illumination non-uniformity, wobbling HIBs were proposed [12-15]: the HIBs axes are rotated to smooth the illumination non-uniformity. However, it was found

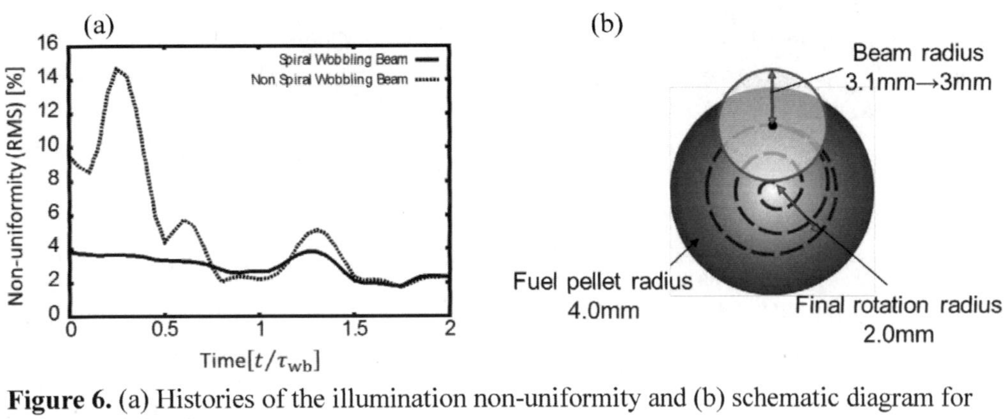


**Figure 6.** (a) Histories of the illumination non-uniformity and (b) schematic diagram for spiral wobbling beam.

that the simple rotation of the HIB axis introduces an unacceptable initial imprint of the HIBs illumination non-uniformity, as shown in Fig. 6(a) (dotted line).

In this Section, we optimize the wobbling beam illumination. We found that the initial imprint is reduced by the spiral wobbling beams (Fig. 6(b)). Figure 6(a) shows the illumination non-uniformity history during the first few rotations. Here $\tau_{wb}$ is the time for one rotation of the wobbling beam axis. The non-spiral wobbling beam has the beam radius of 2.0mm and the beam rotation radius of 3.0mm. For the spiral wobbling beam the beam radius changes from 3.1mm to 3.0mm at $t = 1.3\tau_{wb}$. As shown in Fig. 6(a), the initial imprint of the maximum non-uniformity at the beginning of the irradiation is greatly reduced from 14.7% to 3.56%.

## 5. Conclusions

In this paper, we present the effect of illumination non-uniformity on the target implosion. The target alignment error in a fusion reactor introduces and increases the HIBs illumination non-uniformity. As the illumination non-uniformity increases, the fusion gain is degraded. We found that *dz* should be less than about 130 μm for a sufficient fusion energy output, and that at *dz*<130μm the HIBs illumination non-uniformity was less than 5% [16]. Toward the further smoothing of the HIBs non-uniformity, we also proposed the spiral wobbling beam illumination; the spiral wobbling heavy ion beams would provide a lower illumination non-uniformity. The further studies are required on the beam illumination non-uniformity smoothing in the near future.

## Acknowledgements

This work was partly supported by JSPS, MEXT, CORE (Center for Optical research and Education, Utsunomiya University), ASHULA (ASian core program for High energy density science Using intense LAser photons), and ILE/Osaka University.

## References

[1] Logan B G, Perkins L J and Barnard J J 2008 *Phys. Plasmas* **15** 072701
[2] Miyamoto S and Ogawa M 1995 *J. Plasma Fusion Res.* **71** 951
[3] Arnold R C and Meyer-ter-Vehn J 1987 *Rep. Prog. Phys.* **50** 559
[4] Lindle J D 1995 *Phys. Plasmas* **2** 3933
[5] Petzoldt R W 1998 *Fusion Tech.* **34** 831
[6] Kawata S, Miyazawa K, Kikuchi T and Someya T 2007 *Nucl. Instr. Meth. Phys. Res. A* **577** 332
[7] Miyazawa K, Ogoyski A I, Kawata S, Someya T and Kikuchi T 2005 *Phys. of Plasmas* **12** 122702
[8] Kawata S, Miyazawa K, Ogoyski A I, Kikuchi T, Akasaka Y and Iizuka Y 2008 *J. Phys.* **112** 032028
[9] Someya T, Miyazawa K, Kikuchi T and Kawata S 2006 *Laser Part. Beams* **24** 359-369
[10] Skupsky S and Lee K 1983 *J. Appl. Phys.* **54** 3662
[11] Ogoyski A I, Someya T and Kawata S 2004 *Comput. Phys. Commun.* **157** 160-172
[12] Piritz A R, Temporal M, Lopez Cela J J, Tahir N A and Hoffmann D H H 2003 *Plasma Phys. Control. Fusion* **45** 1733-1745
[13] Piritz A R, Tahir N A, Hoffmann D H H and Temporal M 2003 *Phys. Rev. E* **67** 017501
[14] Basko M M, Schlegel T and Maruhn J 2004 *Phys. Plasmas* **11** 1577
[15] Runge J and Logan B G 2009 *Phys. Plasmas* **16** 133109
[16] Kawata S and Niu K 1984 *J. Phys. Soc. Jpn.* **53** 3416

# Characteristics of extreme ultraviolet emission from high-$Z$ plasmas

**H. Ohashi[1], T. Higashiguchi[1], . Suzuki[1], M. Kawasaki[1], C. Suzuki[2], K. Tomita[3], M. Nishikino[4], S. Fujioka[5], A. Endo[6], B. Li[7], T. Otsuka[7], P. Dunne[7], and G. O'Sullivan[7]**

[1]Utsunomiya University, Department of Advanced Interdisciplinary Sciences, Yoto 7-1-2, Utsunomiya, Japan
[2] ational Institute for Fusion Science, 322-6 Oroshi-cho, Toki, Japan
[3]Interdisciplinary Graduate School of Engineering and Sciences, Kyushu University, 6-1, Kasugakoen, Kasuga, Fukuoka, Japan
[4]Quantum Beam Science Directorate, Japan Atomic Energy Agency, 8-1-7 Umemi-dai, Kizugawa, Kyoto , Japan
[5]Institute of Laser Engineering, Osaka University, 2-6 Yamada-Oka, Suita, Japan
[6]HiLASE Project, Institute of Physics AS, CR, a Slovance 2, 18221 Prague 8, Czech Republic
[7]School of Physics, University College Dublin, Belfield, Dublin 4, Ireland

E-mail: higashi@cc.utsunomiya-u.ac.jp

**Abstract.** We demonstrate the extreme ultraviolet (EUV) and soft x-ray sources in the 2 to 7 nm spectral region related to the beyond EUV (BEUV) uestion at 6.$x$ nm and the water window source based on laser-produced high-$Z$ plasmas. Resonance emission from multiply charged ions merges to produce intense unresolved transition arrays (UTAs), extending below the carbon K edge (4.37 nm). An outline of a microscope design for single-shot live cell imaging is proposed based on high-$Z$ plasma UTA source, coupled to multilayer mirror optics.

## 1. Introduction

Extreme ultraviolet (EUV) lithography by use of a 13.5-nm sources [1] is focused on high-volume production of semiconductor devices. In fact, the fabrication tools for integrated circuits with 0.33NA (numerical aperture), to attain a half-pitch (HP) of 22 nm, have already been developed and are capable of reaching a HP of 16 nm with off-axis illumination techniques [2]. In order to achieve very high resolution HP, a shorter wavelength, which couples to highly reflective multilayer mirrors (MLMs) with a theoretical reflective coefficient of $\sim$ 70%, is a more attractive proposition and explains the present trend in lithography source development [3]. Wavelengths around 6.$x$ nm are especially useful for the final stage beyond the 13.5-nm EUV source [4]. In fact, the 6.$x$-nm beyond EUV (BEUV) emission can be coupled with a Mo/$B_4$C or La/$B_4$C multilayer mirror whose reflectivity is currently 50% at 6.$x$ nm, (theoretical maximum $>$ 70%). The unresolved transition array (UTA) emission exploited in Sn is scalable to shorter wavelengths, and Gd has a similar conversion efficiency to Sn, though at a higher plasma temperature, within a narrow spectral range centered near 6.7 nm.

Before discussing the high-power high-$Z$ plasma source, it is important to summarize the characteristics of efficient UTA light sources used in the 5 to 15 nm region. All are based on

Journal of Physics: Conference Series **688** (2016) 012079 doi:10.1088/1742-6596/688/1/012079

$n = 4 - n = 4$ ($4d - 4f$ and $4p - 4d$) transitions that overlap to generate an intense UTA. For efficient 13.5-nm operation, which corresponds to a photon energy $h\nu \approx 92$ eV, it is important to produce an optimum plasma electron temperature of $30 - 50$ eV. The rare-earth elements of gadolinium (Gd, $Z = 64$) and terbium (Tb, $Z = 65$) produce strong emission near $\lambda = 6.7$ nm ($h\nu \approx 183$ eV) which is maximized at electron temperatures in the $100 - 120$ eV range depending on initial focusing conditions [5, 6]. The spectral behavior of Gd and Tb plasmas is expected to be largely similar to that of Sn plasmas, because in the temperature range of interest, both are dominated by $4d$ open-shell ions. Although the conversion efficiency from the input laser energy to the output in-band EUV emission energy depends on the bandwidth (BW) of the reflection coefficient of the MLM, the maximum CEs have been observed to be higher than 1%. Because it moves to shorter wavelength with increasing atomic number, $Z$, the $n = 4 - n = 4$ UTA is expected to lie in the water window if higher $Z$ elements from $Z = 79$ (Au) to $Z = 83$ (Bi) are used [7]. Higher $Z$ elements such as uranium also emit in the water window but their radioactivity prohibits their use. Much of the previous work on high-$Z$ plasmas has concentrated on the production of quasicontinuum spectra at moderate laser intensities, employing electron temperatures below 100 eV [8].

In this paper, we report the EUV and soft x-ray sources in the 2 to 7 nm spectral region related to the BEUV and the water window source based on laser-produced high-$Z$ plasmas. Resonance emission from multiply charged ions merges to produce intense UTAs, extending below the carbon K edge (4.37 nm). An outline of a microscope design for single-shot live cell imaging is proposed based on high-$Z$ plasma UTA source, coupled to multilayer mirror optics.

## 2. Atomic number dependence on peak wavelengths of TA emission

Because it moves to shorter wavelength with increasing $Z$, the $n = 4 - n = 4$ ($\Delta n = 0$) UTA emission can be used for BEUV source and transmission x-ray microscopy for biological imaging in the water window (see Fig. 1). We have made preliminary studies of the potential of Gd and Tb for BEUV source [5, 6] and Bi as the "water window" soft x-ray source [7].

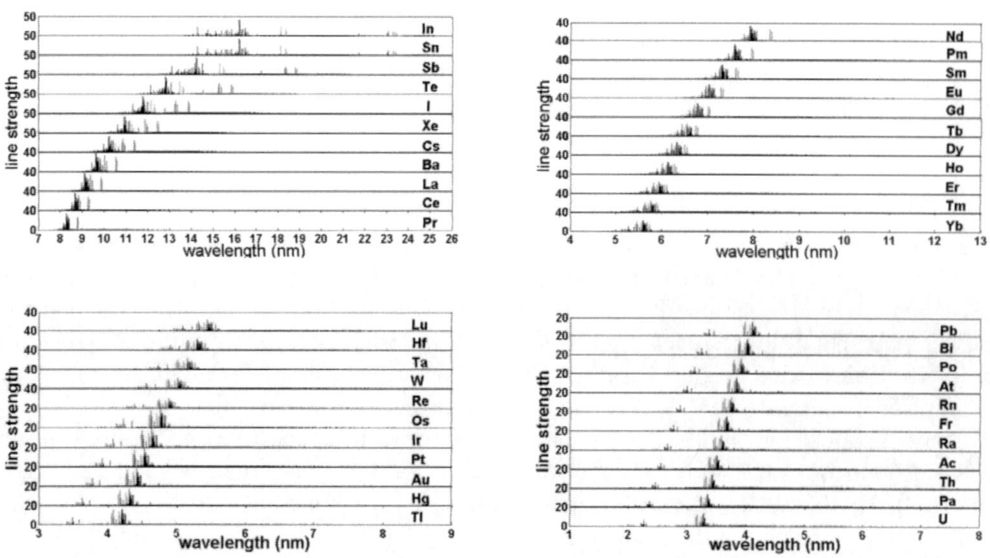


**Figure 1.** Calculated position of $n = 4 - n = 4$ transitions in key ions in elements from indium ($Z = 49$) to uranium ($Z = 92$).

Journal of Physics: Conference Series **688** (2016) 012079 doi:10.1088/1742-6596/688/1/012079

## 3. Results and discussion

The EUV spectra around 3 and 6 nm from high-$Z$ plasmas are shown. Figures 2(a)–2(c) show time-integrated spectra from Au, Pb, and Bi plasmas at a laser intensity of $1 \times 10^{14}$ W/cm$^2$ with a 150-ps pulse duration. Time-integrated EUV spectra between 1 and 6 nm from each element display strong broadband emission near 4 nm, which is mainly due to $n = 4 - n = 4$ transitions from ions with an open $4f$ or $4d$ outermost subshell, together with broadband emission around $2 - 4$ nm due to $n = 4 - n = 5$ transitions from multi-charged state ions with an outermost $4f$ subshell. The latter merge to form a structured feature from which the contributing ion stages may be readily inferred. The intensity of the $n = 4 - n = 4$ UTA emission is higher than that of the $n = 4 - n = 5$ emission. The atomic number spectral dependence is summarized in Fig. 2(d). The predicted photon energy of each experimental peak wavelength was shifted to higher photon energy with increasing atomic number. Neither the emission spectra nor the plasma electron temperatures, however, have been optimized, as shown below. The emission intensity of the $n = 4 - n = 5$ transitions, however, was compared with that of the $n = 4 - n = 4$ UTAs.

One efficient EUV source at 6.$x$ nm is provided by laser-produced plasmas (LPPs) of Gd and Tb. These elements produce strong narrowband emission, which is attributed to a $n = 4 - n = 4$ ($\Delta n = 0$) UTA, at 6.$x$ nm. The spectral behavior of Gd and Tb plasmas is expected to be similar to that of Sn plasmas. Figure 3 shows time-integrated EUV emission spectra from the Nd:YAG LPPs at different laser intensities ranging from $9.7 \times 10^{11}$ to $6.6 \times 10^{12}$ W/cm$^2$. The peak wavelength shifts from 6.7 to 6.8 nm, and is mainly due to $n = 4 - n = 4$ ($\Delta n = 0$) transitions in ions with an open $4f$ or $4d$ outermost subshell. Emission at wavelengths less than 6 nm, increases with increasing laser intensity and according to numerical evaluation, lines in the $\lambda = 2.5 - 6$ nm ($h\nu = 207 - 496$ eV) spectral region originate from Gd ionic charge states between $Gd^{19+}$ and $Gd^{27+}$, and arise from $n = 4 - n = 5$ ($\Delta n = 1$) transitions.

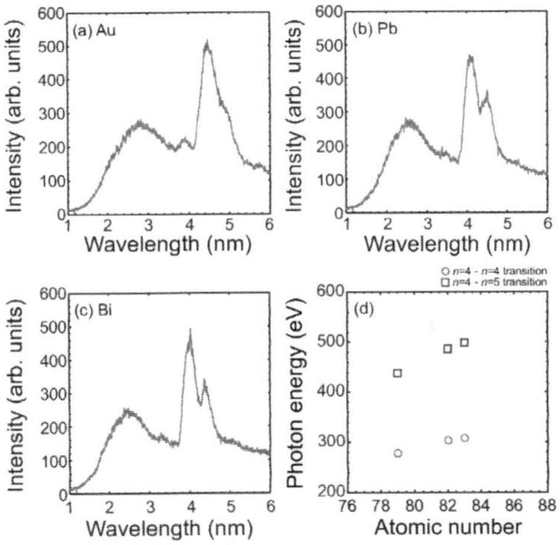


**Figure 2.** Time-integrated spectra from the picosecond-laser-produced high-$Z$ plasmas by use of Au (a), Pb (b), and Bi (c), respectively, and the atomic number dependence on the photon energies of peak emission of the $n = 4 - n = 4$ transition (circles) and the $n = 4 - n = 5$ transition (squares).

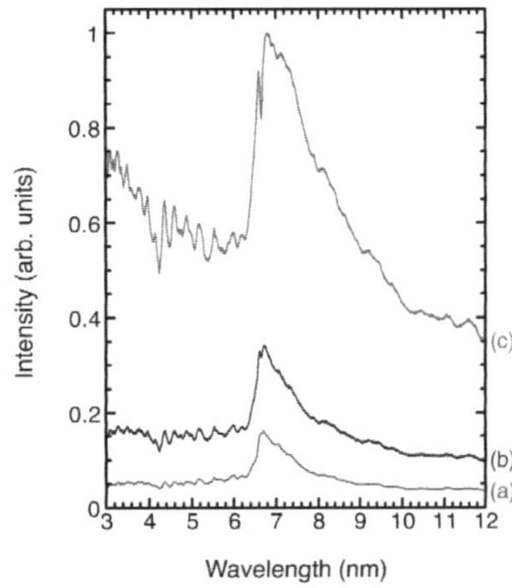


**Figure 3.** Time-integrated EUV emission spectra from the Nd:YAG LPPs at different laser intensities of $9.7 \times 10^{11}$ (a), $2.2 \times 10^{12}$ (b), and $6.6 \times 10^{12}$ W/cm$^2$ (c), respectively. The peak wavelength shifts from 6.7 to 6.8 nm with increasing the laser intensity.

The spatial and temporal evolution of the Gd plasma for 6.$x$-nm emission were studied using the modified 1-dimensional Lagrangian, laser-plasma hydrodynamic simulation code MED103 [9] for a cylindrical solid Gd target. Simulations were performed for a Nd:YAG LPP for laser parameters: $\lambda_L$ = 1064 nm, pulse width, $\tau_L$ = 8.5 ns (FWHM) and peak intensity $2 \times 10^{12}$ W/cm$^2$ incident on a target with a radius of 10 $\mu$m. The electron temperature and density obtained as a function of time and space are presented in Fig. 4. A higher temperature plasma, $T_e$ up to 150 eV, was obtained. Based on a collisional-radiative simulation [10], ionic populations of Rh- to Ag-like are maximized at an electron temperature of 80 – 130 eV. The plasma will thus have strong emission at 6.76 nm when $T_e$ is close to 80 eV [6].

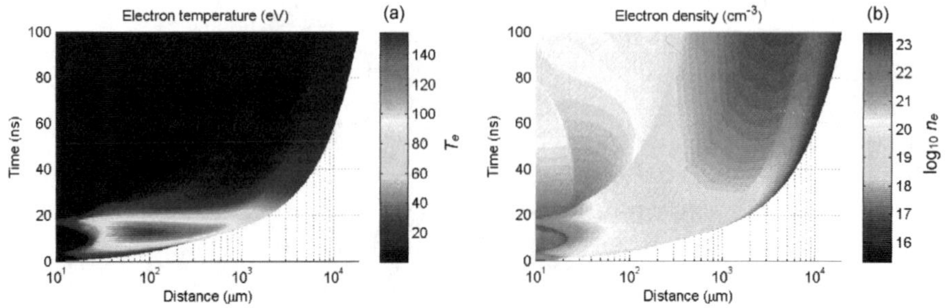


**Figure 4.** Electron temperature, $T_e$ (a) and electron density, $\log_{10} n_e$ (b) distributions as a function of time and space using 1.06-$\mu$m Nd:YAG laser simulated by the modified hydrodynamic code: MED103.

## 4. Summary

We characterized the EUV and soft x-ray emission in the 2 to 7 nm spectral region related to the BEUV at 6.$x$ nm and the water window source based on laser-produced high-$Z$ ($Z = 64 - 83$) plasmas. Resonance emission from multiply charged ions merges to produce intense UTAs, extending below the carbon K edge ($\lambda = 4.37$ nm).

## Acknowledgement

A part of this work was performed under the auspices of ME T (Ministry of Education, Culture, Science and Technology, Japan), and "UU Interdisciplinary Project for *in vivo* Bioimaging and Sencing" from ME T. One of the authors (T.H) also acknowledges support from The Canon Foundation, Research Grant (Basic Research) on TEPCO Memorial Foundation, and Gigaphoton Inc.

## References

[1] George S A, Hou K-C , Takenoshita K, Galvanauskas A and Richardson M C 2007 *Opt. Exp.* **15** 16348
[2] Meiling H, Boeij W, Bornebroek F, Harned , Jong I, Meijer H, Ockwell D, Peeters R, Setten E, Stoeldraijer J, Wagner C, Young S, Kool R, Kürz P and Lowisch M 2012 *Proc. SPIE* **8322** 83221G
[3] Platonov Y, Rodriguez J, Kries M, Louis E, Feigl T and Yulin S 2011 *Proc. 2011 International Workshop on EUV Lithography* (Maui, Hawaii, 13 17 June)
[4] Banine V Y, Koshelev K and Swinkls G H P M 2011 *J. Phys. D* **44** 253001
[5] Otsuka T, Kilbane D, White J, Higashiguchi T, Yugami , Yatagai T, Jiang W, Endo A, Dunne P and O'Sullivan G 2010 *Appl. Phys. Lett.* **97** 111503
[6] Li B, Dunne P, Higashiguchi T, Otsuka T, Yugami , Jiang W, Endo A and O'Sullivan G 2011 *Appl. Phys. Lett.* **99** 231502
[7] Higashiguchi T, Otsuka T, Yugami , Jiang W, Endo A, Li B, Dunne P and O'Sullivan G 2012 *Appl. Phys. Lett.* **100** 014103
[8] Zeng G-M, Takahashi M, Daido H, Kanabe T, Aritome H, akatsuka M and akai S 1990 *J. Appl. Phys.* **67** 3597
[9] Christiansen J P, Ashby D E T F and Roberts K V 1974 *Comput. Phys. Commun.* **7** 271
[10] Colombant D and Tonon G F 1973 *J. Appl. Phys.* **44** 3524

# Modern trends in low-density materials for fusion

**A.S. Orekhov[1], A.A. Akunets[1], L.A. Borisenko[1,2], A.I. Gromov[1], Yu.A. Merkuliev[1], V.G. Pimenov[3], E.E. Sheveleva[3], V.G. Vasiliev[4], N.G. Borisenko[1,5]**

[1] P.N. Lebedev Physical Institute of RAS, Russia
[2] Lomonosov Moscow State University, Russia
[3] N.D. Zelinsky Institute of Organic Chemistry RAS, Russia
[4] A.N. Nesmeyanov Institute of Organoelement Compounds of RAS, Russia
[5] National Research Nuclear University MEPhI (Moscow Engineering Physics Institute)

E-mail: orekhov@sci.lebedev.ru

**Abstract**. Low-density targets continue to yield new experimental data and to put new unsolved questions for driver-plasma experiments. The experiments with such targets are presented in the paper by "Low-density targets that worked in direct and indirect experiments with laser and particle beams" by L.A. Borisenko *et. al.* ibid.
Here we concentrate on nanostructured and aerogel targets' fabrication and characterization. These configurations establish certain standards for contemporary shot experiments.

## 1. Introduction
A variety of techniques is developed and used in order to meet the demand of laser interaction experiments with low-density targets. Different methods are worked out for typical requests below.

Extra fine regular structure is normally ordered to eliminate chaotic instability growth. Definite material structuring is used for plasma instability control or saturation through small-scale structures.

Mass gradients in targets are planned for different designs. Sharp density gradient is required for shock pressure multiplication. Z-gradient could be applied for preheating or energy-release control.

Thin aerogel or nano-snow layers are proved to smooth laser light before target illumination. Thin low-density metal layer on the inside of indirect target raises efficiency of hohlraum.

Well measured parameters and profoundly characterized structures of the target are obligatory for modern laser experiment success. This responsibility is a challenge because the hierarchies of needed parameters vary from experiment to experiment and from facility to facility. Most frequently the fabrication and metrology become unique and inventive.

## 2. Low-density polymer
Structured targets of definite and accurately measured features are requested for laser-plasma experiments and ICF applications, sometimes independently of the solid-matter physics capabilities. The targets' materials have become thin and light-weighted, so numerous problems occur during their characterization. This brings us to development of following instruments:

- better "traditional" diagnostic tools;
- customized software for existing devices;

- specific applications of common equipment;
- verification in the laser experiments in order to simplify the theoretical requirements on target.

Current fabrication process allows some control of the structures in a wide density range. Examples of structures, fabricated independently and in different time are given in Figure 1. The density is known from chemical solution composition.

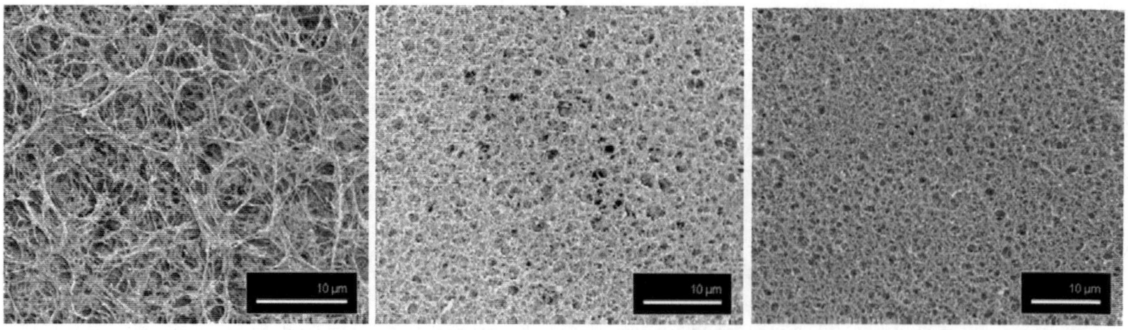


Figure 1. Structure of cellulose triacetate (CTA) aerogel with density 2 mg/cc (left), 5 mg/cc (centre), and 10 mg/cc (right). Scale is 10 µm; solid density is about 1.1 g/cc.

The methods of light materials characterization often result in sample destruction. So we search for stable techniques to produce repeatable fusion targets.

The experiments on high-temperature and high-pressure physics, shock front and implosion studies consume targets with density gradient. Approximately the gradient can be achieved by composing several layers with different densities. For higher efficiency smooth density gradient is needed.

Pressure multiplication in plasma with density increase [1] started our development of the shell target with smooth density gradient 30 mg/cc up to 20 g/cc in a single shell instead of double shell. Several methods are known for one-step production of density gradient samples. Those were performed and reported earlier by N.G. Borisenko *et al.* and W. Nazarov *et al.* [2]. Density profile monitoring during polymerization process indicates certain restrictions. Sharp density gradient (more than 1000 mg/cc/mm) is hard to produce. The supercritically dried divinyl benzene (DVB) aerogel with density gradient up to 400 mg/cc/mm done in University of St. Andrews (UK) was characterized with X-ray tomography (Figure 2).

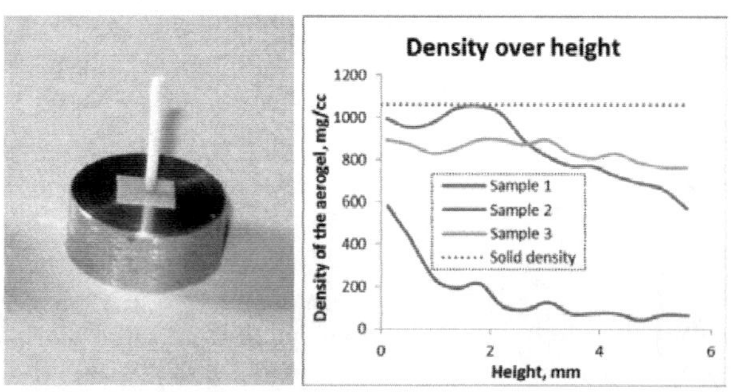


Figure 2. DVB aerogel fixed on tomograph metal mounting (left); dried DVB aerogel density via height for several different examined samples (right).

Sample 1 reaches full solid density at denser part, which looks optically transparent there. The rest two samples are opaque with visible gradual change of turbidity. But x-rays show only one of them (sample 2) displays density gradient, whereas sample 3 has altered structure rather than density over height. The optimization of manufacturing conditions should be worked out. Now we can only use x-rays based selection for needed gradient search.

The autowaves appear in the existing technology route of aerogels with smooth density gradient, reciprocal diffusion of catalyst and monomer are to be analyzed. Opposing diffusion of monomer and catalyst during polymer (samples of polystyrene) or silica gel synthesis/casting causes density modulations in the resulting vertical density gradient. The density waves during synthesizing polymer appear in case of decreasing monomer concentration which stimulates opposing diffusion of monomer and catalyst. The reaction is exothermic, and temperature contributes to acceleration of the reaction rate. During silica gel polymerization the tetraethoxysilane solution produces not only heat but water which intensifies gelation. The acceleration causes sharp decrease over height of monomer concentration and results in lower density in the region with low monomer amount. This process finishes only on the top boundary of the monomer solution.

Different chemical systems have demonstrated modulation in density gradient, so quantitative study of autowaves in gradient gel growth is due to find regimes to eliminate or to damp them.

## 3. Low-density metal

We produce metal nano-snow layers of Bi or Au with 50-300 mg/cc density, pore sizes ~20 µm and solid elements (chains of metal nanoparticles) of ~100 nm diameter.

Subsequent thermal treatment and pressing result in layers density 0.2-4 g/cc, pores being twice or 2.5 times less than predicted, and average particle size increase up to 150-200 nm.

Metal nanoparticles and foam layers are produced inside a special setup according to a widely used technique of metal fog precipitation in the low-pressure buffer gas atmosphere (pure N, Ar or He). Heater evaporates solid metal forming dense atomic metal flow. At 1-5 Torr pressure in the setup chamber (1.5 meter high) metal fog or snow from nanoparticles is formed. Nanoparticle final size depends on pressure and buffer gas composition. Thus using of He results in tinier particles and lower densities, than Ar. The optimum pressure for target densities less than 1/100 of solids is known to be 150-700 Torr. Varying the conditions for fog production (heater temperature, gas composition, and pressure) and precipitation conditions (holder temperature) we can form nanoparticle layers down to 1-0.4% of solid metal density.

Hot melt is evaporated in 3-20 minutes, whereas precipitation lasts for 16-48 hours. Layer inner structures (Figure 3) may be globular, or may form fractal layers. The structure is self-similar in different scales (10-20 µm and 200-400 µm). SEM picture shows pores of 1-10 µm size.

Figure 3. Left to right: cut of Bi snow layer, scale 20 µm; Sn snow layer, scale 2 µm; Cu snow layer, scale 2 µm; Bi snow layer, globules, scale 2 µm.

Parameters of thin low-density metal layer on the holders were examined using a carbon cylinder ("witness" of manufacturing) [3]. The witnesses are situated upright near the working samples during precipitation process, so they have the same low-density metal layer as on the final samples.

Due to symmetric horizontal shape of the witness we know the optical path and measure optical density of the sample on one X-rays photo. With absorption ratio restored from solid calibration wedge we can calculate the density of the sample. The thickness of the target layer is calculated as height of the "cup" on top the witness. We use simple formulas to recover thickness:

$$t = \frac{\rho}{\rho_{solid}} \cdot x ,$$

where $x$ is an absorption thickness in µm, $\rho$ is the density of the layer, $\rho_{solid}$ is the density of pure material. Then

$$I = I_0 \cdot e^{-\mu \cdot t},$$

where $I$ and $I_0$ is the x-ray intensity after and before the sample, $\mu$ is absorption ratio in $\mu m^{-1}$.

Resulting thickness of analysed samples was about 40-150 µm with density varying from 40 to 180 mg/cc for bismuth layers and from 80 to 200 mg/cc for gold ones.

## 4. Low-density polymer and metal composites

Several types of high-Z nanoparticle dopants with different properties may help to explore increasing uniform compression. Steps in density profile ranging from subcritical (about 2 mg/cc) to 100 mg/cc are reported for higher hydrodynamic efficiency of targets. Layers with density gradient are required for equation-of-state (EOS) experiments. Concentration profile of high-Z nanoparticles in aerogel layers is planned for increasing of target efficiency, and for efficient x-ray converters.

Metal densities of less than 100 mg/cc (to as low as 10 mg/cc) are obtained in composites of heavy metal nanoparticles (50-100 nm diameter) inside the polymer network with solid fibers of 0.1 μm size and 3 μm apart. The polymer can contain up to 20% by weight (wt) for Cu nanoparticles and up to 50% (wt.) for Au nanoparticles. Earlier we saw coarser structure of Cu-loaded aerogel [4].

Now the structure of CTA aerogel remains the same (Figure 4) when metal nanoparticle doping is moderate (<30%). In this case inert (plastic) mass and non-uniformities are minimized, when nanoparticles are kept inside the interaction volume for cluster-related laser experiments.

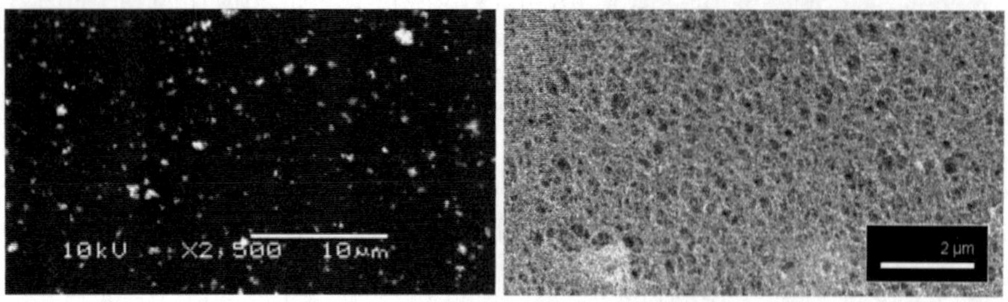


Figure 4. X-ray analysis by self-emission of Cu nanoparticles in CTA film, scale 10 μm (left); structure of 10 mg/cc CTA aerogel with 20% (wt) of Cu nanoparticles, scale 2 μm (right).

## 5. Conclusions

The following distinct trends seem to prevail in low-density target manufacturing techniques.

1. Nanostructuring and uniformity are combined with a wide range of densities requested, from 1-2 mg/cc (gas-like) to more than 100 mg/cc. Polymer foams give way to plastic aerogels – fine uniform materials for easy simulations and comparison with experimental results. Plastic aerogels with quasiperiodic tiny structure are now available for manufacturing within certain range of parameters. Different experimental techniques are used for manufacturing of extremely low and of moderate densities.
2. Metal nano-snow layers are sometimes sintered for better stability. Low-density metals are useful in many respects. Their regularity is reached close to that of plastic aerogels.
3. Combinations and topologic modifications are possible. Composites, metal nanoparticles in plastic matrix are the examples. In fact the diversity of structures now used in laser experiments came to some desired standards, where aerogels and nanostructures remain the preferred form.

## References

[1] D. Batani et al., Phys. Rev. E, 2001, V. 63, 046410
[2] N.G. Borisenko, A.A. Akunets, et al. Fusion science and technology, 2009, V. 55, №4, pp. 477-483.
N. G. Borisenko, W. Nazarov, et al., Journal of Radioanalytical and Nuclear Chemistry, 2014, V. 299, №2, pp. 961-964; DOI 10.1007/s10967-013-2628-2
[3] L.A. Borisenko, A.S. Orekhov, et al., Journal of Radioanalytical and Nuclear Chemistry, 2014, V. 299, №2, pp.955-960; DOI 10.1007/s10967-013-2629-1
I.V. Akimova, N.G. Borisenko, et al., Problems of Atomic Science and Technology, ser. Thermonuclear Fusion, 2012, №2, pp. 122-130 (in Russian).
[4] N.G. Borisenko, et al., Fusion Sciences and Technology, 2006, V. 49, №4, pp. 676-685.

# A novel experimental setup for energy loss and charge state measurements in dense moderately coupled plasma using laser-heated hohlraum targets

**A. Ortner[1,8], D. Schumacher[3], W. Cayzac[1,4], A. Frank[5], M.M. Basko[2], S. Bedacht[1], A. Blazevic[3], S. Faik[6], D. Kraus[1,7], T. Rienecker[6], G. Schaumann[1], An. Tauschwitz[6,8], F. Wagner[1] and M. Roth[1]**

[1]Technische Universität Darmstadt, Darmstadt, Germany
[2]Keldysh Institute of Applied Mathematics, Moscow, Russia
[3]GSI Helmholtzzentrum für Schwerionenforschung GmbH, Darmstadt, Germany
[4]Univ. Bordeaux-CEA-CNRS CELIA UMR 5107, Talence, France
[5]Helmholtzinstitut Jena, Jena, Germany
[6]Goethe-Universität Frankfurt, Frankfurt, Germany
[7]University of California, Berkeley, USA
[8]Helmholtz International Center for FAIR (HIC for FAIR), Frankfurt, Germany

E-mail: a.ortner@gsi.de

**Abstract.** We report on a new experimental setup for ion energy loss measurements in dense moderately coupled plasma which has recently been developed and tested at GSI Darmstadt. A partially ionized, moderately coupled carbon plasma ($n_e \leq 0.8 \cdot 10^{22}$ cm$^{-3}$, $T_e = 15$ eV, $z = 2.5$, $\Gamma = 0.5$) is generated by volumetrical heating of two thin carbon foils with soft X-rays. This plasma is then probed by a bunched heavy ion beam. For that purpose, a special double gold hohlraum target of sub-millimeter size has been developed which efficiently converts intense laser light into thermal radiation and guarantees a gold-free interaction path for the ion beam traversing the carbon plasma. This setup allows to do precise energy loss measurements in non-ideal plasma at the level of 10 percent solid-state density.

## 1. Introduction

The energy deposition of an ion beam in plasmas is a key question for inertial confinement fusion research (ICF), particularly for simulations of $\alpha$-particle heating, for the assessment of heavy ions as drivers (HIF), and for fast ignition concepts with ion beams (FI). In the last three decades a variety of experiments [1-6] have been carried out to get a deeper insight into the physics of ion beam plasma interaction. Most of these experiments have investigated the so-called linear interaction regime, which corresponds to swift ions in an ideal plasma. This ideal non-coupled plasma is characterized by the coupling parameter

$$\Gamma = \frac{e^2}{k_B T_e}(\frac{4\pi}{3}n_e)^{1/3} \ll 0.1$$

In this regime, the stopping power can be well described by perturbative approaches like the standard stopping model (SSM) [7]. But as soon as the plasma becomes non-ideal, these

approaches are no longer valid and strong coupling effects have to be considered like multiple scattering, dynamic screening, bound states and continuum lowering [8]. Up to now only few experimental data exist for the interaction of heavy ions with non-ideal plasma due to the difficulty to create a dense and cold plasma confined for a sufficient time to be probed by an ion beam. One example to exploit this regime is schock-compression of a gas, achieving typical plasma parameters of $n_e \leq 10^{20}$ cm$^{-3}$, $T_e = 1.8$ eV, $z = 0.3$, $\Gamma = 1$ [9].
In this paper we present an alternative approach which can create much higher electron densities up to $10^{22}$ cm$^{-3}$ which are highly relevant for the HIF and FI regimes. A uniform moderately-coupled plasma ($0.1 < \Gamma < 1$) is generated from a solid foil by volumetrical heating with laser-driven hohlraum radiation. This plasma is probed by swift ions ($E = 3.6\,\mathrm{MeV/u}$) to measure the influence of non-ideality of the plasma in the stopping power. In an upcoming experiment, this plasma will be probed with slow ions ($E < 60$ keV/u) to explore the non-linear regime of beam-plasma interaction with a beam-plasma coupling constant $\gamma \geq 0.1$ [10].

## 2. The double hohlraum plasma target

An attractive way to create a uniform plasma state at high densities is using hohlraum cavities of a high-Z material which are heated by intense laser pulses up to radiation temperatures of tens or hundreds of electron volts. These X-rays can volumetrically heat a sample material inside the hohlraum to a hot and dense plasma state which is inertially and thermally confined for several nanoseconds. For the design of a hohlraum target allowing combined experiments with laser and ions beams, two aspects have to be considered. First, the laser light has to be efficiently converted into thermal radiation. Second, the entire interaction path of the ion beam has to be shielded against the inflow of wall material and debris coming from the exploding converter, at least during the measurement time of 4–6 ns. This is essential in order to guarantee an undisturbed energy loss measurement in the pure sample material plasma.
For that purpose, we developed a double hohlraum target consisting of a primary spherical hohlraum (converter hohlraum) in which the laser light is converted into X-rays and further more a secondary cylindrical hohlraum (target hohlraum) where the sample material is placed.

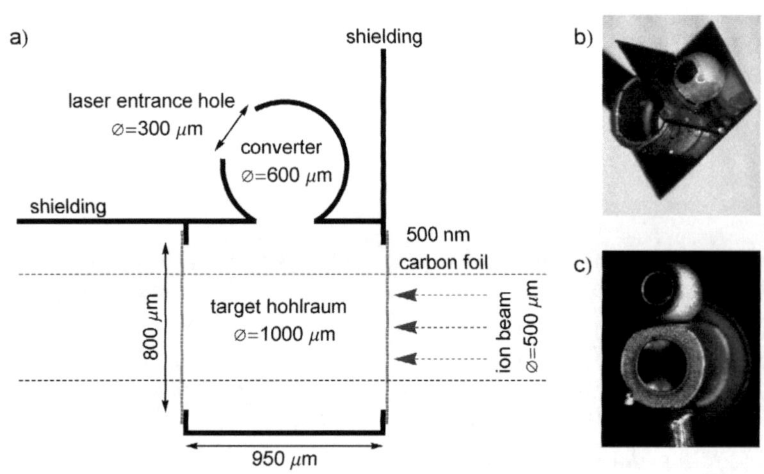


**Figure 1.** a) Geometry of the double hohlraum plasma target. On the right a target b) with and c) without advanced shielding.

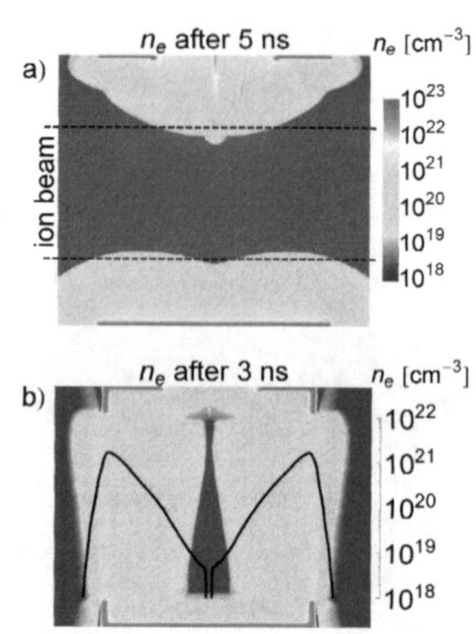


**Figure 2.** 2D hydrodynamic simulations for a) an empty and b) a filled target hohlraum with electron density line out.

The target hohlraum is directly attached to the converter and is heated by the transmitted X-rays from the converter. The design and dimensioning of this sub-millimeter target is shown in Fig. 1. Both hohlraums are made of gold with a wall thickness of 15 µm. As sample material, two thin carbon foils with 100 µg/cm$^2$ areal density are attached at both openings of the cylinder. Extra shielding is added to shield the ion interaction path from debris of the fast expanding converter. The targets are manufactured in the Target Laboratory at Technische Universität Darmstadt. First positive blanks of either brass or steel are formed by micro-chipping and then electroplated with a gold layer. Then, the necessary holes are cut into the gold wall with a laser cutter and the blanks are etched out. The shieldings are directly cut out from 15 µm thick gold foils and the sample material from $\approx$ 500 nm carbon foils. Finally all parts are assembled using a microscope. All components can be fabricated and positioned with an accuracy of $\pm$15 µm.
The converter is driven by a high energy laser beam generating an almost Planckian spectral distribution of X-rays. The hohlraum radiation of both cavities has been measured both time- and spectrally resolved [11] and is in good agreement with theory [13] and simulations [12]. The laser light heats the converter to a temperature of 100 eV and the target hohlraum to 30 eV. Moreover the radiation temperature in the target hohlraum remains above 20 eV for more than 6 ns. This guarantees a homogeneous and constant target heating during the full beam-plasma interaction time. The expansion of the gold into the beam path has been characterized by shadowgraphy measurements and ion probing through an empty hohlraum. As long as the ion bunches are not disturbed or delayed, one can assume that the electron density of the gold plasma on the interaction path is below $10^{16}$ cm$^{-3}$. The insets in Fig. 3 show two shadowgraphy measurements at 4 ns after laser impact without and with shielding where the expanding gold is successfully blocked. The graph in Fig. 3 shows the ion bunch delay with respect to vacuum. After the laser impact at $t = 0$ ns first perturbations occur after 6 ns.
The hydrodynamics of the converter as well as the empty and the filled target hohlraum have been studied extensively by simulations carried out with the RALEF2D code [14]. All simulations are in good agreement with the measurements. For the empty hohlraum they also predict a gold-free time window of about 5 ns, as shown in Fig. 2. Finally the parameters of the carbon plasma are extracted from the simulation of the filled hohlraum and are benchmarked by plasma interferometry measurements. A density plot and a line out of the electron denstity at $t = 3$ ns is given in Fig. 2. In summary, a weakly coupled carbon plasma is generated with $10^{21} < n_e \leq 0.8 \cdot 10^{22}$ cm$^{-3}$, $T_e = 25$ eV and $z = 3.5$ which corresponds to $\Gamma = 0.2$.

## 3. The experimental setup

The experimental setup at the UNILAC accelerator at GSI is shown in Fig. 4. The converter is heated by the high energy laser system PHELIX (150 J in 1.5 ns FWHM at 528 nm) and the generated plasma is probed by a $Ca^{17+}$ ion beam at 3.6 MeV/u. After a time of flight distance (ToF) of 12 m the probe beam can either propagate directly onto a stop detector or be split up by a magnet into its charge states and analysed with a charge state spectrometer. The most critical part of the experiment is the precise alignment of the ion beam since it first has to pass a 500 µm aperture, then the hohlraum cylinder and finally it has to be focussed onto the detectors located 12 m behind the target. The reduction of the beam diameter is necessary to only probe the carbon plasma at the center of the target hohlraum. The radiation temperature in the hohlraum is measured by a streaked X-ray spectrometer through additional 150 µm diagnostic holes in the hohlraum wall [15]. The electron density of the outflowing plasma is measured time-resolved with a Nomarski multiframe interferometer which is also used for shadowgraphy to characterize the hydrodynamic expansion of the gold [15]. The heavy ion beam from the UNILAC accelerator has a bunch frequency of 108 MHz which leads to an ion bunch every 9 ns registered by the detector. Each of these ion bunches has a width of 2.5 ns at FWHM. This pulse length determines the minimum measurement time needed to probe the plasma state and

is the reason for the aforementioned time window of about 5 ns. Custom made CVD-diamond detectors [16] allow to record the ion-beam intensity with a fast oscilloscope over more than 50 µs. The temporal position of each ion bunch can be determined with an accurracy better than 0.1 ns by fitting these signals.
In a first campaign extensive measurements of the energy loss using only the stop detector were recorded. In 2014, a second campaign will be carried out in which the charge state distribution and the energy loss in each charge state channel will be measured. These results will be compared with existing stopping theories and Monte Carlo simulations of the charge state distribution [17].

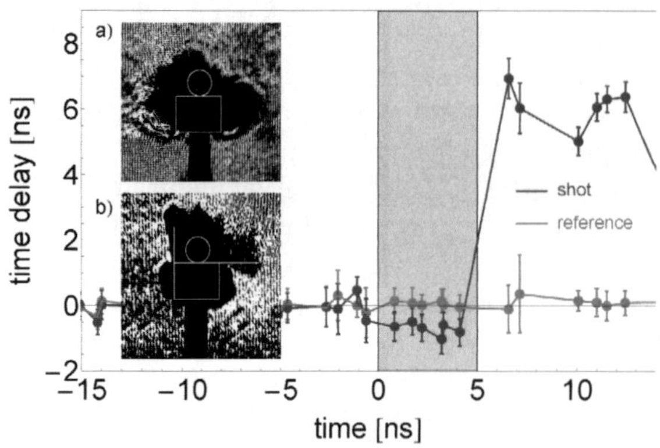


**Figure 3.** ToF delay of the ion bunches probing an empty hohlraum target. The first disruption occurs after 5 ns. The inset shows two shadowgraphies of an unshielded a) and shielded b) target 4 ns after laser impact.

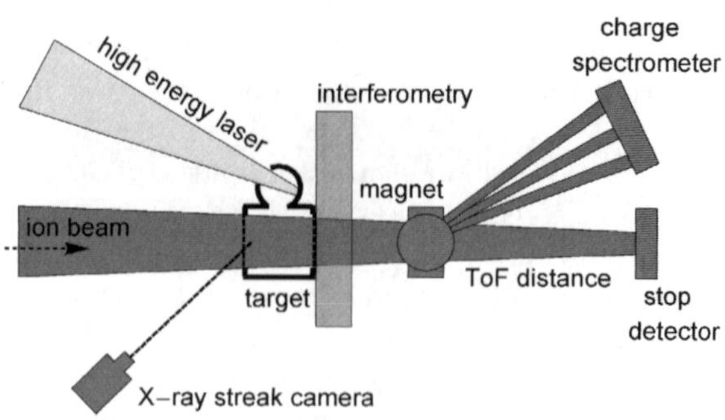


**Figure 4.** Schematics of the experimental setup: The plasma is generated by hohlraum radiation generated in the converter target and is then probed by a bunched ion beam and detected after a 12m ToF distance.

## 4. Acknowledgements

We wish to acknowledge the expert support of the PHELIX and the UNILAC team. This work was supported by the Helmholtz International Center for FAIR.

## 5. References

[1] K.Weyrich, D.H.H.Hoffmann, J.Jacoby, H.Wahl, R.Noll et al., *Nucl. Instr. and Meth. A* **278** 52 (1989)
[2] D.H.H. Hoffmann, K.Weyrich, H.Wahl, D.Gardes, R.Bimbot, C.Fleurier. *Phys. Rev. A* **42**, 2313-2321 (1990)
[3] K.G.Dietrich, D.H.H.Hoffmann, E.Boggasch, J.Jacoby, H.Wahl et al., *Phys. Rev. Lett.* **69** 25, 3623-3626 (1992)
[4] C.Couillaud, R.Deicas, P.Nardin, C.Deutsch, G.Maynard et al., *Phys. Rev. E* **49** 2 1545-1562 (1994)
[5] M.Roth, C.Stöckl, D.O.Gericke, R.Bock, D.H.H.Hoffmann et al. *Europhys. Lett.* **50** 1 28-34 (2000)
[6] A.Frank, A.Blazevic, W.Cayzac, A.Ortner, M.Roth et al., *Phys. Rev. Lett.* **110** 11 115001 (2013)
[7] C.Deutsch, G.Maynard, M.Chabot, D.Gardes et al. *The Open Plasma Physics Journal* **3** 88-115 (2010)
[8] D.O.Gericke, M.Schlanges and T.Bornath *Phys. Rev. E* **65** 3 036406 (2002)
[9] K.Weyrich, H.Wahl, D.H.H.Hoffmann, A.A.Golubev et al. *J. Phys. A: Math. Gen* **39** 17 4749-4754 (2006)
[10] D.O. Gericke, *Laser Part. Beams* **20** 471-474 (2002)
[11] T.Hessling A.Blazevic, A.Frank, D.Kraus, M.Roth, G.Schaumann et al., *Phys. Rev. E* **84** 016412 (2011)
[12] D.Schumacher, PhD thesis, TU Darmstadt, *Untersuchung laserinduzierter Hohlraumstrahlung und Energieverlust von Schwerionen in indirekt geheizten Plasmen* (2012)
[13] G.D.Tsakiris *Physics of Fluids B: Plasma Physics* **4** 9921005 (1992)
[14] A. Ortner, S. Faik, D. Schumacher, M.M. Basko et al. *Nucl. Instr. and Meth. B* **343** 123-131 (2015)
[15] M.Börner, J.Fils, A.Frank and A.Blazevic, *Rev. Sci. Instrum.* **83** 043501 (2012)
[16] W.Cayzac, A.Frank, D.Schumacher, M.Roth et al. *Rev. Sci. Instrum.* **84** 043301 (2013)
[17] A. Ortner, A. Frank, A. Blazevic, M. Roth *Phys. Rev. E* **91** 023104 (2015)

# Precise diagnostics for SG-II facility

Ouyang Xiaoping[1], Yang Lin[1], Zhu Baoqiang[1], Ma Weixin[2], Zhu Jian[2], Zhu Jianqiang[1], and Lin Zunqi[1]

[1] Shanghai Institute of Optics and Fine Machanics, Chinese Academy of Science, China
[2] Shanghai Institute of Laser Plasma, China Academy of Engineering Physics, China

E-mail: oyxp@siom.ac.cn

Abstract. Diagnostics is important for large laser facilities to monitoring laser status and to analyzing physics phenomena on target. SG-II facility in Shanghai is setup in 2001 with eight beams. Each beam's energy is 300J at 351nm, and its pulse width is 1ns. SG-II ninth beam is setup in 2005, its energy is 2700J at 351nm, and its pulse width is 1ns too. In recent years, it is an important international user facility for laser research with precise diagnostics system. With the development of SG-II, a new calorimeter is designed to 400× 400mm$^2$ in order to measure absolute energy up to 6400J at 351nm. Fast photodiode and 12GHz oscilloscope is used to analyze nanosecond pulse shape. In order to analyze petawatt laser pulse, a single shot auto-correlator is designed to measure pulse width between 0.5~18ps. And pulse contrast is obtained by a cross-correlation with fiber-array, which dynamic range is ~$10^6$ when pulse is 1mJ and 10ps. Quality of far-field measurement is 1.1DL for nanosecond pulse ,and that is 1.5DL for picosecond pulse.

## 1. SG-II laser facility

SG-II (ShenGuang-II) laser facility is setup in 2001 with eight beams. Energy of each beam is 300J at 351nm, and pulse width is 1ns. These beams have good focusing ability that 70% energy is encircled in 3DL (3 times Diffraction Limitation).

SG-II ninth beam is setup in 2005. Its energy is 5000J at 1053nm or 2700J at 351nm when pulse width is 1ns. It also can output 300J at 100ps or 30J at 30ps. 70% energy of SG-II ninth beam is encircled in 5DL.

SG-II laser facility is an international user facility for laser research. Its operation rate is 6 shots per day. A highlight experiment is published on Nature Physics [1].

Diagnostics is important for large laser facilities to monitoring laser status and to analyzing physics phenomena on target, such as such as LLNL, OMEGA and GEKKO [2, 3]. In SG-II facility, there are both diagnostics for nanosecond pulse and that for picosecond pulse.

## 2. Diagnostics for nanosecond pulse

Diagnostics for nanosecond pulse can provide laser parameters such as energy, pulse shape and far field.

A calorimeter is used to measure absolute energy of laser pulse. Its aperture is φ 400mm, which is bigger than laser diameterφ 310mm. This calorimeter has a capability to measure 5000J at 1053nm and 351nm. Fast photodiode and 3GHz oscilloscope are used to analyze pulse shape. And Far-field

Journal of Physics: Conference Series **688** (2016) 012082 doi:10.1088/1742-6596/688/1/012082

measurement is used to diagnose focusing ability of laser pulse. Its quality is 1.5DL. This means that 70% energy is encircled in 1.5 times diffraction.

With the development of SG-II, diagnostics for nanosecond pulse have improved its abilities. A new calorimeter is designed to 400×400mm$^2$ in order to measure 16kJ at 1053nm and 6400J at 351nm. Its uncertainty is below 5%. Fast photodiode and 12GHz oscilloscope are used to analyze pulse shape. Far-field measurement and near-field measurement are fit into a package. The quality of far-field measurement is 1.1DL with its view range is 1.7mrad. The resolution of near-field measurement is 0.45mm when beam size is 310×310mm$^2$.

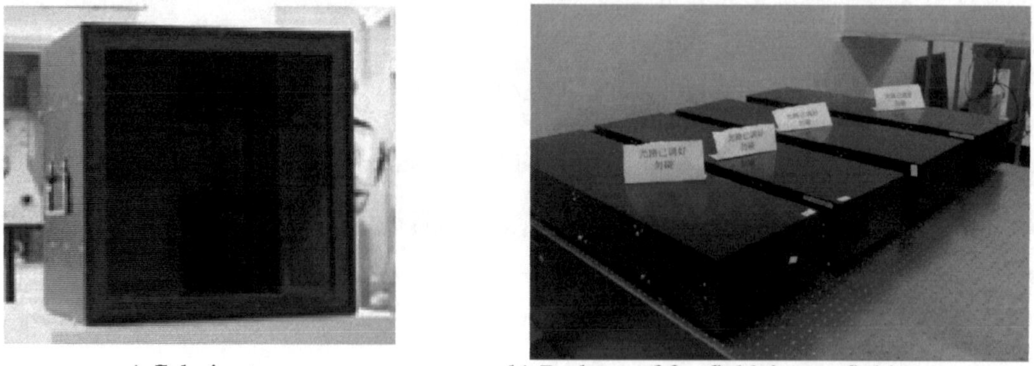

a) Calorimeter b) Package of far-field &near-field measurement

Fig. 1 Calorimeter and package of far-field & near-field measurement

## 3. Diagnostics for picosecond pulse

Diagnostics for picosecond pulse can provide laser parameter such as energy, pulse width, pulse contrast, far-field and spectrum.

Calorimeter with stainless steel shell is designed in order to avoid of leakage and pollution in vacuum compressor. A single shot auto-correlator is designed to measure pulse width between 0.5~18ps. Its resolution is 0.07ps. Tolerance of this auto-correlator is below 12.3% by mirror images and wider size [4]. Pulse contrast is measured by a single shot cross-correlator with fiber-array [5]. Its dynamic range is ~10$^6$ when pulse is 1mJ and 10ps in our experiment. The experiment data on petawatt laser is shown in Fig.2 when energy is 100J, whereas maximum energy of petawatt laser is 377J [6]. Far-field measurement is used to diagnose focusing ability of laser pulse. Its view range is 5mrad, and its quality is 1.5DL. The spectrum of chirped laser is adopted by SP2750 of Princeton Instruments.

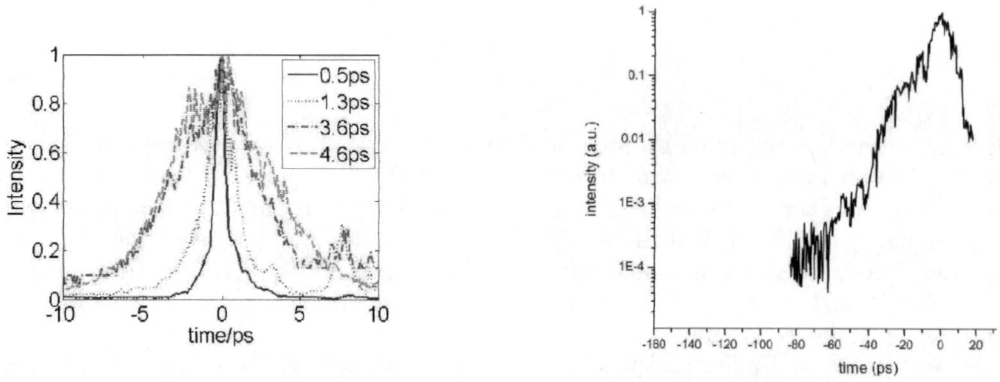


a) Autocorrelation of compressed pulse b) Cross-correlation of pulse contrast

Fig. 2 Pulse width and pulse contrast measurement

The authors thanks very much to Prof. Dai Yaping, Xu Guang, Xie xinglong, Qian Liejia and Len Yuxing, Liu Daizhong.

References

[1] Jiayong Zhong, Yutong Li, Xiaogang Wang, Jiaqi Wang, Quanli Dong, Chijie Xiao, Shoujun Wang, Xun Liu, Lei Zhang, Lin An, Feilu Wang, Jianqiang Zhu, Yuan Gu, Xiantu He, Gang Zhao, Jie Zhang. Modelling loop-top X-ray source and reconnection outflows in solar flares with intense lasers, Nature Physics, 2010, Vol. 6(12): 984~7.

[2] Zuegel, J.D.; Seung-Whan Bahk; Bromage, J.; Dorrer, C.; Earley, R.; Kessler, T.J.; Kruschwitz, B.J.; Morse, S.F.B.; Maywar, D.N.; Oliver, J.B.; Jie Qiao; Rigatti, A.L.; Schmid, A.W.; Shoup, M.J. III; Waxer, L.J.; Kelly, J.H. Novel laser and diagnostic technologies for the OMEGA EP high-energy petawatt laser[J]. Review of Laser Engineering, 2009, 37(6): 437~442.

[3] Yoneyoshi Kitagawa, Hisanori Fujita, Ryosuke Kodama, Hidetsugu Yoshida, Satoshi Matsuo, Takahisa Jitsuno, Tetsuji Kawasaki, Hisao Kitamura, tadashi Kanabe, Shuji Sakabe, Keisuke Shigemori, Noriaki Miyanaga, and Yasukazu Izawa. Prepulse-free petawatt laser for a fast ignitor[J], IEEE Journal of Quantum Electronics, 2004, 40(3): 281~293.

[4] Xiaoping Ouyang, Jingui Ma, Lin Yang, Shunxing Tang, Chong Liu, Yonghua Peng, Liejia Qian, Baoqiang Zhu, Jianqiang Zhu, and Zunqi Lin. Accuracy of single-shot autocorrelation measurements of petawatt laser pulses, Apllied optics, 2012, Vol. 51(18): 3989~3994.

[5] Dongfang Zhang, Liejia Qian, Peng Yuan, Heyuan Zhu, Shuangchun Wen, and Chris Xu. Fiber-array-based detection scheme for single-shot pulse contrast characterization, Optics Letters, 2008, Vol. 33(17): 1969~1971.

[6] Ouyang Xiaoping, Yang Lin, Peng Yonghua, Chen Ming, Ma Jingui, Wang Yongzhi, Tang Shunxing, Liu Chong, Li Hong, Wang Yuyu, Qian Liejia, Zhu Jian, Zhu Baoqiang, Zhu Jianqiang; Analysis on reliability of diagnostics system for picosecond petawatt laser[J]. Chinese Journal of Laser, 2013,40(1): 0108003-1~0108003-8.

# Hot electron spectra on advanced targets in FIREX

**T. Ozaki[1], A. Sunahara[2], H. Shiraga[2], Y. Arikawa[2], S. Fujioka[2], H. Sakagami[1], Z. Zhang[2], H. Nagatomo[2], T. Johzaki[3], T. Namimoto[2], M. Taga[2], S. Kojima[2], Y. Abe[2], K. Ishihara[2] , T. Nagai[2] , S. Sakata[2] , S. Hattori[2], Y. Sakawa[2], H. Nishimura[2] , H. Azechi[2] and FIREX group[2]**

[1]National Institute for Fusion Science, 322-6, Oroshi, Toki, Gifu 509-5292, Japan
[2]Institute of Laser Engineering, Osaka Univ., 2-6, Yamada-Oka, Suita, Osaka, Japan
[3]Grad. of Eng., Hiroshima Univ., 1-4-1, Kagamiyama, Higashi-Hiroshima, Japan

E-mail: ozaki@nifs.ac.jp

**Abstract.** The traditional fast ignition scheme is that a compressed core created by an imploding laser is auxiliary heated and ignited by the hot electrons (produced by a short pulse laser guided through the cone and guided under-dense plasma). However sufficient heating has not be achieved because the hot electron energy is too high and dissipated in the cone, the angular divergence of the hot electron is too large, and the distance from the generation point to the core is too long. Here we clarify the problems of fast ignition by observation of the hot-electron spectra.

## 1. Introduction

Fast ignition (FI)[1,2] will be performed to heat an imploded core which is created by an imploding laser, by hot electrons (optimum energy of 2 MeV)[3]. The hot electrons are produced by the interaction between a chirped ultra-short pulse laser (heating laser) and the pre-formed plasma created by the pre-pulse of the heating laser at the tip of a guiding cone. If the scale length $L$ of the plasma density is long, the effective electron temperature $T_{eff}$ (=slope of the electron spectrum) is too high and the coupling efficiency $\eta_c$ is reduced because the electron passes through the core with small energy deposition. When $T_{eff}$ is too high, the amount of the lower energy electrons is reduced because the laser energy is used for energetic electron production. At the same time, when the scale length L is long, the distance between the core and the hot electron source point: S is long. Therefore the distance between the core and the hot electron generation point, $S$ becomes long. It is one of the key issues that $T_{eff}$ remains low with the higher conversion efficiency from the laser to the hot electrons.

The total coupling efficiency: $\eta$ is the product of the conversion efficiency and the irradiation efficiency $\eta_i$. $\eta_c$ is determined by the conversion efficiency from the laser to the hot electrons, the transparency in cone, the irradiation efficiency $\eta_i$ to the core and the energy deposition in the core. However in the most experiments a little energy of the hot electrons can be deposited because the hot electron energy is too high against an areal density $\rho R$ of the core. It is important to decrease $T_{eff}$ in order to increase $\eta_c$ To obtain a high ion temperature $T_i$, the fuel amount should be tuned.

The hot-electron spectrum can be approximated by two or more Gaussian distributions. Although the higher energy component of the Gaussian spectrum has a small contribution to heating, the lower energy component has been thought to play a heating role. It is difficult to measure the lower component using an electron spectrometer (ESM). The heating by the lower component may not be larger than that we expect. Here we conservatively consider $\eta_c$ and $T_i$ using the spectra with one Gaussian approximation which is determined by ESM[4].

## 2. Comparison of $T_{eff}$ in various targets

The LFEX laser (two beams, 1.053 μm) has an energy from 500 J to 1.6 kJ and the pulse width of 1.5 ps in the experiments. The energy of 82 % is focused at the focal spot size (50%) of 40 μmϕ. Gekko XII (9-12 beams, 250 J/beam, 0.53 μm) is used for implosion of the deuterated polystyrene shell (500 μmϕ, 7 μm-thickness). The neutron yield is measured by the MANDALA and three liquid scintillation detectors. Back ground neutrons caused by γn reaction are remarkably increased over the LFEX energy of 500 J. γn neutrons are shielded by polyethylene blocks. The LFEX laser-plasma interaction position is monitored by the X-ray streak camera.

$T_{eff}$ in plain, cone and integrated targets are compared. LFEX laser has a pre-pulse of about 1 ns with the contrast of about $10^8$. The pre-pulse creates a pre-formed plasma. According to Phukov scaling[5], $T_{eff}$ is determined by $L$ as follows;

$$T_{eff} = \frac{\alpha}{3.546}\left({}^{I}/_{I_{18}}\right)^{\frac{1}{2}}\left(2.465L^{\frac{1}{4}} - 2.223\right), \quad (1)$$

where $I/I_{18}$ is the laser intensity (W/cm$^2$) normalized by $10^{18}$, $\alpha$ is the coefficient factor of 1.5 MeV, respectively. Figure 1 shows the laser intensity dependence of $T_{eff}$. Generally $T_{eff}$ in high-Z material is small because $L$ is short due to the heavy mass. Therefore in low-Z material as a diamond-like-carbon (DLC), we use Au coating. In the conical target, higher $T_{eff}$ is observed than $T_{eff}$ in the plain. The plasmas produced by the irradiation of the inner wall of the cone, gather to the laser axis. As the result, $L$ may be extended.

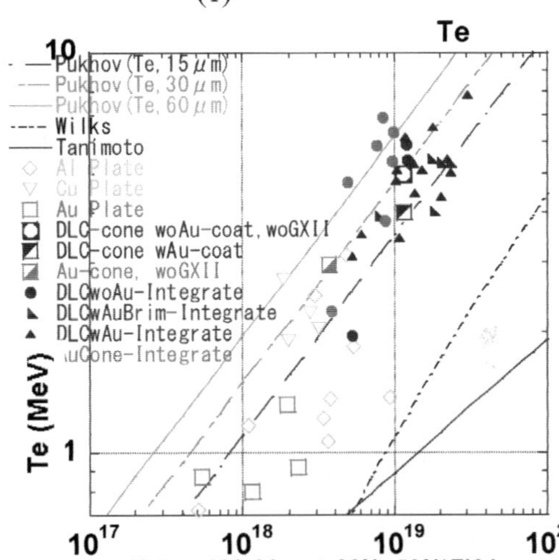


**Figure 1.** $T_{eff}$ in various targets.

In the integrated experiment, we use three different targets, an Au cone shell, a DLC cone shell and a hole-cone shell. In the latest experiment, $T_{eff}$ of 5 keV in the DLC cone shell and 6 keV in Au cone shell can be observed. Those values correspond with $L$ of 15-30 μm according to Phukov scaling. The geometrical effect of the cone may be larger than effects caused through implosion because $T_{eff}$ in integrated targets are almost similar to $T_{eff}$ in cone targets or slightly higher.

## 3. Estimation of the energy deposition in the integrated experiment

The energy deposition in the core can be estimated using the hot electron spectra in the experiment. The conversion efficiency from the laser to the hot electrons (forward direction) can be estimate from the experimental results by Zhang[7] and the computer simulation as a function of the laser intensity. The spectrum observed in Section **2** is used in this calculation. The interaction point between the laser and the hot electrons is assumed to be $L$ (defined in equation (1)) in front of the cone tip. The deposition of the hot electrons in the cone and the core is approximated by the energy deposition in the cold materials of a single electron because the electron temperature of the core plasma is small enough. The classical deposition range $R_a$ (g/cm$^2$) is determined by the hot electron energy $E$ (MeV) as follows;

$R_a = 0.542E - 0.133$, ($E$=0.8-8MeV), (2)

$R_a = 0.407E^{1.38}$, ($E$=0.13-0.8 MeV), (3)

Actually $R_a$ may become short due to the self-magnetic field. When the plasma temperature increases, *RE* may be extended. However those effects are not considered here because the calculation becomes complicated. Instead of this, we introduce the anomalous deposition factor against $R_a$ so as to match the experimental result with the calculation.

The relaxation between the electron and the ion in the core plasma is assumed from the result of the simulation as a function of the mass density. $\rho R$ and the core radius $R$ are obtained to be 0.02 g/cm$^2$ and 27 μm from other experimental results. $T_i$ without auxiliary heating can be estimated to be 0.67 keV from the neutron yield. When the LFEX laser with 612 J heats the imploded core in the Au cone shell target, increasing in $T_i$ of 0.17 keV and in $\eta_c$ of 1.6 % can be obtained[8].

We calculate $\eta_c$ and $T_i$ using the same experimental condition as the GXII-LFEX integrated experiment. If we assume the divergence angle of 62 degrees using reference [9], an anomalous deposition factor can be estimated to be 75%. The anomalous deposition may occur due to the low energy component of two Gaussian distributions of the spectrum, which is not considered here, and the high current effect and the range expansion due to the core heating as mentioned above. At that time, the conversion efficiency of 36%, the cone transparency of 96.5%, $\eta_i$ of 75% and the core deposition of 6.2% can be calculated. Lower $\eta_i$ comes from $S$. In this case, the interaction layer is shifted due to the long $L$. $T_i$ (0.77/0.76keV in experiment/calculation) on the DLC cone shell is smaller than $T_i$ (0.84/0.83keV in experiment/calculation) on the Au cone shell in the experiment. $T_i$ on the DLC cone shell in the calculation corresponds to $T_i$ in the experimental result. By using a low-Z material, the core transparency and the core deposition are slightly improved to be 98.2% and 6.3%, respectively. However $S$ is long because of the thicker tip in the DLC cone shell. $\eta_i$ becomes worse (41%) due to the long $S$.

The laser intensity dependences of $T_i$ and $\eta_c$ have been performed in three different spectra, the average spectrum of the experiment (Standard), the minimum spectrum of the experiment (Improved) and Wilks scaling (Wilks)[6]. If we can suppress $T_{eff}$ to the $T_{eff}$ determined by Wilks scaling, four times or more $T_i$ and $\eta_c$ can be achieved as shown in Figure 2. To achieve Wilks scaling, $L$ should be minimize by the pre-pulse suppression. Therefore $\eta_i$ can be also improved when $T_{eff}$ can be reduced. We survey $T_i$ and $\eta_c$ using realistic spectra obtained in the experiments when $\rho R$ is scanned under the same conditions. There are two ways to scan $\rho R$, variable density with fixed $R$ and variable $R$ with fixed fuel amount as shown in figure 3. When $R$ is fixed, $\eta_i$ determined by $R$ and by $S$, is not changed. If $\rho R$ increases, the fuel amount also increases. Therefore $T_i$ does not increase although $\eta_c$ increases because the deposited energy per one deuteron is not so changed. When the fuel amount is fixed, $R$ should be smaller by increasing $\rho R$. $\eta_c$ does not increase because $\eta_i$ decreases.

In the latest experiment, $\rho R$ is smaller than the most suitable value (0.05 g/cm$^2$) as shown in Figure 3. Therefore if we can achieve higher $\rho R$ by improvement of the implosion, slightly higher $T_i$ can be expected to be 0.86 keV. Indeed higher neutron yield ($T_i$) can be obtained at a higher imploding laser energy, which leads to better implosion. $\eta_c$ is determined by the product of conversion, $\eta_i$ and deposition efficiencies. It is most important to decrease $T_{eff}$ up to Wilks scaling because $T_{eff}$ is related to both $\eta_i$ and the deposition efficiencies. For example, the LFEX laser energy of 3.2 kJ (0.02 g/cm$^2$, 4.4x10$^{18}$W/cm$^2$) is required if $T_{eff}$ can be reduced down to Wilks scaling in order to achieve $T_i$=5 keV, which is a goal of FIREX although 16 kJ is necessary in current experimental scaling.

## 4. Proposal of hole cone hole shell target

The most suitable target for FI can be sought by comparison of the spectra between varied targets using ESM. We compare $T_{eff}$ in the cone irradiated by the LFEX laser with $T_{eff}$ in the plate. Direct heating of the imploded core has been performed by using the hole-cone shell target[10]. $T_{eff}$ does not become high and high neutron yield can be obtained. This means the hole-cone shell target is another candidate of FI. At earlier LFEX injection timing, for example, before the imploding shell is filled by the pre-formed plasma, more effective $\eta_c$ may be expected. At that time, the electron deposition is decreased because the density is still low. However almost the hot electrons can irradiate the core from inside as a cavity even if the beam divergence is large. The relaxation between the electron and the ion is enough although the density is over several g/cm$^3$.

If another hole is drilled in the opposite side of the cone (hole cone hole shell target)[11], $T_{eff}$ may be strongly reduced. The pre-formed plasma cannot be created because the pre-pulse goes through the hole during the imploding phase. If the LFEX laser is injected when the pre-cursor fills the cavity, lower $T_{eff}$ and short interaction distance and higher $\eta_i$ can be expected.

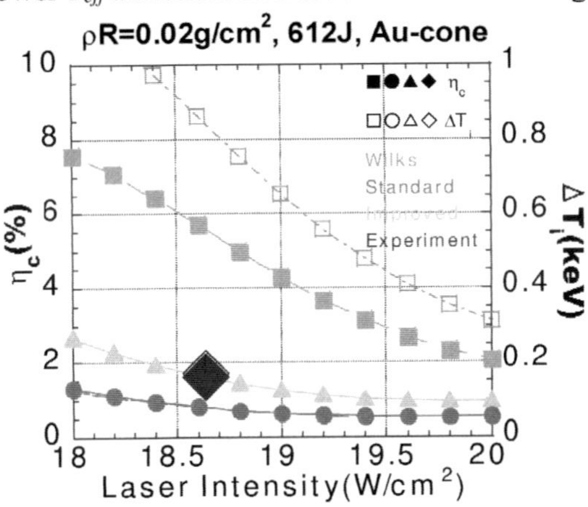


**Figure 2.** $\Delta T_i$ and $\eta_c$ versus Laser intensity.

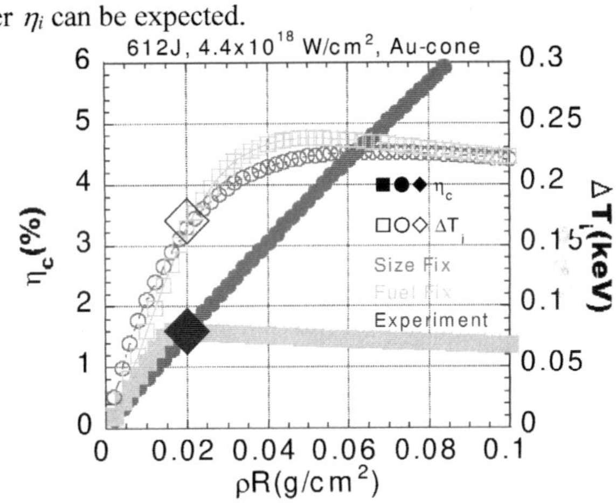


**Figure 3.** $\Delta T_i$ and $\eta_c$ versus $\rho R$.

## 5. Summary

The results in the latest experiment can be explained by the results in a simple calculation using realistic spectra obtained in the experiments. We survey $T_i$ and $\eta_c$ with regard to the LEFX laser intensity and $\rho R$. Higher $T_i$ can be expected by strong implosion. However it is necessary to increase $T_i$ and $\eta_c$ drastically such that $T_{eff}$ should be also decreased drastically. One of the candidates to reduce $T_{eff}$ is the hole-cone-hole shell which does not create the pre-formed plasma until the arrival of the main pulse.

**Acknowledgements**

This work is supported by budgets of NIFS12KUGK059 and the Grant-in-Aid for Scientific Research (24561028).

## 6. References

[1] M. Tabak, et al., POP 1 (1994) 1626.
[2] R. Kodama, et al., Nature 418 (2002) 933-934.
[3] T. Jhozaki, et al., PP Cont. Fus. 51 (2009) 014002.
[4] T. Ozaki, et al., JOP:Conf. Ser.2, 244 (2010) 022056.
[5] A. Pukhov, et al., POP 6 (1999) 2847.
[6] S. Wilks, et al., PRL 69 (1992) 1383.
[7] Z. Zhang, et al., IFSA2013, Nara (Japan).
[8] Y. Arikawa, et al., IFSA2013, Nara (Japan).
[9] T. Yabuuchi, et al., NJOF 11 (2009) 093031.

[10] A. Sunahara, et al., IFSA2013, Nara (Japan).
[11] T. Ozaki, et al., Plasma2013, Warsaw (Poland).

# Laboratory astrophysical collisionless shock experiments on Omega and NIF

**Hye-Sook Park[1,*], J. S. Ross[1], C. M. Huntington[1], F. Fiuza[1], D. Ryutov[1], D. Casey[1], R. P. Drake[2], G. Fiksel[3], D. Froula[3], G. Gregori[4] , N. L. Kugland[5], C. Kuranz[2], M. C. Levy[1,9], C. K. Li[6], J. Meinecke[4], T. Morita[7], R. Petrasso[6], C. Plechaty[1], B. Remington[1], Y. Sakawa[7], A Spitkovsky[8], H. Takabe[7], A. B. Zylstra[6]**

[1] Lawrence Livermore National Lab, 7000 East Ave, Livermore CA, USA
[2] Dept. of Atmospheric, Oceanic, and Space Sciences, Univ. of Michigan, Ann Arbor, MI, USA
[3] Laboratory for Laser Energetics, Rochester, NY, USA
[4] University of Oxford, Oxford, UK
[5] Lam Research Corporation, 4400 Cushing Parkway, Fremont, CA, USA
[6] Massachusetts Institute of Technology, Cambridge, MA, USA
[7] Institute of Laser Engineering, Osaka University, Osaka, Japan
[8] Dept. of Astrophysical Sciences, Princeton University, Princeton, NJ, USA
[9] Dept. of Physics and Astronomy, Rice University, Houston, Texas 77005, USA

*E-mail: park1@llnl.gov

**Abstract.** We are performing scaled astrophysics experiments on Omega and on NIF. Laser driven counter-streaming interpenetrating supersonic plasma flows can be studied to understand astrophysical electromagnetic plasma phenomena in a controlled laboratory setting. In our Omega experiments, the counter-streaming flow plasma state is measured using Thomson scattering diagnostics, demonstrating the plasma flows are indeed super-sonic and in the collisionless regime. We observe a surprising additional electron and ion heating from ion drag force in the double flow experiments that are attributed to the ion drag force and electrostatic instabilities. [1] A proton probe is used to image the electric and magnetic fields. We observe unexpected large, stable and reproducible electromagnetic field structures that arise in the counter-streaming flows [2]. The Biermann battery magnetic field generated near the target plane, advected along the flows, and recompressed near the midplane explains the cause of such self-organizing field structures [3]. A $D^3He$ implosion proton probe image showed very clear filamentary structures; three-dimensional Particle-In-Cell simulations and simulated proton radiography images indicate that these filamentary structures are generated by Weibel instabilities and that the magnetization level (ratio of magnetic energy over kinetic energy in the system) is ~0.01 [4]. These findings have very high astrophysical relevance and significant implications. We expect to observe true collisionless shock formation when we use >100 kJ laser energy on NIF.

## 1. Introduction

Astrophysical collisionless shocks have been of keen interests to the astrophysical community as a mechanism for self-generating magnetic fields and cosmic ray acceleration. A collisionless shock is the condition where the Coulomb mean free path is much larger than the system, yet a shock is formed via plasma instabilities. There are many astrophysical objects, both relativistic and non-relativistic, that show these characteristics such as in Supernova remnants and gamma-ray bursts [5]. When a supernova event occurs, the plasma ejecta stream from the explosion and go through interstellar media, encountering various physical conditions. Among possible outcomes, Weibel instabilities can be created from the momentum anisotropy distribution of the ejecta and ambient media [6]. Conceptually, the Weibel instabilities create self-generated magnetic fields that trap the ions via the Lorenz force, thereby imitating collisions in a usual collisional shocks and leading to collisionless shock formation. The most significant aspect of these instabilities is their conversion from kinetic energy to magnetic energy that can be the mechanism for seed magnetic fields in universe [7]. The signature of the instability is a pattern of current filaments stretched along the axis of symmetry of the flows. The exact understanding of the required physical condition for such occurrence has not been well characterized, and laboratory laser experiments can provide a unique platform to study the electromagnetic Weibel instabilities that occur from high-Mach number plasma flows.

## 2. Experimental set-up

Figure 1 shows our typical laser experimental configuration at the Omega laser facility. Two face-on polyethylene ($CH_2$) plastic foils are illuminated by ~4 kJ of 351 nm laser energy with focal spot diameters of 250 μm on the target surface, generating high-velocity counter-streaming plasma flows. The separation is 8 mm between the targets. We employ Thomson scattering diagnostics to understand the plasma bulk velocity ($v$), electron temperature ($T_e$), ion temperature ($T_i$) and electron density ($n_e$), The Thomson-scattered (TS) light from the 526.5 nm probe studies a plasma volume of 100 μm x 100 μm x 60 μm at the central region of the counter-streaming plasmas. Thomson scattering is measured from the electron feature (collective scattering from electron-plasma waves) and the ion feature (collective scattering from ion-acoustic waves). An example of time composite electron and ion features from the double foil counter-streaming data is shown in the top right panel in Figure 1.

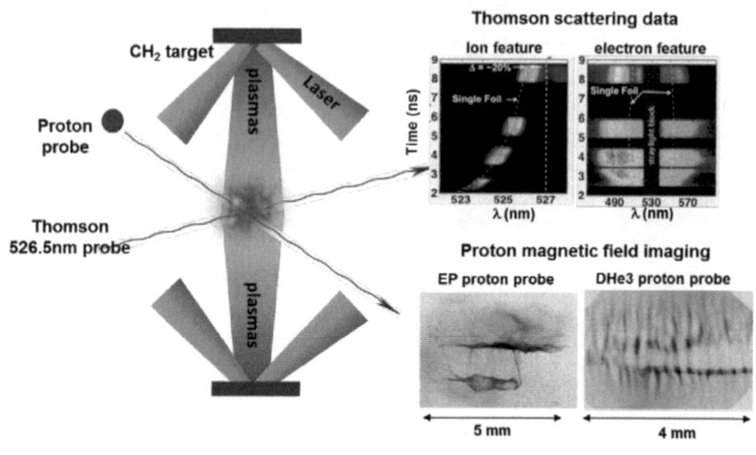


**Figure 1.** Schematic of Omega experiment. We use Thomson scattering to probe the plasma state and proton probe to image the magnetic field structures.

Another key element of our experiment is proton probing that gives understanding of the magnetic field structures. We use 2 different proton sources: 1) by short pulse generated protons; 2) by imploding a capsule filled with deuterium (D) and $^3$helium ($^3$He) fuel. The short pulse generated protons are generated by illuminating a 10 ps pulse of up to 800 J of 1053 nm infrared light onto 40 μm diameter spot for an intensity of ~$2x10^{18}$ W/cm$^2$ on an Au disk target. Electric and magnetic fields deflect the protons which are then recorded on a radiochromic film layered with Al filters to obtain a range of proton energies from 5 to 15 MeV. We also use the 14.7 MeV mono-energetic proton source generated by thermonuclear interaction of D+$^3$He $\rightarrow$ $^4$He+p [8]. For our experiments, the 2 μm thin silica capsules are filled with 18 atm D$^3$He fuel (6 atm of D and 12 atm of $^3$He for equal atomic distribution) and are compressed uniformly by ~9 kJ of laser energy from 18 laser beams. The protons are generated at the peak of compression creating ~$10^8$ protons isotropically lasting ~50 ps with a source size of ~50 μm in diameter. These protons are detected using CR39 film in which the proton tracks are etched and counted to form an image. Examples of these two proton magnetic field imaging are shown in the right-bottom panel in Figure 2.

## 3. Thomson scattering results

Extensive Thomson scattering data were collected for both single-flow and counter-streaming flows. Time-slices of the data from the electron feature are then fitted with the Thomson scattering form-factor allowing a measurement of the electron temperature ($T_e$) and electron density ($n_e$). With constraints from the electron feature fitting, the ion feature can then be used to measure the ion temperature ($T_i$) and bulk plasma flow velocity ($v$). This can be understood from the simplified dispersion relation, $\Delta\lambda = \frac{4\lambda_{probe}}{c}\sin(\theta/2)\sqrt{\frac{ZT_e}{M}+\frac{3T_i}{M}}$. Figure 2 shows the summary of the resulting plasma parameters of $v$, $n_e$, $T_e$ and $T_i$, where single flow is noted in red and the counter-streaming double flows are marked in blue respectively [9].

The bulk flow velocity, Fig 2a, is > 1000 km/s up to 5 ns and is not suppressed for the double flow case, indicating interpenetration. The single flow $n_e$ is ~$5\times10^{18}$ $cm^{-3}$ at maximum and is doubled for the counter-streaming case at $1\times10^{19}$ $cm^{-3}$. This indicates that a shock is not quite formed as we should expect a factor of > 2 increase in electron and ion densities at a shock front. A more surprising discovery is the significant increase in $T_e$ and $T_i$. The $T_e$ was ~100 eV for the single flow whereas the double flow $T_e$ was up to 1 keV increase. The rapid $T_e$ increase at early time is explained by electron-ion collisions from ion slowing-down by drag forces caused by the 'resting' electron gas [10]. Analytic calculation of the electron temperature increase from this model is plotted in black solid line in Fig 2c showing that the model agrees well with the data at early time. Ignoring the electron thermal conduction in this model explains the late-time discrepancy. While this drag-force model explained the $T_e$ increase very well, the ion-ion collisions could not explain the observed $T_i$ increase. When collisional particle-in-cell (PIC) simulations are applied accounting for acoustic two-stream electrostatic instabilities, we are able to reproduce the $T_i$ increase. This instability occurs for $T_e > T_i$ and therefore leads to the heating of the ions to temperatures close to $T_e$ [1]. Our quantitative plasma state measurements suggest that the intra-flow ion and

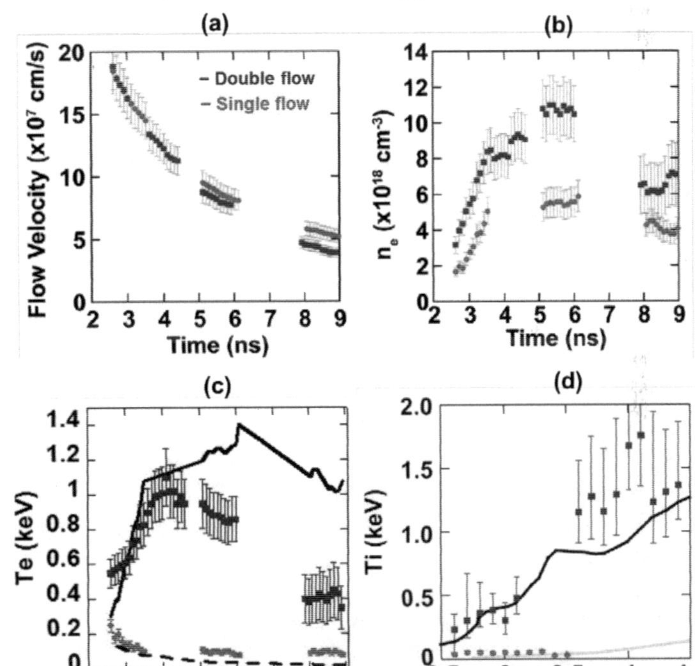


**Figure 2.** Omega Thomson scattering results for both single and double flows. The high increase in $T_e$ and $T_i$ are due to electron drag force and the electrostatic instabilities.

electron collisional effects are important and that inter-flow ion collisions are rare from high velocity flows.

## 4. Observation of self-organizing fields

Using proton probes generated by short pulse lasers on EP, we observe large, stable, reproducible electromagnetic field structures that arise in counter-streaming interpenetrating supersonic plasma flows in the laboratory. Self-organization, whereby energy progressively transfers from smaller to larger scales in an inverse cascade, is widely observed in fluid flows, such as in the nonlinear evolution of multimode Rayleigh-Taylor and Kelvin-Helmholtz instabilities. These surprising structures, predominantly oriented transverse to the primary flow direction, extend for much larger distances than the intrinsic plasma spatial scales, and persist for much longer than the plasma kinetic timescale10. One such example image is shown in Fig 3a. This image is taken at 5.2 ns after the laser, and the sharp co-planar field structures are clearly visible. Their origin is now explained by the magnetic field advection process. Here the Biermann battery magnetic field is generated near the target surface in a toroidal shape, and advects along the electron plasma flows, then recompresses near the midplane [3]. A sketch of this process is shown in Fig 3b The magnetic field strength is evaluated by the caustic analysis [11] by: $B_{cf} = \frac{m_p c}{el}\sqrt{\frac{2W}{m_p}}\frac{d^3}{\sqrt{\pi a^2 b \sin\psi}}$, where $W$ is the proton kinetic energy (8.8 MeV), $b/a$ the aspect ratio (1/20), $l$ is the distance from proton source to object plane (8 mm), $\psi$ is the tilt angle of the plane (5 to 10 degrees). When we apply our experimental measurements and observables, we derive $B_0$=10T. The observed plasma self-organizing magnetic field may play an important role in astrophysics how electromagnetic order emerges from high-velocity plasma flows.

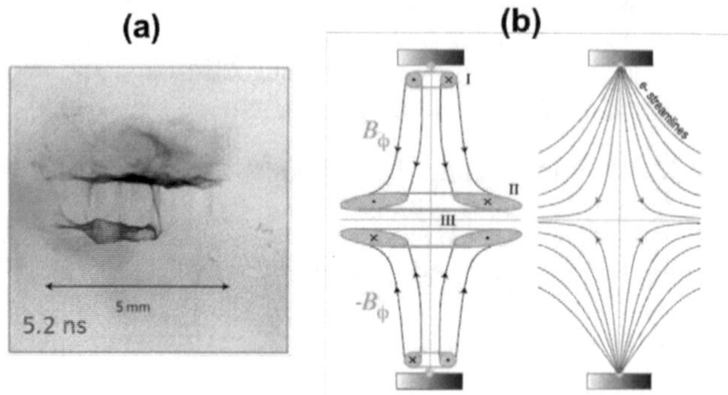


**Figure 3.** Omega EP short pulse generated proton image of counter-streaming plasmas at 5.2 ns after the laser. Highly stable self-organizing planar magnetic field structures are observed. This is from the Biermann battery magnetic field that is generated near the target surface; advects along the electron flows; then recompress in the mid-plane.

## 5. Observation of Weibel filamentation

We recently used a $D^3He$ imploding capsule to create a mono-energetic proton source and imaged the counter-stream double flows on the Omega laser. The result was quite spectacular as shown in Fig 4a where filamentary striation field features are clearly visible along with the co-planar structures. We study whether these filaments could be generated by Weibel instabilities (filamentation is the characteristic morphology for Weibel instabilities.) In order to understand both the Weibel and Biermann battery generated magnetic fields in our proton imaging experimental system, we have conducted detailed 3-dimensional particle-in-cell (3D PIC) simulations with OSIRIS [12] along with proton ray tracing through the electromagnetic field. In addition to the filamentary structure, we also added the Biermann battery field that was described in section 4. The plasma input is also from our experimental measurements as described in section 3: $n_e$=5x10$^{18}$ cm$^{-3}$, $v_e$=$v_i$=1000 km/s, $T_e$=$T_i$=100 eV. The simulation results are shown in Fig 4b showing remarkable resemblance to the data. From the detailed studies of simulation and data, we derive the magnetization, which is defined:

$\sigma = \dfrac{B^2/4\pi}{(\gamma-1)nmc^2}$. This is the quantity that the kinetic energy converts to magnetic energy in the system. Figure 5 shows this magnetization as a function of time. The dotted line is the field strength without pre-existing magnetic field where as the solid line is accounting for the initial field by the Biermann battery. The straight linear line is the theoretical growth rate of the ion Weibel instability. Note that the final total magnetization is 0.01 and the initial Biermann battery field plays little role in the magnetization growth. The Weibel instability filamentation is clearly observed in the laboratory and a significant self-generated magnetization is indicated [4].

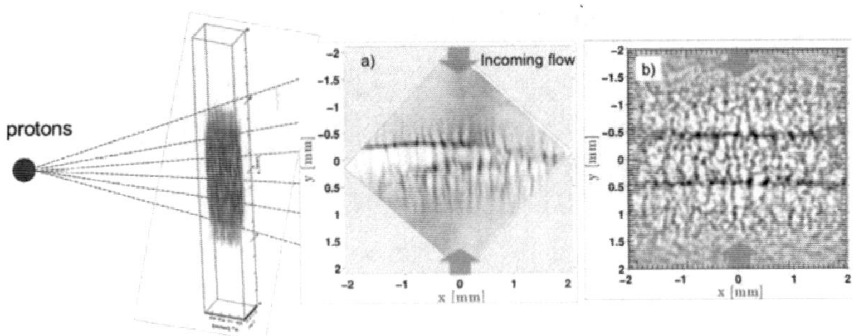


**Figure 4.** Schematic of proton image of D3He3 imploding capsule. (a) is the experimental data where the Weibel filamentary structures are clearly observed; (b) is the 3D Particle-in-cell simulation and the proton ray trajectory from experimental angle. The simulation also added Biermann battery field that shows as the band structure in the middle.

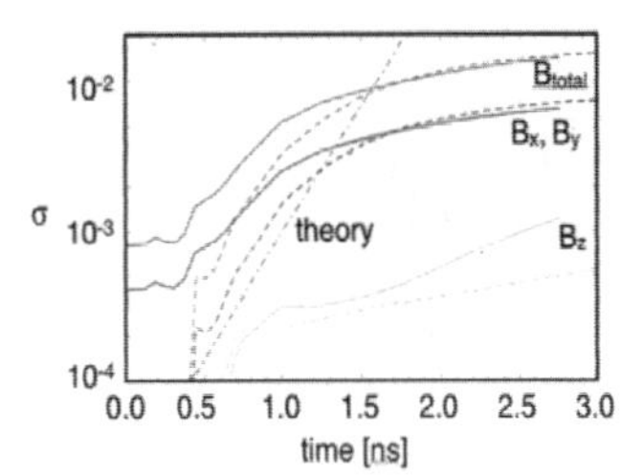


**Figure 5.** Magnetization level derived from the 3D PIC simulation The magnetization is achieved up to 1% level.

## 6. NIF experiments

While the Omega experiments produced a remarkable understanding of counter-streaming plasmas, we note that the Weibel mediated collisionless shock is not fully formed as indicated by the electron density measurements and the field formation in the 3D PIC simulation results. Our NIF experimental configuration will apply ~100 kJ per target and separation of 6~8 mm. The radiation hydro simulation (HYDRA) of our single flow simulation indicate that we will have much higher ne (>$10^{20}$ cm$^{-3}$), $T_e$ (>1 keV) and higher flow velocity (>2000 km/s) (Fig 6). This will produce a well-formed shock as the ion skin depth will be much shorter and a high magnetic Reynolds number will favor producing higher magnetization. Using this flow condition, the 3D PIC simulation (Fig 7) predicts a much higher magnetic field generation of up to ~3 MG (top row) and a shock density of $n/n_0$ ~4. On NIF we will use the D$^3$He generated proton source to image the magnetic field structures and x-ray and neutron spectroscopy to measure the plasma state.

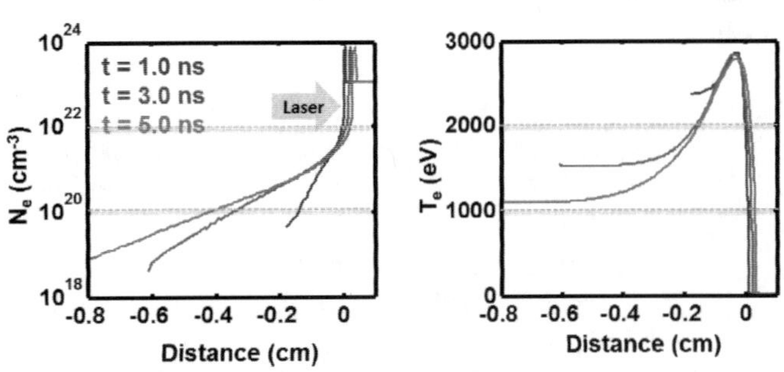


**Figure 6.** HYDRA simulation results for the NIF experimental condition with 100 kJ per target condition. The different color represents the pulse duration. At 5 ns setting, the $N_e$ is ~$10^{20}$ cm$^{-3}$ and the $T_e$ is ~1 keV at 4 mm from the target surface.

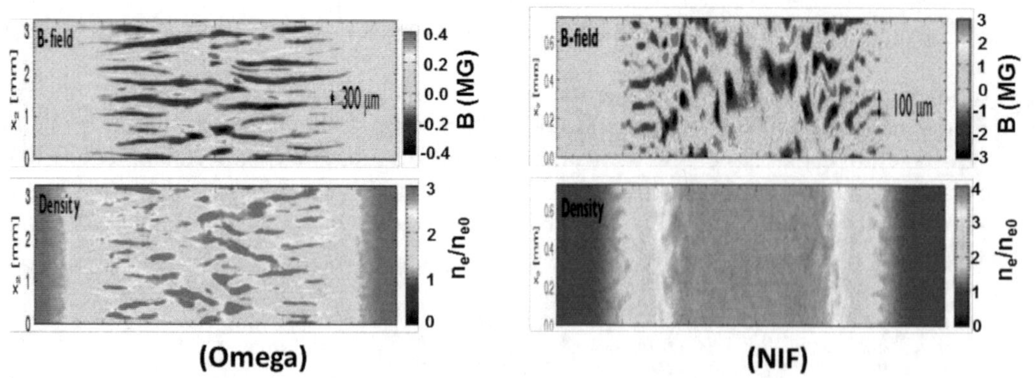


**Figure 7.** 3D PIC simulation results of B-field (top row) and electron density (bottom row) for Omega and NIF cases. The NIF experiments will be able to produce fully formed collisionless shocks ($n_e/n_e0$ ~4) and B field near 3 MG.

In conclusion, laser generated counter-streaming plasma flows have been studied in connection to astrophysical collisionless shocks. We observe that the intra-collisional electron-ion interaction by electron-drag force elevates the electron temperature and electrostatic instabilities upraise the ion temperatures for the double flows. We detect very stable self-organizing field structures that originate from the recompression of the advected Biermann battery magnetic field. The Weibel filamentation is directly imaged for the first time and a magnetization level at 0.01 is derived. The NIF experiment will be able to create true Weibel mediated collisionless shocks.

**Acknowledgements**
This work was performed under the auspices of the Lawrence Livermore National Security, LLC, (LLNS) under Contract No. DE-AC52-07NA27344. Computing support for this work came from an ALCC allocation on Mira and from the LLNL Institutional Computing Grand Challenge program. Additionally, the authors would like to acknowledge the OSIRIS Consortium for the use of the OSIRIS 2.0.

**References**
[1] Ross J S et al. 2013 *Phys. Rev. Let.* vol 110 p 145005
[2] Kugland N et al. 2012 *Nature Physics* vol 8 p 809
[3] Ryutov D D et al. 2013 Phys. Plasmas. Vol 20 p 032703
[4] Huntington C M et al. 2013 *Nature Physics*, submitted
[5] Gruzinov A and Waxman E,1999 *Astrophys. J* vol 511 p 852
[6] Weibel E S 1959 *Phys. Rev. Let.* vol 2 p 83
[7] Medvedev M V et al. 1999 *Astrophys. J* vol 526 p 697; Medvedev M V 2009 *Astrophys. Space Sci* vol 322 p 147
[8] Li C K et al. *Rev. Sci. Inst.* 2006 vol 77 p 10E725
[9] Ross J S et al. 2012 Phys. Plasmas vol 19 p 056501
[10] Ryutov D D et al. 2012 Phys. Plasmas vol 19 p 074501
[11] Kugland N L et al. 2013 Phys. Plasmas vol 20 p 056313
[12] Fonseca R A et al. 2002 Lect. Notes Comp. Sci. 2331 342; Fonseca R A et al. 2008 Plasma Phys. Contr. Fusion vol 50 p 124034

# FM-to-AM conversion measurement for high power nanosecond lasers

**Denis PENNINCKX**

CEA CESTA, 15 avenue des Sablières, CS 60001, 33116 Le Barp, Cedex, France

Denis.Penninckx@cea.fr

**Abstract.** Through numerical simulations we show that the spectral content of amplitude modulations induced by a transfer function converting frequency modulations required for high-power lasers may be very broad. Hence, measurement of FM-to-AM conversion should be first done in the spectral domain to remove unwanted transfer functions at low frequency scale and then in the time domain to obtain an accurate value.

## 1. Introduction

High-power nanosecond lasers like LMJ or NIF require phase modulations both for avoiding stimulated Brillouin scattering in the optics which could cause damage and for beam smoothing of the focal spot. Typical induced spectral broadenings are 200GHz at 1ω, and 600GHz at 3ω for LMJ. Because of any small distortion of the spectrum, part of the phase modulation is converted into unwanted amplitude modulation. This is called FM-to-AM conversion [1]. The spectral content of AM may be narrow (a few gigahertz) or wide (up to the spectral content of the optical pulse, i.e.: 200GHz or 600GHz depending on the wavelength) depending of the type of distortions. FM-to-AM conversion should be reduced as much as possible to avoid laser damage of the optics and changes of the optical waveform reaching the target. Hence, a lot of work has been done during the last ten years to reduce FM-to-AM conversion (by reducing spectral distortions), to reduce the sensitivity to distortions (by changing the phase modulation), and to precompensate for distortions [2]. Only smooth and stable filtering functions may be precompensated. However, to assess the improvement brought by these schemes, FM-to-AM conversion has to be measured with accuracy. The aim of this paper is to give rules for an accurate measurement of FM-to-AM conversion.

## 2. Measuring in the spectral domain

Sprectral broadening is obtained with sinusoidal phase modulations: $e^{imsin(2\pi fmt)}$ where $f_m$=2GHz and m=7 for anti-Brillouin and $f_m$=14.25GHz and m=5 for smoothing. Such modulations induce the power spectral density given in Figure 1. The intensity remains constant in time. FM-to-AM conversions are usually characterized by the α factor equal to $2.(I_{max}-I_{min})/(I_{max}+I_{min})$. It is equal to 0 when there is no FM-to-AM conversion. α should be measured in the time domain. However, it is possible only if the spectral content of AM is narrow, because of the limited bandwidth of single-shot oscilloscopes. Another way of measuring AM is to measure the spectral transfer function and to use a Fourier transform. However, not only the spectral intensity distortions but also the spectral phase distortions should be measured which is quite difficult.

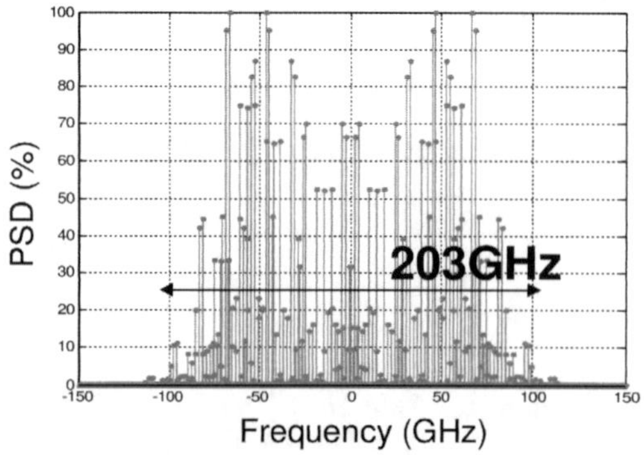


**Figure 1:** Power spectral density of an LMJ pulse.

As a matter of fact, let us consider a two-wave interferometer filter like the one shown in Figure 2. α is equal to 34.4%. We can simulate what could happen to a pulse seeing either only the transfer function in intensity or only the transfer function in phase. It happens that α is still in the same order of magnitude (31,9% for the transfer function in intensity only and 33% in phase only).

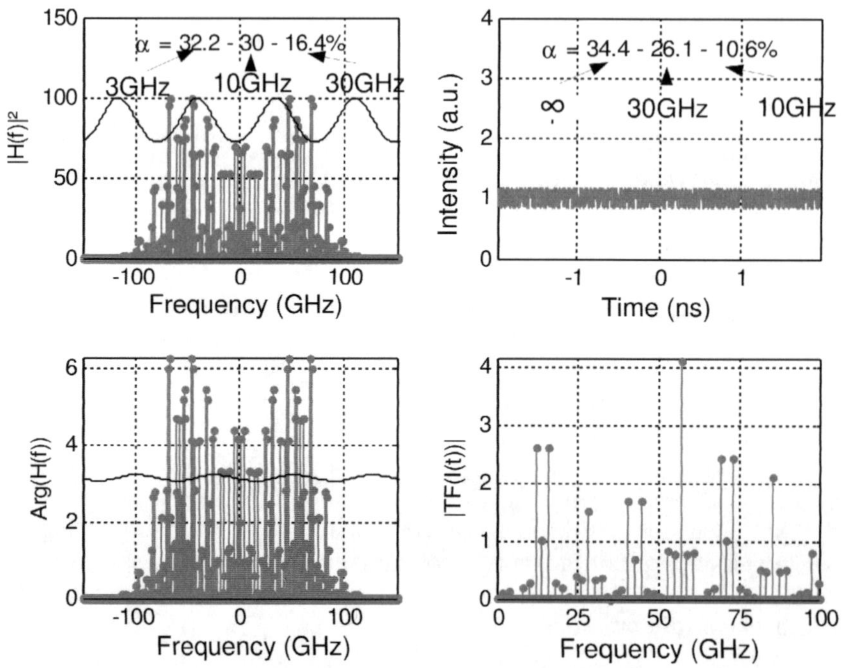


**Figure 2:** FM-to-AM conversion due a two-wave interferometer transfer function. The left part shows the transfer function in intensity (above) and in phase (below). The power spectral density is given at the same frequency scale. On the right, we show the resulting AM modulation and its Fourier transform. The values of α given in the figures correspond to simulated measurements in the spectral domain on the left (by measuring the transfer function in intensity) with a spectral resolution of 3GHz, 10GHz and 30GHz and in the time domain on the right with an infinite, 30GHz or 10GHz bandwidth. The value in the time domain with an infinite bandwidth (34.4%) corresponds to the actual value.

Although in this case, measuring the transfer function in intensity to compute α would give the correct order of magnitude of α, the transfer function in phase cannot be neglected in general since its impact can be as high as the transfer function in intensity. Therefore, measuring the transfer function in intensity to compute α is only possible when the transfer function is supposed to be known and is a dangerous process for the one wanting an accurate value.

## 3. Measuring in the time domain

Thus, a measurement in the time domain seems compulsory. But the measurement bandwidth is limited. Hence, the spectral content of AM should be studied. As a matter of fact, as shown in Figure 3, the level of AM is determined by the amplitude variations of the transfer function (the higher the amplitude variations, the higher the AM level, as can be seen with the vertical green arrows) while the spectral extension of AM is determined by the frequency scale of the transfer function (the lower the frequency scale, the broader the spectral content of AM, as can be seen with the horizontal blue arrows).

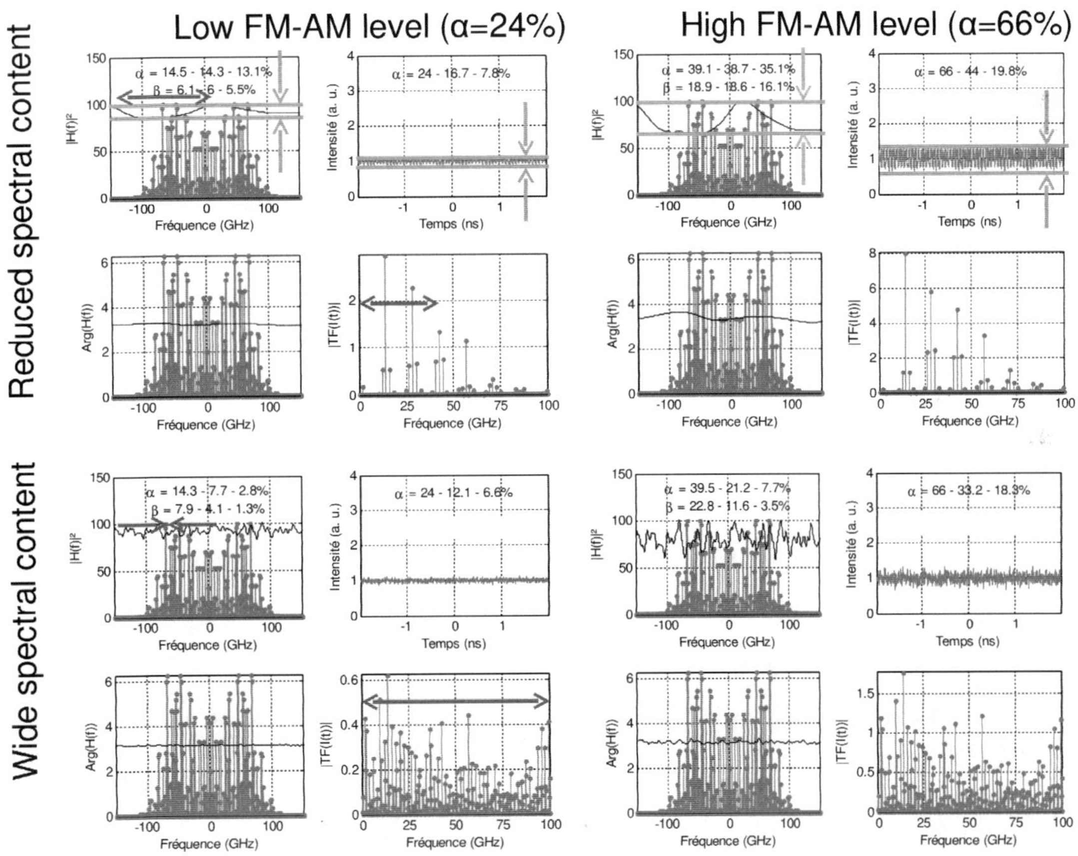


**Figure 3:** FM-to-AM conversion for different transfer functions. Above with a reduced spectral content, below with a wide spectral content. On the left with lower FM-to-AM conversion level. On the right with higher FM-to-AM conversion level.

Spectral distortions may either come from smooth and stable filtering functions (gain narrowing, spectral acceptance for frequency conversion...) or from interferometric filtering due to parasitic effects. Interferometric filtering may have almost any length scale ranging from μm inside coatings to meters between components. These length scales correspond to different times of flight or to different frequency scales as the ones shown in Figure 3. The longer the time of flight, the more rapidly the spectral function changes.

We thus perfomed numerical simulations of α measurements in the time domain with a given oscilloscope bandwidth and we compared the measured values to the actual ones (Figure 4). The transfer functions were randomly drawn with different frequency scale (i.e.: different typical times of flight $\Delta t_c$) but for a given standard deviation σ of the amplitude variation of the transfer function. We can conclude from this figure that for a good correlation between the measured and the actual values, the typical time of flight should be below 1ps.

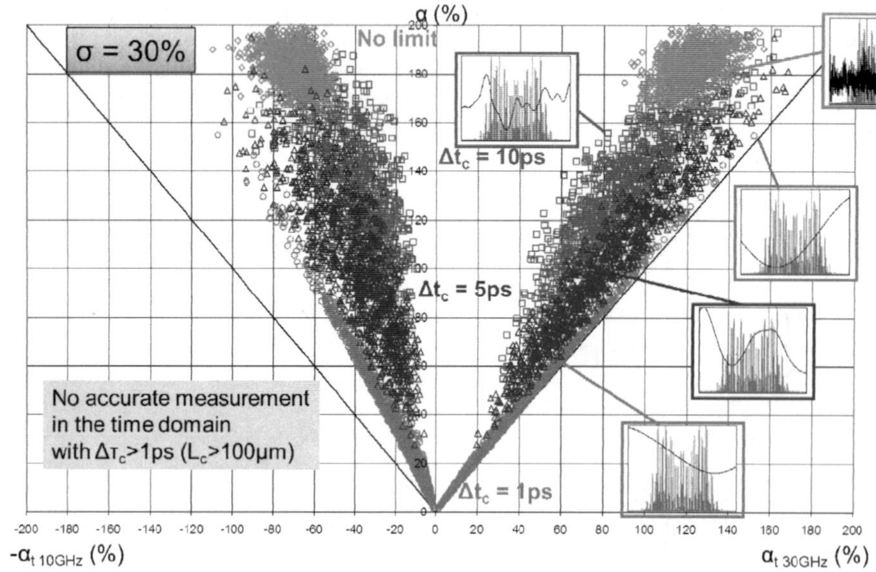


**Figure 4:** Actual FM-to-AM conversion level (vertical axis) versus the measured value when the measurement is performed in the time domain with 30GHz bandwidth (right of the horizontal axis) or 10GHz bandwidth (left of the horizontal axis). An accurate measurement corresponds to a point on the diagonal line and the wider the point cloud the lower the correlation. The different colors correspond to different times of flight. Typical example of transfer functions are given in the insets.

## 4. Conclusion

We have shown that FM-to-AM measurement should be done in two steps:

1) first measuring the spectrum with a high resolution, for instance with a Fabry-Perot interferometer with a reduced phase modulation. This measurement should allow removing you're your set-up parasitic effects with long times of flight by anti-reflection coating, tilting optics...

2) once the times of flight are sufficiently low, typically below 1ps (corresponding to lengths below a few hundreds micrometers), FM-to-AM conversion can be accurately measured in the time domain with an oscillator having at least 10GHz bandwidth, ideally 30GHz. This measurement can be safely used for the compensation of FM-to-AM conversion.

## References

[1] J.E. Rothenberg et al., Proc. SPIE, 1999, 3492, pt. 1-2, pp. 51-61.
[2] S. Vidal et al., Applied Optics, Vol. 51, No. 24, pp. 5818-5825, 08/2012.

# Generation of phase - matched coherent point source in plasma media by propagated X- ray laser seeded beam

**T Pikuz[1,2,*], A Faenov[1,2,*], S Magnitskiy[3], N Nagorskiy[2], M Tanaka[1], M Ishino[1], M Nishikino[1], M Kando[1], Y Kato[4], T Kawachi[1]**

[1] Quantum Science Beam Directorate, Japan Atomic Energy Agency, Umemidai 8-1-7, Kizugawa, Kyoto, 619-0215, Japan
[2]Joint Institute for High Temperatures, Russian Academy of Science, Izhorskaya street 13-2, Moscow, 125412, Russia
[3] Physical department, M.V. Lomonosov Moscow State University, Moscow, Russia
[4]The Graduate School for the Creation of New Photonics Industries, Hamamatsu, Shizuoka, 431-1202, Japan
[*]Institute for Academic Initiatives, Osaka University, 565-0871, Japan

E-mail: pikuz.tatiana@jaea.go.jp

**Abstract**. There is a significant interest in developing the coherent table-top X-ray lasers. Advent of plasma-based transient collisional excitation x-ray laser and particular, injection of coherent seeded beam, especially high-order harmonics, has tremendously improved the spatial coherence of such lasers, what allowed them to be the same widely used as synchrotron sources. Here we report experimental founding of unknown interference structure in a spatial profile of the output beam of the two-stage plasma X-ray laser. That allowed us experimental and theoretical discovering a new phenomenon consisted in a generation of phase-matched coherent point source in a laser plasma media by propagated X-ray laser seeded beam. This phenomenon could extend the applications of such x-ray lasers. For explanation of the observed phenomenon a new method of solving the standard system of Maxwell-Bloch equations has been developed. It was found that the interference pattern in the output laser beam was formed due to an emergence of phase-matched coherent virtual point source in the XRL amplifier and could be treated as the first observation of mirage phenomenon, analogous to the optical mirage, but in X-rays. The obtained results bring new comprehension into the physical nature of amplification of X-ray radiation in laser-induced plasma amplifiers and opening new opportunities for X-ray interferometry, holography and other applications, which requiring multiple rigidly phased sources of coherent radiation.

## 1. Introduction

Development of table-top X-ray lasers (XRL) [1] and their application for investigations of XRL short pulse interaction with matter [2] has an increasing interest during past years. Especially significant progress is achieved in development of multistage plasma-based transient collisional excitation X-ray lasers [3]. Injection of coherent seeded beams [4], in particular high-order harmonics, in plasma-media of X-ray lasers allowed essentially improve the spatial coherence of such lasers and made them more attractive for different applications. Despite of such progress the radiation coherence of modern X-ray lasers is still not ideal and many scientific groups continue to work on improvement both the temporal and spatial coherence of x-ray laser radiation.

In this article we discuss an appearance of stable interference pattern in the output profile of XRL beam, discovered in [5], which allows proposing the generation of phase-matched coherent point source in amplifying plasma media by propagated X-ray laser seeded beam. The properties of such source are explained. An analogous of observed in x-ray spectral range phenomenon to optical mirage is shown.

## II. Experimental set up and results

The experiments have been performed using the soft X-ray laser (SXRL) facility [3] at Kansai Photon Science Institute of Japan Atomic Energy Agency. The output laser pulses were generated with the Ni-like Ag plasma media using the oscillator-amplifier configuration shown schematically in Fig. 1. Thin silver foils for the oscillator and the amplifier were pumped, respectively, with the line-focused travelling wave laser pulses and with the proper timing for the effective amplification. A part of the SXRL beam from the oscillator was amplified in the plasma media of an amplifier. Emission spectrum corresponded to 4d–4p transition of a nickel-like Ag ion at 13.9 nm wavelength. The SXRL pulse after the amplifier had a bandwidth narrower than $10^{-4}$ and pulse duration of ~ 7 ps. The SXRL system worked in 0.1 Hz repetition regime with the typical output energy of the order of 300 nJ.

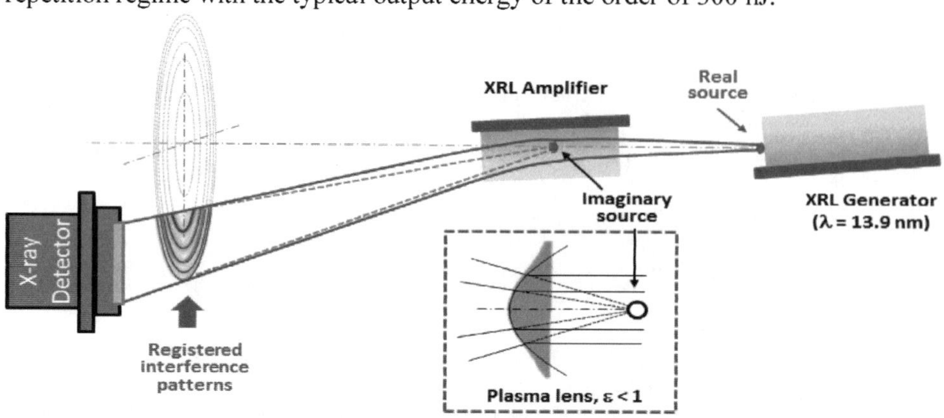


Figure 1. Scheme of the X-ray coherent mirage phenomenon observation caused by generation of imaginary source in amplifier plasma media, which is phase-matched to real source of XRL generator. In inset the focusing properties of plasma as the plasma lens are shown.

In Fig.2 an image of output spatial-intensity distribution of SXRL measured by a high- resolution soft X-ray LiF detector at the distance of ~ 1300 mm is presented. Despite that the image was obtained in 30 consecutive shots a presence of concentric interference fringes in the image is clearly seen and indicates a high stability of the interference pattern formation. That is leading to conclusion that the phenomenon has a fundamental essence. Observation of such interference pattern requires presence of two coherent rigidly phased point sources. The first source is evident. It is formed by the oscillator. The appearance of second source is not obvious. To find location of second source additional experiments have been provided. In these experiments an output intensity distribution of the SXRL beam was measured in a single shot at the distance of 3,314 mm from the amplifier by means of X-ray back illuminated CCD. It was found that in the single shot images the concentric ring structure has high contrast (see Fig. 3) and could be recorded at the limited ranges of relative delays between pumping lasers (( ≤ 10 ps) and relative angles between oscillator and amplifier targets ( ≤ 0.6°).

Interference fringes were observed only in the two-stage laser operation mode, that is, when the generator and amplifier were turned on simultaneously (See Fig.4). The result of comparison of the

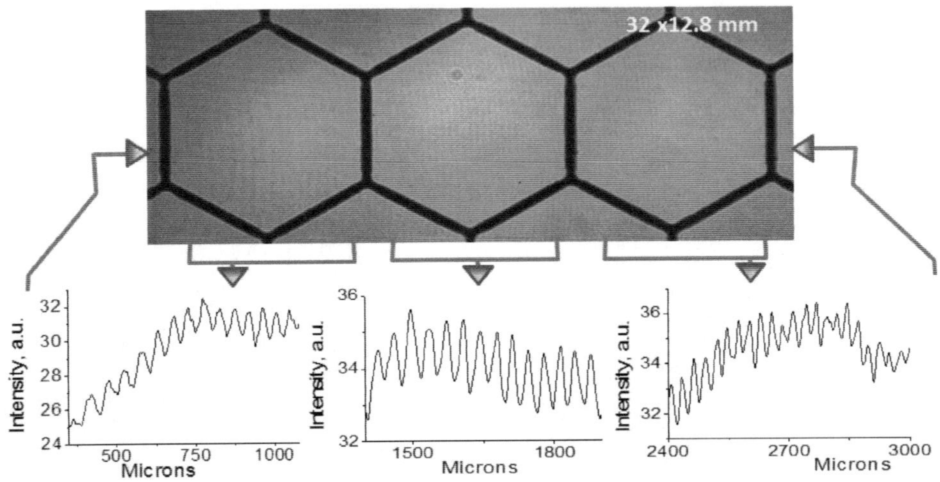


**Figure 2.** Far field intensity distribution of XRL beam measured by the LiF crystal X-ray detector after 30 XRL shots accumulation. Upper image and horizontal traces of intensity distribution (below) show presence of concentric interference fringes practically in the full view of the beam (the hexagonal mesh, placed in the XRL beam before detector, was used as a spatial coordinate's reference).

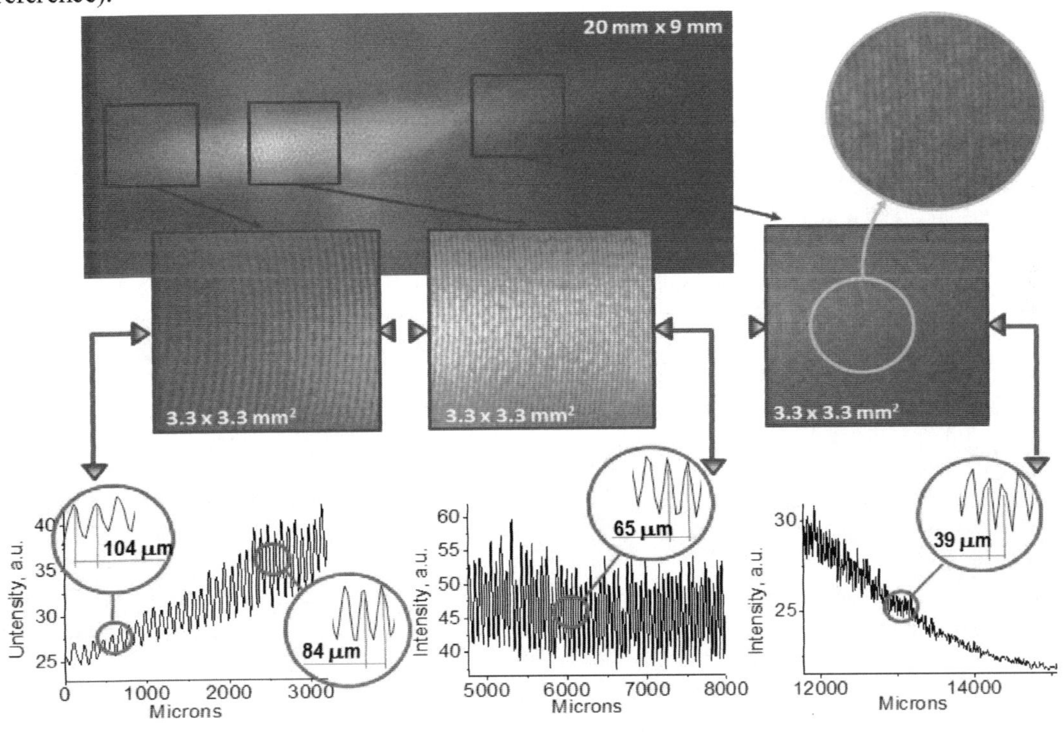


**Figure 3.** Interference pattern in the far field intensity distribution of XRL beam measured by the X-ray CCD (upper). In zoomed parts of image the concentric interference fringes are clearly seen (center). Horizontal traces (below) show gradual decreasing of fringes period from left (closer to position of amplifier target) to right (far away from the target). The brighter area in the full image is the amplified XRL beam.

experimentally obtained interference pattern with the numerically calculated pattern is shown in Fig. 5 and allowed to determine that second point source is located at the distance a = 203 mm between sources in the center of the amplifier plasma.

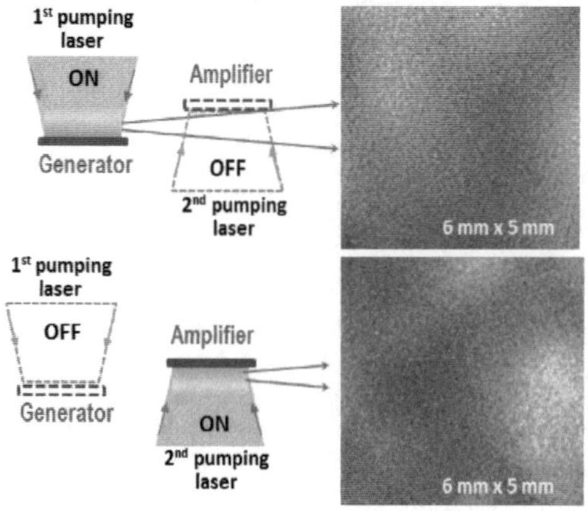


Figure 4. Output intensity distribution of the beam, if only generator is turned ON (upper) or only amplifier is turn ON (below). Interference fringes have not observed.

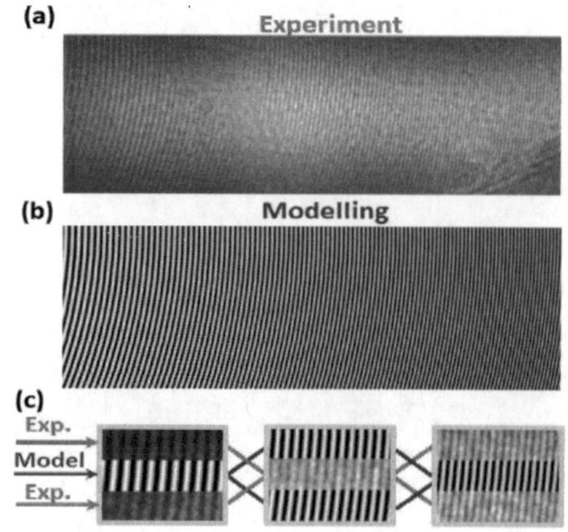


Figure 5. (a) Experimental interference pattern; (b) Modeling of interference pattern for the case a = 203 mm; (c) comparison of experimental and modeled images in correspondent parts.

Provided scrupulous analyses [5] showed that standard Maxwell-Bloch equations, used for description of seeded beams amplification in plasma media, could not directly explain an appearance of two phased almost point sources and observed phenomenon. A new theoretical approach has been developed, which showed that observed effect is similar to formation of visible mirages in optically inhomogeneous atmosphere. The real and imaginary coherent sources could be created in plasmas of generator and amplifier. The difference to visible mirage consists in coherency of X-ray sources and their ability to interfere. The necessary and sufficient conditions for observation of X-ray mirage have been formulated [5]. It was demonstrated that plasma of laser amplifier could modify the x-ray laser wave front in the needed direction and behaves as X-ray optics (see also Fig.1). Moreover the measuring of the interference patterns themselves could be an effective method of diagnostics of X-ray laser plasma media, including the coherence degree of output radiation of such lasers.

**Acknowledgments**
This work is partly supported by JSPS Grant-in-Aid for Scientific Research Kiban (B) No.25289244 and by RFBR-JSPS collaboration program (RFBR Grant 14-02-92107

**References**

[1] Nickles P V et al. 1997 *Phys. Rev. Lett.* **78**, 2748; Dunn J et al. 1998 *Phys. Rev. Lett.* **80**, 2825; Kawachi T et al.2002 *Phys. Rev. A* **66**, 033815; Keenan R et al. 2005 *Phys. Rev. Lett.* **94**, 103901; Wang Y et al. 2004 *Phys. Rev. A* **72**, 053807

[2] Benware B R et al., 1999 *Opt. Lett.* **24**, 1714; Ritucci A et al., 2006 *Opt. Lett.* **31**, 68; Faenov A Ya et al., 2009 *Appl. Phys. Lett.* **94** 231107; Edwards M H et al 2007 *Phys. Rev. Lett.* **99** 195002; Mocek T et al. 2006 *Applied Phys. Lett*. **89** 051501; Norman G et al. 2012 *Journ. Applied Physics* **112** 013104; Ishino M, Faenov A Ya, Tanaka M et al., 2013 *Appl. Phys. A.* **110**, 179

[3] Nishikino M et al. 2009 *Rev. Sci. Instrum.* **80**, 116102

[4] Zeitoun, P. et al. 2004 *Nature* **431**, 426; Al'miev, I. R. et al. 2007 *Phys. Rev. Lett.* **99** 123902; Berrill, M. et al. 2010 *Opt. Lett.* **35**, 2317; Oliva, E. et al. 2012 *Opt. Lett.* **37**, 4341

[5] Magnitskiy S, Nagorskiy N, Faenov A et al., 2013 *Nature Communications* **4** 1936

# Ablation driven by hot electrons in shock ignition

**A R Piriz[1], G Rodriguez Prieto[1], N A Tahir[2] and Y T Zhao[3]**

[1] E.T.S.I. Industriales, Instituto de Investigaciones Energéticas (INEI) and CYTEMA, Universidad de Castilla-La Mancha, 13071 Ciudad Real, Spain
[2] GSI Helmholtzzentrum für Schwerionenforschung, Planckstrasse 1, 64291 Darmstadt, Germany
[3] Institute of Modern Physics, Chinese Academy of Science, 730000 Lanzhou, People's Republic of China

E-mail: roberto.piriz@uclm.es

**Abstract.** An analytical model for the ablation driven by hot electrons is developed. The hot electrons are assumed to carry on the totality of the absorbed laser energy. Efficient energy coupling requires to keep the critical surface sufficiently close to the ablation front. To achieve this goal for high laser intensities a short enough laser wavelength is required. Scaling laws for the ablation pressure and the other relevant magnitudes of the ablation cloud are found in terms of the laser and target parameters.

## 1. Introduction

The shock ignition (SI) approach to inertial confinement fusion (ICF) critically depends on the possibility to generate a very strong shock launched at the stagnation phase of the implosion by a pressure of the order of 1 *Gbar* [1, 2]. In directly driven laser fusion the generation of this shock needs to rise the laser intensity $I$ up to 10 $PW/cm^2$ or more, for a laser wave length $\lambda = 0.35\ \mu m$. At such intensities the laser energy is absorbed in the subcritical region by collisionless processes that produce copious amounts of hot electrons (HE) [3], and it has been argued that they could be beneficial for the generation of the ignitor shock [1, 2]. The main difficulty to address this problem is that a detailed treatment of the HE generation requires to deal with kinetic mechanisms that are difficult to couple with the hydro-codes used for describing the target implosion and the later propagation of the ignitor shock. Therefore, it is necessary to resort to approximations in order to get insight into the role of the HE in SI [2, 4]. In this regard, analytical models can be a valuable complement to the other approaches to the problem.

Here, we present a model based in the evidence available at present from experiments and numerical simulations that show that the effects of the collisionless absorption depend on the irradiance $I\lambda^2$ [5]. Therefore, much of the relevant physics overlaps with early work performed with nanosecond $Nd$ and $CO_2$ lasers of wavelengths $\lambda = 1.06\ \mu m$ and $\lambda = 10.6\ \mu m$, respectively. That work is in agreement with recent PIC simulation results with $\lambda = 0.35\ \mu m$ [3]. Then, consistently with the available evidence we construct our model assuming that the absorbed laser energy is totally transferred to a Maxwellian population of HE with mean temperature $\theta_H$ and particle density $n_H$ that are produced in the neighborhood of the surface with mass density

$\rho_c$ which we will taken as the critical density. Thus, the HE energy flux $q_H$ can be written:

$$q_H = \frac{1}{2}\Phi_H\theta_H \ ; \quad \Phi_H = n_H v_H; \quad v_H = \left(\frac{8\theta_H}{\pi m_e}\right)^{1/2} \quad . \tag{1}$$

As the electrons penetrate into the supercritical plasma the energy flux becomes attenuated by energy loss ($d\theta_H/dr$) and by backscattering ($d\Phi_H/dr$). Since both processes have approximately the same mean free path $\lambda_H$, we have:

$$\frac{dq_H}{dr} = \frac{2q_H}{\lambda_H} \ ; \quad \frac{q_H}{\theta_H^2} \approx const \ ; \quad \lambda_H \approx \frac{Am_p}{Z}\frac{\theta_H^2}{3\pi\rho e^4 ln\Lambda_H} \equiv \frac{\alpha\theta_H^2}{\rho} \quad , \tag{2}$$

where $\rho$ is the mass density, $r$ represents the radial coordinate in spherical geometry and $ln\Lambda_H \approx 10$ is the Coulombian logarithm of the HE.

Then, for studying the ablation driven by HE we will introduce the previous energy deposition equation into the hydrodynamic equation for the energy conservation.

## 2. Ablation model and scaling laws

In spherical geometry we can consider quasi-stationary conditions provided that the transit time of a fluid element for traveling from the ablation surface up to the critical surface is shorter than the characteristic time of variation of the ignitor pulse. Then, the conservation equations for mass, momentum and energy are:

$$\dot{m} = 4\pi\rho v r^2 = const \ ; \quad \rho v\frac{dv}{dr} + \frac{dp}{dr} = 0 \ ; \quad \frac{\dot{m}}{4\pi r^2}\frac{d}{dr}\left(\gamma\epsilon + \frac{v^2}{2}\right) = \frac{dq_H}{dr} \quad , \tag{3}$$

where $v$ is the fluid velocity, $\epsilon = (Z+1)kT/Am_p$ is the specific internal energy ($k$ is the Boltzmann constant, $Z$ and $A$ are the charge and mass numbers, respectively, $m_p$ is the proton mass, $T$ is the electrons and ions temperature), $p = (\gamma-1)\rho\epsilon$ is the pressure, and $\gamma = 5/3$ is the enthalpy coefficient of the ablated plasma.

Eqs.(2) and (3) can be combined to get the following system [6]:

$$\frac{dv}{dr} = \frac{2\gamma(\gamma-1)\epsilon v}{\gamma(\gamma-1)\epsilon - v^2}\left(\frac{q_H}{\alpha\theta_H^2}\frac{1}{\gamma\epsilon v} - \frac{1}{r}\right) \ ; \quad \frac{d\epsilon}{dr} = \frac{2q_H}{\alpha\theta_H^2}\frac{1}{\gamma v} - \frac{v}{\gamma}\frac{dv}{dr} \ ; \quad \frac{dq_H}{dr} = \frac{2q_{Hc}}{\alpha\theta_{Hc}^2}\rho \quad . \tag{4}$$

Eqs.(4) must be solved with the boundary conditions on the ablation surface at $r = r_0$: $q_H(r_0) = \theta_H(r_0) = v(r_0) = \epsilon(r_0) = 0$, and on the absorption surface at $r = r_c$: $\rho(r_c) = \rho_c$, $\theta_H(r_c) = \theta_{Hc}$, and $q_H(r_c) = q_{Hc} = \beta I$, where $\beta$ is the fraction of absorbed laser energy. Since we consider a free expansion, we have to look for the unique transonic solution of the previous equations. Such a solution can be easily obtained by numerical calculation, but for the purpose of finding relationships with the character of scaling laws for the ablation driven by HE it is more convenient to obtain an approximated analytical solution. To that end we first find the asymptotic solution of Eqs.(4) (for $r >> r_0$) and then, we extend it up to the target surface:

$$v \approx \left[\frac{30(\gamma-1)}{7\gamma-5}\frac{q_H r_0}{\alpha\theta_H^2}\right]^{1/3}(\xi-1)^{1/3} \ ; \quad \epsilon \approx \frac{1}{M^2(\gamma-1)}\left[\frac{30(\gamma-1)}{7\gamma-5}\frac{q_H r_0}{\alpha\theta_H^2}\right]^{2/3}(\xi-1)^{2/3} \quad . \tag{5}$$

$$\rho \approx \frac{\dot{m}}{4\pi r_0^2}\left[\frac{7\gamma-5}{30(\gamma-1)}\frac{\alpha\theta_H^2}{q_H}\right]^{1/3}\frac{1}{\xi^2(\xi-1)^{1/3}} \ ; \quad p \approx \frac{\dot{m}}{4\pi M^2 r_0^2}\left[\frac{30(\gamma-1)}{7\gamma-5}\frac{q_H}{\alpha\theta_H^2}\right]^{1/3}\frac{(\xi-1)^{1/3}}{\xi^2}, \tag{6}$$

$$q_H \approx \frac{\dot{m}}{4\pi r_0^2}\left[\frac{2(7\gamma-5)}{30(\gamma-1)}\right]^{2/3}\left(\frac{q_H}{\alpha\theta_H^2}\right)^{1/3} F(\xi) \quad , \tag{7}$$

$$F(\xi) = \frac{(\xi-1)^{2/3}}{\xi} + \frac{1}{\sqrt{3}} tg^{-1}\left[\frac{2(\xi-1)^{1/3}-1}{\sqrt{3}}\right] + \frac{1}{6} log\{\frac{1-(\xi-1)^{1/3}+(\xi-1)^{2/3}}{[(\xi-1)^{1/3}+1]^2}\} + \frac{\pi}{6\sqrt{3}}, \tag{8}$$

where $\xi = r/r_0$. These equations are not applicable for $\xi < 1.06$ but we will use them for radii larger than the isothermal sonic radius $r_s = 1.2 r_0$ ($\xi \geq 1.2$).

Then, from Eqs.(5) to (8) we can find the scaling laws for the magnitudes at $\xi_s = 1.2$ in terms of the laser and target parameters, and of the HE temperature $\theta_{Hc}$ at the critical surface:

$$\rho_s \approx \frac{0.594\, \alpha\theta_{Hc}^2}{F(\xi_c)\ r_0} \ ; \ v_s \approx 0.843\left(\frac{r_0}{\lambda_{Hc}}\right)^{1/3}\left(\frac{q_{Hc}}{\rho_c}\right)^{1/3} \ ; \ \epsilon_s \approx 1.07\left(\frac{r_0}{\lambda_{Hc}}\right)^{2/3}\left(\frac{q_{Hc}}{\rho_c}\right)^{2/3} \quad , \tag{9}$$

$$p_s \approx \frac{0.424}{F(\xi_c)}\left(\frac{\lambda_{Hc}}{r_0}\right)^{1/3} \rho_c^{1/3} q_{Hc}^{2/3} \ ; \ \frac{\lambda_{Hc}}{r_0} \approx 2F(\xi_c)\xi_c^2(\xi_c-1)^{1/3} \quad , \tag{10}$$

where $\lambda_{Hc} = \alpha\theta_{Hc}^2/\rho_c$ is the HE mean free path at the critical surface, $\rho_c = 1.878 \times 10^{-3}(A/Z)/\lambda^2\ g/cm^3$ (with $\lambda$ in $\mu m$), $\xi_c \geq \xi_s = 1.2$, $q_{Hc} = \beta I$. In spherical geometry the laser intensity $I$ at the critical surface depends on the critical radius $r_c$ so that $I = I_0/\xi_c^2$, where $I_0 = W_0/4\pi r_0^2$ is the nominal intensity (on the target surface) and $W_0$ is the laser power.

### *2.1. High intensity regime ($r_c > r_s$)*

The second of the Eqs.(10) gives, in an implicit form, the critical distance $\xi_c$ in terms of the ratio $\lambda_{Hc}/r_0$ and it shows that for $\xi_c \geq \xi_s = 1.2$ it is $\lambda_{Hc}/r_0 \geq 0.75$. For high laser irradiances $I\lambda^2 \geq 10^{15}\ W\mu m^2/cm^2$, we can relate the HE temperature at the critical surface with the laser intensity by using the well known empirical law [7]: $\theta_{Hc} = 7.5 \times 10^{-2}(I\lambda^2)^{1/3}\ eV$, with $I$ in $W/cm^2$, and $\lambda$ in $\mu m$. Then, we can get the HE mean free path at the critical surface and, in particular, the scaling law for the ablation pressure:

$$\frac{\lambda_{Hc}}{r_0} \approx 2.56 \times 10^{-8}\frac{I^{2/3}\lambda^{10/3}}{r_0} \ ; \ p_s \approx 1.28 \times 10^{-7}\left(\frac{\beta^2 Z}{A}\right)^{1/3}\frac{I_0^{8/15} r_0^{1/5}}{\lambda^{4/3}}\ Mbar \quad , \tag{11}$$

where $I$ is in $W/cm^2$, and $\lambda$ and $r_0$ are in $\mu m$.

As we can see from Eq.(11) the scaling of the ablation pressure with the laser intensity $I_0$ is weaker than the well known scaling $p_s \sim I_0^{2/3}$ found at low intensities. From Eq.(10) one can see that this behavior arises from the combination of two opposite effects. From one side the long mean free path of the HE increases the ablation pressure but, on the other side, the effective intensity at the critical surface is reduced as the critical radius $r_c$ increases with the HE mean free path. The second effect dominates in this high intensity regime and the weaker scaling $p_s \sim I_0^{8/15}$ results.

### *2.2. Low intensity regime ($r_c = r_s$)*

For the lower intensities the condition $\lambda_{Hc}/r_0 \geq 0.75$ breaks and Eqs.(9) to (11) are not valid any longer. This limit is achieved when the laser intensity becomes $I \leq 1.6 \times 10^{11}\ r_0^{3/2}/\lambda^5$ ($r_0$ and $\lambda$ in $\mu m$, and $I$ in $W/cm^2$). Since beyond the absorption region density and temperature must be decreasing functions of $r$, the critical surface cannot enters inside the surface with the isothermal sonic radius and it must always be $r_c \geq r_s$ [8]. Therefore, in the lower intensity regime the critical surface must remain attached to the sonic surface so that below the above mentioned limit it is $r_c = r_s \leq 1.2r_0$. Then, $\rho_s = \rho_c$ and from the first of Eqs.(6) we get the critical distance in this lower intensity regime:

$$\xi_c - 1 \approx 0.27\frac{\alpha\theta_{Hc}^2}{\rho_c r_0} \quad or \quad r_c - r_0 \approx 0.27\lambda_{Hc} \quad . \tag{12}$$

In a similar manner, from Eqs.(5) and (6) we re-obtain the scaling laws as follows:

$$\epsilon_s = \frac{v_s^2}{\gamma - 1} \approx 1.3\left(\frac{q_{Hc}}{\rho_c}\right)^{2/3} \quad ; \quad p_s \approx 0.87\ \rho_c^{1/3} q_{Hc}^{2/3} \quad . \tag{13}$$

These are the very well known scalings obtained in planar geometry ablation driven by thermal conduction. In fact, when $r_c = r_s$ they become independent of the particular mechanism driving the ablation, and such a mechanism determines only the position of the critical surface with respect to the ablation front. Therefore, we have:

$$p_s \approx \frac{5 \times 10^{-9}}{\xi_c^{4/3}}\left(\frac{A\beta^2}{Z}\right)^{1/3}\left(\frac{I_0}{\lambda}\right)^{2/3} Mbar \ ; \quad (\xi_c - 1)\xi_c^{4/3} \approx 6.9 \times 10^{-9}\frac{I_0^{2/3}\lambda^{10/3}}{r_0} \quad , \tag{14}$$

($I_0$ is in $W/cm^2$ and $\lambda$ is in $\mu m$) where for obtaining the equation for $\xi_c$ we have used again the empirical law $\theta_{Hc} = 7.5 \times 10^{-2}(I\lambda^2)^{1/3}\ eV$ (for $I\lambda^2 \geq 10^{15}\ W\mu m^2/cm^2$). As an example, we calculate the ablation pressure $p_s$ for the maximum intensity for which Eq.(14) can be applied ($\xi_c = \xi_s = 1.2$, $\lambda_{Hc} = 0.75r_0$), for $\lambda = 0.351\ \mu m$ ($I_0 \approx 4.3 \times 10^{16}\ W/cm^2$), $A/Z = 2.5$, and assuming $\beta = 0.5$. In such a case, we get $p_s \approx 830\ Mbar$. Therefore, it seems that a suitable ignitor shock could be generated by the ablation driven by HE within the regime of lower laser intensities for which we have the best laser/target coupling.

Finally, it is possible to verify the validity of the main model assumptions: a) We can see that the electric field required to maintain the return current is negligible. Thus, it cannot inhibit the inward hot electrons flux. b) The radiation pressure is also negligible in comparison with thermal pressure. Nevertheless, density jumps in the absorption region may still be present. It would place the laser absorption region near the critical surface. c) Quasi-steady state is reached in times ($\sim 10\ ps$) much shorter than the foreseen spike duration ($\geq 100\ ps$). d) The plasma preformed by the compression laser pulse can be neglected except when its characteristic scale-length $l_{Tc}$ is larger than the supra-thermal critical distance $r_c - r_0 \sim \lambda_{Hc}$. The matching condition $l_{Tc} \leq \lambda_{Hc}$ imposes a relationship between the compression pulse intensity $I_{cp}$ and the spike intensity $I$: $I \geq 2.2 \times 10^{-11}(\beta_{cp} I_{cp}\lambda)^2$, in the usual units. In the opposite case, $I$ at the critical surface will be too small, $\theta_{Hc}$ too low, and the supra-thermal ablation cloud may not be assembled.

## AKNOWLEGDEMENTS

This work has been partially supported by the Ministerio de Economía y Competitividad of Spain (ENE2009-09276), by the BMBF of Germany and by the Chinese Academy of Sciences.

### References

[1] R. Betti, C. D. Zhou, K. S. Anderson, L. J. Perkins, W. Theobald, and A. A. Solodov, Phys. Rev. Lett. **98**, 155001 (2007).
[2] L. J. Perkins, R. Betti, K. N. LaFortune, and W. K. Williams, Phys. Rev. Lett. **103**, 045004 (2009).
[3] O. Klimo, S. Weber, V. T. Tikhonchuk, and J. Limpouch, Plasma Phys. Control. Fusion **52**, 055013 (2010).
[4] W. Manheimer, D. Colombant, and A. J. Schmitt, Phys. Plasmas **19**, 056317 (2012).
[5] P. Gibbon and E. Förster, Plasma Phys. Control. Fusion **38**, 769 (1996).
[6] A. R. Piriz, G. Rodriguez Prieto, N. A. Tahir, Y. Zhang, S. D. Liu, and Y. T. Zhao, Phys. Plasmas **19**, 122705 (2012).
[7] S. J. Gitomer and D. B. Henderson, Phys. Fluids **22**, 364 (1979).
[8] C. E. Max, C. F. McKee, and W. C. Mead, Phys. Fluids **23**, 1620 (1980).

# Simulations of proton beam characteristics for ELIMED Beamline

**Jan Psikal[1,2], Jiri Limpouch[1,2], Ondrej Klimo[1,2], Jiri Vyskocil[1,2], Daniele Margarone[2], Georg Korn[2]**

[1] FNSPE, Czech Technical University in Prague, Czech Republic
[2] Institute of Physics, ELI-Beamlines project, ASCR, Prague, Czech Republic

E-mail: jan.psikal@fjfi.cvut.cz

**Abstract.** ELIMED Beamline should demonstrate the capability of laser-based particle accelerators for medical applications, mainly for proton radiotherapy of tumours which requires a sufficient number of accelerated protons with energy about 60 MeV at least. In this contribution, we study the acceleration of protons by laser pulse with parameters accessible for ELIMED Beamline (intensity $\sim 10^{22}$ W/cm$^2$, pulse length $\sim 30$ fs). In our two-dimensional particle-in-cell simulations, we observed higher energies of protons for linear than for circular polarization. Oblique incidence of the laser pulse on target does not seem to be favourable for proton acceleration at such high intensities as the accelerated protons are deflected from target normal axis and their energy and numbers are slightly decreased. The expected numbers of accelerated protons in the energy interval 60 MeV $\pm$ 5% are calculated between $10^9$ and $10^{10}$ per laser shot with estimated proton beam divergence about 20° (FWHM).

## 1. Introduction

With rapidly increasing laser pulse intensities in last decades, ion acceleration by femtosecond laser pulses is of great interest in the present research. Several mechanisms of ion acceleration by short laser pulses have been proposed [1]. The most studied one is target normal sheath acceleration (TNSA) based on laser energy transformation to energetic electrons that penetrate to vacuum and produce there a strong quasi-static electric field that accelerates ions from the target surface [2]. At higher laser intensities (above $10^{20}$ W/cm$^2$), radiation pressure acceleration (RPA) was proposed as an efficient mechanism for generation of quasimonoenergetic ion beams [?]. However, RPA requires to avoid electron heating which can be reduced by using circularly polarized laser beams, but it cannot be removed completely due to two-dimensional effects.

The ELIMED (MEDical applications at ELI) project aims to demonstrate the potential clinical applicability of optically accelerated proton beams [4]. Numerical study of laser-proton acceleration for expected parameters of generated laser pulse is one of important tasks in the preparatory phase of the ELIMED project. In this paper, we summarize the results obtained by means of two-dimensional particle-in-cell simulations. Our goal is to estimate maximum energies of accelerated protons, the number of high-energy protons, and their divergence. We have done computational study of the dependence of proton acceleration on the foil material and thickness, laser polarization, and the incidence angle.

## 2. Simulation method and parameters

Our relativistic colisionless particle-in-cell (PIC) code in two spatial directions and with three velocity components is described in Ref. [5]. In the simulations, we assumed two intensities of laser pulse. The pulse has $\sin^2$ temporal profile and full length about 40 fs (15 laser periods) focused to 3 $\mu$m (FWHM) diameter. In the first case, the pulse reaches maximum intensity equal to $1.4 \times 10^{22}$ W/cm$^2$ (corresponding to 1 PW case in the following text). In the second case, the using of double-plasma mirror is assumed in order to improve the laser pulse contrast, which reduces the intensity of the focused laser beam to $7.2 \times 10^{21}$ W/cm$^2$ (0.5 PW pulse). In most of simulations, the beam is incident normally on the polyethylene ($CH_2$) foil of thickness 200 nm or 1 $\mu$m. The foil is assumed to be ionized to $C^{6+}H_2^+$ plasma with electron density $n_e = 3.5 \times 10^{23}$ cm$^{-3}$ = 200 n$_{ec}$, where $n_{ec}$ is the electron critical density for laser of wavelength 800 nm.

## 3. Results and discussion

### 3.1. Dependence on laser polarization and target composition

Firstly, we calculated the interaction of 1 PW laser pulse, linearly and circularly polarized, with 200 nm thick foil composed of fully hydrogen plasma or $CH_2$ plasma. The proton energy distribution is plotted in Fig. 1 (at time instant 120 fs after the target interaction with maximum laser pulse amplitude). Linear $p$-polarization leads to higher energies and numbers of the most energetic protons compared with circular polarization, as the foil is too thick and focusing too tight for more efficient RPA in the light sail regime [3]. Therefore, we investigated further only the interaction with linearly polarized laser beams.

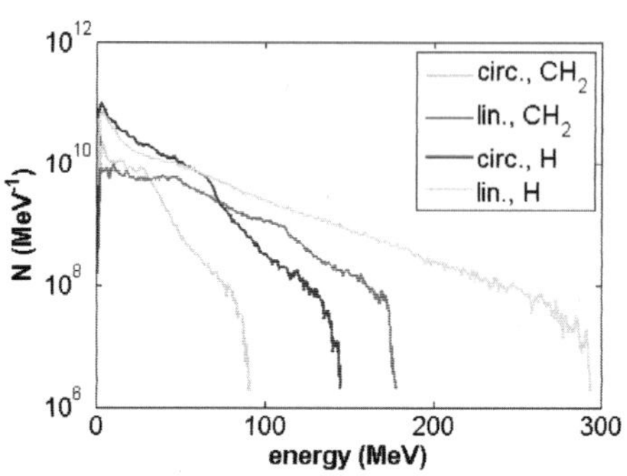


**Figure 1.** Energy spectra of accelerated protons for linear and circular polarization and polyethylene ($CH_2$) and hydrogen (H) foils of thickness 200 nm at laser intensity $1.4 \times 10^{22}$ W/cm$^2$.

It is demonstrated that using pure hydrogen target in the simulations leads to an overestimation of the proton energies in comparison with plastic foils that can be used in experiments. We ascribe this difference to a slower deformation of the target front surface by the radiation pressure when the plasma contains heavier $C^{6+}$ ions. Due to a slower surface deformation, the electrons in the target are heated in a smaller volume. Indeed, the total absorption of the laser pulse energy is about 36% for linear polarization and pure hydrogen target, whereas it is only 25% for polyethylene target. For circularly polarized laser beam, the calculated absorption drops to 21% (hydrogen) and 14% (polyethylene). Due to reduced electron heating and increased radiation pressure on target via ponderomotive force, circularly polarized laser beams can accelerate relatively higher number of protons to lower energies compared with linearly polarized beams for the same absorption rate.

*3.2. Dependence on incidence angle and target thickness*
In difference from lower intensities (below $10^{20}$ W/cm$^2$), oblique incidence (30°) is at intensity $7.2\times10^{21}$ W/cm$^2$ less efficient than the normal incidence for ion acceleration (Fig. 2a) as at these intensities $\vec{j}\times\vec{B}$ heating dominates and the effective laser intensity is decreased for the oblique incidence. The other drawback of the oblique incidence is here preferential oblique direction of the accelerated ions (Fig. 2b). This observation is in agreement with the theoretical and PIC simulation results [6] that predicted the average deflection angle of approximately 5° for protons with kinetic energy 60 MeV. From Fig. 2b, one can also estimate the divergence of high-energy protons, which is about 20° at FWHM. Similar values of the divergence were observed for all simulation runs.

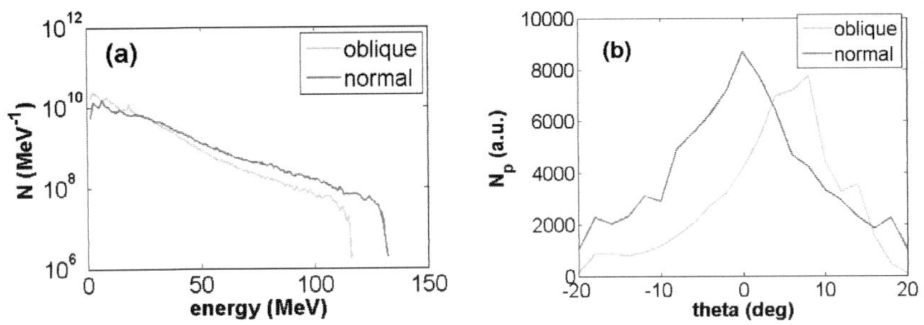


**Figure 2.** Proton spectra (a) and angular distribution of protons with energy >30 MeV (b) for 200 nm thick $CH_2$ foil and laser intensity $7.2\times10^{21}$ W/cm$^2$ for the normal incidence and for the oblique incidence at angle 30°.

Maximum energies and numbers of protons accelerated from thinner (200 nm) and thicker (1 $\mu$m) $CH_2$ foils for normal and oblique incidence are shown in Table 1. Note that the absorption of laser pulse energy is similar for both foil thicknesses and the same incidence angle. It depends strongly on the front target surface (as shown further below) and partly on the incidence angle. The number of protons $N_p$ with energy 57 – 63 MeV, suitable for proton therapy, is estimated from energy transformation efficiency $\eta_{Lp}$ of laser pulse energy to the protons with energy $\varepsilon_p\pm5\%$

$$N_p = \eta_{Lp}\cdot\varepsilon_L/\varepsilon_p, \qquad (1)$$

where $\varepsilon_L$ is laser pulse energy and $\varepsilon_p = 60$ MeV. The laser pulse energy is assumed to be 15 J for 0.5 PW pulse and 30 J for 1 PW pulse.

**Table 1.** Comparison of maximum energies and numbers of protons accelerated from thinner (200 nm) and thicker (1 $\mu$m) planar foils irradiated at normal (0°) and at oblique (30°) incidence by 0.5 PW laser pulse.

| target thickness, angle of incidence | max. energy [MeV] | proton number (57; 63 MeV) |
|---|---|---|
| 200 nm, 0° | 130 | $4.8\times10^9$ |
| 1 $\mu$m, 0° | 85 | $1.4\times10^9$ |
| 200 nm, 30° | 115 | $3.1\times10^9$ |
| 1 $\mu$m, 30° | 75 | $8.0\times10^8$ |

### *3.3. Enhanced proton acceleration by a microstructure on the front side*

For the enhancement of proton energies and numbers due to increased laser pulse energy absorption, microstructures on the front side of thin plastic foils have been already successfully used in the experiments at lower laser intensities about $10^{20}$ W/cm$^2$ [7]. Therefore, we also tested the interaction with these targets at at higher intensities (near to $10^{22}$ W/cm$^2$ for 0.5 PW pulse) in the simulations. The results, described in detail in Ref. [8], are qualitatively similar to simulations and experiments at lower intensity - maximum proton energy is increased about several tens of per cent by using structured front foil surface, whereas the number of accelerated protons can be much higher (almost about one order of magnitude higher) compared with flat foils.

## 4. Conclusions & Perspectives

We have performed numerical simulations of laser-induced ion acceleration for laser intensities of order $10^{22}$ W/cm$^2$ that will be achievable at ELI-Beamlines facility. It is shown that simulations with pure hydrogen targets overestimate the proton energies in comparison with more realistic plasma containing heavier ions. In difference from lower laser intensities, oblique incidence is less efficient than the normal one for the proton acceleration. Linear polarization is preferred to the circular one for the foils of thicknesses $\geq 200$ nm. Maximum energy of accelerated protons is estimated to be about 100 MeV, the number of high-energy (60 MeV $\pm$ 5%) protons suitable for proton therapy should exceed $10^9$ per laser shot, and the calculated proton beam divergence is about 20° (FWHM). The number of high-energy protons can be substantially (several times) increased by introducing a microstructure (deposited nanospheres or grating) on the front surface of a thin foil. Although simulation parameters are chosen to be rather conservative (full pulse duration 40 fs, mostly 0.5 PW laser power instead of 1 PW), 3D simulations are planned for the future to check the validity of our numerical estimates from 2D PIC code.

## Acknowledgement

The authors would like to express their thanks to Prof. S. Kawata from Utsunomiya University (Japan) for his innovative suggestions and to Dr. M. Bussmann and other IT specialists from HZDR Dresden (Germany) for valuable discussions about high performance parallel computing. Partial support by the Czech Science Foundation (projects P205/12/P366 and P205/11/1165), by the Ministry of Education, Youth and Sports of the Czech Republic under projects ELI-Beamlines (CZ.1.05/1.1.00/02.0061) and OPVK (CZ.1.07/2.3.00/20.0087) is gratefully acknowledged.

## References

[1] Macchi A *et al* 2013 *Rev. Mod. Phys.* **85** 751
[2] Wilks S C *et al* 2001 *Phys. Plasmas* **8** 542
[3] Qiao B *et al* 2012 *Phys. Rev. Lett.* **108** 115002
[4] Cirrone G A P *et al* 2013 *Proc. of SPIE* **8779** 87791I
[5] Psikal J *et al* 2006 *Czech. J. Phys.* **56** B515
[6] Morita T *et al* 2008 *Phys. Rev. Lett.* **100** 145001
[7] Margarone D *et al* 2012 *Phys. Rev. Lett.* **109** 234801
[8] Limpouch J *et al* 2013 *Proc. of SPIE* **8780** 878027

# Self-consistent particle-in-cell modelling of short pulse absorption and transport for high energy density physics experiments

**M G Ramsay[1,2], T D Arber[2], N J Sircombe[1,2]**

[1] AWE Aldermaston, Reading RG7 4PR, UK
[2] CFSA, University of Warwick, Coventry CV4 7AL, UK

E-mail: martin.ramsay@awe.co.uk

**Abstract.** In order for detailed, solid density particle-in-cell (PIC) simulations to run within a reasonable time frame, novel approaches to modelling high density material must be employed. For the purposes of modelling high intensity, short pulse laser-plasma interactions, however, these approaches must be consistent with retaining a full PIC model in the low-density laser interaction region. By replacing the standard Maxwell field solver with an electric field update based on a simplified Ohm's law in regions of high electron density, it is possible to access densities at and above solid without being subject to the standard grid and time step constraints. Such a model has recently been implemented in the PIC code EPOCH. We present the initial results of a detailed two-dimensional simulation performed to compare the adapted version of the code with recent experimental results from the Orion laser facility.

## 1. Introduction

To support experimental campaigns on short pulse laser facilities, or develop point designs for fast ignition, laser interaction simulations with peak densities at, or approaching, solid density are required. Simulating high density plasmas places severe constraints on the use of standard particle-in-cell (PIC) techniques to solve Maxwell's equations and the particle equations of motion, limiting the spatial and temporal scales which can be easily modelled. Due to these constraints, detailed modelling of the hot electron transport resulting from a high intensity, short pulse laser interaction is often only possible by sampling the hot electron population and passing this information to a separate, dedicated code [1]. In order to extend the range of densities which can be feasibly modelled, and thus reinforce the link between PIC and hot electron transport codes, an algorithm, based on that developed by Cohen, Kemp & Divol [2], has been implemented in the PIC code EPOCH [3]. This model utilizes a 'hybrid' description for the plasma which replaces the standard Maxwell field solver in regions of high electron density ($\gtrsim 100n_c$) with a field update based on a simplified Ohm's law:

$$\boldsymbol{E} = \eta\left(\boldsymbol{J}_b + \boldsymbol{J}_i\right) - \frac{\boldsymbol{J}_b \times \boldsymbol{B} + \nabla P_b}{en_b},$$

where $\boldsymbol{J}_b = \frac{\nabla\times\boldsymbol{B}}{\mu_0} - \boldsymbol{J}_h - \boldsymbol{J}_i$ is the background electron current density, $\boldsymbol{J}_h$ and $\boldsymbol{J}_i$ are the hot electron and ion current densities, and an isotropic resistivity, $\eta$, and background electron

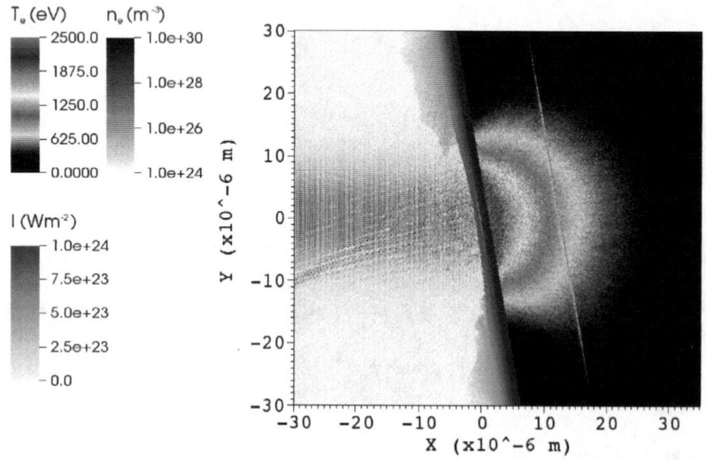


**Figure 1.** Background electron temperature ($T_e$) and density ($n_e$), and the Poynting flux ($I$) of the laser, at t = 1ps (shortly after the peak laser interaction).

pressure, $P_b$, have been assumed. This approach relaxes the grid and time step constraints, permitting simulations to be run at a lower resolution while also dramatically reducing numerical self-heating.

In order to distinguish hot electrons from the background, the electron population is partitioned into two electron species. Background electrons are promoted to the hot electron population if their kinetic energy satisfies the condition: $|\boldsymbol{v}_h| > \alpha\sqrt{\frac{3k_BT_b}{m_e}}$, where $T_b$ is the local background electron temperature and $\alpha$ is a free parameter (typically $\sim 5$).

## 2. Buried layer heating

A detailed 2D simulation was performed using hybrid-EPOCH with the intention of replicating the heating of buried layers observed in recent Orion experiments [4,5] as a means of validating the code against experimental results. A plastic target, containing a thin sample layer of aluminium, was modelled with parameters selected so as to replicate the experimental conditions as closely as possible.

The simulated target consisted of fully-ionised 'plastic' ions, with mass number 6.5 and atomic number 3.5 to represent both hydrogen and carbon ions, at a density of 1.0 g/cc. A 150 nm sample of $Al^{12+}$ at 2.7 g/cc was placed 12 $\mu$m into the plastic. The target was preceded by an exponential density profile of the form $se^{x/L_1} + (1-s)\,e^{x/L_2}$, where $s$ is the ratio of the critical density to the peak electron density in the front plastic layer, $L_1 = 3\,\mu$m and $L_2 = 0.5\,\mu$m. Furthermore, the target was rotated so as to be angled 10° relative to the incident laser pulse.

The laser pulse was modelled as having a Gaussian radial profile with 90% of the energy contained within a 20 $\mu$m diameter spot, a supergaussian temporal profile with full-width-half-maximum (FWHM) of 0.5 ps, total energy of approximately 100 J, and a wavelength of 0.528 $\mu$m.

Approximately $2 \times 10^8$ PIC particles (75% electrons, 5% aluminium ions, 20% 'plastic' ions) were modelled on a mesh of 2600 × 2400 cells covering 65 × 60 $\mu m^2$. The interface between the Maxwell (PIC) and Ohmic (hybrid) field solvers was set between electron densities of $8 \times 10^{28}$ and $9 \times 10^{28}\,m^{-3}$, with the electric field update at intermediate densities being interpolated between the two solutions.

Radial temperature and density gradients were observed within the aluminium layer (see figure 1). In order to obtain a single value for comparison with the equivalent experimental

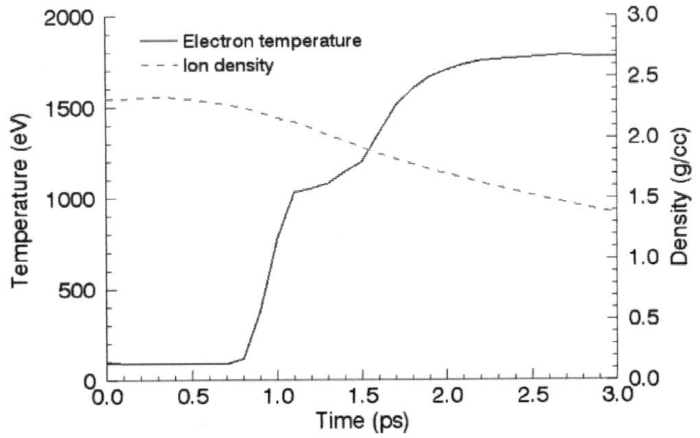


**Figure 2.** Time history of the spatially-averaged electron temperature (blue, solid) and ion mass density (red, dashed) in the aluminium layer.

results, the electron temperature late was allowed to reach a steady state and spatially-averaged. This yielded a predicted electron temperature of 1779 eV, well in excess of the experimentally observed value of 677 eV (see figure 2).

## 3. Line emission comparison

The results of the previous section indicate that the maximum spatially-averaged temperature of the aluminium layer predicted by hybrid-EPOCH compared poorly with the experimental values. However, these values are usually inferred from the K-$\alpha$ radiation emitted by the target. Thus comparing the average of a distribution of temperatures with a single inferred temperature may not be a valid approach. For this reason the atomic kinetics code FLY [6] was used to calculate the expected emission spectrum, in the range 1.8–2.15 keV, for the hybrid-EPOCH target. This energy range encompasses the aluminium He-$\beta$ (1.86 keV) and Ly-$\beta$ (2.05 keV) emission lines, which are generally used as the main temperature and density diagnostics.

The hybrid-EPOCH cells containing aluminium ions were assigned to 100 eV temperature and 0.1 g/cc density bins, covering the ranges 100–6000 eV and 0.5–4.0 g/cc. FLY calculations for each temperature and density bin were performed, and the resultant emission spectra combined to produce a total expected emission spectrum at late time ($t = 3$ ps, $\sim 2$ ps after the peak laser interaction).

Initial comparisons between the emission spectra predicted using temperature and density conditions from the hybrid-EPOCH simulation and those using a single temperature and density indicated that the hybrid-EPOCH-dependent emission could be well characterised by a spectrum corresponding to aluminium at 600 eV and 2.0 g/cc (see figure 3). These values are broadly consistent with the results of similar experiments [4, 7].

## 4. Conclusion

Two dimensional simulations using hybrid-EPOCH (an adapted version of the particle-in-cell (PIC) code EPOCH [3], employing an algorithm for modelling high density plasmas based on that of Cohen *et al.* [2]) have been performed to attempt to validate the code against experimental results [4, 5, 7]. The average temperature of an aluminium layer, embedded 12 $\mu$m into a plastic target, predicted by hybrid-EPOCH exceeds the experimental values. However, if the distribution

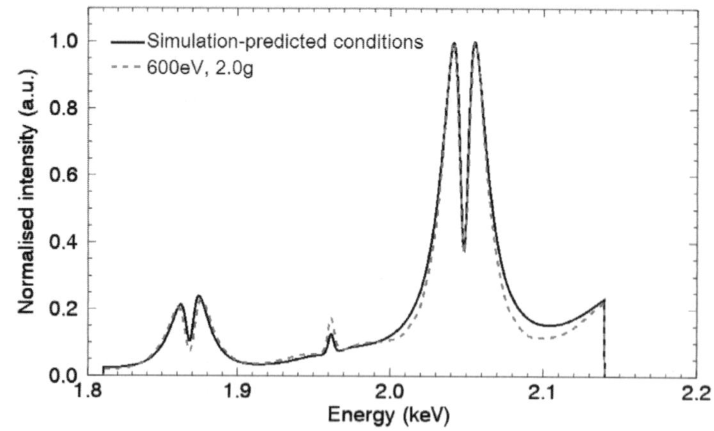


**Figure 3.** FLY-calculated emission spectra using temperature and density conditions predicted by hybrid-EPOCH (black, solid), and a single temperature and density of 600 eV and 2.7 g/cc respectively (red, dashed).

of temperature and density conditions predicted by the code are used by the atomic kinetics code FLY [6] to produce an expected x-ray emission spectrum good agreement is observed with a single spectrum which corresponds to the approximate temperature and density conditions inferred from experiment. Additional simulations with layers buried at differing depths into the plastic targets are currently underway.

Although the observed x-ray emission spectrum from buried layer targets can be characterised by a single temperature and density, the results presented above suggest that there may be radial temperature and density gradients within the layer. Thus the maximum temperature and density within a small portion of the layer may exceed those inferred from the experimental results.

More generally, the above results indicate that hybrid-EPOCH is a useful tool for modelling the initial stages of a short pulse laser experiment. Namely the laser interaction, energetic electron production and transport, return current generation and the associated heating. However, limits to the code's physics models still prevent this from being sufficient for modelling the full range of phenomena associated with short pulse laser experiments. Additional codes are required to accurately model target ablation and shock generation by long pulse beams, radiative processes, and longer timescale ($\sim$ ns) heating and hydrodynamic expansion. Much initial success has been met in linking suitable codes together to encompass these effects [1]. By making use of models such as that detailed in reference [2], and employed by hybrid-EPOCH, PIC codes are able to access the solid density regime, and thus reinforce any linking of PIC and hybrid transport codes.

## References

[1] Sircombe N J, Hughes S J and Ramsay M G 2013 *New Journal of Physics* **15** 025025
[2] Cohen B I, Kemp A J and Divol L 2010 *J. Comput. Phys.* **229** 4591
[3] Arber T D *et al.* *J. Comput. Phys.* Submitted
[4] Hoarty D J *et al.* 2013 *Phys. Rev. Lett.* **110** 265003
[5] Hoarty D J *et al.* 2013 *High Energy Density Physics* **9** 661
[6] Lee R W and Larsen J T 1996 *J. Quant. Spectrosc. Radiat. Transfer* **56** 535
[7] Brown C R D *et al.* 2011 *Phys. Rev. Lett.* **106** 185003

# Hydrodynamic instabilities and mix studies on NIF: predictions, observations, and a path forward

B. A. Remington,[1] L. J. Atherton,[1] L. R. Benedetti,[1] L. Berzak-Hopkins,[1] D. K. Bradley,[1] D. A. Callahan,[1] D.T. Casey,[1] P. M. Celliers,[1] C. J. Cerjan,[1] D. S. Clark,[1] E. L. Dewald,[1] T. R. Dittrich,[1] S. N. Dixit,[1] T. Döppner,[1] D. H. Edgell,[1] M. J. Edwards,[1] R. Epstein,[2] J. Frenje,[3] M. Gatu-Johnson,[3] S. Glenn,[1] S. H. Glenzer,[4] G. Grim,[5] S. W. Haan,[1] B. A. Hammel,[1] A. Hamza,[1] D. Hicks,[1] W. W. Hsing,[1] O. Hurricane,[1] N. Izumi,[1] O. S. Jones,[1] M. H. Key,[1] S. F. Khan,[1] J. D. Kilkenny,[6] J. L. Kline,[5] G. A. Kyrala,[5] O. L. Landen,[1] S. Le Pape,[1] J. D. Lindl,[1] T. Ma,[1] B. J. MacGowan,[1] A. J. Mackinnon,[1] A. G. MacPhee,[1] N. B. Meezan,[1] J. D. Moody,[1] E. I. Moses,[1] A. Nikroo,[6] A. Pak,[1] T. Parham,[1] H.-S. Park,[1] P. K. Patel,[1] R. Petrasso,[3] J. Pino,[1] J.E. Ralph,[1] K. Raman,[1] S. P. Regan,[2] H. F. Robey,[1] J. S. Ross,[1] B. K. Spears,[1] V. A. Smalyuk,[1] P. T. Springer,[1] L. J. Suter,[1] R. Tipton,[1] R. Tommasini,[1] R. P. Town,[1] S. V. Weber[1]

[1]Lawrence Livermore National Laboratory, 7000 East Ave., Livermore, CA 94550 USA
[2]Laboratory for Laser Energetics, University of Rochester, Rochester, New York 14623, USA
[3]Massachusetts Institute of Technology, Cambridge, MA, USA
[4]SLAC National Accelerator Laboratory, 2575 Sand Hill Rd. Menlo Park, CA 94025, USA
[5]Los Alamos National Laboratory, Los Alamos, NM 87545, USA
[6]General Atomics, 3550 General Atomics Court, San Diego, CA, 29186, USA

**Abstract.** The goals of the Mix Campaign are to determine how mix affects performance, locate the "mix cliff", locate the source of the mix, and develop mitigation methods that allow performance to be increased. We have used several different drive pulse shapes and capsule designs in the Mix Campaign, to understand sensitivity to drive peak power, level of coast, rise time to peak power, adiabat, and dopant level in the capsule. Ablator material mixing into the hot spot has been shown conclusively with x-ray spectroscopy. The observed neutron yield drops steeply when the hot spot mix mass becomes too large. The mix appears to be driven by ablation-front Rayleigh-Taylor instabilities. A high foot, higher adiabat drive has a more stable ablation front and has allowed the mix mass in the hot spot to be reduced significantly. Two recent high foot shots achieved neutron yields > $10^{15}$ and measured neutron yield over clean 1D simulation (YOC) > 50%, which was one of the central goals of the Mix Campaign.

## 1 Introduction

The goal of inertial confinement fusion (ICF) is to compress deuterium (D) and tritium (T) to the densities and temperatures required to achieve a burst of D+T → $^4$He+n+17.6 MeV fusion reactions of sufficient intensity that the energy output exceeds the energy input. The indirect-drive approach pursued at the National Ignition Facility (NIF) uses a high power pulsed laser focused into a radiation cavity (hohlraum) to convert to soft x-rays which ablatively drive the implosion of a hollow spherical capsule at the center of the hohlraum. The capsule has a multi-layered ablator (Fig. 1a) and the radiation drive is shaped to launch 3-4 staged shocks to control the adiabat and achieve high compression of the fuel (Fig. 1b). [1-3] It has been known for decades, based on theory [4-11], 2D and 3D simulations, [12-18] and experiments on high power lasers [19-27] that the capsule ablation front in ICF is unstable to the Rayleigh-Taylor (RT) instability and that the ablation process reduces RT growth rates. What is not known sufficiently are the detailed effects RT has on NIF capsule performance. This RT growth, if unchecked, can become large enough to perturb the ablator-fuel and fuel-hot spot interfaces and ultimately leads to ablator material contaminating the hot spot and quenching the burn by radiative cooling. This mix failure mode is often referred to loosely as the "mix cliff", as suggested by the sketch in Fig. 1c. The goals of the Mix Campaign on NIF are to determine if such a mix cliff exists, locate it, determine how steep it is, and find ways to mitigate this mix failure mode. Once effective mitigation methods are found, the plan is to push the more robust capsules harder to test the limits of performance, and find the new mix cliff. This cycle of locating the mix cliff, finding new or more refined mix mitigation methods, then pushing these capsules to their new performance limits will be repeated, as we attempt to enter into the alpha heating ("boot strapping") regime, and push closer to ignition. The remainder of this article describes our progress in this endeavour; recent results are encouraging.

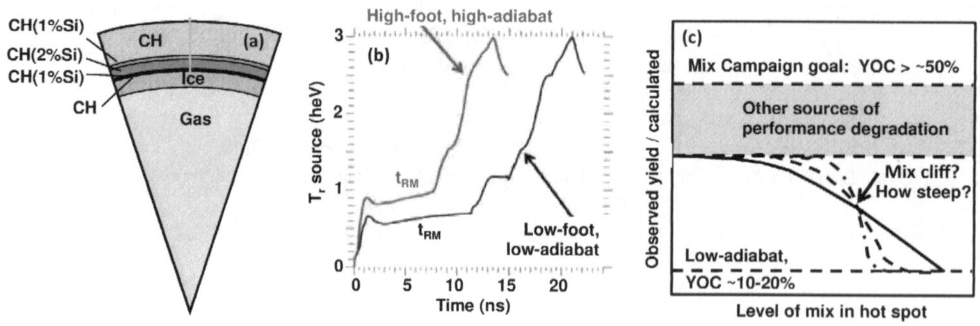


**Figure 1.** (a) Cut-out sketch of a typical capsule used in NIF ignition experiments, corresponding to an outer radius of 1.15 mm. The layers, from the outside in, correspond to 138 μm of CH, 10 μm of CH(1% Si), 35 μm of CH(2% Si), 6 μm of CH(1% Si), 6 μm of CH, and 69 μm of cryogenic DT ice. (b) Drive radiation pulse shapes used for the 4-shock low-foot, low-adiabat (α~1.5) drive (black curve) and the 3-shock high-foot, higher-adiabat (α~2.3) drive (red curve). (c) The basic plan for the Mix Campaign is illustrated in this qualitative sketch. Our expectation at the onset of the Mix Campaign was that as the level of ablator mixed into the hot spot increases, the experimentally observed neutron yield compared to clean 1D simulation (YOC) would decrease. Depending on the steepness, we called this the "mix cliff". To optimize implosions to approach ignition, we need YOC > ~50% (in the absence of alpha heating) to have a tighter correlation between simulations and experiments to allow the wide parameter space to be more fully and efficiently explored.

## 2 Experiments to probe the mix cliff

The first conclusive experimental evidence that there was ablator material penetrating into the DT hot spot came from x-ray emission spectroscopy. In one "tri-doped" DT layered implosion, the innermost 6 μm layer in the ablator (adjacent to the DT layer) had 0.15% by number trace amount of Cu. And the next two layers moving back from the ice had 0.20% trace levels of Ge. In this implosion, Ge He-α emission from the hot spot was seen in the Supersnout-II spectrometer, but no Cu emission from the hot spot was observed, as shown in Fig. 2a. [28] Both the Ge and Cu cold K-edges were observed in the surrounding shell. These experimental observations are consistent with the predictions that ablation-front RT instability carries ablator material from the outside inward into the hot spot, [17] whereas instability growth at the ablator-fuel interface drives much less ablator into the hot spot. In parallel with these spectroscopy experiments, it was noticed that in poorly performing implosions, the neutron yield and ion temperature were low, but the hot spot x-ray emission was high. A very plausible way to explain this is that ablator material, in particular carbon, was mixed into the hot spot and radiatively cooled the hot spot, quenching the burn. This led to the development of the mix model; the carbon mass in the hot spot was inferred from the excess of radiation from the hot spot relative to that expected for clean DT, given the yield, hot spot volume and hot spot temperature. [29] With this mix model in hand, a full ensemble of cryogenic layered implosions can be plotted in terms of DT neutron yield versus experimentally inferred hot spot mix mass, as shown in Fig. 2b. A similar trend between the neutron yield and the hot-spot mix mass was

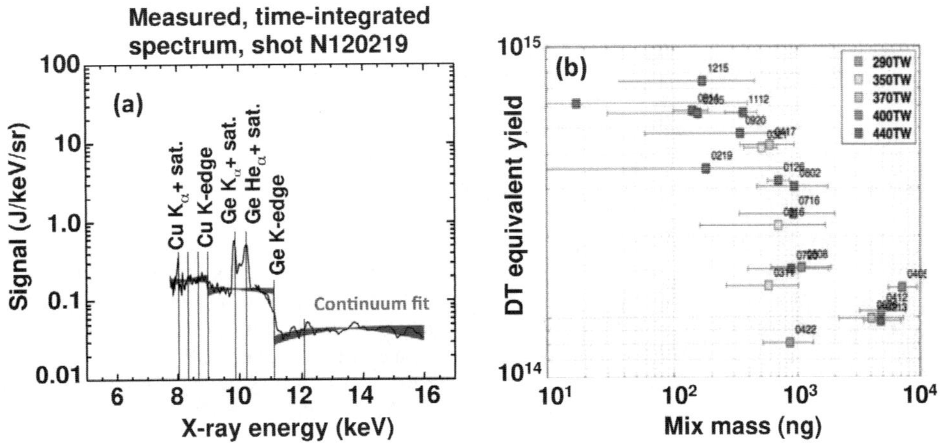


**Fig. 2.** (a) Measured hot-spot x-ray spectrum for an ignition target with a tri-doped ablator (black curve), corresponding to NIF shot N120219. The x-ray continuum from the hot spot transmitted through the compressed shell is modeled (red curve) assuming the x-ray continuum and the shell optical thickness scale with photon energy (hv) as exp(-hv/kT) and $(hv)^{-3}$, respectively. [28] (b) Measured DT neutron yield versus inferred mix mass from the mix model for an ensemble of 21 cryogenic layered DT implosions on NIF. Points are color coded by peak laser power. [29] When the hot spot mix mass exceeds several hundred nanograms, the yield drops steeply, suggesting a "mix cliff".

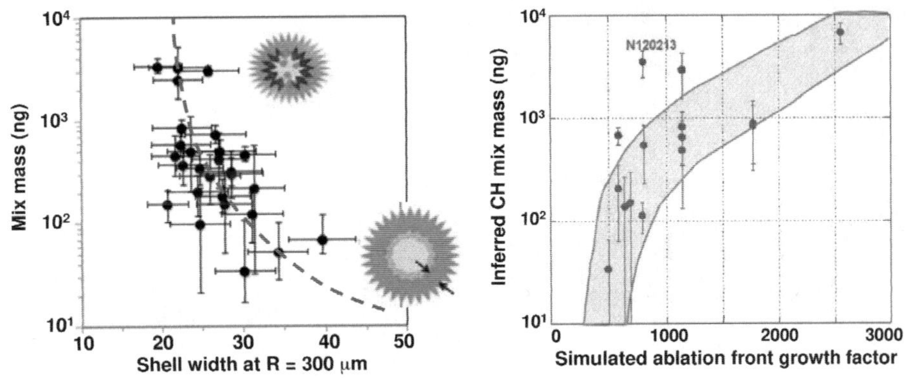


**Fig. 3.** (a) Experimentally inferred hot spot mix mass versus experimentally inferred shell thickness, based on in-flight backlit radiography. [30] (b) Inferred hot spot mix mass versus maximum ablation front growth factors from 2D simulations. [30] The smooth curves in (a) and (b) are only to guide the eye. [30]

observed with the x-ray spectroscopy [28]. It is quite clear from Fig. 2b that when the amount of ablator mixed into the hot spot exceeds several hundred nanograms, neutron yield plummets sharply. The results shown in Fig. 2 all stem from the low-foot, low-adiabat drive shown in Fig. 1b, albeit with some variation in the rise time to peak power (1 ns, 2 ns, and 3 ns). At this point in the Mix Campaign, it was clear that we had located the "mix cliff", and that ablator mix into the hot spot was a strong failure mode that needed to be mitigated. The next section describes our initial work in mitigation.

## 3 Mitigating Rayleigh-Taylor induced hot spot mix

There are several indications that ablation-front RT instability is one of the dominant causes of ablator material being mixed into the hot spot. First, as shown in Fig. 2a, the spectroscopy measurement on the tri-doped capsule showed strong Ge emission from the hot spot, but no emission from Cu. Second, when experimental mix mass is plotted versus experimentally inferred shell width (Fig. 3a), there appears to be a trend that as shell width decreases, mix mass increases. [30] Third, the experimental mix masses tend to increase when plotted against simulated ablation front growth factors, as shown in Fig. 3b. [30] And finally, when mix masses are compared between the nominal capsules (Fig. 1a) and those with increased silicon concentrations, the latter had higher mix masses (not shown). The above four observations are all consistent with ablation front RT being a dominant source of ablator penetrating into the hot spot. Hence, our first mitigation strategy was aimed at a near-term modification that would reduce ablation-front RT growth. We used the same capsule shown in Fig. 1a, but switched to the high-foot, higher-adiabat drive, shown by the red curve in Fig. 1b. Examples of the 2D design simulations comparing the low-foot with high-foot capsule implosions are shown at peak compression in Fig. 4a. [3] The high-foot simulation converges slightly less, with convergence ratios (CR) of 25-30, as opposed to CR ~ 35-40 for the low-foot implosions. The high foot drive creates a higher adiabat ($\alpha$~2.3 for the higher foot pulse versus ~1.5 for lower foot pulse) starting in the very beginning of the drive. The net effect is a higher ablation velocity and larger density gradient scale length, both of which reduce the RT growth rates, and thereby reduce the overall ablator penetration into the hot spot. [5,6,8,9,11]

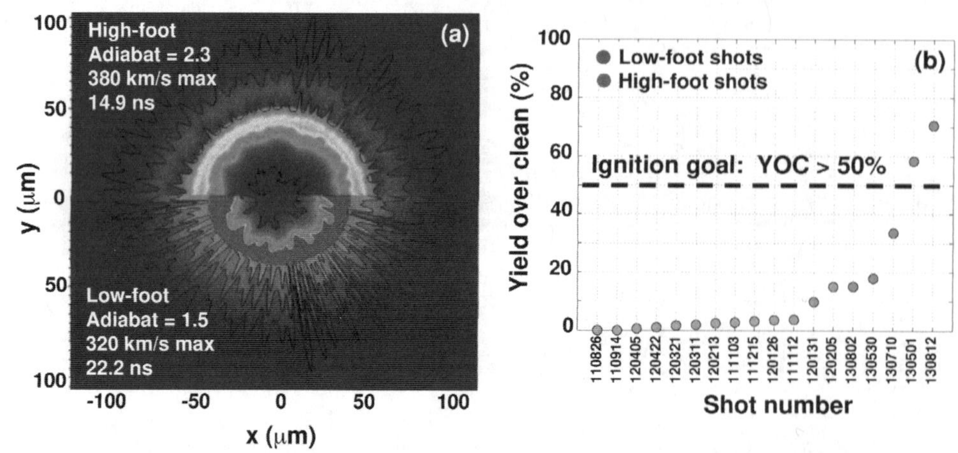


**Fig. 4.** (a) Two dimensional, 100-mode simulations with a spectrum of imposed surface roughness show, in density, the expected instability growth on the capsule. The bottom frame shows the low-foot capsule with adiabat $\alpha = 1.5$ case and the top frame shows the high-foot, adiabat $\alpha = 2.3$ case. The results are at the minimum implosion radius. The high foot simulation converges less and has a lower level of RT induced spikes penetrating into the hot spot. [3] (b) An ensemble of 18 cryogenic layered DT implosions plotted as experimental yield divided by clean 1D simulation. The blue symbols are for the low-foot, low-adiabat series and the green for the new high-foot, high-adiabat series. The low foot implosions have YOC ~ 5-15%, whereas the high foot implosions have YOC ~ 20-70%.

Accordingly, the high-foot simulation shows less RT growth and spikes penetrating into the hot spot at this late stage of the implosion. The experimentally inferred mix masses for these high-foot implosions were low. When the ratio of experimental neutron yield over clean 1D simulation (YOC) is plotted, the high-foot implosions clearly demonstrate higher values, as shown in Fig. 4b. [31] Further, YOC for two of the high foot shots surpassed the 50% threshold, which was one of the central goals of the Mix Campaign, as indicated in Fig. 2b. The goal of achieving YOC > 50% is so that simulations and experiments are more closely correlated, allowing faster optimization of new, higher performance designs. Further, analysis and simulations (not shown) demonstrated that shot N130812 in Fig. 4b produced 50% enhanced nuclear yield due to alpha heating, meaning these high-foot implosions are entering into the regime where alpha heating is becoming significant. [3,31] Alpha dominated burn, in which the yield more than doubles due to the effects of alpha heating, is the next goal being pursued in these mix mitigated high-foot implosions.

## 4 Conclusion and future directions

In summary, the Mix Campaign conducted an extensive series of experiments. A steep mix cliff was found, using a number of diagnostic techniques. In particular, x-ray spectroscopy with tri-doped capsules, and the mix model based on comparing continuum x-ray yield to nuclear yield allowed the mix cliff to be mapped out, for the low-adiabat drives. Our mitigation strategy to date has focused on reducing ablation-front RT instability growth, using a higher adiabat design. This has allowed implosions with YOC > 50% and produced the first preliminary indications of the effects of alpha heating. This high YOC due to higher adiabat comes at the expense of reduced fuel areal density, however. Future work will test the limits of the high-foot design, and will also consider additional mitigation techniques such as alternate ablators, [32,33] as we attempt to push performance further into alpha heating and start to approach the ignition regime.

## 5 Acknowledgements

This work was performed under the auspices of the U.S. Department of Energy by Lawrence Livermore National Laboratory under Contract DE-AC52-07NA27344.

## References

[1] S.W. Haan et al., "Point design targets, specifications, and requirements for the 2010 ignition campaign on the National Ignition Facility", PoP 18, 051001 (2011).
[2] M.J. Edwards et al., "Progress towards ignition on the National Ignition Facility", PoP 20, 070501 (2013).
[3] T.R. Dittrich et al., "Physics Design Basis for a High-Foot/High-Adiabat Inertial Confinement Fusion Capsule for the NIF", PRL, submitted (2013).
[4] J.D. Lindl and W.C. Mead, "Two-dimensional simulation of fluid instability in laser-fusion pellets, " PRL 34, 1273 (1975).
[5] H. Takabe, K. Mima, L. Montierth, and R. L. Morse, "Selfconsistent growth rate of the Rayleigh-Taylor instability in an ablatively accelerating plasma," Phys. Fluids 28, 3676 (1985);
[6] D. H. Munro, "Analytic solutions for Rayleigh-Taylor growth rates in smooth density gradients," Phys. Rev. A 38, 1433 (1988).
[7] S.W. Haan, " Onset of nonlinear saturation for Rayleigh-Taylor growth in the presence of a full spectrum of modes," PRA 39, 5812 (1989).
[8] S.W. Haan, "Weakly nonlinear hydrodynamic instabilities in inertial fusion," Phys. Fluids B 3, 2349 (1991).
[9] M. Tabak, D. H. Munro, and J. D. Lindl, "Hydrodynamic stability and the direct drive approach to laser fusion," Phys. Fluids B 2, 1007 (1990).

[10] D. Shvarts et al., "Nonlinear evolution of multimode Rayleigh-Taylor instability in two and three dimensions," PoP 2, 2465 (1995).
[11] R. Betti, V.N. Goncharov, R.L. McCrory, P. Sorotokin, and C.P. Verdon, "Selfconsistent stability analysis of ablation fronts in inertial confinement fusion," PoP 3, 2122 (1996).
[12] R. L. McCrory, L. Montierth, R. L. Morse, and C. P Verdon, "Nonlinear evolution of ablation-driven Rayleigh-Taylor instability," Phys. Rev. Lett. 46, 336 (1981).
[13] C. P. Verdon, R. L. McCrory, R. L. Morse, G. R. Baker, D. I. Meiron, and S. A. Orszag, "Nonlinear effects of multifrequency hydrodynamic instabilities on ablatively accelerated thin shells," Phys. Fluids 25, 1653 (1982).
[14] H Sakagami and K. Nishihara, "Rayleigh-Taylor instability on the pusher-fuel intact surface of stagnating targets," Phys. Fluids B 2, 2715 (1990).
[15] S.V. Weber et al., "Modeling of Nova indirect drive Rayleigh-Taylor experiments," PoP 1, 3652 (1994).
[16] M.M. Marinak et al., "Three-dimensional single mode Rayleigh-Taylor experiments on Nova," PRL 75, 3677 (1995).
[17] B.A. Hammel et al., "High-mode Rayleigh-Taylor growth in NIF ignition capsules", HEDP 6, 171 (2010) ; "Diagnosing and controlling mix in NIF implosion experiments," PoP 18, 056310 (2011).
[18] D.S. Clark et al., "Short-wavelength and three-dimensional instability evolution in National Ignition Facility ignition capsule designs",PoP 18, 082701 (2011).
[19] J.S. Wark, J.D. Kilkenny, A.J. Cole, M.H. Key, and P. T. Rumsby, "Observations of the Rayleigh-Taylor instability in laser imploded microballoons," Appl. Phys. Lett. 48, 969 (1987).
[20] J. D. Kilkenny, "Experimental results on hydrodynamic instabilities in laser-accelerated planar packages," Phys. Fluids B 2, 1400 (1990).
[21] B.A. Remington et al., "Single-mode and multimode Rayleigh-Taylor experiments on Nova," PoP 2, 241 (1995).
[22] K.S. Budil et al., "Experimental comparison of classical versus ablative Rayleigh-Taylor instability," PRL 76, 4536 (1996).
[23] H. Azechi et al., "Direct-drive hydrodynamic instability experiments on the Gekko XII laser," PoP 4, 4079 (1997).
[24] S.G. Glendinning et al., "Ablation front Rayleigh-Taylor growth experiments in spherically convergent geometry," PoP 7, 2033 (2000).
[25] V.A. Smalyuk et al., "Evolution of shell nonuniformities near peak compression of a spherical implosion," PRL 87, 155002 (2001).
[26] V.A. Smalyuk et al., "Rayleigh-Taylor growth stabilization in direct-drive plastic targets at laser intensities of $\sim 10^{15}$ W/cm$^2$," PRL 101, 025002 (2008).
[27] C. Cherfils et al., "Convergent Rayleigh-Taylor experiments on the Nova laser," PRL 83, 5507 (1999).
[28] S.P. Regan et al., "Hot-spot mix in ignition-scale inertial confinement fusion targets", PRL 111, 045001 (2013).
[29] T. Ma et al., "Onset of Hydrodynamic Mix in High-Velocity, Highly Compressed Inertial Confinement Fusion Implosions", PRL 111, 085004 (2013).
[30] T. Döppner et al., "Mix in cryogenic DT layered implosions on the National Ignition Facility," PoP, in preparation (2013).
[31] H.S. Park et al., "High-adiabat, high-foot, low-mix cryogenic DT layered capsule implosion experiments on the National Ignition Facility", PRL, submitted (Oct. 2013).
[32] D.T. Casey et al., "High performance, indirect-drive implosions using beryllium ablators at Omega," PoP, in preparation (2013).
[33] J.S. Ross et al., "High-density carbon capsule experiments on NIF," PRL, in preparation (2013).

Journal of Physics: Conference Series **688** (2016) 012091 doi:10.1088/1742-6596/688/1/012091

# K-shell spectroscopic diagnosis of suprathermal electrons at fusion-relevant environmental conditions

**O Renner[1], M Šmíd[1,2], D Khaghani[3,4] and F B Rosmej[3,4]**

[1] Institute of Physics, Academy of Sciences CR, 18221 Prague, Czech Republic
[2] Czech Technical University in Prague, FNSPE, 11519 Prague, Czech Republic
[3] Sorbonne Universités, Pierre et Marie Curie, 75252 Paris Cedex 05, France
[4] LULI, École Polytechnique, CEA, CNRS, 91128 Palaiseau, France

E-mail: renner@fzu.cz

**Abstract.** Fast-electron-induced changes in the ion charge distribution lead to a substantial modification of radiative properties of x-ray emitting systems. We report Cu K-shell spectra emitted from 1.5-μm-thick Cu foils irradiated by laser intensities up to $5\times10^{16}$ W/cm$^2$ using the fundamental and frequency tripled radiation of the Prague iodine laser system PALS, i.e., at conditions characteristic for laser spikes anticipated in the shock ignition scheme. The emission of the lower-ionization-stage ions observed as a fine structure merged with the Cu K$\alpha_{1,2}$ doublet and spectral lines corresponding to transitions in He-like to Ne-like Cu atoms were identified. The spectra analysis based on a combination of MULTI2D hydrodynamic modelling of the plasma evolution and its emissivity synthesized by collisional-radiative code FLYCHK provides information on distribution of hot electron fractions at low plasma temperatures. The procedure used for interpretation of experimental data contributes to a development of alternative methods for diagnosis of suprathermal electrons.

## 1. Introduction

Generation of fast electrons accompanying high-intensity laser-matter interaction, their transport and energy deposition in a near-solid density material with a varied degree of ionization represent one of the key issues for realization of diverse schemes of inertial confinement fusion [1]. Non-Maxwellian electrons strongly affect properties of x-ray emitting plasma systems, in particular their radiative losses, spectral line intensities, ratios, and shapes (including appearance of non-diagram lines and satellites accompanying parent transitions) [2]. Advanced methods of x-ray imaging spectroscopy thus offer a direct experimental technique for studying the distribution of hot electrons in plasmas.

The purpose of the reported research is to contribute to a development of suprathermal electron diagnostics based on a detailed study of K-shell emission of mid-$Z$ elements. The conditions of the experiment with Cu foils irradiated by fundamental and frequency tripled radiation of the iodine laser PALS, i.e., at variable coupling parameter $I\lambda^2$ (where $I$ and $\lambda$ are the intensity and wavelength of the laser radiation), are close to those characterizing laser spikes in the shock ignition fusion scheme [3].

In general, fast electrons are generated due to combined effects of stimulated parametric processes in the lower density plasma and resonance absorption near the plasma critical surface. A survey of recent spectroscopic experiments studying the hot-electron-energy deposition and target-heating effects via Cu K-shell emission can be found in papers [4,5]. In contrast to these studies, we present x-ray spectra covering a broader range of photon energies with considerably higher spectral resolution.

The observed fine structure is interpreted in terms of suprathermal electron effects. We compare the measured data with synthetic spectra produced using MULTI2D hydrodynamic modeling of the plasma evolution [6] post-processed by collisional-radiative code FLYCHK [7] and discuss implications for hot electron diagnosis.

## 2. PALS experiment

The experiment was carried out using the first ($\lambda_1 = 1.315$ μm) and third ($\lambda_3 = 0.438$ μm) harmonics of the PALS laser [8]. The radiation with maximum energy of 450 J, pulse duration of 0.25 ns and focal spot radius of 35 μm was incident normal to the surface of a 1.5-μm-thick Cu foil. The maximum intensity of $4.7\times10^{16}$ W/cm$^2$ delivered on the target using the first laser harmonic (1ω) corresponds to the coupling parameter $I\lambda^2 = 8\times10^{16}$ Wμm$^2$/cm$^2$.

The spectral emission was observed tangentially to the target surface by using a x-ray spectrometer with a crystal of quartz (223) spherically bent to a radius of 150 mm. The spectrometer was aligned to cover the photon energy range of 7.9 – 8.5 keV and provided 1D spatial resolution of 14 μm in the direction of the target normal. The spectra were recorded on x-ray film Kodak Industrex AA400, scanned with a table-top scanner, calibrated with respect to experimental geometry and tabulated Cu K$\alpha_{1,2}$ transitions (8047.78 and 8027.83 eV, respectively) and recalculated to intensity scale using the film characteristic curve.

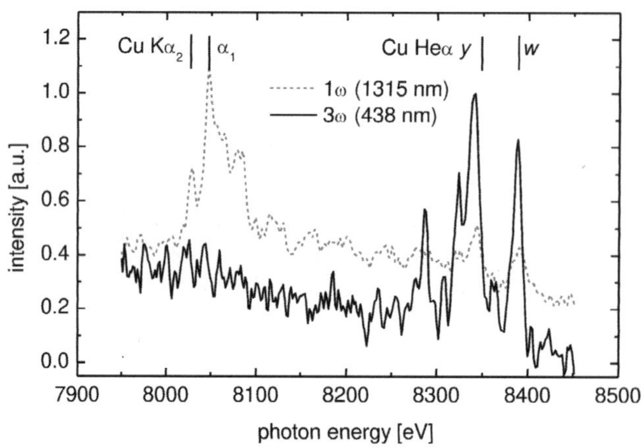


**Figure 1.** Comparison of Cu K-shell spectra recorded using 1ω and 3ω laser radiation.

The single shot spectra shown in figure 1 demonstrate clearly the effect of the laser coupling parameter. Whereas at 3ω the Cu *K*-shell emission is governed by resonance line transitions in highly ionized atoms and their satellites (the so-called quasi-optical transitions), the 1ω laser interaction results in an increased emission of inner-shell electronic transitions in weakly ionized atoms. In accordance with this, the strongest transitions observed in 3ω-generated spectra were identified as resonance *w* (with photon energy 8391 eV) and intercombination *y* (8347 eV) lines of He-like Cu and their Li-like satellites. In contrast, the experimental spectra recorded using the 1ω laser are dominated by x-ray emission near K$\alpha_1$ (8047.8 eV) and K$\alpha_2$ (8027.8 eV) transitions in single-ionized Cu II atoms overlapped by Cu III till Cu XX (Ne-like) emission. Complementary spectral structure observed between Kα and Heα lines belongs to emission from Li- till F-like copper ions; the line identification is provided in the next section.

The presence of these transitions in x-ray spectra generally indicates an occurrence of fast (suprathermal) electrons with the energy above the K-shell ionization limit of the atom in a given ionization state. Further we shall concentrate to the analysis of the 1ω-generated spectrum and to its diagnostic potential with respect to the determination of hot electron population.

Journal of Physics: Conference Series **688** (2016) 012091 doi:10.1088/1742-6596/688/1/012091

## 3. Theoretical analysis

The importance of hot electrons for K-shell spectra formation is indicated in figure 2. The shown spectra were synthesized using the FLYCHK code [7] for the Cu plasma density of 0.894 g/cm$^3$, size of 70 μm (equal to the focal spot dimension and relevant for incorporation of reabsorption effects), two characteristic bulk plasma temperatures (200 and 1600 eV) and variable hot electron fractions $f_{HE} = 0$, $10^{-4}$, $10^{-3}$, and $10^{-2}$. Whereas at higher plasma temperature the effect of hot electrons is more or less negligible (except the growth of the He-like resonance $w$ and Kβ line intensities at highest $f_{HE}$), the presence of hot electrons at lower temperatures is of paramount importance for the plasma K-shell emissivity. The shape of the emitted spectra varies in accordance with figure 2a but still more important is the variation of their intensities. To achieve the same maximum intensities for spectra emitted at $f_{HE} = 0$, $10^{-4}$, $10^{-3}$, and $10^{-2}$, the code-predicted data had to be multiplied by a scaling factor of $3\times10^{12}$, 90, 9, and 1, respectively. In other words, the presence of hot electrons is decisive for observability of K-shell emission from lower ionization stages.

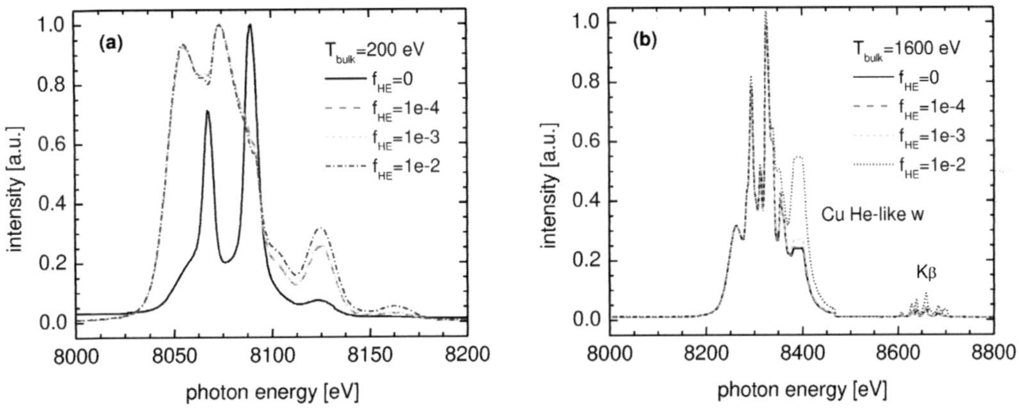


**Figure 2.** Spectra synthesized by CR code FLYCHK demonstrate the effect of hot electron fraction on Cu K-shell emission at bulk plasma temperature of 200 eV (a) and 1600 eV (b).

Previous investigation of x-ray emission from Cu targets irradiated by intense laser beams [4,5] demonstrated that the time-integrated K-shell spectra cannot be satisfactorily modeled by considering only a single set of relevant plasma parameters (temperature, density, size and hot electron fraction). To comply with this conclusion, the plasma evolution was simulated by using the lagrangian hydrodynamic code MULTI2D [6] assuming 2D cylindrical geometry. The 1.5-μm-thick Cu foil subdivided into 30 cells in axial and 50 in radial direction was irradiated by a gaussian laser beam. To simplify the production of the final spectrum, instead of time integration over the full spatial plasma distribution the output data were interpolated onto a fixed eulerian grid and the evolution of the temperature and density at the intersection of the laser axis with the target surface was extracted.

These parameters interpolated onto 20 time steps were used for production of synthetic spectra via collisional-radiative code FLYCHK [7]. At each step, a set of simulations was performed for hot electron densities varied in the range of $10^{16}$ - $10^{22}$ cm$^{-3}$. According to the Beg scaling [9], the energy of suprathermal electrons was set to 94 keV.

The final spectra fitting was done by using an optimization code which varies the time-evolution of $f_{HE}$ in order to obtain the best agreement of the integrated spectrum to the experimental one. In successive iterations, the code checks whether the modification of $f_{HE}$ within any step leads to a better fit with the experimental spectra. After performing the desired modification, the code proceeds to search for the global optimum.

The comparison of the best fit theoretical spectrum with the experiment is presented in figure 3a. Since the signal to noise ratio in the measured single shot spectra was relatively low, the experimental data shown in this figure represent a sum of 6 individual spectral records taken at identic conditions.

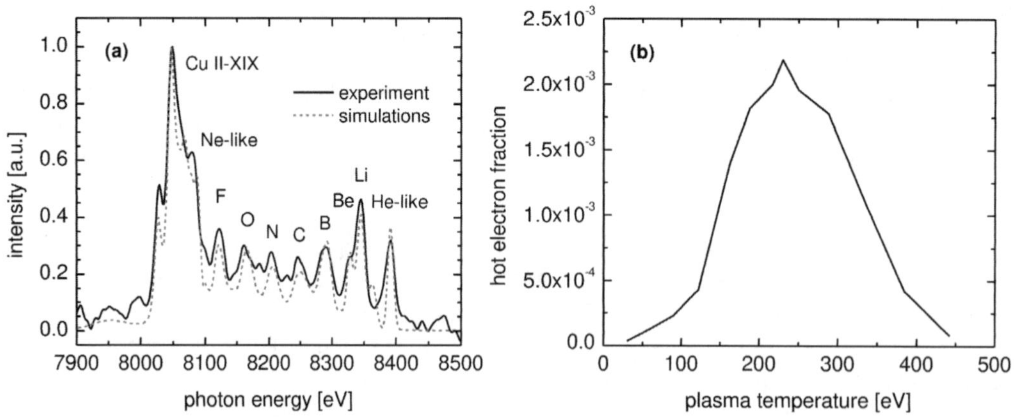


**Figure 3.** Comparison of experimental and synthetic spectra (a) and corresponding hot electron fractions found for the stage of low-temperature plasma emission (b).

The overall agreement of both spectra is very good. The marked peaks characterizing the principal emission of He- to Ne-like copper ions are well resolved and clearly identified. The best-fit fractions of suprathermal electrons corresponding to the plasma states in the range of bulk temperatures up to 500 eV are depicted in figure 3b. The found values of $f_{HE}$ span between $1\times10^{-4}$ - $2.2\times10^{-3}$ and reach the maximum at approximately 230 eV. Outside the depicted temperature range, the values of $f_{HE}$ cannot be univocally determined (as indicated for higher temperatures in figure 2b).

The fitting of the spectra in the photon energy range between Cu Kα and Heα emission can consequently be used for assessment of hot electron population in colder plasmas. The validation of this conjecture by PIC code simulations is under the progress.

## 4. Conclusion

By using a combination of MULTI2D hydrodynamics and FLYCHK collisional-radiative code, we have developed a model for interpretation of K-shell emission spectra emitted from Cu foil irradiated by the intense laser beam. Despite simplifications used in modeling the plasma evolution and its emissivity, we believe that the procedure used in spectra interpretation provides a viable method for determination of hot electron fraction in low temperature plasmas.

## Acknowledgments

This research was supported by Academy of Sciences of the Czech Republic Programme of internal support of international collaboration, project M100101208, and by MSMT projects LM2010014, CZ.1.07/2.3.00/20.0279 and CZ.1.05/1.1.00/02.0061.

## References

[1] Guskov S Yu 2013 *Plasma Phys Reports* **39** 1
[2] Rosmej F B 2012 *Highly Charged Ions* eds R Hutton, Y Zou (NY: Taylor & Francis) p 267
[3] Betti R, Zhou C D, Anderson K S et al 2007 *Phys Rev Lett* **98** 155001
[4] Nilson P M, Theobald W, MilehamC et al 2011 *Phys Plasmas* **18** 042702
[5] Huntington C M, Kuranz C C, Malamud G et al 2012 *Rev Sci Instr* **83** 10E114
[6] Ramis R, Meyer-ter-Vehn J and Ramírez J 2009 *Comput Phys Commun* **180** 977
[7] Chung H K, Chen M H, Morgan W L et al 2005 *High Energy Density Phys* **1** 3
[8] Jungwirth K, Cejnarova A, Juha L et al 2001 *Phys Plasmas* **8** 2495
[9] Beg F N, Bell A R, Dangor A E et al 1997 *Phys Plasmas* **4** 447

# Advances in shock timing experiments on the National Ignition Facility

H F Robey[1], P M Celliers[1], J D Moody[1], J Sater[1], T Parham[1], B Kozioziemski[1], R Dylla-Spears[1], J S Ross[1], S LePape[1], J E Ralph[1], M Hohenberger[2], E L Dewald[1], L Berzak Hopkins[1], J J Kroll[1], B E Yoxall[1], A V Hamza[1], T R Boehly[2], A Nikroo[3], O L Landen[1] and M J Edwards[1]

[1]Lawrence Livermore National Laboratory, 7000 East Ave., Livermore, CA 94550 USA
[2]Laboratory for Laser Energetics, Univ. of Rochester, Rochester, New York 14623, USA
[3]General Atomics, 3550 General Atomics Court, San Diego, CA, 29186, USA

**Abstract.** Recent advances in shock timing experiments and analysis techniques now enable shock measurements to be performed in cryogenic deuterium-tritium (DT) ice layered capsule implosions on the National Ignition Facility (NIF). Previous measurements of shock timing in inertial confinement fusion (ICF) implosions were performed in surrogate targets, where the solid DT ice shell and central DT gas were replaced with a continuous liquid deuterium (D2) fill. These previous experiments pose two surrogacy issues: a material surrogacy due to the difference of species (D2 vs. DT) and densities of the materials used and a geometric surrogacy due to presence of an additional interface (ice/gas) previously absent in the liquid-filled targets. This report presents experimental data and a new analysis method for validating the assumptions underlying this surrogate technique.

## 1. Introduction

Inertial confinement capsule implosions on the National Ignition Facility (NIF) are performed using the indirect-drive concept [1], where the energy of a temporally shaped laser pulse from 192 NIF beams is converted to thermal x rays in a cylindrical high-Z enclosure referred to as a hohlraum. The x-ray radiation uniformly ablates the surface of a low-Z (CH) spherical ablator shell (r = 1.1 mm), which surrounds a spherical cryogenic layer of solid DT fuel (70 $\mu$m thickness, $\rho = 0.255$ g/cm$^3$) and central low-density (0.3 mg/cm$^3$) DT gas fill. The hohlraum x-ray drive launches a series of four shocks to spherically compress the ablator shell and fuel, thereby creating the high density and temperature plasma conditions required to initiate DT fusion reactions in the hot, central fuel core.

To experimentally diagnose and adjust the strength and timing of the shocks, a modified hohlraum geometry (known as the "keyhole" target) is used. A gold diagnostic cone is added, penetrating both the hohlraum and capsule walls, to allow for the direct measurement of shock propagation in the capsule interior. Following the technique first described in [2], the capsule and Au cone are filled with liquid deuterium (D2) at $\rho = 0.170$ g/cm$^3$ (replacing the DT ice and gas of an ignition capsule), and the VISAR (Velocity Interferometer System for Any Reflector) diagnostic [3, 4] is used to optically measure the velocity vs. time of the highly reflective leading shock front [5]. Initial experiments demonstrating the shock measurement technique and its application to directly driven implosions on the OMEGA Laser Facility are reported in [6, 7], and results for indirectly-driven implosions on NIF and initial tuning results for NIF ignition pulses are given in [8, 9].

The use of liquid D2 in the tuning targets presents two surrogacy issues: 1.) a material surrogacy (single species D2 vs. two-species DT) and 2.) a geometric surrogacy (the presence or absence of an additional ice / gas interface in the ignition target geometry). Reference [10] presented the initial data assessing these surrogacy assumptions for the first three shocks. Here, a more extensive discussion of the modified analysis procedures is given together with an extension of the experimental data to include the 4[th] shock.

To quantify differences in shock strengths between liquid D2 tuning targets and layered DT ice ignition targets, the keyhole target was modified to allow for the growth of DT ice layers. Figure 1(a) shows an x-ray radiograph taken through the LEH of a "thick" DT ice layer grown on NIF (shot N121101). The tip of the Au cone and the inner surface of the Si-doped CH ablator are clearly seen. The gap between these two surfaces was 400 $\mu$m. The interface between the DT ice and the equilibrium DT vapor is visualized by x-ray refraction. The uniformity of the layer thickness is shown in Figure 1(b), where the image of Figure 1(a) is "un-folded" to visualize the ice layer thickness relative to the ablator inner surface. The nominal thickness goal of this layer was 300 $\mu$m, but considerable variation of the thickness is seen.

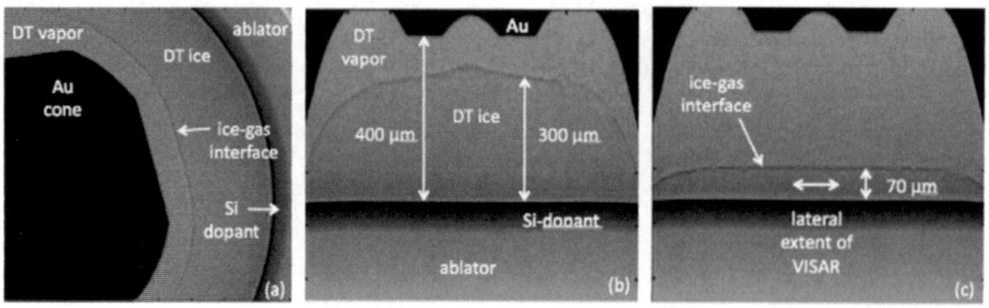


**Figure 1.** (a) X-ray radiograph of the capsule interior (shot N121101) showing the tip of the Au diagnostic cone, the Si-doped CH ablator, and the contour of a 300 $\mu$m thick DT ice layer. (b) Un-wrapped view of the DT ice layer of (a) showing ice thickness variation relative to the ablator inner surface. (c) Un-wrapped view of a 70 $\mu$m thick DT ice layer (shot N121108) showing reduced ice thickness variation.

To study the effect of multiple shock release and recompression events from the DT ice to gas, layers of thickness corresponding to those used in ignition targets (nominal value 70 $\mu$m [11]) were grown as well. Figure 1(c) shows an example of such a layer (N121108). The thickness uniformity is considerably improved. This is important, as successive shocks will transit the full extent of this layer. The lateral extent of the VISAR measurements at the ablator inner surface is ~350 $\mu$m, and the uniformity of the layer thickness over this field-of-view was measured to be better than 1 $\mu$m.

## 2. Geometric surrogacy between continuous liquid D2 and DT ice layer

To explore the geometric surrogacy issue, we now look at the results from thin DT ice layers. Figure 2 shows a comparison of streaked VISAR interferometer data for (a) N121112 (liquid D2) and (b) N121108 (70 $\mu$m DT ice layer). Time runs from left to right, and lateral (bottom-to-top) motion of the interference fringes is directly proportional to the shock velocity with fringe motion upward indicating an accelerating shock. Fringe motion is due to reflections from the leading shock front in the fuel (D2 or DT), and discontinuities in the fringe positions indicate the arrival time of shocks. The first of these, labeled "1" and seen at ~13.5 ns, is the time at which the 1[st] shock breaks out of the CH ablator into the fuel. Subsequent discontinuities indicate the time at which the increasingly stronger 2[nd], 3[rd], and 4[th] shocks overtake or merge with preceding shocks.

In Figure 2(a), all 4 shocks are seen, with the arrival time of shocks 2, 3 and 4 close to that desired for optimal ignition tuning. Figure 2(b) shows several new features. The VISAR interferometer signal abruptly ceases at $t \approx 17.5$ ns, as the 1[st] shock passes through the DT ice-gas interface. The resulting rarefaction reflected from this interface generates pressure gradients in the expanding DT ice. As will be quantified later, the reduced pressure in the expanding ice drops below that required for reflectivity of the VISAR laser ($\lambda$ = 660 nm) [5]. As the 2[nd] and 3[rd] shocks traverse this decompressing ice region, they briefly recompress it to the pressure level required for VISAR reflection. This is seen in Figure 2(b) at t ≈ 18.1 and 19.2 ns as transient returns of VISAR reflectivity. The weaker signal from 19.2 to 20.5 ns is thermal emission from the stronger 3rd shock; here the VISAR is functioning as a Streaked Optical Pyrometer (SOP) [12, 13], recording thermal emission within the band-pass of the VISAR optical system.

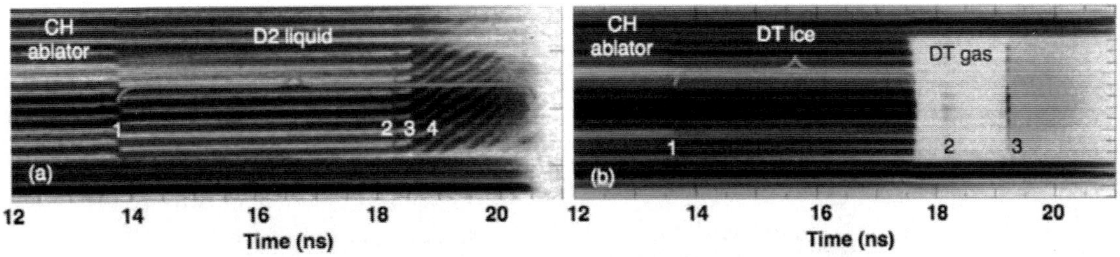


**Figure 2.** (a) VISAR streaked interferometer images for (a) a standard liquid D2-filled keyhole tuning target and (b) a 70 $\mu$m thick DT ice layer. Shock mergers 2 and 3 in (b) occur in the expanding DT gas region.

To analyze the data beyond the time of shock breakout, we use the reported VISAR reflectivity measurements of Celliers [5], where it was shown that VISAR reflectivity in liquid D2 exhibits a threshold response in both electron density ($Ne > Ne_{crit}$) and pressure ($P > 0.5$ Mbar). A metric for VISAR reflectivity from the simulations is proposed as the pressure at critical density. Figure 3 shows contour plots from a 1-D HYDRA [14] simulation to illustrate this. Figure 3(a) is a contour plot of pressure as a function of Lagrangian radius and time. The trajectories of the 4 shocks are superposed on this contour plot. One can clearly see a repeating sequence of shock compression and release events at the inner edge of the DT ice layer with a series of rarefaction waves propagating from the ice/gas interface upward toward the ablator. Figure 3(b) shows the corresponding contour plot of electron number density. The critical density for VISAR absorption or reflectivity is $Ne_{crit} = 2.6e21$ / $cm^3$. In Figure 3(b), the VISAR critical density is located very near the inner edge of the DT ice layer.

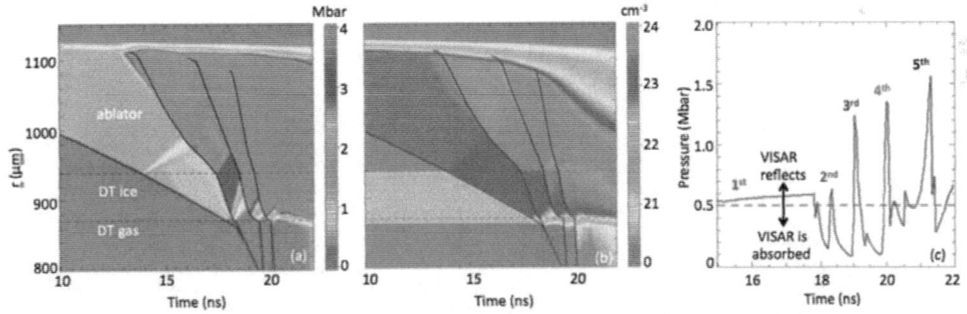


**Figure 3.** (a) Simulated contours of (a) pressure and (b) electron density plotted as radius vs. time for N130517. (c) Simulated pressure at critical density showing predicted time history of VISAR reflection and absorption.

The physical picture of the VISAR features labeled 2 and 3 in Figure 2(b) is now becoming clearer. These are not the locations of sequential shock mergers, but rather the time at which successive shocks recompress the releasing DT ice material to a pressure above the 0.5 Mbar threshold for VISAR reflectivity. Figure 3(c) plots this metric, the pressure at critical density, from the contour plots of Figures 3(a, b). For the transit of the first shock, this metric shows a continuously increasing pressure of 0.5-0.6 Mbar due to the slightly accelerating 1st shock as it transits the ice layer. Upon shock breakout near 17.7 ns, the pressure abruptly drops as a rarefaction fan is reflected back into the ice layer. At ~18.2 ns, the 2nd shock recompresses this expanding DT and briefly brings the pressure above the 0.5 Mbar threshold for reflectivity. The cycle then repeats, and the same process happens for the recompression caused by the 2$^{nd}$, 3$^{rd}$ and 4$^{th}$ shocks. As Figures 3(a-c) show, a 5$^{th}$ shock is also predicted to occur.

Figure 4 applies this new VISAR analysis method to two DT layered keyhole shots to compare the transient reflection features observed after 1$^{st}$ shock breakout with predictions from simulation. Figure 4(a) shows a comparison for shot N121108 (previously discussed in Figure 2(b)). The VISAR streaked interferometer data is shown together with a lineout of the intensity taken through the vertical center of the data. The intensity is high for the first shock as it transits the DT ice, and then drops to a rather low value upon breakout. The two recompression features are seen in the lineout as transient increases in signal intensity. The simulated pressure at critical density is shown on the right-hand side of the Figure. The timing of 1$^{st}$ shock breakout, and 2$^{nd}$ and 3$^{rd}$ shock recompression is in excellent agreement with the data.

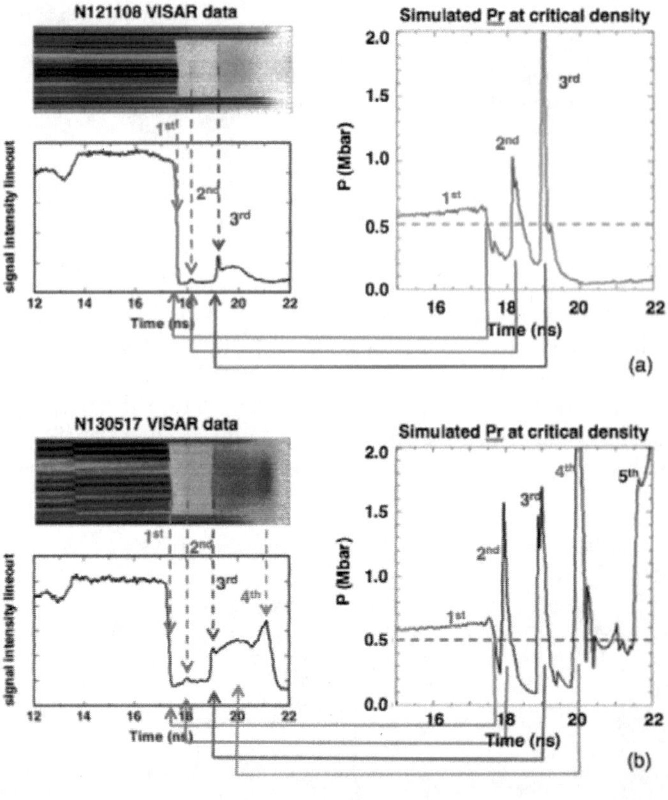


**Figure 4.** Comparison of measured vs. simulated shock timing for (a) a 3-shock and (b) a 4-shock system.

Figure 4(b) shows a second example of a similar 70 $\mu$m DT layer (N130517), but now with the 4$^{th}$ laser pulse included. The data now shows a 4$^{th}$ shock feature beginning at ~20.6 and peaking at 21 ns, whereas no such feature was observed in the VISAR data of Figure 4(a). The timing of 1$^{st}$ shock breakout and 2$^{nd}$ and 3$^{rd}$ shock recompressions is again in excellent agreement with simulation. The experimentally observed 4$^{th}$ shock feature, by contrast, is observed to be 600 ps later than in simulation.

## 3. Effect of electron preheat on 4$^{th}$ shock timing

There are several possible explanations for the observed discrepancy between simulated and observed 4$^{th}$ shock timing. Any mechanism that causes an additional expansion of the ice layer, for example, would delay the appearance of the 4$^{th}$ shock at the observed critical density for VISAR reflectivity.

Possible mechanisms for such an effect include the DT equation of state (EOS) release behavior, mix at the DT ice/gas interface, x-ray preheat, or supra-thermal (or hot) electron preheat. Since the data and simulations seem to agree for the first three shocks, we need a mechanism that is specific to the timing of the 4$^{th}$ shock. Both EOS release and mix at the ice/gas interface will begin at first shock breakout, so these seem less likely. The current simulations routinely include M-band preheat, which is calibrated by Dante measurements [15], so that leaves hot electron preheat as perhaps the most likely mechanism.

Figure 5 shows the Filter-Fluorescer Experiment (FFLEX) [19] measurements for shot N130517 (DT layer of Figure 4(b)), which are used to infer the hot-electron preheat in NIF hohlraums. FFLEX on NIF consists of 10 time-resolved channels that measure hard x-ray Bremsstrahlung emission over a wide energy range. The FFLEX data are well fit with a two temperature (T1, T2) Maxwellian distribution of supra-thermal electrons. Figure 5(a) shows the preheat power, defined as the power (GW) in the Maxwellian tail for electron energies > 170 keV. As discussed in [11], hot electrons in this energy range can penetrate the ablator and directly heat the DT fuel. Even though there is considerably more preheat energy in the higher temperature (T2) component, it occurs too late to have a significant effect on the 4$^{th}$ shock timing that is observed by VISAR. Inclusion of the T2 component in the simulations has little effect on the observed time of the 4$^{th}$ shock of Figure 4(b).

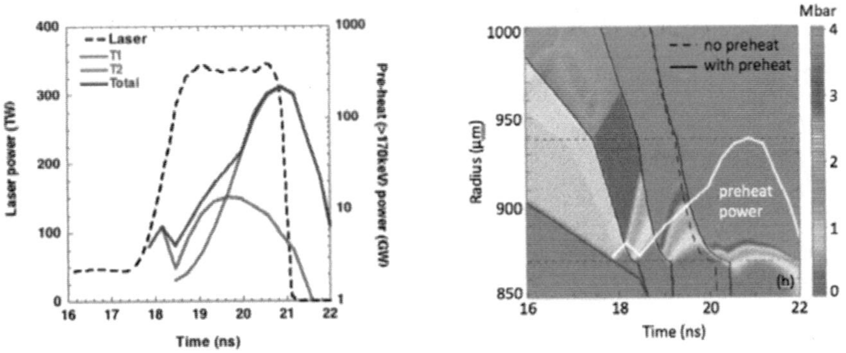


**Figure 5.** (a) Comparison of measured laser power history for N130517 (black dashed) vs. electron preheat (>170 keV) power. (b) Simulated pressure contours (radius vs. time) with shock trajectories super-posed. Solid black line shows shock trajectories with preheat. Dashed line shows shock trajectories without preheat. White curve is the total (T1 + T2) electron preheat power history.

Inclusion of the lower-temperature (T1) component, however, does have an effect. The supra-thermal electron preheat (both T1 and T2) is included in the HYDRA simulations using the non-local electron transport option, which is an extension of the method of [17]. Figure 5(b) shows the effect of this added preheat on the shock trajectories. The solid (dashed) black lines show the simulated shock trajectories with (without) preheat. The preheat power history is over-plotted with the white curve. The first three shocks are unchanged, but the 4$^{th}$ shock is delayed by approximately 300 ps, which is about half the discrepancy between measurement and simulation observed in Figure 4(b).

Figure 6(a) plots the pressure at critical density from three simulations: one with no electron preheat (black), one with electron preheat from FFLEX measurements (red), and one with twice the FFLEX measured power (blue). The predicted timing of the first three shocks is unchanged, but the timing of the 4$^{th}$ shock is delayed by 300 and 600 ps for the 1X and 2X preheat cases, respectively. This suggests that we need a slightly higher level of electron preheat than that measured by FFLEX to explain the VISAR observations. This could result, for example, from an anisotropic distribution of the electron flux. Note also in Figure 6 that the originally predicted 5$^{th}$ shock feature (black) at ~21.3 ns is completely absent in the simulations with preheat. This agrees with the observations of Figure 4(b), which show only one additional shock feature beyond the 3$^{rd}$ shock, not two.

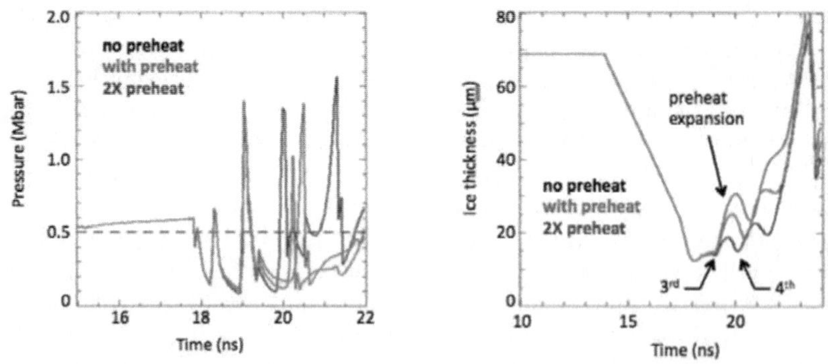


**Figure 6.** (a) Comparison of simulated pressure at critical density with no electron preheat (black) and with 1X and 2X electron preheat (red and blue, respectively). (b) Temporal evolution of the simulated thickness of the DT ice layer with no electron preheat (black) vs. with 1X and 2X electron preheat (red and blue, respectively).

The delay of the 4th shock timing shown in Figure 6(a) can be further understood by looking at the preheat expansion of the DT ice layer as shown in Figure 6(b). The DT ice layer was originally 68 $\mu$m thick. The first three shocks compress the layer by nearly a factor of 6 to ~ 12 $\mu$m. With electron preheat added, the layer expands by an additional 7-14 $\mu$m during the decompression that occurs between the 3rd and 4th shocks. The 4th shock then re-compresses and exits the ice layer at 20.1 ns (no preheat, black) vs. 20.4 and 20.7 ns (with 1X or 2X preheat), respectively. This preheat induced expansion of the ice layer at 4th shock breakout is a significant fraction (~50%) of the original 3-shock compressed layer thickness and continues to affect the implosion beyond the timing of the 4th shock. Further details and discussion of the implications of this preheat induced expansion on the fuel shell areal density (ρR) and the pressure and density in the hot-spot center are given in [18].

## 5. Acknowledgements
This work was performed under the auspices of the U.S. Department of Energy by Lawrence Livermore National Laboratory under Contract DE-AC52-07NA27344.

## 6. References
[1] Lindl J D *et al* 2004 *Phys. Plasmas* **11** 339
[2] Munro D H *et al* 2001 *Phys. Plasmas* **8** 2245
[3] Barker L M and Hollenbach R E 1972 *J. Appl. Phys.* **43** 4669
[4] Celliers P M *et al* 2004 *Rev. Sci. Instrum.* **75** 4916
[5] Celliers P M *et al* 2000 *Phys. Rev. Lett.* **84** (24) 5564
[6] Boehly T R *et al* 2009 *Phys. Plasmas* **16** 056302
[7] Boehly T R *et al* 2011 *Phys. Rev. Lett.* **106** 195005
[8] Robey H F *et al* 2012 *Phys. Plasmas* **19** 042706
[9] Robey H F *et al* 2012 *Phys. Rev. Lett.* **108** 215004
[10] Robey H F *et al* 2013 *Phys. Rev. Lett.* **111** 065003
[11] Haan S W *et al* 2011 *Phys. Plasmas* **18** 051001
[12] Trainor R J *et al* 1979 *Phys. Rev. Lett.* **42** 1154
[13] Miller J E *et al* 2007 *Rev. Sci. Instrum.* **78** 034903
[14] Marinak M M *et al* 2001 *Phys. Plasmas* **8** 2275
[15] Dewald E L *et al* 2004 *Rev. Sci. Instrum.* **75** 3759
[16] Dewald E L *et al* 2010 *Rev. Sci. Instrum.* **81** 10D938
[17] Schurtz G P, Nicolai P D, and Busquet M 2000 *Phys. Plasmas* **7** 4238
[18] Robey H F *et al* 2013 submitted to *Phys. Plasmas*

Journal of Physics: Conference Series **688** (2016) 012093 doi:10.1088/1742-6596/688/1/012093

# XFEL resonant photo-pumping of dense plasmas and dynamic evolution of autoionizing core hole states

**F B Rosmej[1,2], A Moinard[1,2], O Renner[3], E Galtier[4*], J J Lee[4], B Nagler[4], P A Heimann[4], W Schlotter[4], J J Turner[4], R W Lee[4], M Makita[5], D Riley[5], J Seely[6]**

[1] *Sorbonne Universités, Pierre and Marie Curie, UMR7605, LULI, case 128, 4 place Jussieu, F-75252 Paris Cedex 05, France*

[2] *LULI, Ecole Polytechnique, CEA, CNRS, Physique Atomique dans les Plasmas Denses, Palaiseau, France*

[3] *Institute of Physics, Academy of Sciences CR, Prague, Czech Republic*

[4] *SLAC Accelerator National Laboratory, California, USA*

[5] *Queen's University of Belfast, Centre for Plasma Physics, UK*

[6] *Naval Research Laboratory, Washington, DC, USA*

* part of the work was done being a post-doc at LULI-PAPD

E-mail: frank.rosmej@upmc.fr

**Abstract.** Similarly to the case of LIF (Laser-Induced Fluorescence), an equally revolutionary impact to science is expected from resonant X-ray photo-pumping. It will particularly contribute to a progress in high energy density science: pumped core hole states create X-ray transitions that can escape dense matter on a 10 fs-time scale without essential photo-absorption, thus providing a unique possibility to study matter under extreme conditions. In the first proof of principle experiment at the X-ray Free Electron Laser LCLS at SCLAC [Seely, J., Rosmej, F.B., Shepherd, R., Riley, D., Lee, R.W. *Proposal to Perform the 1st High Energy Density Plasma Spectroscopic Pump/Probe Experiment*", approved LCLS proposal L332 (2010)] we have successfully pumped inner-shell X-ray transitions in dense plasmas. The plasma was generated with a YAG laser irradiating solid Al and Mg targets attached to a rotating cylinder. In parallel to the optical laser beam, the XFEL was focused into the plasma plume at different delay times and pump energies. Pumped X-ray transitions have been observed with a spherically bent crystal spectrometer coupled to a Princeton CCD. By using this experimental configuration, we have simultaneously achieved extremely high spectral ($\lambda/\delta\lambda \approx 5000$) and spatial resolution ($\delta x \approx 70$ μm) while maintaining high luminosity and a large spectral range covered (6.90 - 8.35 Å). By precisely measuring the variations in spectra emitted from plasma under action of XFEL radiation, we have successfully demonstrated transient X-ray pumping in a dense plasma.

## 1. Introduction

Laser Induced Fluorescence "LIF" [see, e.g. 1] is a well-known method of laser spectroscopy in the optical wavelength range that has received worldwide attention, e.g., in physics, biology, medicine, and technology. Its physical principle is based on the stimulated photo-absorption of an electric dipole transition by a laser photon that matches the energy difference of the upper and lower atomic states (i.e., fulfills conditions of the resonant photo-pumping). The pumped upper states can decay via spontaneous radiative emission, the so-called fluorescence radiation. Here, we report on the laser-induced fluorescence in the X-ray spectral range that could be realized only with the emergence of the $4^{th}$ generation of light sources, X-ray Free Electron Lasers. LCLS at SLAC in United States [2] has been the first installation available for the scientific community. Although the principle of the X-ray LIF can also be approximated by an effective two level system, the X-ray spectral range shows several important differences. First, radiative transition probabilities scale as $Z^4$ (where Z is the effective charge of the ion) and transition energy as $Z^2$, i.e., the laser intensity requested to efficiently pump X-ray transitions scales as $I \propto Z^6$ [3]. Second, in most of the cases X-ray pumping creates core hole states that are autoionizing. To induce effective changes in X-ray energy levels, X-ray pumping must therefore be realized in a time interval corresponding to the characteristic time scale of autoionization. As autoionization rates are very large (characteristic time scales are of the order of 1-100 fs), efficient pumping in the X-ray range is a great challenge: extremely high X-ray photon intensities are requested on a 10 fs time scale to "beat the Auger clock". Moreover, X-ray fluorescence may be seriously reduced by the non-favorable branching ratio $A/(A+\Gamma)$ where A is the spontaneous X-ray transition probability and $\Gamma$ the autoionization rate. We note that these constraints are also strongly related to the production of "hollow" ions in intense radiation fields [3-5], a subject that became recently also of interest in the high-energy laser community [6].

## 2. Experimental setup

Figure 1 (left) shows the scheme of the experimental setup. A dense plasma was created by a YAG laser ($\lambda$=800 nm, E=24 mJ, $\tau$=120 ps, $I \approx 4 \cdot 10^{12}$ W/cm$^2$) irradiating solid aluminium and magnesium

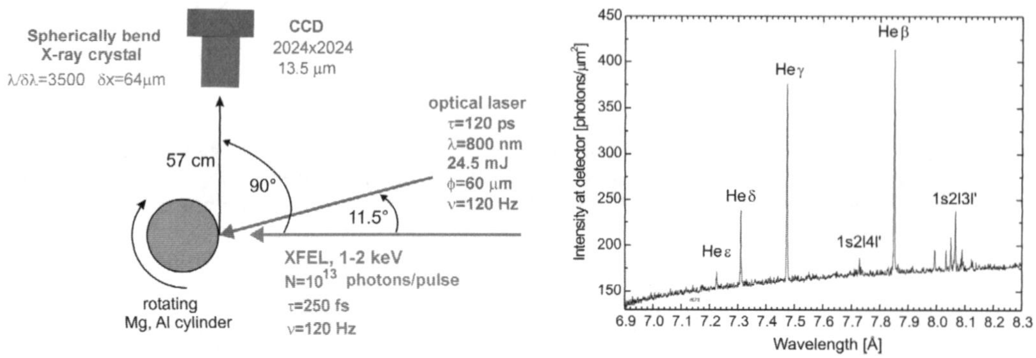


**Figure 1:** Schematic of the experimental setup (left) and emission spectrum of Mg (right)

that was mounted on a rotating cylinder (to provide fresh target surfaces for each shot at repetition frequency of 120 Hz). At 11.5° relative to the optical laser beam, the XFEL (about $10^{13}$ photons/shot, pulse duration 250 fs) was focused to a diameter of 20 μm in the dense plasma plume at different delay times and X-ray pump energies. Pumped X-ray transitions have been observed with a Bragg crystal coupled to a Princeton CCD (2048x2048 pixels). The spherically bent crystal of mica (004) set at the distance of 57 cm from the plasma source provided a magnification factor of 0.21 and a high spectral resolution at the level of $\lambda/\delta\lambda \approx 3500$ while maintaining a large spectral range (6.90 - 8.3 Å). The spatial resolution in a direction perpendicular to the expanding plasma (about 64 μm) was essentially determined by the CCD pixel size of 13.5 μm . The crystal was protected with a 2 μm thick polycarbonate ($C_{16}H_{14}O_3$) foil (effectively resulting in 4 μm thick foil due to transmission of incoming and outgoing rays), the CCD was protected with a 1 μm polypropylene ($C_3H_6$) coated with 0.1 μm aluminum from each side to supress the visible light. Spectra have been corrected for crystal reflectivity, CCD response and filter transmission.

The right part of Fig. 1 shows the Mg spectrum obtained when irradiating the rotating Mg cylinder by optical laser only, i.e., without XFEL interaction. The spectrum shows the He-like resonance line series ($He_\beta$ = 1s3p $^1P_1$ - 1s$^2$ $^1S_0$, $He_\gamma$ = 1s4p $^1P_1$ - 1s$^2$ $^1S_0$, $He_\delta$ = 1s4p $^1P_1$ - 1s$^2$ $^1S_0$, $He_\varepsilon$ = 1s5p $^1P_1$ - 1s$^2$ $^1S_0$,) and associated dielectronic statellite emission, $He_\beta$ - satellites 1s2l3l' - 1s$^2$2l'' and $He_\gamma$ - satellites 1s2l4l' - 1s$^2$2l''. The spectrum demonstrates excellent quality and resolution of the implemented X-ray diagnostics: a) high signal to noise ratio, b) large spectral coverage 6.9-8.3 Å, c) extremely high spectral resolution - note that even single LSJ-split configurations in the $He_\beta$-satellites 1s2l3l'-1s$^2$2l'' are resolved (for further details on $He_\beta$ - satellites see [7,8]).

### 3. Resonant X-ray pumping

Figure 2 shows the X-ray spectral distributions observed at solid Al targets. The upper spectrum was obtained when irradiating the Al cylinder with the optical laser only. The strongest emission is the He-like resonance line $He_\alpha$ = W = 1s2p $^1P_1$ - 1s$^2$ $^1S_0$ and the intercombination line Y = 1s2p $^3P_1$ - 1s$^2$ $^1S_0$. Li-like satellite emission 1s2l2l'-1s$^2$2l'' (indicated by 1s2l2l':=$K^1L^2$ in Fig. 2) is rather weak. The spectral distributions indicated with He...C designate pumping of the $He_\alpha$ transition (pump photon energy set to 1597 eV), pumping of the Li-like configuration $K^1L^2$ (pumping $K^2L^1+h\nu_{XFEL}\rightarrow K^1L^2$ at

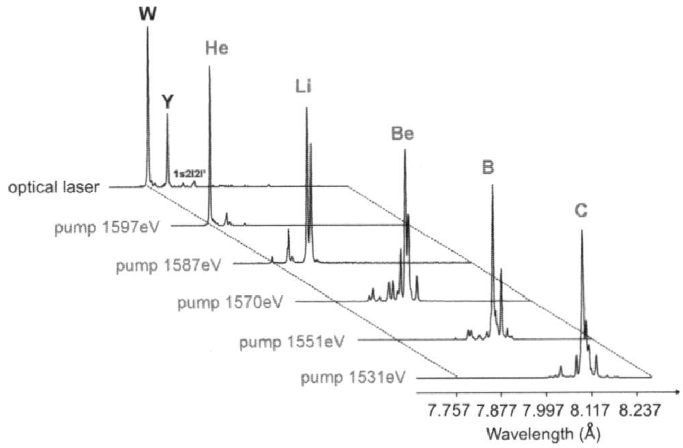


**Figure 2:** X-ray spectral distribution of Al pumped at different XFEL energies.

1587 eV), Be-like configuration $K^1L^3$ (pumping $K^2L^2+h\nu_{XFEL}\rightarrow K^1L^3$ at 1570 eV), B-like configuration $K^1L^4$ (pumping $K^2L^3+h\nu_{XFEL}\rightarrow K^1L^4$ at 1551 eV) and C-like configuration $K^1L^5$ (pumping $K^2L^4+h\nu_{XFEL}\rightarrow K^1L^5$ at 1531 eV).

The pumping at 1587 eV (curve labeled as "Li" in Fig. 2) is of particular interest. This photon energy corresponds to the transitions 1s$^2$2l + $h\nu_{XFEL}$ $\rightarrow$ 1s2l2l'. The pumped states are therefore the autoionizing Li-like levels of the type 1s2l2l'. Although the spontaneous electric dipole radiative transitions that originate from these states are of very low intensity for the case without pumping (see upper spectrum with transitions labelled 1s2l2l' in Fig. 2), the X-ray pump creates an impressive intensity increase. It is important to emphasize that emission lines, being entirely absent in the dense plasma produced by the optical laser only (upper spectrum in Fig. 2, emission labelled "Be", "B" and "C"), have been pumped to intensities that exceed those of the resonance line W (usually the most intensive X-ray emission line) by a factor of about 100. This is an extremely important demonstration of X-ray pumping: ion charge state distributions of dense matter can be probed via X-ray transitions that have been inaccessible up to present due insufficient signal/noise, background level. This is particular important for high energy density science: one of the most challenging plasma regimes are strongly coupled plasmas where the coupling parameter is proportional to $n^{1/3}/T$ with "n" being the density and "T" the temperature. High coupling parameters request therefore high density and low temperatures, but low temperatures usually prohibit X-ray transitions.

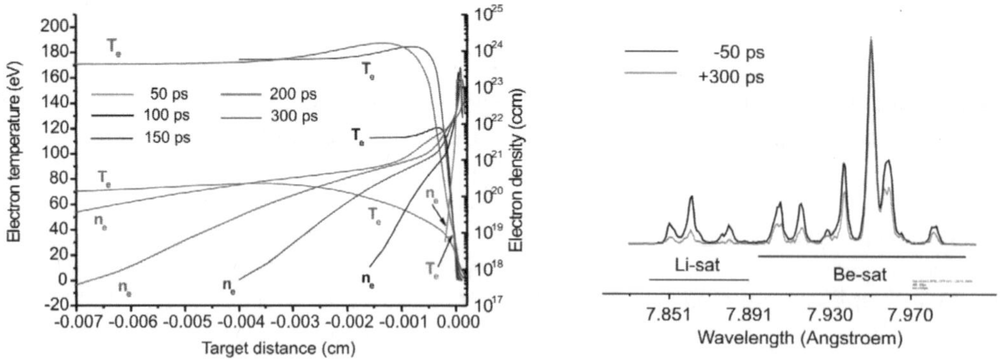


**Figure 3:** MULTI-fs simulations of the optical YAG laser (left) and transient photopumping of Be-like configurations $K^1L^3$ (right).

Our final concern has been the demonstration that X-ray pumping also provides a possibility to study the evolution of transient dense matter. Figure 3 shows its successful demonstration via a time dependent pumping of the Be-like configurations (right part of Fig. 3) which was realized by setting different delay times between the maximum of the optical laser pulse and the XFEL onset. As can be seen from the spectra, the spectral distribution changes in time, thus reflecting the dense plasma evolution. The left part of Fig. 3 {switch left- and right-hand part of Fig. 3?} shows the MULTI-fs simulations [9] of aluminum plasmas created by the optical YAG laser that provides insight in the transient evolution of temperature and density for different target distances.

## 4. Conclusion

In the first proof of principle experiment at the LCLS X-ray Free Electron Laser, we have successfully demonstrated transient X-ray pumping of the dense plasma. The application of curved Bragg crystal X-ray spectroscopic technique provided high-resolution spectral distributions that permitted interpretation that goes much beyond the hydrogenic approximations (via resolving even LSJ-split levels in complex autoionizing configurations that permits in turn to separate the different excitation channels). These results open new horizons in X-ray pumping/X-ray fluorescence XFEL physics.

## 5. References

[1] Demtröder W 2008 *Laser spectroscopy* Springer.
[2] https://portal.slac.stanford.edu/sites/lcls_public/Pages/Default.aspx
[3] Rosmej F B 2012 *Exotic states of high density matter driven by intense XUV/X-ray Free Electron Lasers"* in "Free Electron Laser", InTech2012, p. 187, ed. S. Varró, ISBN 978-953-51-0279-3.
[4] Rosmej F B and Lee R W 2007 *Europhysics Letters* **77** 24001.
[5] Rosmej F B, Lee R W, Schneider D H G 2007 *High Energy Density Physics* **3** 218.
[6] Colgan J, Abdallah J Jr, Faenov A Ya, Pikuz S A, Wagenaars E, Booth N, Culfa O, Dance RJ, Evans RG, Gray R J, Kaempfer T, Lancaster K L, McKenna P, Rossall A L, Skobelev I Yu, Schulze K S, Uschmann, I Zhidkov A G, Woolsey N C 2013 *Phys. Rev. Lett.* **110** 125001.
[7] Rosmej F B, Calisti A, Talin B, Stamm R, Hoffmann D H H, Süß W, Geißel M, Faenov A Ya, Pikuz T A 2001 *JQSRT* **71** 639.
[8] Skobelev I Yu, Faenov A Ya, Pikuz T A, Magunov A I, Flora F, Bollanti S, DiLazzaro P, Murra D, Reale A, Reale L, Tomassetti G, Ritucci A, Petrocelli G, Martellucci S, Lisi N, Rosmej F B 2002 *JETP* **95** 421.
[9] Ramis R, Eidmann K, Meyer-ter-Vehn J, Hüller S 2012 *Computer Phys. Comm.* **183** 637.

Journal of Physics: Conference Series **688** (2016) 012094 doi:10.1088/1742-6596/688/1/012094

# A bright neutron source driven by relativistic transparency of solids

**M. Roth[1,2], D. Jung[2], K. Falk[2], N. Guler[2], O. Deppert[1], M. Devlin[2], A. Favalli[2], J. Fernandez[2], D. C. Gautier[2], M. Geissel[3], R. Haight[2], C. E. Hamilton[2], B. M. Hegelich[4], R. P. Johnson[2], A. Kleinschmidt[1], F. Merrill[2], G. Schaumann[1], K. Schoenberg[2], M. Schollmeier[3], T. Shimada[2], T. Taddeucci[2], J. L. Tybo[2], F. Wagner[1], S. A. Wender[2], C. H. Wilde[2], G. A. Wurden[2]** *Technische*

[1]*Universität Darmstadt, D-64289 Darmstadt, Germany*

[2] *Los Alamos National Laboratory, Los Alamos, NM 87545, USA*

[3]*Sandia National Laboratories, Albuquerque, NM 87185, USA*

[4]*University of Texas, Austin, TX 78712-1068, USA*

*E-mail: markus.roth@physik.tu-darmstadt.de*

**Abstract**. Neutrons are a unique tool to alter and diagnose material properties and excite nuclear reactions with a large field of applications. It has been stated over the last years, that there is a growing need for intense, pulsed neutron sources, either fast or moderated neutrons for the scientific community. Accelerator based spallation sources provide unprecedented neutron fluxes, but could be complemented by novel sources with higher peak brightness that are more compact. Lasers offer the prospect of generating a very compact neutron source of high peak brightness that could be linked to other facilities more easily.

We present experimental results on the first short pulse laser driven neutron source powerful enough for applications in radiography. For the first time an acceleration mechanism (BOA) based on the concept of relativistic transparency has been used to generate neutrons. This mechanism not only provides much higher particle energies, but also accelerated the entire target volume, thereby circumventing the need for complicated target treatment and no longer limited to protons as an intense ion source. As a consequence we have demonstrated a new record in laser-neutron production, not only in numbers, but also in energy and directionality based on an intense deuteron beam. The beam contained, for the first time, neutrons with energies in excess of 100 MeV and showed pronounced directionality, which makes then extremely useful for a variety of applications.

The results also address a larger community as it paves the way to use short pulse lasers as a neutron source. They can open up neutron research to a broad academic community including material science, biology, medicine and high energy density physics as laser systems become more easily available to universities and therefore can complement large scale facilities like reactors or particle accelerators. We believe that this has the potential to increase the user community for neutron research largely.

## 1. Situation

In nuclear-, material-, biology[1] - and related sciences[23], neutrons have been used to modify and diagnose matter. Their unique properties make them a useful tool to explore areas, where other diagnostics fail. Moreover neutrons produced either by spallation or nuclear reactions (i.e. giant resonance excitation) can serve to mimic reactions usually found in nuclear reactors and therefore help gain a better understanding of aging and stability of present and future nuclear systems. More important as usually the samples to be tested are small in size a compact system is preferential as the total number of neutrons and thereby the required shielding can be minimized. So in general, accelerator based sources and reactors have been used to provide neutrons of high fluxes. Nevertheless, such a scheme is far away from flexible and there are many challenges to face. On the other hand, there is a growing need for flexible, high brightness sources that can be utilized at smaller laboratories[4], which only could be provided by a laser-accelerated ultra-fast particle source as driver.

As one application in mind, neutron radiography offers complementary information to x-ray diagnostics, especially because of the largely different cross sections for different Z- materials. It therefore can serve as a novel diagnostic, for example in high energy density physics, active interrogation of sensitive materials and laboratory planetary sciences. Ultra intense lasers have been developed to a point, where laser driven particle sources, especially pulsed, intense proton sources, could potentially serve as a laser driven, compact neutron source.
The general concept of a laser driven neutron source is, simplified, to exchange the ten to hundreds of meters long conventional accelerator by approximately 600 μm of laser acceleration and to run an intense burst of ions, e.g. deuterons (pitcher) into a converter material (catcher) to release a directed pulse of energetic neutrons with a pulse duration of less than a nanosecond.

## 2. Laser ion generation

Laser particle acceleration is an important part of today's high energy and laser physics research. This young field of research has been opened by the advent of the *chirped pulse amplification* (CPA)[5], which allows laser peak powers up to the petawatt level[6]. Todays laser facilities are able to investigate relativistic laser-matter interaction, or laser-plasma interaction, which is the basis for efficient particle acceleration with lasers. Laser-ion acceleration is primarily driven by relativistic electrons, generated by the interaction of an ultra-intense laser beam with a solid target or a gas jet. The laser peak intensity thereby has to exceed $10^{18}$ $W/cm^2$ to immediately accelerate electrons close to the speed of light. Today's lasers are able to reach maximum intensities of more than $10^{21} W/cm^2$ [7]. The dominant ion acceleration starts off the rear, non-irradiated surface by a rapid charge separation. This mechanism is known as *Target Normal Sheath Acceleration* [8] and has been investigated for about a decade. For proton energies in the few MeV range the cross sections for neutron evaporation by (p,xn) reactions is quite large and conversion efficiencies of $10^{-4}$ have been measured in first experiments. Ions accelerated by the TNSA mechanism have an exponential energy distribution with an upper cut-off energy determined by the maximum of the accelerating electric potential. So far protons have been accelerated up to roughly 60MeV [9] and heavier ions (Carbon) up to 5 MeV/u [10], while it hast to be noted that since the discovery of these ion beams 1999, no significant progress has been made in terms of maximum particle energy.

Recent advances at TRIDENT show a very efficient conversion efficiency using the Break Out Afterburner (BOA)[11] mechanism and show peak energies approaching 200 MeV. The BOA mechanism only requires $10^{20} W/cm^2$ and linear polarization. Therefore, this mechanism can be used using present laser facilities having pulse length of 500 fs and energies of about 100 J on target. The main difference between TNSA and BOA is the de-coupling of the ion acceleration from the driving laser field due to the thickness of the target. In contrast, for the BOA mechanism the electrons that are accelerating the ions are still interacting with the laser field.

With compact, laser ion acceleration, energies the regime of spallation will become accessible in the near future and consequently the conversion efficiency will rise by several orders of magnitude. Regardless of the production mechanism, the neutron pulse length resembles the initial proton beam pulse length and can be in the picosecond regime and due to the focusability of laser driven proton beams the source size can be made very compact. Apart from the target area, none of the components require shielding and the sample to be exposed to the neutron beam can be mounted very close to the source, gaining a large factor simply by the solid angle.

## 3. Experiments

We have been able to perform the first laser driven neutron source based on the BOA regime, powerful enough for a first test on fast neutron radiography. These results have been recently published in Physical Review Letters[12] and highlighted in numerous journals (like e.g. NATURE[13], La Recherche, Physicsworld.com etc). We could demonstrate the production of beamed neutrons ranging from a few MeV all up to 150 MeV dependent on the laser parameter and could provide a first radiograph of dense objects[14]. We used an F/1.5 off-axis parabolic mirror to typically focus 80 J of 1.053 $\mu$m laser light in a 600 fs pulse. The on-target focus has been measured to be around 3 $\mu$m in radius ($1/e^2$-condition, containing >60% of the laser energy) with a peak intensity of 1 x $10^{21}$ W/cm$^2$. The laser pulse duration and beam parameters were carefully recorded during the whole campaign. Thin, freestanding, plastic (CH2) and deuterized plastic (CD2) foils with thicknesses from 200 nm to 3.2 $\mu$m were used to generate proton and deuteron beams. The high temporal contrast of the Trident laser with $10^{-7}$ at -4 ps (ratio of preceding laser irradiation compared to the peak intensity)[15] enables interaction of the laser with a highly overdense target even for nm-scaled targets and hence very efficient acceleration of deuterons through the BOA mechanism. The proton and deutron beams were deposited in a Beryllium converter, providing a high cross section for neutron production and at the same time minimizing generation of unwanted high-energy bremsstrahlungs photons.

We also have implemented and calibrated numerous diagnostic systems to detect the ion and neutron beams in these laser experiments. We have a calibration curve for BTI Bubble detectors[16] extended to higher neutron energies, shielded, used and calibrated neutron time of flight detectors (nToF), successfully implemented nuclear activation techniques and fielded a suite of additional neutron diagnostics ($3He$ detectors, scintillating fibers, $^{10}B$ detectors) to short pulse laser experiments. Thus, in recent experiments we demonstrated the full characterization of the laser driven neutron field.

## 4. Results

In the main part of the experiment, we generated proton beams in excess of 135 MeV and deuteron beams above 180 MeV total particle energy. These ion beams were dumped in a beryllium converter. The converter was placed at distances of 5 mm up to 5 cm behind the driver target to control the size of the neutron source. The lateral dimensions of the converter were adjusted accordingly to capture the full ion beam, which had an average cone half-angle of about 30° (see Ref. 14); the length of the converter was typically 5cm, sufficiently long to stop protons and deuterons with energies of 135 MeV and 180 MeV, respectively.

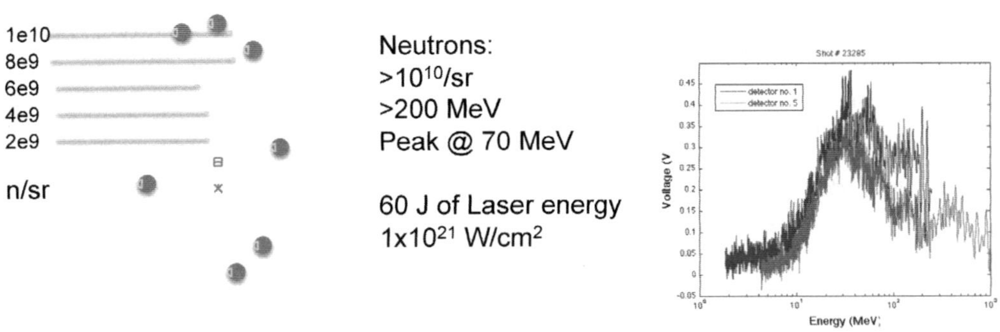


Figure 1: Neutron results: A clear, forward peaked energy distribution of the neutrons is visible, when using sub-micron CD targets and the BOA mechanism. The neutron energies extend to above 100 MeV peaking at 50-70 MeV.

The neutron beam distribution measured with the bubble detectors showed two components: an isotropic component from (p,n) and (d,n)-reactions and a peaked forward component from deuteron breakup-reactions (see Fig. 1). The neutron yield in the forward direction was one order of magnitude higher with up to 1.2 x $10^{10}$ n/sr. All other directions measured much lower neutron fluxes of up to $4 \times 10^9$ n/sr. The nTOF-detectors detected the same forward directed neutron distribution as the bubble detectors. The ones placed orthogonal to the laser beam and parallel to its counter propagation direction still measured the peak neutron flux at 8 MeV with maximum energies of up to 25 MeV. However, in the direction of the driving ion beam, the neutron flux peaked at energies of up to 50 MeV with maximum energies of 100 MeV to 150 MeV. A typical spectrum is shown in Fig. 1 (blue and green lines from two detectors behind each other). The high maximum neutron energies imply that other neutron generating mechanisms are present such as pre-compound reactions[17].

## 5. Summary

We experimentally demonstrated a high power, laser-driven neutron source. The driving ion beam consisting of either only protons from CH targets or protons and deuterons from CD targets with maximum energies of 135 MeV and 178 MeV, respectively. The beam was dumped in a Be-converter for the generation of neutrons via (p,n), (p,2n), and (d,n) reactions and via deuteron break-up reactions. The neutron beam has been characterized with a set of diagnostics consisting of a dozen bubble detectors, 3 nTOF detectors, and a neutron imager. With CD targets of 600 nm thickness, where efficient acceleration of deuterons takes place in the BOA regime, the neutron distribution has a strong, forward directed, component with $1.2 \times 10^{10}$ n/sr and maximum neutron energies of up to 150 MeV. This constitutes two orders of magnitude higher flux and 6 to 8 times higher maximum neutron energies than previously reported by other groups. The ability to generate dense MeV-neutron beams with lasers also opens the field of neutron and neutron-assisted research to a much broader community, including small-scale laboratories as well as universities and also paves the way to new compact and cost-efficient neutron-based applications.

## 6. Acknowledgements

We gratefully acknowledge the expert support of the Trident laser team and LANSCE in conducting the experiments. This work was supported by LANL Laboratory Directed Research and Development (LDRD), the DOE Office of Fusion Energy Sciences (OFES), BMBF 06Da9044I, HIC4FAIR, the University of Darmstadt, DFG, and Sandia. Sandia is a multiprogram laboratory operated by Sandia Corporation, a Lockheed Martin Company, for the National Nuclear Security Administration under Contract No. DE-AC04-94AL85000. Work was performed under the auspices of the U.S. Department of Energy by the Los Alamos National Security, LLC, Los Alamos National Laboratory.

## References

[1] S. Krueger, et al., Langmuir 17, 511 (2001).
[2] N. Kardjilov, etal., J. Neutron Res. 14, 29 (2006).
[3] A. Buffler, Radiat. Phys. Chem. 71, 853 (2004).
[4] I. Anderson, et al., Report of a technical meeting held in Vienna, 18–21 May 2004 (IAEA,2005).
[5] D. Strickland and G. Mourou, , Optics Communications 56, 219 (1985)
[6] M.D. Perry et al., Opt. Letters 24, 160 (1999)
[7] S.-W. Bahk, et al., Opt. Lett. 29, 24 (2004)
[8] S. C. Wilks, et al., Phys. Plasmas 8, 542 (2001)
[9] R. A. Snavely, et al., Phys. Rev. Lett. 85, 2945 (2000)
[10] B. M. Hegelich, et al., Phys. Plasmas 12, 056314 (2005)
[11] L.Yin, B.J. Albright, B.M. Hegelich, and J.C. Fernandez, Laser and Particle Beams *24*, 291-298 (2006)
[12] M. Roth, et al., Phys. Rev. Lett., 110, p. 044802 (2013)
[13] NATURE research Highlights, *A Tabletop neutron source*, Nature 494, 9 2013
[14] D. Jung, et al., Physics of Plasmas 20, 056706 (2013)
[15] S. H. Batha, et al., Rev. Sci. Instrum. 79, 10F305 (2008).
[16] Bubble Technology Industries Inc. *calibration of BTI BD-PND neutron detectors*. Information note (1996)
[17] A. De, S. Ray, and S. K. Ghosh, Phys. Rev. C 37, 2441 (1988).

# Direct drive targets for the megajoule facility UFL-2M

**V B Rozanov[1,3], S Yu Gus'kov[1], G A Vergunova[1], N N Demchenko[1], R V Stepanov[1], I Ya Doskoch[1], R A Yakhin[1] and N V Zmitrenko[2]**
[1] P.N. Lebedev Physical Institute of RAS,
53 Leninskij Prospekt, Moscow 119991, Russia
[2] M.V. Keldysh Institute of Applied Mathematics of RAS,
4 Miusskaya Sq., Moscow 125047, Russia

E-mail: rozanov@sci.lebedev.ru

**Abstract.** Development of direct drive target schemes for the megajoule facility is a topical problem of up-to-date inertial fusion physics. The choice of possible schemes and solutions depends essentially on the irradiation conditions. The installations both running (NIF) and under construction (LMJ) are destined to the 3$\omega$ irradiation in PDD (polar direct drive) configuration. The UFL-2M installation that is under construction is based on 2$\omega$ irradiation and a symmetrical scheme of direct drive target irradiation. Under these conditions possible schemes for direct drive targets demonstrating the ignition and the achievement of gain $G$=10÷20 are considered in this report. At the same time, the possibilities are analyzed for the target compression and ignition with a reliability reserve at the conditions that can deviate from the standard ones, and if our understanding of the physics of the processes is not completely adequate to the physics of the real processes.

## 1. Introduction

In 2012 a decision has been made to build up a megajoule laser facility UFL-2M at the Russian Federal Nuclear Centre, All-Russian Research Institute of Experimental Physics, VNIIEF [1]. One of the problems connected with the creation and further application of the facility concerns the development of direct drive targets, since the design of the targets strongly depends on the parameters of the facility. The parameters of the facility specified up to the present moment are listed in table 1.

**Table 1.** Basic characteristics of the UFL-2M facility.

| Main dimensions | Parameters of the irradiation system | Irradiation symmetry |
|---|---|---|
| Size – 322.5×67 $m^2$ | Number of laser beams – 192 | 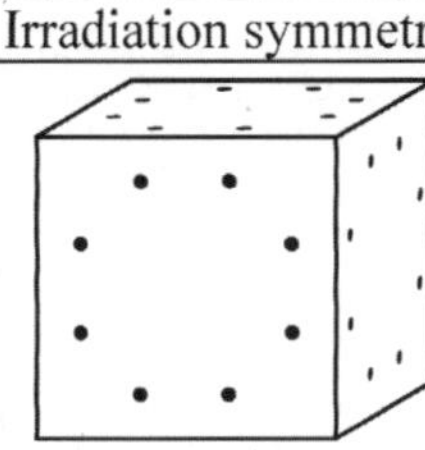 |
| Laser bay length – 130 m | Laser beam cross-section – 400×400 $mm^2$ | |
| Laser bay height – 17.6 m | Laser operating wavelength $\lambda_0$ – 1053 nm | |
| Interaction chamber – ∅10 m | Target irradiation wavelength $\lambda$ – 527 nm | |
| Chamber bay height – 34 m | Output laser energy on $\lambda_0$ – 4.6 MJ | |
| Clean rooms area – 16800 $m^2$ | Laser energy on the target – 2.8 MJ | |

An important difference in choosing the target design for UFL-2M, as compared to the available NIF and LMJ facilities, lies in the following: 2$\omega$ ($\lambda$=0.53 μm) is taken as the working (main) wavelength and the system and geometry of irradiation are close to spherically symmetrical (in

[3] To whom any correspondence should be addressed.

contrast to the NIF and LMJ, where the irradiation comes from the poles). The main feature of irradiation geometry in UFL-2M facility is also shown in table 1: the irradiation over the planes of the cube – each of the 6 planes contains 8 modules consisting of 4 beams. It is also of importance that for $\lambda$=0.53 μm one can expect within the target area higher energy of the laser pulses (about 2.8 MJ).

The problem of direct drive targets for megajoule pulses has been considered repeatedly, and, at first sight, seems to be easy. However, the experience gained from the National Ignition Campaign at the NIF laser shows that the problem of target design is a difficult one, and the choice of target parameters (size, mass, composition of the layers) needs serious approach. One of the solutions may be based on choosing the design, which would be low-sensitive to the irradiation and compression conditions, and our possible ignorance of some details of the processes taking place in the targets.

The present paper covers the results obtained from the analysis and 1D simulations of direct drive targets under the conditions of UFL-2M facility. The simulations have been made with the help of DIANA-code [2]. The choice of direct drive target scheme is based on the analysis of calculation results for direct drive targets under conditions of the NIF facility [3].

## 2. The choice of target parameters

Irradiation of the target by a smooth growing pulse from [3] (see figure 1) has been considered. It is assumed that 1.48 MJ is absorbed (for $2\omega$ and $3\omega$) at the critical surface, and later we'll define the laser pulse energy under which the absorbed energy is 1.48 MJ. In the table 2, the comparison of $2\omega$ and $3\omega$ irradiation regimes is made: target dynamics was simulated with account for the α-particle energy deposition and without it (i.e. a pure gasdynamic compression case). The results show that moderate decrease in $\rho R$ and $\eta$ for $2\omega$ as compared to $3\omega$ (by 30-40%) leads to a significant drop in $G$.

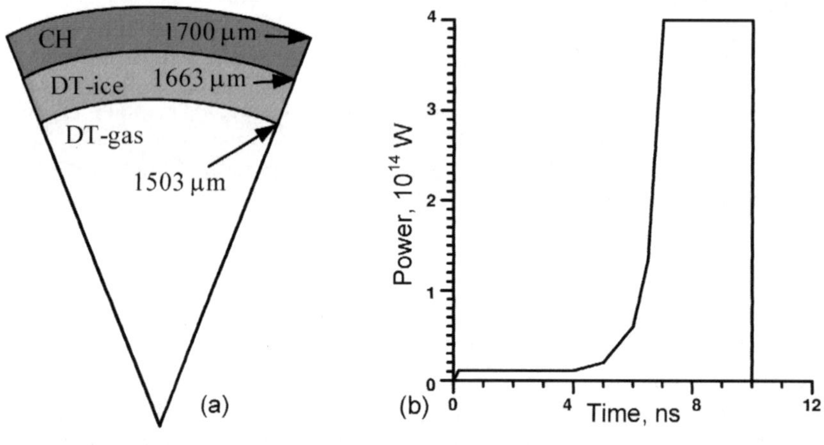


**Figure 1.** Baseline direct drive target design (a) and pulse (b).

**Table 2.** The comparison of $2\omega$ and $3\omega$ irradiation regimes. Here $G$ is gain, $\eta$ - hydrodynamic efficiency, $E_{DT}$ - energy of DT (gas + ice layers), and $W_\alpha \cdot E_{DT}$ product yields the energy of α-particles.

| α-particle energy deposition is switched on | | α-particle energy deposition is switched off | |
|---|---|---|---|
| $2\omega$ | $3\omega$ | $2\omega$ | $3\omega$ |
| $G = 2.73$ | $G = 38.7$ | $W_\alpha = 0.722$ | $W_\alpha = 1.59$ |
| $\rho R = 0.719$ g/cm$^2$ | $\rho R = 1.109$ g/cm$^2$ | $\rho R = 0.815$ g/cm$^2$ | $\rho R = 1.128$ g/cm$^2$ |
| $\eta = 7.39\%$ | $\eta = 9.54\%$ | $\eta = 7.39\%$ | $\eta = 9.54\%$ |
| | | $E_{DT} = 0.087$ MJ | $E_{DT} = 0.132$ MJ |

Additionally, the calculations have been performed for various descriptions of the EOS of CH and DT contained in targets: Thomas-Fermi (TF) approximation, empiric cold curve + equilibrium ionization, empiric cold curve + non-equilibrium ionization, empiric cold curve + TF approximation. For $2\omega$ the sensitivity range is $G$=1.28÷18.1, and $G$=27.7÷44.2 for $3\omega$. The target with higher critical density (shorter laser wavelength, $\lambda$=0.35 μm) is less sensitive to the lack of knowledge of EOS details.

The simulations have shown that the target given in figure 1a turns to be too heavy (CH + DT-ice mass is approximately 2670 μg) for the irradiation at the wavelength $\lambda$=0.53 μm. In the calculations with the target of the mass reduced to about 2150 μg (the proportions given in figure 1a being conserved) one observes a maximum energy yield $G$=18, $G$=10 for $m$=1850 μg, and $G$=13 for $m$=2400 μg. The illustration of the optimized target compression and burn dynamics is given in figure 2. The dimensions of this modified baseline target are shown in figure 3. It is seen in figure 2a that thermonuclear burn grows from 10.9 to 11.1 ns, i.e. it starts before the moment of maximum compression and affects the collapse dynamics, while the fuel remains relatively "cold" during implosion, as it is indicated by the behaviour of the adiabat (figure 2b).

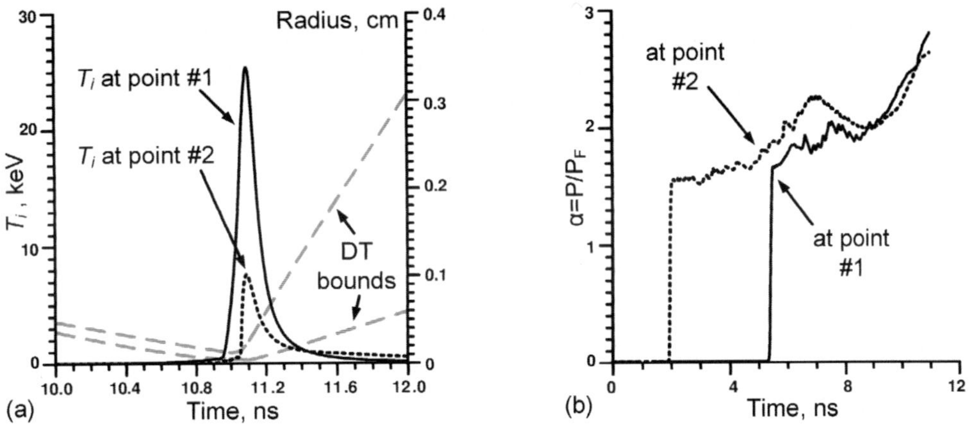


**Figure 2.** The behaviour of the modified baseline target: (a) - R-t diagrams of the DT layer bounds (dashed grey lines) and time dependence of the ion temperature $T_i$ (black lines); (b) – time dependence of the adiabat $\alpha$. The observation points are taken at mass coordinates of 20% (#1, solid black lines) and 90% (#2, dotted black lines) from the centre of the target.

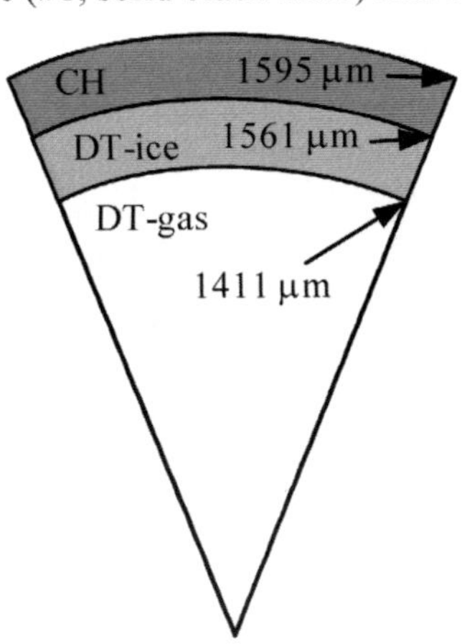


**Figure 3.** The sketch of the modified baseline target.

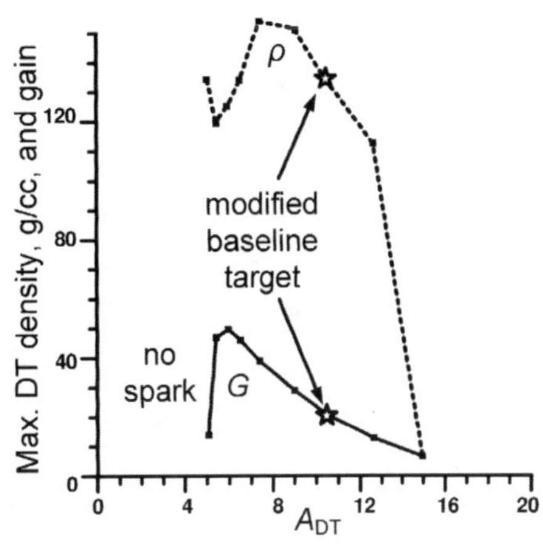


**Figure 4.** The dependence of gain $G$ (solid line) and maximum fuel density $\rho$ (dotted line) on the DT layer initial aspect ratio.

## 3. The sensitivity of modified baseline target to different variations of experimental conditions

As far as different models of EOS are concerned, the range of $G$ variations turned to be moderate, $G$~8÷21. The laser pulse variation being $E_L$=1.38÷1.58 MJ (pulse duration, $\tau_L$=10 ns) $G$~18÷22,

$\rho R$=0.85÷0.96 g/cm$^2$, hydrodynamic efficiency $\eta$=6.83÷6.4%. Under the change of pulse duration $\tau_L$=9÷11 ns ($E_L$=1.48 MJ) the thermonuclear yield changes insignificantly $G$~21÷18, $\rho R$=0.9 g/cm$^2$, $\eta$=7.2÷6%. The presence of an area (possibly areas) of the result ($G$) weak sensitivity within the space of the EOS, $E_L$ and $\tau_L$ parameters is an interesting fact. The possibilities of utilization of this fact are to be still evaluated. The calculations for different aspect ratios of DT layer (at the given target mass) have revealed more sharp dependences; they are shown in figure 4. For the aspect ratio $A_{DT}$=5.96 the prediction gives $G$=49.7, and $G$=13.9 for $A_{DT}$=5.06. This decrease is due to the decrease in the DT-gas temperature and the worsening of ignition conditions. Under large values $A_{DT}$>12 the compression and the DT layer density decrease. A possible conclusion here may be as follows: when choosing the target design one should avoid the variants, which make use of maximal values of thermonuclear gain.

## 4. Conclusions

- For UFL-2M conditions different variants of 1D optimization give $G$=4÷45.
- In our opinion special attention should be paid to the search for the "rough" target design that would be low-sensitive to probable deviations from "standard" condition and possible drawback in the description of physical processes in ICF target.

## Acknowledgments
This work was partially supported by the RFBR grant #11-02-01305-a.

## References

[1] Garanin S G, Bel'kov S A and Bondarenko S V 2012 Conception of UFL-2M laser facilitiy *39$^{th}$ Zvenigorod Int. Conf. on Plasma Physics and Controlled Fusion (Zvenigorod, Russia, 6–10 February 2012)* R-12

Bel'kov S A, Garanin S G and Shagalkin Yu V 2013 The UFL-2M facility: first steps on the way to its construction *40$^{th}$ Zvenigorod Int. Conf. on Plasma Physics and Controlled Fusion (Zvenigorod, Russia, 11-15 February 2013)* R-10

[2] Zmitrenko N V, Karpov V Ya, Fadeev A P and Shpatakovskaya G V 1982 *Voprosy Atomnoy Nauki i Tekhniki. Seriya: Metodiki i Programmy Chislennogo Resheniya Zadach Matematicheskoy Fiziki* **2** 38 (in Russian)

[3] 2012 Cryogenic Deuterium and Deuterium–Tritium Direct-Drive Implosions on OMEGA *Laboratory for Laser Energetics Annual Report* (Rochester: University of Rochester) p 72

# Integrated simulations for ion beam assisted fast ignition

**H Sakagami[1], T Johzaki[2], A Sunahara[3] and H Nagatomo[4]**

[1] Fundamental Physics Simulation Research Division, National Institute for Fusion Science, 322-6 Oroshi-cho, Toki 509-5292, Japan
[2] Graduate school of Engineering, Hiroshima University, 1-3-2 Kagamiyama, Higashi-Hiroshima 739-8511, Japan
[3] Institute for Laser Technology, 2-6 Yamada-oka, Suita 565-0871, Japan
[4] Institute of Laser Engineering, Osaka University, 2-6 Yamada-oka, Suita 565-0871, Japan

sakagami.hitoshi@nifs.ac.jp

**Abstract**. Although the energy conversion efficiency from the heating laser to fast electrons is high, the coupling efficiency from fast electrons to the core is estimated to be very low due to large divergence angle of fast electrons in fast ignition experiments at ILE, Osaka University. To mitigate this problem, a plastic thin film or low-density foam, which can generate not only proton ($H^+$) but also carbon ($C^{6+}$) beams, is combined with currently used cone-guided targets and additional core heating by ions is expected. According to integrated simulations, it is found that these ion beams can enhance the core heating by 20~60% and it shows a possibility of ion beam assisted fast ignition.

## 1. Introduction

A fuel target is imploded by long-pulse implosion lasers and its compressed core is heated by a short-pulse ultrahigh-intense laser in the fast ignition scheme. Incorporated fast ignition experiments have started at ILE, Osaka University to demonstrate that the compressed core could be heated up to 5 keV using Au cone-guided targets under the FIREX project. First series of the incorporated experiments was performed in 2009, and only 30-fold enhancement in neutron yield, which was ~ 1/30 smaller than that in 2002 experiments, was achieved, and lower energy coupling from the heating laser to the imploded core was suspected.

2D PIC simulations indicate that even the coupling efficiency from the heating laser to fast electrons is generally high (>40 %) but the divergence angle of fast electrons is large (~90 degree), and it results in low coupling efficiency from fast electrons to the core [1]. The Weibel instability, which is induced by a fast electron flow and its counter stream of background electrons, generates quasi-static magnetic fields, and they grow up to more than hundred Mega gauss, large enough to scatter several MeV electrons within 1 μm. Furthermore, filamentations of the heating laser induce deformation of the plasma surface, and it could enhance the fast electron divergence. To mitigate this critical issue, advanced target designs, in which external-compressed and/or self-generated magnetic fields are expected to guide fast electrons to the core, are proposed.

On the other hand, high-energetic well-collimated proton beams were observed in many laser-thin foil experiments [2,3], and they can be used to additionally heat the core in cooperation with fast electrons. As protons are supposed to come from contaminations on the target surface, it's unclear and uncontrollable. Thus two types of plastic (CH) targets to generate not only proton ($H^+$) beam but also carbon ($C^{6+}$) beam are introduced. One type is a thin film target where ions are accelerated by the sheath field at the rear surface, and the other is a low-density foam target where ions are accelerated by the Ponderomotive force at the front surface [4]. Ion beam characteristics are investigated by 2D PIC simulations and core heating properties are evaluated by integrated simulations [5].

## 2. Target design

We put the CH as ion beam generator close to the cone tip in order to shorten the distance between the generation point and the imploded core. This design can reduce core-arrival time lags due to different ion energies, and we can accept ion beams with wide energy range. It can also reduce time lags between fast electrons and ions, and it is easy to introduce into FIREX experiments combining with currently used cone-guided targets. As the heating laser is designed to have 10 ps pulse length and the cone tip is placed 50 μm away from the implosion center, we assume that ion beams should propagate 60 μm and 5 ps time lag are accepted. So ions with the speed faster than 0.04 of the light speed can be accepted, and it means that $H^+$ with the energy higher than 0.7 MeV and $C^{6+}$ higher than 9 MeV can be used to heat the core.

There are two well-known ion acceleration mechanisms. One is the sheath field acceleration on the rear surface of the target and the other is the Ponderomotive force acceleration on the front surface of the target. To adapt the sheath field acceleration, we introduce the thin film target as shown in figure 1 (a). The Au cone plasma (Z=30, A=197, $20n_{cr}$, 60 degree open angle, 10 μm tip width) is introduced and the CH thin film (Z=6, A=12, $17.14n_{cr}$ and Z=1, A=1, $2.86n_{cr}$, 3 μm thickness) is placed 5 μm away from the cone tip surface. The low-density $SiO_2$ foam plasma (Z=10, A=20, $0.5n_{cr}$) is also placed in front of the CH thin film to enhance fast electron generation [5]. When the sheath field is induced, dense background electrons in the Au plasma are pulled into the CH thin film and it results in the sheath field being weakened. Therefore 0.5 μm gaps between the Au and the CH thin film are introduced to prevent the electron flow. For the Ponderomotive force acceleration, we introduce the low-density foam target as shown in figure 1 (b). The same Au cone plasma is also introduced and the low-density CH foam (Z=6, A=12, $8.57n_{cr}$ and Z=1, A=1, $1.43n_{cr}$, 8 μm thickness) is placed in front of the cone tip surface.

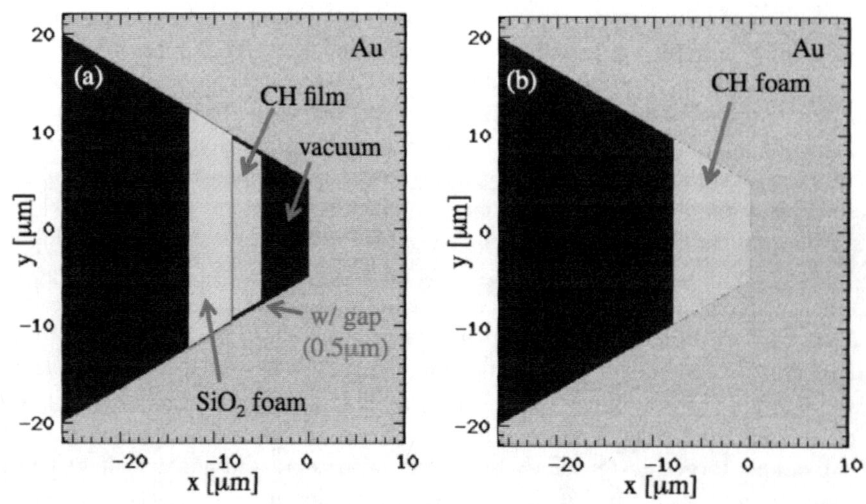


**Figure 1.** Targets for ion beam assisted fast ignition with (a) the sheath field acceleration and (b) the Ponderomotive force acceleration.

### 3. Simulation results

The heating laser is set to $\lambda_L$=1.06 μm, $I_L=10^{20}$ W/cm$^2$, $\tau_{rise}=\tau_{fall}$=50 fs, $\tau_{flat}$=400 fs and $\phi_{FWHM}$=10 μm Gaussian. In the case of the low-density foam target, super-Gaussian laser ($I_L=5.333\times10^{19}$ W/cm$^2$, $\phi_{FWHM}$=20 μm, $\alpha$=100), whose total energy is same as the Gaussian laser, is also irradiated to aim small divergence of ion beams. Fast electrons and ions are observed at 1 μm behind of the cone tip surface with 30 (±15) μm width. To ignore a circulation of fast electrons, we introduce an artificial cooling region (1 μm width), in which fast electrons are gradually cooled down to the initial temperature, behind the observation region, top and bottom regions of the Au plasma.

Time evolutions of fast ion beam intensities and time averaged fast ion energy spectra with and without the gap are shown in figure 2 (a) and (b), respectively. As sheath fields are not weakened due to existence of the gap because the thin CH film is isolated from background Au electrons, ions are much more accelerated and the ion beam intensity increase in the case with the gap.

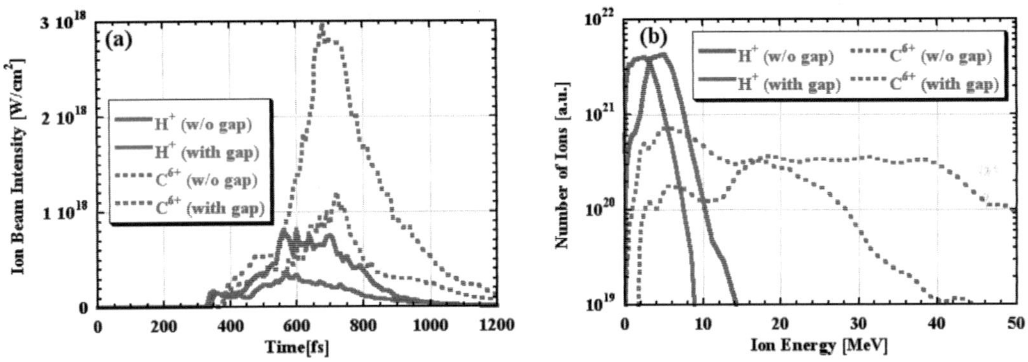


**Figure 2.** Fast ion ($H^+$:solid, $C^{6+}$:dash) properties. (a) time evolutions of fast ion beam intensities and (b) time averaged fast ion energy spectra with (red) and without (blue) the gap.

Time evolutions of fast ion beam intensities and time averaged fast ion energy spectra in the case of Gaussian and super-Gaussian lasers are shown in figure 3 (a) and (b), respectively. It is confirmed that $C^{6+}$ beam divergence is improved with the super-Gaussian laser. As the total laser energy of both lasers are same, ion beam intensities and spectra are not so affected by the laser profile.

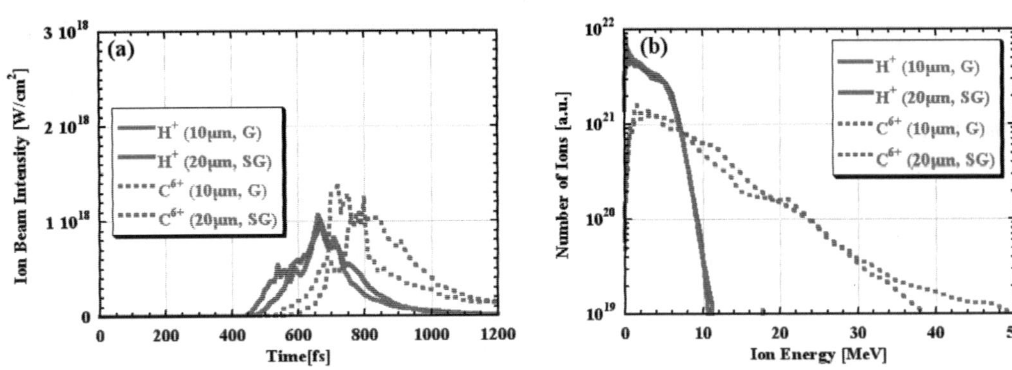


**Figure 3.** Fast ion ($H^+$:solid, $C^{6+}$:dash) properties. (a) time evolutions of fast ion beam intensities and (b) time averaged fast ion energy spectra in the case of Gaussian (blue) and super-Gaussian (red) lasers.

Journal of Physics: Conference Series **688** (2016) 012096 doi:10.1088/1742-6596/688/1/012096

## 4. Integrated simulations

Using the time-dependent profiles of fast electrons and ions, which are observed in $0 < y < 3$ μm in 2D PIC code, into 1D RFP-Hydro code, we carried out integrated simulations to evaluate the core heating properties [6], including 7 μm transport in the Au cone tip. Time evolutions of averaged core electron temperatures are shown in Figure 4 for all 4 cases. As electrons finish to heat the core but ions do not reach the core yet due to slow speed at 2 ps, electrons play a dominant role in core heating. After that, ions finally reach and heat the core instead of electrons. Thus temperature increment after 2 ps is dominantly caused by ions. In the case of the sheath field acceleration target with the gap, enhancement to temperature increment by ions is larger than that without the gap. On the other hand, the propagation of fast electrons is disturbed by the same sheath field and electron heating is reduced. So there is a tradeoff between ion and electron contributions, and the target without the gap results in better core heating under current conditions.

We summarize enhancement by ions for average core electron temperatures in table 1. Putting the CH thin film or low-density foam to the cone-guided target can successfully generate $H^+$ and $C^{6+}$ beams and enhance the core heating by 20~60%. But structured targets are not well optimized at this moment, and it just shows potential for additional heating. Core heating simulations were performed by 1D RFP-Hydro code, and 2D Fokker-Planck simulations are needed to clarify ion beam assisted fast ignition.

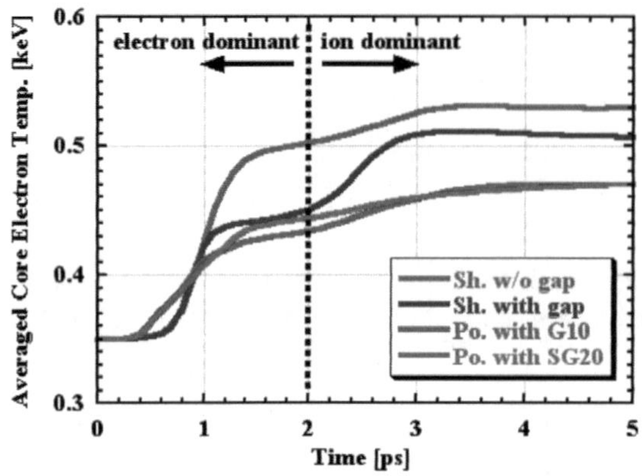


**Figure 4.** Time evolutions of averaged core electron temperatures for all 4 cases.

**Table 1.** Summary of enhancement by ions for average core electron temperatures.

| | | Sh. w/o gap | Sh. with gap | Po. with G10 | Po. with SG20 |
|---|---|---|---|---|---|
| Temperature increment keV | e only | 0.15 | 0.10 | 0.08 | 0.09 |
| | $e+H^++C^{6+}$ | 0.18 | 0.16 | 0.12 | 0.12 |
| Enhancement by ions % | | 20 | 60 | 50 | 33 |

**Acknowledgments**

This work was partially supported by JSPS Grant-in-Aid for Scientific Research (C)(25400539).

**References**

[1] Johzaki T, Nagatomo H, Sunahara A, Cai H B, Sakagami H, Nakao Y and Mima K 2011 *Nucl. Fusion* **51** 073022
[2] Hatchett S P, Brown C G, Cowan T E, et al. 2000 *Phys. Plasmas* **7** 2076
[3] Snavely R A, Key M H, Hatchett S P, et al. 2000 *Phys. Rev. Lett.* **85** 2945
[4] Wilks S C, Kruer W L, Tabak M and Langdon A B 1992 *Phys. Rev. Lett.* **69** 1383
[5] Sakagami H, Johzaki T, Nagatomo H and Mima K 2009 *Nucl. Fusion* **49** 075026
[6] Johzaki T, Sakagami H, Nagatomo H and Mima K 2007 *Laser Part. Beams* **25** 621

# A scaled experiment to study energy dissipation process during longitudinal compression of charged particle beams

**Y Sakai[1], M Nakajima[1], J Hasegawa[1], T Kikuchi[2] and K Horioka[1]**

[1]Department of Energy Sciences, Tokyo Institute of Technology, Japan
[2]Department of Nuclear System Safety Engineering, Nagaoka University of Technology, Japan

E-mail: sakai.y.ba@m.titech.ac.jp

**Abstract.** Beam behavior during longitudinal bunch compression of charged particles was investigated using a compact simulator device based on electron beams. Beam current waveforms and bunch compression ratios were measured as a function of the initial beam current. We found that the current waveform became blunt and the compression ratio degraded at higher beam currents. These results indicate that space-charge fields dissipate the kinetic energy of beam particles.

## 1. Introduction

In heavy-ion fusion (HIF), the beam power must be increased to TW level. To achieve such an extremely high beam power, longitudinal beam compression has been adopted in the final stage of the drivers[1][2]. However, this bunching process is disturbed by strong space-charge fields. In particular, abrupt beam compression with a large velocity tilt may cause beam instabilities through the collective motions of beam particles and may substantially dissipate beam kinetic energy into thermal energy (random motion of particles), thereby degrading beam quality.

In general, a nonstationary beam mismatched to the external focusing field has free energy [3]. Thermodynamically, a nonstationary beam relaxes to its final equilibrium state as defined by the Maxwell-Boltzmann velocity distribution while dissipating free energy by long-time particle interactions through intrabeam scattering and/or space-charge fields. The quality and profile of resulting beam can be predicted on the basis of external focusing potential and beam self-potential [4][5]. However, the HIF final bunching stage employs an abrupt one-pass beam manipulation with linac, and beams have no time to completely dissipate any free energy induced by the strong velocity tilt. Thus, it is very hard to predict the final beam quality and profile, both which strongly depend on energy dissipation processes governed by collective effects [6].

To investigate the dissipation processes during longitudinal beam compression, we propose a scaled-down experiment using a compact and flexible simulator based on electron beams. Because of the small inertia and large specific charge of electrons, GeV-kA heavy-ion beams (typical in HIF) can be scaled down to keV-mA electron beams. In fact, a beam velocity $\beta$ and a longitudinal perveance $K_L$ equivalent to HIF beam parameters are readily available with electron beams (see Table 1). Moreover, electron motions can be restricted in the longitudinal direction by applying a longitudinal magnetic field. If beam compression is performed one-dimensionally,

**Table 1.** Typical beam parameters for heavy ion and electron beams. To obtain these values we assumed a bunch duration of 100 ns and a geometry factor $g$ of 5.7 are assumed.

| Beam species | q/m [C/kg] | Voltage [V] | Current [A] | $\beta$ $(v_z/c)$ | $K_L$ [mm] |
|---|---|---|---|---|---|
| HIB ($Pb^+$) | $4.67\times10^5$ | $1\sim10\times10^6$ | $0.1\sim10\times10^3$ | 0.10~0.32 | 0.038~38 |
| Electron | $1.75\times10^{11}$ | $2.7\sim27\times10^3$ | $0.27\sim27\times10^{-3}$ | 0.10~0.32 | 0.038~38 |

the current waveform of the bunched beam directly reveal the distribution of beam particles in longitudinal phase space. Thus, we have the prospect of discussing the particle dynamics by comparing the time evolution of beam current waveforms with those obtained from numerical simulations.

So far, we have constructed a compact electron-beam-based simulator for longitudinal compression experiments and investigated the effects of the initial velocity spread on the longitudinal focusing ability in a parameter region dominated by emittance [7]. This paper presents recent progress in our on-going study.

## 2. Experimental configuration and beam parameters

### *2.1. Experimental setup*

Figure 1 shows a schematic diagram of the simulator device. The device consists of a hot-cathode electron gun, an induction voltage adder, a solenoidal transport line, and a current probe. In the experiment, first, a continuos electron beam is injected into the induction cavity. Next, the voltage adder, composed of five units, applies a time-varying voltage to the beam in order to induce a head-to-tail velocity tilt. Then, the beam is compressed quasi-one-dimensionally in the longitudinal direction as it drifts toward a focal point through the transport line with a solenoidal magnetic field of 5.7 mT.

As a modulation voltage (longitudinal focusing potential) for drift compression, we define an ideal voltage waveform by,

$$V_{mod}(t) = \frac{m_e}{2e}\frac{1}{(\sqrt{\frac{m_e}{2eV_o}} + \frac{T-t}{L_f})^2} - V_0, \tag{1}$$

here $V_0 = 2.8$ kV is the extracting voltage of the electron gun, $T = 80$ ns and $L_f = 1.92$ m are the initial pulse duration of the modulated beam, and the focal length from the endpoint of the modulation gap, respectively. With this ideal waveform, we can apply a linear focusing force in the longitudinal direction; that is, all electrons in the bunch are gathered at the focal point. Figure 2 shows the ideal and experimental voltage waveforms. To generate the ideal modulation voltage waveform, the voltage adder was carefully tuned [8].

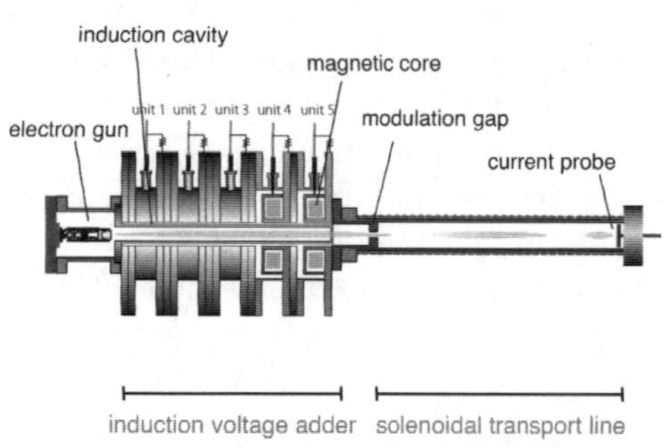


**Figure 1.** Schematic diagram of the simulator device.

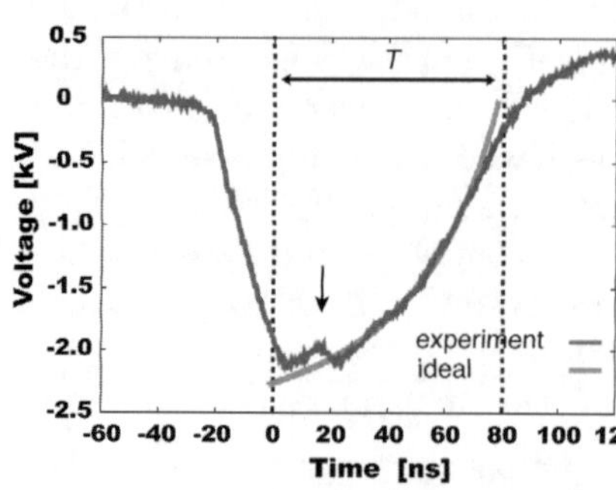


**Figure 2.** Ideal and experimental modulation voltage waveforms.

**Table 2.** Relevant parameters for longitudinal beam dynamics in this experiment. Here, bracket denotes the index of the space-charge effect for the 20-times compressed case.

| $I_0$ [mA] | g-factor | $\epsilon_{zz'_0}$ [m-rad] | $K_L$ [mm] | $K_L Z_m/\epsilon^2_{zz'_0}$ | $(K_L Z_m/\epsilon^2_{zz'_0})$ |
|---|---|---|---|---|---|
| 0.1∼1 | 5.7 | $2.37\times10^{-2}$ | 0.109∼1.09 | 0.49∼4.9 | 9.8∼98 |

*2.2. Analytical estimation of beam parameters*

Longitudinal envelope equation [9] can provide a general description for the longitudinal behavior of the beam, so we used that equation to estimate beam parameters in our simulator [8].

For this study, the beam was extracted from the 2.8-kV electron gun with currents from 100 $\mu$A to 1 mA. Assuming the beam longitudinal temperature, $T_L$, is equivalent to the cathode temperature, we can use $k_b T_L = 0.1$ eV. Values for the major beam parameters are listed in Table 2. Regarding the value for $K_L Z_m/\epsilon^2_{zz'}$, which is an index of the strength of the space-charge force on beam dynamics, the bunched beam in our simulator covers a wide parameter space from emittance-dominated to space-charge-dominated regions; $K_L Z_m/\epsilon^2_{zz'_0} \gg 1$, where $\epsilon_{zz'_0}$ is the initial emittance. Detailed descriptions are available in [8] and [10].

*2.3. Estimation of beam dynamics by numerical simulation*

In this study, a simplified one-dimensional electrostatic particle-in-cell simulation was carried out to examine the particle dynamics. In the calculation, we used the experimental modulation voltage shown in Fig. 2. We assumed that the space-charge field can be calculated with the g-factor model [9][10] and adopted it instead of the usual Poisson solver.

## 3. Results and discussions

Figure 3 shows typical current waveforms of compressed beams normalized by initial beam currents $I_0 = 152$, 581, or 1050 $\mu$A, respectively. In this experiment, the beam current was measured at $L = 1.95$ m, which was 3 cm downstream from the nominal focal point at $L_f = 1.92$ m. For $I_0 = 152$ $\mu$A, the figure shows two peaks in addition to the first main peak. These peaks correspond to locally compressed components caused by the error in the modulation voltage shown with the arrow in Fig. 2. However, by increasing the initial beam current, these peaks were gradually smoothed; they completely disappeared at $I_0 = 1050$ $\mu$A. In addition, the main peak heights decreased with increasing $I_0$. Figure 4 shows the dependence of the peak compression ratio on the initial beam current. Here the compression ratio is defined by the main peak height of the normalized current waveforme. As shown in Fig. 4, the ratio monotonically decreased with increasing beam current.

Figures 5(a) and 5(b) show numerical simulation results for the time evolution of beam particles in longitudinal phase space for $I_0 = 152$ and 1050 $\mu$A, respectively. In Fig. 5(a) and

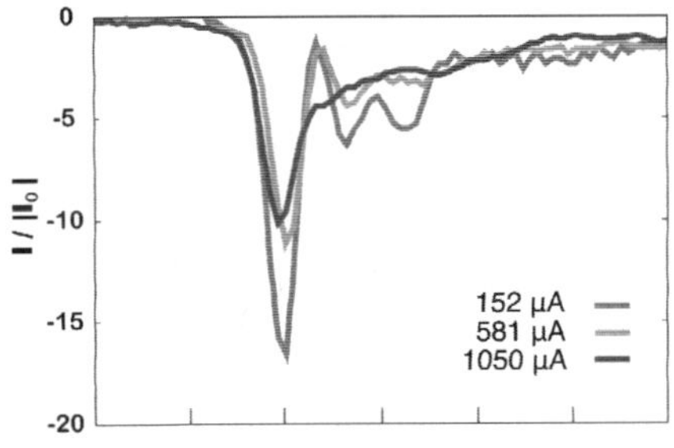


**Figure 3.** Normalized compressed beam current waveforms.

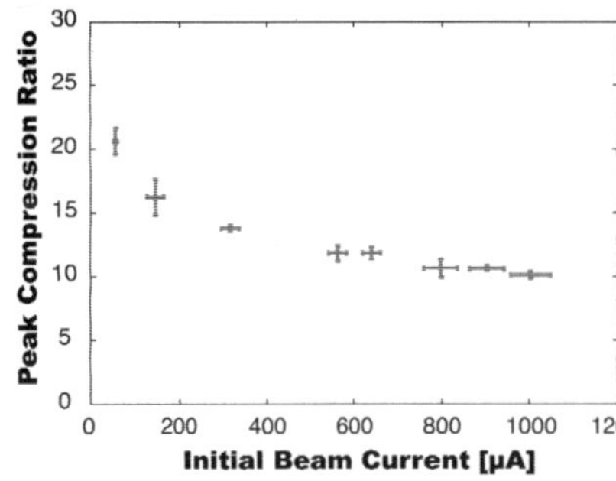


**Figure 4.** Effect of initial beam current on peak compression ratio.

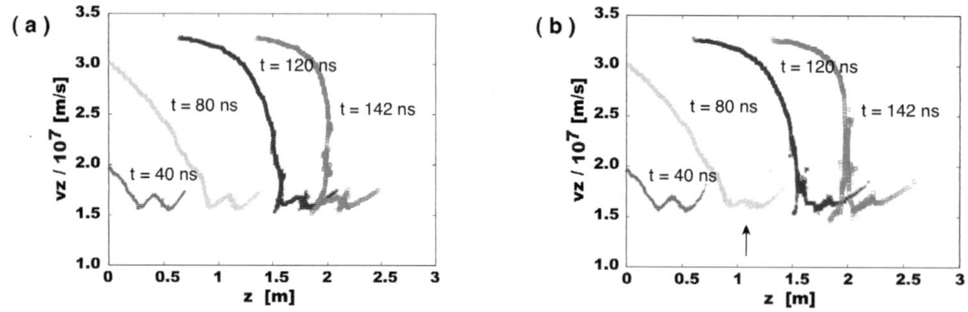


**Figure 5.** Numerically simulated particle dynamics in phase space (a) for $I_0 = 152\ \mu$A and (b) for $I_0 = 1050\ \mu$A.

5(b), $t = 0$ ns marks the time when the beam modulation started and $t = 142$ ns is the time of maximum bunch compression predicted for the ideal case. Figures. 5(a) and 5(b) both show that a twisted structure in the head of the beam bunch. This structure is due to the fluctuations in the longitudinal focusing force induced by the modulation voltage (see Fig. 2), which correspond to the two peaks observed in the experiment. In Fig. 5(a), the distribution gradually rotates with time and becomes spatially focused while closely maintaining the initial velocity tilts. Interestingly, at the higher beam current in Fig. 5(b), the distributions contain additional modulations. These modulations seem to be localized at reflection points of the distributions (arrow in the figure), which correspond to nonlinearly focused beam components. This phenomenon indicates that nonlinearly focused components induce additional perturbations in the beam that accelerate relaxation of the velocity distribution. As shown in Figs. 3 and 4, the current waveform and compression ratio obviously depend on the initial beam current. To discuss details of dissipation processes occurring in our bunching experiments, the g-factor model may not be most suitable because it assumes a uniform charge density in an ellipsoidal beam bunch. Thus, in our future study, two-dimensional electrostatic PIC code will be used instead.

## 4. Conclusions

A compact electron-beam-based simulator was used to conduct bunch compression experiments in parameter space ranging from emittance-dominated to space-charge-dominated. We found that the current waveforms of bunched beams depend on the initial beam current. In addition, beam particle dynamics were numerically simulated, and the results also supported the particle diffusion in the phase space. These results indicate that relaxation of the free energy affects the longitudinal focusing dynamics.

## References

[1] R. O. Bangerter, Fusion Eng.Des. 44 71 (1998)
[2] E. P. Lee and J. Hovingh, AIP Conference Proceedings, No249, PT.2 1713-1724 (1992)
[3] M. Reiser, J. Apple. Phys. 70, 1919 (1991)
[4] N. Brown and M. Reiser, Phys. Plasmas 2 (3) (1995)
[5] B. Beaudoin, I. Haber, R. A. Kishek et al., Phys. plasmas, 18 013104 (2011)
[6] S. M. Lund, D. P. Grote, R. C. Davidson, Nucl. Instr. and Meth. A 544 (2005) 472-480
[7] A. Nakayama, Y. Sakai, Y. Miyazaki et al., "Longitudinal bunch compression study with induction voltage modulator", Proc. IFSA-2011 (in press)
[8] Y. Sakai, M. Nakajima, J. Hasegawa et al., "A scaled experiment to study dynamics during longitudinal compression of intense charged particle beams", Nucl. Instr. and Meth. A (available online)
[9] M. Reiser, Theory and Design of Charged Particle Beams (Wiley, New York, 1994)
[10] T. Kikuchi et al., "Numerical Analysis of Beam Dynamics for Final Pulse Compression Simulated by Compact Electron Beam Device for Heavy Ion Inertial Fusion", PLASMA 2011, Kanazawa, Japan, 23P165-B (2011)

Journal of Physics: Conference Series **688** (2016) 012098 doi:10.1088/1742-6596/688/1/012098

# Thomson scattering measurement of a collimated plasma jet generated by a high-power laser system

**T. Ishikawa[1], Y. Sakawa[2], T. Morita[2], Y. Yamaura[1], Y. Kuramitsu[3], T. Moritaka[2], T. Sano[2], R. Shimoda[4], K. Tomita[4], K. Uchino[4], S. Matsukiyo[4], A. Mizuta[5], N. Ohnishi[6], R. Crowston[7], N. Woolsey[7], H. Doyle[8], G. Gregori[8], M. Koenig[9], C. Michaut[10], A. Pelka[11], D. Yuan[12], Y. Li[12], K. Zhang[13], J. Zhong[13], F. Wang[13], and H. Takabe[1,2]**

[1]Graduate School of Sciences, Osaka University, 1-1 Machikaneyama, Toyonaka, Osaka, Japan
[2]Institute of Laser Engineering, Osaka University, Japan
[3]National Central University, Taiwan
[4]Interdisciplinary Graduate School of Engineering Sciences, Kyushu University, Japan
[5]RIKEN, Computational Astrophysics Lab., Japan
[6]Department of Aerospace Engineering, Tohoku University, Japan
[7]Department of Physics, University of York, UK
[8]Department of Physics, Oxford University, UK
[9]LULI, Ecole Polytechnique, France
[10]LUTH, Observatoire de Paris, France
[11]Helmholtz-Zentrum Dresden-Rossendorf, Germany
[12]Institute of Physics, Chinese Academy of Sciences, China
[13]Institution National Astrophysical Observatories, CAS, China

E-mail: ishikawa-t@ile.osaka-u.ac.jp

**Abstract.** One of the important and interesting problems in astrophysics and plasma physics is collimation of plasma jets. The collimation mechanism, which causes a plasma flow to propagate a long distance, has not been understood in detail. We have been investigating a model experiment to simulate astrophysical plasma jets with an external magnetic field [Nishio *et al.*, *EPJ. Web of Conferences* **59**, 15005 (2013)]. The experiment was performed by using Gekko XII HIPER laser system at Institute of Laser Engineering, Osaka University. We shot CH plane targets (3 mm $\times$ 3 mm $\times$ 10 $\mu$m) and observed rear-side plasma flows. A collimated plasma flow or plasma jet was generated by separating focal spots of laser beams. In this report, we measured plasma jet structure without an external magnetic field with shadowgraphy, and simultaneously measured the local parameters of the plasma jet, i.e., electron density, electron and ion temperatures, charge state, and drift velocity, with collective Thomson scattering.

## 1. Introduction

Astrophysical jets are as well-collimated outflows from compact objects generated by the dynamic interactions in an accretion disc. The collimation mechanism, which causes a plasma flow to propagate a long distance, has not been understood in detail. There are three effects to contribute to the collimation of astrophysical jets: radiative cooling, interstellar medium, and magnetic fields. The experiments on plasma jets have been conducted with high-power laser facilities[1-4]. In the previous report from our group, Nishio *et al.*[5] compared the plasma jets with and without an eternal perpendicular magnetic field (0.2 $\sim$ 0.3 T). In this paper we investigate plasma jet formation without an external magnetic field with collective Thomson scattering (CTS) measurements. We shot a plastic (CH) target plane with two high-power laser beams and generated plasma jet structure. We measured the structure with optical diagnostics, and simultaneously measured the local parameters of the plasma jet (electron density $n_e$, electron and ion temperatures $T_e$, $T_i$, charge state $< Z >$, and drift velocity $v_d$) with CTS. Moreover we evaluated error bars of the obtained plasma parameters.

## 2. Experiment

The experiments were performed with Gekko XII HIPER laser system at the Institute of Laser Engineering, Osaka University. Figure 1 shows a schematic drawing of the experimental setup. A CH target was irradiated by two beams with the energy of $\sim$120 J/beam at a wavelength of 351 nm with a pulse duration of 500 ps. A plasma flow was observed at the rear side of the plane. To produce a non-uniform plasma flow, the CH plane target was irradiated by laser beams focused at different positions[6]. The distance between two focal spots at the target surface was $\sim$260 $\mu$m and the focal spot size was 300 $\mu$m. We measured the plasma jet structure with shadowgraphy using an ICCD camera (200 ps gate width), and simultaneously measured the local parameters of the plasma jet. The wavelength and the pulse width of the probe beam were 532 nm and $\sim$10 ns, respectively. Figure 1(c) shows the target chamber and schematics of diagnosis. We used an X-ray pinhole camera to observe the focal spot separations. The Thomson scattering probe (a wavelength of 532 nm and the energy of $\sim$0.3 J in $\sim$8 ns) was focused at $x = 0, y = 0, z = 3.0$ mm. The incident angle of the Thomson probe with respect to the HIPER laser was 45 degrees, and the scattered light was detected at 90 degrees from the probe laser by using a triple-grating spectrometer (TGS)[7] and a 2 ns-gate ICCD camera as shown in Fig. 1(c).

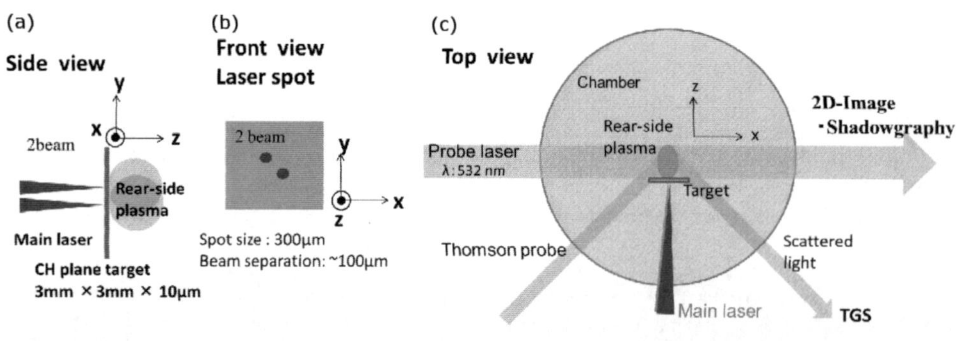


**Figure 1.** (a) Side and (b) front views showing the configuration of main laser beams and a target. (c) Top view of the target chamber and schematics of the diagnostics.

## 3. Results

Figure 2(a) shows a shadowgraphy image at 40 ns from the main laser timing. The rear-side plasma expands in the target normal direction, and two plasma jets appear. Figure 2(b) shows the CTS spectrum at the same shot and timing as Fig. 2(a). Figure 2(c) shows the spectra derived from Fig. 2(b) at $z = 3.0$ mm from target rear-surface. The wavelength of the probe laser is blocked at the TGS as shown with a shaded area [Fig. 2(c)]. The Thomson scattering power is expressed by using the spectral density function S(k) with two species $i$ ($i$ = H for hydrogen and $i$ = C for carbon). In Fig. 2(c), the experimental data(dots) are fitted with $S(k)$[8](red line). Here, $T_i = T_H = T_C, v_d = v_H = v_C = v_e, n_H = n_C$ are assumed. Electron density $n_e$ is obtained from the total scattered power calibrated by Rayleigh scattering[9]. The average charge state of carbon $< Z_C >=< Z >$ is derived self-consistently to satisfy the collisional-radiative model[10] with $< Z_H >= 1$. We neglect the scattering from the electron component and consider only the ion-acoustic feature because the electron component is small in the wavelength of $\lambda = 531 - 533$ nm. The plasma parameters, $n_e$, $T_e$, $T_i$, $< Z >$, and $v_d$ are obtained from these spectra at each position[Fig. 3]. The electron densities at $x = 0.3$, $z = 3.3$ and $x = 2.4$, $z = 5.4$ mm are larger than those around them. $T_e$ and $T_i$ at each position are $15 - 25$ eV. They are nearly constant for $x = 0 - 4.1$, $z = 3.0 - 6.1$ mm at 40 ns. The charge state at each position is 4. The drift velocity increases as a function of the distance along $z$-axis. A typical sensitivity of the fit is estimated by increasing (green lines) and decreasing (blue lines) a single parameter and find a new best-fit result for the spectrum obtained at $x = 0, z = 3.0$ mm [Fig. 2(c)] as shown in Figs. 4(a)-4(d). As a result, the sensitivity for $n_e$, $T_e$, $T_i$, and $v_d$ are $-15 \sim +30\%, \pm15\%, -35 \sim +45\%$, and $\pm5\%$, respectively. Moreover we calculated the sensitivity of $< Z >$ from those of $n_e$ and $T_e$, as a result, it was $\pm3\%$. We assumed that these sensitivities are independent of the position.

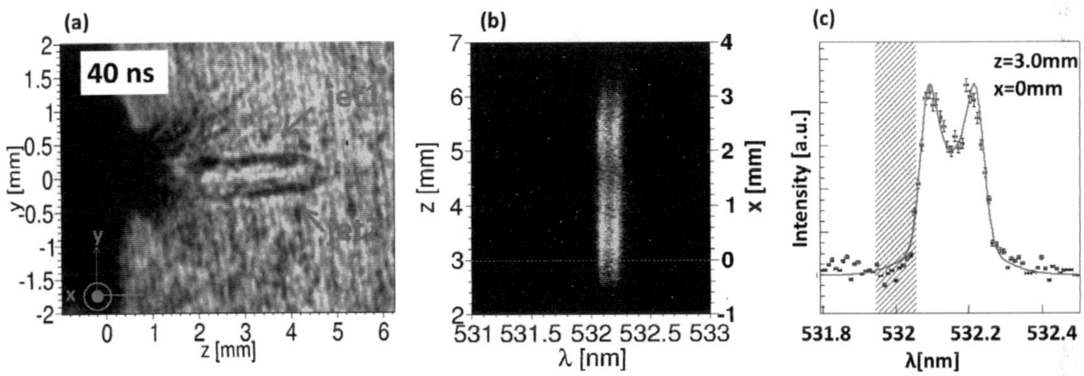


**Figure 2.** (a) The shadowgraphy image at 40 ns from the main laser. Total laser energy was 248 J. (b) CTS data at the same shot and timing (40 ns) as in (a). (c) A CTS spectrum at $x = 0, z = 3.0$ mm from target (dotted lines in (b)). The experimental data are shown in dots and the best fit for the spectrum is shown in a red line.

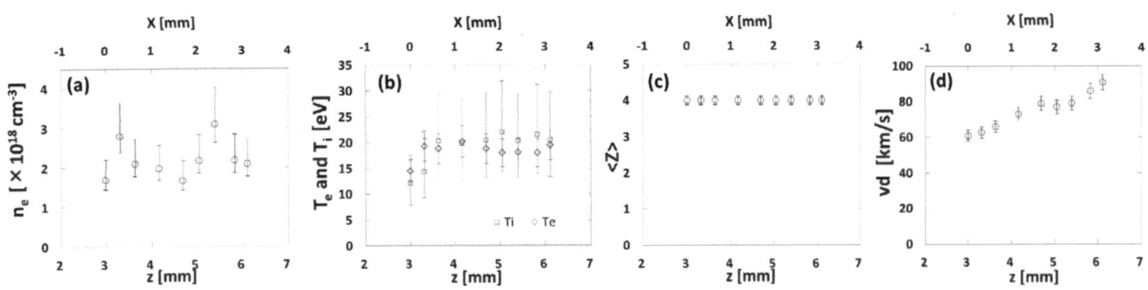


**Figure 3.** The obtained (a)$n_e$, (b) $T_e$ and $T_i$, (c) $< Z >$, and (d) $v_d$ as a function of position $z$.

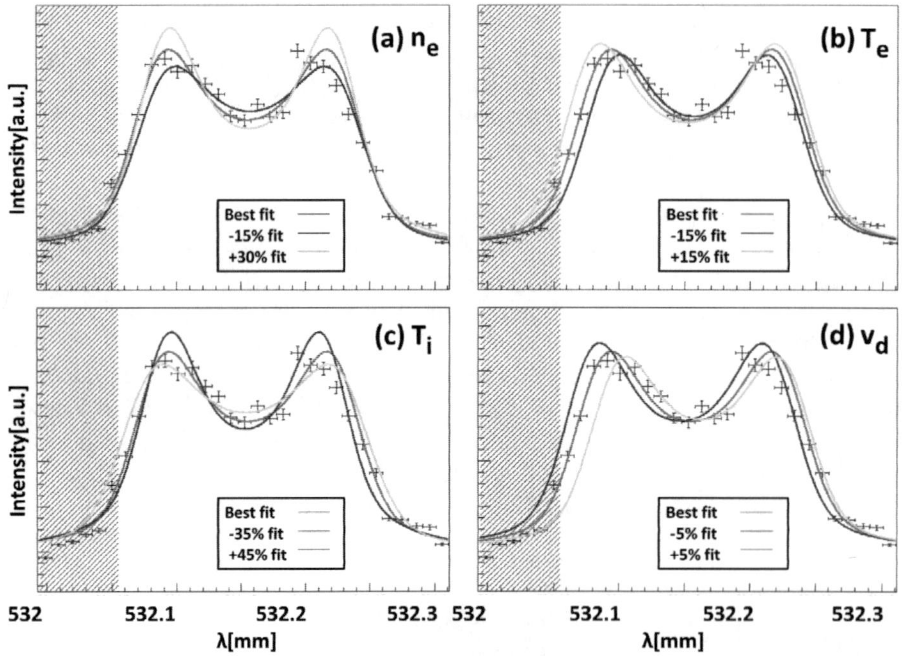


**Figure 4.** Sensitivities to the fit at $z = 3.0$ mm for (a) $n_e$, (b) $T_e$, (c) $T_i$, and (d) $v_d$. Red lines are the best-fit results. Green and blue lines are fitting lines when a single parameter is increased and decreased, respectively, by -15 and +30% ($n_e$), ±15% ($T_e$), -35 and +45% ($T_i$), and ±5% ($v_d$) from the best-fit results.

## 4. Conclusion

We generated a rear-side plasma jet by irradiating a CH plane target with high-power laser beams. We measured the plasma jet with optical diagnosis and collective Thomson Scattering, simultaneously. The electron densities at $x = 0.3$, $z = 3.3$ and $x = 2.4$, $z = 5.4$ mm are higher than those around them.

## Acknowledgments

The authors acknowledge the staff at the Institute of Laser Engineering, Osaka University.

## References

[1] K. Shigemori *et al.*, *Phys. Rev.* E **67**, 8838-41 (2000)
[2] Ph. Nicolai *et al.*, *Phys. Plasmas.* **15**, 082701 (2008)
[3] A. Faenov, V. Dyakin, A. Magunov, T. Pikuz and Skobelev, *Phys. Scripta.* **53**, 591-596 (1996)
[4] A. N. Mostovych, B. H. Ripin and J. A. Stamper, *Phys. Rev. Lett.* **62**, 2837-2840 (1989)
[5] K. Nishio *et al.*, *EPJ. Web of Conferences* **59**, 15005 (2013)
[6] Y. Kuramitsu *et al.*, *Astrophys Space Sci.* **336**, 269-272 (2011)
[7] S. Hassaballa *et al.*, *IEEE. Trans. Plasma Sci.* **32**,127 (2004)
[8] J. Sheffield, D. Froula, S. H. Glenzer and N. C. Luhmann, Jr, *Plasma Scattering of Electromagnetic Radiation: Theory and Measurement Techniques* 2nd ed. (Academic Press Inc.) (2010)
[9] T. Morita *et al.*, *Phys. Plasmas.* **20**, 092115 (2013)
[10] H.-K. Chung *et al.*, *High Energy Density Physics* **1**, 3-12 (2005)

# Atomic processes and equation of state of high Z plasmas for EUV sources and their effects on the spatial and temporal evolution of the plasmas

**Akira Sasaki[1], Atushi Sunahara[2], Hiroyuki Furukawa[2], Katsunobu Nishihara[3], Takeshi Nishikawa[4], and Fumihiro Koike[5]**

[1] Quantum Beam Science Directorate, Japan Atomic Energy Agency, 8-1-7 Umemidai, Kizugawa-shi, Kyoto, 619-0215, Japan
[2] Institute for Laser Technology, 1-8-4 Utsubohonmachi, ishi-ku, Osaka, 565-0871, Japan
[3] Institute of Laser Engineering, Osaka University, 2-6 Yamadaoka, Suita, Osaka, 565-0871, Japan
[4] Department of Electrical and Electronic Engineering, Okayama University, 1-1 aka 1-chome, Tsushima, Okayama 700-8530, Japan
[5] Physics Laboratory, School of Medicine, Kitasato University, 1-15-1 Kitasato, Sagamihara, Kanagawa 228-8555, Japan

E-mail: a_sasaki@osa.att.ne.jp

**Abstract.** Laser-produced plasma (LPP) extreme ultraviolet (EUV) light sources have been intensively investigated due to potential application to next-generation semiconductor technology. Current studies focus on the atomic processes and hydrodynamics of plasmas to develop shorter wavelength sources at $\lambda = 6.x$ nm as well as to improve the conversion efficiency (CE) of $\lambda = 13.5$ nm sources. This paper examines the atomic processes of mid-z elements, which are potential candidates for $\lambda = 6.x$ nm source using n=3-3 transitions. Furthermore, a method to calculate the hydrodynamics of the plasmas in terms of the initial interaction between a relatively weak prepulse laser is presented.

## 1. Atomic processes of mid-z elements for shorter wavelength sources

Extreme ultraviolet (EUV) light sources at $\lambda = 6.x$ nm have attracted attention as a mean to further reduce the process rule of lithographic technology at wavelengths where multilayer optics with a high reflectivity is available. The 4d-4f transition from near Pd-like ions, which is used for $\lambda = 13.5$ nm sources with Sn plasmas, is scalable to shorter wavelengths by increasing the atomic number of the target material. Gd and Tb plasmas show strong emission in the $\lambda = 6.x$ nm band [1]–[3].

The 4d-4f transition exhibits an unresolved transition array (UTA) with a broad structure [4], from which strong emission can be obtained. However, for sources in the $\lambda = 6.x$ nm band, the spectral width is too broad compared to the 0.5% bandwidth of the multilayer optics. Moreover, Gd and Tb sources should require a high temperature plasma of $T_e \approx 100$ eV, a pumping laser with a one order of magnitude greater intensity than those used for Sn sources may be necessary. Thus, an alternative source material may be better suited to realize efficient sources.

In this study, we investigate the atomic energy levels and transition probabilities using the HULLAC code [5] for a variety of atomic elements over a wide range of charge states. Figure

1 shows the wavelengths of various transitions from Mo ions, where the numbers indicate the principal quantum numbers of the lower and upper states of the transition. It is shown that Mo can be used as a source for different wavelengths as the ionization proceeds. The figure also suggests that the n=3-3 transition from ions with $\approx 20$ of Mo may be a potential source at $\lambda$ = 6.x nm.

Figure 2 shows the average wavelength of 3d-4f and 3p-3d transitions. The wavelength of the 3d-4f transition depends strongly on the ion charge. In the case of Kr, the emission near the $\lambda$ = 6.x nm band can only be obtained from Co-like ion. On the other hand, the wavelength of the 3p-3d transition is almost constant from a Co-like to K-like ions, exhibiting a trend similar to those of the 4d-4f and 4p-4d transitions of Sn.

Figure 3 shows the calculated emission spectrum from the Mo plasma. The emission from a plasma sphere with a radius of 0.3 mm is calculated assuming that the plasma is in a local thermal equilibrium (LTE) state [7]. The strong peak at $\lambda \approx 6.9$ nm arises from the 3p-3d (j=3/2-1/2) transition of Co-like Mo. The estimated spectral efficiency, which is defined by the ratio of the emission in 0.5% bandwidth relative to the total emission, is 1%. It has been shown that the calculated wavelengths have an uncertainty due to the effect of the configuration interaction. For this line, the calculated wavelength is 6.8711 nm, which is slightly shorter than the value in the NIST database (6.9596 nm) [6]. The emission line should be chosen using its accurate wavelength and then the co-optimization with the multilayer should be carried out.

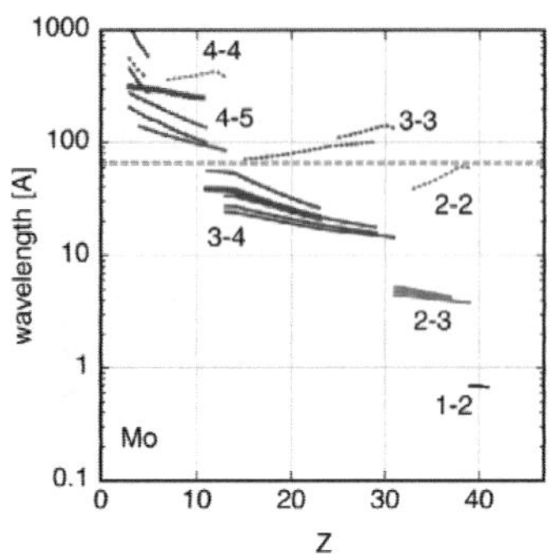


**Figure 1.** Average wavelengths of various transitions of Mo ions.

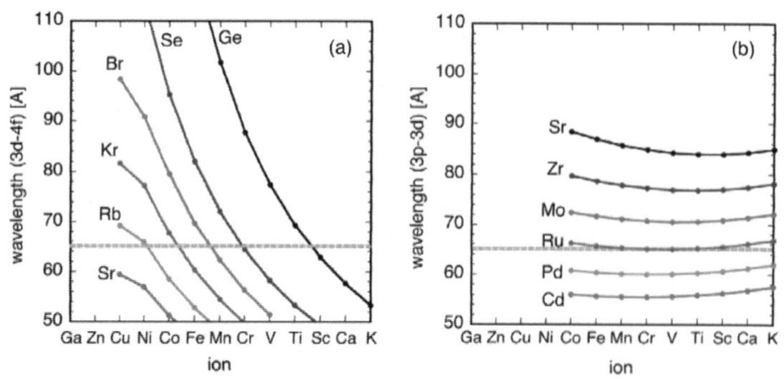


**Figure 2.** Average wavelengths of (a) 3d-4f transition of Cu-like to K-like ion of Ge to Sr, and (b) 3p-3d transition of Co-like to K-like ion of Sr to Cd.

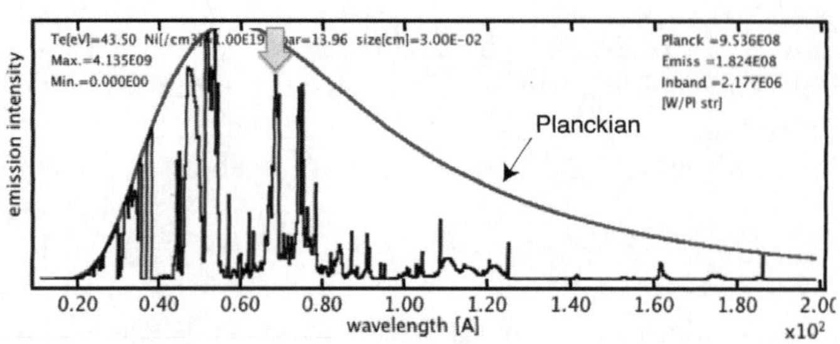


**Figure 3.** Calculated emission spectrum from Mo plasma at $T_e$ = 43 eV and $n$ = $10^{19}$ cm$^{-3}$. Arrow denotes the 3p-3d (j=3/2-1/2) transition of Co-like Mo.

## 2. Modeling the development of the spatial structure of a laser irradiated Sn droplet

During the development of EUV sources at $\lambda = 13.5$ nm, a mass-limited target such as a small droplet of Sn with a radius of few 10 $\mu$m has been found to be useful to minimize the emission of debris. However, such a droplet is smaller than the spot size of the laser, resulting in an inefficient absorption of the laser energy. To increase the efficiency and EUV power from a plasma, a longer scale length of the plasma is useful. Therefore, double-pulse irradiation is employed, where the first laser pulse produces a plasma with a scale length of $\approx 100$ $\mu$m and the second laser pulse heats the plasma to a temperature sufficient for the EUV emission, as illustrated in fig. 4 [8]. Because the average density of the mist is reduced three orders of magnitude compared to the solid density, the power balance model suggests that a CE of more than 4% is feasible [9]. To further optimize of EUV sources, the initial interaction between the relatively weak laser pulse and the solid target as well as subsequent fragmentation of the target material are important.

Anisimov and Khokhlov have suggested mechanisms for the instabilities associated with the melting, evaporation, and condensation processes during the interaction between a relatively weak laser pulse and the target material [10]. We have investigated modeling methods of such phase transition processes and their effects on the structure of the plasma. Figure 5 shows the Van-der-Waals equation of state of Sn [11]. Sn has a two-phase region below the critical temperature of $T_c = 7297$ K in which the liquid and gas phases of Sn coexist with an arbitrary arrangement.

Because the material in a cell is assumed to be in a single state in conventional hydrodynamics simulations, the average state is considered even for a two-phase region. In this study, as shown in fig. 6, the cell is divided to two once the temperature and volume of the cell reach the liquid-gas two-phase region. The volume fraction and mass of each cell, which correspond to the liquid and gas phases, is determined according to the Maxwell construction assuming that the total volume is conserved, but the arrangement of each region is determined using a probability. This method allows the spontaneous formation of the plasma structure due to phase transition processes to be simulated. As an example, fig. 6 also shows the result of a test calculation for the fragmentation of a liquid droplet during isothermal expansion.

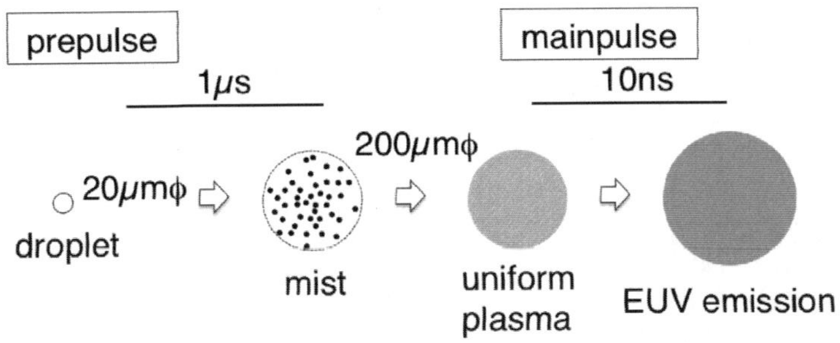


**Figure 4.** Schematic diagram of the excitation of a Sn droplet target.

## 3. Summary

We show the results of the recent studies on the atomic processes and hydrodynamics of plasmas used as EUV sources. We propose a method to simulate the hydrodynamics of the initial interaction between a laser and the matter as well as subsequent particle formation. In the future, we intend to theoretically and experimentally validate our method and apply it to the optimization of EUV sources.

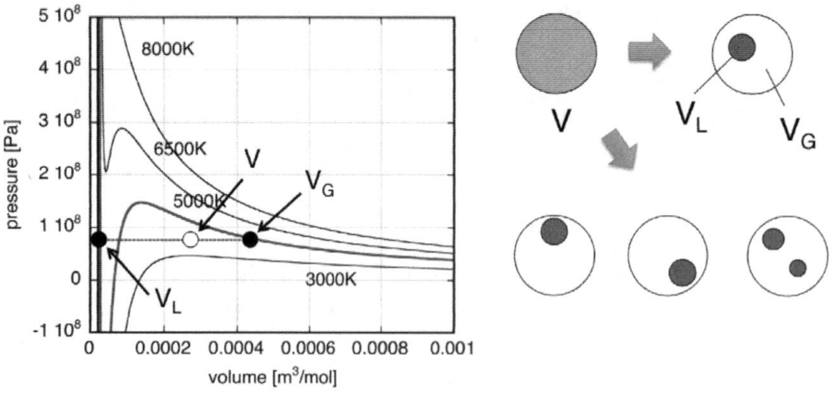


**Figure 5.** Van-der-Waals equation of state for Sn, and schematic diagram of the possible arrangements in the liquid and gas phase regions.

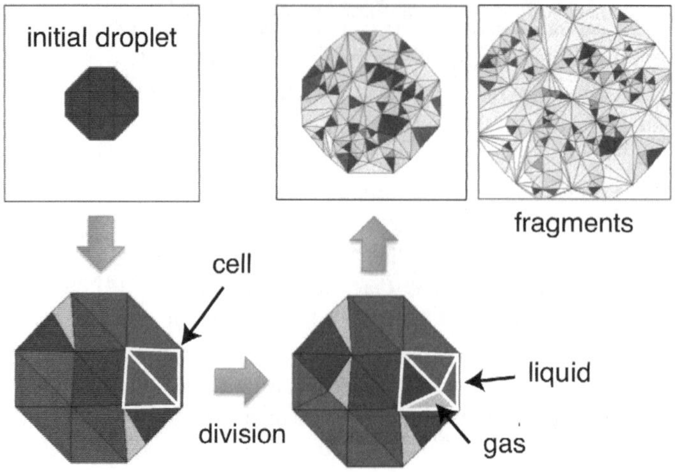


**Figure 6.** Modeling of the liquid-gas phase transition using cell division. When the density and temperature of the cell reach those of the two phase region, the cell is divided into two cells, where one is in the liquid phase and the other is in the gas phase.

## Acknowledgement
The present work is supported by a Grant-in-Aid for Scientific Research (B) No. 23340185 from the Japan Society for the Promotion of Science (JSPS).

## References
[1] Otsuka T et al. 2010 *Appl. Phys. Lett.* **97** 111503.
[2] Otsuka T et al. 2010 *Appl. Phys. Lett.* **97** 231503.
[3] Higashiguchi T et al. 2011 *Appl. Phys. Lett.* **99**, 191502.
[4] O'Sullivan G and Carroll P K 1981 *J. Opt. Soc. Am.* **A71** 227.
[5] Bar-Shalom et al. 2001 *J. Quant. Spectrosc. Radiat. Transf.* **71** 169.
[6] http://www.nist.gov/pml/data/asd.cfm
[7] Murakami M 1990 *J. X-ray Sci. & Tech.* **2** 127.
[8] Endo A *Proceedings of the 2012 EUV source workshop* can be found at http://www.euvlitho.com/
[9] isihihara K et al. 2008 Phys. Plasmas **15** 056708.
[10] Anisimov S I and Khokhlov V K *Instabilities in laser-matter interaction* 1995 CRC Press, Tokyo.
[11] Young D A and Alder D J 1971 *Phys. Rev.* **A3** 364.

# Evaluation Method for Thermal Conductivity in Warm Dense Matter by using Ruby Fluorescence Probe

**Takuya Takahashi[1], Yoshimasa Kawaguchi[1], Takumi Ohuchi[1], Kazumasa Takahashi[1], Toru Sasaki[1*], Takashi Kikuchi[2], Tsukasa Aso[1], and Nob. Harada[1]**

[1]Department of Electrical Engineering, Nagaoka University of Technology, Kamitomioka 1603-1, Nagaoka, Niigata, Japan

[2]Department of Nuclear System Safety Engineering, Nagaoka University of Technology, Kamitomioka 1603-1, Nagaoka, Niigata, Japan

E-mail: sasakit@vos.nagaokaut.ac.jp

**Abstract.** We have proposed a concept of experimentally estimating thermal conductivity in warm dense matter from the ruby fluorescence. To reduce the dimension of the system, a cylindrically arranged sample tamped by the ruby capillary has been considered. From the estimated ruby temperature, in which is simulated by the time-dependent thermal diffusion in equation, the ruby fluorescence can be obtained from 0.5 mm to 0.6 mm. The results indicated that the low density regime as $\rho/\rho_s < 0.004$ is possible to evaluate the ruby fluorescence.

## 1. Introduction

Magnetic confinement fusion (MCF) such as international thermonuclear experimental reactor (ITER) and DEMO is expected to use a divertor made of tungsten, which is the highest melting temperature in the material [1]. The tungsten divertor is exposed by the highly energetic particles from the core plasmas. The heat flux from the highly energetic particles on the divertor in ITER is estimated to be 10 MW/m$^2$ at steady state and over 1000 MW/m$^2$ at plasma disruption and mitigated edge localized modes (ELMs) [2, 3, 4]. Up to now, all of MCF systems cannot reach the heat flux expected during ITER ELMs. To evaluate the heat flux during ITER ELMs, the several experiments have been demonstrated by using electron[5] or plasma guns[6, 7], or intense laser aiming at reproducing relevant heat fluxes[8]. These experiments have mainly evaluated metallurgical characteristics of tungsten divertor. Predicting performances of tungsten divertor in MCF, we should analyze not only the metallurgical properties but also the thermophysical properties of ablated tungsten. The ablated tungsten is a warm dense matter and its properties should be evaluated.

The physics of warm dense matter (WDM) are interested in transport coefficients. The transport coefficients have been evaluated by quantum molecular dynamics[9, 10], statistical modeling[11, 12, 13], and *ab initio* simulations[14]. Recently, in order to evaluate the transport properties in WDM, the deviation of the Lorenz number have been pointed out for dense hydrogen plasma from theoretical estimations[14]. The Lorenz number $L$ is the ratio between

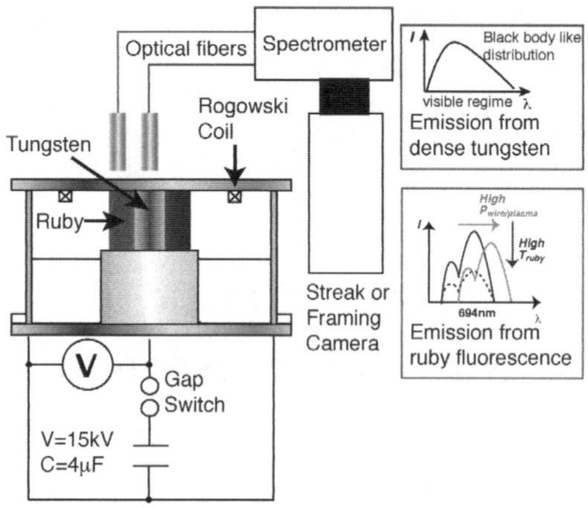


**Figure 1.** Typical schematic diagram of a direct measurement method for the dense tungsten plasma by using ruby fluorescence probe.

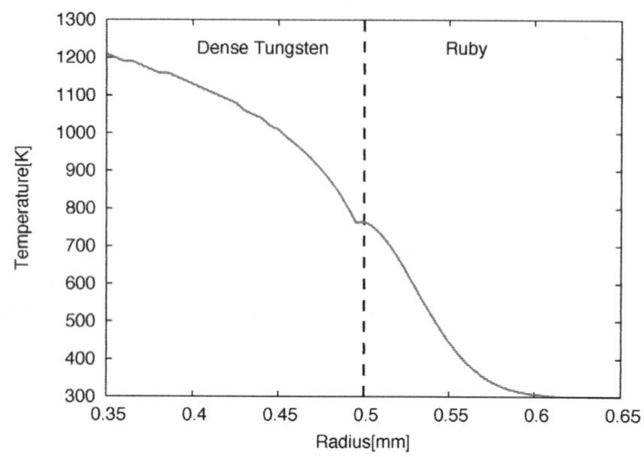


**Figure 2.** Radial temperature distribution profile at 50 $\mu$s.

electrical and thermal conductivities divided by the temperature, $L = \frac{e^2}{k_B T}\frac{\kappa}{\sigma}$, which, $e$ is the elementary charge, $k_B$ is the Boltzmann constant, $T$ is the temperature, $\kappa$ is the thermal conductivity, and $\sigma$ is the electrical conductivity. From the previous semi-empirical evaluation, the dense thermal conductivity evolution is similar up to the degenerate regime. Thus, the thermal conductivity for low density regime should evaluate the thermal conductivity from experimental observations.

In this paper, we propose a concept of experimentally estimating thermal conductivity in warm dense matter from the ruby fluorescence. To estimate the region of the ruby fluorescence, the temperature of the ruby by using one-dimensional numerical simulation is evaluated. To confirm the timing to generate the warm dense tungsten, we demonstrate an exploding wire discharge in vacuum.

## 2. Concept of experimental observation of the thermal conductivity in dense tungsten plasma as a method of ruby fluorescence probe

We consider the direct measurement method by using ruby fluorescence probe. The ruby is usually use the pressure gauge for high pressure experiments as a diamond anvil cell method. Mao [15] has been developed a formula of the fluorescence shift as a function of pressure.The lifetime of ruby fluorescence depends on the ruby temperature [16]. Thus, the ruby fluorescence is expected to use a probe for the evaluating ruby temperature.

To obtain the thermal conductivity in dense tungsten plasma, the sense of the measurement method is following equations.

$$Q = -\kappa \nabla T, \quad (1)$$

$$\frac{\partial T}{\partial t} = \frac{1}{\rho C_{\mathrm{v}}} \nabla \cdot \kappa \nabla T, \quad (2)$$

where, $Q$ is the heat flux, $T$ is the temperature, $\rho$ is the density, and $C_{\mathrm{v}}$ is the specific heat. To solve the equation, we should understand the physical properties of each medium. To evaluate the specific heat of dense tungsten plasma, the pulsed-power discharge technique is compatible due to the correctly evaluation of the input energy and the temperature for dense tungsten

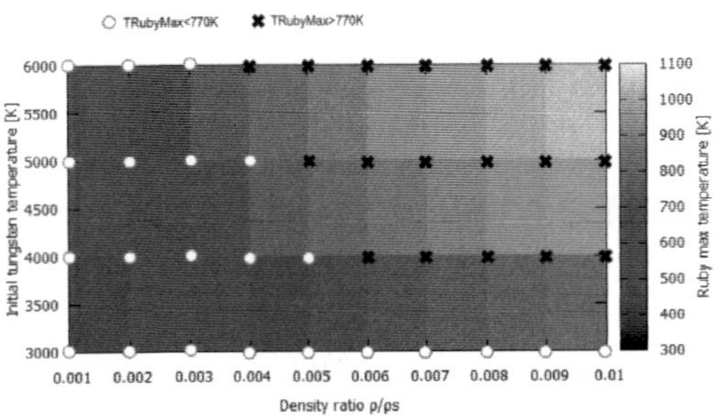


**Figure 3.** Maximum temperature of ruby capillary as a function of initial temperature and density of tungsten plasma.

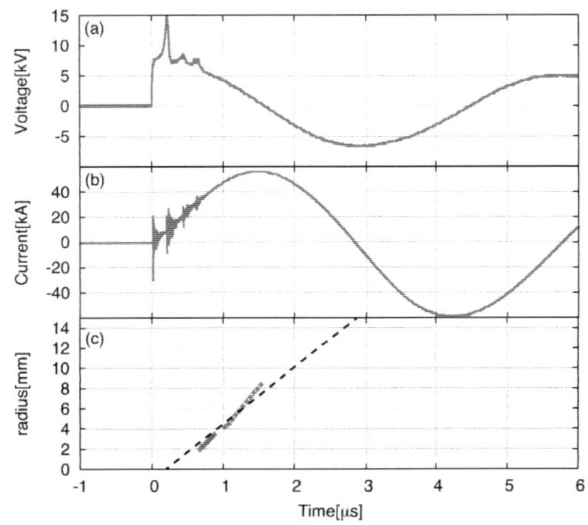


**Figure 4.** Time-evolution of exploding wire for tungsten in the vacuum. (a) voltage, (b) current, and (c) expanding radius.

plasma. From the time-resolved spectroscopic measurements of both the sample and the ruby fluorescence, we determine the temperature of dense tungsten plasma and that of ruby capillary. The time evolution of input power for the dense tungsten plasma is directly estimated by the input voltage and the current. Therefore, the specific heat of the dense tungsten plasma as a function of that temperature decide the time evolution of input power.

Figure 1 shows the typical image of a direct measurement method for the dense tungsten plasma by using ruby fluorescence probe. The dense tungsten plasma is generated by the pulsed-power discharge. The ruby capillary as a tamper can be sustain the high pressure up to 3 GPa, which is enough to conserve the dense tungsten plasma without hydrodynamic behavior.

The ruby fluorescence decreases the increasing temperature of ruby. Therefore, in order to evaluate the achieving temperature of ruby, we have estimated by a numerical simulation of the thermal diffusion equation. The hydrodynamic behavior is neglected due to the confined dense tungsten plasma by the ruby capillary. The initial condition of the numerical simulation is set to be 5000 K for the dense tungsten plasma and 300 K for the ruby capillary. The radii of the dense tungsten plasma and the ruby capillary are 0.5 mm and 2.0 mm, respectively. The thermodynamic properties and the thermal conductivity for both mediums are respected to be Refs. [17, 18, 19].

Figure 2 shows the radial temperature distribution profile at 50 $\mu$s. Evaluated ruby temperature is approximately from 800 K at the boundary between the ruby and the dense tungsten plasma to 300 K at 0.6 mm in radius. It reveals that the ruby fluorescence can be obtained from 0.5 mm to 0.6 mm. The ruby fluorescence are possible to observe the spectroscopic measurements. Figure 3 shows a maximum temperature of ruby capillary as a function of initial temperature and density of tungsten plasma. The results indicated that the low density regime as $\rho/\rho_s < 0.004$ is possible to evaluate the ruby fluorescence. Thus, the thermal conductivity in low density regime can be evaluated by this method.

## 3. Experimental results and discussion

To evaluate the thermal conductivity from the ruby fluorescence, we estimate a behavior of the exploding wire for tungsten in vacuum. To confirm the behavior of the exploding wire for tungsten, we have measured a time-evolution of the exploding wire for tungsten with $\phi = 0.2$ mm in the vacuum as shown in Fig. 4. From the time-evolution of the voltage and the current, the time of vaporization is 220 ns from the beginning of discharge. The time of vaporization ensures the comparison of between the experimental observation and the numerical estimation. After the vaporization of wire, the wire/plasma radius which is observed by the streak camera expands as a function of time. The expansion velocity of the wire/plasma is estimated to be $5 \times 10^3$ m/s. Assuming 1 mm of the interior diameter of the ruby capillary, the propagation time of the tungsten plasma is estimated to be about 100 ns for filled in the capillary. The density distribution of plasma is made uniform through the propagation of rarefaction (sound) wave; in our experimental condition, the sound wave in the plasma propagates within a hundred nanoseconds. Thus, the thermal conductivity for the dense tungsten plasma is possible to evaluate after 500 ns from the beginning of discharge.

## 4. Conclusions

We have proposed a concept of experimentally estimating thermal conductivity in warm dense matter from the ruby fluorescence. To reduce the dimension of the system, a cylindrically arranged sample tamped by the ruby capillary has been considered. From the estimated ruby temperature, in which is simulated by the time-dependent thermal diffusion in equation, the ruby fluorescence can be obtained from 0.5 mm to 0.6 mm. From the numerical evaluation, The results indicated that the low density regime as $\rho/\rho_s < 0.004$ is possible to evaluate the ruby fluorescence. the ruby fluorescence probe are possible to evaluate the thermal conductivity in dense plasma.

## References

[1] INTERNATIONAL ATOMIC ENERGY AGENCY *Summary of the ITER Final Design Report* (ITER EDA DOCUMENTATION SERIES No. 22)
[2] Pitts R A, Kukushkin A, Loarte A, Martin A, Merola M, Kessel C E, Komarov V and Shimada M 2009 *Physica Scripta* **T138** 014001
[3] Hiwatari R, Okano K, Asaoka Y, Shinya K and Ogawa Y 2005 *Nuclear Fusion* **45** 96
[4] Loarte A, Lipschultz B, Kukushkin A S, Matthews G F, Stangeby P C, Asakura N, Counsell G F, Federici G, Kallenbach A, Krieger K, Mahdavi A, Philipps V, Reiter D, Roth J, Strachan J, Whyte D, Doerner R, Eich T, Fundamenski W, Herrmann A, Fenstermacher M, Ghendrih P, Groth M, Kirschner A, Konoshima S, LaBombard B, Lang P, Leonard A W, Monier-Garbet P, Neu R, Pacher H, Pegourie B, Pitts R A, Takamura S, Terry J and Tsitrone E 2007 *Nuclear Fusion* **47** S203
[5] Hirai T, Brezinsek S, Kuehnlein W, Linke J and Sergienko G 2004 *Physica Scripta* **T111** 163
[6] Temmerman G D, Zielinski J J, van der Meiden H, Melissen W and Rapp J 2010 *Applied Physics Letters* **97** 081502
[7] Garkusha I E, Arkhipov N I, Klimov N S, Makhlaj V A, Safronov V M, Landman I and Tereshin V I 2009 *Physica Scripta* **T138** 014054
[8] Kajita S, Ohno N, Takamura S, Sakaguchi W and Nishijima D 2007 *Applied Physics Letters* **91** 261501
[9] Desjarlais M P, Kress J D and Colins L A 2002 *Physical Review E* **66** 025401
[10] Clerouin J, Renaudin P, Laudernet Y, Noiret P and Desjarlais M P 2005 *Physical Review B* **71** 064203
[11] Redmer R 1999 *Physical Review E* **59** 1073
[12] Kuhlbrodt S and Redmer R 2000 *Physical Review E* **62** 7191
[13] Kuhlbrodt S, Holst B and Redmer R 2005 *Contributions to Plasma Physics* **45** 73
[14] Holst B, French M and Redmer R *Physical Review B* **83** 235120
[15] Mao H K and Bell P M 1986 *Journal of Geophysical Research* **91** 4673
[16] Lakowicz J R (ed) 1994 *Topics in Fluorescence Spectroscopy: Volume 4: Probe Design and Chemical Sensing* (Springer)
[17] Ho C Y, Powell R W and Liley P E 1972 *Journal of Physical and Chemical Reference Data* **1** 279
[18] Ho C Y, Powell R W and Liley P E *Thermal Conductivity of the Elements: A Comprehensive Review* (*Journal of Physical and Chemical Reference Data. Supplements* vol 3) (American Institute of Physics)
[19] Linstrom P and Mallard W Nist chemistry webbook, nist standard reference database number 69

# Study on pulsed-discharge devices by using pulse-forming-network modules toward intense X-ray source

**Nobuyuki Anzai**, Daiki Takewaki, Fumitaka Tachinami, Kazumasa Takahashi, Toru Sasaki*, Tsukasa Aso, Takashi Kikuchi, and Nob. Harada**

Department of Electrical Engineering, Nagaoka University of Technology, Kamitomioka 1603-1, Nagaoka, Niigata, Japan

E-mail: *sasakit@vos.nagaokaut.ac.jp, **s125046@stn.nagaokaut.ac.jp

**Abstract.** A pulsed-power generator with a high rate of current rise was studied toward generating intense X-ray source from an X-pinch plasmas. The pulsed-power generator consists of 48 pulse-forming-network (PFN) modules with a three-stage of LC ladder circuit. To evaluate the rate of current rise for the pulsed-power generator, we demonstrated the short circuit experiments with low operation voltage. The rate of current rise depends on the number of PFN modules due to the decrease of inductance of PFN. The rate of current rise for 48 PFN modules at 10 kV of an operation voltage is estimated to be 0.1 kA/ns. To predict the rate of current rise for the requirement to obtain the intense X-ray from the X-pinch, the circuit simulation was demonstrated. The results indicated that the operation voltage requires over 70 kV for the rate of current rise of 1 kA/ns.

## 1. Introduction

An X-ray source generated by a compact pulsed-power device is an important tool for observing an interior of dense plasma such as warm dense matter (WDM), a lithography, and so on [1, 2, 3, 4, 5]. An X-pinch plasma [6, 7, 8], which is several thin wires crossed a load, is one of the intense X-ray source with the small source size. When the pulsed-power discharge occurs, the metallic thin wires are ablated and the large currents concentrate in the crossing point. The Lorentz force drives the implosion of the ablated wires, due to the interaction of the large currents and the generated magnetic field. However, to obtain the X-ray source for the X-pinch plasma, a rate of current rise of the pulsed-power generator is required to be 1 kA/ns, i.e. 100 kA of current and 100 ns of current rising time [9, 10, 11, 12]. To achieve the required high rate of current rise for generating X-pinch plasma, the pulsed-power generators have the pulse compression system and/or the relatively high voltage system [9, 13, 14]. However, the huge pulsed-power generators are difficult to use the applications such as the plasma diagnostics. To realize the tabletop X-pinch system, the inductance and the operation voltage are key issue to reduce the size of the system.

To generate X-ray source by using X-pinch system, a pulse forming network (PFN) module was used, having the advantages of holding high peak currents, ease of fabrication, and so on [15]. From a previous study [5, 16, 17], we have optimized the circuit topology of the PFN

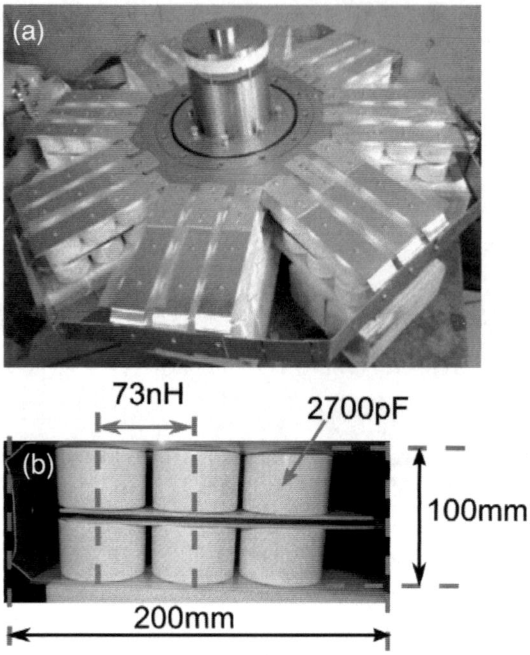


**Figure 1.** Photograph of X-ray source obtained by the X-pinch driven by the PFN modules. (a) overview of the X-pinch device, (b) modules of PFN.

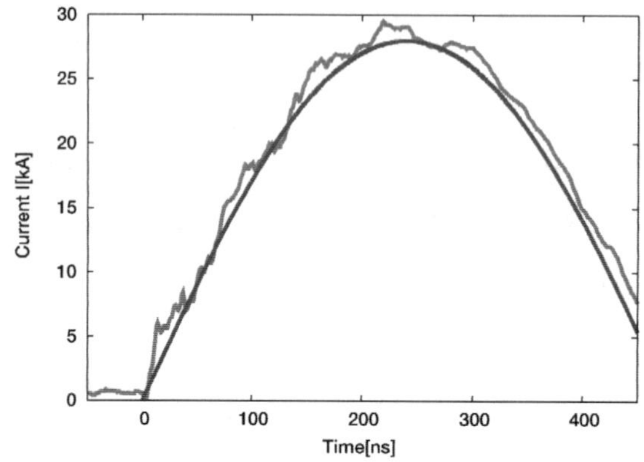


**Figure 2.** Typical current waveform at 10 kV of the operation voltage and the obtained fitting current waveform from the circuit simulation by ATP-EMTP[18].

module through circuit simulations. The results indicated that a module as three stage LC-ladder of PFN was suitable for the X-pinch light source. The number of module has also been estimated to be 48 with the 25 nH of the circuit inductance [17]. However, from the previous experimental results, the circuit inductance of the X-pinch system was estimated to be 60 nH. Thus, we should understand the X-pinch system to obtain 1 kA/ns of the rate of current rise.

In this study, we have evaluated the rate of current rise of the X-pinch system by using a PFN module from experimental observation and numerical estimation. To understand the behavior of the rate of current rise, the dependence on the number of PFN modules and the required operation voltage are evaluated.

## 2. Experimental Setup

Figure 1 is a photograph of X-ray source obtained by the X-pinch driven by the PFN modules. In order to reduce the inductance of the X-pinch system, the PFN modules were coaxially arranged in octagon plates, in which 48 PFN modules can be connected to drive the X-pinch. The PFN module as shown in Fig. 1(b) consists of a three-stage of LC ladder circuit. To obtain the high rate of current rise and the sufficient input energy for X-pinch, the capacitance and the inductance are considered. The inductance of the PFN module is determined by the size of capacitor. To reduce the inductance, the smaller ceramic and high-voltage capacitor is selected to be 2700 pF. The conductors of the PFN module makes the inductance. The inductance of the PFN modules are estimated to be 73 nH from the theoretical evaluation. The discharge gap switch, which is a self-breakdown due to the applied voltage, consists of the center of X-pinch system as shown in Fig. 1(a). From the previous experimental evaluation, the total inductance of the X-pinch system is estimated to be 60 nH.

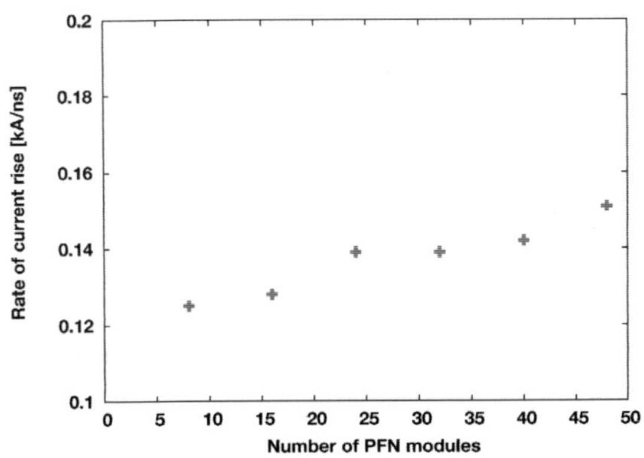


**Figure 3.** Rate of current rise as a function of the number of PFN modules with 10 kV of the operation voltage.

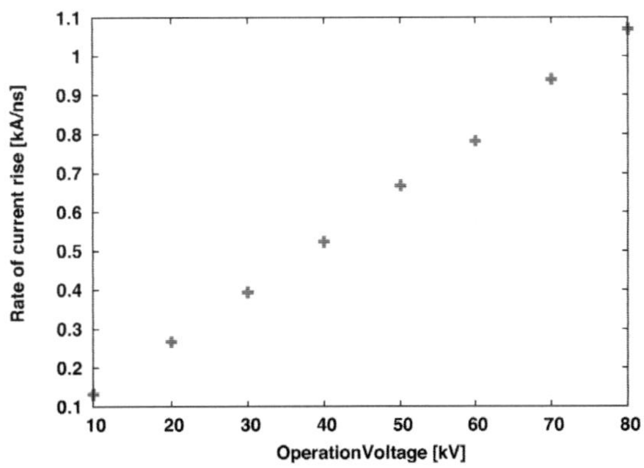


**Figure 4.** Numerical evaluation of the rate of current rise as a function of the required operation voltage with 48 of PFN modules.

Figure 2 shows the typical current waveform at 10 kV of the operation voltage and the obtained fitting current waveform from the circuit simulation by the Alternative Transients Program-Electromagnetic Transients Program (ATP-EMTP) [18]. The experimentally obtained peak current and rise time were estimated to be 30 kA and 200 ns, respectively. From these evaluations, the rate of current rise is estimated to be 0.15 kA/ns. To understand the X-pinch system, we evaluate the rate of current rise dependence on the number of PFN modules and the required operation voltage.

## 3. Experimental results and Discussions

Figure 3 is the rate of current rise as a function of the number of PFN modules with 10 kV of the operation voltage. The PFN modules for the X-pinch system arranges the co-axial symmetry for each experiments. The result indicates that the number of PFN modules depends on the rate of current rise. The improvement of the rate of current rise for the increase of the PFN modules is the reduction of inductance due to the parallel connection of PFN modules. Additionally, the rate of current rise increases from the peak current because of the increase of the capacitance of PFN modules.

Figure 4 is the numerical evaluation of the rate of current rise as a function of the required operation voltage with 48 of PFN modules. The results indicate that the increase of the operation voltage is almost linearly the increase of the rate of current rise. To obtain the X-ray from the X-pinch, the required operation voltage for ensured to the rate of current rise is estimated to be about 70 kV. Therefore, we should take care the insulation of the X-pinch system.

From the these results, to construct the table-top X-pinch system, we should consider to reduce the inductance of the X-pinch system, that is, the operation voltage decreases. Thus, the structure of the gap switch and the size of the X-pinch system will be considered to decrease the inductance of the X-pinch system.

## 4. Concluding Remarks

The pulsed-power generator with a high rate of current rise was studied toward generating intense X-ray source from an X-pinch plasmas. The pulsed-power generator consists of 48 PFN modules with a three-stage of LC ladder circuit. To evaluate the rate of current rise for the pulsed-power generator, we demonstrated the short circuit experiments with low operation voltage. The rate of current rise depends on the number of PFN modules due to the decrease of inductance of

PFN. The rate of current rise for 48 PFN modules at 10 kV of an operation voltage is estimated to be 0.1 kA/ns. To predict the rate of current rise for the requirement to obtain the intense X-ray from the X-pinch, the circuit simulation was demonstrated. The results indicated that the operation voltage requires over 70 kV for the rate of current rise of 1 kA/ns.

To demonstrate the table-top X-pinch system, we will consider the structure of the gap switch and the size of the X-pinch system to reduce the operation voltage.

## Acknowledgements

This work was partly supported by Grant-in-Aid for Challenging Exploratory Research from Japan Society for the Promotion of Science (24656184) and by the Murata Science Foundation.

## References

[1] Higashiguchi T, Otsuka T, Yugami N, Jiang W, Endo A, Li B, Dunne P and O'Sullivan G 2012 *Applied Physics Letters* **100** 014103
[2] Harvey-Thompson A J, Lebedev S V, Patankar S, Bland S N, Burdiak G, Chittenden J P, Colaitis A, De Grouchy P, Doyle H W, Hall G N, Khoory E, Hohenberger M, Pickworth L, Suzuki-Vidal F, Smith R A, Skidmore J, Suttle L and Swadling G F 2012 *Phys. Rev. Lett.* **108**(14) 145002
[3] Knapp P F, Pikuz S A, Shelkovenko T A, Hammer D A and Hansen S B 2012 *Physics of Plasmas* **19** 056302
[4] Stygar W A, Ives H C, Fehl D L, Cuneo M E, Mazarakis M G, Bailey J E, Bennett G R, Bliss D E, Chandler G A, Leeper R J, Matzen M K, McDaniel D H, McGurn J S, McKenney J L, Mix L P, Muron D J, Porter J L, Ramirez J J, Ruggles L E, Seamen J F, Simpson W W, Speas C S, Spielman R B, Struve K W, Torres J A, Vesey R A, Wagoner T C, Gilliland T L, Horry M L, Jobe D O, Lazier S E, Mills J A, Mulville T D, Pyle J H, Romero T M, Seamen J J and Smelser R M 2004 *Physical Review E* **69** 046403
[5] Sasaki T, Miki Y, Tachinami F, Saito H, Takahashi T, Anzai N, Kikuchi T, Aso T and Harada N 2013 *Nuclear Instruments and Methods in Physics Research Section A: Accelerators, Spectrometers, Detectors and Associated Equipment* –
[6] Beg F N, Krushelnick K, Lichtsteiner P, Meakins A, Kennedy A, Kajumba N, Burt G and Dangor A E 2003 *Applied Physics Letters* **82** 4602
[7] Shelkovenko T A, Pikuz S A, Sinars D B, Chandler K M and Hammer D A 2002 *Physics of Plasmas* **9** 2165
[8] Shelkovenko T A, Sinars D B, Pikuz S A and Hammer D A 2001 *Physics of Plasmas* **8** 1305
[9] Wu J, Sun T P, Wu G, Wang L P, Han J J, Li M, Cong P T, Qiu A C and Lv M 2011 *Physics of Plasmas* **18** 052702
[10] Kharlov A V, Kovalchuk B M and Zorin V B 2006 *Review of Scientific Instruments* **77** 123501
[11] Shelkovenko T A, Pikuz S A, Song B M, Chandler K M, Mitchell M D, Hammer D A, Ivanenkov G V, Mingaleev A R and Romanova V M 2005 *Physics of Plasmas* **12** 033102
[12] Aranchuk L E and Larour J 2009 *IEEE TRANSACTIONS ON PLASMA SCIENCE* **37** 575
[13] Mitchell I, Bayley J M, Chittenden J, Worley J F, Dangor A, Haines M and Choi P 1996 *Review of Scientific Instruments* **67** 1533–1541
[14] Greenly J B, Douglas J D, Hammer D A, Kusse B R, Glidden S C and Sanders H D 2008 *Review of Scientific Instruments* **79** 073501
[15] Caballero L and Smith P 2009 *Plasma Science, IEEE Transactions on* **37** 1948–1953
[16] Miyamoto T, Tachinami F, Sasaki T, Kikuchi T and Harada N 2011 *The Seventh Conference on Inertial Fusion Sciences and Applications (IFSA2011)* p We_107
[17] Tachinami F, Anzai N, Sasaki T, Kikuchi T and Harada N 2013 *Nuclear Instruments and Methods in Physics Research Section A: Accelerators, Spectrometers, Detectors and Associated Equipment* –
[18] SCOTT-MEYER W 1984 *ATP Rule Book* (Bonneville Power administration)

Journal of Physics: Conference Series **688** (2016) 012102 doi:10.1088/1742-6596/688/1/012102

# Study on Exploding Wire Compression for Evaluating Electrical Conductivity in Warm-Dense Diamond-Like-Carbon

**Toru Sasaki[1], Kazumasa Takahashi[1], Takahiro Kudo[1], Takashi Kikuchi[1,2], Tsukasa Aso[1], Nob. Harada[1], Shinsuke Fujioka[3], and Kazuhiko Horioka[4]**

[1] Department of Electrical Engineering, Nagaoka University of Technology, Kamitomioka 1603-1, Nagaoka, Niigata, Japan
[2] Department of Nuclear Safety System Engineering, Nagaoka University of Technology, Kamitomioka 1603-1, Nagaoka, Niigata, Japan
[3] Institute of Laser Engineering, Osaka University, Yamadaoka 2-6, Suita, Osaka, Japan
[4] Department of Energy Sciences, Tokyo Institute of Technology, Nagatsuta 4259, Yokohama, Kanagawa, Japan

E-mail: `sasakit@vos.nagaokaut.ac.jp`

**Abstract.** To improve a coupling efficiency for the fast ignition scheme of the inertial confinement fusion, fast electron behaviors as a function of an electrical conductivity are required. To evaluate the electrical conductivity for low-Z materials as a diamond-like-carbon (DLC), we have proposed a concept to investigate the properties of warm dense matter (WDM) by using pulsed-power discharges. The concept of the evaluation of DLC for WDM is a shock compression driven by an exploding wire discharge with confined by a rigid capillary. The qualitatively evaluation of the electrical conductivity for the WDM DLC requires a small electrical conductivity of the exploding wire. To analyze the electrical conductivity of exploding wire, we have demonstrated an exploding wire discharge in water for gold. The results indicated that the electrical conductivity of WDM gold for 5000 K of temperature has an insulator regime. It means that the shock compression driven by the exploding wire discharge with confined by the rigid capillary is applied for the evaluation of electrical conductivity for WDM DLC.

## 1. Introduction

Critical to fast ignition [1] is the transport of the laser-generated fast electrons and their associated heating of compressed DT fuel. The coupling efficiency of laser energy to these fast electrons and the energy deposited in the fuel should be improved the cone materials. Improving coupling efficiency, we should consider the behaviors of unstoppable, un-ejected, and diverging fast electrons. From the numerical results, the low-Z cone is expected to be the improvement of coupling efficiency, because the fast electrons are scattered by the large Coulomb potential of highly charged ions in high-Z cone wall [2]. On the other hand, the transport of fast electrons in the cone depends on the electrical conductivity in warm dense matter (WDM) state [3].

From above evaluations, the diamond-like-carbon (DLC)[4] cone, which is one of the low-Z cone, is promised to increase the coupling efficiency due to the redaction of stopping power in cone compared to the high-Z cone. The properties of DLC respects the diamond and the

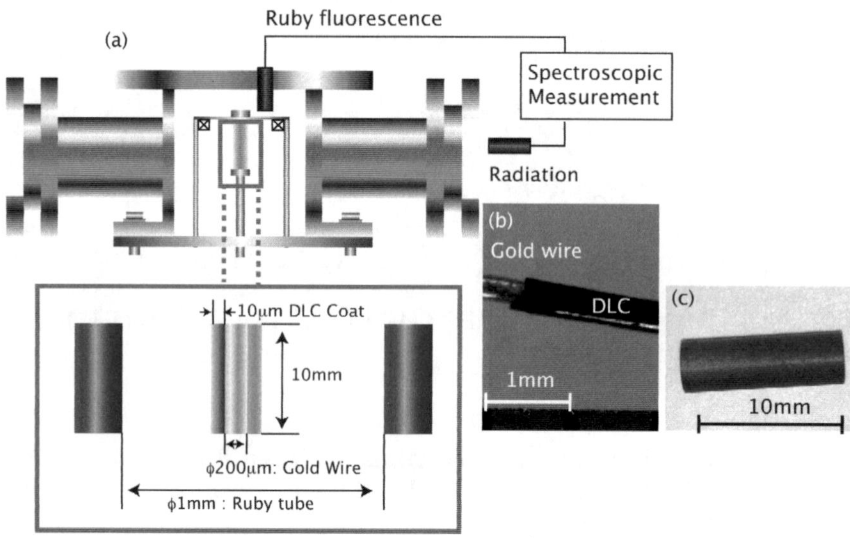


**Figure 1.** Proposal images for shock compression driven by an exploding wire discharge with confined by a rigid capillary. (a) Side view of experimental setup, (b) DLC coated on the gold wire, (c) Ruby tube as a rigid capillary.

graphite, however, materials in WDM state are in a complex area. The DLC is also made by the plasma-enhanced chemical vapor deposition on the substrate. Therefore, the properties of vapor DLC is difficult to obtain the conventional method.

To evaluate the electrical conductivity in DLC WDM, we propose a concept to investigate the WDM properties of insulator by using pulsed-power discharges. The concept of the evaluation of electrical conductivity for WDM DLC is a shock compression driven by an exploding wire discharge with confined by a rigid capillary. The qualitatively evaluation of the electrical conductivity for the WDM DLC requires a small electrical conductivity of the exploding wire. To analyze the electrical conductivity of exploding wire, we have demonstrated an exploding wire discharge in water for gold.

## 2. Proposed method

The WDM generation by using pulsed-power discharge is qualitatively evaluation of the electrical conductivity and the other plasma parameters. However, the pulsed-power discharge is difficult to make the WDM for insulator. To evaluate the electrical conductivity for DLC WDM, the shock compression driven by an exploding wire discharge with confined by a rigid capillary is considered as shown in Fig. 1. The exploding wire has a huge ablation pressure approximately a few GPa. Thus, the pressure of exploding wire drives the shock heating for the insulator as the DLC membrane which is coated on the wire. The heated DLC membrane state is observed by the ruby fluorescence for the pressure and the emission spectrum for the temperature. As shown in Fig. 1 (b), we can fabricate the DLC membrane coated on the wire for gold.

The electrical conductivity is measured by the time evolution of voltage-current waveforms. Therefore, the electrical conductivity for the exploding wire should be small compared to that for DLC WDM. The electrical conductivity with the heated diamond at solid density have been reported to be over $10^4$ S/m with 1 eV of temperature [3]. It means that the exploding wire should be less than $10^4$ S/m. To confirm the electrical conductivity of the exploding wire, we have evaluated the electrical conductivity of WDM.

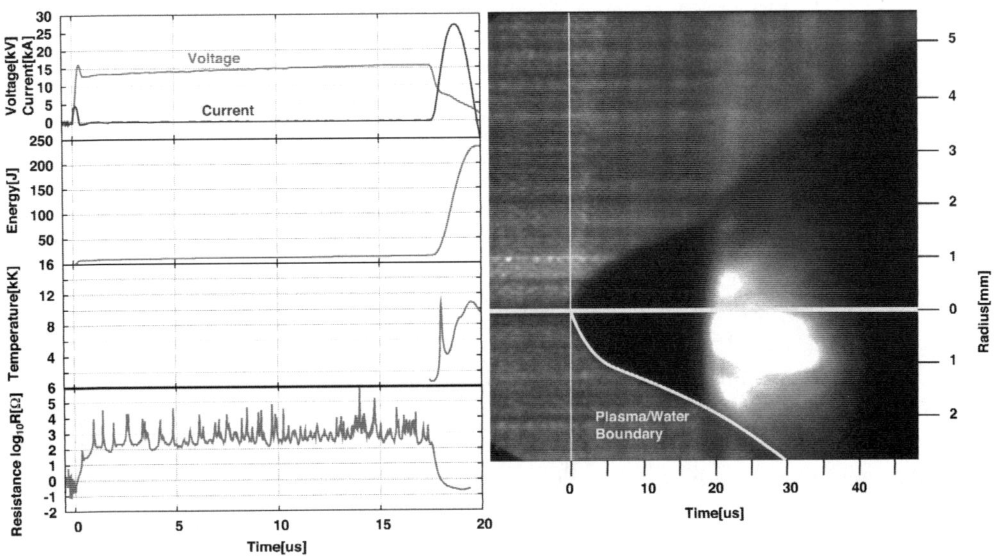


**Figure 2.** Typical waveforms, input energy history, temperature evolution, resistance evolution, and evolution of wire/plasma boundary of gold wire having 24 mm length with 100 $\mu$m in diameter.

## 3. Experimental results and Discussions

To evaluate the electrical conductivity for the exploding wire as the function of density and temperature, we have demonstrated by the exploding wire in water. We used thin gold wires with 50 - 100 $\mu$m in diameter and 18 - 38 mm in length as the load [5, 6, 7]. From previous results as Ref. [5], we have found the high resistance for copper and silver in warm dense regime. Therefore, the smaller electrical conductivity in the exploding wire should be confirmed to the several materials.

Figure 2 shows typical time evolutions of the voltage-current waveform, the input energy, the temperature, the resistance, and the hydrodynamic behavior for gold wire with 100 $\mu$m in diameter and 18 mm in length. The wire/plasma boundary and its temperature strongly depend on the input energy history. We can see that the resistance of wire/plasma respected by the voltage-current waveforms is estimated to be $10^3$ $\Omega$ from the beginning of discharge to 20 $\mu$s. The resistance of wire/plasma constitutes the resistance of water. The density of wire/plasma is proportional to the wire/plasma radius. Thus, the exploding wire of gold in dense regime has a highly resistance.

Figure 3 shows the electrical conductivities as a function of density at temperature of 5000 K $\pm$ 10 % for gold. Theoretical conductivities based on the models of Spitzer [8], Lee-More [9], and Ichimaru [10] at 5000 K are also shown in Fig. 3. From the previous section, we discuss that the electrical conductivity for the exploding wire should be small compared to that for DLC WDM. The results indicated that the electrical conductivity for observing DLC WDM is possible to measure the exploding wire for gold with well-defined density control, in which ranges from $0.1\rho_s$ to $0.001\rho_s$. It means that the shock compression driven by the exploding wire discharge with confined by the rigid capillary is applied for the evaluation of electrical conductivity for WDM DLC.

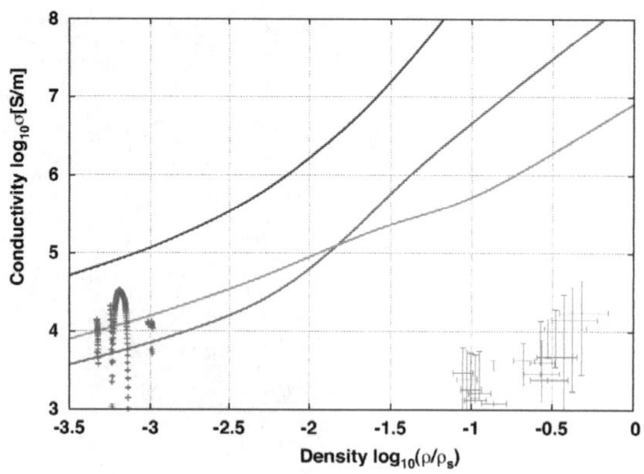


**Figure 3.** Electrical conductivity of gold as a function of density at 5000 ± 10%. The red line reveals the Ichimaru conductivity model, the green line denotes the Lee-More conductivity model, and the blue line indicates the Spitzer conductivity model.

## 4. Concluding Remarks

To evaluate the electrical conductivity for low-Z materials as a DLC, we have proposed a concept to investigate the properties of WDM by using pulsed-power discharges. The concept of the evaluation of DLC for WDM is a shock compression driven by an exploding wire discharge with confined by a rigid capillary. The qualitatively evaluation of the electrical conductivity for the WDM DLC requires a small electrical conductivity of the exploding wire. To analyze the electrical conductivity of exploding wire, we have demonstrated an exploding wire discharge in water for gold. The results indicated that the electrical conductivity of WDM gold for 5000 K of temperature has an insulator regime. It means that the shock compression driven by the exploding wire discharge with confined by the rigid capillary is applied for the evaluation of electrical conductivity for WDM DLC.

## Acknowledgements

This work was partly supported by the NIFS Collaboration Research programs (NIFS13KUGK073).

## References

[1] Kodama R, *et. al.*, 2001 *Nature* **412** 798
[2] Johzaki T, *et. al.*, 2011 *Nuclear Fusion* **51** 073022
[3] McKenna P, *et. al.*, 2011 *Physical Review Letters* **106** 185004
[4] Fujioka S, *et. al.*, 2012 *Plasma Physics and Controlled Fusion* **54** 124042
[5] Sasaki T, *et. al.*, 2006 *Laser and Particle Beams* **24** 371
[6] Sasaki T, *et. al.*, 2008 *Journal of Physics: Conference Series* **112** 042026
[7] Sasaki T, *et. al.*, 2010 *Physics of Plasmas* **17** 084501
[8] Spitzer Jr L and Härm R 1953 *Physical Review* **89** 977
[9] Lee Y T and More R M 1984 *Physics of Fluids* **27** 1273
[10] Kitamura H and Ichimaru S 1995 *Physical Review E* **51** 6004

# Study on AC-DC Electrical Conductivities in Warm Dense Matter Generated by Pulsed-power Discharge with Isochoric Vessel

**Toru Sasaki***, **Takumi Ohuchi, Takuya Takahashi, Yoshinari Kawaguchi, Hirotaka Saito, Yasutoshi Miki, Kazumasa Takahashi, Takashi Kikuchi, Tsukasa Aso, and Nob. Harada**

Department of Electrical Engineering, Nagaoka University of Technology, Kamitomioka 1603-1, Nagaoka, Niigata, Japan

E-mail: *sasakit@vos.nagaokaut.ac.jp

**Abstract.** To observe AC and DC electrical conductivity in warm dense matter (WDM), we have demonstrated to apply the spectroscopic ellipsometry for a pulsed-power discharge with isochoric vessel. At 10 $\mu$s from the beginning of discharge, the generated parameters by using pulsed-power discharge with isochoric vessel are 0.1 $\rho_s$ ($\rho_s$: solid density) of density and 4000 K of temperature, respectively. The DC electrical conductivity for above parameters is estimated to be $10^4$ S/m. In order to measure the AC electrical conductivity, we have developed a four-detector spectroscopic ellipsometer with a multichannel spectrometer. The multichannel spectrometer, in which consists of a 16-channel photodiode array, a two-stages voltage adder, and a flat diffraction grating, has 10 MHz of the frequency response with covered visible spectrum. For applying the four-detector spectroscopic ellipsometer, we observe the each observation signal evolves the polarized behavior compared to the ratio as $I_1/I_2$.

## 1. Introduction

Warm dense matter (WDM) is one of the states of matter, in which is neither the solid nor the plasma. The related physical phenomena in WDM are the implosion process of fuel pellet in inertial confinement fusion [1], the formation of giant planets (e.g. Jupiter) [2], and so on. The WDM is unclear physical properties such as the equation of state and the transport properties of both electrons and ions. To understand the physical properties, the quantum molecular dynamics or ab-initio simulations have predicted the AC-DC electrical conductivity and the equation of state in WDM regime [3, 4]. Desjarlais [5] has been predicted the low density WDM in disagreement with the Drude model. Yoneda, *et. al.* [6] has experimentally demonstrated the ultrashort-pulse ellipsometric pump-probe experiments and the pump-probe transmission experiments, respectively. However, these experimental results have obtained the single wavelength of probe light. Thus, the experimental data is lacking to evaluate the numerical results.

Recently, the isochoric heating of foamed metal using pulsed power discharge as a making technique of warm dense matter have been developed by Amano, *et. al.*[7]. Because of usable transparent tamper as a sapphire, the method is possible to observe in WDM with well-defined density-temperature. The lifetime of generated WDM in this method is longer than that of laser

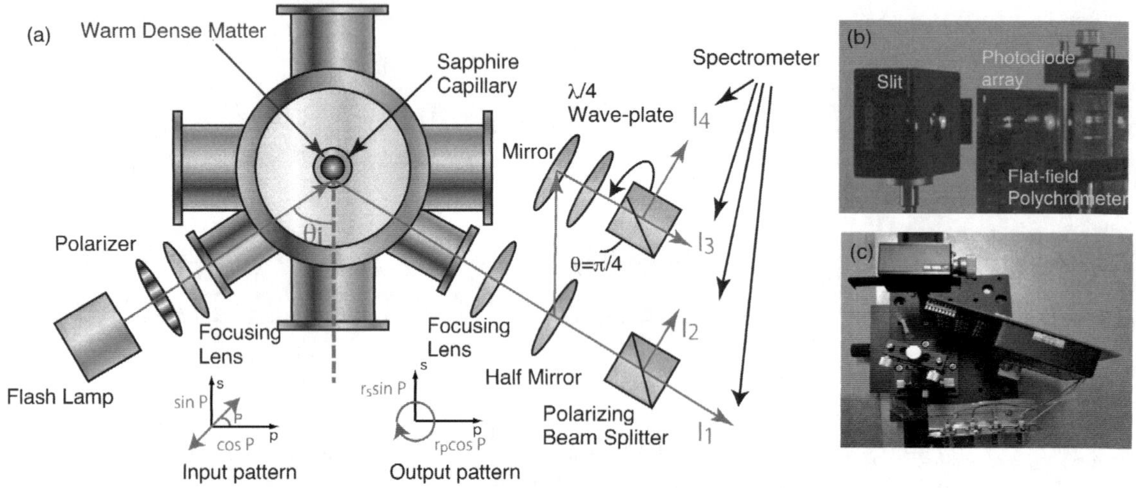


**Figure 1.** Experimental setup for observing AC and DC electrical conductivities in WDM by using pulsed-power discharge with isochoric vessel. (a)schematic diagram for observing AC conductivity, (b) photograph of spectrometer, and (c) topview of spectrometer.

experiments. Because of prolonged timescale, a spectroscopic ellipsometry is possible to apply the method.

The spectroscopic ellipsometry is generally used as the condense matter physics [8]. To observe the AC electrical conductivity as an optical properties, the ellipsometer requires four sensitive spectrometers, in which observe the reflectivity and the phase for each polarization. In case of WDM generated by the pulsed-power discharges, the typical timescale of emission from WDM is several tens of microseconds [7]. The spectrometer with fast response is required to analyze the optical properties observed by the spectroscopic ellipsometry.

To observe AC and DC electrical conductivity in warm dense matter (WDM), we have demonstrated to apply the spectroscopic ellipsometry for a pulsed-power discharge with isochoric vessel. In order to measure the AC electrical conductivity, we have developed a four-detector spectroscopic ellipsometer with a multichannel spectrometer and observed the polarization parameters.

## 2. Experimental Setup

Figure 1 shows an experimental setup for observing AC and DC electrical conductivities in WDM by using pulsed-power discharge with isochoric vessel [7].To achieve isochoric heating, copper metallic foam with pore sizes ranging from 50 to 600 $\mu$m and with $\sim$ 90 % porosity was packed into a hollow sapphire capillary ($\phi$ 5 mm $\times$ 10 mm), in which can sustain up to 4 GPa. To avoid surface creepage on the sapphire, the interior pressure of chamber is less than $10^{-3}$ Pa by using turbo-molecular and rotary pumps. Low-inductance capacitors (3 or 6 $\times$ 1.87$\mu$F) were charged up to 15 kV to ensure vaporization of the foam. Time evolutions of current $I(t)$ and voltage $V(t)$ in the foam and plasma were measured by a Rogowski coil and by high-voltage probes.

To observe the AC electrical conductivity on WDM, a four-detector spectroscopic ellipsometer which is used to determine reflectivity and phase for each polarization is developed. The four-detector spectroscopic ellipsometer consists of a flash lamp, a polarizer, beam polarizers, and spectrometers. The flush lamp as an probe light for observing generated by Xe gas filled pulsed-discharge with 20 $\mu$s of duration. The radiation temperature of flush lamp is estimated to be

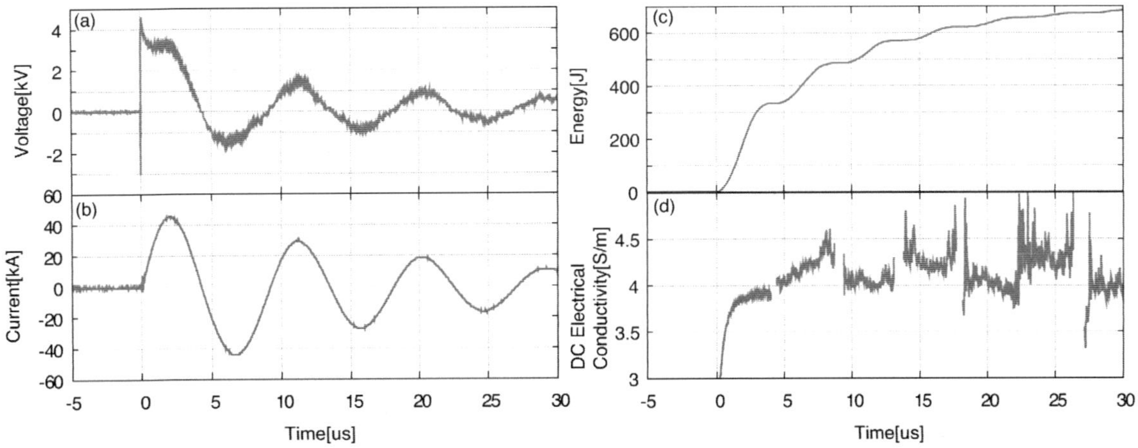


**Figure 2.** Typical time-evolutions of (a) volatage, (b) current, (c) input energy, and DC electrical conductivity for copper with $0.1\rho_s$ of density.

7000 K with the continuum spectrum. The flush lamp is triggered by a delay pulser after the onset of the discharge as generating WDM. The light from flush lamp polarized at 45 degree, in which means equivalent p- and s-polarized components. The probe light arrives at 60 degree as $\theta_i$ from the target normal.

The four-detector spectroscopic ellipsometer has two components denoted $I_1/I_2 \propto |r_p|^2/|r_s|^2$ and $(I_3 - I_4) / (I_3 + I_4) \propto |r_s||r_p| \sin\delta / (|r_s|^2 + |r_p|^2)$. To observe the spectroscopic information, spectrometers have been constructed as shown in Fig 1 (b) and (c). The spectrometer consists of a slit, a flat-field polychrometer, and detectors made with 16ch photodiode array having voltage adders. The voltage adders merge the detection signals from 16ch of the signals of photodiode to 4ch signals. The merged signals means the regime of observation wavelength corresponded to 340 - 370 nm, 370 - 530 nm, 530 - 680 nm, and 680 - 800 nm. Each signal is observed by the fast data logger. The time resolution of the 16ch photodiode array having voltage adders is estimated to be 10 MHz from the sinusoidal LED signals. The detectors are calibrated using data from cold copper.

## 3. Experimental results and Discussions

Figure 2 shows typical time-evolutions of voltage, current, input energy and DC electrical conductivity for copper foam with $0.1\rho_s$ of density. The voltage waveform eliminates the inductive voltage in the circuit. As shown in Figs. 2 (a) and (b), the peak voltage and current are 3 kV and 45 kA, respectively. The oscillation frequency is estimated to be 125 kHz. The input energy achieves 400 J at 10 $\mu$s from the beginning of discharge. The copper foam with $0.1\rho_s$ of density ensure the vaporization at 5 $\mu$s from the beginning of discharge. From spectroscopic observations, the temperature of copper is estimated to be about 4000 K. As shown in Fig. 2 (d), the DC electrical conductivity increases at the beginning of discharge because of the decrease of effective cross section for copper foam compared to the solid copper. After the ablation of copper foam, the DC electrical conductivity is estimated to be $10^4$ S/m.

Figure 3 shows typical time-evolutions of detection signals obtained by the four-detector spectroscopic ellipsometer. The onset time as $t = 0$ indicates the triggered timing of the flash lamp. The results indicates that the each observation signal evolves the polarized behavior compared to the ratio as $I_1/I_2$ after the onset time of the flash lamp when the timing for WDM generation corresponds to 8 $\mu$s. However, the observation signals also evolve an oscillating behaviors before the onset time of the flash lamp. The oscillation frequency of the observation

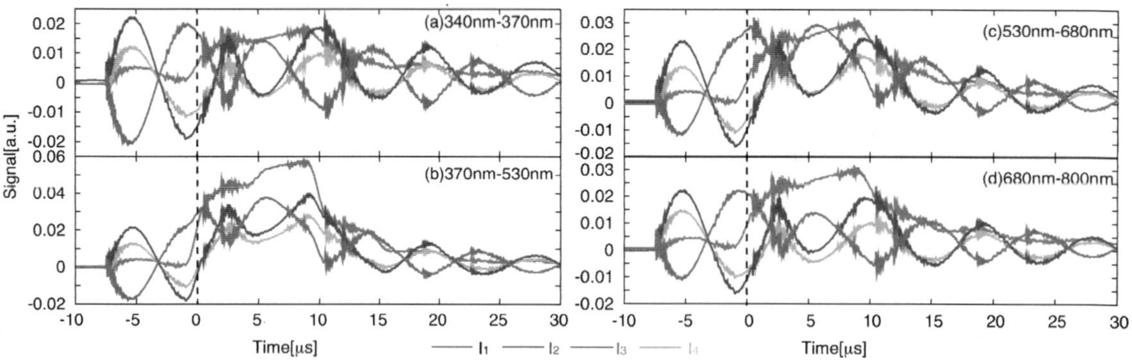


**Figure 3.** Typical time-evolutions of detection signals obtained by the four-detector spectroscopic ellipsometer, (a)340-370 nm, (b)370-530 nm, (c)530-680 nm, and (d)680-800 nm. The onset time as $t = 0$ indicates the triggered timing of the flash lamp.

signal before the onset time of the flash lamp corresponds to the oscillation frequency of the pulsed-power discharge for WDM generation. It means that the observation signals integrated the noise from the pulsed-power discharge for WDM generation. To confirm the AC electrical conductivity by using pulsed-power discharge, we should care the noise from the pulsed-power discharge for WDM generation and the intensity of the probe light.

## 4. Concluding Remarks

To observe AC and DC electrical conductivity in WDM, we have demonstrated to apply the spectroscopic ellipsometry for a pulsed-power discharge with isochoric vessel. At 10 $\mu$s from the beginning of discharge, the generated parameters by using pulsed-power discharge with isochoric vessel are 0.1 $\rho_s$ of density and 4000 K of temperature, respectively. The DC electrical conductivity for above parameters is estimated to be $10^4$ S/m. In order to measure the AC electrical conductivity, we have developed a four-detector spectroscopic ellipsometer with a multichannel spectrometer. The multichannel spectrometer, in which consists of a 16-channel photodiode array, a two-stages voltage adder, and a flat diffraction grating, has 10 MHz of the frequency response with covered visible spectrum. For applying the four-detector spectroscopic ellipsometer, we observe the each observation signal evolves the polarized behavior compared to the ratio as $I_1/I_2$. To confirm the AC electrical conductivity by using pulsed-power discharge, we should care the noise from the pulsed-power discharge for WDM generation and the intensity of the probe light. To observe the quantitatively AC electrical conductivity in WDM, the improvement of the light source and the detector will be considered.

## References

[1] Lindl J 1995 *Physics of Plasma* **2** 3933
[2] Davidson R C (ed) 2003 *Frontiers in High Energy Density Physics* (NATIONAL ACADEMIES)
[3] Mazevet S, Desjarlais M P, Collins L A, Kress J D and Magee N H 2005 *Physical Review E* **71** 016409
[4] Dharma-wardana M W C 2006 *Physical Review E* **73** 036401
[5] Desjarlais M P, Kress J D and Colins L A 2002 *Physical Review E* **66** 025401
[6] Yoneda H, Morikami H, Ueda K and More R M 2003 *Physical Review Letters* **91** 075004
[7] Amano Y, Miki Y, Sasaki T, Kikuchi T and Harada N 2012 *Review of Scientific Instruments* **83** 085107
[8] Collett E 1980 *Surf. Sci.* **96** 156

# DPSSL pumped 20-TW Ti:sapphire laser system for DD fusion experiment

**T Sekine[1], Y Hatano[1], Y Takeuchi[1], and T Kawashima[1]**

[1]Development Bureau Hamamatsu Photonics, K.K., 1820, Kurematsu-cho, Nishi-ku, Hamamatsu, Shizuoka 431-1202 Japan

E-mail: t-sekine@crl.hpk.co.jp

**Abstract.** A diode-pumped solid-state laser (DPSSL) pumped 20-TW output Ti:sapphire laser system has been developed. A diode-pumped Nd:glass laser with output energy of 12.7 J in 527 nm was used as a pump source for a 20-TW Ti:sapphire amplifier. A CeLiB6O10 nonlinear optical crystal was used as a frequency doubler of the Nd:glass DPSSL[1]. Figure 1 shows typical output pulse energy of the 20-TW amplifier as a function of pumping energy and a near field pattern. A 1.65 J pulse energy was obtained by 4.5 J pump energy. The amplified seed pulse is compressed to typically 60 fs as shown in Fig. 1 by a vacuumed pulse compressor with 80% of transmissivity. Encircled energy ratio, into a circled with 8 m diameter area, of far field pattern focused by off-axis parabolic mirror with F# of 3 is numerically evaluated to 40% at TW class output condition. Then focal intensity would reach to $10^{18}$W/cm$^2$. This all-DPSSL system contributes for stable and continual investigation of laser induced plasma experiment. We have succeeded continual and high efficient generation of DD fusion neutron from CD nano-particles by cluster fusion scheme using the 20-TW laser. A yield of ~105 neutrons per shot was stably observed during continuous 100 shots with repetition rate of 0.1Hz.

## 1. Introduction

After inventing of chirped-pulse-amplification (CPA) scheme[1], new technologies such as spectrum phase controlling, dielectric multilayer-coating for broadband wavelength and fabrication of high-quality large aperture laser material have been developed. Recently, 100 TW to 1 PW class laser system with laboratory scale have been commercialized. In these reasons, researches of electron or ion acceleration and neutron generation from laser induced plasma have been grown[2].These researches expect creation of some new industrial applications for medical, analysis and energy fields. For practical realization of generated particles-beams from laser induced plasma, stable characteristics are required for laser system. And low-maintenance and low running cost for laser system are also required. For stable operation, current pumping source of flash-lamp for green laser have to be replaced to laser-diode (LD) because LD has stable characteristics, high electrical efficiency and long lifetime. We have constructed all diode-pumped solid-state laser (DPSSL) pumped Ti:sapphire laser system. This system has been completely designed by us. And as a first demonstration in the world, a 20 TW output power have been performed. In this paper, one of our investigations of neutron generation using the 20-TW system has been presented.

## 2. Configuration of a DPSSL pumped 20 TW laser system

Figure 1 shows a schematic diagram of DPSSL pumped 20-TW Ti:sapphire system (MATSU-I). This system consist of Ti:sapphire oscillator, pulse stretcher, acousto-optic programmable dispersive filter, regenerative amplifier, 1st main amplifier for 2 TW output, 2nd main amplifier for 20 TW output and vacuumed pulse compressor. The regenerative amplifier and the 1st amplifier are pumped by two diode-pumped Nd:YAG lasers. Output energy are typically 250 mJ in 532 nm at 10Hz for each system. 2nd amplifier pumped by a diode-pumped Nd:glass laser (KURE-I). The 2-TW at 10Hz system is independently operated by programmable logic controller. The 20-TW system is operated at 1Hz limited by plasma experiment.

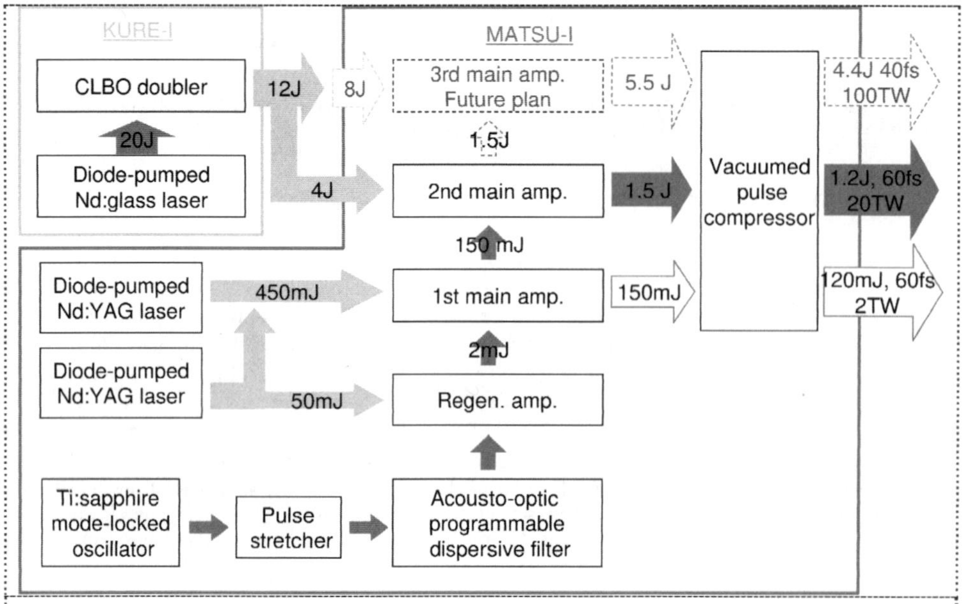


**Figure 1.** Configuration of DPSSL pumped 20 TW laser system.

## 3. Ti:sapphire laser system (MATSU-I)

The Ti:sapphire oscillator is mode-locked by semiconductor saturable absorber mirror with output characteristics of 33 fs pulse duration at 83 MHz repetition rate, 100 mW average power. Chirped pulse expanded by the pulse stretcher is amplified to several mJ by Ti:sapphire regenerative amplifier[6]. Ti:sapphire crystal with diameter of 12 mm and 30 mm are use for the 1st and the 2nd amplifier, respectively. In the 1st amplifier, seed pulse output from the regenerative amplifier is amplified to 150 mJ with 6 pass amplification by 450 mJ pump energy. In the 2nd amplifier, a 150 mJ of seed pulse is boosted-up to 1.65 J by 4 pass amplification with 5 J pumping energy. Figure 3 shows pulse energy of seed pulse at the 2nd amplifier as a function of pumping energy. Plot shows an experimental result and solid line shows a calculated result. Experimental result is good agreement with calculation. And also, Fig. 2 shows a NFP of output pulse of the 2nd amplifier. After that, beam aperture is expanded to factor of 3 by image relaying vacuumed telescope. The seed pulse is compressed to typical 60 fs by the vacuumed pulse compressor with 80% of transmissivity. Pulse duration measured by SPIDER method is shown in Fig. 2. In this experimental result, a 22 TW output have been obtained with 1.32 J with 60 fs at 1Hz. Encircled energy ratio, into a circled with 8 μm diameter area, of far field pattern focused by off-axis parabolic mirror with F# of 3 is numerically evaluated to 40% at low pulse energy condition. Transmissivity from end of the compressor to the focal point is about 70%. Then focal intensity would reach to $10^{18}$W/cm$^2$. Table 1 shows a typical characteristics of MATSU-I system.

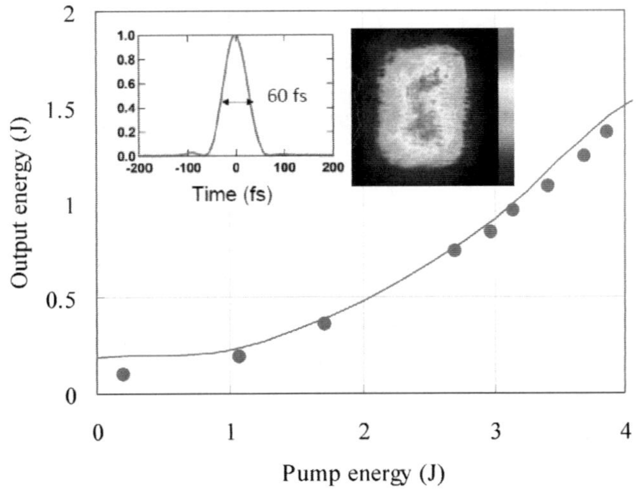


**Figure 2.** Output characteristics of TI:sapprhire laser.

**Table 1.** Typical specifications of 20 TW laser system.

| Parameter | Value |
|---|---|
| Peak power | 20 TW |
| Pulse energy on target | 1.2J |
| Pulse duration | 60fs |
| Focusability | 40% in 8 μm diameter |
| Focus intensity | 4 x $10^{18}$ W/cm$^2$ |

**4. Investigation of high-intensity laser applications**

We have succeeded continual and high efficient generation of DD fusion neutron from CD nano-particles by cluster fusion scheme using 20-TW MATSU-I laser. Figure 3 shows typical time-of flight(TOF) signal of neutron and neutron yields. A neutron TOF signal was clearly measured. A yield of ~$10^5$ neutrons per shot was stably observed during continuous 100 shots with repetition rate of about 0.1 Hz. Then the repetition rate is limited by performance of vacuum instruments. Generation of high energy deuterium ions with energy up to 700 keV was also observed. In this coulomb explosion physical mechanism, neutron yield is depends on focal intensity of irradiating laser. Therefore stable and high throughput high-intensity laser is required for investigation of new industrial applications.

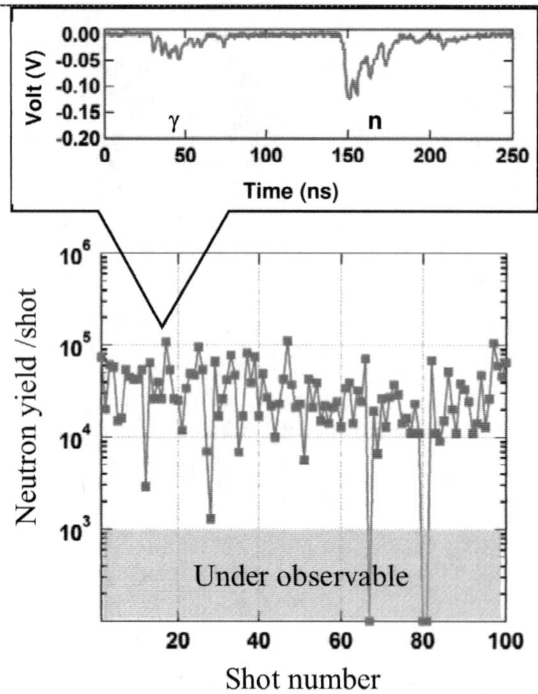


**Figure 3.** Figure with short caption (caption centred).

## 5. Conclusion

A DPSSL pumped 20-TW output Ti:sapphire laser system has been developed. Diode-pumped Nd:glass laser with output energy of 12.7 J in 527 nm was used as a pump source. Typical output characteristics with peak power of 1.2 J, pulse width of 60 fs at 1 Hz repetition were achieved. Schematic diagram shown in Fig. 1 and calculation curve shown in Fig. 3 indicate that a 5.5 J of amplification would be obtained by adding 3rd Ti:sapphire main amplifier with pumping of residual green-energy from KURE-I. In addition, the vacuumed pulse compressor is designed for maximum input pulse energy of 30 J. Development of all-DPSSL 100-TW to 1-PW system for investigation of new industrial application would be expected in near future.

*Reference*

[1] D. Strickland and G. Mourou, "Compression of amplified chirped optical pulses," Opt. Comm. 56 (6), 447–449 (1985).
[2] E. Esarey, et al., "Physics of laser-driven plasma-based electron accelerators," Rev. Mod. Phys. 81 (3), 1229–1285 (2009)
[3] R. Yasuhara, et al., "213 W average power of 2.4 GW pulsed thermally controlled Nd:glass zigzag slab laser with a stimulated Brillouin scattering mirror," Opt. Lett. 33 1711-1713 (2008).
[4] T. Sekine, et al., "84 dB amplification, 0.46 J in a 10 Hz output diode-pumped Nd:YLF ring amplifier with phase-conjugated wavefront corrector," Opt. Exp. 18, 13927-13934 (2010).
[5] T. Sekine, et al., "A 12.5 J high efficient second harmonics generation from CsLiB6O10 nonlinear crystal by diode-pumped Nd:glass laser", Opt. Exp. Submitted (2012).
[6] S. Matsuoka, et al., "1 TW, 10 Hz and 0.1 TW, 1 kHz All-Solid-State Femtosecond Lasers" Las. Eng., 34 (9), 610-615 (2006)

# Dressed Charge of Electron by Radiation Reaction

**Keita SETO[1*], Sen Zhang[2], James KOGA[3], Toseo Moritaka[1], Kensuke Homma[4,5], Yoshihide Nakamiya[4], Hideo NAGATOMO[1] and Kunioki MIMA[1,6]**
[1]Institute of Laser Engineering, Osaka University, Suita, Osaka, Japan.
[2]Okayama Institute for Quantum Physics, Kita-ku, Okayama, Japan.
[3]Quantum Beam Science Directorate, Japan Atomic Energy Agency, Kizugawa, Kyoto, Japan.
[4]TheGraduate School of Science, Hiroshima University, Higashihiroshima, Hiroshima, Japan
[5]International Center for Zetta-Exawatt Science and Technology, Ecole Polytechnique, Route de Saclay, Palaiseau Cedex, France
[6]TheGraduate School for the Creation of New Photonics Industries, Hamamatsu, Shizuoka, Japan.

* E-mail: keita.seto@eli-np.ro [1]

**Abstract.** With the progress of ultraintense short pulse laser technologies, the maximum intensity of the lasers has reached the order of $10^{22}$W/cm$^2$. Now, several institutes are aiming at higher intensities of over $10^{24\sim}$W/cm$^2$. It is expected that these lasers can investigate the regime of the ultra-intense field effects like electron-positron pair creation and annihilation, these high-intense laser fields enters into the non-linear QED regime. However, when an electron interacts with lasers with intensities over $10^{22}$W/cm$^2$, it has been predicted that the radiation reaction effect becomes significant. The strong radiation field can induce the fluctuation of QED vacuum (vacuum polarization or photon-photon scatterings). Though the propagator for fields is deformed by it, some kind of a polarized charge exists as the dress. In this paper, we will discuss about this dress of an electron generated by radiation in the description of classical physics. It leads to the avoidance of the mathematical difficulty of the radiation reaction problems.

## 1. Introduction

Recently, the plans of the construction of ultra-high intense lasers are becoming more concrete and one can hear some interesting plans for investigating new physics [1]. This new regime includes the QED region, like electron-positron pair creation/annihilation [2]. These processes are regarded as being generated by the laser intensities of $10^{24\sim}$W/cm$^2$. However, there is an important physical process before these, radiation reaction [3] which is the main problem addressed in this paper. Radiation reaction is a very basic process. When an electron is accelerated, it emits the Liénard-Wiechert electromagnetic field. If this radiation energy loss becomes significant, the electron feels an additional force via radiation mechanism. This feedback from radiation is called radiation reaction [4~6]. This radiation reaction remains the mathematical dificulty, namely the "run-away" effect [6]. In this paper, we discuss this radiation reaction in the QED vacuum [7,8] with the stabilization of the run-away effect [9-10, 13]. Finally, the theory leads to the "dressed charge" via quantum vacuum fluctuations, the fermionic 1-loop.

[1] Current affiliation: Extreme Light Infrastructure – Nuclear Physics (ELI-NP) / Horia Hulubei National Institute for R&D in Physics and Nuclear Engineering (IFIN-HH), 30 Reactorului St., Bucharest-Magurele, jud. Ilfov, P.O.B. MG-6, RO-077125, Romania

## 2. Radiation Reaction in Quantum Vacuum

What is reason why we need to consider QED vacuum fluctuations [7,8]? The answer is, the original theory of radiation reaction is equal to the model of an electron. H. A. Lorentz considered that an electron has the charge distribution of a spherical surface, each charge elements interact with others via the Liénard-Wiechert field [5, see Fig. 1]. When the elements are co-moving, the Liénard-Wiechert force has directivity. This field we call the "retarded field". P. A. M. Dirac suggested that the radiation field has not only the retarded field $F_{\mathrm{ret}}{}^{\mu\nu}$, but also an advanced field $F_{\mathrm{adv}}{}^{\mu\nu}$.

$$F_{\mathrm{LAD}}{}^{\mu\nu} = \left.\frac{F_{\mathrm{ret}}{}^{\mu\nu} - F_{\mathrm{adv}}{}^{\mu\nu}}{2}\right|_{x=x(\tau)} = -\frac{m_0\tau_0}{ec^2}\left(\frac{d^2w^\mu}{d\tau^2}w^\nu - \frac{d^2w^\nu}{d\tau^2}w^\mu\right) \tag{1}$$

This field on the electron's point $F_{\mathrm{LAD}}{}^{\mu\nu}$ is the field for radiation reaction (in this paper, we call it the LAD field). It should be regarded as the self interaction of the electron. This LAD field generates an additional acceleration and the equation of motion becomes,

$$m_0\frac{d}{d\tau}w^\mu = -eF_{\mathrm{ex}}{}^{\mu\nu}w_\nu - eF_{\mathrm{LAD}}{}^{\mu\nu}w_\nu . \tag{2}$$

This is the standard model of radiation reaction, named the Lorentz-Abraham-Dirac (LAD) equation [6]. But this equation has the mathematical difficulty of the exponential divergence $dw^\mu/d\tau = C^\mu(\tau)\times \exp(\tau/\tau_0)\to\infty$ during the small period, so called the "run-away" solution [6], the method for the stabilization has been required.

In QED, it allows the following processes. (1) An electron emits the photon field, (2) emitted photons enters the fermionic 1-loop (the sea of a "virtual" pair). (3) Then, this "virtual" pair production vanishes and goes back to the photons. The sea of the fermion 1-loop is regarded as a medium for light. This medium is called "QED vacuum" which is all around us according to QED. Since this effect leads the photon-photon scatterings, the propagator for fields should be modified and the permittivity of vacuum should be changed like $\varepsilon_0 \mapsto \varepsilon$. Therefore, we have to consider the correspondence between $F = (\mathbf{E},\mathbf{B})$ acting on an electron and $F_{\mathrm{LAD}} = (\mathbf{D},\mathbf{H})$ which we observe via the vacuum polarization. Now, the field $F^{\mu\nu}$ makes the field of the QED vacuum fluctuation $M^{\mu\nu}$ (Fig. 2). For putting the symbol $\langle\bullet|\circ\rangle$ as $\langle A|B\rangle \equiv A_{\mu\nu}B^{\mu\nu}$, the vacuum fluctuation (the corrected behavior of propagating fields) is derived from the first order of the Heisenberg-Euler Lagrangian;

$$L = -\frac{1}{4\mu_0}\langle F|F\rangle + \frac{\alpha^2\hbar^3\varepsilon_0{}^2}{360m_0{}^4c}[4\langle F|F\rangle^2 + 7\langle F|{*F}\rangle^2] \tag{3}$$

$$M^{\mu\nu} = -\frac{\alpha^2\hbar^3\varepsilon_0{}^2}{45m_0{}^4c^2}[4\langle F|F\rangle F^{\mu\nu} + 7\langle F|{*F}\rangle * F^{\mu\nu}] \tag{4}$$

J. Schwinger demonstrated the Larmor formula from the LAD field $F_{\mathrm{LAD}}{}^{\mu\nu}$ [11] and showed,

$$\partial_\mu F_{\mathrm{LAD}}{}^{\mu\nu}(x) = 0 . \tag{5}$$

From this Maxwell equation, $F_{\mathrm{LAD}}{}^{\mu\nu}$ is a source-less field which is observed at the outer position of an electron. Therefore, the LAD field is a superposition field of the bare field $F^{\mu\nu}$ and QED vacuum fluctuation $M^{\mu\nu}$,

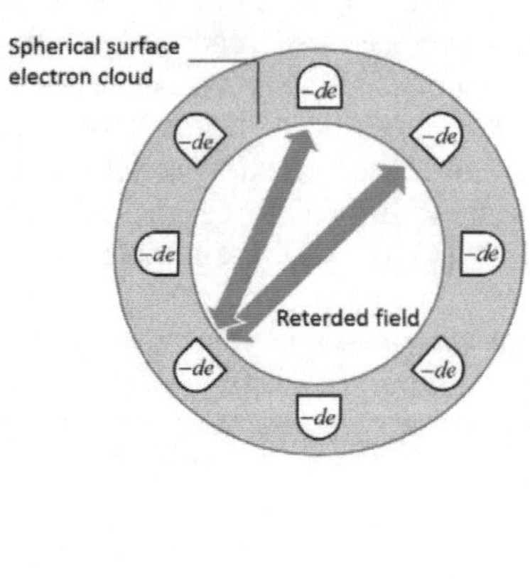


Fig.1 The electron model by Lorentz. The charge elements are distributed on a spherical surface. Each element interacts with others via the Liénard-Wiechert field (the retarded field).

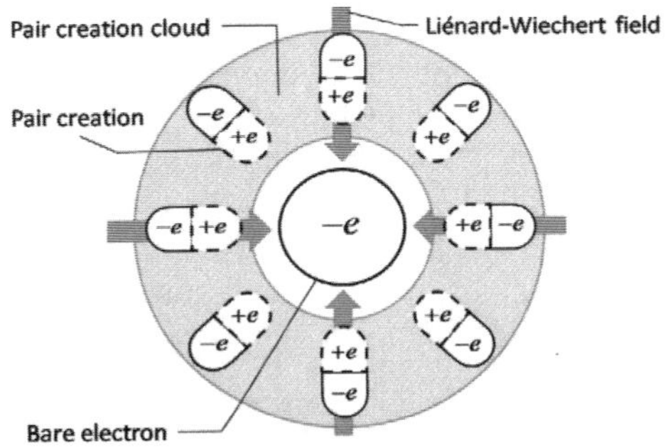


Fig.2 The new model of the electron. The bare electron dresses the QED vacuum fluctuation as a charge. This fluctuation comes from photon-photon scattering which is the diagram of the fermion 1-loop with two incoming photons and two outgoing photon. This dress has a total positive charge.

$$F^{\mu\nu} + \frac{1}{\varepsilon_0 c} M^{\mu\nu} = F_{\mathrm{LAD}}{}^{\mu\nu}. \tag{6}$$

Transforming Eq. (6) with the definition of (4) and using standard perturbation, the bare field becomes

$$F^{\mu\nu} = \frac{1}{1-\eta\langle F_{\mathrm{LAD}} \,|F_{\mathrm{LAD}}\rangle} F_{\mathrm{LAD}}{}^{\mu\nu}. \tag{7}$$

($\eta = 4\alpha^2\hbar^3\varepsilon_0 / 45 m_0{}^4 c^3$, $F_{\mathrm{LAD}\,\alpha\beta} F_{\mathrm{LAD}}{}^{\alpha\beta} \le 0$). If we take the limit of $\hbar \to 0 \Rightarrow \eta \to 0$, then $F^{\mu\nu} \to F_{\mathrm{LAD}}{}^{\mu\nu}$. From the above discussion, our new equation of motion becomes,

$$m_0 \frac{d}{d\tau} w^\mu = -\frac{e}{1-\eta\langle F_{\mathrm{LAD}} \,|F_{\mathrm{LAD}}\rangle} (F_{\mathrm{ex}}^{\mu\nu} + F_{\mathrm{LAD}}{}^{\mu\nu}) w_\nu. \tag{8}$$

We can demonstrate this equation is stable due to the factor of $1/[1-4\alpha^2\hbar^3\varepsilon_0 / 45 m_0{}^4 c^3 \, (F_{\mathrm{LAD}\,\alpha\beta} F_{\mathrm{LAD}}{}^{\alpha\beta})]$, and Eq.(8) can avoid run-away (the detail of demonstration is in Ref. [9] or more strictly in Ref.[10]).

## 3. Dress of Charge via Radiation Reaction

In this section, we focus on the term of the radiation reaction force in Eq. (8).

$$f_{\mathrm{reaction}}{}^{\mu} = -\frac{e}{1-\eta\langle F_{\mathrm{LAD}} \,|F_{\mathrm{LAD}}\rangle} F_{\mathrm{LAD}}{}^{\mu\nu} w_\nu \tag{9}$$

This formula comes from two parts. The first part is the LAD radiation reaction force $f_{\mathrm{LAD}}{}^{\mu} = -eF_{\mathrm{LAD}}{}^{\mu\nu} w_\nu$ and the second is the additional term.

$$f_{\mathrm{reaction}}{}^{\mu} = f_{\mathrm{LAD}}{}^{\mu} - \frac{e\eta\langle F_{\mathrm{LAD}} \,|F_{\mathrm{LAD}}\rangle}{1-\eta\langle F_{\mathrm{LAD}} \,|F_{\mathrm{LAD}}\rangle} F_{\mathrm{LAD}}{}^{\mu\nu} w_\nu \tag{10}$$

Now, we regard the LAD force as the field which is generated by a bare electron. We consider the extra term in Eq. (10) as, coming from the fermionic 1-loop. Here, $-e = -e_{\mathrm{bare}}$ and $\delta e_{\mathrm{dress}}(\tau) = -e\eta\langle F_{\mathrm{LAD}} \,|F_{\mathrm{LAD}}\rangle / [1-\eta\langle F_{\mathrm{LAD}} \,|F_{\mathrm{LAD}}\rangle] \ge 0$, the radiation reaction force becomes,

$$f_{\mathrm{reaction}}{}^{\mu} = [-e_{\mathrm{bare}} + \delta e_{\mathrm{dress}}(\tau)] F_{\mathrm{LAD}}{}^{\mu\nu} w_\nu. \tag{11}$$

Sometime we use the word of the "coupling" for the charge since it couples the dynamics between a particle and fields. The new formula of radiation reaction arranges this coupling by the additional charge of $\delta e_{\mathrm{dress}}$ depending on the LAD field $F_{\mathrm{LAD}}$, we can consider it as the "dress of charge" by radiation. Actually, the Hisenberg-Euler Lagrangian renormalizes the propagation of radiation fields, we could incorporate a modern physics scheme into classical physics. From this fact, the dress is essentially equivalent to the stabilization of radiation reaction [9-10,13].

## 4. Summary

Radiation reaction is one of the important process in ultra-high intense laser - high energy electron interactions. The standard model for it, the LAD model was developed as an electron model which is a pure solution of the Maxwell equation. In this paper, we improved the electron model by using QED vacuum. The equation of motion becomes Eq. (8). This equation is stable and can avoid the run-away problem [9]. For these reasons, our equation (8) is a successor of Lorentz, Abraham and Dirac.

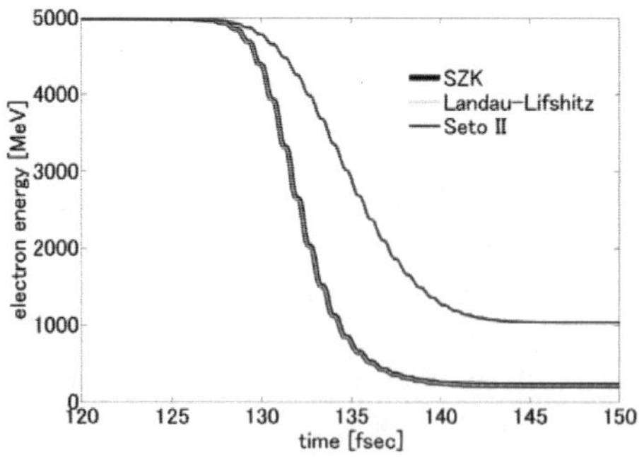


Fig.3 The time evolution of an electron's energy. It simulates the head on collision between a 5GeV electron and the intense laser ( $5\text{PW} - 1\times 10^{22}\ \text{W/cm}^2$ $\lambda = 0.82\mu\text{m}$ , $\tau_{\text{pulse}} = 22\text{fs}$ ). We used the SZK Eq.(8) [9], the Landau-Lifshitz (LL) equation [12] and Seto II Eqs.(12-13) [13]. SZK converges to LL, it mean LL includes the main effects of QED vacuum fluctuation for the stability. The difference of the final energy derive the fact that Seto II equips QED radiation coupling. It will be carried out in ELI-NP [14].

The essence of our method comes from the dress of charge (the part of the coupling with the dependence of the LAD field $F_{\text{LAD}}$ ), $\delta e_{\text{dress}}$ by QED vacuum fluctuations, the fermionic 1-loop. This method will become new reference for laser-plasma simulations. Recently, our group succeeded the next model with QED modified radiation from an electron. Defining $\mathfrak{F}_{\text{hom}} = F_{\text{ex}} + F_{\text{Mod-LAD}}$,

$$dm(x)\frac{dw^{\mu}}{d\tau} = -d\mathfrak{C}^{\mu\nu}{}_{\alpha\beta}(x)\mathfrak{F}_{\text{hom}}{}^{\alpha\beta}w_{\nu} \ , \tag{12}$$

$$\frac{d\mathfrak{C}^{\mu\nu\alpha\beta}}{dm} = \frac{e}{m_0}\frac{(1-\eta\langle\mathfrak{F}_{\text{hom}}|\mathfrak{F}_{\text{hom}}\rangle)g^{\mu\alpha}g^{\nu\beta} + \eta\langle\mathfrak{F}_{\text{hom}}|{}^*\mathfrak{F}_{\text{hom}}\rangle\times\frac{1}{2!}\varepsilon^{\mu\nu\alpha\beta}}{(1-\eta\langle\mathfrak{F}_{\text{hom}}|\mathfrak{F}_{\text{hom}}\rangle)^2 + (\eta\langle\mathfrak{F}_{\text{hom}}|{}^*\mathfrak{F}_{\text{hom}}\rangle)^2} \ . \tag{13}$$

By using this equation, we can find more active effect of an electron's anisotropic coupling $d\mathfrak{C}^{\mu\nu\alpha\beta}/dm$ by checking the time evolution of an electron's energy in Fig.3. The detail of this model is in Ref.[13].

## 5. Acknowledgements

This work is partly supported under the auspices of the Japanese Ministry of Education, Culture, Sports, Science and Technology (MEXT) project on "Promotion of relativistic nuclear physics with ultra-intense laser" and also supported by Extreme Light Infrastructure – Nuclear Physics (ELI-NP) – Phase I, a project co-financed by the Romanian Government and the European Union through the European Regional Development Fund

## References

[1] Report on the Grand Challenges Meeting, 27-28 April 2009, Paris.
[2] V. Yanovsky, V. Chvykov, G. Kalinchenko, P. Rousseau, T. Planchon, T. Matsuoka, A. Maksimchuk, J. Nees, G. Cheriaux, G. Mourou, and K. Krushelnick, Opt. Express 16, 2109 (2008).
[3] A. Zhidkov, J. Koga, A. Sasaki and M. Uesaka, Phys. Rev. Lett. 88, 185002 (2002).
[4] M. Abraham, *Theorie der Elektrizität: Elektromagnetische Theorie der Strahlung*, (Teubner, Leipzig, 1905).
[5] H. A. Lorentz, *The Theory of Electrons and Its Applications to the Phenomena of Light and Radiant Heat, A Course of Lectures Delivered in Columbia Univ., New York, in March and April 1906*, 2nd edition (Leipzig, New York, 1916).
[6] P. A. M. Dirac, Proc. Roy Soc. A **167**, 148 (1938).
[7] W. Heisenberg and H. Euler, Z. Phys. **98**, 714 (1936).
[8] J. Schwinger, Phys. Rev. **82**, 664 (1951).
[9] K. Seto,S. Zhang, J. Koga, H. Nagatomo, M. Nakai, K. Mima, Prog. Theor. Exp. Phys. **2014**, 043A01 (2014).
[10] K. Seto, Prog. Theor. Exp. Phys. **2015**, 023A01 (2015).
[11] J. Schwinger, Phys. Rev. **75**, 1912 (1949).
[12] L. D. Landau and E. M. Lifshitz, *The Classical theory of fields* (Pergamon, New York, 1994).
[13] K. Seto, arXiv: 1502.05319v3 (2015).
[14] O. Teşileanu, D. Ursescu, R. Dabu, and N. V. Zamfir, Journal of Physics: Conference series 420, 012157 (2013), and RA5-TDR (Technical Design Report) in ELI-NP.

# Laser-induced synthesis and decay of Tritium under exposure of solid targets in heavy water

E.V. Barmina[1], S.F. Timashev[2,3], and G.A. Shafeev[1,3]

[1] Wave Research Center of A.M. Prokhorov General Physics Institute of the Russian Academy of Sciences, Moscow, Russian Federation

[2] L.Ya. Karpov Institute of Physical Chemistry, Moscow, Russian Federation

[3] National Research Nuclear University MEPhI, Moscow, Russian Federation

**Abstract**

The processes of laser-assisted synthesis of Tritium nuclei and their laser-induced decay in cold plasma in the vicinity of solid targets (Au, Ti, Se, etc.) immersed into heavy water are experimentally realized at peak laser intensity of $10^{10}$-$10^{13}$ W/cm$^2$. Initial stages of Tritium synthesis and their laser-induced beta-decay are interpreted on the basis of non-elastic interaction of plasma electrons having kinetic energy of 5-10 eV with nuclei of Deuterium and Tritium, respectively.

PACS: 42.62.–b; 61.46.+w; 78.66.–w

Recent theoretical work shows the capability of laser radiation to directly excite nuclear levels of energy [1]. However, exciting a nuclear transition would require an X-ray or gamma-ray lasers with an intensity greater than $10^{20}$ Watts/cm$^2$. At the same time, laser generated photons can indirectly influence the nuclear levels and alter or induce nuclear reactions. Laser beams at intensity level of $10^{18}$ W/cm$^2$ are capable of inducing nuclear transformations under laser exposure of solid targets in vacuum. This can be achieved at pico- and femtosecond laser pulse durations [2-4]. The yield of $10^3$ alpha particles per pulse has been detected at peak laser intensity on the target of $2 \times 10^{18}$ W/cm$^2$ [5]. In accordance with [5], the mechanism of initiation of a neutron-less fusion reaction is due to the acceleration of electrons and protons in the plasma separated by strong laser field.

New possibilities of laser initiation of nuclear reactions have been demonstrated at peak laser intensity levels of $10^{10}$-$10^{13}$ W/cm$^2$ [6-9]. The crucial role in laser-induced acceleration of nuclear decays belongs to nanoparticles (NPs) and nanostructures (NS) on the target immersed into liquid that interact with laser beam in presence of unstable isotopes. In fact, the possibility of initiation of nuclear transformations in cold laser-induced plasma with electron temperature of 5-10 eV has been shown. Relatively low laser intensity of $10^{12}$-$10^{13}$ W/cm$^2$ can be enhanced both on NPs and NS by a factor of $10^6$-$10^8$ due to plasmon resonance of charge carriers in the nano-entities presented in the solution. In this way the electric field of laser radiation becomes comparable to intra-atomic field and may cause the distortion of electronic shells.

Tritium is another unstable isotope with half-life of 12.2 years. It would be interesting to induce its beta-decay by exposure to laser radiation, as it was done previously for other nuclides. In this work we demonstrate for the first time the nuclear synthesis and decay of Tritium under laser exposure of various targets in $D_2O$ [15].

Different targets were used for exposure of bulk metal targets in a heavy water with some tritium content ($D_2O$ + $\xi$DOT), where $\xi$ stands for molar fraction of $D_2O$ with T. The content of T was measured according to its beta-activity with the use of a Perkin-Elmer beta-spectrometer Tri-Carb 3110TR Liquid Scintillation Analyzer with fluorescent dye ULTIMA GOLD™ cocktail. The accuracy of measurements was 1%. Deuterium content was 98.8 %. Two laser sources were used for exposure of various targets in heavy water. The first source was a Cu vapor laser, pulse duration of 15 ns, wavelength of 510.6 nm, and repetition rate of 15 kHz. The estimated peak power on the target was $10^{10}$ W/cm$^2$. The second laser source was a Nd:YAG laser Duetto™, Timebandwidth, with pulse duration of 10 ps and repetition rate of 50 kHz. Three harmonics of this laser, first at 1064 nm, second at 532 nm, and third at 355 nm were used for target exposure. The estimated peak power on the target was $10^{13}$ W/cm$^2$, $5\times10^{11}$ W/cm$^2$, and $2\times10^{11}$ W/cm$^2$ at 1064, 532, and 355 nm, respectively.

Laser exposure of various targets placed into a glass cell with liquid was carried out by focusing laser beam on the target with spherical lens of 5 cm focal distance. Laser beam was focused on the target from bottom through a window made of fused silica to a spot on the target of 50 μm in diameter. The targets had the shape of rods with diameter about 1 mm. The sides of the targets were electrically isolated from the solution. Typically, 2 ml of heavy water was placed inside a glass cell cooled by flowing water. The inner diameter of the glass cell was 1 cm. In another set of experiments the metallic target was cathodically biased with respect to another electrode made of Platinum wire 0.5 mm in diameter. The applied voltage was 25 V DC. To ensure the conductivity of the solution, 15 mg of metallic Na were added to 5 ml of $D_2O$ thus producing conducting solution of NaOD.

In the first set of experiments the ablation of targets was carried out in 99.9% pure $D_2O$ (supplied by Euriso-top) without Tritium content. In this case the content of DOT molecules in $D_2O$ was at the background level of the beta-spectrometer, which corresponds to $\xi = 3.54\cdot10^{-14}$. In one set of experiments the voltage was not applied to the target, in the second set of experiments the ablation was carried out simultaneously with electrolysis. Application of voltage to electrodes is accompanied by reduction of Deuterium (and Tritium) on the target as bubbles. Isolation of the sides of target from the solution confined the emission of reduced gas to ablated area.

Typical exposure time of the targets was 1 hour. The target material is dispersed into surrounding liquid in the shape of nanoparticles. A deep crater is formed in the target surface after laser exposure. The inner surface of the crater and the target surface around it are decorated with dense array ($4 \times 10^9$ cm$^{-2}$) of nanostructures with average lateral size of 100 nm (Fig. 1). Nanoparticles ablated from the target form a colloidal solution. Their presence in the beam path hinders the exposure of the target especially at the wavelength of the third harmonics (355 nm) due to strong absorption of laser radiation.

Table 1. Laser exposure of targets in pure $D_2O$ with addition of NaOD ($\xi = 3.54 \cdot 10^{-14}$).

| № | Material | Wavelength of laser beam, nm | Intensity of laser radiation, W/cm$^2$ and pulse width | Time of irradiation, min | Electrolysis | Activity, Ci/l | Change with respect to initial level |
|---|---|---|---|---|---|---|---|
| 1 | Initial $D_2O$ | - | - | - | - | $8.64\times10^{-8}$ $\xi=3.54\cdot10^{-14}$ | - |
| 2 | Au | 532 | $7\times10^{11}$ 10 ps | 60 | + | $4.06\times10^{-4}$ $\xi=1.66\cdot10^{-10}$ | Increase $4.7\times10^3$ |
| 3 | Au | 1064 | $10^{13}$ 10 ps | 32 | - | $1.62\times10^{-6}$ $\xi=6.63\cdot10^{-13}$ | Increase by factor 18.8 |
| 4 | Ti | 532 | $7\times10^{11}$ | 62 | + | $4.74\times10^{-4}$ $\xi=1.94\cdot10^{-10}$ | Increase $5.5\times10^3$ |
| 5 | Ti | 1064 | $10^{13}$ | 90 | - | $1.57\times10^{-6}$ $\xi=6.42\cdot10^{-13}$ | Increase by factor 18.2 |
| 6 | Ti | 355 | $2\times10^{11}$ | 61 | + | $9.62\times10^{-8}$ $\xi=3.94\times10^{-14}$ | Increase by factor 1.11 |
| 7 | Pd | 355 | $2\times10^{11}$ | 75 | + | $2.46\times10^{-6}$ $\xi=10^{-12}$ | Increase by factor 28.2 |
| 8 | Pd | 1064 | $10^{13}$ | 62 | + | $2.31\times10^{-6}$ $\xi=9.47\times10^{-13}$ | Increase by factor 26.7 |

It was found that laser exposure of various targets in $D_2O$ leads to formation of Tritium, and its content depends on several experimental parameters. The results are presented in Table 1. Highlighted samples were exposed under cathodic bias. Laser ablation of targets without electrolysis leads to the increase of Tritium activity to the level which is about 20 times above the background ($10^{-7}$ Curie/l). The synthesis of Tritium is even more pronounced if the ablated target is cathodically shifted. Ablation of either Ti or Au targets with electrolysis results in synthesis of Tritium to the level, which is $10^3$ times above the background. The largest change of activity is observed at laser wavelength of 1064 nm. The optical density of the solution at the wavelength of the third harmonics of a Nd:YAG laser (355 nm) is high, especially after generation of some amount of nanoparticles, so the intensity of the laser beam on the target is much lower than at 1064 nm.

The following set of experiments was carried out in heavy water with initial Tritium content. Laser ablation of bulk targets for 40-60 min leads to generation of nanoparticles (NPs). In case of Au and Se targets these NPs are made of Au and Se, respectively [16, 17]. Both these colloids absorb in the visible range of spectrum and are reddish in appearance. The solution without NPs (or with small remaining fraction of NPs) was tested for Tritium content about one week after laser exposure. Tritium content in the initial $D_2O$ was also measured in the same set of samples. Then in another set of experiments the generated colloids were exposed to laser radiation in absence of the target. Tritium content increases by a factor of 1.9 after ablation of a bulk Se target with 10 ps pulses of the second harmonics of a Nd:YAG laser. Further laser exposure of Se NPs solution does not alter this increased level of Tritium activity. Again, exposure of the colloidal solution to radiation of a Cu vapor laser leads to decrease of T activity by a factor of 6.7. Qualitatively similar results are obtained for NPs of Au (Table 2).One can see that activity of Tritium (its content) is higher by a factor of 1.8 than in the initial $D_2O$ right after laser generation of Au NPs with picosecond laser pulses. Further laser exposure of the colloidal solution alone does not alter much the activity of T. On the contrary, small decrease of T activity is observed in case of exposure of Se colloidal solution with 15 ns pulses of a Cu vapor laser.

The exposure of different targets (Au, Ag, Be) to radiation of a 15 ns laser pulses has small effect on the activity of Tritium. The exception is exposure of Se target to this radiation that markedly reduces T content in the exposed solution (see Table 2).

Laser exposure of targets that are cathodically biased in $D_2O$ with relatively high DOT content leads to some decrease of Tritium activity. These results are presented in Table 2.

Table 2. Decay of Tritium activity during laser exposure of different targets with electrolysis.

| No | Material | Kind of liquid | Wavelength of laser beam, nm | Intensity of laser radiation, W/cm$^2$ | Time of irradiation, min | Activity, Ci/l |
|---|---|---|---|---|---|---|
| 1 | Initial $D_2O$+ DOT | $D_2O$+ DOT | - | - | - | $1.22\times10^{-3}$ ($\xi = 0.5\cdot10^{-9}$) |
| 2 | Au | NaOD+DOT | 532 | $7\times10^{11}$ | 60 | $1.12\times10^{-3}$ |
| 3 | Au | NaOD+DOT | 1064 | $10^{13}$ | 30 | $9.03\times10^{-4}$ |
| 4 | Ti | NaOD+DOT | 532 | $7\times10^{11}$ | 60 | $1.11\times10^{-3}$ |
| 5 | Ti | NaOD+DOT | 1064 | $10^{13}$ | 90 | $3.35\times10^{-4}$ |
| 6 | Pd | NaOD+DOT | 1064 | $10^{13}$ | 90 | $1.22\times10^{-3}$ |
| 7 | Ti | $Li_3C_6H_5O_7$+ DOT | 1064 | $10^{13}$ | 54 | $6.35\times10^{-4}$ |

The final value of activity after laser exposure is around $10^{-3}$ Ci/l. The variation of activity caused by laser ablation is well pronounced even at high level of initial Tritium activity. As it follows from data presented in Table 1, at low initial Tritium content the increase of Tritium content after laser exposure is observed. One may suggest that this result is the competition between laser-induced decay of Tritium and its formation.

The efficiency of nuclear processes occurring in the course of heavy water electrolysis can depend on the character of roughness of the electrode surfaces on a nanometer scale, the "spikiness" parameters [18, 19] in particular. Indeed, it is precisely in the regions of the sharpest surface relief alterations that high electric field strengths making for the acceleration of electrons and high mechanical stresses depressing the activation barriers for electrochemical processes can both get realized. This parameter is out of control in most experiments with electrolysis of heavy water. On the contrary, laser ablation of metallic targets by sub-nanosecond laser pulses leads to formation of self-organized nanostructures (NS) on the target. The average size and density of NS depends on laser fluence on the target and target material. Therefore, laser ablation technique used in the present work provides guaranteed "spikiness" of the target that may generate high electric fields in the vicinity of the ablated target. Local electric field near the tips of NS under cathodic bias can amount to $10^9$ V/m, and they may serve as efficient emitters of high energy electrons.

The observed phenomena of T synthesis cannot be ascribed to nuclear fusion in its classical sense. There should be a low-energy channel that may alter the interaction of deuterons. One may suggest the following scheme of reactions [22, 23]. If the original nucleus is stable (for the sake of definiteness, let it be a deuteron $d^+$) and the kinetic energy $E_e$ of the electron is too low for any nuclear process involving the breakdown of the original nucleus to take place, the electron-nucleus interaction can result in the emission of a neutrino ν (an irreversible process!) and formation in the nucleus of an intermediate vector boson $W^-$ as a virtual particle (the electron cannot directly interact with a quark). Thereafter there forms a d quark, as a result of interaction between the virtual vector boson $W^-$ and one of the u quarks of a nuclear proton, and the latter converts into a virtual neutron. No actual neutron can form in the nucleus. True, if this were the case, the nucleus being formed would decay, because a two-neutron system is unstable [24]. But for such decay into two neutrons to take place, the mass of the nucleus being formed, which is equal to the sum of the masses of the deuteron and electron, is insufficient. (In the case under consideration, the energy deficit $\Delta Q$ is $(m_p + m_e - 2m_n)c^2 \approx -3.01$ MeV, where $c$ is the velocity of light in vacuum and $m_p$, $m_e$, and $m_n$ are the masses of the proton, electron, and neutron, respectively.) The conclusions drawn in [24] that the existence of a nuclear stable dineutron is impossible does not relate to the nucleus being introduced here with its mass less than the total mass of two neutrons, despite its zero electric charge, baryon number equal to 2, and zero leptonic charge. We will refer to such a nucleus as $\beta$-dineutron and represent the sequence of the processes being discussed as follows:

$$e^-_{he} + d^+ \rightarrow {}^2n_{isu} + \tilde{\nu}\,, \qquad (1)$$

$$^2n_{isu} \rightarrow d^+ + e^- + \tilde{\nu}\,. \qquad (2)$$

In these expressions, the subscript "he" on the notation of the initial electron points to the initiative character of interaction between the electron and deuteron in the formation of the $\beta$-dineutron $^2n_{isu}$ in the in-shake-up state (the latter being indicated by the subscript "isu"). Inasmuch as the threshold of such an "inelastic scattering" of the electron by the deuteron is solely determined by the doubled rest mass of the neutrino and amounts to less than 1 eV, the electron kinetic energies of $E_e \sim 10$ eV will apparently be adequate for one to observe such processes. This means that the mass of the $\beta$-dineutron practically coincides, as noted above, with the mass of the deuterium atom. The Triton in the experimental conditions of present work may be synthesized according to the following equation:

$$d^+ + {}^2n_{isu} \rightarrow t^+ + n + Q(3.25 MeV)\,. \qquad (3)$$

In turn, Tritium nucleus formed in the cold laser plasma may also interact with electrons. According to [22, 23], decay of Tritium may be accelerated ("e-catalysis") owing to this interaction and formation of an intermediate product β-trineutron $^{3}n_{isu}$:

$$t^{+} + e^{-} \rightarrow {}_{0}^{3}n_{isu} + \nu \rightarrow {}_{2}^{3}He^{2+} + 2e^{-} + \nu + 2\tilde{\nu} + Q(0.019MeV), \quad (4)$$

All experimental results presented above can be interpreted on the basis of the processes (1), (3), and (4). Indeed, processes (1)-(3) dominate at low initial Tritium content in $D_2O$ ($10^{-7}$ Curie/l) and presence of electrons having kinetic energy $E_e$ ~ 5-10 eV in cold plasma. As a result, Tritium content increases by 3 orders of magnitude. On the other hand, predominant β-decay of Triton takes place under relatively high initial Tritium beta-activity ~ $10^{-3}$ Curie/l.

The laser-induced decay of Tritium is a competing process that determines the stationary level of Tritium during laser exposure. Note that significant decrease of Tritium content occurs during 1 hour laser exposure, which is many orders of magnitude shorter than life time of this nuclide.

Thus, laser-assisted synthesis of Tritium from $D_2O$ has been realized. The synthesis proceeds under exposure of various solid targets with laser pulses. There is definite threshold of the intensity and duration of laser pulses, the synthesis takes place only with short (10 ps) laser pulses and peak intensities exceeding $10^{11}$ W/cm$^2$. Tritium is synthesized without any cathodic bias on the target though the application of this bias largely increases the rate of its synthesis. The increase of Tritium content requires careful check. In particular, long-term measurements of fluorescence are required to confirm the decrease of generated Tritium with its characteristic half-life of 12.2 years.

References

1. Wong, A. Grigoriu, J. Roslund, T.-S. Ho, and H. Rabitz, *Phys. Rev.* A 84, 053429 (2011).

2. Andreev A.V., Gordienko V.M., Dykhne A.M., Saveliev A.B., Tkalya E.V., 1997, *JETP Lett.* **66** (5) 312.

3. Andreev A.V.; Volkov R.V.; Gordienko V.M.; Dykhne A.M.; Mikheev P.M.; Tkalya E.V.; Chutko O.V.; Shashkov A.A.; 1999, *JETP Lett.* 69 (5) 343.

4. Schwoerer H., Gibbon P., Düsterer S., Behrens R., Ziener C., Reich C., Sauerbrey R. *Phys. Rev. Lett.*, 86 (11), 2317 (2001).

5. V. S. Belyaev, A. P. Matafonov, V. I. Vinogradov, V. P. Krainov, V. S. Lisitsa, A. S. Roussetsky, G. N. Ignatyev, V. P. Andrianov, *Phys. Rev. E* 72 (2005), 026406.

6. A.V. Simakin, G.A. Shafeev, arXiv:0906.4268; A.V. Simakin, G.A. Shafeev, *Physics of Wave Phenomena,* 16 (4) (2008) 268-274; A.V. Simakin, G.A. Shafeev, *Journal of Optoelectronics and Advanced Materials*, (2010) 12(3), 432 – 436.

7. A.V. Simakin, G.A. Shafeev, *Quantum Electronics* 41 (7) 614 – 618 (2011).

8. A.V. Simakin, G.A. Shafeev, *Physics of Wave Phenomena* (2011) 19(1), 30–38.

9. G.A. Shafeev, Laser-induced nuclear decays in Uranium isotopes, in: Uranium: Characteristics, Occurrence and Human Exposure, Editors: Alik Ya. Vasiliev and Mikhail Sidorov, ISBN: 978-1-62081-207-5, Novapublishers Inc, New York, 2012, p.p. 117-153.

10. Volkovich A.G., Govorun A.P., Gulyaev A.A. Zhukov S.V., Kuznetsov V.L., Rukhadze A.A., Steblevsky A.V., Urutskoev L.I., Lebedev Bulletin of the Lebedev Physics Institute, 2002, No 8, 45 – 50.

11. G.A. Shafeev, A.V. Simakin, F. Bozon-Verduraz, M. Robert, *Applied Surface Science* **254** (2007) 1022–1026.

12. Fleishmann M., Pons S. and Hawkins M. Electrochemically induced nuclear fusion of deuterium, *J. Electroanal. Chem.* 1989. V. 261. P. 301-308.

13. Iwamura Y., Itoh T., Gotoh N., Toyoda I. Detection of anomalous elements, X-ray, and excess heat in a $D_2$-Pd system and its interpretation by the electron-induced nuclear reaction model, *Fusion Technology*, 1998. V. 33. P. 476-492.

14. Takahashi A. Progress in condensed matter nuclear science. In: Condensed Matter Nuclear Science. Proc. 12th Intern. Conf. on Cold Fusion, Yokohamacity, Japan. 27 Nov-2 Dec 2005. World Scientific Publishing Co. Pte. Ltd.: Singapore, 1-25 p.

15. E.V. Barmina, P.G. Kuzmin, S.F. Timashev, G.A. Shafeev, Laser-induced synthesis and decay of Tritium under exposure of solid targets in heavy water, http://arxiv.org/abs/1306.0830.

16. P.V. Kazakevich, A.V. Simakin, V.V. Voronov, and G.A. Shafeev, *Applied Surface Science,* **252** (2006) 4373–4380.

17. P.G. Kuzmin, G.A. Shafeev, V.V. Voronov, R.V. Raspopov, E.A. Arianova, E.N. Trushina, I.V. Gmoshinsky and S.A. Khotimchenko, *Quantum Electronics*, **42**, (2012) 1042.

18. Timashev S.F., Polyakov Yu.S., Lakeev S.G., Misurkin P.I., Danilov A.I. Fundamentals of fluctuation metrology, *Russian Journal Physical Chemistry* A. 2010. V. 84. No 10, 1807-1825.

19. Mirsaidov U., Timashev S.F., Polyakov Yu.S., Misurkin P.I., Polyakov S.V., Musaev I. *Analytical method for parameterizing the random profile components of nanosurfaces imaged by atomic force microscopy* // Analyst. 2011. V. 136. N 3, P. 570-576

20. S. Lau Truong, G. Levi, F. Bozon-Verduraz, A.V. Petrovskaya, A.V. Simakin, G.A. Shafeev, *Applied Surface Science* **254** (2007) 1236–1239.

21. E.V. Barmina, S. Lau Truong, F. Bozon-Verduraz, G. Levi, A.V. Simakin, G.A. Shafeev, , *Quantum Electronics*, 40 (2010) 346-348.

22. Timashev S. F., V. I. Muromtsev V.I., Akovantseva A.A. Nuclear Processes Initiated by Electrons , Russian Journal of Physical Chemistry A, 2013, Vol. 87, No. 6, pp. 1063–1069.

23. Timashev S.F. "Beta-dineutron" as a radioactive element and its possible role in nuclear synthesis processes (in press).

24. Zel'dovich Ya.B., Gol'danskii V.I., Baz' A.I. Systematics of the lightest nuclei, Physics – Uspekhi. Advances in Physical Sciences. 1965. V. 8. N 2. P. 177–201.

# Mitigation of Laser Imprinting with Diamond Ablator for Direct-Drive Inertial Confinement Fusion Targets

**K Shigemori[1], H Kato[1], M Nakai[1], R Hosogi[2], T Sakaiya[2], H Terasaki[2], S Fujioka[1], A Sunahara[3] , and H Azechi[1]**

[1]Institute of Laser Engineering, Osaka University, 2-6 Yamada-Oka, Suita, Osaka, 565-0871 Japan
[2]Department of Earth and Planetary Science, Osaka University1-1 Machikaneyama-Cho, Toyonaka, Osaka, 560-0043 Japan
[3]Institute for Laser Technology, 2-6 Yamada-Oka, Suita, Osaka, 565-0871 Japan

E-mail:shige@ile.osaka-u.ac.jp

**Abstract.** We propose a novel scheme to mitigate the initial perturbation imprinting due to irradiation non-uniformity. Diamond was potential candidate for the ablator material for ICF targets due to its stiffness. The stiffness is very important parameter for imprint mitigation because the laser imprint is primary as a function of pressure perturbation due to lase irradiation non-uniformity. We measured the imprint amplitude of diamond foils and plastic (CH) foils. The experimental data suggest the initial imprinting is drastically mitigated for the diamond foils.

## 1. Introduction

Hydrodynamic instabilities are still the most crucial issue on inertial confinement fusion (ICF) targets. Among the instabilities, Rayleigh-Taylor instability (RTI) gives the largest growth factor during the implosion process because the perturbations on the target surface grow exponentially with time [1]. Initial perturbations on the target surface are due mainly to target fabrication and to imprinting by laser irradiation non-uniformity. For direct-drive ICF targets, the initial imprinting is the most important issue to seed the RTI. Previous experimental and theoretical studies revealed the detailed physics of the laser imprinting [2-5].
The imprinting is modelled as imprint efficiency, which is the equivalent perturbation amplitude on the surface normalized by the laser intensity modulation. The imprint efficiency is defined by the imcompressive model as [5]:

$$\eta_{\mathrm{imp}} = \frac{1}{3}\frac{kd_0}{\gamma + 1\{1 + 0.2(kd_0)^2\}}F(k, D_{\mathrm{sb}}, d_0) \tag{1}$$

where $k$ is wave number of the perturbation, $d_0$ is initial target thickness, $\gamma$ is specific heat ratio, and $D_{sb}$ is standoff distance. $F$ is the smoothing factor of the imprint efficiency, and approximated to be $2/k^2D_{sb}^2$. This suggests that, in order to reduce the imprint efficiency, thermal smoothing is very controllable with several technique. Previous investigation for mitigation of laser imprinting were carried out with x-ray pre-irradiation [6], picket pulse [7], low density foam target [8], high-Z dopant [9], and so forth. These technique are very effective for small wavelength perturbation, however in many cases, the adiabat of the target is increased prior to shock transition due to the process of enhancing the standoff distance.

Other possible mitigation scheme is to change the target stiffness $\gamma$. We here focus on diamond, which is the hardest material, to be a candidate for the direct drive ablator because the diamond is low-Z material, easy to handle, and possible to fabricate to hollow shell. The diamond ablator has been employed as a possible candidate for indirect-drive NIF ignition target [11]. Diamond also has very large Hugonit elastic limit (HEL: ~200 GPa) [10], which means the compression below the HEL is very small. For example, The effective $\gamma$ below the HEL at 100 GPa is more than 20 whereas the $\gamma$ for CH is 2. This fact suggests that very strong mitigation is probable for the stiffer material. Thus we carried out a proof of principle experiment in order to demonstrate the effect on imprinting mitigation in the aspect of the material stiffness..

## 2. Experiments

Experiments were done on GEKKO-XII laser facility at Institute of Laser Engineering. The setup and overall experimental conditions were similar to previous hydrodynamic instability experiments [12, 13]. We observed the perturbation growth of the accelerated foils due to the RTI, in which the initial perturbation was seeded by a spatially modulated foot pulse laser. Schematic view of the experimental setup is shown in Figure 1.

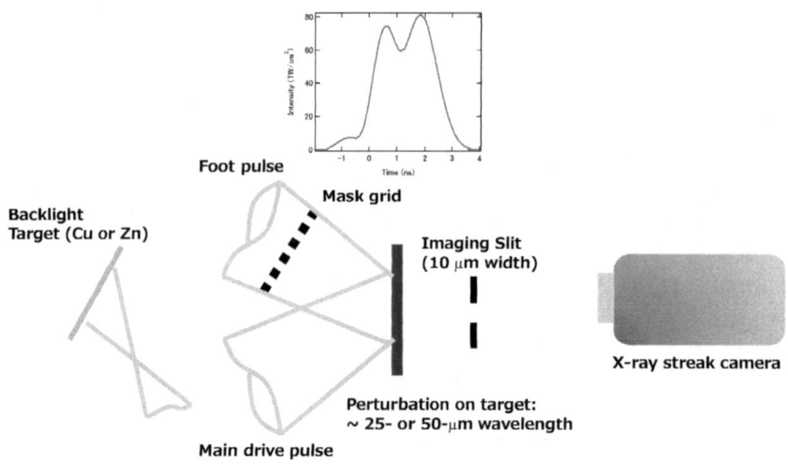


**Figure 1.** Experimental set up and typical pulse shape

Target foils were irradiated by three beams of the GEKKO-XII laser ($\lambda$: 0.527 μm). The pulse duration of each beam we 1.3 ns. Each pulse with time delay which made a foot pulse followed a main drive pulse. The typical stacked pulse shape is shown in Figure 1. Spot diameter on the target was 300 μm. We put a grid mask for the imprint beam just in front of the irradiated foil in order to make a spatial pattern on the foil. The period of the grids were 50 and 100 μm. The main drive beam followed the foot pulse for amplification of the imprinted surface by the RTI because the imprint amplitude is usually too small to detect. Intensities of the foot pulse and the main pulse were $4 \times 10^{12}$ W/cm$^2$ and $8 \times 10^{13}$ W/cm$^2$, respectively. The backlighter beam was irradiated ~1.0 ns after the onset of the main drive pule when the imprint perturbation was amplified due to the RTI.

The targets on the measurements were single crystal Type-Ib diamond foils with thickness of ~ 10 μm. The orientation of the target surface was (000). We also employed polystyrene (CH) foils with thickness of 25 μm in order to obtain reference data.

Imprint amplitude is evaluated by measuring areal density perturbation of the laser-irradiated foil. The real density was measured by a conventional face-on backlighting technique. Copper or zinc foils were employed for generation of 1~1.5 keV x-rays. The backlit x-ray was imaged by a slit (10 μm-width) onto a photocathode of an x-ray streak camera. For the analysis of the areal density perturbation, the spatial resolution of the whole diagnostic system was separately measured by taking a grid backlit image. Temporal resolution was ~ 100 ps. We also measured the target trajectories for both targets with side-on x-ray backlighting in order to determine the target acceleration.

## 3. Experimental results

Figure 2 shows raw streaked images by the face-on backlighting measurements. Thee tie origin ($t$=0) is the onset of the main drive pulse. In the measurements, we observed later timing of the main pulse by changing the time delay of the backlighting laser pulse. Significant modulation of the x-ray intensity is seen for the CH foil whereas there is no clear modulation for the diamond foil. The experimental results were analysed to obtain the temporal evolution of the areal density perturbation. In the analysis process, we took into account the spatial resolution function and the mass absorption ratio of the x-ray backlight x-ray for Cu and Zn.

Figure 3 shows the analysed temporal evolution of areal density perturbation for both targets. The amplitudes shown in Fig. 3 is sums of the Fourier amplitude for the wavelength of 45 – 70 μm because the perturbations on the experiments have a lot of wave numbers although the grid mask is with the single mode perturbation. This is probably due to diffraction by the grid, "hole closure" by plasmas, and other reasons. The areal-density perturbation for the CH grows with time whereas no clear growth is found for the diamond target. The dotted line in Fig. 3 is the detection threshold of the measurement which is determined by the white spectrum of the Fourier amplitude. That means the areal density perturbation for the diamond foil is nearly the noise level.

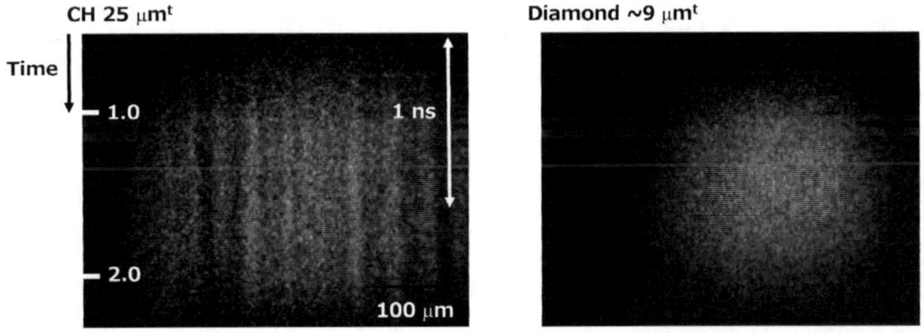


**Figure 2.** Raw streaked backlit images for CH and diamond foils

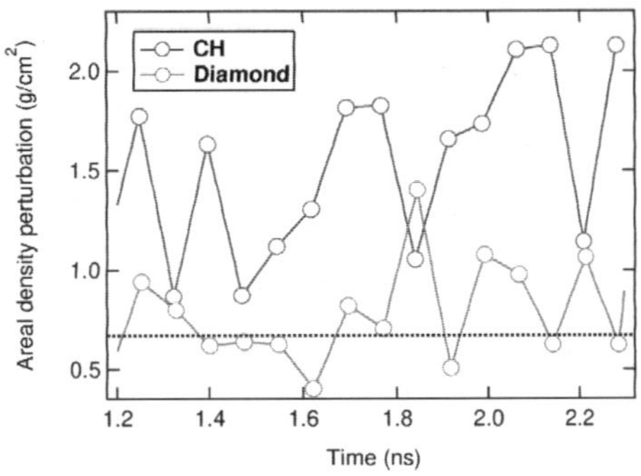


**Figure 3.** Areal density perturbation or CH and diamond foils for wavelength of 45 -70 μm. Dotted line is detection threshold for the diagnostic system

## 4. Discussions and Conclusion

The measured imprinting is apparently reduced for the diamond ablator. In this experimental condition, there are few differences between the diamond and the polystyrene targets. This fact could conclude that the imprint mitigation for the diamond is due to its stiffness.

Although we observed the significant mitigation of the imprinting, imprint efficiency should be quantitatively obtained. We are planning to carry out next experiment to use grid at far field position in order to produce a single mode imprint beam pattern to characterize the imprint efficiency. Also we are going to compare with the imprint model with one-dimensional code for further understanding of the imprint mitigation, and to apply for the optimization of diamond ICF targets.

## Acknowledgements

This work was performed under the joint research project of the Institute of Laser Engineering, Osaka University. The authors would like to acknowledge the dedicated technical support by the staff at the GEKKO-XII facility for the laser operation, target fabrication, and plasma diagnostics. This work was supported by JSPS KAKENHI Grant Number 23340175.

## References

[1] Nuckolls J, Wood L, Thiessen A, Zimmerman G 1972 *Nature* **239** 139

[2] Kalantar D H, Key M H, Da Silva L B, Glendinning S G, Ramington B A, Wolfrum E, Kim N S, Neely D, Zhang J, Wark j S, Demir A, Lin J, Smith R, Tallents G L, Lewis C L S, MacPhee A, Warwick J, Knauer J. P 1997 *Phys. Plasmas* **4** 1985

[3] Wolfrum E, Wark J, Zhang Z, Kalantar D H, Key M H, Remington B A, Weber S W, Neely D, Rose S, Warwick J, MacPhee A, Lewis C L S, Demir A, Lin J, Smith R, Tallent R L 1998 *Phys. Plasmas* **5** 227

[4] V. N. Goncharov, S. Skupsky, T. R. Boehly, J. P. Knauer, P. McKenty, V. A. Smalyuk, R. P. J. Town1, O. V. Gotchev, R. Betti and D. D. Meyerhofer 2000 *Phys. Plasmas* **7** 2062

[5] Nakai M, Azechi H, Miyanaga N, Shigemori K, Kawasaki T, Nagaya T, Nishikino M, Kanai Y, Ohnuki D, Nishimura H, Shiraga H, Maegawa O, Ishizaki R, Nishihara K, Takabe H, Yamanaka T 2002 *Phys. Plasmas* **9** 1734

[6] Nishikino M, Shiraga H, Miyanaga N, Ohnishi N, Shigemori K, Fujioka S, Nakai M, Nishimura H, Azechi H, Mima K, Takabe H, Nakai S, and Yamanaka T 2002 *Phys. Plasmas* **9** 1381

[7] Metzler N, Velikovich A L, Schmitt A J, and Gardner J H 2002 *Phys. Plasmas* **9** 5050

[8] Smalyuk V A, Goncharov V N, Anderson K S, Betti R, Craxton R S , Delettrez J A, Meyerhofer D D, Regan S P and Sangster T C 2007 *Phys. Plasmas* **14** 032702

[9] Hu S X, Fiksel G, Goncharov V N, Skupsky S, Meyerhofer D D, and Smalyuk V A 2012 *Phys. Rev. Lett.* **108** 195003

[10] McWilliams R S, Eggert J H, Hicks, D G, Bradley D K, Celliers P M, Spaulding D K, Boehly, T R, Collins, G W, Jeanloz R 2010 *Phys. Rev. B* **81** 014111

[11] Biener J, Ho D D, Wild C, Woerner E, Biener M M, El-dasher B S, Hicks D G. Eggert J H, Celliers P M, Collins G W, Teslich Koziozіemski B J, Haan S W, Hamza A V 2009 *Nucl Fusion.* 2009 **49** 112001

[12] Shigemori K, Azechi ., Nakai M, Honda M, Meguro K, Miyanaga N, Takabe H. Mima K 1997 *Phys. Rev. Lett.* **78** 250

[13] Azechi H , Nakai M, Shigemori K, Miyanaga N, Shiraga H, Nishimura H, Honda M, Ishizaki R, Wouchuk, J G, Takabe H, Nishihara K, Mima K, Nishiguchi A, Endo T 1997 *Phys. Plasmas* **4** 4079

# Probe beam-free detection of terahertz wave by electroluminescence induced by intense THz pulse

**J. Shin[1], Z. Jin[2], Y. Nosaka[1], T. Nakazawa[1], and R. Kodama[1,2,3]**

[1] Graduate School of Engineering, Osaka University, 2-1, Yamadaoka, Suita, Osaka 565-0871, Japan
[2] Photon Pioneers Center, Osaka University, 2-1, Yamadaoka, Suita, Osaka 565-0871, Japan
[3] Institute of Laser Engineering, Osaka University, 2-6 Yamadaoka, Suita, Osaka 565-0871, Japan

E-mail: jhshin@ef.eie.eng.osaka-u.ac.jp

**Abstract.** Recently, a table-top fs laser system can generate MW terahertz (THz) pulse with its electric field higher than 100 kV/cm can be generated by several schemes. Such a strong THz field can directly drive electrons inside various materials. Here, we demonstrated a direct THz electric field detection method by measuring the electroluminescence induced by intense THz pulse inside commonly available light emitting diode. An intense THz wave obtained by the two-color laser scheme was focused onto LED along with an external DC bias to induce luminescence which we found proportional to the amplitude of the incident THz field. The scheme can be useful to realize a low-cost, probe-free THz detection and imaging system.

## 1. Introduction

Since the diagnostics of temporal profile of terahertz (THz) wave was demonstrated [1, 2], THz radiation has gained attention from a broad range of engineering and research areas including security, biomedical imaging, material physics and etc [3, 4]. Those *coherent* methods, electro-optical sampling or photoconductive switching, measure coherent portion of THz radiation, providing phase information of THz waves. However, they are difficult to setup, because they require a probe beam and we need to align both the probe beam and the THz wave precisely so that they are synchronized and spatially overlapped. On the contrary, incoherent detectors, such as bolometer and Golay cell, do not require a probe beam, but provide only THz power from all heat (incoherent) sources as well as the coherent sources. For that reason, the incoherent detectors have lower signal-to-noise. Still, they are convenient to use.

With recent research efforts pursuing strong THz radiation sources, MW THz pulse with an electric field higher than 100 kV/cm can be obtained with a table-top fs laser system by several methods such as two-color laser scheme [5] and tilted-pulse-front scheme [6]. Such a strong THz field can directly drive the electrons in various materials such as nitrogen gas plasma [7] and quantum well semiconductor [8], and the effects have been observed through luminescence phenomena.

Here, we demonstrate a direct THz electric field detection method by measuring the electroluminescence (EL) induced by intense THz pulse inside common light emitting diode (LED), as illustrated in the Fig. 1. The scheme has the advantage of both the coherent and incoherent method; it can be classified under coherent method, but does not require a probe beam to operate. Because it makes use of the conventional LED technology, it may be useful for realizing a low-cost, probe-beam-free THz detection and imaging system.

Journal of Physics: Conference Series **688** (2016) 012108 doi:10.1088/1742-6596/688/1/012108

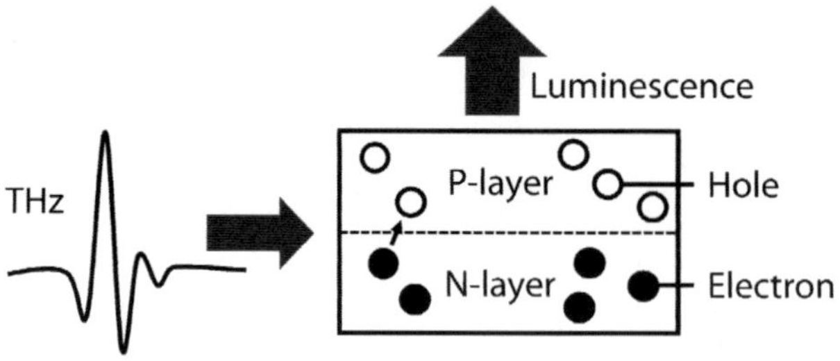


**Figure 1.** Electroluminescence driven by THz field.

## 2. Experiments

Figure 2a shows the setup of experiments performed using the P-cube laser facility at the Graduate School of Engineering, Osaka University. It was adjusted to deliver 30 mJ of energy in $\sim$100 fs long pulse with its spectrum centered on 800 nm. It was negatively chirped for optimal generation of THz wave. The laser was focused into ambient air by an f/20 lens. A 100-μm thick $\beta$-barium borate (BBO) crystal behind the focal lens generated the second harmonic ($2\omega$) pulse by the type-I second harmonic generation from the fundamental laser pulse ($\omega$). A polymer filter was put to allow transmit only the THz wave generated from the plasma channel, while blocking the residual $\omega$ and $2\omega$ pulses. The THz pulse was then focused by an off-axis parabola (OAP) onto the LED chip at an incident angle of 45 degree. A GaAlAs LED manufactured by *Toshiba* (Model No. *TLN1108*) was used in our experiments, which emits near-infrared light with its central wavelength at 870 nm. The emitted luminescence is collected to a gated intensified CCD camera. To reduce the noise, the CCD gate was opened 5 ns before the arrival of the THz wave and closed after 50 ns. The entire LED emission measurement system is carefully enclosed and sealed in a black box with a long-pass THz filter window to ensure that no scattering light from the laser and plasma channel can arrive at the camera. The function generator *DG535* (*Stanford Research Systems, Inc.*) provided the supporting DC bias (discussed in the Appx.) for the LED and the synchronized trigger signal for the CCD gate.

The generated THz wave had a single-cycle profile with a period of 1 ps, as shown in Fig. 2b, which is measured by a single-shot THz time-domain spectroscopy using a pair of reflective echelon [9]. The maximum THz energy of about 5 μJ was measured with our calibrated Golay Cell, corresponding to an electric field of $\sim$ 0.77 MV/cm at the focal spot.

Fig. 3 shows the typical LED emission enhancement patterns with different supporting DC bias

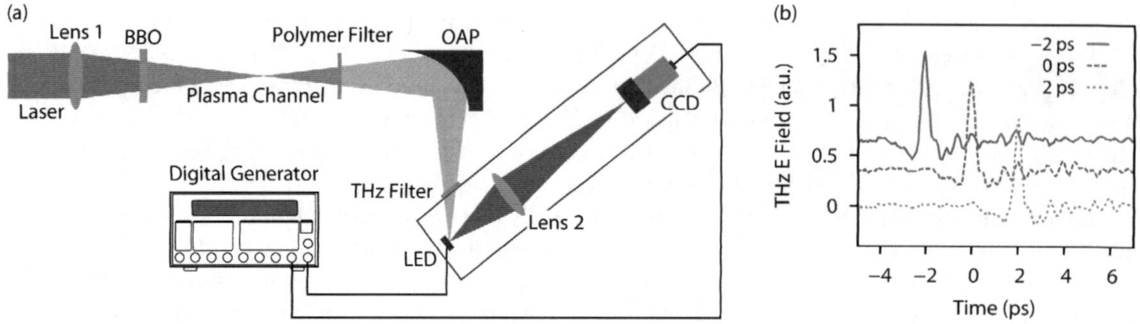


**Figure 2.** (a) Experimental setup. (b) THz waveforms measured by a single-shot time-domain spectroscopy method with different time delays of probe beam.

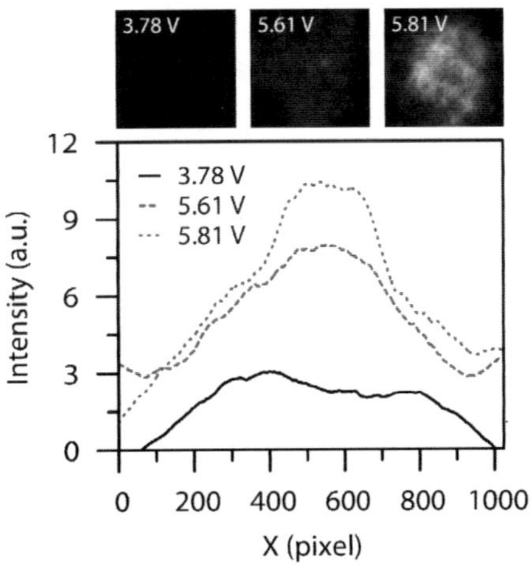


**Figure 3.** Intensity profiles of the EL induced by 0.77 MV/cm THz pulses accompanied by supporting DC biases. The upper insets show the CCD images of the enhanced emission pattern from LED chip.

voltages of 3.78 V, 5.61 V and 5.81 V, with insets showing the CCD images of the LED chip after subtracting the image without incident THz pulse as the background. When the bias is lower than 3.78 V, the EL induced by THz field is too weak so that the counts on CCD are close to the noise level.

Fig. 4 shows that different peak amplitude of the incident THz pulse changes the amount of EL enhancement proportionally. It is also clear that the EL enhancement becomes larger with higher bias voltages, which is consistent with the measurements using ns pulses as shown in the Appendix.

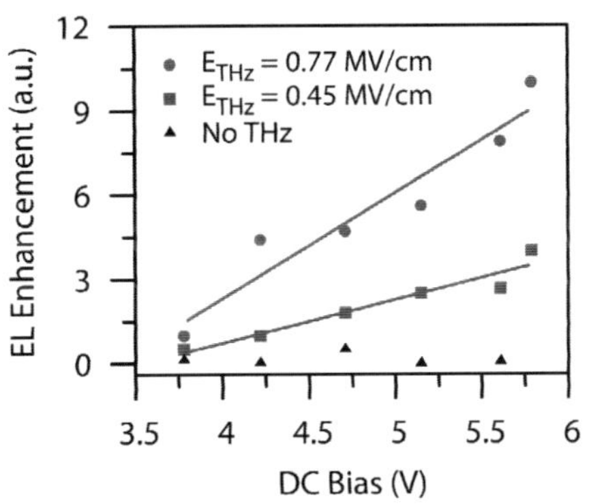


**Figure 4.** Enhancement of EL against the given external DC bias.

## 3. Conclusion

We showed that a direct THz electric field detection is possible by measuring the EL induced by THz pulses inside commercially available LED. This method is coherent with its signal sensitive to the electric field of the THz wave, but has an advantage over other coherent methods that it does not require a probe beam. Because our results rely upon the well-established LED technology, it may be useful for realizing a low-cost, probe-beam-free THz imaging system.

## Appendix

The response of commercially available LEDs for electric signals shorter than $\sim$ ns are not well investigated. Since the THz pulse from two-color scheme usually has a single-cycle waveform with a duration of ps level, it is necessary to investigate the LED's response to a short voltage pulse. We used *DG535*, a digital pulse generator to apply drive pulses with various time durations from μs to 5 ns to the LED chip. The emission of the LED is collected and measured by a 16 bit CCD camera. Fig. A1 shows the response of the LED. If the pulse duration is longer than 200 ns the response is linear regardless of the support bias. With shorter drive pulses, the emission intensity drops non-linearly and reaches to the noise level as the pulse duration becomes shorter than 25 ns which is comparable to the typical rise and fall time (15 ns) of this LED.

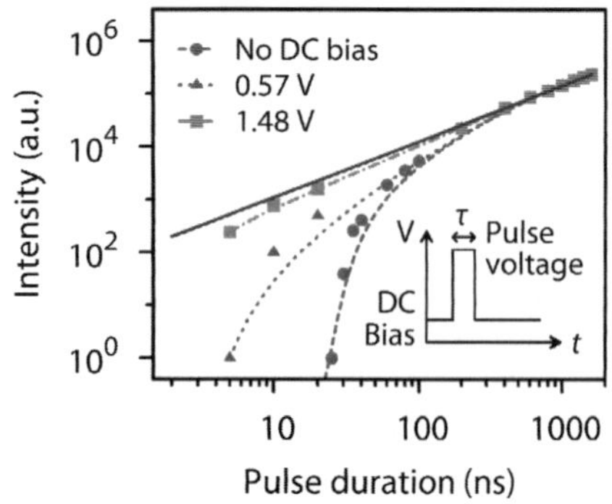


**Figure A1.** Pulse response of the LED. The black solid line shows a linear relation between the emission intensity and pulse duration. The red circles, blue triangles, and green squares show the measured results with an supporting DC bias of 0 V, 0.57 V and 1.48 V, respectively.

The pulse duration limit, however, can be overcome by applying a supporting DC bias, as indicated in the inset of Fig. A1. The effect of the pulse voltage is investigated by comparing the difference between the emission intensity with and without the drive pulse. The pulse response gets more close to linear decrease (black solid line) with the support of the bias.

It is noted that 5 ns, which is the limit of our pulse generator, is still much longer than the THz pulse duration. However, this measurement suggests that the EL induced by short THz pulses can be detected with a sufficient supporting bias.

## References

[1] Zhang X C, Jin Y and Ma X F 1992 *Applied Physics Letters* **61** 2764
[2] Auston D H, Cheung K P and Smith P R 1984 *Applied Physics Letters* **45** 284
[3] Tonouchi M 2007 *Nature Photonics* **1** 97–105
[4] Ferguson B and Zhang X C 2002 *Nature materials* **1** 26–33
[5] Kim K Y, Taylor A J, Glownia J H and Rodriguez G 2008 *Nature Photonics* **2** 605–609
[6] Hebling J, Yeh K L, Hoffmann M C, Bartal B and Nelson K A 2008 *Journal of the Optical Society of America B* **25** B6
[7] Liu J, Dai J, Chin S L and Zhang X C 2010 *Nature Photonics* **4** 627–631
[8] Hirori H, Shinokita K, Shirai M, Tani S, Kadoya Y and Tanaka K 2011 *Nature communications* **2** 594
[9] Kim K Y, Yellampalle B, Taylor A J, Rodriguez G and Glownia J H 2007 *Optics letters* **32** 1968–70

Journal of Physics: Conference Series **688** (2016) 012109 doi:10.1088/1742-6596/688/1/012109

# Dependence of Ablative Rayleigh–Taylor Instability on High-Z Dopant Concentration

**Takashi Shiroto[1], Naofumi Ohnishi[1], Atsushi Sunahara[2], and Shinsuke Fujioka[3]**

[1]Department of Aerospace Engineering, Tohoku University, Sendai 980-8579, Japan
[2]Institute for Laser Technology, Osaka 550-0044, Japan
[3]Institute of Laser Engineering, Osaka University, Suita 565-0871, Japan

E-mail: `tshiroto@rhd.mech.tohoku.ac.jp`

**Abstract.** We conducted two-dimensional simulations of inertial confinement fusion targets to evaluate effects of high-Z doping on implosion hydrodynamics. It was found that an ablation structure drastically changes with concentration of dopant material. We also confirmed that even a lightly-doped target can suppress Rayleigh–Taylor instability on short wavelength, while a long-wavelength perturbation is difficult to be suppressed with any dopant concentration. The high-Z doping is thus only effective for a spherical implosion with high-mode perturbations.

## 1. Introduction

Rayleigh–Taylor instability (RTI) at ablation front in an accelerated inertial confinement fusion (ICF) target is still a critical issue to achieve spherical implosion even though continuous efforts have been devoted to obtain inertial fusion energy. A linear growth rate of the ablative RTI $\gamma$ is obtained from the linear stability theory [1, 2];

$$\gamma = \alpha\sqrt{\frac{kg}{1+kL}} - \beta k v_a, \tag{1}$$

where $k$ is the wavenumber of the perturbation, $g$ is the acceleration, $L$ is the density scale length, $v_a$ is the ablation velocity, and $\alpha$ and $\beta$ are the constants that depends on the ablation structure. This formula suggests the ablative RTI is suppressed especially at short wavelength thanks to mass ablation flow or finite scale length.

A concept of a high-Z doped target has been proposed to suppress the ablative RTI [3]. In the high-Z doped target, mean-free-path difference between X-rays and thermal electrons results in a double ablation structure. The linear growth rate is reduced at radiation-driven ablation (RA) front, and the ablative RTI is completely suppressed at electron-driven ablation (EA) front. However, high-Z doping may be ineffective for high-density implosion because high-Z material tends to deprive kinetic energy from the ICF target by radiation energy loss. This fact awakes us to existence of the optimum amount of high-Z doping to achieve high-density implosion.

In this study, we first conducted one-dimensional (1D) simulations of planar brominated targets to explore a transition of the ablation structure with varying the fraction of dopant. Then dispersion relation of the ablative RTI was investigated with two-dimensional (2D) simulations of the planar targets. Finally, comparison of axisymmetric implosion simulations clarified implosion performances of the brominated fuel shells.

## 2. Numerical methods

We conducted numerical simulations with 2D Eulerian radiation hydrodynamics (RHD) code, RAICHO, and its axisymmetric version (one-fluid, two-temperature model) [4]. They solve the Euler equations with some energy sources with realistic equations of state (EOS) for ions and electrons. The energy sources are determined by processes of laser absorption, thermal conduction, and radiative transfer.

The governing equations are discretized with cell-centered finite volume method. The numerical flux at the cell interface is estimated by AUSM-DV scheme [5] with 2nd-order MUSCL method [6]. The hydrodynamic part is integrated by the 1st-oder explicit Euler method. The thermal transfer part is integrated by the 1st-order implicit Euler method with Bi-CGSafe linear solver [7] preconditioned by incomplete LU factorization. The EOS for ions is based on the Cowan model and that for electrons is based on the Thomas–Fermi model [8]. The laser absorption of inverse-bremsstrahlung is calculated by 1D ray-tracing by neglecting the refraction of the laser rays. The thermal conduction is computed by the diffusion approximation with Spitzer–Härm conductivity [9] and the flux limiter of 0.1. The X-ray radiative transfer is formulated by the multi-group diffusion approximation with the variable Eddington factor and the collisional radiative equilibrium opacity (32 energy groups).

## 3. Results and discussions

### *3.1. Double ablation structure*

We first conducted 1D simulations of the planar targets with varying the fraction of bromine. The simulation domain is described in Figure 1(a). The initial target thickness was set to 16 $\mu$m and irradiated by laser. The input laser was a flat-topped pulse with maximum intensity of $1 \times 10^{14}$ W/cm$^2$ and 0.53-$\mu$m wavelength.

Figure 2 shows the density profiles of each brominated target at 3.0 ns after the beginning of the laser irradiation. Legends indicate the fraction of bromine in composition. The peak density of the target becomes lower as the fraction of bromine increases for the lightly-doped cases (less than 0.5%). In contrast, the heavily-doped target (more than 0.5%) behaves oppositely. This phenomenon is caused by the difference of X-ray absorption between the lightly-doped region and heavily-doped region. In the case of the lightly-doped target, the self-emitted X-rays are absorbed at high-density region and depress the peak density. On the other hand, in the case of the heavily-doped target, most of X-rays are absorbed at low-density region, forming a plateau

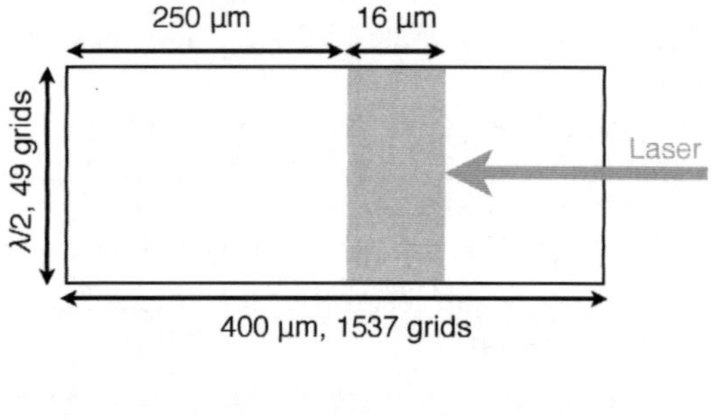


(a) planar target

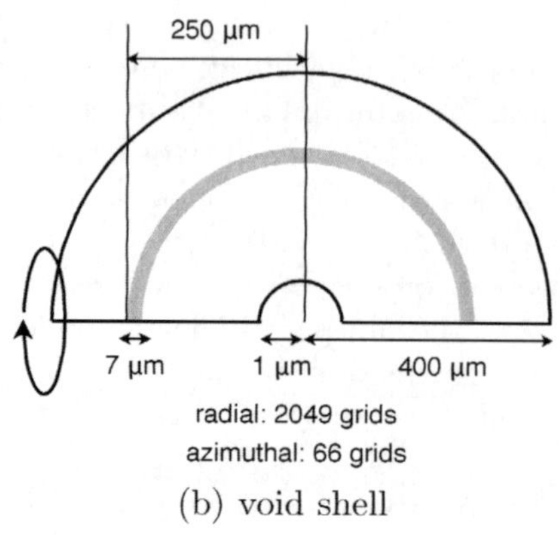


(b) void shell

**Figure 1.** Computational domains and conditions.

region, and then the peak density is kept high. Hence, there is a fraction of high-Z to minimize the peak density.

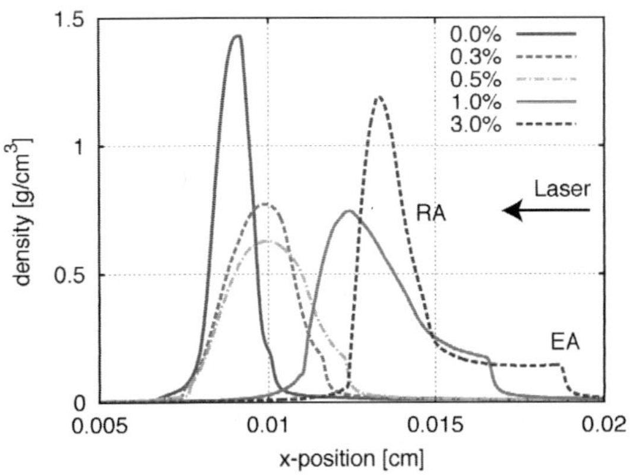


**Figure 2.** Density profiles with different doping fractions at the acceleration phase ($t = 3.0$ ns).

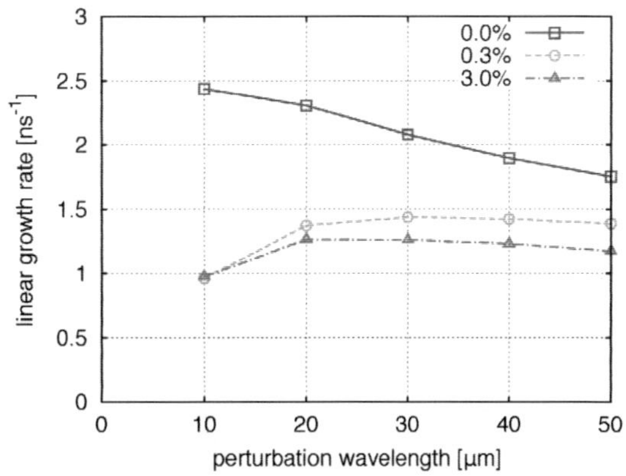


**Figure 3.** Linear growth rates for the ablative RTI at the acceleration phase ($t = 2.4 - 3.0$ ns).

### *3.2. Ablative RTI*

The dispersion relation of the ablative RTI was examined with 2D simulations of the planar targets. The initial target thickness was set to 16 $\mu$m and irradiated by laser. The laser condition was the same as the 1D cases.

Figure 3 shows the dispersion relation for the ablative RTI at the acceleration phase. The fractions appearing on the legends correspond to the targets used in GEKKO XII experiments [10]. The 0.3%-doped target reduces the linear growth rate of the ablative RTI at the short wavelength ($\simeq$ 10 $\mu$m) as well as the 3.0%-doped one does since the RA front gradient of the former is gentler than that of the latter. However, the 0.3%-doped target does not suppress the ablative RTI so as the 3.0%-doped target does in the intermediate wavelength (30 – 50 $\mu$m) because the ablation velocity becomes lower as the fraction of bromine decreases. Thus, the lightly-doped target is effective only for the short wavelength in the suppression of the ablative RTI.

### *3.3. Spherical implosion*

Axisymmetric implosion simulations were conducted to evaluate implosion performances of the brominated void shells. The simulation domain is described in Figure 1(b). The initial target diameter was set to 500 $\mu$m, and the initial target thickness was 7 $\mu$m. The input laser was a flat-topped pulse with maximum power of 1.5 TW, 0.53-$\mu$m wavelength, and 2-ns pulse duration.

Figures 4 and 5 depict density contours with low-mode ($l = 6$) perturbation and those with high-mode ($l = 32$) perturbation, respectively. The lower-left presents the density of the non-

**Table 1.** Maximum compression density of the brominated shells.

| $l$ | 6 | | | 32 | | |
|---|---|---|---|---|---|---|
| % of Br | 0.0 | 0.3 | 3.0 | 0.0 | 0.3 | 3.0 |
| $\rho_{max}$[g/cm$^3$] | 47.1 | 40.4 | 33.4 | 13.4 | 15.9 | 17.0 |

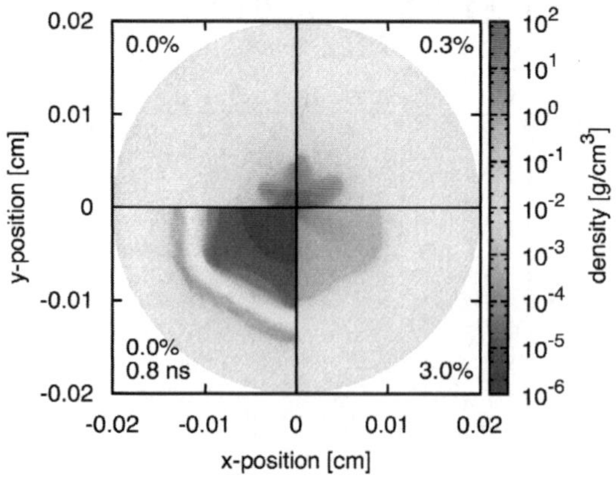


**Figure 4.** Density contours with low-mode perturbation ($l = 6$) at the maximum compression.

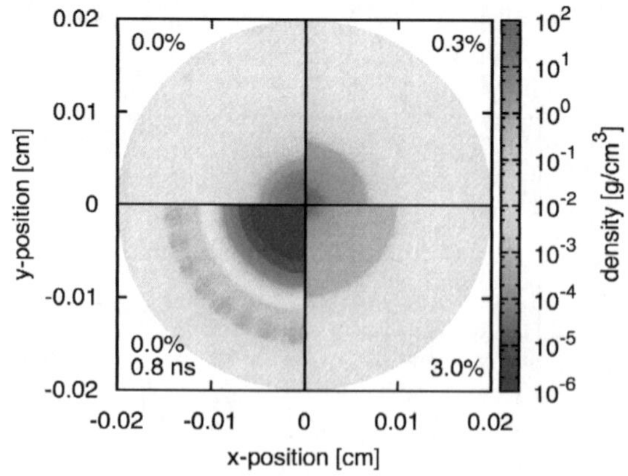


**Figure 5.** Density contours with high-mode perturbation ($l = 32$) at the maximum compression.

doped target at the acceleration phase (0.8 ns), the others are for the brominated targets (0.0%, 0.3%, 3.0%) at the maximum compression. Table 1 presents the maximum compression density of the brominated targets. The maximum compression density with low-mode perturbation becomes smaller with decrease of the target momentum due to the radiation loss. Furthermore, the high-Z doping cannot suppress the low-mode perturbation so much. On the other hand, in the case with high-mode perturbation, the maximum compression density becomes greater with increase of the bromine concentration. It may come from the fact that the high-Z doping spoils the nonlinearity of the ablative RTI and prevents a break of the target. As a consequence, the high-Z doping is only effective for suppression of high-mode perturbations.

## 4. Conclusions

We conducted two-dimensional RHD simulations to confirm the effect of high-Z doping in laser-accelerated targets. The ablation structure seems to change depending on the fraction of the high-Z dopant. A 0.3-mol% doped target can suppress Rayleigh–Taylor instability on the short wavelength. In a spherical implosion, we may conclude that the high-Z doping is only effective for high-mode perturbations.

## Acknowledgments

TS would like to thank the computer staff at Institute of Laser Engineering and Cybermedia Center, Osaka University for conducting the simulations on NEC SX-8R.

## References

[1] Takabe H et al. 1985 *Phys. Plasmas* **28** 3676
[2] Budil K S et al. 1996 *Phys. Rev. Lett.* **76** 4536
[3] Fujioka S et al. 2004 *Phys. Rev. Lett.* **92** 195001
[4] Ohnishi N, Nishikino M and Sasaki A 2006 *J. Phys. IV* France **133** 1193
[5] Wada Y and Liou M S 1994 *AIAA Paper* 94-0083
[6] van Leer B 1979 *J. Comp. Phys.* **32** 101
[7] Fujino S, Fujiwara M and Yoshida M 2006 *Trans. of JSCES* **8** 145 (in Japanese)
[8] More R M et al. 1988 *Phys. Fluids* **31** 3059
[9] Spitzer L and Härm R 1953 *Phys. Rev.* **89** 977
[10] Fujioka S *private communication*

# Beryllium ignition target design for indirect drive NIF experiments

**A N Simakov[1], D C Wilson[1], S A Yi[1], J L Kline[1], J D Salmonson[2], D S Clark[2], J L Milovich[2] and M M Marinak[2]**

[1] Los Alamos National Laboratory, Los Alamos NM, USA
[2] Lawrence Livermore National Laboratory, Livermore CA, USA

E-mail: simakov@lanl.gov

**Abstract.** Beryllium (Be) ablator offers multiple advantages over carbon based ablators for indirectly driven NIF ICF ignition targets. These are higher mass ablation rate, ablation pressure and ablation velocity, lower capsule albedo, and higher thermal conductivity at cryogenic temperatures. Such advantages can be used to improve the target robustness and performance. While previous NIF Be target designs exist, they were obtained a long time ago and do not incorporate the latest improved physical understanding and models based upon NIF experiments. Herein, we propose a new NIF Be ignition target design at 1.45 MJ, 430 TW that takes all this knowledge into account.

## 1. Advantages of Be ablators and previous NIF Be target designs

National Ignition Facility (NIF) is working hard towards demonstrating thermonuclear ignition and burn of millimeter size, indirectly driven, inertially confined, cryogenic deuterium-tritium (DT) capsules in a gold hohlraum. So far, NIF has primarily employed plastic-ablator (CH) capsules with germanium (Ge) or silicon (Si) dopant, although recently a number of experiments with high density carbon (HDC or "diamond") ablators have been also carried out. Herein, we propose an alternative NIF target that employs copper (Cu) doped beryllium-ablator (Be) capsules.

Be ablator offers multiple advantages over carbon-based ablators [1, 2]. Such advantages can be utilized, in particular, to significantly improve capsule robustness to instabilities [1]. In fact, overall, Be is arguably the best ablator there exists. The most important Be advantages, as demonstrated [2] both experimentally and computationally, are the mass ablation rate $\dot{M}_a$ [mg/(cm$^2$ ns)], ablation pressure $P_a = C_s\dot{M}_a$ [Mbar] and ablation velocity $V_a = \dot{M}_a/\rho$ [$\mu$m/ns], which are significantly larger than those of CH and HDC. Here, $C_s$ is the sound speed in the ablated plasma and $\rho$ is the ablator mass density. Dependences of $\dot{M}_a$, $P_a$ and $V_a$ on ablator material properties can be understood by considering propagation of a thermal wave that carries heat into the ablator by the radiation heat conduction mechanism [3] and are given by

$$\dot{M}_a \propto \sqrt{\frac{A_i}{1+Z_i}\frac{T_r^3}{\kappa_R(\rho,T_r)}},\quad P_a \propto \frac{T_r^2}{\sqrt{\kappa_R(\rho,T_r)}},\quad V_a \propto \sqrt{\frac{A_i}{1+Z_i}\frac{T_r^3}{\rho^2\kappa_R(\rho,T_r)}}. \qquad (1)$$

Here, $A_i$ and $Z_i$ are the charge and atomic numbers of the ablator ions, $T_r$ is the radiation temperature and $\kappa_R$ is the Rosseland opacity. Aside from lithium, a reactive metal, and large-grained lithium hydride, Be has the lowest opacity and thus highest $\dot{M}_a$, $P_a$ and possibly $V_a$ of all ablator materials at relevant hohlraum temperatures $T_r \sim 200 - 300$ eV. This is illustrated in Figs. 1 and 2 for Be, CH and HDC. In particular [2], $\dot{M}_a^{Be}/\dot{M}_a^{CH+0.6\%Ge} \approx 2.1$ and $\dot{M}_a^{Be}/\dot{M}_a^{HDC} \approx 1.5$. Higher Be $\dot{M}_a$ results in a higher $V_a$ and better ablative stabilization of the ablation front Rayleigh-Taylor instabilities. Indeed, employing $\rho_{Be} = 1.845$, $\rho_{HDC} = 3.51$ and $\rho_{CH} = 1.06$ g/cm$^3$ gives $V_a^{Be}/V_a^{HDC} \approx 2.89$ and $V_a^{Be}/V_a^{CH} \approx 1.24$. In addition, lower Be opacity results in lower capsule albedo and thereby higher absorbed energy, a several-percent effect. Higher Be thermal conductivity at cryogenic temperatures results in smoother DT ice layer than with CH. Finally, Be capsule fabrication is fairly well understood [4].

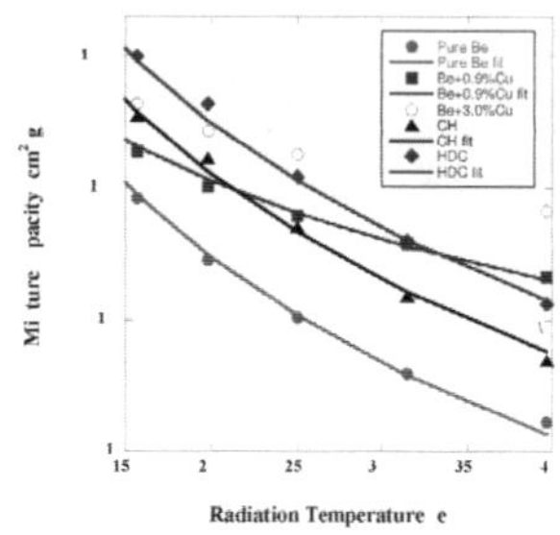

**Figure 1.** Rosseland opacities for Be, HDC and CH at normal densities.

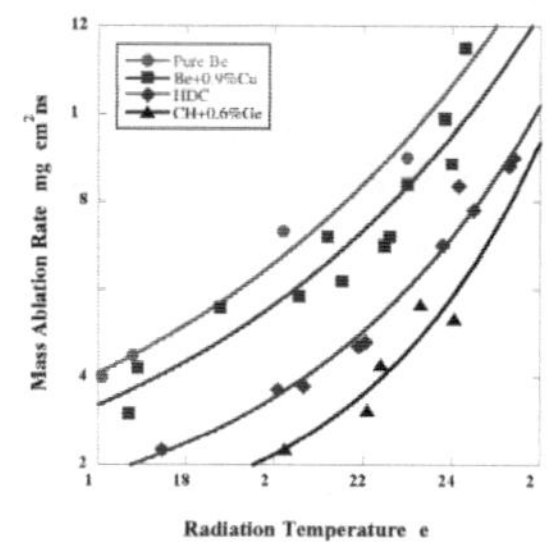

**Figure 2.** Measured $\dot{M}_a$ for Be, HDC and CH (lines are least-squares fits $\propto T_r^3$) [2].

NIF Be targets have been designed, but not used, before. Examples are Rev. 3 [5] and Rev. 5 [6] targets. However, since these designs were created, it became clear that hohlraum energetics is less favorable than earlier thought and that design calculations have to assume about 15-17% of energy back-scatter [7] as well as arbitrarily reduce the main laser pulse power by extra 15% [8]. In addition, previous Be designs did not employ the recently formulated "high-flux" (HF) hohlraum model [9] that (i) improves the atomic physics model of the hohlraum wall by employing DCA rather than XSN atomic configuration model; and (ii) increases the electron heat flux limiter $f$ from 0.05 to 0.15.

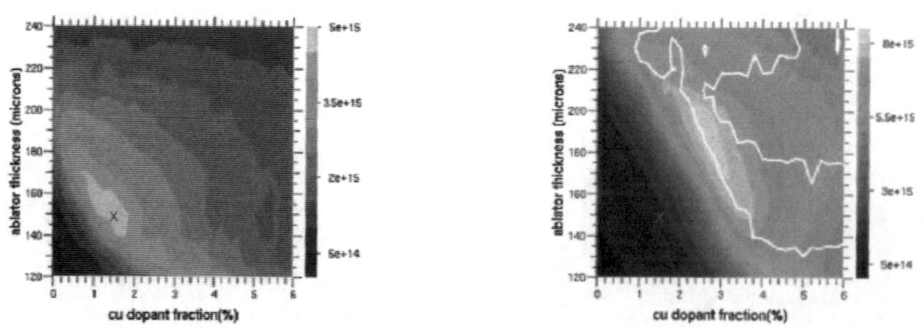


**Figure 3.** Neutron yield contours without $\alpha$-particle energy deposition for a Be capsule for the "old" (left) and HF (right) hohlraum model. Red cross indicates an updated Rev. 5 design. The white contour indicates neutron yield of $1.7 \times 10^{18}$ with the $\alpha$ deposition.

Figure 3 demonstrates effects of the HF model on neutron yield of a NIF Be ignition capsule with $\alpha$-particle energy deposition neglected, as obtained by scanning the (ablator thickness, maximum Cu dopant) parameter space. For the scan, we used a fixed ice-ablator interface radius of 945 $\mu$m, a DT ice layer thickness of 69 $\mu$m, and the standard dopant pyramid structure. The capsule was driven by a frequency dependent source (FDS) obtained from an integrated simulation with the standard NIF 5.75-mm gold hohlraum and a tuned laser pulse's energy of 1.2 MJ and maximum power of 415 TW. The right figure used an FDS obtained with the HF model and the left an FDS obtained with the XSN model and $f = 0.05$ (the "old" model). The red cross indicates an updated Rev. 5 NIF Be target design (147 $\mu$m of ablator and 1.5% of Cu dopant) obtained with the "old" model, which is clearly far from optimal according to the HF model. Thus, it is clear that existing NIF Be designs are outdated and a new design is necessary.

## 2. Proposed new Be target design

For a new Be ignition target, we chose to employ the standard NIF gold hohlraum with a diameter of 5.75 mm and the laser entrance hole (LEH) diameter of 3.15 mm. The helium fill gas density of 1.6 mg/cm$^3$ was selected to moderate the hohlraum wall and the capsule blow-off motion for optimal laser penetration but not to lead to hydro-coupling. We also elected to use a conservative, four shock laser pulse with total delivered energy of 1.45 MJ and maximum delivered power of 430 TW, with assumed 15% of backscatter and the 0.85 main pulse power multiplier. We tuned the pulse's foot in the standard way to optimize shock strength and timing for ensuring low-entropy implosion. An important constraint was to ensure that the first shock melted the ablator by creating pressure in excess of 2.8 Mbar, so that the Be polycrystalline structure did not introduce hydrodynamic perturbations. We had to determine the pulse's time dependent cone fractions as well as beam pointing positions to ensure the implosion symmetry. Finally, we had to optimize the capsule. We chose to follow the updated Rev. 5 design by using the same DT ice thickness of 69 $\mu$m, ice-ablator interface radius of 945 $\mu$m, and dopant pyramid structure (i.e., going outside from the ice-ablator interface, 5 $\mu$m of undoped Be, 5 $\mu$m of doped Be at 40% of the maximum dopant level, 18 $\mu$m of doped Be at the maximum dopant level and 5 $\mu$m of doped Be at 40% of the maximum dopant level). However, we had to optimize the overall ablator thickness and the maximum Cu dopant level by using 2D integrated target simulations.

The optimal ablator thickness and dopant level were obtained by scanning the parameter space (130 $\mu$m < thickness < 200 $\mu$m, 1 % < dopant < 5%). The design candidates were selected by requiring a reasonably high yield and peak velocity, keeping remaining mass at the 1-2% level, and staying below the Cu solubility in Be limit of ~3%. Some of the results are shown in Table 1. Eventually, we selected the ablator thickness of 155 $\mu$m and dopant level of 2%. Instability simulations indicate that this dopant level is most robust.

**Table 1.** Yield (MJ)/Peak no-burn velocity ($\mu$m/ns)/Remaining ablator mass (%).

| Abl./Dop. | 1% | 2% | 3% | 4% |
|---|---|---|---|---|
| 180 $\mu$m | | | 15.5/338/5.5 | |
| 175 $\mu$m | | 14.4/353/3.1 | 17.8/343/5 | |
| 170 $\mu$m | | 14.5/357/3.6 | 18.0/346/4.5 | |
| 165 $\mu$m | | 15.0/363/2.5 | 18.6/349/4 | |
| 160 $\mu$m | | 16.0/373/2.3 | 16.4/354/3.5 | |
| 155 $\mu$m | 0.4/403/0 | **14.8/379/2** | 18.4/359/3.5 | 19.6/344/4.5 |
| 150 $\mu$m | | 13.5/384/1.2 | 18.4/363/3 | |
| 145 $\mu$m | | 10.6/393/1 | 16.4/368/2.5 | |
| 140 $\mu$m | | 0.2/405/0 | 13.1/379/1.8 | |

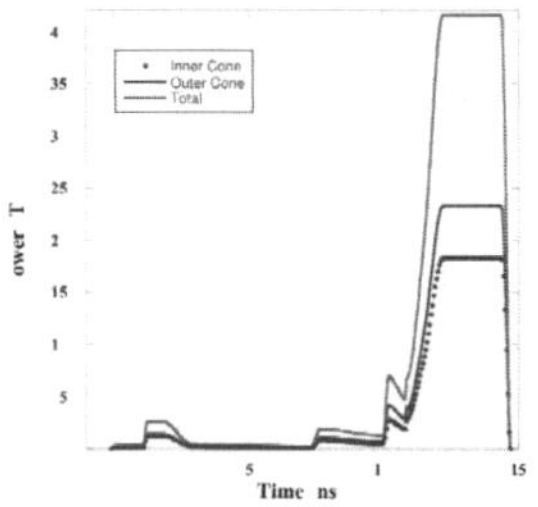


**Figure 4.** Delivered laser power.

For the selected hohlraum and capsule parameters, the time-dependent delivered laser power is shown in Fig. 4. The absorbed power is obtained by removing 15% for backscatter and degrading the main pulse by another 15%. The required for symmetry cone fractions for the picket, trough, second, third and fourth pulses are 0.5, 0.5, 0.43, 0.4 and 0.44, respectively. The main pulse cone fraction is easily achievable with cross-beam power transfer.

The capsule robustness is demonstrated in Fig. 5, which shows the yield degradation with an increase in the capsule interface roughness. The roughness multiplier is used to artificially increase the specified roughness of all the interfaces. The dominant instability source is the ice roughness. The green line shows the results for the specified Rev. 3 Be design roughness (with ice shown in green in Fig. 6), while the red corresponds to a recently measured Be capsule roughness (assuming CH ice roughness shown in red in Fig. 6) and indicates that the capsule can still produce a significant yield with all interfaces twice as rough as readily achievable.

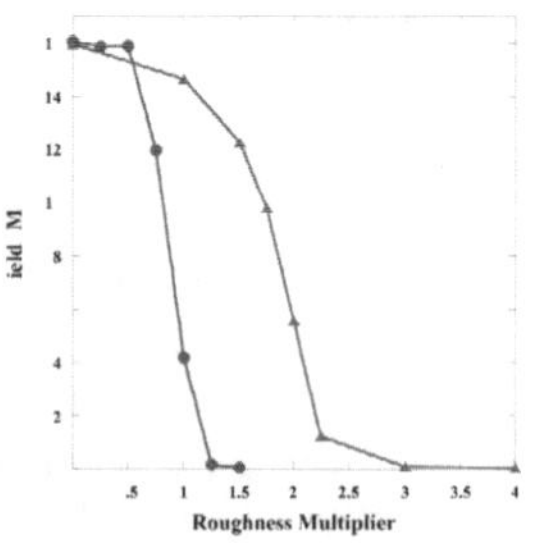


**Figure 5.** Yield degradation vs. interface roughness for specified NIF Rev. 3 Be target (in green) and recent surface roughness measurements (in red).

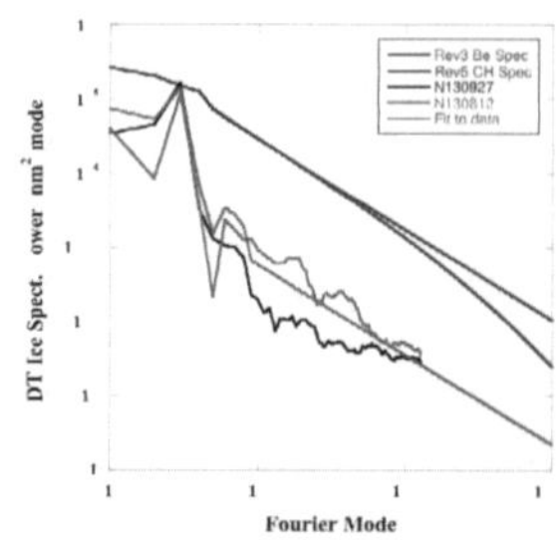


**Figure 6.** DT ice roughness for Rev. 3 Be (green) and Rev. 5 CH (blue) designs, recent CH capsule measurements (black and magenta) and a fit to the last two (red).

## ACKNOWLEDGEMENT

We wish to thank B. J. Kozioziemski, S. W. Haan and D. A. Callahan from LLNL for multiple enlightening discussions. This work was performed at Los Alamos National Laboratory, operated by Los Alamos National Security, LLC for the U.S. Department of Energy under contract DE-AC52-06NA25396.

## References

[1] Wilson D C *et al* 1998 *Phys. Plasmas* **5** 1953
[2] Olson R E *et al* 2011 *Phys. Plasmas* **18** 032706
[3] Zel'dovich Ya B and Raizer Yu P 2002 *Physics of Shock Waves and High-Temperature Hydrodynamic Phenomena* (Mineola, NY: Dover Publications)
[4] Huang H *et al* 2012 *Bull. Am. Phys. Soc.* **57 no. 12** 113
[5] Haan S W *et al* 2009 *Fusion Sci. Technol.* **55** 227
[6] Haan S W *et al* 2011 *Phys. Plasmas* **18** 051001
[7] Hinkel D E *et al* 2011 *Phys. Plasmas* **18** 056312
[8] Jones O S *et al* 2012 *Phys. Plasmas* **19** 056315
[9] Town R P J *et al* 2011 *Phys. Plasmas* **18** 056302

# Integrated modelling framework for short pulse high energy density physics experiments

**N. J. Sircombe, S. J. Hughes and M. G. Ramsay**

AWE, Aldermaston, Reading, RG7 4PR, UK.

E-mail: nathan.sircombe@awe.co.uk

**Abstract**. Modelling experimental campaigns on the Orion laser at AWE, and developing a viable point-design for fast ignition (FI), calls for a multi-scale approach; a complete description of the problem would require an extensive range of physics which cannot realistically be included in a single code. For modelling the laser-plasma interaction (LPI) we need a fine mesh which can capture the dispersion of electromagnetic waves, and a kinetic model for each plasma species. In the dense material of the bulk target, away from the LPI region, collisional physics dominates. The transport of hot particles generated by the action of the laser is dependent on their slowing and stopping in the dense material and their need to draw a return current. These effects will heat the target, which in turn influences transport. On longer timescales, the hydrodynamic response of the target will begin to play a role as the pressure generated from isochoric heating begins to take effect. Recent effort at AWE [1] has focussed on the development of an integrated code suite based on: the particle in cell code EPOCH, to model LPI; the Monte-Carlo electron transport code THOR, to model the onward transport of hot electrons; and the radiation hydrodynamics code CORVUS, to model the hydrodynamic response of the target. We outline the methodology adopted, elucidate on the advantages of a robustly integrated code suite compared to a single code approach, demonstrate the integrated code suite's application to modelling the heating of buried layers on Orion, and assess the potential of such experiments for the validation of modelling capability in advance of more ambitious HEDP experiments, as a step towards a predictive modelling capability for FI.

## 1. Integrated Modelling

The use of short pulse lasers to heat materials to high temperatures in order to study their properties has been a key part of the laser programme at AWE for some time – both on the HELEN laser [2, 3] and, more recently, the more energetic Orion laser [4, 5]. The compression of material by long pulses prior to heating offers an opportunity to study a broad range of temperatures and densities in the laboratory. A complete description of an Orion experiment requires an extensive range of physics, including: accurate treatment of Maxwell's equations and kinetic effects where the laser is absorbed; collisional effects for the onward transport of the fast particles generated, including return current effects; and hydrodynamic response in the dense target, using an appropriate equation of state (EOS).

We have tools available to model these key aspects: EPOCH – a massively parallel, relativistic electromagnetic particle in cell (PIC) code for modelling laser-plasma interaction; THOR – a Monte-Carlo fast electron transport code for modelling hot electron energy deposition in dense plasmas; CORVUS – an ALE radiation hydrodynamics code, with an extensive suite of physics packages available, for modelling hydrodynamic compression, laser pre-pulse, and target evolution following the short pulse energy deposition.

Since combining all of the required physics in one code is not practical, or tractable with existing (and anticipated) HPC, we link our kinetic, transport and hydro-codes together (see figure 1) to provide an integrated modelling capability. Our aim is to prepare a robust tool for high energy density physics (HEDP) research that will allow as much of the relevant physics to be modelled as possible.

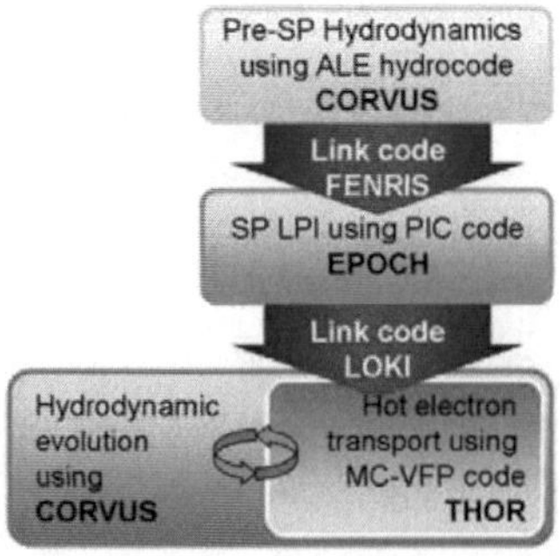


**Figure 1.** Integrated modelling framework. Hydrodynamic effects prior to the arrival of the short pulse are modelled in hydro-code CORVUS, which is linked to the PIC code EPOCH using the link code FENRIS. Hot electrons from EPOCH are fed back into CORVUS via the MC-VFP code THOR, using the link code LOKI, and their onward transport is captured by sub-cycling THOR within CORVUS

We use our modelling suite to study the heating of aluminium microdots buried 4 microns from the front of 10 micron diamond foils. Initially we consider only the action of the short pulse, before demonstrating the capability to model both compression and heating. The challenge of modelling the combination of both long and short pulse LPI, with fast electron transport through dense material and long time-scale hydrodynamic motion, has much in common with that of fast ignition (FI). Experiments on systems such as Orion offer an opportunity to refine and test modelling capabilities required for developing FI point designs in the future.

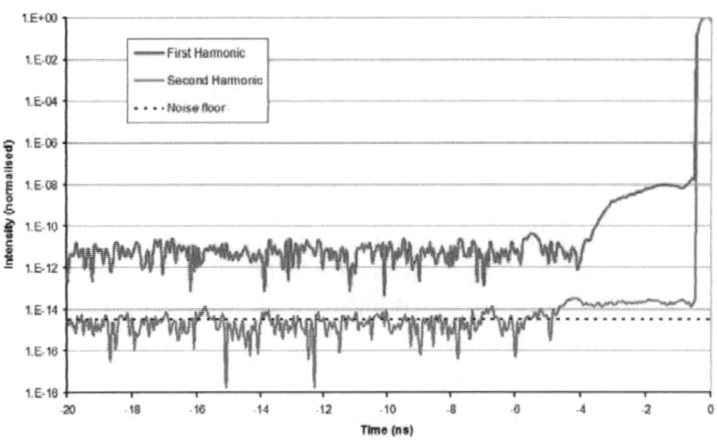


**Figure 2.** Measured pre-pulse from the first and second harmonic short-pulse beams on Orion.

## 2. The Orion Pre-pulse

Each of Orion's two petawatt beamlines can deliver a peak intensity of $\sim 10^{21}$Wcm$^{-2}$ and in excess of 500J in the first harmonic ("red" light). One beamline can be converted to deliver in excess of 100J in the second harmonic ("green" light), using a sub aperture 3mm thick KDP crystal. Measurements taken using a two-photodiode pre-pulse monitor (see figure 2) indicate a 3-4 ns pedestal [6] due to parametric fluorescence in the OPA. Results indicate a contrast of approximately $10^8$ in the first harmonic and $10^{14}$ for the second harmonic. The laser package in CORVUS was configured to deliver the recorded red pre-pulse to target, see figure 3. CORVUS density and temperature profiles are then mapped onto EPOCH's Eulerian mesh using the link code FENRIS.

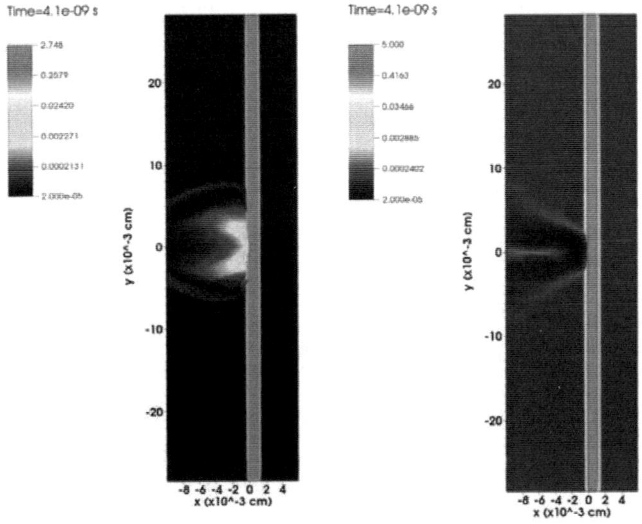


**Figure 3.** Density profiles for plastic (left) and diamond (right) foils, with buried Al microdots, following the pre-pulse.

## 3. Laser plasma interaction modelling

EPOCH was used to model the interaction of the main pulse with the plasma profile generated from the pre-pulse modelling performed in CORVUS. A 500J, red, 0.5ps (full width half maximum) pulse was focussed to a 10μm spot and angled at 20° to target normal. The hot electron spectra generated are captured by a particle probe placed at solid density, behind the laser interaction region. The link code LOKI generates a time-resolved source distribution function in ($y$, $E$, ) from the probe data which is passed to the transport code THOR.

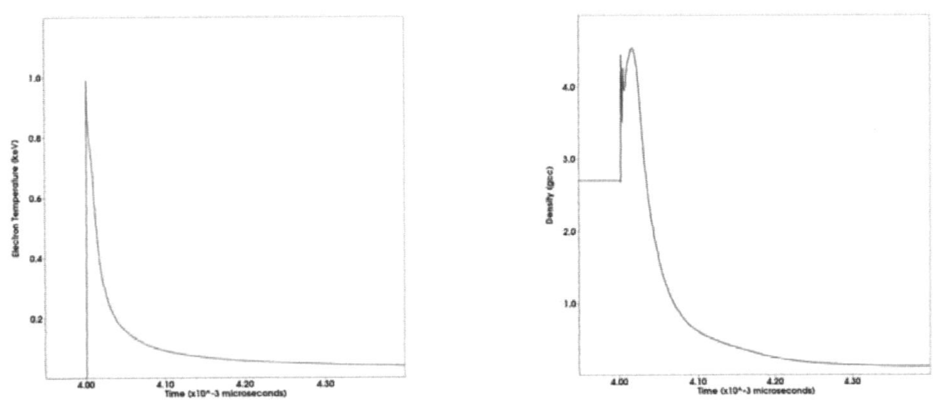


**Figure 4.** Temperature (left) and density (right) histories for a single cell at the centre of the aluminium sample.

## 4. Transport modelling

CORVUS modelling, using THOR for hot electron transport, shows the Al sample reaching a temperature close to 1 keV at a density in excess of 4 g/cc (see figure 4). There is little bulk hydrodynamic motion during the heating phase. However, capturing the long timescale evolution of the target as the heated material expands necessitates the integration of a robust fast electron transport model within a hydrodynamic framework.

## 5. Long pulse compression

The rear surface of the diamond buried-microdot target was illuminated by 500J of third harmonic ("blue" light) long-pulse energy over ~1ns. Ablation drives a shock into the target which compresses the Al (see figure 5), achieving a peak density of ~7g/cc, falling to ~5.5 g/cc by the time the shock reaches the front of the target. Using THOR, sub-cycling within CORVUS, with an EPOCH generated source distribution to model the effect of short pulse heating, starting at 4ns, close to peak density, temperatures of ~800eV at densities ~7g/cc can be achieved, with higher densities likely with further optimisation.

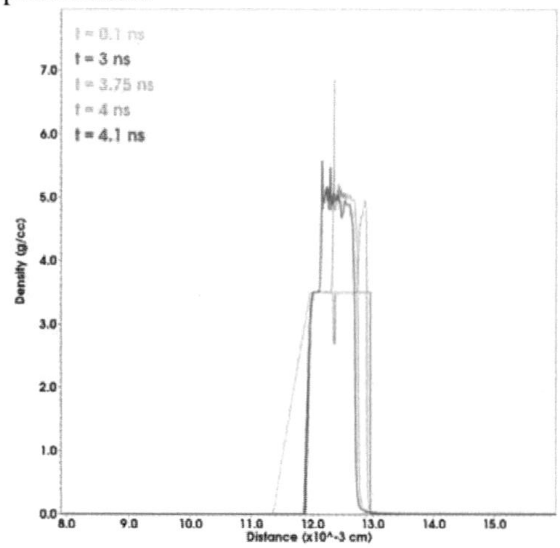


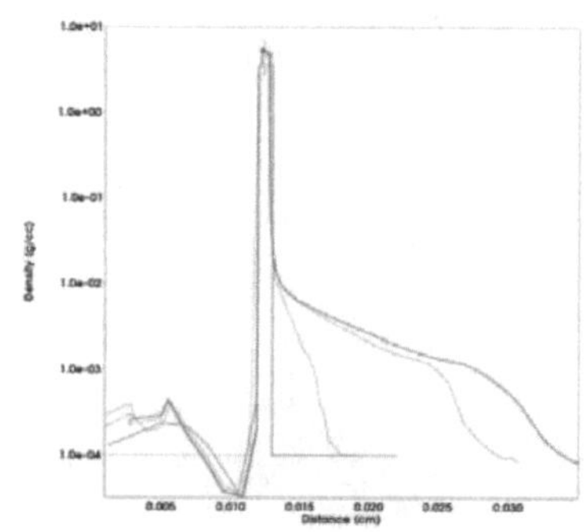

**Figure 5.** Linear (left) and log (above) density line-outs through the target at various times showing the generation of a pre-plasma at the front, a long scale length plasma at the rear, and a shock travelling through the target which compresses the sample.

## 6. Discussion

We have shown how an integrated modelling framework can be employed to simulate short and long pulse irradiated targets with an application to HEDP experiments. The use of explicit PIC sources in a hybrid code captures the distribution in energy and divergence angle, as well as the spatial and temporal evolution of the hot electron population, which is difficult to capture with a parameterized analytic form.

Coupling transport and hydrodynamic algorithms also confirmed that, over the timeframe of the short-pulse interaction, an isochoric approximation is often sufficient, as little hydrodynamic motion takes place. This means that a considerable quantity of modelling can be carried out using just THOR, drawing on the same set of EPOCH simulations for the hot electron source. This provides a flexible tool for designing and interpreting experiments which can be refined with further PIC simulations and augmented with bulk hydrodynamic motion as required. The isochoric approximation is less appropriate if pulses get longer and more energetic, as would be the case for many FI concepts and for HEDP experiments which utilize long-pulse beamlines, it is here that the integration of THOR into CORVUS comes to the fore.

The capability outlined here can already be applied to help design and interpret experiments; the efficacy of this approach will improve over time as HEDP campaigns progress. Continuing development of both the constituent codes and a refinement of the links between them, underwritten with new experimental data from facilities such as Orion will ensure that these nascent capabilities develop into a practical, predictive modelling tool which can help the design and interpretation of ever more ambitious HEDP experiments in the coming years.

## References

[1] N J Sircombe, S J Hughes and M G Ramsay, New J. Phys.15 025025 (2013).
[2] D J Hoarty et al. HEDP, 1, 3 (2007).
[3] C R D Brown et al. Phys. Rev. Lett. 5, 106 (2011).
[4] D J Hoarty et al. Phys. Rev. Lett. 6, 110 (2013).
[5] D J Hoarty et al. HEDP 9, 661 (2013).
[6] D Hillier et al. Applied Optics, 52, 4258, (2013).

Journal of Physics: Conference Series **688** (2016) 012112 doi:10.1088/1742-6596/688/1/012112

# Control and ultra-short pulse generation in stimulated Raman backscattering

**Miloš M. Škorić[1,2,3], Ljubomir Nikolić[4], Ljupčo Hadžievski[1]**
**Dimitri Batani[3], Seiji Ishiguro[2] and Kunioki Mima[5]**

[1]Vinča Institute of Nuclear Sciences, University of Belgrade, 11001 Belgrade, Serbia
[2]National Institute for Fusion Science, Graduate University for Advanced Studies, Japan
[3]Université Bordeaux, CEA, CNRS, CELIA, F-33405 Talence, France
[4]Geomagnetic Laboratory, NRCan, Ottawa, ON, K1A OY3, Canada
[5]Institute of Laser Engineering, Osaka University, Japan

E-mail: `mskoric@vinca.rs`

**Abstract.** A stimulated Raman scattering (SRS) is a great concern for laser fusion, causing an energy and symmetry loss and target preheating. In particular, for dominating backward-SRS (BRS), a complexity of nonlinear saturation was revealed. Recently, to develop lasers at multi-exawatts and beyond, relevant to high-energy physics, a proposed Raman amplification based on BRS met a restrictive operation window. Here, we stress BRS generics that due to a nonlinear frequency shift (relativistic/trapping effect), it nonlinearly saturates with intermittent pulsations. A "break up" of Manley-Rowe invariants explains a non-steady saturation. Further, a coherent pulsation BRS regime is proposed for femto-sec pulse generation in a thin exploding foil plasma, with parameter scalings investigated by analytics and particle simulations.

## 1. Introduction

Nonlinear laser plasma instabilites are useful paradigms of complex phenomena. In laser fusion, backward stimulated Raman scattering (BRS) in an underdense plasma results in a reduced coupling of laser energy to the target [1-3]. Recently, novel ultra-intense lasers beyond multi-exawatts, when standard CPA fails, were discussed for high-energy physics. Raman amplification scheme [4], with a counter-propagating pico-second laser pump and a femto-second resonant seed, was proposed. Still, simulations and experiments found a narrow parameter window to avoid parasitic instabilities (Forward-SRS, relativistic MI, etc.) and pulse destruction already at moderate intensities [5]. Firstly, in a standard 3-coupled mode model of BRS we outline the condition for an absolute instability [2, 6], that is readily satisfied in a uniform plasma which amplifies large Raman signals from a background noise. Even at low intensity, $I \sim 10^{14}$ $W/cm^2$, $\sim$10 microns long, 0.1 of a critical density ($N_0$), absolute BRS dominates. As a generic feature, due to a nonlinear frequency shift (in EPW) [7-9], instead to a steady-state via pump depletion, BRS saturates through intermittent pulsations with reflectivity bursts and spectral broadening [2, 6]. The absolute BRS sets in for the interaction length $L_0$ shorter than, both, the plasma length $L$ and absorption length $L_a$. The fact that, both, interaction and absorption length depend on plasma dynamics, contributes to an overall Raman complexity. Finally, large coherent pulsation regime is proposed for efficient femto-sec optical pulse generation by BRS in thin foil plasmas [9], with scalings studied by analytics and particle simulations.

## 2. Model

Stimulated Raman backscattering in a plasma is a paradigm of a three-wave parametric instability (3WI) whereby a strong electromagnetic-laser light (pump) wave decays into an electron plasma wave (EPW) and a backscattered light wave downshifted in frequency. The coupled 3WI process obeys a resonant matching condition for frequencies and wavenumbers of three waves ($\omega_0 = \omega_1 + \omega_2$, $\mathbf{k}_0 = \mathbf{k}_1 + \mathbf{k}_2$). We model BRS as a resonant parametric coupling of 3WI for $a_i(x,t)\exp[i(k_i x - \omega_i t)]$, in a weakly varying envelope approximation [1,2,6]

$$\frac{\partial a_0}{\partial t} + V_0 \frac{\partial a_0}{\partial x} = -M_0 a_e a_s, \tag{1}$$

$$\frac{\partial a_s}{\partial t} - V_s \frac{\partial a_s}{\partial x} = M_s a_0^* a_e, \tag{2}$$

$$\frac{\partial a_e}{\partial t} + V_e \frac{\partial a_e}{\partial x} + \Gamma_e a_e - i\delta \left|a_e\right|^2 a_e = M_e a_0^* a_s, \tag{3}$$

where $V_i > 0$ are the group velocities, $\Gamma_e$ is a damping rate for EPW ($\Gamma_0 = \Gamma_s = 0$ for light waves is used), $M_i > 0$ are the coupling coefficients and $a_i$ are the wave amplitudes, where $i = 0,\ s,\ e$, stand for the pump, backscattered wave and EPW, respectively. A self-modal cubic nonlinearity in (3) is a generic nonlinear phase detuning (shift) due to relativistic/trapped electrons effect, $\sim \delta\left|a_2\right|^2$; e.g. a relativistic shift as $\delta = 3\omega_{pe}^3/16c^2 n_0^2 k_2^2$, [7-9]. With standard boundary conditions, $a_0(0,t) = E_0$, $a_s(L,t) = a_e(0,t) = 0$, BRS is absolute instability for $L/L_0 > \pi/2$, where $L_0 = (V_s V_e)^{1/2}/\gamma_0$ is the interaction length and $\gamma_0 = E_0(M_e M_s)^{1/2}$ is the uniform growth rate. If EPW damping, $\Gamma_e \neq 0$, it defines the absorption length $L_a = V_e/\Gamma_e$ with Raman backscattering becoming absolute under an extra condition, $L_0/L_a < 2$, [1-2].

### *2.1. Nonstationarity and "Break-up" of Manley-Rowe invariants*

Assuming the steady state ($\partial/\partial t \to 0$), conserved quantities (well known Manley-Rowe invariants, [1]) are calculated from (1-3), as [2, 9], (subscripts 1, 2 replace $s$, $e$, respectively)

$$m_0 = V_0 n_0(x) - V_1 n_1(x) = const., \tag{4}$$

$$m_1 = V_0 n_0(x) + V_2 n_2(x) = const., \tag{5}$$

$$K(x) = A_0 A_1 A_2 \sin\phi - \frac{\delta}{4} A_2^4 = const. \tag{6}$$

with $n_i(x) = A_i(x)^2$, $i = 0,1,2$, and $a_i(x,t) = A_i(x,t)\,e^{i\phi_i(x,t)}$, where $A_i$ and $\phi_i$, are the amplitude and phase of the wave, with the total phase shift given as $\phi = \phi_0 - \phi_1 - \phi_2$.
For boundary conditions, $n_0(0) = 1, n_1(L) = 0, n_2(0) = 0$, the third invariant gives $K(0) = 0$. However, as at the rear boundary $x = L$, in general, $A_2(L) \neq 0$, implying $K(L) \neq 0$; which breaks the invariance condition, i.e, $K(x) \neq const.$; hence, contradicting a basic assumption of the steady state. This simple argument by Škorić et al., [2, 9], predicts an onset of BRS complexity due to nonlinear phase detuning; as presently readily found in simulations and experiments on nonlinear BRS in laser plasmas [3, 8]. We also note above relevance to proposed Raman compression and amplification schemes, found to be limited to a weakly nonlinear regime [4-5].

## 3. Self-organization and route to complexity

The reflectivity $R$, which designates a fraction of incident laser intensity reflected backward [1,6]

$$R = \frac{V_0\left|a_1(0)\right|^2}{V_1\left|a_0(0)\right|^2}; \tag{7}$$

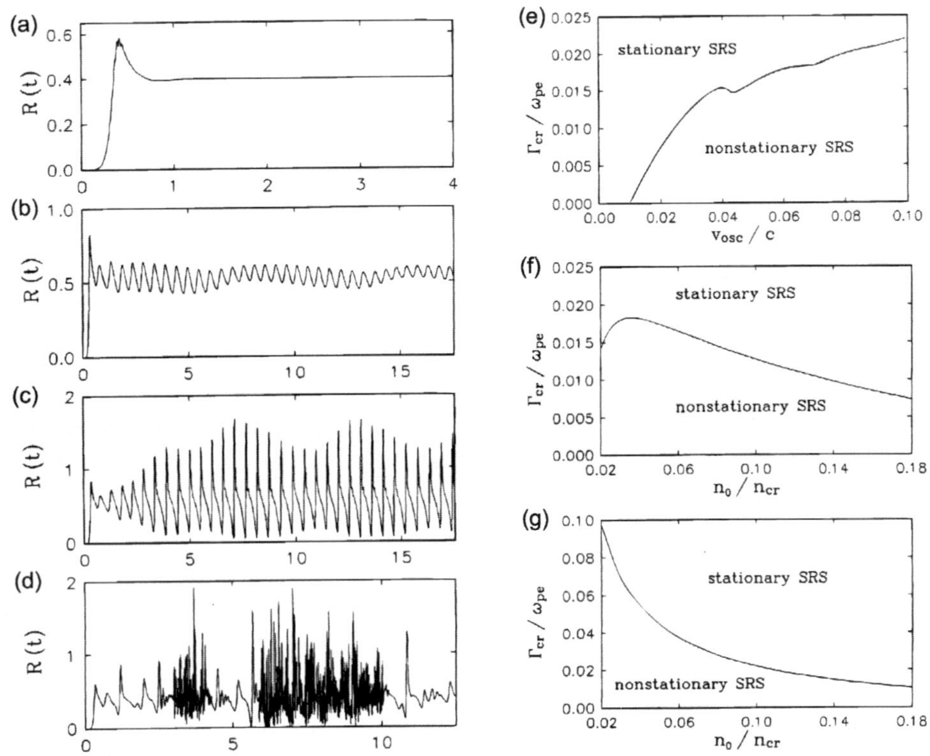


**Figure 1.** Raman complexity in time ($10^3/\omega_0$) versus EPW damping rate $\Gamma/\omega_{pe}$: (a) 0.06, (b) 0.02, (c) 0.018, (d) 0.012, for $\beta_0 = 0.1$, $N_0 = 0.1n_{cr}$, $L = 100c/\omega_0$, $T_e = 1keV$ and Bifurcation diagrams for onset of complexity (e) $(\Gamma - \beta_0)$ and $(\Gamma - N_0)$ for (f) $\beta_0 = 0.03$, (g) $\beta_0 = 0.1$

has maximum normalized to unity in the stationary case. To solve the system (1-3), we choose realistic boundary conditions which satisfy the condition for absolute instability. A series of numerical simulations in space-time are run for system (laser and plasma) parameters related to hohlraum plasmas $N_0 = 0.1n_{cr}$, $T_e$ =1 keV, $L = 100c/\omega_0$ and varying $\Gamma_e/\omega_{pe}$. As laser pump $\beta_0$ (ratio of electron quiver velocity to speed of light) increases beyond absolute instability threshold, self-organized saturated states follow a quasi-periodic route to intermittency $FP \to P \to QP \to I \to C$, where $FP$ stands for unimodal fixed point, $P$ for periodic, $QP$ for quasiperiodic, $I$ for intermittent, and $C$ for chaos, as shown in Fig. 1, [2, 6]. Intensity dependent intermittent pulsations and spectral broadening are inherent in BRS complexity as observed in recent experiments and simulations [3, 8]. Still, frequency broadenings measured are typically within a few percent of $\omega_0$ and justify a weak-coupling model (e.g. Figs 2-6 [6]). We stress a generic role of nonlinear detuning term in (3), as if absent ($\delta$=0), BRS goes to a steady-state failing to recover complex Raman features. Realistically, for large BRS, due to hot electrons from kinetic EPW and bulk plasma heating, interaction length $L_0$ and absorption length $L_a$ change in time. Generally, a growth of both, the bulk temperature and effective damping (collisional, Landau) increase the absolute treshold and suppress the BRS instability. To illustrate, in Fig. 1 we show characteristic lengths effect on BRS dynamics and bifuraction diagrams for an onset of complexity. Raman supression and transition from nonstationary to steady-state ($FP$), indicated by a borderline (Fig. $1e - g$), follows a dissipation increase.

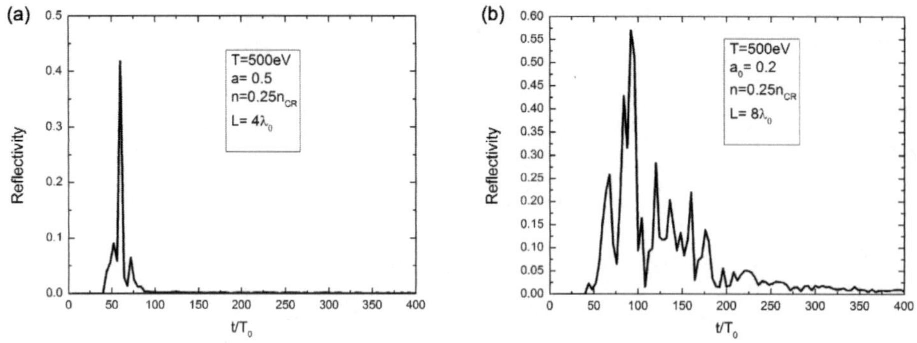


**Figure 2.** (a) Particle simulation of Raman compression into coherent fs-pulse in thin hydrogen foil plasma ($T \sim 10$ fs); (b) Unsuitable parameters choice results in incoherent Raman pulsations.

## 4. Femto-sec pulse generation by Raman backscattering in thin foil plasma

Here, we explore a new method for generation of ultra-short, fs-range optical pulses [9-10]. The scheme exploits a quasi-periodic coherent regime of BRS, as above, in laser interaction with a thin foil plasma. In an underdense exploding foil plasma, beyond the threshold for absolute BRS, reflectivity saturates with large pulsations [8-10]. For laser intensity in $10^{17}$ W/cm$^2$ range, hot electrons can rapidly halt the Raman instability and produce a single coherent pulse. By a proper choice of laser/ plasma parameters, an ultrashort Raman backscatter pulse of few laser periods, i.e. in 10 fs range is generated; as illustrated by relativistic particle simulatation results in Fig. 2. Therefore, one possibly finds a simple, robust and flexible scheme for proliferation of energetic optical fs-pulses by using the ps-scale laser driver system, now widely available.

## 5. Conclusion

We outlined a condition for absolute instability in stimulated Raman backscattering and crucial role of a nonlinear phase detuning in a route to complexity. Nonlinear saturation via intermittent pulsations is common, defined by the ratio between three parameters: interaction, plasma and absorption length, respectively. Accordingly, effective control of self-organized states, including kinetic BRS suppression and ultra-short coherent pulse generation appears to be feasible.

## Acknowledgments

Supported by the Ministry of Education & Science of Serbia, project III 45010. The IdEX visiting scholars program of Université Bordeaaux and COST1208 Action supports are acknowledged by MMŠ. We thank Dr. D. Hinkel, Dr. L. Yin and Dr. V. Tichonchuk for valuable discussions.

## References

[1] Forslund D W, Kindel J M and Lindman E L 1975 *Phys. Fluids* **18** 1002
[2] Kono M and Škorić M M 2010 *Nonlinear Physics of Plasmas* (Berlin: Springer) ch 12
[3] Hinkell D *et al* 2011 *Phys. Plasmas* **18** 056312
[4] Mourou G A *et al* 2012 *Optics Comm.* **285** 720
[5] Trines R M G M., Fiuza F, Bingham R *et al* 2011 *Nature Physics* **7** 87
[6] Škorić M M *et al* 1996 *Phys. Rev. E* **53** 4056; ibid. 1994 *AIP Conf. Proc.* **318** 380
[7] Kruer W L 1990 *Phys. Scr.* T**30** 5; Bingham R 1990 ibid. T**30** 24
[8] Yin L *et al* 2013 *Phys. Plasmas* **20** 012702; Berger R L *et al* 2013 *Phys. Plasmas* **20** 032107
[9] Škorić M M *et al* 1997 *AIP Conf. Proc.* **406** 381; ibid. 2009 **1188** 15
[10] Škorić M M, Nikolić Lj and Ishiguro S 2013 *J. Plasma Phys.* **79** 1003

# Hydrodynamic growth and mix experiments at National Ignition Facility

V: A. Smalyuk[1], J. Caggiano[1], D. Casey[1], C. Cerjan[1], D. S. Clark[1], J. Edwards[1], G. Grim[2], S. W. Haan[1], B. A. Hammel[1], A. Hamza[1], W. Hsing[1], O. Hurricane[1], J. Kilkenny[3], J. Kline[2], J. Knauer[4], O. Landen[1], J. McNaney[1], M. Mintz[1], A. Nikroo[3], T. Parham[1], H.-S. Park[1], J. Pino[1], K. Raman[1], B. A. Remington[1], H. F. Robey[1], D. Rowley[1], R. Tipton[1], S. Weber[1], and C. Yeamans[1]

[1]Lawrence Livermore National Laboratory, NIF Directorate, Livermore, CA 94550, USA

[2]Los Alamos Na National Laboratory, Los Alamos, NM 97544, USA

[3]General Atomics, San Diego, CA 92186, USA

[4]Laboratory for Laser Energetics, University of Rochester, Rochester, NY 14623, USA

**Abstract.** Hydrodynamic growth and its effects on implosion performance and mix were studied at the National Ignition Facility (NIF). Spherical shells with pre-imposed 2D modulations were used to measure Rayleigh-Taylor (RT) instability growth in the acceleration phase of implosions using in-flight x-ray radiography. In addition, implosion performance and mix have been studied at peak compression using plastic shells filled with tritium gas and imbedding localized CD diagnostic layer in various locations in the ablator. Neutron yield and ion temperature of the DT fusion reactions were used as a measure of shell-gas mix, while neutron yield of the TT fusion reaction was used as a measure of implosion performance. The results have indicated that the low-mode hydrodynamic instabilities due to surface roughness were the primary culprits to yield degradation, with atomic ablator-gas mix playing a secondary role.

## 1. Introduction

Hydrodynamic instabilities and mix play a central role in the performance degradation; any drive asymmetries and surface imperfections are amplified by the hydrodynamic instabilities during implosion resulting in a distorted shell with reduced hot-spot temperature, volume, and pressure [1]. In addition, small-scale turbulence seeded by these instabilities can mix ablator material into the DT hot spot [1]. The presence of mixed ablator material was correlated with reduced experimental yields and temperatures in high-compression layered DT implosions [2], but it has not been directly measured. Layered DT implosions have previously been modeled using 2D simulations intended to capture performance degradation due to instabilities and drive asymmetries [3], but lacking a model to predict atomic ablator-fuel mix [4]. These simulations over-predicted the yields by a factors from ~5 to ~30 for high-compression implosions. As a way to explain the measured performance, 2D simulations used large, un-physical multipliers (up to 3-5x) on the capsule surface roughness to bring simulated yields down to the measured levels [3]. This prompted a hypothesis that the instability growth factors were larger than in 2D simulations. Since the discrepancies were especially pronounced in implosions with significant inferred ablator-fuel mix [2], it also suggested another hypothesis that atomic mix was a major contributor to yield degradation because there were no models of atomic mix used in 2D simulations.

Two sets of experiments were designed to test hypotheses about lower growth factors used in simulations, and the atomic mix being the dominant factor in performance degradation.

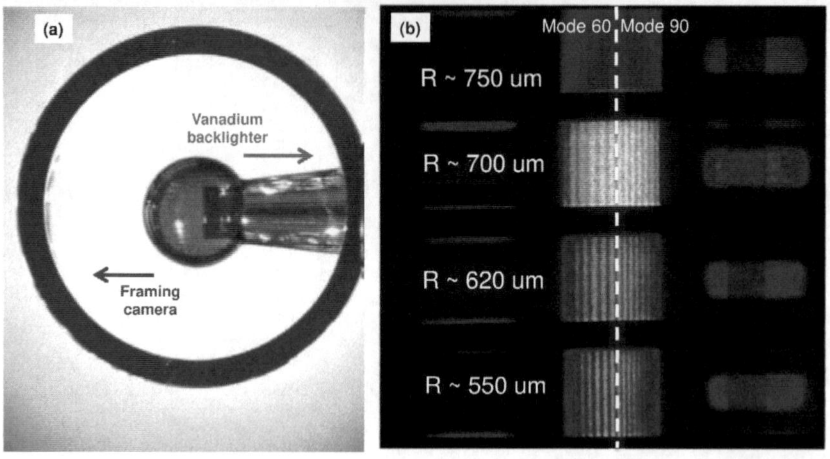


**Fig. 1.** (a) Image of the capsule and the gold cone. (b) Measured capsule radiographs captured on a framing camera. The central four images of the growing capsule modulations were formed using 20-μm wide slit at shell radii of 750 μm, 700 μm, 620 μm, and 550 μm. The capsule had pre-imposed, side-by-side, 2D sinusoidal modulations at mode numbers of 60 and 90.

The ablation-front RT instability experiments are described in Sec. 2; while the atomic-mix experiments near peak compression are discussed in Sec. 3. The results are summarised in Sec. 4.

## 2. Ablation-front Rayleigh-Taylor instability experiments

Rayleigh-Taylor growth experiments were designed to measure instability growth factors at the acceleration phase of spherical implosions at National Ignition Facility [5]. Figure 1(a) shows an image of the capsule and the gold cone used in these experiments. The capsules had pre-imposed, 2D sinusoidal modulations at three wavelengths, 240 μm (mode 30), 120 μm (mode 60), and 80 μm (mode 90). The initial modulation amplitudes were in the range from 0.4 μm to 1.7 μm. The gold cone provided a possibility for the backlighter x-rays to pass through a single wall of the shell, enabling high-quality radiographs of the growing modulations. The experiments were conducted with shaped, ignition-like pulses, with total laser energies of ~1.3 MJ and peak x-ray radiation temperatures of ~280 eV. The pulses were the energy-scaled version of the drive used in layered cryogenic DT implosions. The nominal 209-μm thick plastic capsules with nominal 1120-μm outer diameters had the same Si-doped layers, as used in other similar implosions experiments. Figure 1(b) shows measured capsule x-ray radiographs captured on a framing camera. The central four images of growing capsule modulations were formed using 20-μm wide slit, while images on right and left sides of the slit images were formed with 20-μm and 50-μm pinholes. The temporal resolution of the framing camera was 100 ps, while spatial resolution of the slit images was 20 μm. The images were formed using ~5.4 keV x-rays generated by a vanadium backlighter. The measurements were conducted for convergence ratios up to ~2, when the shell radius was decreased down to ~500 μm in the implosions.

Figure 2 shows measured optical-depth growth factors for the modes 30, 60, and 90 as a function of mode number. The results are compared with simulation predictions at shell radius of 620 μm. While the measured growth factor at mode 30 was close to that predicted, the measured growth factors at modes 60 and 90 were larger than predicted by factors of ~1.3 and ~2.5, respectively. These results demonstrated that ablation-front growth factors were under-predicted in the simulations used to model the performance of layered DT implosions. They also indicate that lower predicted growth factors used in 2D simulations can partially explain the need for large, un-physical multipliers on the capsule surface roughness to bring simulated yields down to the measured levels.

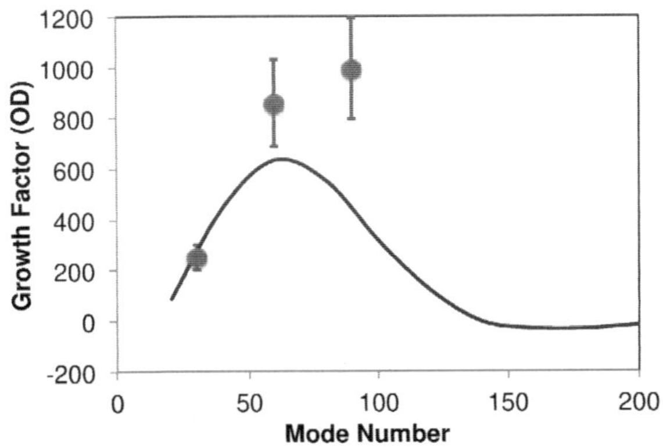


**Fig. 2.** Measured (circles) and simulated (solid curve) optical-depth growth factors as a function of mode number at shell radius of ~620 μm.

## 3. Atomic-mix experiments in the deceleration phase

Figure 3(a) shows the capsule schematic and the laser pulse used in the atomic-mix spherical implosion experiments. Plastic shells had a nominal 209-μm thickness and 2280-μm-initial outer diameter. Si-doped layers were included to reduce preheat of the inner CH ablator from 2-4 keV M-band emission from the Au hohlraum wall. The capsules included a CD layer with 4.0-μm thickness, placed at either the inner shell surface, or offset by up to 8.0 μm from the inner surface by the CH-only layers. The shells were filled with tritium gas that included a small contamination of deuterium gas of 0.1% by atom fraction. All implosions used a laser pulse with peak power of ~435 TW and total laser energy of ~1.5 MJ.

In implosions without CD layers (labeled "CH capsules"), the measured TT and DT yields along with DT ion temperatures probed the conditions in the central part of the core. In implosions with CD layers, the TT yields were similar to those in the CH capsules, while DT yields were higher (up to ~6 times), and DT ion temperatures were lower (~ 2.0 keV vs 3.4 keV). The lower measured temperature supported the hypothesis that the DT neutrons were primarily generated in the colder region where D and T were atomically mixed. As the recession of the CD layers from the inner surface increased, the measured DT yields decreased, indicating that much of the plastic mixed into the gas came from a region close to the inner surface.

The experimental results were compared with 2D simulations using the code ARES [6]. To capture large wavelength low-mode (l <~ 100) instabilities, simulations were performed using an angular resolution of 1/8 degree, with imposed surface roughness at unstable interfaces. The K-L mix model [4] was included to capture the turbulent regime and the effects of mix at scales smaller than the computational grid. The free parameter in this method was the initial turbulent mixing length, $L_0$, set at all unstable interfaces.

As shown in Fig.3, the simulations match well the whole set of experiments. For this, a multiplier of 3 times on the nominal outer surface roughness was needed to match the measured conditions in the central hot core, as determined by the TT yield. The measured outer surface roughness in these experiments varied from 0.5 to almost 2 times the nominal value used in our simulations; thus the 3x nominal values needed to match the TT yield represent up to a factor of ~6 above the measured roughness. This need for a multiplier is consistent with previous 2D simulations of high-compression layered DT implosions [3]. To explain the conditions in outer colder core (DT yield and ion temperature), a low roughness parameter of $L_0$=0.1 nm was used in the K-L mix model. By varying $L_0$ in the K-L model, the DT yield in the mix region was affected, while the TT yield in the central hot-spot region was not particularly sensitive to $L_0$. On the other hand, varying a multiplier on the outer-surface roughness and low-mode hydrodynamic instabilities had a strong effect on TT yield in the central part of the hot spot, while having no significant effect on the DT mix yield. These results indicated that the low-mode hydrodynamic instabilities due to surface non-uniformities were the primary reason to yield degradation, with atomic ablator-gas mix playing a secondary role.

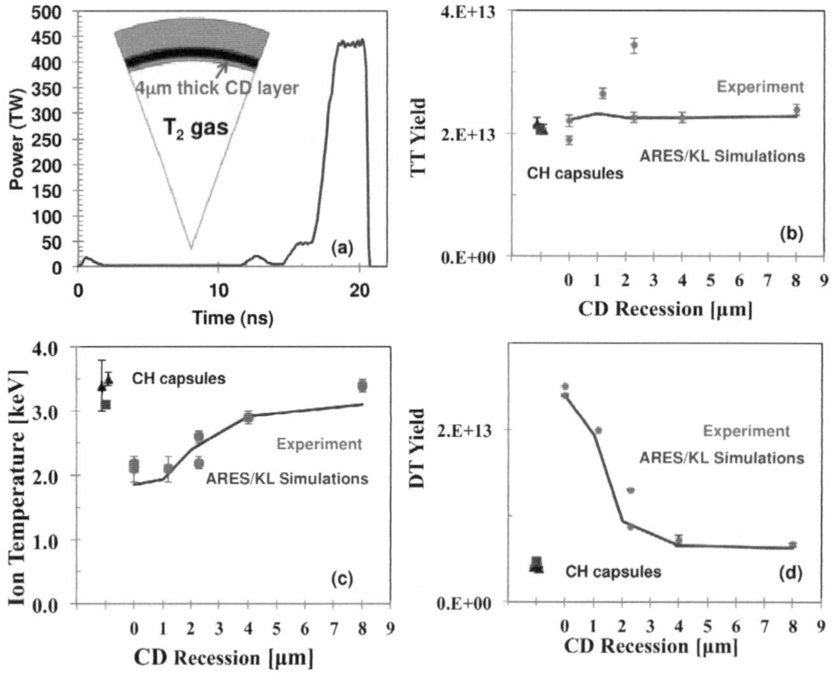


**Fig. 3.** (a) Pulse shape used in experiments. Schematic of the capsule with a 4-μm thick CD layer, placed either at inner shell surface, or recessed up to 8 μm form the inner surface by CH layers. Measured and simulated (b) TT neutron yield, (c) ion temperature, and (d) DT neutron yield, as a function of recession depth of the CD layer, and for CH capsules without CD layer. Results of 2D ARES simulations including K-L mix model are shown with solid curves. The squares represent simulations of the CH capsules without CD layers.

The implosion results were also consistent with higher measured growth factors at ablation front, compared to those predicted in simulations, as shown in the previous Section.

## 4. Conclusions

The results of ablation-front, Rayleigh-Taylor instability experiments have shown that modulation growth was under-predicted in 2D simulations at low mode numbers of 60 and 90, while it was close to predictions at the mode number of 30. The results of atomic-mix experiments using plastic shells with CD layers have indicated that the low-mode hydrodynamic instabilities due to surface roughness were the primary culprits to yield degradation, with atomic ablator-gas mix playing a secondary role. This work was performed under the auspices of the U.S. Department of Energy by Lawrence Livermore National Laboratory under Contract DE-AC52-07NA27344.

## References

[1] S. W. Haan et al., Phys. Plasmas **16**, 051001 (2011)
[2] T. Ma et al., Phys. Rev. Lett. **111**, 085004 (2013)

[3] D. S. Clark et al., Phys. Plasmas **20**, 056318 (2013)
[4] G. Dimonte and R. Tipton, Phys. Fluids **18**, 085101 (2006)
[5] E. I. Moses et al., Phys. Plasmas **16**, 041006 (2009)
[6] R. M. Darlington et al., Comput. Phys. Commun. **135**, 58 (2001)

Journal of Physics: Conference Series **688** (2016) 012114 doi:10.1088/1742-6596/688/1/012114

# Direct heating of imploded plasma in the fast ignition

**Atsushi Sunahara**
Institute for Laser Technology, 2-6 Yamadaoka, Suita, Osaka 565-0871, JAPAN
E-mail: suna@ile.osaka-u.ac.jp

**Tomoyuki Johzaki**
Hiroshima University1-4-1 Kagamiyama, Higashi-Hiroshima, 739-8527, JAPAN

**Hideo Nagatomo, Kunioki Mima, Hiroyuki Shiraga, Hiroshi Azechi**
Institute of Laser Engineering, Osaka University, 2-6 Yamadaoka, Suita, Osaka 565-0871, JAPAN

**Yohitaga Mori, and Yoneyoshi Kitagawa**
The graduate school for the creation of new photonics industries, 1955-1 Kurematsu-cho, Nishiku, Hamamatsu, Shizuoka, 431-1202, JAPAN

**Abstract.** We propose the direct heating of an imploded plasma core by ultra-intense lasers in inertial confinement fusion, to increase the heating coupling efficiency. In this scheme, both fast-electrons and fast-ions heat the plasma core. Experiments using this direct heating scheme has been carried out at GXII and LFEX laser facility at Osaka Univeristy. To model this direct heating scheme, we developed the 1D simulation model and carried out simulations using the experimental conditions. Comparison between results of the simulation and the experimental observations validates the simulation model. We show that even in the unoptimized experimental conditions used in simulations, our calculations show that the maximum temperature, 1.6 keV, of the CD plasma.

## 1. Introduction

In the fast ignition scheme of laser fusion, the laser-accelerated particles are used to heat an imploded plasma core. In this connection, we have been attempting to increase the energy coupling from the heating lasers to the plasma core by various methods such as generating fast electrons with improved high-contrast of the heating lasers, usage of magnetic fields, and optimization of the cone shape. For further increase of the energy coupling, we are currently trying to use laser-accelerated ions, in addition to electrons, for heating plasma. When deuterated carbon (CD) plasma is directly irradiated by an ultra-intense laser, carbon and deuteron ions and electrons are all accelerated. Estimates based on the hole-boring model by S. C. Wilks [1], give that the kinetic energy of the ions reaches to the order of MeV by the ultra intense laser, and its conversion efficiency from the laser to the fast ions is of the order of 1%. This ion acceleration is confirmed by PIC simulation using the FIREX condition. Sakagami et al.,[2]. In this paper, we show the numerical analysis of ion acceleration and its energy deposition to the plasma. Since we have already conducted the first experiment based on this idea using GXII and LFEX laser facility at Osaka University, we compare the numerical results with the experimental data. The comparison shows good agreement between the two. In Section 2, we present our numerical model, and in Section 3, the comparison of our calculations with the experimental observations, respectively. The conclusions are given in Section 4.

## 2. Heating model

We estimate heating of the imploded core by both fast-electrons and fast-ions. First, an ultra-intense laser irradiates and generates fast electrons and ions. In calculation of the conversion efficiency from laser to the electrons, we assume that the energy distribution of the fast electrons is given by exp(-E/T), where E is the electron energy in MeV and T the slope temperature in MeV, respectively. Also we assume that the total conversion fraction from the laser energy to the forward fast electrons is 0.4. We calculate the slope temperature using two models, i.e., Wilks model [1] and Haines model [3]. Transport of fast-electrons is calculated by the ray-tracing method using the stopping range in fully ionized CD plasma. We use the conical calculation volume, to take into account of the divergence angle of the fast electrons. In the present work, we set the 35 deg of the half-angle for the conical volume. The energy deposition of the fast electrons to the plasma is calculated using the drag heating and ohmic heating. In the drag heating, we used the stopping range of the fast electrons [4], and in the ohmic heating $\eta j^2$ at each point, respectively. Here $j$ and $\eta$ are the bulk electric current (equal to the fast electron current) and the resistivity, respectively. We use Spitzer resistivity $\eta$ [5]. Subsequent to direct irradiation of the ultra-intense laser with the fully ionized CD plasma, $C^{6+}$ and $D^+$ ions as well as electrons are accelerated. The fast electron are accelerated by the ponderomotive force of the laser. This leads to charge separation within the Debye length, and the ions in this range are accelerated by the electric field caused by the charge separation. This ion acceleration is studied by S. C. Wilks using PIC simulation of the hole-boring and ion acceleration [1]. He also calculated the velocity of accelerated ions using the momentum balance between the laser and the ions. We use the Wilks model in estimation of the kinetic energy and the conversion efficiency of the accelerated $C^{6+}$ and $D^+$ ions. We also assume that these fast ions have the monotonic energy and the particle numbers given by Wilks and they are injected at the boundary of the simulation volume. The divergence half-angle of the fast ions is set to 20 deg. Transport of each ion particle is calculated based on the stopping range model of Mehlhone [6], where the stopping power is composed of three different processes; the nuclei, the valence electrons and the free electrons. From deceleration of the ion particles, we calculate the energy deposition of the fast ions to the plasmas.

The energy equations of our model are;

$$\rho c_{vi} \frac{DT_i}{Dt} = -Q_{ei}, \tag{1}$$

$$\rho c_{ve} \frac{DT_e}{Dt} = +Q_{ei} + \nabla \cdot (\kappa_e \nabla T_e) + S_e - S_r. \tag{2}$$

In the above, $T_i$, $T_e$ are the ion and electron temperatures, respectively. $Q_{ei}$ is the electron-ion temperature relaxation given by the Spitzer model5], and $S_e$ is the heating term based on the energy deposition of the fast electron and the ions. $\kappa_e$ is the electron thermal conduction with the flux-limited diffusion model with Spitzer conductivity. $c_{vi}(\rho, T_i)$, $c_{ve}(\rho, T_e)$ are the electron and ion specific heats, respectively. They are set to $(3/2)k_B$ and $(3/2)Zk_B$, respectively, where Z is the averaged atomic number, $Z = 3.5$ for CD plasmas, and $\rho$ is the density of the core plasma, respectively. $k_B$ is the Boltzmann constant. $S_r$ is the radiation loss term, calculated from bremsstrahlung from the calculated volume.

## 3. Results

We calculated the temperature of the core plasma, base on the model described in Section 2, using the experimental conditions[7]. In the experiment, the ultra-intense laser LFEX pulse duration is 1.5ps, and Spot diameter was $60\mu m$.. The energy of LFEX laser was 400J, and 613J, respectively. We have conducted two shots in the LFEX and GXII laser facility, Osaka University. From the 2D hydrodynamic simulations, we estimated the core density to be $2g/cm^3$, and the initial core temperature $0.8keV$, respectively. We use these values for the initial core CD plasma conditions in our calculation. We calculated the heating of fast-ion and fast-electrons to the compressed CD core plasma when the heating laser energy is 613J. In Fig. 1, we show the temperature profiles for the conditions (a) both fast-ions and fast-electrons contributions are included, and (b) only the fast-electrons considered. In Condition (a), calculated values of the conversion efficiency from laser to fast $C^{6+}$ ions, and the estimated kinetic energy is 1%, and $8MeV$, respectively. The fast deuteron has the energy and conversion efficiency which are 1/6 of those of the fast $C^{6+}$, since the mass of deuteron is 1/6 of carbon mass. In this figure, we see that the contribution of the fast-ions, mainly $C^{6+}$, to the core heating is significant compared to that of the fast electrons. Here Wilks morel was used to estimate the slope temperature of fast electrons.

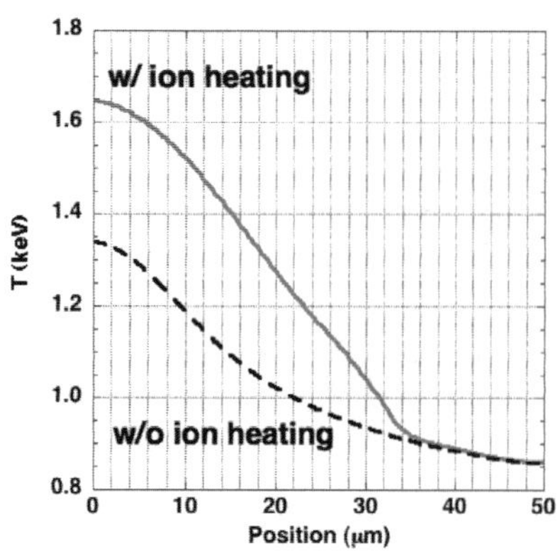


**Figure 1.** Temperature profile of heated core CD plasma with (a) fast ions and fast electrons at 12.8 ps (solid line), and (b) fast electron alone at 9.6ps (dashed line).

In order to compare our model calculation with the experimental observation, We also calculated the neutron yield of thermal fusion and the beam fusion, respectively, using the calculated temperature profile, and assuming that the core sustain duration time is given by dividing the core radius by the sound velocity at the maximum temperature at each computation mesh. In Fig. 2, we compare our calculated thermal- and beam - neutrons with the corresponding experimental data. The calculated curves for the beam and the thermal neutron are in reasonable agreement with the experimental data. Therefore, we conclude that our model simulates well the actual experiments.

Next, we calculated the neutron weighted core temperature using the slope temperatures of Wilks [1] and Haines [2], as a function of the heating laser energy. The heating laser spot size and the duration time are set to be $60\mu m$ and $1.5ps$, respectively. Figure 3 shows that the core temperature increases with the heating laser energy, and 5 keV core temperature is achievable at the laser intensity of $10^{21}W/cm^2$ corresponding thelaser energy of 30 kJ with 1.5ps duration and $60\mu m$ spot diameter. Here, both Wilks and Haines models are used to estimate the fast-electron slope temperature.

This is still not in the optimized conditions for achieving the ignition temperature, since the conditions such as the core density and the laser spot and the duration time are those given by the current experimental conditions and not optimized. However, even in this unoptimized conditions, the above calculated results are good and demonstrate that our new scheme of using

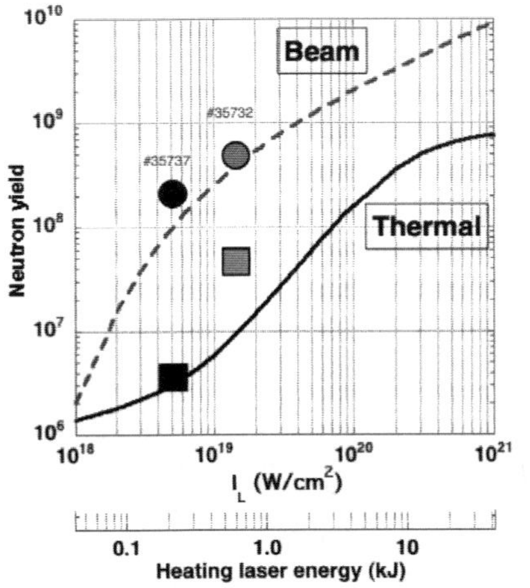


**Figure 2.** Calculated yield of thermal DD-neutrons shown by solid line, and beam-DD neutrons shown by dashed line. Circles are experimental beam DD yield. Squares are experimental thermal DD yeild.

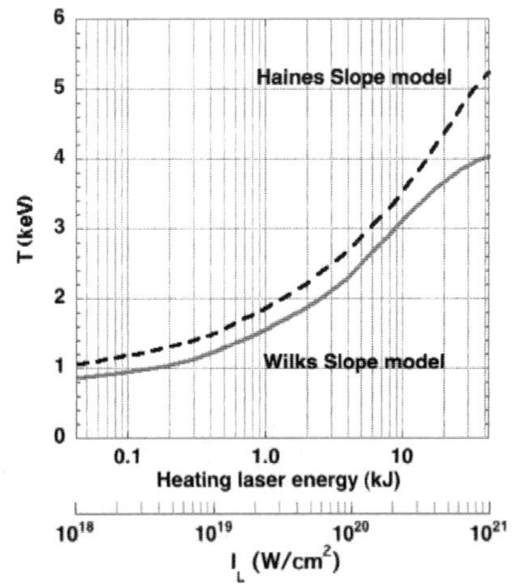


**Figure 3.** Calculated neutron weighted ion temperatures with (a) Wilks model shown by solid line, and (b) Haines model shown by dashed line.

both the fast ions and the fast electrons in heating the core is a very promising one. and the ion assist in heating the core is not small compared to that of fast electrons.

## 4. Conclusion

We calculated heating of a compressed CD plasma core using an ultra-intense laser in a new fast ignition scheme. Our scheme simultaneously includes both fast-electron and ions. By a model based calculation using the conditions of the previous experiment, and we confirmed that our model is valid by comparing calculated results with the measurements. Based on this model, we calculated that the maximum temperature of the CD plasma obtained in the present example calculation is 1.6keV. At the laser intensity of $10^{21}W/cm^2$, and the 1.5ps duration laser pulse, calculation shows that the 5keV core temperature is achievable, even though the conditions are not optimized. In the future work, we will optimize the relevant conditions to achieve higher ignition temperatures in this scheme.

## Acknowledgments

This research was partially supported by the JSPS. Also, it is partially supported by the collaborative research program, Institute of Laser Engineering. The authors are grateful for the support of the computer room of ILE and the Cyber-Media Center at Osaka University.

## References

[1] S. C. Wilks, Phys. Rev. Lett. **69**, 1383 (1992).
[2] H. Sakagami et al., Nuclear fusion **49**, 075026 (2009).
[3] M. G. Haines, M. S. Wei, F. N. Beg, and R. B. Stephens, Phys. Rev. Lett. **102**, 045008 (2009).
[4] A. A. Solodov, Phys. Plasmas **15**, 042707 (2008).
[5] L. Spitzer Jr., *Physics of Fully Ionized Gases*, Interscience (Wiley), New York, (1962).
[6] T. A. Mehlhorn, J. Appl. Phys. **52**, 6522 (1981).
[7] Y. Kitagawa et al., submitted to the proceedings to this IFSA2013 conference.

# Illumination non-uniformity of spirally wobbling beam in heavy ion fusion

**T. Suzuki[1], K. Noguchi[1], T. Kurosaki[1], D. Barada[1], S. Kawata[1], Y. Y. Ma[1], and A. I. Ogoyski[2]**

[1] Utsunomiya University, Utsunomiya, Japan
[2] Varna Technical University, Bulgaria

E-mail: t092335@gmail.com, kwt@cc.utsunomiya-u.ac.jp

**Abstract.** In inertial confinement fusion, the driver beam illumination non-uniformity leads a degradation of fusion energy output. The illumination non-uniformity allowed is less than a few percent in inertial fusion target implosion. Heavy ion beam (HIB) accelerator provides a capability to oscillate a beam axis with a high frequency. The wobbling beams may provide a new method to reduce or smooth the beam illumination non-uniformity. In this paper the HIBs wobbling illumination scheme was optimized.

## 1. Introduction

In inertial confinement fusion, the driver beam illumination non-uniformity leads a degradation of fusion energy output. Therefore, it is important to reduce the illumination non-uniformity [1]. The beam illumination non-uniformity allowed is less than a few percent in inertial fusion target implosion [2-5].

In heavy ion inertial fusion (HIF) the heavy ion beam (HIB) has preferable features, and the HIB axis is precisely controlled with a high frequency [5-9]. The HIBs illumination non-uniformity would be mitigated by the wobbling beam motion, that is, the HIB axis oscillation or rotation [5, 6]. In this study, the illumination scheme is optimized based on the spirally wobbling HIBs.

In direct drive inertial fusion, lasers or HIBs are used as an energy driver. The energy efficiency of the HIB generation is high [10-12]. In this paper we study about the illumination non-uniformity of the spirally wobbling HIBs in direct drive inertial fusion.

In HIF a fuel target is irradiated by HIBs, when the fuel target is injected and aligned at the center of the fusion reactor [13]. In an actual fusion reactor a fuel target alignment error may happen. The target alignment error induces the HIBs illumination non-uniformity.

## 2. HIB illumination non-uniformity

In this study, we employ ($Pb^+$) ion HIBs with the mean energy of 8GeV. The HIB temperature is 100MeV and the HIB transverse distribution is the Gaussian profile. The beam radius at the entrance of a fusion reactor is 35mm and the radius of a fusion reactor is 3m. We employ an Al monolayer pellet target structure with a 4.0mm external radius. The 32-HIBs positions are given as presented in [14]. The HIBs illumination non-uniformity is evaluated by the global *rms*, including also the Bragg peak effect in the energy deposition profile in the target radial direction. In this study, one HIB is divided into many beamlets, and the precise energy deposition is computed [2, 15-16].

## 3. Spiral wobbling beam

So far, we have found that the growth of the Rayleigh-Taylor instability would be mitigated well by a continuously vibrating non-uniformity acceleration field with a small amplitude compared with that of the averaged acceleration [5, 6]. It is realized by using a wobbling beam. Figure 1 shows a schematic diagram for the wobbling beam. However, in our previous work [5] we found that at the initial stage of the wobbling HIBs illumination the illumination non-uniformity becomes huge and cannot be accepted for a stable fuel target implosion.

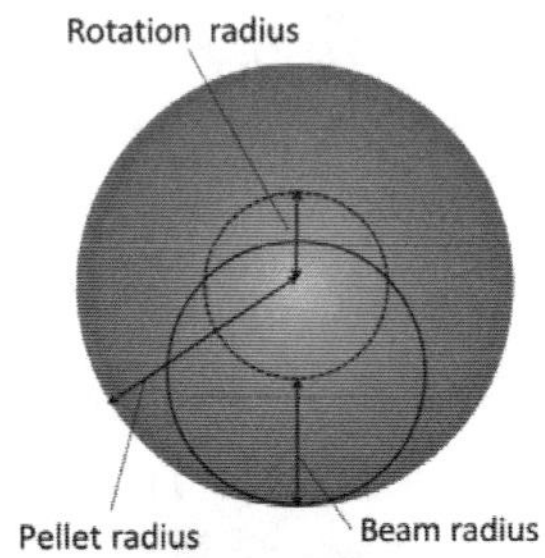


**Figure 1.** Schematic diagram for spiral wobbling beam

Then this problem on the initial imprint of the rotating HIBs illumination was solved by spiral wobbling HIBs as shown in Fig. 2 [6]. When the spirally wobbling beams in Fig. 2 are used, the initial imprint of the non-uniformity at the beginning of the irradiation is greatly reduced from 14% to 4%. In our study, for the spiral wobbling beam the beam radius changes from 3.1mm to 3.0mm at $t = 1.3\tau_{wb}$. Here $\tau_{wb}$ is the time for one rotation of the wobbling beam axis. The beam rotation radius becomes 2.0mm at $t = 2.0\tau_{wb}$. After that, the beam rotation radius is 2.0mm. In this study, we employ the spiral wobbling beam for the HIBs illumination non-uniformity study.

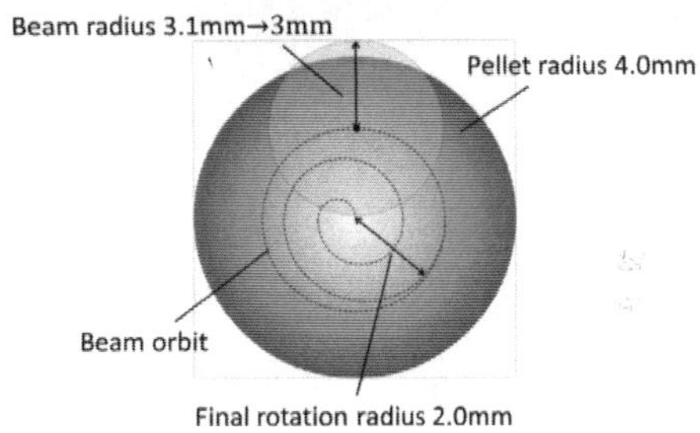


**Figure 2.** Schematic diagram for spiral wobbling beam

Figure 3 shows the amplitude of the mode $(n, m) = (2, 0)$ vs. time, and Figure 4 presents the spectrum of the mode (2, 0) in its frequency space. Here $(n,\ m)$ are the polar and azimuthal mode numbers, and $S_n^m$ is the amplitude of the spectrum, respectively. If the deposition energy distributed is perfectly spherically symmetric, the amplitude of the spectrum is 1.0 in the mode $(n, m) = (0, 0)$ in our study. For this reason, the amplitude of the mode $(n, m) = (0, 0)$ becomes large. As a result, the amplitude of spectrum mode $(n, m) = (2, 0)$ is largest and the mode $(n, m) = (2, 0)$ is dominant throughout the HIBs illumination. In Fig. 3 the time is normalized by the wobbling beam axis rotation time $\tau_{wb}$. In Figure $f_{wb}$ shows the wobbling HIBs rotation frequency. The result in Figure 4 demonstrates that the small non-uniformity of the HIBs energy deposition has the oscillation with the same frequency and the double frequency with the wobbling HIBs oscillation frequency of $f_{wb}$

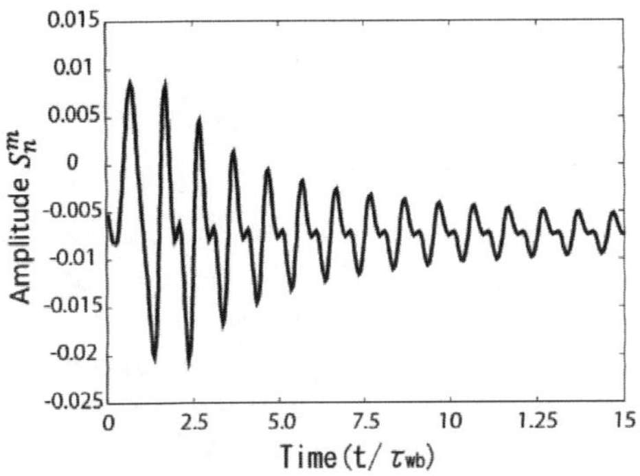


**Figure 3.** The mode (2, 0) amplitude of HIBs non- uniformity vs. time

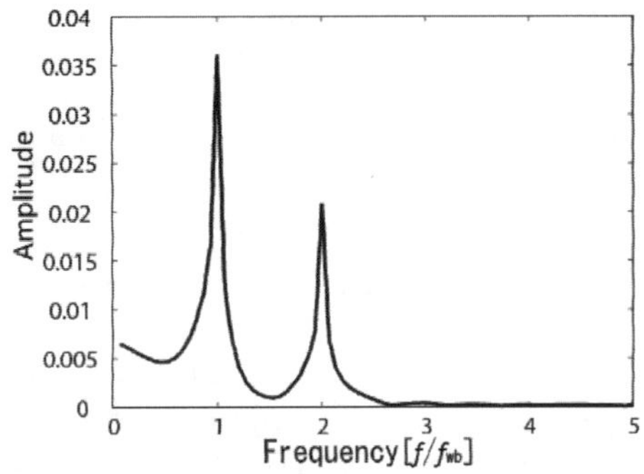


**Figure 4.** Spectrum of the mode (2, 0) in its frequency space

## 4. Optimization of illumination scheme

In this section we optimize the illumination scheme for the spirally wobbling HIBs. The irradiation arrangement of the HIBs is divided into the upper and lower three layers. We changed the angle of $\Delta\theta_1$, $\Delta\theta_2$ and $\Delta\theta_3$ as shown in Fig. 5. We found that the maximum non-uniformity is reduced well, when $(\Delta\theta_1, \Delta\theta_2, \Delta\theta_3) = (0, 0.2, 0.4)$deg or $(\Delta\theta_1, \Delta\theta_2, \Delta\theta_3) = (0.2, 0.2, 0.4)$deg. Figure 6 shows the illumination non-uniformity histories in these cases.

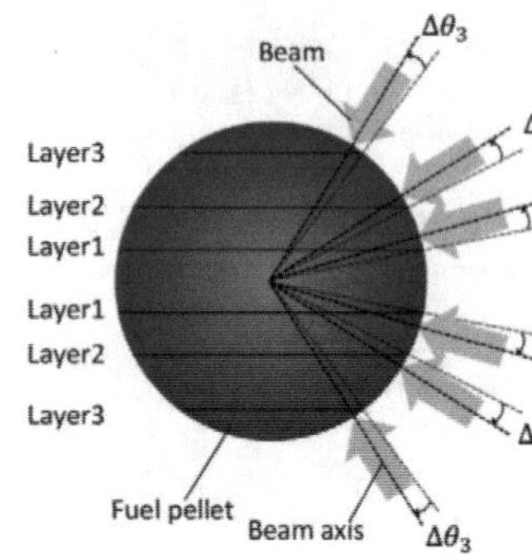


**Figure 5.** Schematic diagram for the definition of $\boldsymbol{\Delta\theta}$

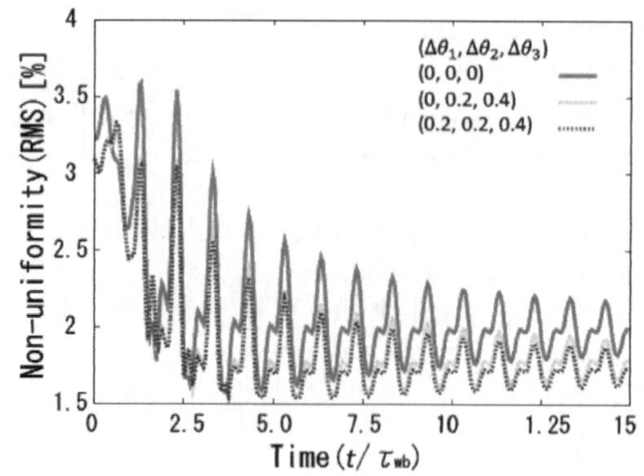


**Figure 6.** Histories of the illumination non-uniformity

Next, the fuel target alignment error is evaluated in a fusion reactor. In an actual inertial confinement fusion reactor as shown in Fig. 7, the target alignment error may appear. Figure 8 shows the maximum illumination non-uniformities vs. the fuel target displacement $dz$. By optimizing the beam illumination scheme, the HIBs illumination non-uniformity is reduced further.

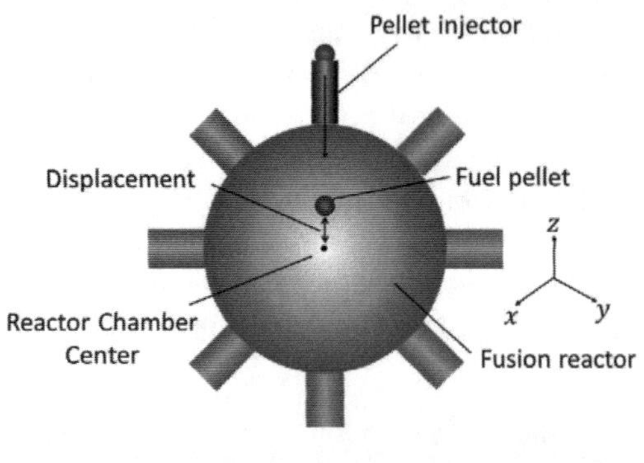


**Figure 7.** Target alignment error in a reactor

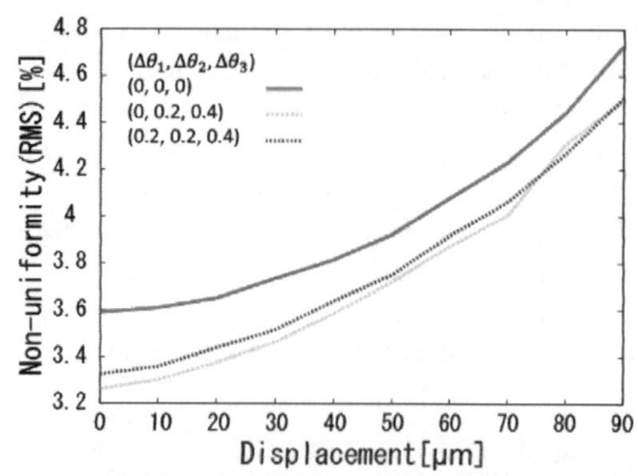


**Figure 8.** Mximal non-uniformity vs. pellet displacement

Finally, we evaluate the spectra in order to analyze the oscillation of the HIBs illumination non-uniformity. To this end, we decompose the illumination non-uniformity by the spherical harmonic spectral using the 3D precise HIBs deposition energy distribution [15]. Figure 9 shows the amplitude histories of the spherical harmonic spectra. In Fig. 9, it is confirmed that the oscillation amplitude is reduced by the HIBs optimized illumination scheme. Figure 10 presents the amplitude of the mode $(n, m) = (2, 0)$ in its frequency space. We found that the oscillation frequency is synchronized well with the wobbling frequency of the illumination HIBs.

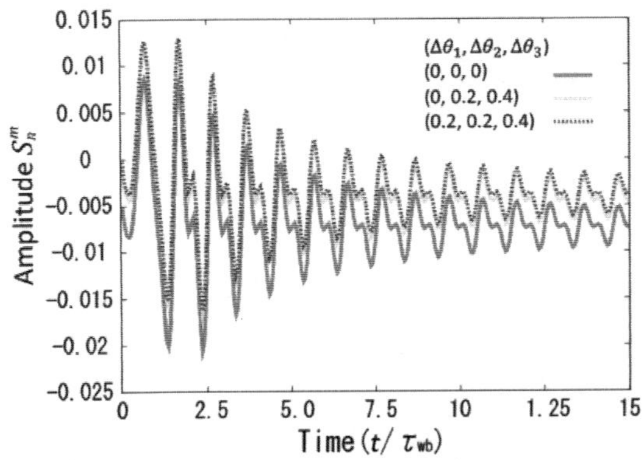

**Figure 9.** Histories of the spherical harmonic

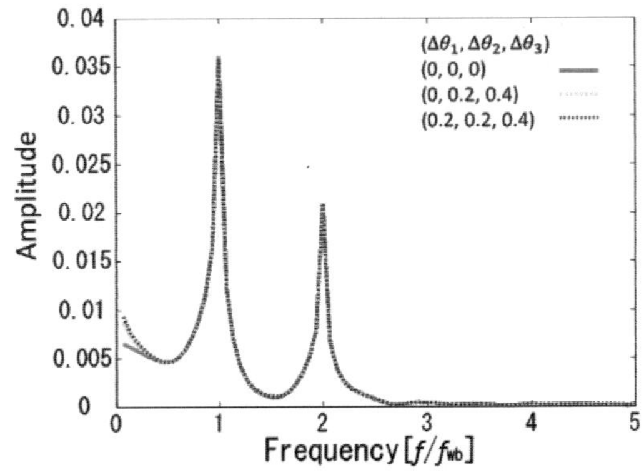

**Figure 10.** Frequency spectrum

## 5. Conclusions

In this study, we optimized the HIBs illumination scheme and evaluated a fuel target alignment error. The illumination non-uniformity is reduced by the optimization of the HIBs irradiation angles to the target fuel. In addition, we found that the frequency spectrum of the HIBs illumination non-uniformity is synchronized with the rotation frequency of the wobbling beams. This result may work to reduce also the growth of the Rayleigh-Taylor instability originated from the HIBs illumination non-uniformity [5, 6].

## Acknowledgments

This work was partly supported by JSPS, MEXT, CORE (Center for Optical research and Education, Utsunomiya University), ASian core program for High energy density science Using intense LAser photons: ASHULA, and ILE/Osaka University.

## References

[1] Logan B G, Perkins L J and Barnard J2008 *J. Phys. Plasmas* **15** 072701
[2] Miyazawa K, Ogoyski, Kawata S and Someya T 2005 *Phys. Plasmas* **12** 122702
[3] Kawata S and Niu K 1984 *J. Phys. Soc. Jpn.* **53**, 3416
[4] Sasaki J and Nakamura T, 2001 Jpn. *J. Appl. Phys.* **40** 968
[5] Kawata S, Iizuka Y, Kodera Y, Ogoyski A I and Kikuchi T 2009 *Nucl. Instr. Meth. A* **606** 152
[6] Kawata S 2012 *Phys. Plasmas* **19** 024503
[7] Basko M M, Schlegel T and Maruhn *J* 2004 *Phys. Plasmas* **11** 1577
[8] Piriz A R, Tahir N A, Hoffmann D H H and Temporal M 2003 *Phys. Rev. E* **67** 017501
[9] Piriz A R, Temporal M, Lopez Cela J J, Tehir N A and Hoffmann D H H 2003 *Plasma Phys. Control. Fusion* **45** 1733-1745
[10] Miyamoto S and Ogawa M 1995 *J. Plasma Fusion Res.* **71** 951
[11] Arnold R C and Meyer-ter-Vehn 1987 *J. Rep. Prog. Phys.* **50** 559
[12] Lindle J D 1995 *Phys. Plasmas* **2** 3933
[13] Petzoldt R W 1998 *Fusion Tech.* **34** 831
[14] Skupsky S and Lee K 1983 *J. Appl. Phys.* **54** 3662
[15] Ogoyski A I, Kawata S and Popov P H 2010 *Compt. Phys. Commun.* **181** 1332
[16] T. Someya, Ogoyski A I, Kawata S and Sasaki T 2004 *Phys. Rev. ST Accel. Beams* 7 044701

# Imploded Plasma Heating by Irradiation of Heating Laser through a Cone with a Hole for Fast Ignition

**M Taga, H Shiraga, S Fujioka, and H Azechi**

Institute of Laser Engineering, Osaka University, Suita, Osaka 565-0871, Japan

E-mail: taga-m@ile.osaka-u.ac.jp

**Abstract**

It is of great importance for the fast ignition research to investigate the heating properties of the imploded core plasma by injection of the heating laser. The open-end cone was introduced recently. An expanding self-emission of x-ray from the core plasma near the cone tip was observed after the heating laser irradiation through the cone. It indicates that the core plasma was heated by the heating laser.

## 1. Introduction

The process of the proposed fast ignition scheme is to create an imploded core plasma by Gekko-XII laser having twelve beams which were symmetrically arranged and to heat it by LFEX high intensity peta watt laser. We are conducing experiment using a deuterated polystyrene shell target with an Au cone for guiding LFEX laser. The fast electrons generated at the cone tip irradiated by the LFEX laser are expected to heat the core plasma. [1-4] So far, a closed-tip cone was used. Recently, we introduced an open-end cone, and x-ray self-emission from the heated core plasma was observed.

## 2. X-ray imaging diagnostics

Since the typical size and life of the imploded core plasma are typically of order of 100 μm and 100 ps, respectively, we used an x-ray streak camera (XSC) with spatial and temporal resolutions of 10 μm and 10 ps, respectively, for observation of the core plasma dynamics (Fig. 1). The data were taken as a part of multi-imaging x-ray streak camera (MIXS). [5-7]

The time of the LFEX laser irradiation relative to the x-ray self-emission history from the core plasma was specified with an accuracy of 7 ps by using the non-imaged hard x-ray signal caused by hot electron generation with LFEX laser irradiation of the target as shown in Fig. 2. Yellow arrow in Fig. 2 indicates the arrival time of the LFEX laser. This non-imaged signal acts as a noise for the self-emission image from the core plasma. By removing this noise signal assuming uniform irradiation on the cathode, we could obtain the net self-emission history of the core plasma.

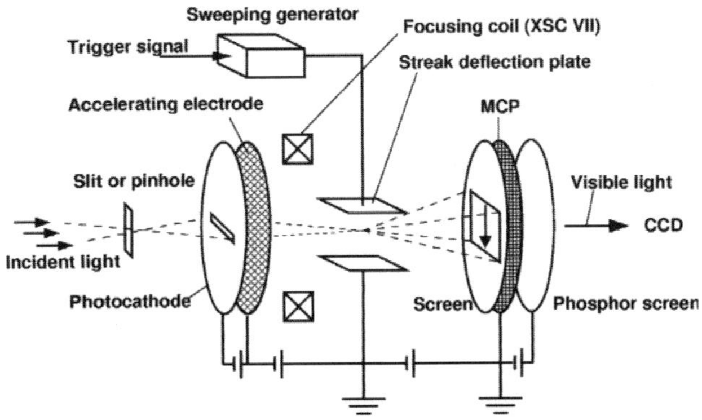


Fig. 1 The schematic drawing of the x-ray streak camera.

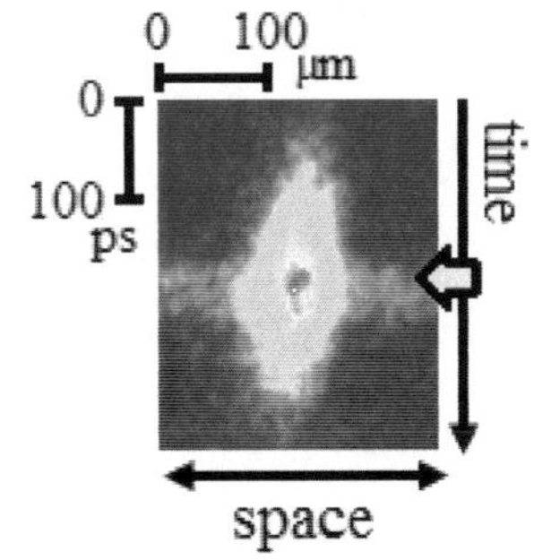


Fig. 2 The temporally resolved self-emission image and non-imaged signal.

## 3. Experiments

Fast ignition experiment was performed by using a deuterated polystyrene shell (diameter = 505 μm, thickness = 7.5 μm) with an open-end Au cone (opening angle = 45 degree, diameter of the hole = 33.7 μm, distance between shell center and cone tip = 47μm). The shell target was irradiated with Gekko-XII laser (8 beams, 2240 J / 1.3 ns, 0.53 μm) and the LFEX laser (615 J / 1.5 ps, 1.05 μm) was injected through the cone. We needed 9 beams to make an uniform core plasma, but one beam was missing due to a failure in the laser system. The cone was used to keep the LFEX laser path to be vacuum, while the hole at the tip allows the LFEX beam to directly irradiate the imploded core plasma. Because the diameter of the hole is approximately the same as the diameter of heating laser spot, the electrons around the hole are expected to be heated and to heat the core plasma without being blocked at the tip.

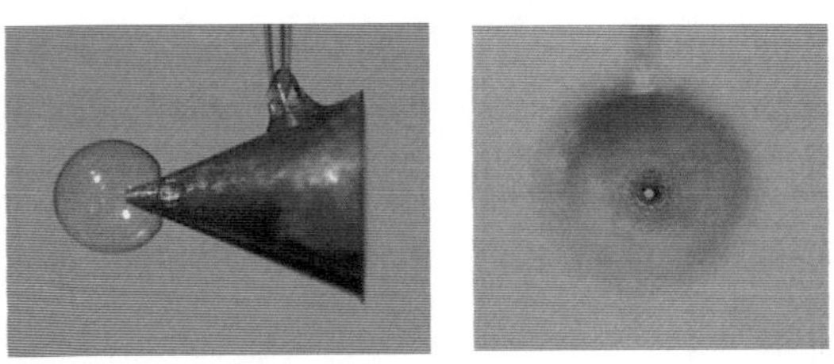

Fig. 3 The shell target with the open-end cone.

## 4. Experimental Results

The image obtained by the X-ray streak camera (Fig. 4 (b)) shows that the imploded core plasma was moving towards the cone. (The white triangle indicates the initial target shell center and the black triangle points the cone tip in Fig. 4.) This is reasonable due to the absence of the drive beams near the cone. Furthermore, because one beam was missing, the core plasma became nonuniform. The self-emission from the core plasma fell down from about 350 ps (Fig. 4 (c)). After LFEX laser irradiated at about 390 ps, the emission restarted, especially near the cone tip. This emission went up until 420 ps, well after the LFEX beam irradiation, which indicates that it is a self-emission from the core.

The region of this restarted emission was found to be gradually expanding over the core plasma from the cone tip with a velocity of about $2.5\times10^{8}$ cm/s. This result indicates that the imploded core plasma was heated by the LFEX laser irradiation, and the heated region expanded to the whole core plasma.

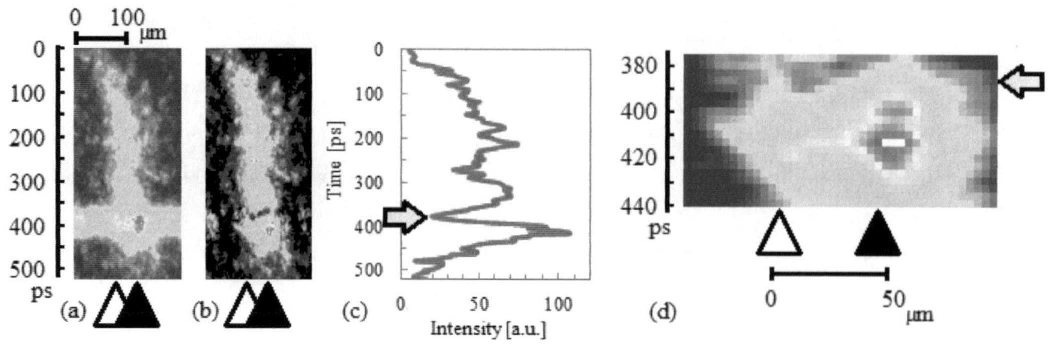


Fig. 4 (a) The self-emission image of the core plasma and the non-imaged signal of high-energy X-rays due to the LFEX laser irradiation. (b) The self-emission image after removal of the non-imaged signal. (c) The history of spatially integrated net signal in Fig. 4(b). Only signal within full-width-half-maximum of the spatial profile at each time is integrated. (d) The enlarged view of the restarted self-emission region after smoothing processing of the image. (0 ps indicates the starting time of the self-emission of the core plasma.)

The similar x-ray self-emissions near the tip of the cone without the hole were obtained in a separate experiment. However, those occurred in both cases with and without heating laser as shown in Fig. 5. These emissions started before the time of the peak x-ray emission from the core, continued for more than 100 ps and did not expand. Thus it can be concluded that these were effect by the imploding laser or the imploded plasma hitting the closed tip of the core. The image shown in Fig. 4 is different from those.

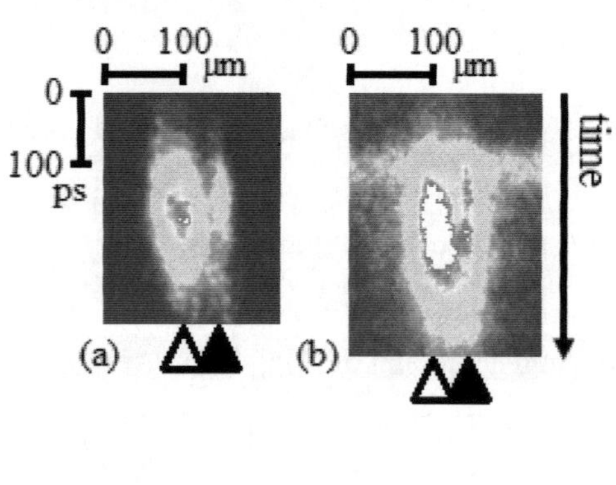


Fig. 5 (a) The obtained emission without LFEX laser in 2010. (b) The obtained emission with LFEX laser in 2012.

Clear difference between open-end cone and standard cone in x-ray self-emission feature from the core plasma near the core tip may indicate effectiveness of the open-end cone in the core plasma heating. Details of the core and cone dynamics will be investigated by using various types of the cone, for example, diameter of the open-end, core material, and so forth. A cone made of 10 μm-thick diamond-like carbon (DLC) is such a candidate, and is ready to use for our experiment. X-rays emitted inside a DLC cone can be easily observed from outside since this DLC cone is almost transparent to 3-5 keV x-rays. It is expected that we can observe the plasma heating features inside the cone by using such a DLC cone.

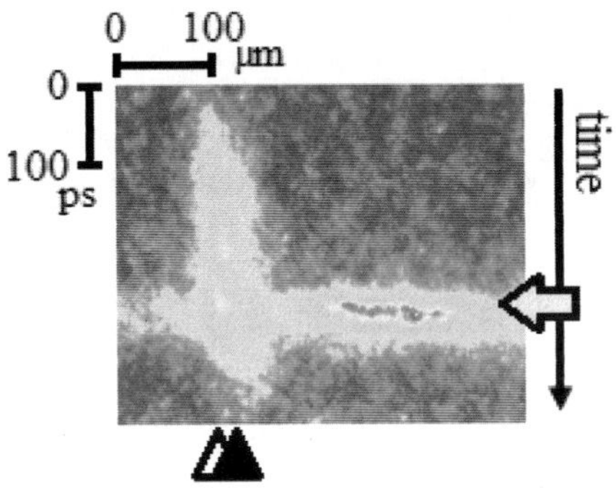


Fig. 6 The self-emission image of the core and the emission in the DLC cone.

## 5. Conclusions

Temporally resolved x-ray self-emission images around the heating beam injection time were obtained for a fast ignition target with an open-end cone. It is indicated that the core plasma was directly heated by the heating laser injection. We will continue the experiment to confirm utility of the open-end cone and examine the heating mechanism of the core plasma by LFEX laser.

## References

[1] Shiraga H *et al* 2011 *Plasma Phys. Control. Fusion* **53** 124029

[2] Shiraga H *et al* 2012 *High Energy Density Physics* **8-3** pp 227-230

[3] Fujioka S *et al* 2012 *Plasma Phys. Control. Fusion* **54** 124042

[4] Azechi H *et al* 2009 *Nucl. Fusion* **49-10** 104024

[5] Shiraga H *et al* 1995 *Rev. Sci. Instrum.* **66-1** pp 722-724

[6] Shiraga H *et al* 1997 *Rev. Sci. Instrum.* **68-1** pp 745-749

[7] Shiraga H *et al* 1997 *Rev. Sci. Instrum.* **68-1** pp 828-830

# Suppression of beam merging and hosing instabilities in magnetized fast ignition fusion

**Toshihiro Taguchi, Thomas M. Antonsen[1], Jr. and Kunioki Mima[2]**

Department of Electric and Electronic Engineering, Setsunan University, Osaka, Japan
[1]Institute for Research in Electronics and Applied Physics, University of Maryland, USA
[2]Graduate School for the Creation of New Photonics Industries, Hamamatsu, Japan

E-mail: taguchi@ele.setsunan.ac.jp

**Abstract.** The magnetized two-stream instability has been investigated in linear and nonlinear regimes. Using a linear analysis of the Vlasov equation in two dimensional wave number space, we studied the competition between modes with wave numbers parallel and transverse to the beam direction. In the analysis, it is found that a near transverse mode is still unstable even for large electron temperature. The nonlinear analysis is performed using a hybrid simulation code. When a sufficiently strong magnetic field is applied along the beam direction, the instabilities are well suppressed and the electron stream becomes laminar. When the magnetic field strength is not large enough, however, electron flow stagnates and the total number of beam electrons penetrating the plasma is largely reduced.

## 1. Introduction

The fast ignition scheme is a promising approach for the inertial confinement fusion program. In the scheme, irradiation of a sufficiently compressed target by an ultra intense laser pulse assists the fuel to burn. As a result, the requirement on the energy of a compression laser is very much reduced. In the fast ignition scheme, relativistic electrons transmit energy from the critical surface of the laser to the compressed fuel. The electrons are generated at the critical surface, and hence the density of the electron beam is of the order of the critical density given by the laser wavelength. As a result, a large electric current flows in an overdense plasma.

When such a large electron flow penetrates the plasma, a return electric current is induced in the background plasma in order to maintain current neutrality. Such a two-stream state is unstable to two kinds of perturbation. One is longitudinal, in which the wave number is parallel to the beam propagation direction. For this mode, an electrostatic field grows as well as a density modulation. The other mode is transverse, in which the wave number is perpendicular to the beam propagation direction. For this mode, known as the Weibel instability, current separation takes place and a quasi-static magnetic field grows. The magnetic field deflects the electrons broadening their spread in angles. The wide angular spread of the electron beam reduces the total number of electrons reaching the fuel region and degrades the efficiency of energy deposition.

In order to suppress the broadening of the angular spread, a group at the Institute of Laser Engineering [1] proposed the application of a strong magnetic field. In their experiments, an intense laser pulse irradiates a small capacitor and drives a large electric current through a one-turn coil connecting two capacitor plates. As a result, a large magnetic field, which is on order of 1 kT, is generated inside the coil. When such a large magnetic field is applied along the beam propagation direction, the angular spread can be suppressed in two ways. One is by guiding the electrons along the magnetic field. The strong magnetic field restricts transverse motion, and the beam propagates along the magnetic field lines.

The other mechanism is suppression of the Weibel instability. The dispersion relation of a two-stream plasma is modified by a strong magnetic field, and unstable modes are stabilized. Due to these effects, the magnetic field increases the number of electrons reaching the fuel, and enhances the fusion burn rate. The scheme is named magnetized fast ignition (MFI) [2]. In this paper, we will describe the linear and nonlinear analysis of the magnetized two stream instability for a two dimensional configuration and discuss its advantages for the fast ignition scheme.

## 2. Linear dispersion relation of the two-stream instability in a strong magnetic field

In order to investigate the stability of the two-stream state in a strong magnetic field, we calculate the linear dispersion relation using Vlasov theory. In the weak instability regime, kinetic effects are an important factor of the analysis [3]. Figure 1 shows that the maximum growth rate of the Weibel instability as a function of the external magnetic field. In the figure, the results of three approaches, fluid analysis (a dashed line), Vlasov analysis (solid lines) and PIC (Particle In Cell) simulation (solid circles), are shown for different electron beam temperatures. The other parameters, such as background electron temperature (5 keV) and the density ratio of beam to background electrons (1:9) are fixed in all analyses. The drift velocity of the beam is set to be $0.9c$ and the ions are immobile so that only electrons engage in the instability. In the figure, the electron cyclotron frequency $\omega_c$ and growth rate $\gamma$ are normalized by the electron plasma frequency $\omega_{pe}$, which is evaluated using the background plasma density. In the figure, $T_h$ is the beam electron temperature.

As shown in the figure, the growth rate decreases with increasing applied magnetic field and no unstable modes exist above a certain value of the magnetic field, which depends on the electron temperature. The result indicates that the Weibel instability is stabilized in a sufficiently strong magnetic field. The figure also shows a difference between fluid model and the Vlasov model in comparison with the PIC results. The Vlasov results well predict the threshold magnetic field found in the PIC simulation, while the fluid analysis underestimates the threshold value. Furthermore, the fluid analysis cannot predict the existence of unstable modes for high electron temperature cases ($T_h \geq$ 50keV).

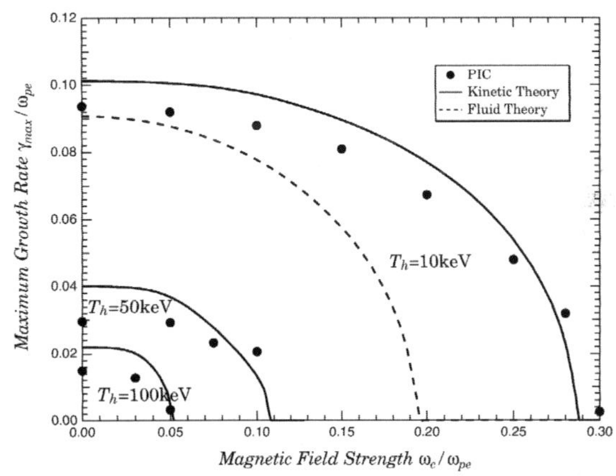


**Figure 1.** Dependence of maximum growth rate of the Weibel instability upon external magnetic field strength

In a real configuration, it is necessary to include both longitudinal and transverse wavenumbers. We next calculate the dispersion relation including these and obtain growth rates in two-dimensional wave vector space. Figures 2 (a)-(c) show the results. In these results, the beam electron temperature is set to be 10 keV and the other parameters are the same as in Fig.1. Figures 2 (a), (b) and (c) are for different external magnetic field strengths. In these figures, $k_x$ is the longitudinal wave number component and $k_y$ is the transverse wave number component. As shown in the figures, the growth rate of the transverse modes depends on the strength of the external magnetic field, while the longitudinal modes are not affected as much by the field.

When the external field is weak, the transverse mode grows as well as the longitudinal modes. As the strength of the field increases, the transverse modes are stabilized but the longitudinal modes are still unstable for the parameters considered.

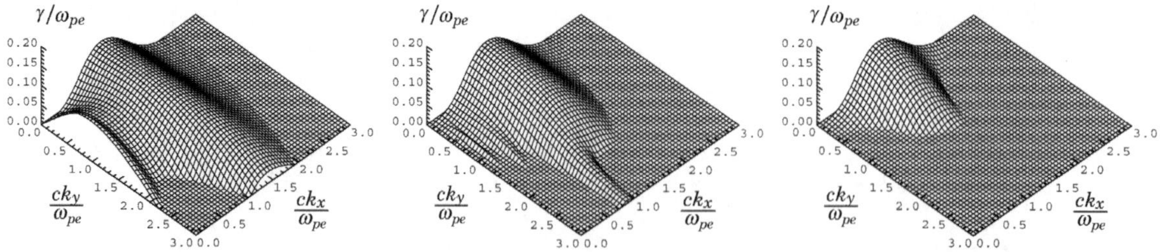


(a) $\frac{\omega_c}{\omega_{pe}} = 0.01$ (b) $\frac{\omega_c}{\omega_{pe}} = 0.5$ (c) $\frac{\omega_c}{\omega_{pe}} = 1$

**Figure 2.** 2D growth rate profiles for the two-stream state for different external magnetic field strengths: $k_x$ and $k_y$ are the longitudinal and transverse components of the wave vector.

When the flow velocity of the electron beam approaches the speed of light, the longitudinal perturbation is suppressed because of the relativistic mass effect. As a result, the growth rate of the transverse modes becomes larger than that of the longitudinal modes. Figure 3 shows an example. In this figure, $\omega_c/\omega_{pe}$=0.5, the flow velocity is 0.99$c$ and the temperature of beam electrons is 150keV. Since the vertical scale of Fig.3 is much smaller than that of Figs.2, it should be noted that the growth rate of Fig.3 is much smaller than Fig.2.

Since the beam temperature is high, there is no unstable mode along the $k_y$ axis ($k_x$=0 line). Nevertheless, there are near transverse modes whose growth rate is much larger than the longitudinal unstable modes (at around $k_x \approx$1.0).

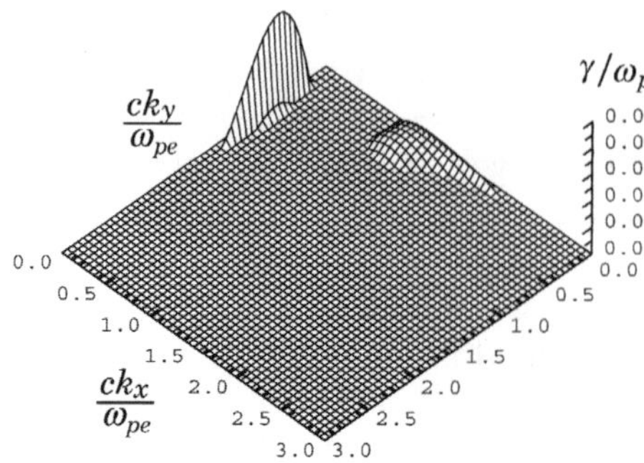


**Figure 3.** 2D growth rate profile for the two-stream state for the case of large flow velocity and large beam electron temperature.

## 3. Nonlinear hybrid simulation

In order to analyze the nonlinear evolution of the beam instability, we performed several runs of a hybrid simulation, which treats the background plasma as a fluid and the beam electrons as particles as in a PIC simulation. Typical results are shown in Figs.4(a)-(c), which are two dimensional false color images displaying the beam electron density. Figures 4 (a), (b) and (c) correspond to the cases of the normalized magnetic field $\omega_c/\omega_{pe}$=0, 0.5 and 1, respectively. The beam velocity is 0.99$c$ and the initial beam temperature is 150keV. The left and right boundaries are open, while the top and the bottom boundaries are periodic. The beam electrons are continuously injected in the vicinity of the left boundary as their density is set to be 1/10 of the background plasma density. The length is scaled in $\mu$m and the background plasma density is taken to be $10^{22}$ 1/cm$^3$.

When there is no magnetic field (Fig.4(a)), the growth rate is large and the beam breaks up into filaments due to the transverse instability. Each filament carries net electric current as the background electrons evacuate the filaments to maintain charge neutrality. The current filaments then attract each other due to the magnetic force and they tend to collide with each other. As a result, the current filaments change direction randomly. This behavior results in the large angular spread of the electrons, and hence the number of electrons reaching to the core plasma will be small.

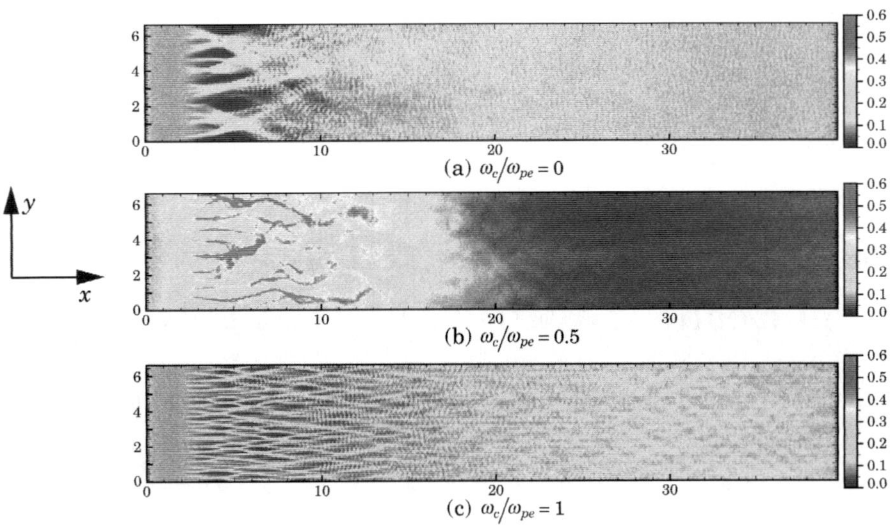


**Figure 4.** Beam electron densities for different external magnetic field strengths at $t$=310fs. The density is normalized to the background plasma density.

On the other hand, when the magnetic field is strong enough (Fig.4(c)), the beam flows quasi-laminarly, although fine filament structures still remain. One reason for the quasi-laminar flow is the guiding effect on the filament of the strong magnetic field. As a result, a sufficiently strong magnetic field has the capability to suppress the increase of the angular spread due to both the reduction of the growth rate of the beam instability and the guiding effect on the filaments.

However, electron flow stagnates for a medium strength magnetic field. As shown in Fig.4(b) ($\omega_c/\omega_{pe}$=0.5), the electrons accumulate near $x \approx 10\mu$m and the stagnation of the electron flow prevents the propagation of trailing beam electrons. The beam stagnation may occur because the external magnetic field is perturbed by the kink instability and reconnection of the magnetic fields lead to formation of magnetic islands. As a result, the number of electrons reaching the fuel will be largely reduced.

## 4. Conclusion

We have analyzed beam instability in both linear and nonlinear regimes. In the linear stage, it is found that a kinetic analysis is necessary to describe the weak instabilities. When the magnetic field is strong enough, the pure Weibel instability, which only has transverse wave vector, is suppressed, but near transverse unstable modes still exist as well as longitudinal modes.

According to the nonlinear analysis, it is found that application of sufficiently strong magnetic field is effective in suppressing the broadening of the angular spread of the beam electrons and increases the number of electrons reaching to core. However, the application of an insufficiently strong magnetic field causes stagnation of the electron flow. The cause of this stagnation remains for future study.

## References

[1] S. Fujioka et al., Sci. Rep., Vol.3, 01170 (2013).
[2] K. Mima et al., Proceedings of IFSA 2013, to be published.
[3] B. Hao et al., Phys. Rev. E, Vol 80, 006402 (2009).

Journal of Physics: Conference Series **688** (2016) 012118 doi:10.1088/1742-6596/688/1/012118

# High Energy Density Physics Research Using Intense Heavy Ion Beam at FAIR: The HEDgeHOB Program

**N.A. Tahir[1], A. Shutov[2], A.R. Piriz[3], C. Deutsch[4] and Th. Stöhlker[1]**

[1] GSI Helmholtzzentrum für Schwerionenforschung, 64291 Darmstadt, Germany
[2] Instutute of Problems of Chemical Physics, Chernogolovka, Moscow Region, Russia
[3] E.T.S.I. Industriales, Universidad de Castilla-La Mancha, 13071 Ciudad Real, Spain
[4] LPGP, Universite Paris–Sud, 91405 Orsay, France

E-mail: n.tahir@gsi.de

**Abstract.** International project, Facility for Antiprotons and Ion Research (FAIR), has entered in its construction phase at Darmstadt. It is expected that the new powerful heavy ion synchrotron, SIS100 will deliver a strongly bunched intense beam of energetic uranium ions that will provide the scientists with an efficient and novel tool to research High Energy Density (HED) Physics in the laboratory. Over the past 15 years, substantial theoretical work has been done to design numerous experiments that can be done at this facility in this field. This work has resulted in an extensive scientific proposal named HEDgeHOB, that includes experiment proposals addressing various aspects of HED matter, for example, planetary physics, equation of state, hydrodynamic instabilities and others. In this paper we present a summary of this work.

## 1. Introduction

Due to its wide applications to numerous branches of basic and applied physics, High Energy Density (HED) physics is one of the most active research fields worldwide. Static as well as dynamic schemes have been employed over the past many decades for this purpose. The most popular static technique involves compression of a small sample of material in a diamond anvil cell to generate HED states. Dynamic schemes, on the other hand, rely on shock compression to induce HED stases in matter. Traditionally, shocks are produced by high power chemical explosives, gas guns, high power lasers and magnetic devices. Recent developments and expected progress in the near future in the technology of strongly bunched, well focused and energetic high intensity ion beams have provided the scientists with a novel, very efficient tool to study HED physics [1, 2].

It is to be noted that one may generate HED matter using two alternative dynamic methods, namely isochoric heating and shock compression of the material. It is also important to note that particle beams offer great flexibility as they can be used in both type of schemes whereas all other drivers are only suitable for either of the two methods.

The facility for Antiprotons and Ion Research (FAIR) at Darmstadt, has entered into construction phase that includes the building of a new very powerful heavy ion synchrotron, SIS100. According to the design parameters, the SIS100 beam will comprise of $5 \times 10^{11}$ uranium ions that will be delivered in 50 ns long bunch. A flexible particle energy range (400 eV/u – 2.7 GeV/u) will be available while the beam can be focused to a spot size characterized with a

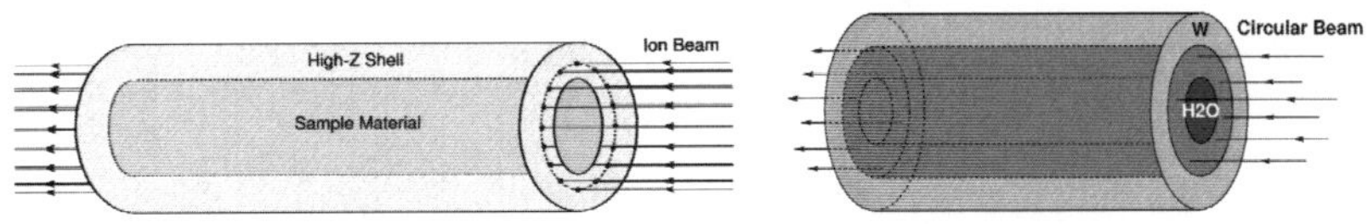


(a) (b)

**Figure 1.** LAPLAS experimental scheme, (a) driven by annular focal spot; (b) driven by a circular focal spot.

full width at half maximum (FWHM) = 1 mm. Provisions have also been made to generate an annular focal spot to do compression experiments.

Over the past 15 years, substantial theoretical work that includes sophisticated two– and three–dimensional numerical simulations as well as analytic modeling, has been done to design HED physics experiments at FAIR. This work has resulted in a very wide range experimental scientific program named, HEDgeHOB for FAIR. A brief overview of this program is presented in this paper.

## 2. High Energy Density Physics and Planetary Science Research at FAIR

HEDgeHOB includes different experiment proposals that consider isochoric heating technique as well as shock compression mechanism to generate samples of HED matter in the laboratory. The former technique is employed in the HIHEX (Heavy Ion Heating and EXpansion) experiment that has been designed to study the Equation of State (EOS) properties of HED matter [2]. The latter technique has been used in the LAPLAS (LAboratory PLAnatery Science) experiment design [3, 4] as well as for hydrodynamic instability studies [5]. In this section we present recent theoretical studies using the LAPLAS scheme.

The two configurations of the LAPLAS experiment are shown in Fig. 1(a) and (b), respectively. In the first case the target is driven by a hollow beam with an annular focal spot, whereas is the latter scheme, a circular focal spot is used to implode the target. In case of an annular focal spot, the sample material is not directly heated by the beam because the energy is deposited in the surrounding shell that leads to very high pressure. This high pressure launches a strong shock radially inwards that reflects at the cylinder axis and then re-verbrates between the axis and the boundary between the sample and the outer shell, thereby leading to a low–entropy compression. Simulations have shown that using the SIS100 beam parameters, hydrogen can be compressed to about 2 – 3 g/cm$^3$ having temperature in the range of 10000 – 20000 K and pressure up to about 30 MBar. These are the conditions that are expected to exist in the interior of Jupiter and other hydrogen rich exo-planets. In case of a circular focal spot, on the other hand, the sample material is also directly heated by the beam. However, due to the much higher pressure in the surrounding shell, the sample is still strongly compressed, although the final temperature in this case is higher. Further details about compression of hydrogen can be found in [6].

Considering both LAPLAS configurations, detailed two–dimensional hydrodynamic simulations have been carried out to study compression of water which is abundant in Uranus and Neptune as well as in similar water rich exoplanets. Figs. 2(a) – (c), show the density, temperature and pressure vs target radius in the compressed water region using $10^{11}$, $3 \times 10^{11}$ and $5 \times 10^{11}$ ions per bunch, respectively, for an annular focal spot. It is seen that the physical conditions in the sample material correspond to the theoretically predicted [7] superionic phase of water. In Fig. 2(d) we show the regions of the phase diagram of water that can be accessed

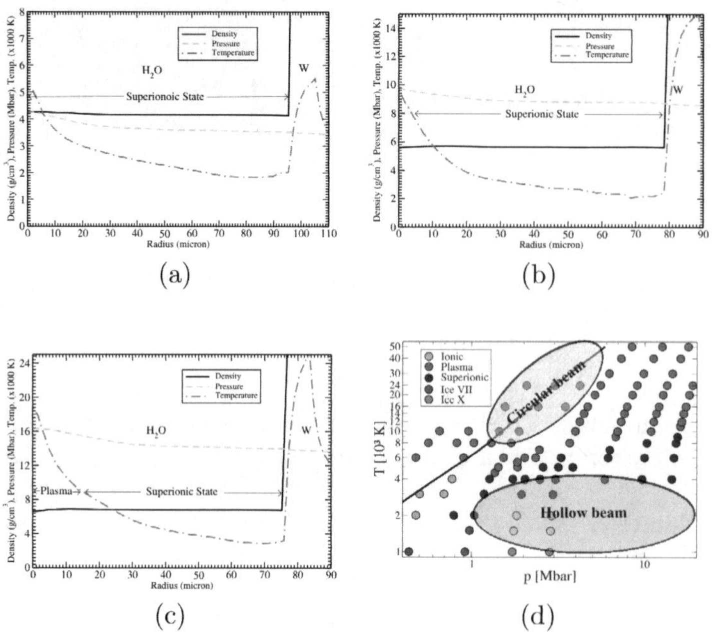


**Figure 2.** (a) $\rho$, T and P vs radius in compressed water using $10^{11}$ ions per bunch; (b) using $3 \times 10^{11}$ ions per bunch; (c) using $5 \times 10^{11}$ ions per bunch and (d) shows regions of water phase diagram [7] that can be accessed by the two LAPLAS schemes respectively.

at the FAIR facility in the LAPLAS experiments. It is seen that the superionic phase can only be accessed using an annular focal spot because in case of a circular focal spot, water is directly heated by the beam and the final temperature is higher. In this case one can generate plasma state of water. Further details can be found in [4].

Equation of state of iron under extreme conditions of density and pressure is also of great relevance to planetary physics as the earth core is made of solid iron. The discovery of huge earthlike extrasolar solid planets, named "superearths" or "exoearths" have made study of this problem much more important. The pressure in the earth's solid core is believed to be about 3 MBar, whereas it is expected that the pressure in the core of larger extrasolar earthlike planets could be as high as 10 MBar. In order to generate these extreme conditions, we have done simulations of compression of solid iron in the LAPLAS scheme considering an annular focal spot using the SIS100 uranium beam. The results are presented in Figs. 3(a) – (e). Fig. 3(a) shows the temperature, pressure and density vs radius in the compressed iron for a beam intensity of $10^{11}$ ions per bunch. It is seen that iron is compressed to 50 % higher than the solid density while a pressure of 2 MBar is achieved. The temperature is relatively low (650 K). In Fig. 3(b) we plot the same variables as in Fig. 3(a), but using a beam intensity of $2 \times 10^{11}$ ions per bunch. In this case the compression is 75 % higher than the solid density while the pressure is around 4 MBar and the temperature is about 900 K. Fig. 3(c) shows that using a beam intensity of $3 \times 10^{11}$ uranium ions per bunch, the compression is 87 % above the solid density, the temperature is about 1000 K whereas the pressure is above 5 MBar. Figs. 3(d) and (e) show that using higher intensities, iron can be compressed to twice the solid density and pressure as high as 8 MBar can be achieved. This shows that LAPLAS experiments can generate wide range of extreme physical conditions in iron relevant to the structure of a variety of solid earthlike planets. It is also imortant to study LAPLAS implosion of iron using a circular

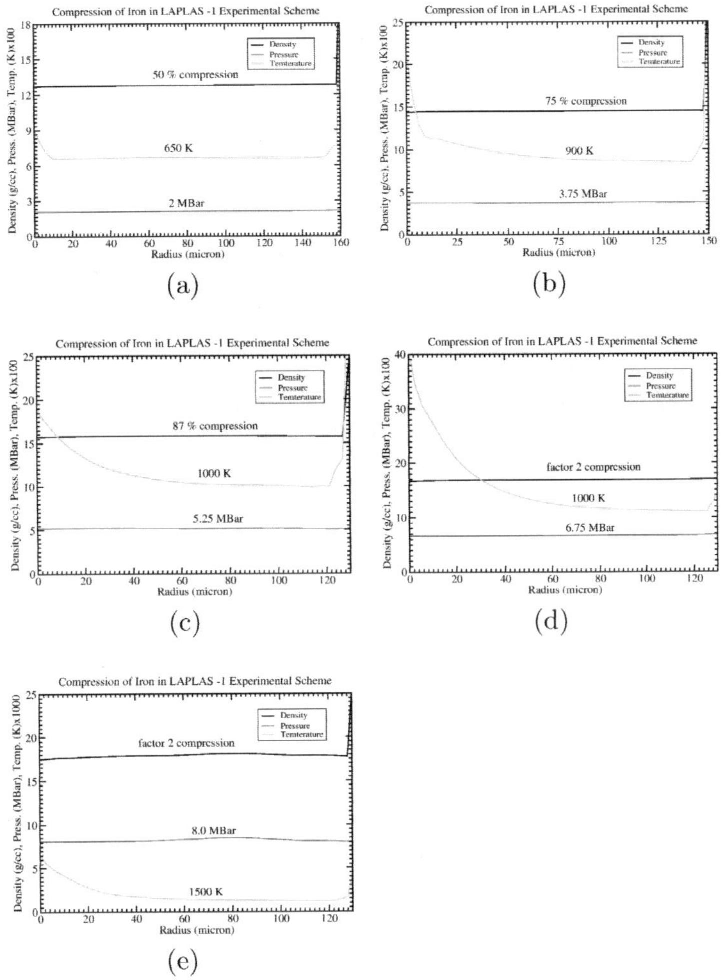


**Figure 3.** $\rho$, T and P vs radius in compressed iron (a) using $10^{11}$ ions per bunch; (b) using $2 \times 10^{11}$ ions per bunch; (c) using $3 \times 10^{11}$ ions per bunch; (d) using $4 \times 10^{11}$ ions per bunch and (e) using $5 \times 10^{11}$ ions per bunch

focal spot in order to generate higher temperatures. This work is in progress.

## References

[1] N.A. Tahir et al., Phys. Rev. E. **61**, 1975 (2000).
[2] N.A. Tahir et al., Phys. Rev. Lett. **95**, 035001 (2005).
[3] N.A. Tahir et al., Phys. Rev. E **63**, 016402 (2001).
[4] N.A. Tahir et al., New J. Phys. **12**, 073022 (2010).
[5] N.A. Tahir et al., Phys. Plasmas **18**, 032704 (2011).
[6] N.A. Tahir et al., High Energy Density Phys. **2**, 21 (2006).
[7] M. French et al., Phys. Rev. B **82**, 174108 (2010).

# Shock Formation by Plasma Filaments of Microwave Discharge under Atmospheric Pressure

**Masayuki Takahashi and Naofumi Ohnishi**
Department of Aerospace Engineering, Tohoku University, Japan
E-mail: mtakahashi@rhd.mech.tohoku.ac.jp

**Abstract.** A one-dimensional compressible fluid calculation was coupled with a finite-difference time-domain code and a particle-in-cell code with collision to reproduce propagation of electromagnetic wave, ionization process of plasma, and shock wave formation in atmospheric microwave discharge. Plasma filaments are driven toward the microwave source at 1 atm, and the distance between each filament is one-fifth of the wavelength of the incident microwave. The strong shock wave is generated due to the high plasma density at the atmospheric pressure. A simple analysis of the microwave propagation into the plasma shows that cut-off density of the microwave becomes smaller with the pressure decrease in a collisional plasma. At the lower pressure, the smaller density plasma is obtained with a diffusive pattern because of the smaller cut-off density and the larger diffusion effect. In contrast with the 1-atm case, the weak shock wave is generated at a rarefied condition, which lowers performance of microwave thruster.

## 1. Introduction

Significant advancement for space transportation is needed to utilize an outer space as academic study, space travel, military defense, weather observation and so on. Recently, microwave propulsion system has been investigated to launch a small satellite with saving operation cost by high-power energy transmission from a microwave oscillator established on a ground. The irradiated microwave is focused by a parabolic mirror installed on the vehicle, and the intense microwave generates a plasma with sustaining a strong shock wave which supplies an impulsive thrust through interactions with the rocket thruster. In past experiments of the plasma formation by the microwave [1–3], filamentary arrays of the plasma were observed in contrast with a bulk plasma obtained in a laser breakdown [4]. The plasma filaments propagate toward the microwave source at km/s-order velocity, and the distance between each filament is about one-quarter of the incident microwave wavelength. However, at the lower pressure, the plasma cannot form the filamentary structure, and the measured impulse of the microwave rocket drastically drops down [2, 3]. Detailed mechanism of the thrust decrease is not identified by the past experiments; therefore, the plasma and shock generations by the microwave discharge should be examined.

In this study, we have developed a one-dimensional integrated simulator by combining a finite-difference time-domain (FDTD) code, a particle-in-cell with Monte Calro collision (PIC-MCC) code, and a computational fluid dynamics (CFD) code to assess the shock wave and plasma formations by the microwave discharge. We then numerically assess the ambient pressure effects for the plasma and shock wave structures, and specify the mechanism of the thrust decrease at the lower pressure.

## 2. Numerical Methods

### *2.1. Interaction between electromagnetic field and plasma*

The Maxwell equations are solved by the FDTD method for simulating the microwave propagation. The PIC-MCC method is employed to reproduce the plasma generation in the microwave discharge. In the PIC module, the equation of motion for the charged particles is integrated to examine the interaction between the charged particles and the electromagnetic field assessed by the FDTD simulation. The current density is calculated on the PIC module, and it is fed back to the FDTD module. Also, in the MCC module with a reactive system of $N_2$, $N_2^+$, $N_2^*$, N, $e^-$, the electron-neutral elastic collision, the electron impact ionization, the electron impact excitation to the first excited state, the neutral dissociation, and the ion-neutral elastic collision processes are considered for a pure nitrogen gas.

### *2.2. Shock wave calculation*

Because the strong shock wave is induced by the microwave discharge, it is necessary to assess the shock wave propagation by coupling the compressible fluid calculation with the plasma simulation. An energy absorption rate by the plasma is estimated from the PIC-MCC simulation, and the absorbed energy is deposited as a source term of a one-dimensional Euler equation to drive the shock wave. For solving the Euler equation, a discretization is performed by a cell-centered finite volume manner. The AUSM-DV method [5] is employed for the numerical flux with the second-order MUSCL method. The temporal integration is done by the first-order Euler explicit method.

## 3. Simulation Condition

The initial plasma spot which has the Gaussian distribution is set on a computational domain for supposing the first gas breakdown, and the microwave is irradiated from the left boundary to reproduce the microwave-plasma interaction simply. The maximum electric field intensity is 5 MV/m, the microwave frequency is 110 GHz. The simulation domain has the length of $5\lambda$ ($\lambda$ is the microwave wavelength), and the number of grid points is 600.

## 4. Results and Discussions

### *4.1. Plasma structures at 1 atm and 0.006 atm*

At 1 atm, the plasma filaments are obtained by the microwave incident, and the distance between each filament is about $\lambda/5$, showing well agreement with the experimental and numerical studies [1–3, 6–8] (Fig. 1(a)). The incident microwave is reflected by a over-critical plasma (a critical density of a collisionless plasma: $n_c = 1.5 \times 10^{14}$ cm$^3$), and the reflected wave generates the quasi-standing wave in front of the plasma, which causes the discrete structure of the plasma filaments (Fig. 1(b)). On the other hand, the clear structure is not obtained at 0.006 atm because of the larger diffusion effect of the electron due to the lower ambient pressure. At 0.006 atm, the maximum density of the plasma becomes smaller than the atmospheric case because the microwave screening density of the plasma decreases with pressure dropping unlike the collisionless plasma in which the critical density is constant. In addition, for the lower pressure, the propagation speed of the plasma front becomes larger than the 1-atm case due to the larger diffusion effect. The estimated propagation speed is order of 1000 km/s at 0.006 atm, while the propagation speed is assessed as order of 10 km/s at 1 atm.

### *4.2. Analytical prediction of cut-off density*

The microwave cannot penetrate into the plasma slab if the plasma density exceeds the cut-off density, and the number density stops to increase because the ionization reaction is suppressed in the plasma. So, the maximum density of the plasma has a relation with the microwave cut-off

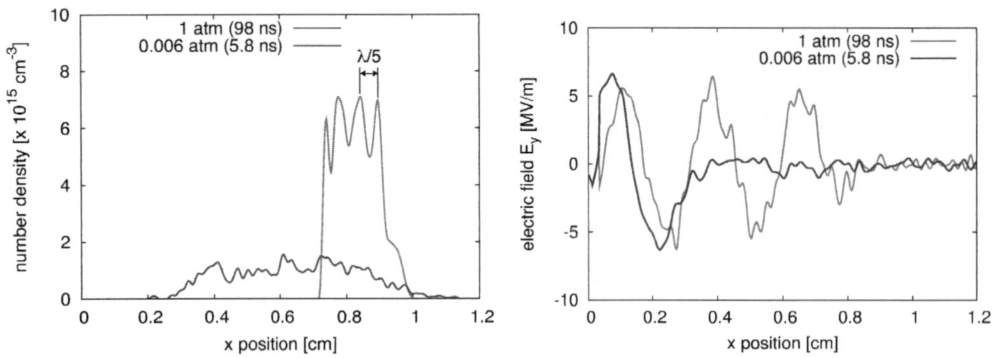


(a) Electron number density (b) Electric field

**Figure 1.** Electron number density and electric field at 1 atm and 0.006 atm.

density. Because the critical density $n_c$ for the microwave cut-off is given by $n_c = \frac{\epsilon_0 m_e \omega^2}{e^2}$ in the collisionless plasma ($\omega^2 >> \nu_m^2$), there is no pressure dependence for it, where $m_e$ is the electron mass , $\omega$ is the incident wave frequency, and $e$ is the elementary charge, $\nu_m$ is the elastic collision frequency for electron-neutral collision. However, in the collisional plasma, because the cut-off density changes with the pressure, the maximum density of the plasma slab depends on the ambient pressure. We analytically estimate the cut-off density in the atmospheric plasma to assess the maximum density dependence on the ambient pressure.

For the collisional case, we introduce a wave number to assess the microwave penetration into the plasma. The complex wave number $k_x$ has a dispersion relation in the collisional plasma;

$$k_x = \sqrt{\eta + \xi i}, \tag{1}$$

where $i$ is the imaginary unit, $\eta \equiv \frac{\omega^2}{c^2}(1 - \frac{\omega_p^2}{\omega^2+\nu_m^2})$, and $\zeta \equiv \frac{\omega^2}{c^2}(\frac{\omega_p^2}{\omega^2+\nu_m^2}\frac{\nu_m}{\omega})$ [9]. Equation (1) can be separated to a real part $k_{\mathrm{Re}}$ and a imaginary part $k_{\mathrm{Im}}$, that is, $k_{\mathrm{Re}} = \sqrt{\frac{\sqrt{\eta^2+\xi^2}+\eta}{2}}$ and $k_{\mathrm{Im}} = \sqrt{\frac{\sqrt{\eta^2+\xi^2}-\eta}{2}}$. Because the microwave damping by the plasma becomes effective if $k_{\mathrm{Im}} > k_{\mathrm{Re}}$, we obtain the simple relation using the plasma frequency $\omega_p^2 = \frac{ne^2}{\epsilon_0 m_e}$ and the critical density of the collisionless plasma $n_c = \frac{\epsilon_0 m_e \omega^2}{e^2}$;

$$n > \frac{\omega^2 + \nu_m^2}{\omega^2} n_c \equiv n_c'. \tag{2}$$

where $n_c'$ is the cut-off number density for the collisional case. The cut-off density $n_c'$ depends on the ambient pressure $p$ because an elastic collision frequency $\nu_m$ is the function of the pressure $p$. Since the elastic collision frequency $\nu_m$ is typically in proportional to the ambient pressure, the cut-off density $n_c'$ decreases with the pressure lowering, which causes the dropping of the maximum density at the rarefied condition.

Figure 2 shows the maximum density estimated by our simulation and the theoretical cut-off density $n_c'$ with the elastic collision frequency of $\nu_m = 5.3 \times 10^9 p\ \mathrm{s}^{-1}\mathrm{torr}^{-1}$ [6]. We can obtain a rough agreement for the pressure dependence between the simulation result and the theoretical prediction, but the theoretical line cannot explain a peak at 0.5 atm of our simulation result. The maximum density becomes larger than the cut-off density $n_c'$ because $n_c'$ just indicates that the plasma slab begins to reflect the incident microwave. Detailed analysis is needed to precisely explain the maximum density dependence on the ambient pressure in the microwave discharge.

### *4.3. Shock wave structures at 1 atm and 0.006 atm*
In the microwave propulsion, the shock wave structure induced by the microwave discharge is important for estimating the thrust performance. By combining the FDTD code, the PIC-MCC code, and the CFD code, we can estimate the shock wave structure driven by the intense microwave. At 1 atm, the strong shock wave is generated by the dense plasma due to the larger cut-off density (Fig. 3), resulting in the large thrust in the microwave propulsion. Plateau high pressure remains behind the shock wave like a detonation structure [3] because an energy absorption region at the ionization front moves toward the microwave source with the plasma propagation. On the other hand, at 0.006 atm, the shock wave pressure becomes smaller than the 1-atm case because the plasma density is smaller by the lower cut-off density. This weak shock wave causes the performance dropping of the microwave propulsion at the rarefied condition.

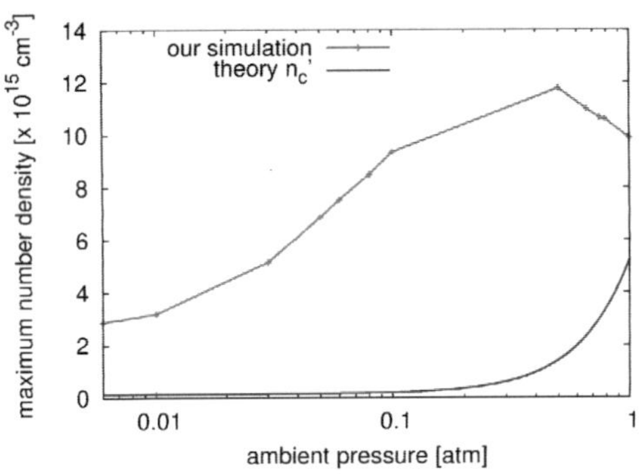


**Figure 2.** Maximum density calculated by our simulation and theoretical cut-off density.

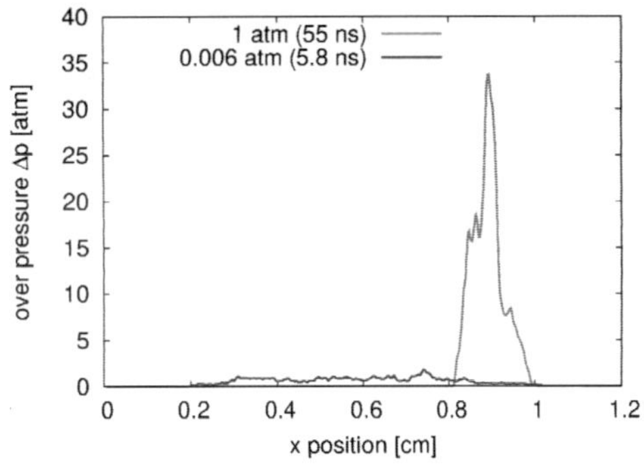


**Figure 3.** Over pressure profiles at 1 atm and 0.006 atm (over pressure: $\Delta p = p - p_0$, where $p$ is pressure and $p_0$ is ambient pressure).

## 5. Conclusion
To simulate the plasma and shock wave formations induced by the microwave discharge, we have developed a coupling simulator constructed by the FDTD module, the PIC-MCC module, and the CFD module. The filamentary arrays are driven toward the microwave source at the atmospheric pressure, and the distance between each filament is one-fifth of the microwave wavelength. Also, the strong shock wave is obtained due to the high plasma density at 1 atm. By a simple estimation of the microwave penetration, we presented that the cut-off density of the collisional plasma becomes smaller with the ambient pressure lowering. At lower pressure, the rarefied and diffusive plasma is obtained because of the smaller cut-off density and the larger diffusion effect. The weak shock wave is generated at the lower pressure, which causes the performance deterioration in the rarefied environment for the microwave propulsion.

## References
[1] Hidaka Y. et al. 2008 *Physical Review Letters* **100** 035003.
[2] Oda Y. and Komurasaki K. 2006 *Journal of Applied Physics* **100** 113307.
[3] Oda Y. 2008 *Doctor Thesis.*
[4] Mori K. et al. 2004 *Journal of Applied Physics* **95** (11) 5979–5983.
[5] Wada Y. and Liou M. S. 1994 AIAA Paper 94-0083.
[6] Chaudhury B. et al. 2010 *Physics of Plasmas* **17** 123505.
[7] Boeuf J. P. et al. 2010 *Physical Review Letters* **104** 015002.
[8] Nam K. S. and Verboncoeur P. J. 2009 : *Physical Review Letters* **103** 055004.
[9] Raizer Y. P. 1991 : *Gas Discharge Physics*, Springer publisher.

8th International Conference on Inertial Fusion Sciences and Applications (IFSA 2013) IOP Publishing
Journal of Physics: Conference Series **688** (2016) 012120 doi:10.1088/1742-6596/688/1/012120

# Emission of High-Energy Ions in the SHOTGUN III Divergent Gas-Puff Z-Pinch Experiment

**Keiichi TAKAS GI[1], Masayuki IWATA[2], and Mineyuki NISHIO[3]**
[1]Institute of Quantum Science, ihon University, Tokyo 101-8308, Japan
[2]Graduate School of Science and Technology, ihon University, Tokyo 101-8308, Japan
[3]Anan ational College of Technology, Tokushima 774-0017, Japan

E-mail: takasugi@phys.cst.nihon-u.ac.jp

**Abstract.** Ion pinhole measurements of high-energy ions were conducted on the divergent gas-puff z-pinch plasma. Two types of ions, 1.7 2.5 MeV and 0.1 0.7 MeV, were observed. The former was observed only on the axis. The latter showed uite different characteristics between positive and negative discharges. These ions were considered to be accelerated by inductive electric field generated by the pinch.

## 1. Introduction

The gas-puff z pinch is a repetitive and efficient system for producing high energy-density plasma [1]. The plasma focus can repetitively converge the plasma to a small spot. The divergent gas-puff z pinch has been devised for the realization of efficient radiation source combining both advantages [2]. In the experiment of the divergent gas-puff z-pinch, hard x-ray of 150 – 200 keV, which has not been observed in the conventional gas-puff z pinch, was observed near the surface of anode. This is considered to be caused by electron beam that occurred by the pinch.

Owing to the existence of electron beam, the generation of ion beam has also been expected simultaneously. High energy ions have been observed in plasma focus, and the mechanism of the ion acceleration has been attributed to electromagnetic induction [3]. The ion measurement using a pinhole has been conducted in this system, and a coaxial structure of ion distribution has been observed [4]. The divergent gas-puff z pinch is similar to the plasma focus in that it is a nonuniform system in the axial direction.

Ion measurement using Thomson parabola energy analyzer has been conducted on the Ar divergent gas-puff z-pinch plasma [5]. The emissions of singly to triply ionized Ar ions with the energy of MeV-order have been identified in the axial direction. Experiment on the ion generation relating to the direction of discharge current has been conducted using the bipolar charging power supply. And the generation of these ions has been identified in the axial direction independent of the current direction [6].

In this research, the ion pinhole measurement was conducted to understand the mechanism of high-energy ion generation in the divergent gas-puff z-pinch discharge. And the energy-dependent spatial distribution of ions was measured using a filtered pinhole camera. The relation between the ion distribution and the direction of the current was investigated using the bipolar charging power supply.

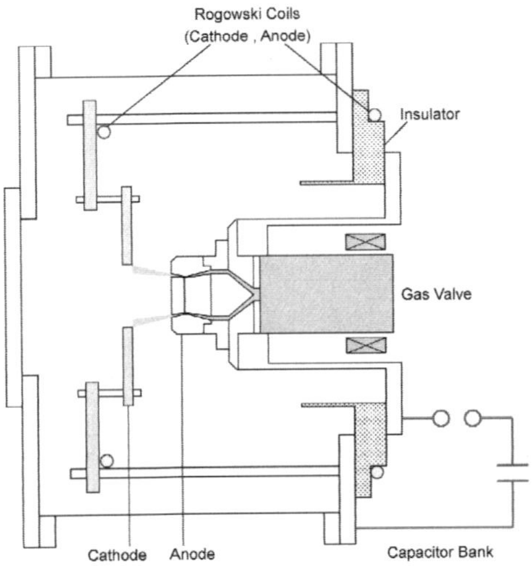


**Figure 1.** Schematic diagram of the SHOTGUN III divergent gas-puff z-pinch device.

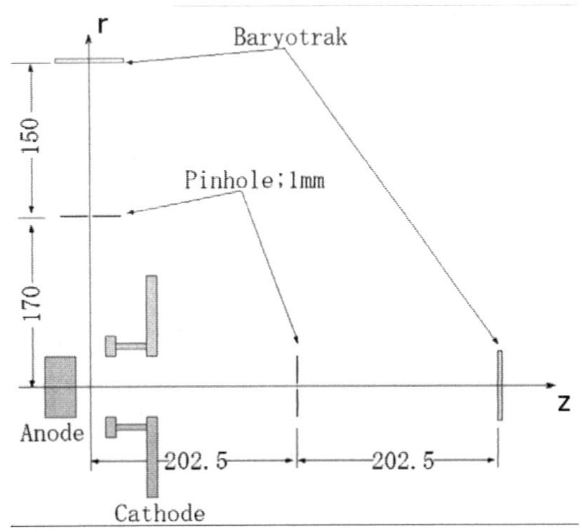


**Figure 2.** Arrangement of the ion pinhole measurement system.

## 2. Experimental setup

The experiment was carried out on the SHOTGUN-III device at Nihon University (Fig. 1). The energy storage section consists of 40 kV, 12 $\mu$F capacitor bank. The advantage of the device is that the bank can be charged to either positive or negative, so that the dependence of polarity of the current can be examined.

Gas-puff was performed by a high-speed gas valve and an annular divergent Laval nozzle mounted on the center conductor. The angle of divergence was 10 degrees. The nozzle diameter was 30 mm, and the inner diameter of opposing electrode was 60 mm. The distance between the electrodes was 30 mm. Ar was used as the operating gas, and the plenum pressure of the gas valve was 5 atm.

In order to investigate the spatial distribution of high-energy ions emitted from the pinch plasma, ion pinhole measurement was conducted. The ion pinhole cameras were placed both axially and radially, as shown in Fig. 2. The diameter of the pinhole was 1 mm. As the distance between the center of electrodes and the pinhole was nearly equal to that between the pinhole and the detector, almost equal size image should be obtained. Track detector Baryotrak-P (CR-39) was used for detecting ions. It is sensitive to ions above 100 keV, and electrons are not detectable [3]. In this experiment each detector was cut as 67 $\times$ 67 mm.

A high-speed camera with an image intensifier (Hamamatsu V3347U) was used for the observation of radial motion of the plasma.

## 3. Divergent gas-puff z-pinch discharge

In the typical operation of divergent gas-puff z pinch, peak discharge current appeared at about 2 $\mu$s after the start of discharge. The current was about 100 kA at the charging voltage of 20 kV. A current dip was observed just after this moment. This indicated the increase of circuit inductance due to the pinch, and the generation of soft x-ray was also observed.

Figure 3 shows visible framing photographs at the charging voltage of 20 kV. Each exposure time was 100 ns, and the photographs were taken at 1.3, 1.8, 2.0, 2.1 and 2.3 $\mu$s after the start of discharge, respectively. At 1.3 $\mu$s the shape of the plasma was similar to that of the initial gas, and the motion of the plasma was little. At 1.8 $\mu$s the plasma shrinked near the anode. The

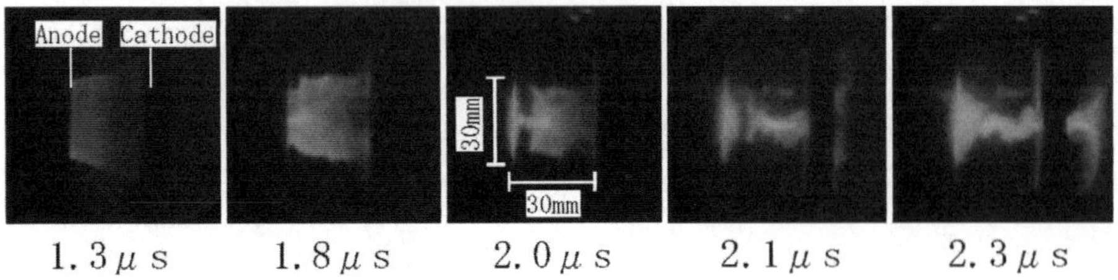


**Figure 3.** Framing photographs of divergent gas-puff z-pinch plasma at the charging voltage of 20 kV.

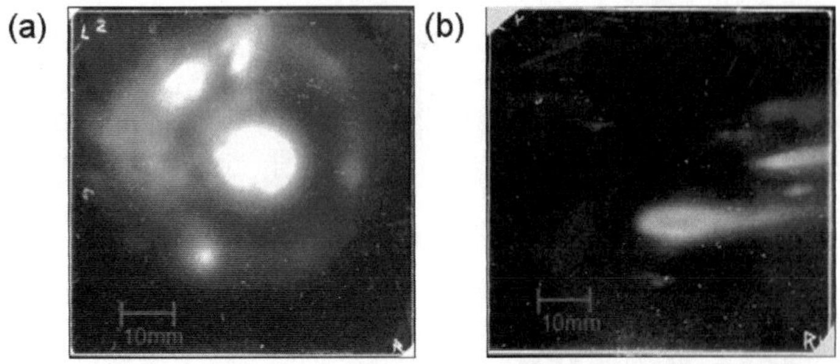


**Figure 4.** Ion tracks of (a) axial and (b) radial detectors at the charging voltage of 20 kV.

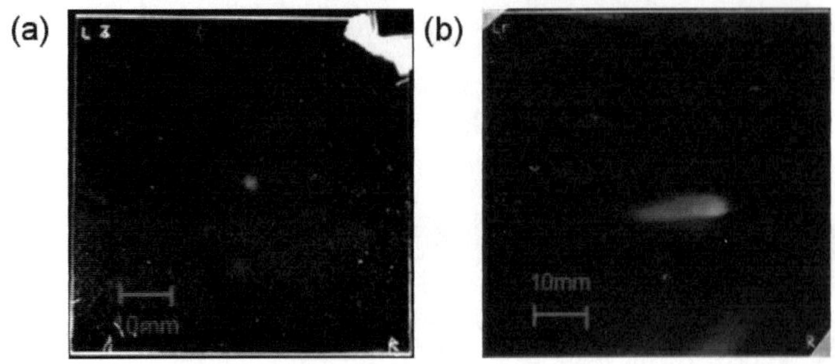


**Figure 5.** Ion tracks of (a) axial and (b) radial detectors at the charging voltage of -20 kV.

Rayleigh-Taylor instability developed over the plasma column. At 2.0 $\mu$s the plasma shrinked strongly just in front of the anode and collapsed on the axis. At 2.1 $\mu$s the shrink took place over the plasma column, and the plasma was expanding slightly. At 2.3 $\mu$s a helical MHD (kink) instability developed, and the plasma began to expand.

The peak current and the pinch time were quite similar at the charging voltage of -20 kV. These evidences indicate that there is no big difference between the positive and the negative discharges in the discharge path and the sequence of radial motion.

## 4. Ion pinhole measurement

At first, the experiment was carried out at the charging voltage of 20 kV. Figure 4 shows the ion tracks in the axial and radial directions. Double coaxial ion tracks were obtained by the axial detector. The diameter of the outer circle was about 40 mm, and the diameter of the center circle was about 16 mm. And the tracks extending to the right was observed in the radial detector.

Next, the experiment was carried out at -20 kV. Figure 5 shows the ion tracks to the axial and radial directions. A small spot of ion tracks was observed in the axial direction, which was quite different from positive discharge. And the tracks extending to the left was observed in the radial detector. Crucial difference appeared in the ion tracks by the difference of the polarity of

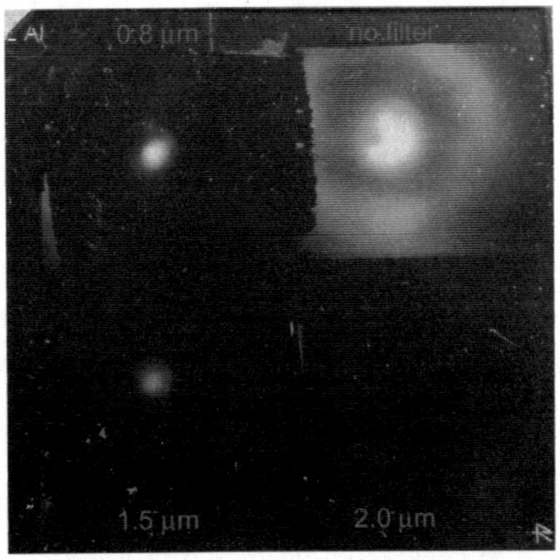


**Figure 6.** Ion tracks of 4-hole pinhole camera with filters. The charging voltage was 20 kV.

the storage system.

Then filtered measurement with 4-hole pinhole camera was conducted. The charging voltage was 20 kV. Figure 6 shows the ion tracks with 0.8, 1.5, 2.0 $\mu$m Al filters and without a filter. The tracks without a filter show coaxial structure. Spot images of ion track were observed with 0.8 $\mu$m and 1.5 $\mu$m filters. No ion tracks was observed with 2.0 $\mu$m filter. The size of the spot was about 2 mm, which was almost the same as that of discharge with negative power supply.

## 5. Summary and discussion

Ar ions which pass through 0.8 $\mu$m, 1.5 $\mu$m and 2.0 $\mu$m Al foils have the energy exceeding 0.7 MeV, 1.7 MeV and 2.5 MeV, respectively. There are two types of high energy ions. The first ions have tracks of coaxial structure, which cannot come over 0.8 $\mu$m filter. The energy of these ions is between 0.1 and 0.7 MeV.

The second ions have tracks of a spot on the axis, which can come over 1.5 $\mu$m filter. The energy of these ions is between 1.7 and 2.5 MeV. As the spread of ion tracks in a spot is small, the ions were thought to have come from the pinch plasma which was observed at 2.0 $\mu$s in Fig. 3. These ions were considered to be the same as the ions which were detected by Thomson parabola analyzer [6]. Although the filtered measurement was not made for the negative discharge, the high-energy component should have been included in the observed spot in Fig. 5 (a). Those high-energy ions were possibly generated by some acceleration mechanism independent of the current direction.

The ions which form the coaxial structure were observed only in the positive discharge. The diameter of the center circle was about 16 mm, and it is considerably larger than the pinch plasma. The ions near the anode surface were considered to be accelerated by inductive electric field which occurred at the moment of pinch. The diameter of the outer circle was about 40 mm, and it was beyond the plasma diameter. The ions in this part came through the outside of the plasma, and the orbits were bent by the magnetic field. The direction of Lorentz force acting on the ions was to converge them toward the cathode. The ion tracks observed in the radial detectors had similar tendency to them.

## References

[1] Shiloh J, A. Fisher A and Rostoker 1978 *Phys. Rev. Lett.* **40**, 515
[2] Takasugi K and Kiuchi E 2007 *Plasma Fusion Res.* **2**, 036
[3] Moser A, Sadowski M, Herold H and Schmidt H 1982 *J. Appl. Phys.* **53**, 2959
[4] Bertalot L et.al. 1980 *Phys. Lett.* **79A**, 389
[5] ishio M, Takasugi K and Sakuma H 2011 *NIFS-PROC* **87**, 87
[6] ishio M, Sakuma H and Takasugi K 2011 *Plasma Fusion Res.* **6**, 1201009

Journal of Physics: Conference Series **688** (2016) 012121 doi:10.1088/1742-6596/688/1/012121

# Concept for Generation of Warm Dense Matter of Insulator due to Flyer Impact Accelerated by Electron Beam Irradiation using Intense Pulsed Power Generator

**Fumihiro TAMURA [1], Ryota HAYASHI [1], Takahiro KUDO [1], Arata WATABE [1], Kenji KASHINE [2], Akira TOKUCHI [1,3], Takashi KIKUCHI [1], Kazumasa TAKAHASHI [1], Toru SASAKI [1], Tsukasa ASO [1], Nob. HARADA [1], and Weihua JIANG [1]**

1) Nagaoka University of Technology, Japan
2) Yuge National College of Maritime Technology, Japan
3) Pulsed Power Japan Laboratory Ltd., Japan

E-mail: tamura2316@stn.nagaokaut.ac.jp

**Abstract.** We have proposed a concept for generation method of warm dense matter (WDM) by using flyer impact accelerated by intense electron beams. To generate the warm dense matter by using flyer impact, the output current of electron beams generated by the intense pulsed-power generator as ETIGO-II is evaluated. The results denote that the beam current and the pulse duration are 11 kA and 50 ns, respectively. The achievable parameters of WDM by using flyer impact are estimated by the simplified model. It indicated that the sample temperature achieves the provided electron beams with lower conversion efficiency.

## 1. Introduction

Properties of insulators in warm dense matter (WDM) are interested in an optimization of cone material for the fast ignition [1], evaluating interiors of giant planets [2], and so on. For example, diamond-like-carbon (DLC) is expected as one of effective cone materials, and the accurate properties in WDM are required to increase the coupling efficiency.

The WDM region is in dense ($10^{21}$ - $10^{24}$m$^{-3}$) and moderate temperature ($10^3$ - $10^5$K). To understand the properties of WDM, effects of ion-ion correlations and degenerate electrons are analyzed by studying the dependence of equation of state (EOS). To generate WDM for conductive materials, several experimental approaches by using pulsed-power discharge have been demonstrated [3-6]. However, it is difficult to generate the WDM from insulators by the method, because insulators are quite low conductivity. To generate WDM of insulator needs high energy density and another method besides the isochoric heating of Joule heating. Thus, we consider a different approach for generation of WDM from insulator with well-defined plasma parameters. To solve the problems, we focus on the flyer impact method by using intense pulsed-power system. The method of flyer impact by using an intense ion beams has been demonstrated [7,8]. The results indicated that the flyer generated from an aluminum foil was 7.7km/s in the velocity and 13GPa in the ablation pressure at ion beam energy density of 2kJ/cm$^2$. Thus, the method of flyer impact provides the heating insulator samples and the well-defined material state from the shock relations.

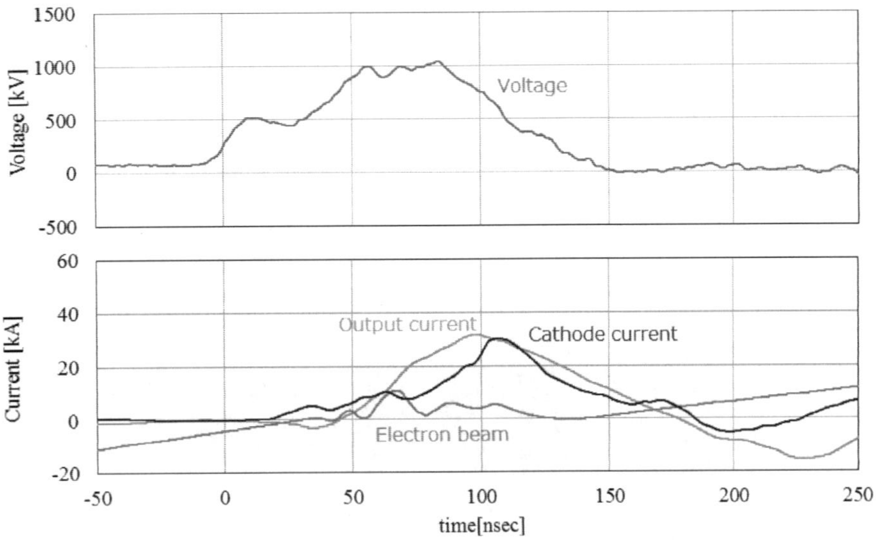


**Figure 1.** The typical output waveforms from ETIGO-II with electron beam diode.

In order to consider the method of flyer impact, we propose the flyer acceleration and impact on an insulator target with intense electron beam by the intense pulsed power supply "ETIGO-II"[9].

## 2. Intense pulsed power generator "ETIGO-II" as experimental setup

The normal output parameters from ETIGO-II in the case of impedance matching at a load are 1MV, 590kA, 50nsec (FWHM), respectively. In this experiment, the load as an electron beam diode consists of a cathode made of a SUS plate having $\phi$ 105 mm in diameter and an anode made of a SUS mesh having $\phi$ 210 mm in diameter. The output voltage of pulse-transmission line in ETIGO-II was measured using a capacitive voltage divider located at upstream of the electron beam diode. The output current of pulse-transmission line in ETIGO-II was measured by pick-up coil located at between the voltage divider and the electron beam diode. The output currents at the cathode and the electron beam current were measured using Rogowski-coils.

Figure 1 shows typical output waveforms of ETIGO-II with electron beam diode. The output voltage is 1 MV at the peak with 50 ns of the pulse duration. The output current of pulse-transmission line in ETIGO-II is 30 kA. The output current of cathode is also 30 kA with 30 ns delayed from the current of pulse-transmission line. The electron beam current is 11 kA at the peak with the duration of 50 ns.

## 3. Evaluation of sample temperature in WDM for insulator

To evaluate the EOS of WDM for insulator, we consider the achievable parameter by using flyer impact. From the beam-flyer conversion efficiency obtained by the previous experiments as the intense ion beams, we estimate the achievable temperature of WDM sample for silicate.

Figure 2 shows the processes of energy conversion by using flyer impact. The electron beams heats the flyer with the ablation. The flyer collides and heats the silicate sample. These processes have several energy conversions. The increase of temperature $\Delta T$ for the silicate is estimated by the following equation,

$$\Delta T = \frac{E_s}{\rho_s C_{vs} S_s L_s} = \frac{\eta_{e-f}\eta_{f-s}E_e}{\rho_s C_{vs} S_s L_s}, \quad (1)$$

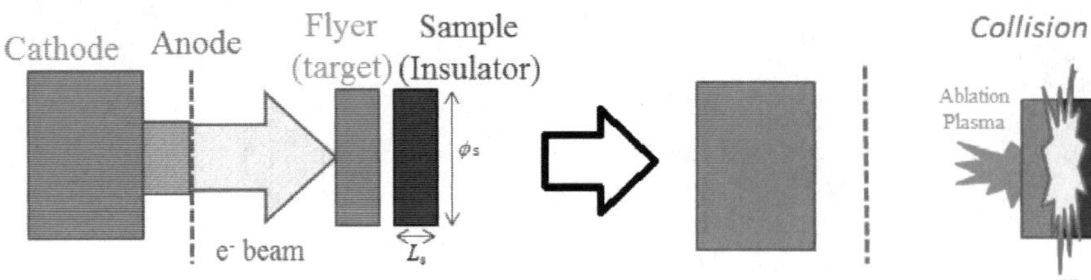


**Figure 2.** Processes of energy conversion by using flyer impact

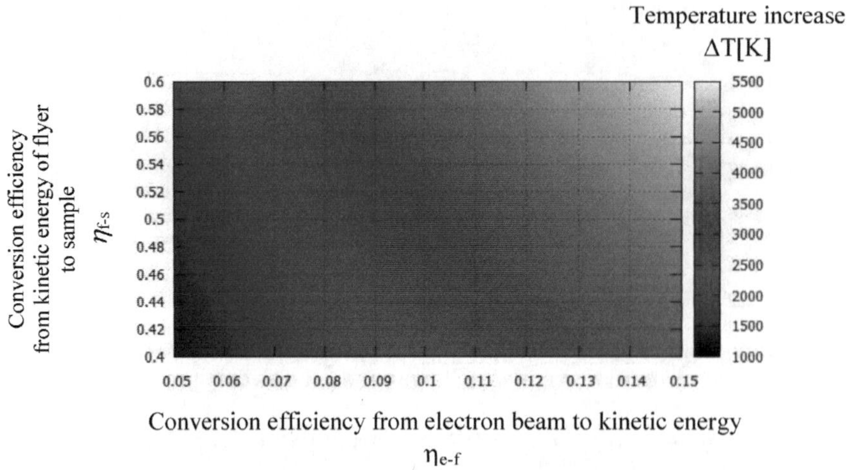


**Figure 3.** Achievable temperature of the sample as a function of the conversion efficiency $\eta_{e\text{-}f}$ from the electron beam energy to the kinetic energy of flyer and the transmission efficiency from the kinetic energy of flyer to the sample.

where, $E_s$ is the input energy from the electron beams, $\eta_{e\text{-}f}$ is the conversion efficiency from the electron beam energy to the kinetic energy of flyer, $\eta_{f\text{-}s}$ is the transmission efficiency from the kinetic energy of flyer to the sample, $\rho_s$ is the sample density, $C_{vs}$ is the specific heat of the sample, $S_s$ is the cross section of the sample, and $L_s$ is the heating depth of the sample. The input energy from the electron beams is proportional to the particle energy as 1 MeV, the electron beam current as 11 kA, and the duration of the electron beam as 50 ns. The conversion efficiency $\eta_{e\text{-}f}$ assumes around 10 % from the previous experiments as the intense ion beams. The cross section and the heating depth are determined by the flyer cross section. The properties of silicate are used density $\rho_s$=3.28kg/m$^3$ and the specific heat $C_{vs}$=1152 J/K/kg [2].

Figure 3 shows the achievable temperature of the sample as a function of the conversion efficiency $\eta_{e\text{-}f}$ from the electron beam energy to the kinetic energy of flyer and the transmission efficiency from the kinetic energy of flyer to the sample. The initial sample temperature is 300 K for these estimations. The sample temperature $T_s$ is over 3000 K at $\eta_{e\text{-}f}$=0.1 and $\eta_{f\text{-}s}$ = 0.5. It means that the sample temperature and the input energy are enough to generate WDM at lower efficiency of conversion from the electron beam energy to the sample heating.

## 4. Design for flyer experiment to generate WDM of insulator

Figure 4 shows the design of the load for flyer acceleration. The flyer target is located on the focal point of the spherical electrodes because of focussing electron beam and irradiating intense electron beam. Figure 5 shows the measurement systems for the flyer and the sample. The electron beam current is measured by the Rogowski-coil to estimate the input power to flyer. Flyer velocity as the flyer energy is measured by time of flight method using optical fibers with photo diodes. The sample temperature and propagation shock velocity trace are observed with the spectroscopic observation and VISAR.

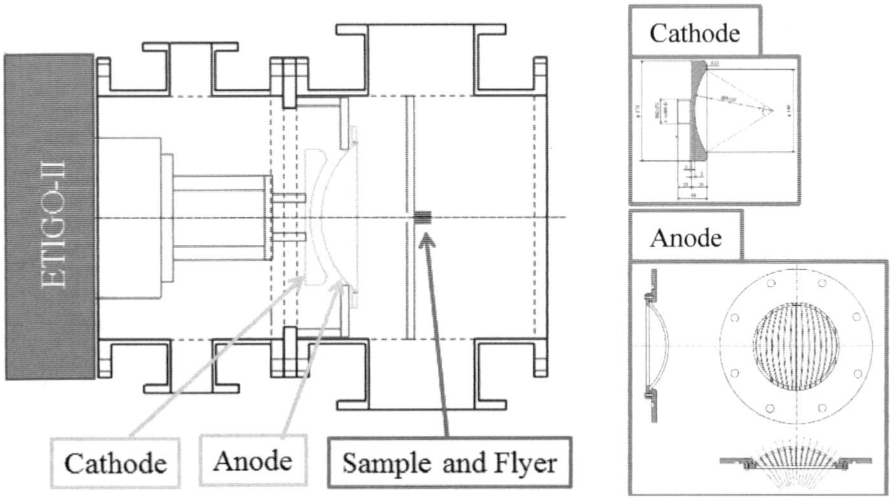


**Figure 4.** Chamber and electrode design for flyer acceleration by electron beam with ETIGO-II.

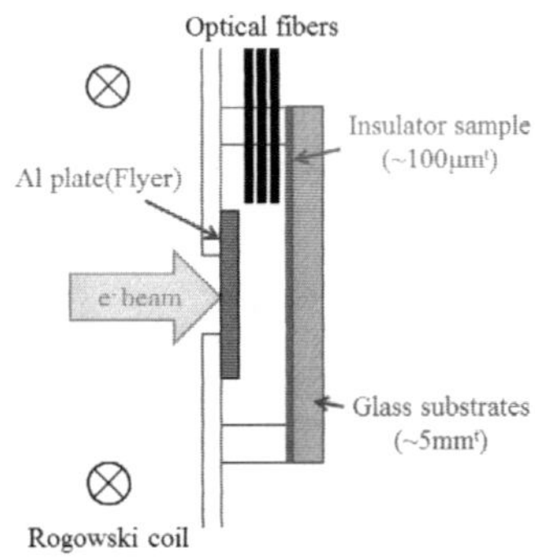


**Figure 5.** Measurement systems for flyer acceleration.

## 5. Summary

We have proposed a concept for generation method of WDM by using flyer impact accelerated by the intense electron beams. To generate the WDM by using flyer impact, the output current of electron beams generated by the intense pulsed-power generator as ETIGO-II is evaluated. The results denoted that the beam current and the pulse width are 11 kA and 30 ns, respectively. The achievable parameters of WDM by using the flyer impact are estimated by the simplified model. It indicated that the sample temperature achieves the provided electron beams with low conversion efficiency. The design of the flyer impact driven by the intense electron beams is considered. We will make the system and demonstrate the experiment of generating WDM for insulator.

## References

[1] Fujioka S et al 2012 Plasma Physics Controlled Fusion **54** 124042
[2] Kurosawa K et al 2012 J. Geophys. Res. **117** E04007
[3] Amano Y et al 2012 Rev. Sci. Instrum. **83** 085107
[4] Renaudin P et al 2002 Phys. Rev. Lett. **88** 215001
[5] Sasaki T et al 2010 Phys. Plasmas **17** 084501
[6] DeSilva A W and Katsouros J D 1998 Phys. Rev. E **57** 5945
[7] Kashine K et al 2002 Jpn. J. Appl. Phys. **41** 4014
[8] Kashine K 2013 2nd IGCN SO-6
[9] Jiang W et al 1993 Jpn. J. Appl. Phys. **32** L752

# Spectroscopic measurements of ablation plasma generated with laser-driven intense extreme ultraviolet (EUV) light

**N Tanaka[1], K Hane[1], H Shikata[2], M Masuda[1], K Nagatomi[2], A Sunahara[3], M Yoshida[2], S Fujioka[1], and H Nishimura[1]**

[1] Institute of Laser Engineering, Osaka University, Yamadaoka 2-6, Suita, Osaka, 565-0871, Japan

[2] Faculty of Science and Engineering, Kinki Univestiy, Kowakae 3-4-1, Higashi-Osaka, Osaka, 577-8502, Japan

[3] Institute for Laser Technology, Yamadaoka 2-6, Suita, Osaka, 565-0871, Japan

E-mail: tanaka-n@ile.osaka-u.ac.jp

**Abstract.** Material ablation by a focused Extreme ultraviolet (EUV) light is studied by comparing expanding ion properties and plasma parameters with laser ablation. The kinetic energy distributions of expanding ions from EUV and laser ablation showed different spectra implying different geometries of plasma expansion. The calculation results of plasma parameters showed that EUV energy is mostly deposited in high electron density region close to the solid density, while laser energy is deposited in low energy density region. Plasma parameters experimentally obtained from visible spectra did not show noticeable difference between EUV and laser ablation due to the corresponding low cut off density.

## 1. Introduction

Extreme ultraviolet (EUV or XUV) is extremely short-wavelength radiation ranging from a few to a few tens of nm, and many types of radiation sources have been widely developed. Although one of the most well known applications of this radiation is EUV photolithography, recent progress on EUV sources can lead to many other possibilities in the industry, such as material processing and material analysis. One of the possible applications is micro- and nano-machining using material ablation by focused EUV [1,2].

Since there are significant differences between EUV and conventional laser in terms of the wavelength and photon energy, heating mechanism and ablation physics of EUV ablation are expected to be quite different from that of laser ablation. Dynamics of material ablation strongly depends on the wavelength of the incident radiation or laser, especially with a pulse length in nano-second order. Ablation plasma is formed in pico-second order after the front end of the laser reaches the sample surface. This means that the rest of laser enters the plasma. In the plasma, the laser light does not penetrate further at a certain electron density called the critical density $n_c$ written as follows,

$$n_c = \frac{\omega^2 m_e \varepsilon_0}{e^2} = \frac{4\pi^2 c^2 m_e \varepsilon_0}{\lambda^2 e^2}\ , \qquad (1)$$

where $\omega$ is the angular frequency of the incident radiation or laser, $m_e$ is the electron mass, $\varepsilon_0$ is the vacuum permittivity, $c$ is the speed of light, and $\lambda$ is the wavelength of the incident radiation or laser. The critical density corresponding to the EUV wavelength range exceeds the solid density (e.g. $6.1 \times 10^{24}$ cm$^{-3}$ for 13.5 nm EUV), while that corresponds to conventional laser wavelength is 1-2 orders of magnitude smaller than the solid density (e.g. $1.1 \times 10^{21}$ cm$^{-3}$ for 1064 nm Nd:YAG laser). One hand, the dominant heating process of laser ablation is the collisional energy absorption by electrons in low-density plasma followed by thermal heating of the materials by electrons. On the other hand, EUV can penetrate through the plasma and reach the sample surface, and deposit its energy into the material directly. Furthermore, orbital electrons would be directly excited by EUV photons with energy about 100 eV, e.g., photo-ionization. These possible differences in plasma heating mechanisms between EUV and lasers could result in different plasma properties. Thus, an effective approach for understanding the EUV ablation physics is the study of the ablation plasmas and its comparison with conventional laser ablation. In this paper, plasma characteristics of EUV and laser ablations are shown. The expanding ion properties and plasma parameters were studied both experimentally and by calculation.

## 2. Experimental setup

A laser produced plasma (LPP) EUV source was utilised in the experiments. A drive laser (Nd: YAG, 1064 nm) was focused onto a solid Xe target. The EUV light emitted from the Xe surface was collected by a toroidal elliptical total reflection mirror with a gold coated reflection surface. For the laser ablation experiments, a Nd:YAG laser (1064 nm) was directly focused onto the sample. The focused EUV or laser intensity was in the order of $10^9$ W/cm$^{-2}$, and the spot size was 150 $\mu$m. Silicon and aluminium were chosen as the ablation samples because they have L absorption edges in Xe EUV wavelength range.

Two different measurements were conducted. One was time-of-flight (TOF) measurement of expansion ions by using a charge collector and the other was spectroscopy of the ablation plasma. TOF is a popular and useful technique for studies of flowing particles with velocities, such as particle beams or particle emissions. It is also suitable for the study of ablation plasma, because particle expansion is a key for understanding the ablation dynamics [3]. Time evolution of the ion current flowing into the charge collector was measured at the sample surface normal. Photoemission spectra from the ablation plasma were measured using a spectrometer in the visible range (350-700 nm) located at the 90-degree side view of the sample surface.

## 3. Ion expansion from EUV and laser ablations

TOF measurement of expanding ions from aluminium sample was carried out. The charge and mass are estimated to be $Al^+$ for the analysis of the TOF signals. The kinetic energy distributions are shown in Fig. 1. The TOF spectra of ion current flowing into the charge collector are also shown in the figures. The two kinetic energy distributions show completely different spectra: laser ablation have a convex spectrum and EUV ablation have a slightly concave spectrum. The shapes of spectra can be explained by an analytical model that includes the dimension of plasma expansion [4].

$$\frac{dN}{dE} = \frac{E/E_0^{(\alpha-2)/2} \exp(-E/E_0)}{\Gamma_{(\alpha/2)}}, \tag{2}$$

where $E_0$ is the plasma temperature in eV, and $E$ is the ion kinetic energy, $\alpha$ is the geometry of plasma expansion. $\alpha = 1$ to 3 corresponds to planar, cylinder, and spherical geometries, respectively. The plasma expansion is one dimensional in all coordinate systems. Plots of eq. 2 form convex ($\alpha = 1$), linear ($\alpha = 2$), and concave ($\alpha = 3$) spectra in semilogarithmic plot. The

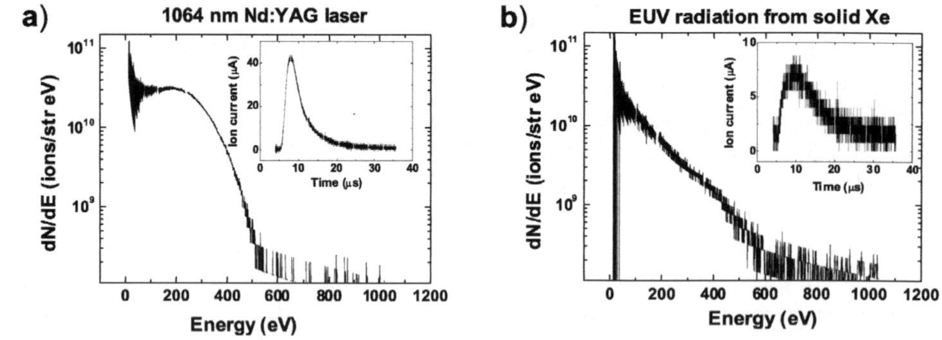


**Figure 1.** Kinetic energy spectra of the expanding ions measured by the charge collector. Small graphs show the TOF spectra. a) A convex ion energy spectrum was observed from laser ablation plasma, and b) a slightly concave spectrum was observed from EUV ablation plasma.

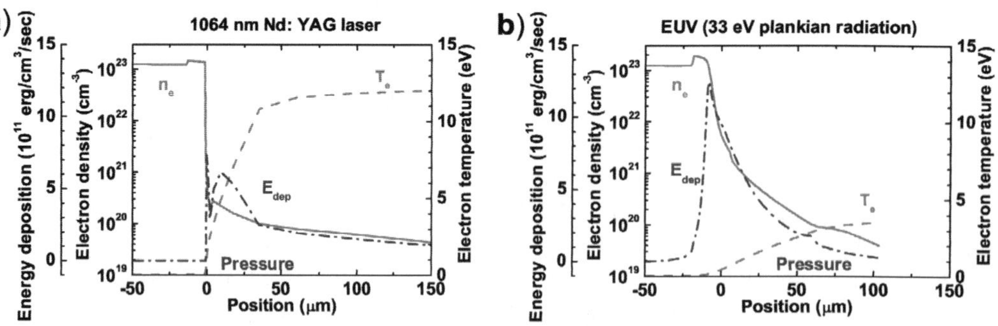


**Figure 2.** Spatial distributions of electron density, electron temperature, energy deposition and pressure along the sample surface normal simulated by the STAR-1D code. Both a) laser and b) EUV ablation of CH sample were calculated.

concave ion spectrum of EUV ablation implies that the plasma expansion is likely more planar expansion, which is one-dimensional expansion along the sample surface normal.

## 4. Comparison of plasma parameters

Figure 2 shows spatial distributions of the electron density, electron temperature, and energy deposition along the sample normal calculated by a one-dimensional radiation hydrodynamic code, STAR-1D [5], for both laser and EUV ablations of CH sample. The input laser or EUV parameters are in the same order of magnitude as that in experiments. A 33 eV Plankian radiation spectrum with a dilution factor of 0.03 was used to simulate the incident EUV spectrum from Xe target. Corresponding EUV intensity on the sample was $5 \times 10^9$ W/cm$^{-2}$. Plasma parameters at the peak of Gaussian pulse of the incident laser or radiation are shown. EUV energy deposition have a peak inside the original surface position at 0 $\mu$m, where the electron density is in the order of $10^{22}$ cm$^{-3}$, while laser energy is mostly deposited in lower density region. Electron temperature in EUV ablation plasma is about 3 times lower than that in laser ablation plasma. It implies that most of EUV photon energy is deposited into particles in high-density region and consequently energy deposition into individual particle is small.

Spectroscopic measurements in visible range (350-700 nm) were conducted for Si ablation plasmas. Most of peaks of the spectra correspond to photo-emissions from $Si^{+}$ or $Si^{2+}$ ions. Figure 3 shows the electron densities derived from the Stark broadening of the spectra, and the

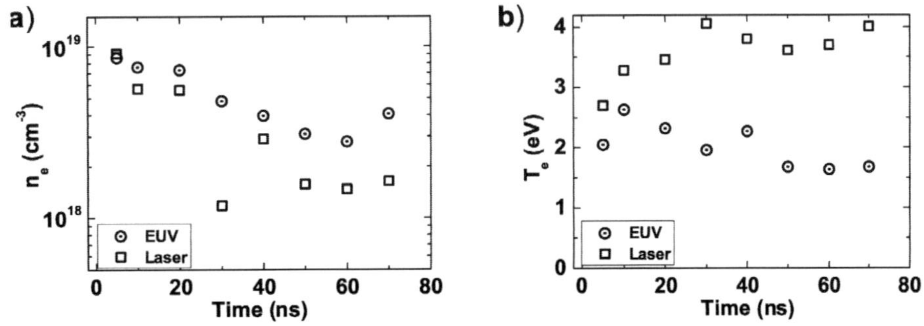


**Figure 3.** Plasma parameters, a) electron densities and b) electron temperatures derived from the time-resolved visible spectra of laser ablation Si plasma (open square plots) and EUV ablation plasma (dotted open circle plots).

electron temperatures by using Boltzmann plot for both EUV and laser ablation plasmas. The plasma densities are almost $10^{19}$ cm$^{-3}$ during irradiation (0-10 ns), and they gradually decrease in both EUV and laser ablations. Slightly higher plasma temperature ($\sim$1 eV) is observed in laser ablation compared to EUV ablation during irradiation (Fig. 3 b). The difference becomes larger as the time implying difference in cooling mechanism between two radiation sources. Noticeable difference between laser and EUV was not observed. One possible reason would be the wavelength range. Visible light from high density plasma is cut off due to the critical density and information of core plasma may not have been observed in this set of experiments. Time and specially resolved spectroscopy of shorter wavelength range such as VUV or soft X-ray would provide detailed information of plasma parameters.

## 5. Conclusion

Characteristics of EUV ablation of materials were studied by comparing expanding ion spectra and plasma parameters with that of laser ablation. Energy distributions of expanding ions from EUV and laser ablation show different ion spectra. An analytical model was applied for analysis. It has been found that the plasma expansion of EUV ablation is likely one dimensional along the sample surface normal. Simulation results of spatial distributions of plasma parameters showed clear differences between EUV and laser ablation plasma. EUV energy was mostly deposited into high-density region inside the original surface position resulting in lower electron temperature than that of laser ablation. The plasma parameters were also derived from experimental results of visible spectra. No significant difference was observed in visible wavelength range, and spectroscopy in shorter wavelength range would provide detailed plasma information.

## Acknowledgments

This work was supported by JSPS Grant-in-Aid for Young Scientists (B) number 25800303, Grant-in-Aid for Scientific Research (B) number 22340172, and MEXT Project for Creation of Research Platforms and Sharing of Advanced Research Infrastructure "Opening up a new photonics industry through high-intensity lasers".

## References

[1] Barkusky F, Bayer A, Mann K 2011 *Appl. Phys. A* **105** 17.
[2] Makimura T, Miyamoto H, Uchida S, Niino H and and Murakami K 2007 *J. Phys. Conf. Ser.* **59** 279.
[3] Pelah I 1976 *Physics Letters* **59A** 348.
[4] Murakami M, Kang Y -G, Nishihara K, Fujioka S, Nishimura H 2005 *Phys. Plasmas* **12** 062706.
[5] Sunahara A and Tanaka K A 2010 *Fusion Engineering and Design* **85** 935.

# Plasma physics effects on thermonuclear burn rate in the presence of hydrodynamic mix

**Xian-Zhu Tang, Zehua Guo, Grigory Kagan, Christopher McDevitt, and Bhuvana Srinivasan**

Theoretical Division, Los Alamos National Laboratory, Los Alamos, NM 87545

E-mail: xtang@lanl.gov

**Abstract.** Hydrodynamic mix can significantly degrade thermonuclear burn rate in an inertial confinement fusion (ICF) target. Successful mitigation requires a detailed understanding of the physical mechanisms by which mix affects burn. Here we summarize the roles of three distinct plasma physics effects on burn rate. The first is the well-known effect of enhanced thermal energy loss from the hot spot and the mitigating role of self-generated or externally-applied magnetic field. The second is the fuel ion separation via inter-species ion diffusion driven by the powerful thermodynamic forces exacerbated by mix during the implosion process. The third is the fusion reactivity modification by fast ion transport in a mix-dominated ICF target, where hot plasma is intermingled with cold fuel.

## 1. Introduction

Fusion yield in an inertial confinement fusion experiment primarily comes from the reaction between deuterium (D) and tritium (T) for their significantly greater fusion cross section in the multi-keV energy range. The thermonuclear burn rate, or the fusion power density, is [1]

$$p_f \equiv n_D n_T \langle \sigma v \rangle E_f, \tag{1}$$

with $n_{D,T}$ the number densities of D and T, $\langle \sigma v \rangle \equiv (n_D n_T)^{-1} \int \sigma |\vec{v}_D - \vec{v}_T| f_D f_T d\vec{v}_D d\vec{v}_T$ the fusion reactivity integral, and $E_f$ the energy release per fusion reaction. With the cross section $\sigma$ and energy release $E_f$ determined by nuclear physics, the DT fusion yield rate is affected by plasma physics through the product of $n_D$ and $n_T$, and the normalized ion velocity distribution functions $f_{D,T}/n_{D,T}$ in the fusion reactivity integral. In a dense ICF plasma, $f$ is close to be a Maxwellian $f^M$, so one can write $f = f^M + \delta f$ with $\delta f = f - f^M$ small in magnitude compared with $f^M$. The fusion reactivity integral can then be written in two parts,

$$\langle \sigma v \rangle = \underbrace{\frac{1}{n_D n_T} \int \sigma |\vec{v}_D - \vec{v}_T| f_D^M f_T^M d\vec{v}_D d\vec{v}_T}_{\widehat{\langle \sigma v \rangle}} + \underbrace{\frac{1}{n_D n_T} \int \sigma |\vec{v}_D - \vec{v}_T| (\delta f_D f_T + f_D \delta f_T) d\vec{v}_D d\vec{v}_T}_{\widetilde{\langle \sigma v \rangle}}, \tag{2}$$

where $\widehat{\langle \sigma v \rangle}(T_i)$ is determined by ion temperature $T_i$ alone, and $\widetilde{\langle \sigma v \rangle}$ is a function of the deviation of the distribution function from a Maxwellian ($\delta f_{D,T}$). Distinctly different plasma transport physics enter these three yield factors: $n_D n_T$, $\widehat{\langle \sigma v \rangle}$, and $\widetilde{\langle \sigma v \rangle}$.

## 2. Effect of mix on yield via hot spot thermal energy loss – $\widehat{\langle\sigma v\rangle}$

Hydrodynamic mix can significantly degrade the fusion yield rate in ICF experiments. The most-recognized effect is the cooling of the hot spot due to mix-enhanced thermal loss by electron thermal conduction. Since the fusion reactivity $\widehat{\langle\sigma v\rangle}$ is a strong function of $T_i$, temperature drop in the multi-keV range can lead to a drastic reduction in $\widehat{\langle\sigma v\rangle}$. The way that mix enhances thermal energy loss from the hot spot is through a much enlarged hot/cold interface $\Sigma$, which enters the energy exchanged between hot (gas) and cold (ice) plasma as, $Q = \int_\Sigma \kappa_{\perp e}\nabla T_e \cdot d\mathbf{S}$. The rapid (exponential in early stage) growth in $\Sigma$ can have a large impact on $Q$. The enhanced hot spot thermal energy loss produces (1) a lower $T_e$ and hence $T_i$ via electron-ion collisional equilibration in the hot spot, and (2) additional ice mass that is heated up and becomes part of a more massive but cooler hot spot. The trade-off between additional hot spot mass and lowered hot spot temperature can be illustrated with a simplified model that assumes a spatially uniform hot spot, where the volume integrated fusion yield rate is

$$\mathcal{P}_f = \int p_f d\mathbf{x} = \frac{4\pi R_{hs}^3}{3}\frac{p_D p_T}{T^2}\langle\sigma v\rangle E_f \approx \frac{4\pi R_{hs}^3}{3}\frac{p_D p_T}{T^2}\widehat{\langle\sigma v\rangle} E_f. \tag{3}$$

In the usual ICF range of $T_i = 1 - 8$ keV, $\widehat{\langle\sigma v\rangle} \propto T^4$, so the total yield rate of a hot spot with given radius ($R_{hs}$) and pressure ($p_D$ and $p_T$) scales as $T^2$. Interestingly enough, if one is able to drive a hot spot to $T_i > 8$ keV but $< 25$ keV, the reactivity tapers off as $\widehat{\langle\sigma v\rangle} \propto T^2$, so the temperature dependence of $\mathcal{P}_f$ cancels out in leading order.

Mitigation of mix-enhanced hot spot thermal energy loss, i.e. reducing $Q$, can be realized by decreasing the electron thermal conductivity $\kappa_{\perp e}$ across the mix interface and/or limiting the dynamical growth in $\Sigma$. Magnetization of electrons is an effective way of reducing $\kappa_{\perp e}$ across the magnetic field. For this to work in an ICF target, the magnetic field needs not only to be strong enough to magnetize the thermal electron, but also to align with the mix interface. It turns out that both self-generated and externally-applied magnetic fields can satisfy that, but by different physical mechanisms. By self-generation, we mean that there is no seed magnetic field, so the Biermann's battery effect [2], which is attributed to a cross gradient of electron density and temperature, is responsible for a sufficiently large magnetic field that wraps around the bubble and spike of the mix interface [3, 4]. Although it is the electrons that drive the Biermann term, the magnetic field self-generation is closely linked to the ion fluid vorticity [5, 6]. To be more specific, the ion vorticity associated with the interpenetration of hot and cold plasmas under hydrodynamic mix is where the Biermann-battery-induced magnetic flux bundles concentrate. Since the vorticity roll-up is at the mix interface where the flow shear is the largest, the induced magnetic field is aligned with the mix interface. An externally applied seed magnetic field offers valuable control for the field magnitude and hence the level of electron magnetization. Hydrodynamic mix can efficiently amplify the field strength by the conventional magnetohydrodynamic dynamo mechanism, also known as the stretch-and-fold dynamo action. Even more importantly, the stretching action is most active where the flow shear is large, namely at the mix interface, and hence it rapidly aligns the amplified field with the mix interface [6], which is typical of a chaotic dynamo [7]. The fact that a mix-interface-aligned magnetic field not only reduces thermal loss but also stabilizes the short wavelength Rayleigh-Taylor modes [6, 8] suggests an intriguing strategy to control hydrodynamic mix by external magnetic field, which can be effective even though the magnetic pressure remains orders of magnitude smaller than the kinetic energy density in the implosion flow $\rho v^2/2$ and the plasma thermal pressure [6].

## 3. Effect of mix on yield via fuel ion separation – $n_D n_T$

For an assembled target with a specific $n_e = n_D + n_T$, the burn rate is maximized when $n_D = n_T$ everywhere in the burn region. This optimal fuel arrangement is usually satisfied initially by design in both the gas fill and the ice fuel layer, but the implosion dynamics can drive D and T separation [9, 10] even given the initial condition that $n_D/n_T$ is a global constant. The underlying physics is inter-species ion diffusion driven by thermodynamic cross terms, namely the ion pressure gradient (baro-diffusion), the ambipolar electric field (electro-diffusion), and the ion and electron temperature gradients (thermo-diffusion). This effect can be quantified in a two-ion component (DT) plasma by following the time evolution of the D relative mass concentration, $c \equiv m_D n_D/\rho$ with $\rho = m_D n_D + m_T n_T$, $\rho\frac{\partial c}{\partial t} + \rho\vec{u}\cdot\nabla c + \nabla\cdot\vec{i} = 0$, where $\vec{u}$ is the plasma flow velocity and the inter-species ion diffusive flux is given by [11, 12, 13]

$$\vec{i} = -\rho D\left[\nabla c + \overbrace{k_p\nabla\log p_i}^{\text{barodiffusion}} + \underbrace{(ek_E/k_BT_i)\,\nabla\Phi}_{\text{electrodiffusion}} + \overbrace{k_T^{(i)}\nabla\log T_i}^{\text{ion thermo-difusion}} + \underbrace{k_T^{(e)}\nabla\log T_e}_{\text{electron thermo-difusion}}\right]. \quad (4)$$

The severity of fuel ion separation is determined by two factors: (1) the magnitude and sign of various relative diffusion coefficients ($k_p, k_E, k_T^{(i)}, k_T^{(e)}$), and (2) the gradient length scales of ion pressure $p_i$, plasma ambipolar potential $\Phi$, ion and electron temperature. For given two species of ions, $k_p, k_E, k_T^{(i)}$, and $k_T^{(e)}$ are known functions of $c$ only [11, 13, 14]. In a DT mixture, $k_p$ and $k_E$ are positive and of similar value in the range of (0,0.11). The ion thermo-difusion coefficient $k_T^{(i)}$ is also positive and larger than both $k_p$ and $k_E$. The electron thermo-diffusion coefficient $k_T^{(e)}$ is negative and of a magnitude slightly smaller than $k_p$ and $k_E$. All of the four relative diffusion coefficients achieve maximum value at $c = 0.5 - 0.6$ [14].

In a spherically symmetric implosion, the gradient length scales are on the order of the hot spot radius $R_{hs}$. When hydrodynamic mix is present, hot plasma and cold fuel interpenetrate, forming bubbles and spikes that later develop into a turbulent cascade toward smaller structures. The gradient length scales for $p_i, \Phi, T_i, T_e$ are then set by the width of the hot plasma filament $l_{\text{mix}}$, which can be far smaller than the nominal hot spot radius $R_{hs}$. This rapid scale reduction by hydrodynamic mix aggravates the inter-species ion diffusion and hence DT separation in the hot plasma region. It should be noted that DT fuel ion separation is mostly a problem in the hot plasma where the ion temperature is high. This can be understood from the temperature scaling of the classical DT inter-ion diffusivity, $D \propto T^{5/2}$. The ice fuel layer suffers little species separation from DT inter-diffusion, until it is heated up into the hot plasma.

## 4. Effect of mix on yield via fusion reactivity changes by fast ion transport – $\widetilde{\langle\sigma v\rangle}$

The third and last factor in the fusion burn rate $p_f$ is $\widetilde{\langle\sigma v\rangle}$ which accounts for the deviation of the ion distribution from a Maxwellian. In a collisional plasma anticipated for a cryogenic ICF target, $\delta f_{D,T}$ and hence $\widetilde{\langle\sigma v\rangle}$ should be negligibly small unless the plasma is close to a hot/cold interface. The exception near the interface is due to the well-known Knudsen layer effect [15, 16] which comes about because the mean-free-path of a tail ion scales as $E^2$, so the fast ions from the hot plasma side tend to suffer free-streaming loss into the cold but dense fuel layer. Since fast ions in the Gamow peak (at $E_g \sim 3 - 4T_i$ for $T_i = 10$ keV, and even larger $E_g/T_i$ for lower $T_i$) contribute the most to $\langle\sigma v\rangle$, the tail ion loss can significantly degrade the fusion reactivity in the hot spot Knudsen layer next to the hot/cold interface. In other words, $\widetilde{\langle\sigma v\rangle}$ can be negative and has a magnitude that is a substantial fraction of $\widehat{\langle\sigma v\rangle}$ in a Knudsen layer of width $L_{Kn} = (E_g/T_i)^2\lambda_i$, where $\lambda_i$ is the mean-free-path of a thermal ion [17].

Similar to the fuel ion separation issue, the Knudsen reactivity reduction is less of a concern if the implosion is spherically symmetry for an ignition target in which $R_{hs} \gg L_{Kn} > \lambda_i$, since the volume affected $\sim 4\pi R_{hs}^2 L_{Kn}$ is small compared with the hot spot volume $(4/3)\pi R_{hs}^3$. Hydrodynamic mix can fundamentally alter this picture, as the interface area $\Sigma$ can grow much larger than $4\pi R_{hs}^2$ as mix progresses. The degree of mix can be characterized by a mix zone width $L_{\mathrm{mix}}$ and the typical width $l_{\mathrm{mix}}$ of the small scale hot plasma filament in the mix zone. For $L_{\mathrm{mix}} \sim R_{hs}$ and $l_{\mathrm{mix}} \sim L_{Kn}$, the Knudsen layer effect on fusion reactivity would be present throughout the hot spot. A subtlety arises for cryogenic targets where the escaping fast ions from hot plasma can fuse with the cold DT fuel layer, so while $\widetilde{\langle\sigma v\rangle}$ is negative on the hot plasma side of the mix interface, it can be positive on the other side [17]. This so-called inverse Knudsen layer effect on fusion reactivity in the cold region can substantially recover the lost reactivity in the hot region if the hot plasma has $T_i > 8-10$ keV and the cold fuel reaches $T_i = 1.5$ keV. If the hot plasma is of sub-ignition temperature, say $T_i = 1-2$ keV, the Knudsen layer reactivity reduction is the dominant mechanism. This is quantified by the solution of Fokker-Planck equation for a hot plasma surrounded by cold fuel, taking into account pitch angle scattering, slowing down, and energy scattering against background ions and electrons, and the ambipolar electric field [18, 19].

## 5. Summary

Hydrodynamic mix is intrinsically an interfacial phenomenon which can impact fusion yield rate through (1) enhanced thermal loss, (2) fuel ion separation, and (3) modification of fusion reactivity. These bring a number of plasma physics issues to the forefront in order to understand the impact of mix on burn. Among them, we have investigated (1) the subtle role of magnetic field in mitigating thermal loss and controlling mix itself; (2) the fundamental inter-species diffusive transport theory that underlies ion species separation by thermodynamic forces; and (3) the competition between Knudsen and inverse Knudsen effect on fusion reactivity in a mix-dominated cryogenic ICF target.

## Acknowledgments

We wish to thank numerous useful discussions with Steve Batha, John Benage, Bob Chrien, Jim Cooley, Hans Herrmann, Nelson Hoffman, Scott Hsu, Ray Leeper, Kim Molvig, Jas Mercer-Smith, and many colleagues at MIT, Univ. of Rochester, and LLNL. This work was supported by U.S. Department of Energy via the LANL LRDR program.

## References

[1] Atzeni S and Meyer-Ter-Vehn J 2004 *The physics of inertial fusion* (Oxford University Press, Inc.).
[2] Mima K, Tajima T, and Leboeuf J N 1978 Phys. Rev. Lett. **41** 1715-1719.
[3] Srinivasan B, Dimonte G, and Tang X Z 2012 Phys. Rev. Lett. **108** 165002.
[4] Srinivasan B and Tang X Z 2012 Phys. Plasmas **19** 082703.
[5] Kulsrud R M, Cen R, Ostriker J P, and Ryu D 1997 Ap. J. **480** 481-491.
[6] Srinivasan B and Tang X Z 2013 Phys. Plasmas **20** 056307.
[7] Tang X Z and Boozer A H 2000 Phys. Plasmas **7** 1113.
[8] Chandrasekhar S 1961 *Hydrodynamic and hydromagnetic stability* (Oxford University Press, Inc.).
[9] Amendt P, Landen O L, Robey H F, Li C K, and Petrasso R D 2010 Phys. Rev. Lett. **105** 115005.
[10] Casey D T, Frenje J A, Johnson M G, Manuel M J E, Rinderknecht H G, Sinenian N, Séguin G H, Li C K, Petrasso R D 2012 Phys. Rev. Lett. **108** 075002.
[11] Kagan G and Tang X Z 2012 Phys. Plasmas **19** 082709.
[12] Amendt P, Bellei C, and Wilks S C 2012 Phys. Rev. Lett. **109** 075002.
[13] Kagan G and Tang X Z 2012 Phys. Rev. Lett. **109** 269501.
[14] Kagan G and Tang X Z 2013 Thermo-diffusion in inertially confined plasmas *submitted for publication.*
[15] Henderson D B 1974 Phys. Rev. Lett. **33** 1142-1144.
[16] Molvig K, Hoffman N M, Albright B J, Nelson E M, and Webster B 2012 Phys. Rev. Lett. **109** 095001.
[17] Tang X Z, McDevitt C J, Guo Z, and Berk H 2013 Fusion yield recovery by escaping hot-spot fast ions in the neighboring fuel layer *submitted for publication.*
[18] Tang X Z, Berk H, Guo Z, and McDevitt C J 2013 Reduced Fokker-Planck models for fast particle distribution across a transition layer of disparate plasma temperatures *preprint* LA-UR-13-27676.
[19] McDevitt C J, Tang X Z, Guo Z, and Berk H 2013 A comparative study of tail ion distribution with reduced Fokker-Planck models *preprint* LA-UR-13-27598.

Journal of Physics: Conference Series **688** (2016) 012124 doi:10.1088/1742-6596/688/1/012124

# Development of position measurement unit for flying inertial fusion energy target

**R Tsuji[1], T Endo[2], H Yoshida[3] and T Norimatsu[4]**

[1] Faculty of Engineering, Ibaraki Univ., Hitachi, 316-8511, Japan
[2] Faculty of Engineering, Hiroshima Univ., Higashi-Hiroshima, 739-8527, Japan
[3] Faculty of Engineering, Gifu Univ., Gifu, 501-1193, Japan
[4] Institute of Laser Engineering, Osaka Univ., Suita, 565-0871, Japan
E-mail: tsuji@mx.ibaraki.ac.jp

**Abstract.** We have reported the present status in the development of a position measurement unit (PMU) for a flying inertial fusion energy (IFE) target. The PMU, which uses Arago spot phenomena, is designed to have a measurement accuracy smaller than 1 μm. By employing divergent, pulsed orthogonal laser beam illumination, we can measure the time and the target position at the pulsed illumination. The two-dimensional Arago spot image is compressed into one-dimensional image by a cylindrical lens for real-time processing. The PMU are set along the injection path of the flying target. The local positions of the target in each PMU are transferred to the controller and analysed to calculate the target trajectory. Two methods are presented to calculate the arrival time and the arrival position of the target at the reactor centre.

## 1. Introduction

In a direct drive inertial fusion energy (IFE) power reactor, a spherical fuel target is injected into a reactor chamber and irradiated by the driver laser beams. The target must be shot by the driver laser beams with an accuracy of less than 20 μm [1]. To calculate the arrival time and the arrival position of the flying IFE target at the reactor centre, we must measure and analyse the time data and the position data of a flying IFE target in real time.

In Japan, a demonstration experiment of the target injection is under way [2]. In the experiment, the target with a diameter of 3~5 mm with (or without) a cone will be injected at a speed of 100±1 m/s. Our goals are (1) to make a position measurement unit (PMU) with a position measurement accuracy of 1 μm and (2) to make a system that can calculate the trajectory of a flying IFE target in real time.

In this paper, we report the present status of the development of the position measurement unit and propose the methods to calculate the target trajectory.

## 2. Principle of position measurement and the conceptual design of the PMU

Position measurement methods using an Arago spot have been proposed to measure the position of a flying IFE target [3, 4, 5]. An Arago spot is a bright spot which appears at the centre of a shadow of a spherical object, as shown in figure 1. This is because the diffracted waves from the edge of a spherical object interfere constructively on the central axis. We demonstrated that a position measurement accuracy of less than 1 μm can be achieved for a stationary ball bearing [5].

Two orthogonal laser beam illuminations on the target gives a three-dimensional position of the flying target, $(X, Y, Z)$, as shown in figure. 2 [6]. If a pulsed laser with a pulse width smaller than 10 ns is used, then the flight distance of the target at laser illumination is smaller than 1 μm for a target velocity of 100 m/s. Use of a pulsed laser and a clock gives the time, $T$, of the laser illumination. Our PMU employs the Arago spot formed by the pulsed, orthogonal laser beam. A divergent beam was also employed [6]. Divergent beams magnify the shadow of the target and magnify the displacement of the target i.e. displacement of the Arago spot. It enabled accurate measurement of the Arago spot at a large distance.

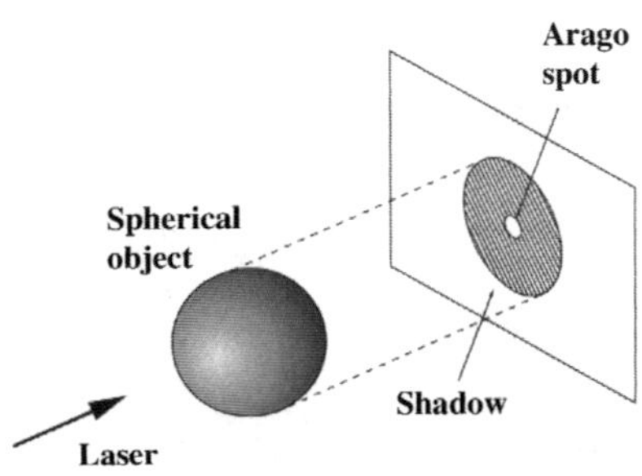


**Figure 1.** The Arago spot.

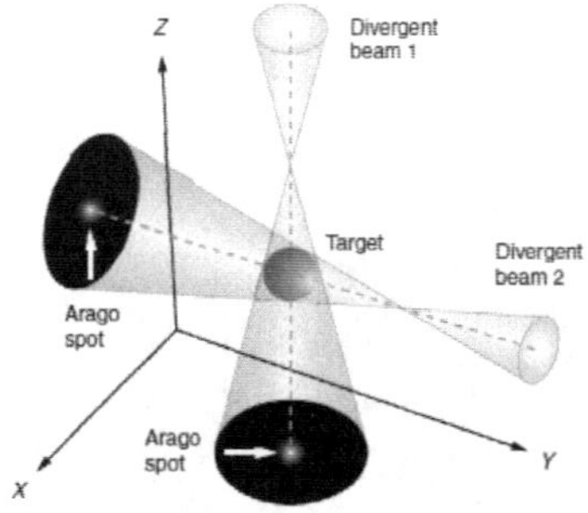


**Figure 2.** Three-dimensional position measurement.

A data reduction technique is essential for real-time data processing. Figure 3 shows the shadow and Arago spot. Two-dimensional imaging data has ~ MB (1000 × 1000 pixels). For simplicity of the discussion of the data reduction, it is assumed that the intensity distribution of the Arago spot can be approximated by the Gaussian function. Consider two-dimensional Gaussian function, as

$$F(x, y) = \exp(-x^2 - y^2) = \exp(-x^2)\exp(-y^2) . \quad (1)$$

Integrating (1), a one-dimensional Gaussian function can be obtained, as

$$G(x) = \int \exp(-x^2 - y^2)\mathrm{d}y = C\exp(-x^2) . \quad (2)$$

The above integration can be done physically by the cylindrical lens. Functions (1) and (2) have a maximum at $x = 0$. Figure 4 shows the compression of the Arago spot image by the cylindrical lens [7].

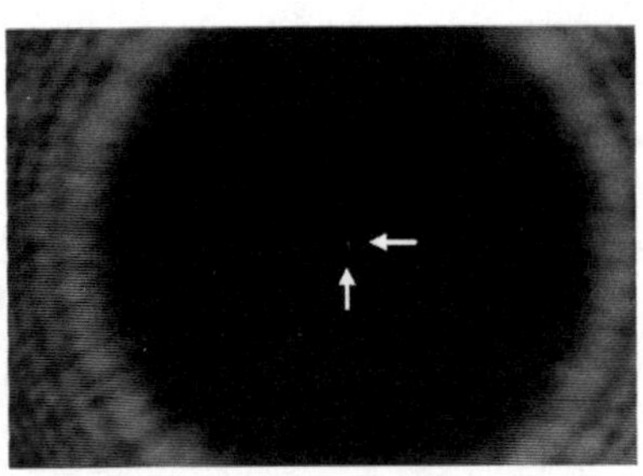

**Figure 3.** Shadow and the Arago spot (central point).

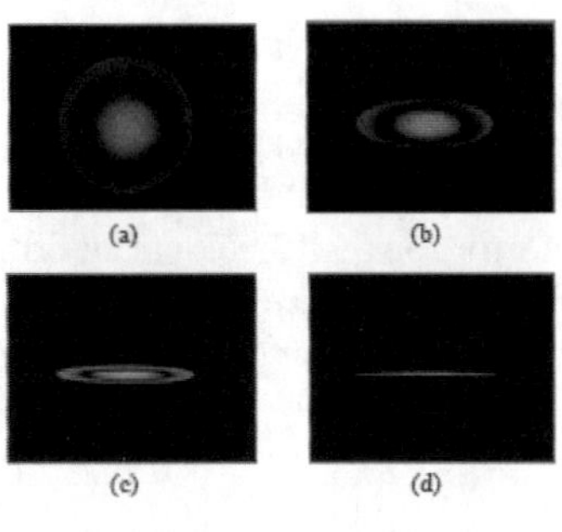


**Figure 4.** Compression of the Arago spot.

One-dimensional imaging data has ~ kB (1000 pixels), and it is easy to analyse one-dimensional data in real time. A linear sensor is placed at the focal point of the cylindrical lens, as shown in figure 5.

The conceptual schematic diagram of PMU is presented in figure 6. When a flying target crosses the PMU, a trigger signal is transmitted to the pulsed laser. A laser pulse is emitted and divided by the beam splitter and a flying target is irradiated by the orthogonal laser beams, which are reflected by the mirror. The Arago spot image is divided by the beam splitter and compressed by the cylindrical lens.

Journal of Physics: Conference Series **688** (2016) 012124 doi:10.1088/1742-6596/688/1/012124

The compressed Arago spot image is measured by the linear CMOS (or CCD) sensor. Measured position data are then transferred to the controller. And a part of the laser pulse is detected by the photo sensor to measure the time of illumination. The controller collects the time data and the position data in each PMU and calculates the arrival time and the arrival position of the target at the reactor centre. Some key technologies were tested but system integration of the PMU is under development.

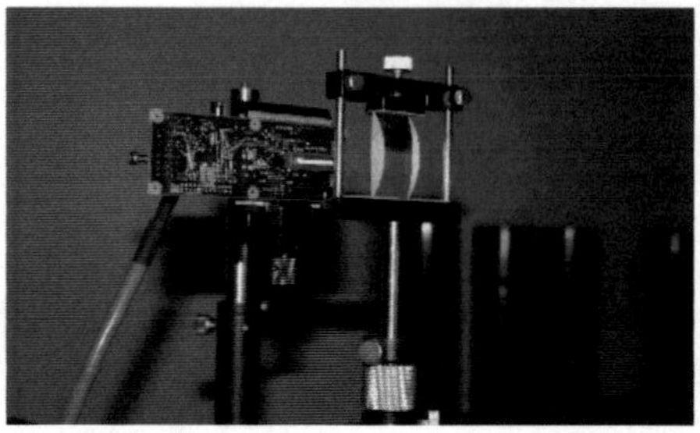

**Figure 5.** Cylindrical lens and sensor.

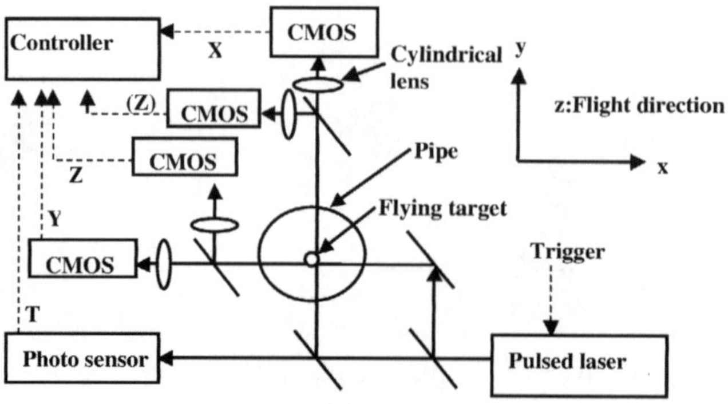


**Figure 6.** Schematic diagram of PMU.

## 3. Setting of the PMU

PMUs are set along the target injection path, as shown in figure 7. In the reactor chamber, the position of the target can be measured by the illumination of the pulsed divergent laser beam, as shown in figure 8. The Arago spot of the target is detected outside of the reactor chamber.

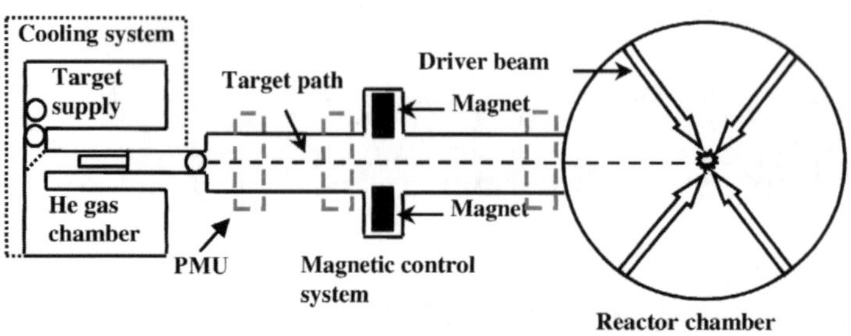


**Figure 7.** Setting of the PMUs (broken box).

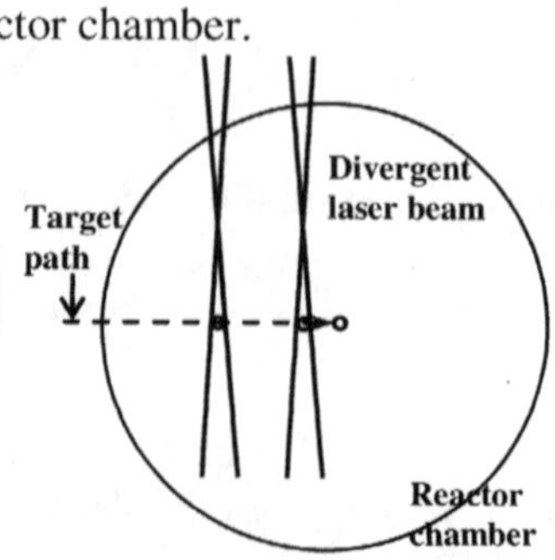


**Figure 8.** Measurement in reactor.

## 4. Determination of target trajectory

The PMU measures the local position of the target in each PMU with an accuracy of less than 1 μm. Is it possible to convert the local coordinates into a global coordinates? If all of the local coordinate centres of the PMUs are set at a scheduled position, as shown in figure 9(a), then the task of the data conversion is easy. However, it is difficult to set and keep PMUs at the scheduled position.

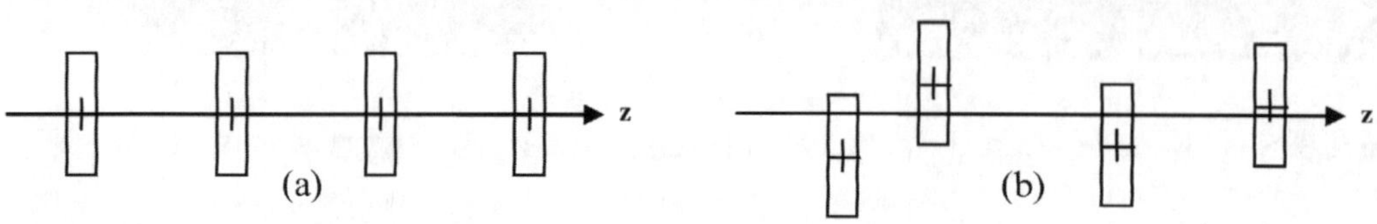


**Figure 9.** (a) Left: PMU setting with no error. (b) Right: Actual PMU setting with setting errors.

The actual setting is shown in figure 9(b). Even in this case, the setting error of PMU can be compensated as follows [8]. To show the principle of compensation simply, consider the target motion with a constant velocity. First, the z-axis is defined as the line between the centres of the first PMU and the last PMU, as shown in the solid line in figure 10(a). The target trajectory is represented as the dotted line. Assuming a constant velocity, the ratio of the time interval between PMUs equals the ratio of the distance between PMUs from the equation $Vdt = dZ$. Therefore, the z-coordinate of the centre of the PMU, $dZ_i$, is determined. If the data ($dZ_i$, $Y_i$) are plotted, a zigzag line appears, as in figure 10(b). Comparing the dotted line with the zigzag line, the setting error $dY_i$ can be determined.

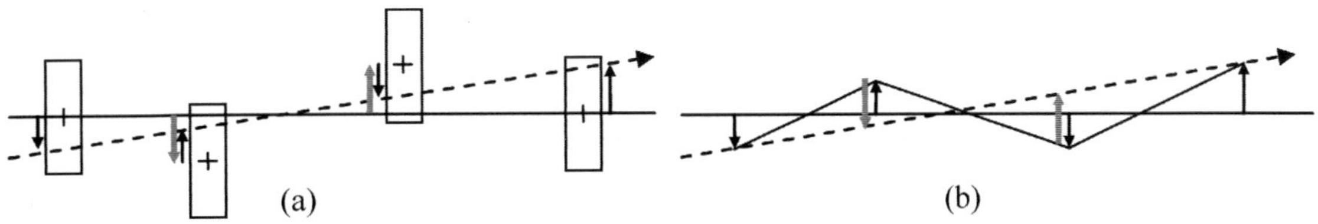


**Figure 10.** (a) Left: Actual setting of PMUs (box). The cross in the box represents the centre of the local coordinate system. The solid line represents the z-axis. The dotted line represents constant velocity motion. The thick arrow represents setting error $dY_i$. The thin arrow represents local position data $Y_i$. (b) Right: Plotting data.

### *4.1. Trajectory determination using equation of motion*

First, a heavy reference target is injected. The data ($T_i$, $X_i$, $Y_i$, $Z_i$) for the reference target in the i-th PMU are obtained. Comparing the data with a parabolic trajectory, the compensations ($dX_i$, $dY_i$, $dZ_i$) are determined. Then, the IFE target is injected. The local data ($T_{it}$, $X_{it}$, $Y_{it}$, $Z_{it}$) of the injected target are obtained and these data are converted into global data ($T_{it}$, $X_{it}+dX_i$, $Y_{it}+dY_i$, $Z_{it}+dZ_i$). In cases of no air resistance (in case of a dry wall reactor, HIB reactor, or target with cone), assuming the parabolic trajectory, the arrival time and the arrival position at the reactor centre can be determined [8]. In case where air resistance exists (in case of a spherical target in wet wall reactor), giving the pressure data and solving the equation of motion with air resistance, the arrival time and the arrival position can be determined, as well.

### *4.2. Trajectory determination as an optimization problem*

First, a few tens of IFE targets are injected sequentially and the data ($T_{ij}$, $X_{ij}$, $Y_{ij}$, $Z_{ij}$) for the j-th target in the i-th PMU are obtained. Approximate forms for the arrival time and the arrival position are constructed by minimizing the error at the reactor centre (optimization problem). Then, the IFE target is injected and the local data ($T_{it}$, $X_{it}$, $Y_{it}$, $Z_{it}$) of the injected target are obtained. Substituting these data into the approximate forms, the arrival time and the arrival position is determined.

## Acknowledgments

This work was supported by JSPS KAKENHI Grant Number 24561020.

## References

[1] Goodin D, Gibson C, Petzoldt R, Siegel N, Thompson L, Nobile A, Besenbruch G and Schultz K 2002 *Fusion Eng. Des.* **60** 27-36
[2] Norimatsu T, Endo T, Yoshida H, Tsuji R, Kajimura Y, Nagata M, Matsumoto M and Azechi H Release of Real Size, Fast Ignition Target from Sabot: IFSA 2013 Nara (Nara, Japan, 8-13 Sept. 2013) O.Th_C10
[3] Petzoldt R, Alexander N, Carlson L, Flint G, Goodin D, Spalding J and Tillack M 2007 *Fusion Sci. Technol.* **52** 454-58
[4] Carlson L, Tillack M, Lorentz T, Spalding J, Alexander N, Flint G, Goodin D and Petzoldt R 2007 *Fusion Sci. Technol.* **52** 478-82
[5] Saruta K and Tsuji R 2007 *Jpn. J. Appl. Phys.* **46** 6000-06
[6] Saruta K and Tsuji R 2008 *Jpn. J. Appl. Phys.* **47** 1742-44
[7] Sakauchi H and Tsuji R 2009 *Plasma Fusion Res.* **4** S1012
[8] Tsuji R, Endo T, Yoshida H, Matsumoto M and Norimatsu T 2012 Trajectory Calculation of Injected Inertial Fusion Energy Target: Conference of Inertial Fusion Energy CIFE 2012 (Yokohama, Japan, 25-27 Apr. 2012) CIFE5-3

# Efficient neutron generation from solid-nanoparticle explosions driven by DPSSL-pumped high-repetition rate femtosecond laser pulse

**T Watari, K Matsukado, T Sekine, Y Takeuchi, Y Hatano, R Yoshimura, N Satoh, K Nishihara, M Takagi and T Kawashima**
Hamamatsu Photonics, K.K., 1820 Kurematsu, Hamamatsu 431-1202 Japan

E-mail: takeshi.watari@lsr.hpk.co.jp

**Abstract.** We propose novel neutron source using high-intensity laser based on the cluster fusion scheme. We developed DPSSL-pumped high-repetition-rate 20-TW laser system and solid nanoparticle target for neutron generation demonstration. In our neutron generation experiment, high-energy deuterons were generated from coulomb explosion of CD solid-nanoparticles and neutrons were generated by DD fusion reaction. Efficient and stable neutron generation was obtained by irradiating an intense femtosecond laser pulse of $>2\times10^{18}$ W/cm$^2$. A yield of $\sim10^5$ neutrons per shot was stably observed during 0.1-1 Hz continuous operation.

## 1. Introduction

Recent developments of high-intensity laser enable us to evolve a new type neutron source. A number of experiments [1-5] demonstrated the possibility of laser-driven fusion using pure $D_2$ or $CD_4$ clusters. In these works, multi-keV deuterium ions were generated by Coulomb explosion of a few nanometres clusters. Since the cross-section of deuterium-deuterium (DD) reaction reaches its maximum at 1750 keV with center-of-mass system, much higher ion energy is required for efficient neutron generation.

Features of the Coulomb explosion are that ion energy distribution function is proportional to the square root of its energy and the maximum energy is proportional to its radius [6,7]. Larger particles result in higher ion energy, although an intense laser irradiation is required for expelling electrons from the larger particles. We fabricated the solid density deuterated-polystyrene (CD) nanoparicles of 250 nm in diameter to obtain high ion energy of ~1 MeV. Stable neutron generation was demonstrated by irradiating 20-TW diode-pumped-solid-state-laser (DPSSL) to these solid nanoparticles.

## 2. Experimental setup

### *2.1. Laser system*

In our laboratory, we have developed a DPSSL-pumped high-repetition-rate 20-TW laser (MATSU-1) and a pump laser KURE-1, a laser-diode-pumped Nd:glass laser system [8]. The KURE-1 is single longitudinal and transverse mode, nano-second pulse width, 10Hz repetition rate, diode-pumped green laser system based on thermally-edge-controlled zigzag slab amplifier techniques. Maximum output energy is 21.4 J and 12.7 J in 1053 nm and 527 nm, respectively. MATSU-1 consist of Ti:sapphire oscillator, pulse stretcher, acousto-optic programmable dispersive filter, regenerative amplifier, 1st main amplifier for 2 TW output, 2nd main amplifier for 20 TW output and vacuumed pulse compressor. The regenerative amplifier and the 1st amplifier are pumped by two diode-pumped Nd:YAG lasers. In the 1st amplifier, seed pulse output from the regenerative amplifier is amplified to 150 mJ with 6 pass amplification by 450 mJ pump energy. In the 2nd amplifier, seed pulse is boosted-up to 1.5 J by 4 pass amplification with 5 J pumping energy from KURE-1 laser. The seed pulse is compressed to typical 60 fs by the vacuumed pulse compressor with 80% of transmissivity. MATSU-1 generate 1.2 J ($\lambda$ = 800 nm) with a pulse width 60 fs at 10 Hz. DPSSL is extremely stable at a longer life compared with conventional flash-lamp-pumped laser. In the focus spot of the MATSU-1 laser, 40 percent of the energy was focused into 8 μm diameter spot. Considering the laser transmission efficiency, the laser intensity on the target was ~$4\times10^{18}$ W/cm$^2$. The repetition rate was 0.1-1 Hz due to the limitation of data acquisition.

*2.2. Target and injection system*

We developed CD nanoparticles used as a target instead of gas clusters [9]. Figure 1(a) shows the SEM image of the nanoparticles which average size was 280 nm. The average size can be controlled with high precision between 150-700 nm. We, therefore, expect to give a certain amount of selectivity to the ion energy obtained by Coulomb explosion of these nanoparticles. Figure 1(b) shows the schematic view of the nanoparticle injection system. MATSU-1 enters from the direction perpendicular to the paper. Injection timing and number density of the nanoparticles were measured by the Mie scattering light intensity using He-Ne laser probe. The nanoparticles are injected to the MATSU-1 focus point at the center of interaction chamber. In this experiment, typical number density of injected nanoparticles was ~$10^9$ cm$^{-3}$.

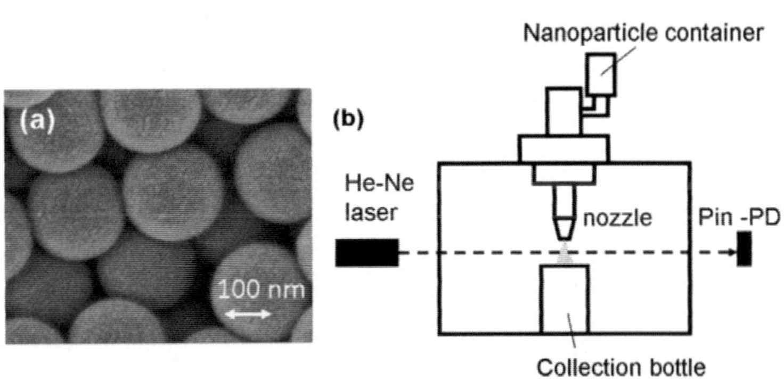


Figure 1. (a) SEM image of CD nanoparticles which average size was 250 nm. (b) Schematic view of the nanoparticle injection system.

*2.3. Target and injection system*

We have developed a Multichannel Time-of-flight system with Kinetic energy Analyzing Device (MT-KAD) to observe the ions of multispecies generated from relativistic laser plasma [10]. In this experiment, we used MT-KAD system to observe the ions generated from Coulomb explosion of CD nanoparticles. This system is composed of a ten-channel scintillation detector array and a permanent magnet that generates a magnetic field of ~0.09 T. The magnetic field and time-of-flight (TOF) information enables us to distinguish protons, deuterons, and stripped carbons. The layout of our detectors is shown in Fig. 2. The MT-KAD system constructed in the detector chamber and observed ions accelerated forward. Neutron detector, consisting of a plastic scintillator (BC420, 180 mm diameter, 100 mm long) coupled to a photomultiplier tube (H2431-50, Hamamatsu Photonics K.K.) was located outside the interaction chamber at distance of 140 cm from the laser focus point.

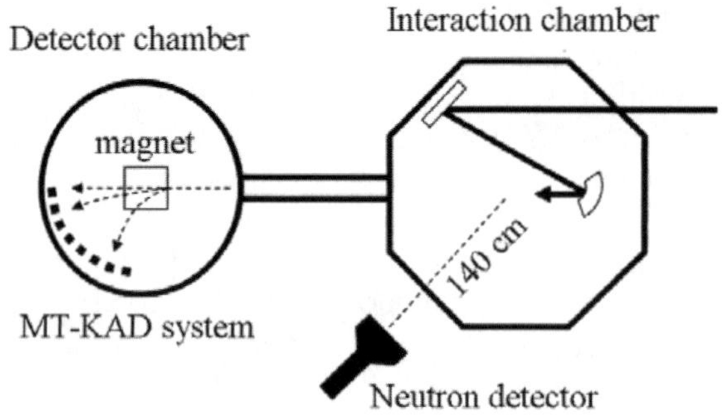


Figure 2. The layout of the detectors

**3. Experimental setup**

At first, we measured the deuterons generated by Coulomb explosion. Figure 3 shows the typical ion TOF signal of a single shot with CD nanoparticles. These signals were obtained by three of ten-channel detectors in detector chamber. Generation of high energy deuterons with energy up to 1.5 MeV was observed. On the other hand, no deuteron signal was observed in CH nanoparticles but only high-energy proton (up to 1.4 MeV) was observed. These observations clearly indicate that high-energy ions were generated by Coulomb explosions of solid-nanoparticles.

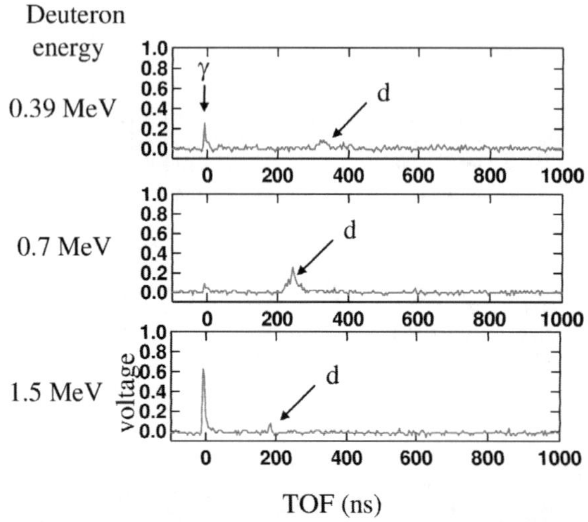


Figure 3. The typical ion TOF signal of a single shot

In the neutron generation experiment, we used a cylindrical CD plane, called "neutron booster", in the interaction chamber. It was arranged around a nanoparticle injection nozzle to enhance the fusion reactions (See Fig.4(a)). High energy deuterium ions generated by Coulomb explosion might hit the plane and increase DD fusion reactions.

A typical signal of a neutron detector in the neutron generation experiment is shown in Fig.4(b). The signals due to the γ-ray and neutron bursts are easily resolvable. The relative delay with > 60 ns corresponds to the sum of the traveling time of deuterium inside the booster and the TOF of 2.45-MeV neutron to the detector. Assuming the neutron isotropic emission, the neutron yield was ~$10^5$ per shot. Such an efficient neutron generation was stably observed during continuous 100 shots (See Fig. 4(c)).

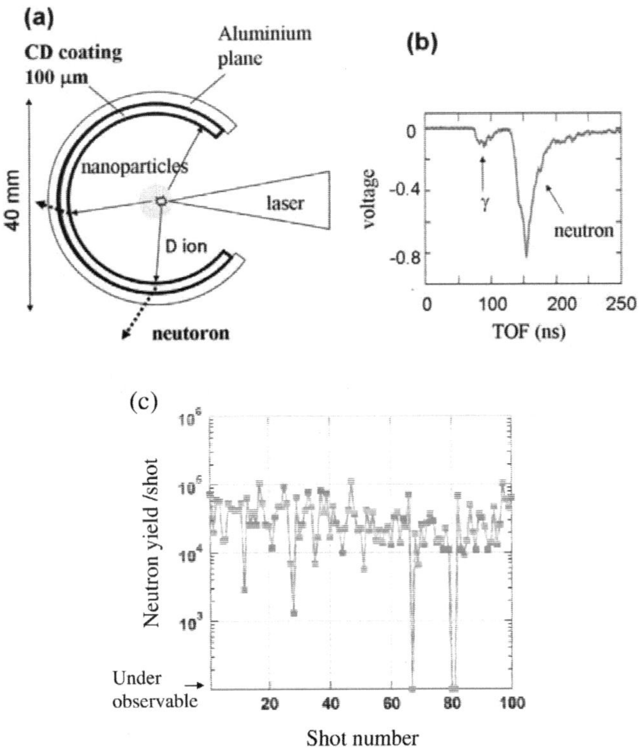


Figure 4. (a) Schematic view of the "neutron booster". (b) Single shot neutron TOF signal. (c) The history of neutron generation during continuous 100 shots.

## 4. Conclusion

Neutron generation experiment with 20-TW LD-pumped laser and CD nanoparticle was demonstrated. Energetic deuterium ions with energy up to 700 keV were observed. A yield of ~$10^5$ neutrons per shot was stably observed during 100 shots of 0.1-1 Hz continuous operation. This indicates that it may be possible to further enhance the neutron generation with an increase in laser energy. This demonstration will be expected to bring the progress of novel neutron source, which is characterized by a pulsed emission and point source, for many applications.

## References

[1] Ditmire T et al. 1999 Nature **398** 489
[2] Ditmire T et al. 2000 Phys. Plasmas **7** 1993
[3] Madison K W et al. 2003 J. Opt. Soc. Am. B **20** 113
[4] Madison K W et al. 2004 Phys. Rev. A **70** 053201
[5] Hartke R et al. 2005 Nucl. Instrum. Methods Phys. Res. A **540** 464
[6] Nishihara K et al. 2001 Nucl. Instrum. Methods Phys. Res. A **464** 98
[7] Nishihara K et al. 2013 Laser ion acceleration and neutron source in short-pulse solid-nanoparticle interaction *This proceeding* (to be published)
[8] Yasuhara R et al. 2008 Opt. Lett. **33** 1711
[9] Tsukigase A, H. Nakamura 2006 Kobunshi Ronbunshu, **63** 266
[10] Matsukado K et al. 2010 Rev. Sci. Instrum. **81** 023304

# Relativistic intensity laser interactions with low-density plasmas

**L. Willingale[1], P. M. Nilson[2], C. Zulick[1], H. Chen[3], R. S. Craxton[2], J. Cobble[4], A. Maksimchuk[1], P. A. Norreys[5], T. C. Sangster[2], R. H. H. Scott[5], and C. Stoeckl[2]**

[1] Center for Ultrafast Optical Science, University of Michigan, 2200 Bonisteel Boulevard, Ann Arbor, Michigan 48109 USA
[2] University of Rochester - Laboratory for Laser Energetics, Rochester, New York 14623, USA
[3] Lawrence Livermore National Laboratory, Livermore, California 94551, USA
[4] Los Alamos National Laboratory, New Mexico, USA
[5] Central Laser Facility, Rutherford Appleton Laboratory, Oxfordshire, UK

E-mail: wlouise@umich.edu

**Abstract.** We perform relativistic-intensity laser experiments using the Omega EP laser to investigate channeling phenomena and particle acceleration in underdense plasmas. A fundamental understanding of these processes is of importance to the hole-boring fast ignition scheme for inertial confinement fusion. Proton probing was used to image the electromagnetic fields formed as the Omega EP laser pulse generated a channel through underdense plasma. Filamentation of the channel was observed, followed by self-correction into a single channel. The channel radius as a function of time was found to be in reasonable agreement with momentum-conserving snowplough models.

## 1. Introduction

The highly non-linear interaction of a high-intensity laser propagating through an underdense plasma leads to ponderomotive and relativistic self-focusing [1, 2], filamentation [3, 4], scattering [5], soliton formation [6, 7], channel formation [8, 9] and in extreme conditions complete cavitation [4]. As the laser pulse propagates through an underdense plasma, the ponderomotive force acts to expel the electrons from the regions of highest intensity. A channel, depleted of electrons, forms on the laser propagation axis. The ions, having a much smaller charge-to-mass ratio, are not affected directly by the laser fields, but will move due to the charge separation electric field. The laser sweeps out an approximately cylindrical region, producing a radial electric fields, that accelerates ions in the transverse direction in a Coulomb explosion [10, 11]. The creation of a channel may be used to access the dense fuel core for fast ignition inertial confinement fusion [12], where a laser is used to channel through millimeter-scale underdense plasma [13]. If the ponderomotive force is strong enough, a shock front can propagate out radially, potentially accelerating the ions to higher energies [14] and forming a blast wave [15]. The channel can facilitate Direct Laser Acceleration of the electrons to many times the ponderomotive energy [16, 17, 18, 19] and lead to bright x-ray generation [20].

This experiment combined a relatively long pulse length ($\sim$ 10 ps) with a high power ($\approx$ 90 TW), approaching the expected requirements for fast ignition [13]. Proton probing [21]

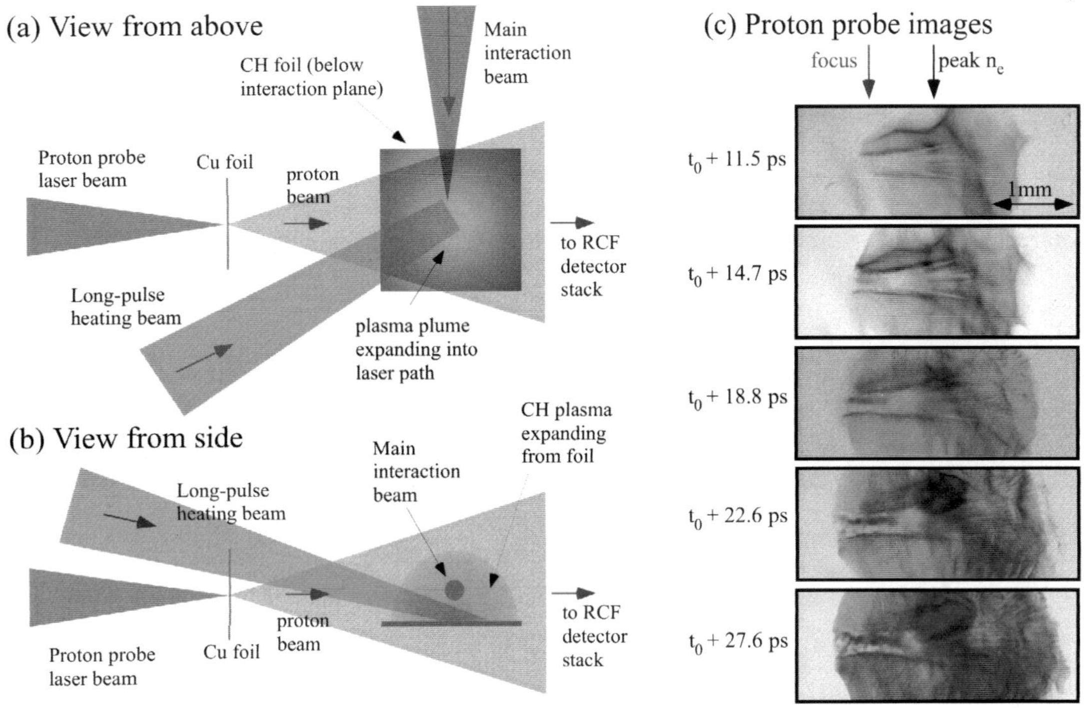


**Figure 1.** Schematic of the experimental configuration for the proton radiography of the underdense plasma interaction: (a) the view from above and (b) the view from the side. (c) Proton probe images showing a sequence times with the laser pulse traveling from left to right. The contrast of the images has been adjusted to show the features.

was used to image channel formation through underdense plasma [22, 23]. This paper presents additional data from the experiments described in references [19, 24], where the evolution of the electromagnetic fields in the channel was measured using proton radiography. Specifically, we consider the channel radius as a function of time.

## 2. Experimental set-up

The experiment was performed at the Laboratory for Laser Energetics using the Omega EP laser system. A single long-pulse laser was used to create an underdense plasma target. Two short-pulse beams provided the main interaction pulse and a pulse to generate a proton probe beam. A schematic showing the experimental configuration is shown in figure 1 with views both (a) from above and (b) from the side. The long-pulse beam (2.5 ns duration, wavelength of 351 nm ($3\omega$), $1160 \pm 110$ J energy, 800 $\mu$m diameter focal spot) generated a plasma plume from a planar plastic (CH) target. The two-dimensional hydrodynamic code SAGE [25] determined the plasma electron density profile along the main interaction pulse propagation path to be approximately gaussian (half-width = 650 $\mu$m) with a peak plasma density of $n_e = (5 \pm 1) \times 10^{19}$ cm$^{-3}$ (5% of the critical density). The main interaction pulse ($8.2 \pm 0.4$ ps, $743 \pm 11$ J, giving a pulse power of $\approx$ 90 TW) was focused by an $f/2$ off-axis parabola to a spot where 80% of the energy was contained in a radius of 22 $\mu$m. The focal plane was at a density of $n_e = 2.5 \times 10^{19}$ Wcm$^{-2}$, this was 0.8 mm before the peak density. The peak vacuum intensity was $(2.8 \pm 0.7) \times 10^{19}$ Wcm$^{-2}$, corresponding to a peak normalized vector potential of $a_0 = 4.7 \pm 0.6$. For creating the proton radiography beam, a third pulse ($\approx$ 0.9 ps, $\approx$ 236 J, wavelength of 1.053 $\mu$m) was focused

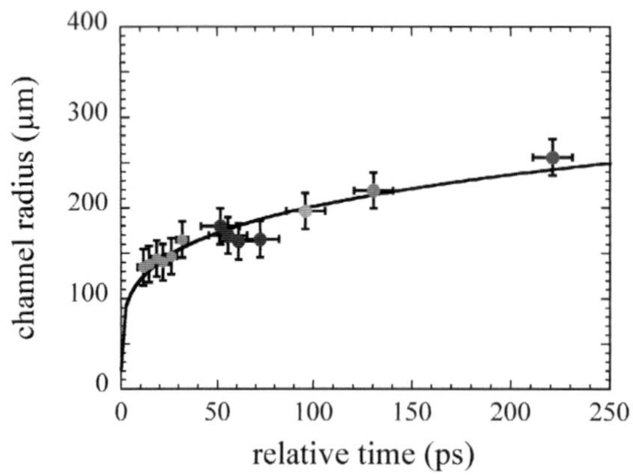


**Figure 2.** The channel radius as a function of time with the different color markers indicating the different shots used. The line has the function $R = r_0 + \beta t^\alpha$.

onto a 50 $\mu$m thick copper foil. Target normal sheath acceleration [26] produced an excellent quality probe beam possessing qualities such as a small virtual source size ($\sim$ $\mu$m) [27] and good laminarity [28] providing good spatial resolution. The proton generation time was on the order of the laser pulse length ($\sim$ 1 ps) providing good temporal resolution [21]. Since the time it takes the protons to travel from the source foil to the interaction is dependent upon the proton energy, different radiochromic film (RCF) layers will be imaging different times in the interaction, creating an image sequence. To observe different times in the channel evolution, the timing between the main interaction pulse and the proton generating pulse was changed. Further details of the set-up can be found in references [19, 24].

## 3. Experimental results

Figure 1 (c) shows proton probe images of the channel formation during the interaction. The time, $t_0$, is defined as when the leading edge of the main interaction pulse reached half-maximum intensity at focus. The strong radial electric fields within the central region deflect the proton probe out of the channel. Examples of channel images from later times are presented in reference [24]. The channel radii from several different shots were analyzed at a position of 300 $\mu$m after the focal plane, and are plotted in figure 2. The different shots are illustrated as different marker colors. The fit $R = r_0 + \beta t^\alpha$ was made to the experimental data as shown in figure 2, where $r_0 = 22$ $\mu$m was assumed to be the focal spot radius and the factors $\beta = 54.6 \pm 5.5$ and $\alpha = 0.26 \pm 0.02$ were determined. Momentum-conservation snowplough models predict $\alpha = 1/3$ for a spherical geometry and $\alpha = 2/5$ for a cylindrical geometry [6]. Therefore the channel expansion was in reasonable agreement with the expected scaling.

In the earliest image, the laser has already propagated a significant distance from the left of the image. But further in to the plasma the channel had broken up into a number of filaments, suggesting that the laser had broken up into several individual filaments [3, 4]. The short focal length parabola ($f/2$) and the imperfect focal spot would assist filamentation. Even though the front of the laser pulse may break up and create a number of individual channels, the pulse was long enough that the excess plasma forming the filamentary channels was expelled by the subsequent part of the laser pulse. Therefore the filaments merge and self-correct into a single channel. This self-correction of the channel has been previously observed in simulations [13]. The filaments merged at a speed of $\approx 0.1c$. A dark region in the center of the channel appears after the initial stages of channel formation and progresses through the channel at a speed of

$\approx 0.12c$, and this likely indicates a region of field inversion inside the channel [22].

## 4. Summary

In conclusion, proton probing was used to image the interaction of a high-energy, high-power laser pulse through millimeter-scale plasma with a peak density of $5 \times 10^{19}$ cm$^{-3}$. The radial channel expansion driven by the ponderomotive potential was in reasonable agreement with the momentum conservation snowplough models [6]. The laser undergoes filamentation, leading to the formation of a number of individual channels. However, providing the laser duration was sustained, the channel eventually self-corrects into a single hollow structure. The finding that filamentation instabilities can be corrected is promising for hole-boring fast ignition schemes.

## Acknowledgments

The authors gratefully acknowledge technical assistance from the staff of the Laboratory for Laser Energetics. This material is based upon work supported by the Department of Energy National Nuclear Security Administration under Award Number DE-NA0002028. Work performed by LLNL under the auspices of U.S. DOE under contract DE-AC52-07NA27344.

## References

[1] Sprangle P, Tang C-M, and Esarey E 1987 *IEEE Trans. Plas. Sci.* **PS-15**, 145
[2] Borisov A B, Borovskiy A V, Korobkin V V, Prokhorov A M, Rhodes C K, and Shiryaev O B 1990 *Phys. Rev. Lett.* **65**, 1753
[3] Najmudin Z, Krushelnick K, Tatarakis M, Clark E L, Danson C N, Malka V, Neely D, Santala M I K, Dangor A E 2003 *Phys. Plas.* **10**, 438
[4] Nilson P M, *et al.* 2010 *New J. Phys.* **12**, 045014
[5] Forslund D W, Kindel J M, and Lindman E L 1975 *Phys. Fluids* **18**, 1002
[6] Naumova N M, Bulanov S V, Esirkepov T Zh, Farina D, Nishihara K, Pegoraro F, Ruhl H, and Sakharov A S 2001 *Phys. Rev. Lett.* **87**, 185004
[7] Romagnani L, *et al.* 2010 *Phys. Rev. Lett.* **105**, 175002
[8] Borghesi M, MacKinnon A J, Barringer L, Gallard R, Gizzi L A, Meyer C, and Willi O 1997 *Phys. Rev. Lett.* **78**, 879
[9] Sarkisov G S, Bychenkov Yu V, Novikov V N, Tikhonchuk V T, Maksimchuk A, Chen S-Y, Wagner R, Mourou G, and Umstadter D 1999 *Phys. Rev. E* **59** 7042
[10] Burnett N H and Enright G D 1990 *IEEE J. Quant. Elec.* **26**, 1797
[11] Krushelnick K *et al.* 1999 *Phys. Rev. Lett.* **83**, 737
[12] Tabak M, Hammer J, Glinsky M E, Kruer W L, Wilks S C, Woodworth J, Campbell E M, Perry M D, and Mason R J 1994 *Phys. Plas.* **1**, 1626
[13] Li G, Yan R, Ren C, Wang T-L, Tonge J, and Mori W B 2008 *Phys. Rev. Lett.* **100**, 125002
[14] Wei M-S *et al.* 2004 *Phys. Rev. Lett.* **93**, 155003
[15] Nilson P M *et al.* 2009 *Phys. Rev. Lett.* **103**, 255001
[16] Pukhov A, Sheng Z M, and Meyer-ter-Vehn J 1998 *Phys. Rev. Lett.* **5**, 1880
[17] Mangles S P D *et al.* 2005 *Phys. Rev. Lett.* **94**, 245001
[18] Arefiev A V, Breizmann B N, Schollmeier M, and Khudik V N 2012 *Phys. Rev. Lett.* **108**, 145004
[19] Willingale L *et al.* 2013 *New J. Phys.* **15**, 025023
[20] Kneip S *et al.* 2008 *Phys. Rev. Lett.* **100**, 105006
[21] Borghesi M, Schiavi A, Campbell D H, Haines M G, Willi O, Mackinnon A J, Patel P, Galimberti M and Gizzi L A 2003 *Rev. Sci. Inst.* **74**, 1688
[22] Kar S *et al.* 2007 *New J. Phys.* **9**, 402
[23] Sarri G *et al.* 2010 *Phys. Plas.* **17**, 113303
[24] Willingale L *et al.* 2011 *Phys. Rev. Lett.* **106**, 105002
[25] Craxton R S, and McCrory R L 1984 *J. Appl. Phys.* **56**, 108
[26] Hatchett S P *et al.* 2000 *Phys. Plas.* **7**, 2076
[27] Borghesi M, Mackinnon A J, Campbell D H, Hicks D G, Kar S, Patel P K, Price D, Romagnani L, Schiavi A, and Willi O 2004 *Phys. Rev. Lett.* **92** 055003
[28] Cowen T E, *et al.* 2004 *Phys. Rev. Lett.* **92**, 204801
[29] Bulanov S V, Pegoraro F 2002 *Phys. Rev. E* **65**, 066405

# Effects of the irradiation of a finite number of laser beams on the implosion of a cone-guided target

**T Yanagawa[1], H Sakagami[2], H Nagatomo[3], A Sunahara[4]**

[1]Department of Physics, Nagoya University, Furo-cho, Chikusa-ku, Nagoya, Aichi 464-8602, Japan

[2]Fundamental Physics Simulation Division, National Institute for Fusion Science, Oroshi-cho, Toki, Gifu 509-5292, Japan

[3]Institute of Laser Engineering, Osaka University, Suita, Osaka 565-0871, Japan

[4]Institute for Laser Technology, Suita, Osaka 565-0871, Japan

E-mail: yanagawa.takumi@nifs.ac.jp

**Abstract**. In direct drive laser fusion, the non-uniformity of the laser absorption on the target surface caused by the irradiation of a finite number of laser beams is a sever problem. GekkoXII laser at Osaka University has twelve laser beams and is irradiated to the target with a dodecahedron orientation, in which the distribution of the laser absorption on the target surface becomes non-uniform. Furthermore, in the case of a cone-guided target, the laser irradiation orientation is more limited. In this paper, we conducted implosion simulations of the cone-guided target based on GekkoXII irradiation orientation and compared the case of using the twelve beams and nine beams where the three beams irradiating the cone region are cut. The implosion simulations were conducted by a three-dimensional pure hydro code.

## 1. Introduction

In laser fusion, a fuel capsule is irradiated by many high intense lasers and imploded to so high density and temperature that nuclear fusion reactions occur. To achieve both high density and temperature at the same instant is very difficult and requires the high uniformity of the implosion. Thus, in fast ignition (FI) scheme [1], that is one of the schemes to achieve laser fusion, the implosion process is used only for the compression of the target, and then, the target is heated by an external source. In Fast Ignition Realization EXperiment (FIREX) at Osaka University, a hollow Au cone is inserted to the target and a heating laser is irradiated into the cone [2]. In this scheme, the laser light interacts the tip of the Au cone, which generates many high energetic electrons and then, these high energetic electrons heat the compressed fuel core. A FI experiment at 2002 showed 1000-fold increase in neutron yield compared with the case without heating, which also showed that the temperature is raised by the heating laser [3].

The fuel core is efficiently heated by using the cone-guided target, but there are some problems in the implosion process due to the presence of the cone. One problem is about the irradiation orientation of the implosion laser. The laser orientation is restricted so that the cone may not be irradiated and ablated. In FIREX, GekkoXII is used for the implosion, which has twelve laser beams designed to irradiate into the target with a dodecahedron orientation, but in many case, the three beams that have the possibility to irradiate the cone surface are not used (called as nine-beam implosion) [4]. In the case of the nine-beam implosion, the target is imploded more asymmetrically, which leads to the inefficient compression of the fuel.

So far many implosion experiments and simulations of the cone-guided target have been conducted, where the position of the fuel core formed by the asymmetric implosion, the interaction with the Au cone and core, and the collapse of the cone are investigated [5-7]. In this paper, focusing on the effects of the irradiation orientation of GekkoXII, the density decrease and implosion dynamics in the case of the nine-beam implosion are investigated and discussed comparing with a twelve-beam implosion case that uses all laser beams of GekkoXII; it is more symmetric implosion than the nine-beam case.

## 2. Simulation conditions

In order to investigate the effects of GekkoXII irradiation orientation on the cone-guided implosion, we used a pure hydro code IMPACT-3D [8]. IMPACT-3D is Eulerian code and solves compressible and inviscid hydro equations in three-dimensional Cartesian coordinates. The target is assumed as a simple two-layer target that consists of a shell and fuel as shown in figure 1. The inner radius and outer radius are 130 μm and 180 μm respectively. The cone is set as the full opening angle is 30 degree, the apex is located on the target center, and the distance between the tip and the target center is 20 μm. The cone is treated as a solid boundary in this simulation, which means that the cone never deformed. To compute the fluid with the cone boundary embedded, immersed boundary (IB) method is introduced into IMPACT-3D. The IB method was developed by Peskin and has received many attentions in computational fluid dynamics [9, 10]. This method allows computing flow around the embedded boundary without using the boundary fitted coordinate.

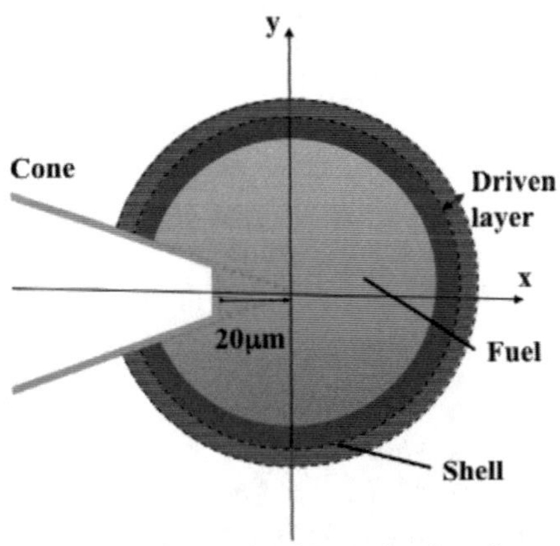


Figure 1. The schematic of the target. The inner radius and outer radius are 130 μm and 180 μm respectively. The cone is set as the full opening angle is 30 degree, the apex is located on the target center, and the distance from the tip to the target center is 20 μm. The shaded area in the shell is a driven layer.

IMPACT-3D is pure hydro code that does not include laser absorption. Therefore, the implosion is mimicked by pressure drive. A high-pressure region of 20 μm width called as driven layer (*see also figure 1*) is imposed in the shell, which accelerates the shell inward. Adding the pressure distribution on the driven layer, we simulate the cone-guided implosion with the GekkoXII irradiation orientation. The pressure distribution is defined as proportional to the laser absorption distribution that is computed by three-dimensional raytracing, in which we assume the target as plasma and absorption process as inverse bremsstrahlung. The laser beam diameter is the same size to the target diameter, the f-number is infinite, and the profile in space is uniform. Figure 2(a) and 2(b) are the pressure distribution of the driven layer in the case of the twelve-beam and nine-beam without the cone respectively. The cone is inserted into the "cone area" as shown in figure 2. The initial density of the fuel and shell are 1 g/cc and 45 g/cc, and the pressure of the fuel and shell except the driven layer are 1 Mbar. This initial profile is set so that the radial convergence ratio is achieved as approximately 3.5.

Journal of Physics: Conference Series **688** (2016) 012127 doi:10.1088/1742-6596/688/1/012127

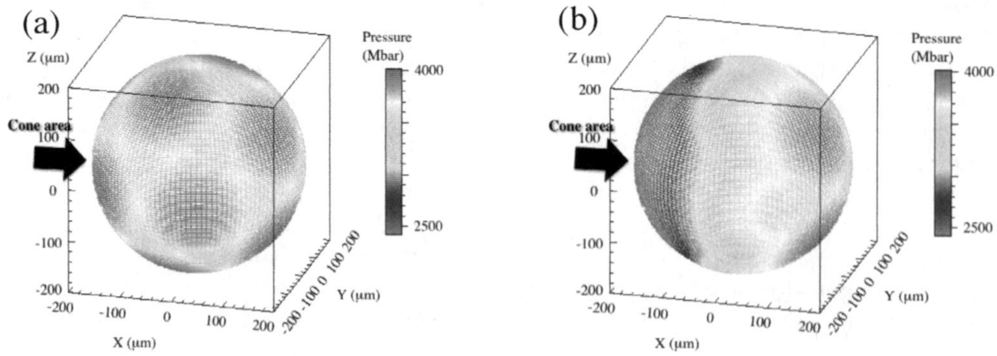


Figure 2. The pressure distribution on the driven layer.
(a) Twelve-beam case. (b) Nine-beam case.

## 3. Results and discussion

Figure 3(a) and 3(b) show the density iso-surface corresponding to the fuel-shell interface near the maximum compression time in the case of the twelve-beam and nine-beam respectively. It is found that the irregularity of the target surface caused by the irradiation non-uniformity of the implosion laser is emphasized by Rayleigh-Taylor (RT) instability in the twelve-beam case. On the other hand, in the case of the nine-beam, the target is crushed in toward the cone, which means that the fuel flow toward the cone side occurs. Figure 4 shows the time evolution of the gravity point of the fuel region. The red and blue lines are in the case of the twelve-beam and nine-beam respectively. In the case of the nine-beam, the gravity point moves toward the minus direction of the x-axis, which is corresponding to the direction toward the cone side, unlike the twelve-beam case where the gravity point keeps at the same position. It is found that the strong fuel flow toward the cone side occurs in the case of the nine-beam. In the iso-surface figure of the nine-beam case, a tumor like shape appears in front of the cone tip. This tumor like shape was formed even in the case without the cone. Furtherer analysis is required to explain the mechanism to form such a tumor. However, the tumor size is very small and it has little effect on the implosion.

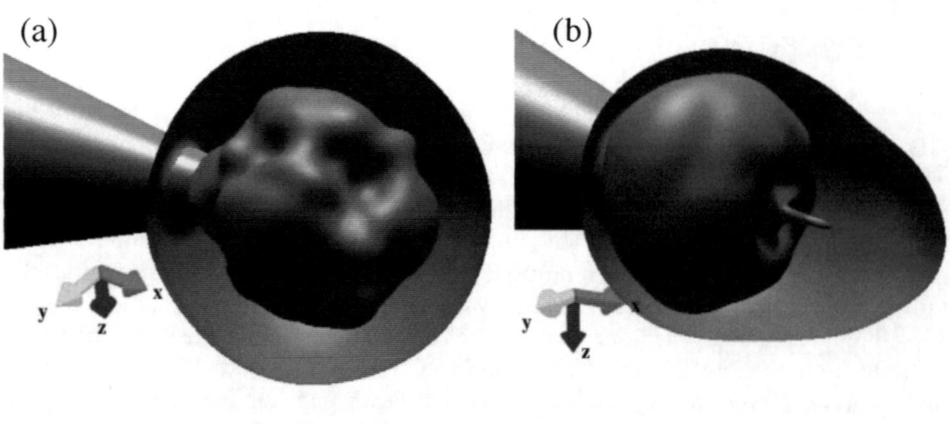


Figure 3. The density iso-surface corresponding to the fuel-shell interface.
(a) Twelve-beam case. (b) Nine-beam case.

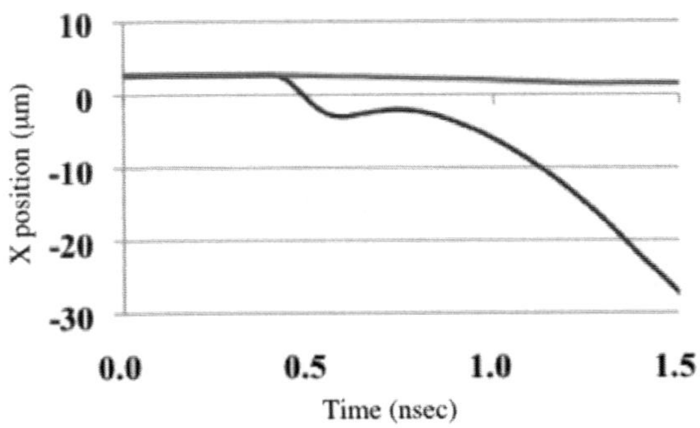


Figure 4. The time evolution of the gravity point of the fuel region. The red and blue lines indicate the twelve-beam and nine-beam case respectively.

Figure 5 shows the time evolution of the average density in the fuel region with the trajectory of the fuel-shell interface in the case of a spherical implosion that is simulated as the pressure of the driven layer is defined by the average pressure of the driven layer of the twelve-beam case. The black, red, and blue lines indicate the density of the spherical, twelve-beam and nine-beam case respectively. The dashed line indicates the trajectory of the fuel-shell interface in the case of the spherical implosion. Comparing the twelve-beam and the nine-beam case, the maximum density of the fuel region decreases by approximately 30 % in the case of the nine-beam. On the other hand, the density decrease between the twelve-beam and spherical case is very small, that is approximately 1%. Therefore, it is considered that the density decrease is due to the strong asymmetric flow.

In this simulation, the cone was treated as the solid boundary. In reality, it is important matter whether the cone tip could tolerate such a strong asymmetric flow due to the GekkoXII nine-beam implosion. In next step, it is required to estimate the effect of the asymmetric implosion on the cone tip.

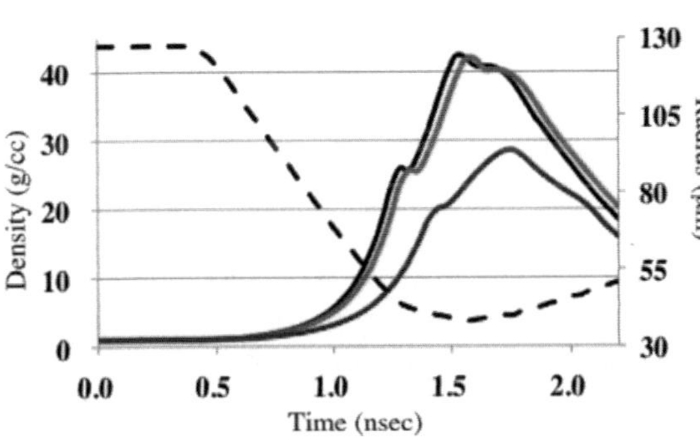


Figure 5. The time evolution of the average density in the fuel region and the trajectory of the fuel-shell interface of the spherical implosion case. The black, red and blue lines indicate the spherical, twelve-beam and nine-beam case. The dashed line indicates the trajectory.

## References

[1] Tabak M, *et al* 1994 *Phys. Plasmas* **1** 1626–34
[2] Azechi H, *et al* 2006 *Plasma Phys. Control. Fusion* **48** B267–75
[3] Kodama R, *et al* 2002 *Nature* **418** 933
[4] Shiraga H, *et al* 2011 *Plasma Phys. Control. Fusion* **53** 124029
[5] Stephens R B, *et al* 2005 *Phys. Plasmas* **12** 056312
[6] Nagatomo H, *et al* 2009 *Nucl. Fusion* **49** 075028
[7] Theobald W, *et al* 2011 *Phys. Plasmas* **18** 056305
[8] Sakagami H and Nishihara K 1990 *Phys. Rev. Lett.* **65** 432–35
[9] Peskin C S 1972 *J. Comput. Phys.* **10** 252–71
[10] Mittal R and Iaccarino G 2005 *Annu. Rev. Fluid Mech.* **37** 239–61

# 10-TW high-contrast double CPA laser system for ion acceleration

**A. ogo[*,1], H. Sakaki[1], . Fukuda[1], S. inno[1], M Kanasaki[1,2], H. Kiriyama[1], T. Shimomura[1], K. Ogura[1], M.Mori[1], P. R. Bolton[1], and K. Kondo[1]**

[1]Kansai Photon Science Institute, Japan Atomic Energy Agency, 8-1-7 Umemidai, Kizugawa, Kyoto 619-0215, Japan
[2]Graduate School of Maritime Sciences, Kobe University, Kobe 658-0022, Japan

**Abstract.** We demonstrate an insertable pulse cleaning module (IPCM) that is connected to a commercial CPA Ti:sapphire laser system and reduces the background of amplified spontaneous emission (ASE), when the output energy and repetition rate of the original laser system is completely preserved. The ASE temporal contrast is suppressed by three orders of magnitude on the benefit of double saturable absorbers involved in the module. Our method can reform old conventional lasers into high-contrast advanced systems.

## 1. Introduction

Advances in the technology of ultraintense chirped-pulse-amplification (CPA) [1] lasers are propelling studies of the laser-driven particle acceleration to relativistic energies. Especially in the case of ion acceleration, a laser pulse with intensity well exceeding $10^{18}$ W/cm$^2$ irradiates a solid foil target, when the laser-solid interaction can be complicated by the presence of an amplified spontaneous emission (ASE) at a leading edge of the high-intensity laser pulse.

Several techniques have been developed to reduce the ASE of high-intensity laser pulses: saturable absorber (SA) [2, 3], double CPA laser system [4], optical parametric chirped pulse amplification (OPCPA) combined with SA [5], or cross polarized wave generation ( PW) [6, 7]. In these methods, the ASE suppression was accomplished by revising the laser system itself or developing a completely new laser. To enhance the temporal contrast in lasers of current architecture, plasma mirror [8, 9] can be installed after the final compressor of the system. This approach has the advantage that no further ASE can arise after the plasma mirror. However, it leads to energy losses that can no longer be compensated for. In addition, the plasma mirror is a fundamentally destructive device and thus is hard to accomplish with high-repetition-rate lasers. It was reported [9] that the total reflectivity of their double plasma mirror is around 50% and the total number of shots available was limited to 2000 with a maximum repetition rate of 1 Hz.

In the present work, we propose a novel method to enhance the temporal contrast of conventional laser systems, making NO change on the output energy, the repetition rate, and the current architecture of the laser: insertable pulse cleaning module (IPCM) with double SAs. In the IPCM beamline, a pulse compressor is placed uppestream of a stretcher, in inverse with CPA lasers. Connecting IPCM to the intermediate stage of a single CPA laser, the system works

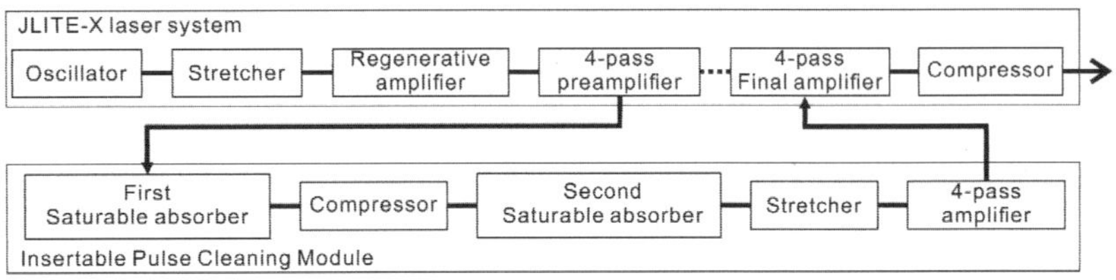


**Figure 1.** The laser configuation of Insertable Pulse Cleaning Module (IPCM) connected to JLITE- laser system.

as a double CPA laser in whole, when SAs involved in IPCM are expected to suppress the ASE drastically. In previous works , SAs were used for a femtosecond (fs) pulses of 1-10 $\mu$J [2, 5] or stretched pulses of 1 mJ [3]. The present work makes a first attempt to inject a few-10s-mJ laser into SA to filter the ASE without causing any damage on the devise. Thereby, the IPCM with SAs can be placed further downstream on the laser system line. This makes it possible to filter the ASE generated in the upperstream components including a regenerative amplifier and a muti-pass amplifier, in our case. In addition, the IPCM has a muti-pass amplifier independently to compensate the energy of pulses decreased in the process of ASE-filtering with SAs. Therefore, the laser system completely preserves the final output energy after the installation of IPCM, although the laser contrast is enhanced.

## 2. Experimental setup

Development of IPCM has been performed at JAEA using JLITE- [10], which was built based on a commercial Ti:sapphire fs laser system produced by Continuum in 2003. The setup is shown in Fig. 1. The front end consists of a Ti:sapphire oscillator that delivers 12-fs pulses and a pulse stretcher based on Öffner triplet [11]. The laser pulse stretched to $\sim$1 ns with a positeve charp is injected into a regenerative amplifier and reaches $\sim$2-mJ energy. Then, the pulse is amplified 4 times through a Ti:sapphire crystal pumped with a 532-nm Nd:YAG laser (Continuum, Surelite II-10) and obtains the energy of $\sim$30 mJ. In the original JLITE- system, the laser pulse is subsequently introduced into the final 4-pass amplifier, the crystal of which is pumped with two 600-mJ, 532-nm Nd:YAG lasers (Continuum, Powerlite 9010), and reaches $\sim$500-mJ energy.

In this development, we connect the IPCM as a bypass line between the 4-pass preamplifier and the final 4-pass amplifier, mentioned above. The IPCM consists of the first SA, a grating compressor, the second SA, a stretcher and another 4-pass amplifier. The stretched pulse from the 4-pass preamplifier encounters the first SA (CVI Melles Griot, RG-850) having a thickness of 1 mm. Then, the ASE component lying prior to the 1-ns stretched pulse is expected to be absorbed. Consequently, the first SA works as a pulse cleaner in the time range farer than 0.5 ns from the pulse time center. A similar method is also seen in [5, 3]. Suppressing ASE in the ns range is crucially required by solid-target experiments, because the hydrodynamic expansion of preplasma can progress with a typical velocity of $\sim$1-10 $\mu$m/ns [12], resulting in a serious deformation of the target in the ns time scale. The pulse is then temporally compressed to $\sim$50 fs by the grating compressor and passes through the second SA (CVI Melles Griot, RG-850) of 2 mm in thickness, when the ASE pedestal lying just before the rising edge of the fs main puse is absorbed. In other words, the second SA works up to the time range of picosecond (ps) prior to the main pulse peak.

The input beam energies for the first and the second SAs are $\sim$30 mJ and $\sim$10 mJ, respectively. In previous works [5, 2], SAs were used for fs pulses with the input energies of

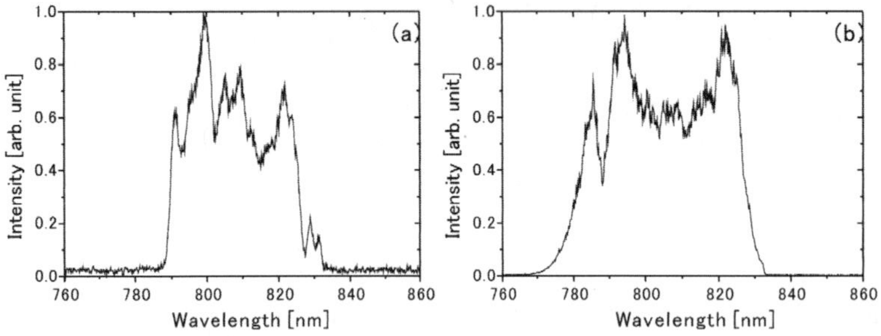


**Figure 2.** Spectra obtained with (a) and without (b) the IPCM.

1-10 $\mu$J that is $10^4$-times smaller than the present case. In our work, by setting the input energy fluence to be ~30 mJ/cm$^2$ and ~10 mJ/cm$^2$ for the first and second SAs, respectively, the laser pulse is transmitted without any damage. The energy transmittance observed is 70% and 40% for the first and the second SAs, respectively. In our knowledge, the present work is the first attempt to inject a few 10s mJ class laser pulses into SAs for the purpose of suppressing ASE.

The laser pulse cleaned by the double SAs is again stretched by the other Öffner-triple stretcher, when the beam diameter is set to be 5 mm in order to induce a laser pulse having a few mJ energy. The stretched laser undergoes 4-pass amplification to be compensated for the energy lost in the cleaning procedure. Therefore, the laser pulse completely recovers its energy at the entrance of the IPCM ~30 mJ, even though its ASE is drastically suppressed, and arrives at the final 4-pass amplifier.

## 3. Temporal profile measurements

Previous works [3, 5] reported that after passing through a saturable absdorber laser pulses often suffer from spectral narrowing or reduction on the component having shorter wavelength. This unfavorable effect is attributed to the characteristics of the SA that the transmittance rapidly increases with the wavelength around 800 nm. It is known [5] that OPCPA pre-amplifier can compensate for the narrowed bandwidth by adjusting pulse timing and phase-matching conditions. In the present system, using no OPCPA, we have successfully minimize the spectral narrowing by decreasing the amplification on longer-wavelength component at the regenerative amplifier [13]. In JLITE- system, a 5-$\mu$m-thick pellicle is used as an etalon and placed in the resonance cavity of the regenerative amplifier. The incidence angle of the intracavity beam is adjusted to attenuate the longer-wavelength component via Fresnel reflection effect on the both surface of the pellicle. Figute 2(a) shows the adjusted spectra obtained after the final amplifier. The bandwidth is 35 nm (FWHM), which is not suffering significant spectral narrowing in comparison with the case without the IPCM, 45 nm (FWHM), shown in Fig. 2(b).

To demonstrate the compressability of pulses undergoing the pulse cleaning with the IPCM, the temporal profile of the laser main pulse is measured after the final compressor with SHG-FROG spectrogram. Figure 3(a) shows a typical SHG-FROG result obtained with the IPCM, when the laser system is working at full power. The duration of the re-compressed pulse is determined to be 45.7 fs (FWHM) in average with a standard deviation of 6.8 fs, which is comparable with the result 40 fs [10] obtained without the IPCM.

The temporal profile including ASE is observed by a high dynamic range third-order cross correlator (Amplitude Technologies, Sequoia) having a temporal measurement range of 570 ps. The measurement is performed at full power mode with a repetition rate of 10 Hz. Figure 3(b) shows the trace of the pulse cleaned by the IPCM in a black line, where the leading edge of the main pulse is drastically suppressed in intensity. The contrast ratio between the intensities of the main pulse and ASE is determined to be $1 \times 10^9$ and $2.5 \times 10^8$ at 500 ps and 150 ps prior to the main-pulse arrival, respectively. It is worth noting that one cannot find any residual temporal

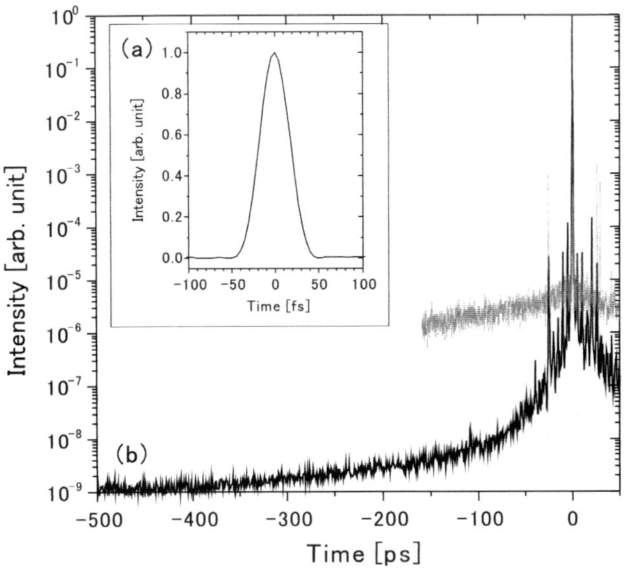


**Figure 3.** Temporal profiles measured at full power mode. (a) A temporal profile around the main peak acquired with SHG-FROG spectrogram. (b) Third-order cross correlation traces of the pulses from the final compressor obtained with (solid line) and without (dashed grey line) the IPCM.

distortion, including a prepulse having fs duration, at the leading edge ranging from -500 ps to -50 ps. This characteristics is clearly beneficial to realize preplasma-free interaction with a target. On the other hand, without the IPCM, the intensity contrast ratio is $5 \times 10^5$ at -150 ps. Evidently, our IPCM have achieved the ASE suppression with three-orders of magnitude. Here, we have to notice that the first SA works on the leading edge of the stretched laser pulse having 1-ns duration. Hence, the ASE suppression seen in the range closer than -500 ps is achieved predominantly with the second SA, which cleans the pulse up to the rising edge of the re-compressed pulse. Therefore, the contrast ratio is probably expected to be improved further more one or two orders of magnitude (resulting in $10^{10}$-$10^{11}$ contrast) at the range < -500 ps, not seen in Fig. 3(b), with the benefit of the first SA.

## Acknowledgment

This work was supported by a Scientific Research (C) No. 25420911 commissioned by the Ministry of Education, Culture, Sports, Science and Technology, Japan (ME T) and the Funding Program for Next Generation World-Leading Researchers (NE T Program) from the Japan Society for the Promotion of Science (JSPS).

## References

[1] D. Strickland and G. Mourou, Optics Comm. **56**, 219-221 (1985).
[2] J. Itatani *et al.*, Opt. Commun. **148**, 70-74 (1998).
[3] S. Fourmaux *et al.*, Opt. Express **19**, 8486-8497 (2011).
[4] M. P. Kalashnikov *et al.*, Opt. Lett. **30**, 923.925 (2005).
[5] H. Kiriyama *et al.*, Opt. Lett. **37**, 3363-3366 (2012).
[6] G. I. Petrov *et al.*, Opt. Lett. **26**, 355.357 (2001).
[7] V. Chvykov *et al.*, Opt. Lett. **31**, 1456.1458 (2006).
[8] G. Doumy *et al.*, Phys. Rev. E **69**, 026402 (2004)
[9] C. Thaury *et al.*, at. Phys. **3**, 424-429 (2007).
[10] M. Mor *et al.*, Phys. Rev. ST Accel. Beams **12**, 082801 (2009).
[11] G. Cheriaux *et al.*, Opt. Lett., **21**, 414-416 (1996).
[12] A. Yogo *et al.*, Phys. Rev. E **77**, 016401 (2008).
[13] C. P. J. Barty *et al.*, Opt. Lett. **21**, 668-670 (1996).

Journal of Physics: Conference Series **688** (2016) 012129 doi:10.1088/1742-6596/688/1/012129

# Laser-plasma experiments to study super high-energy phenomena during extreme compression of the Earth's magnetosphere by Coronal Mass Ejections *

**Yu P Zakharov, A G Ponomarenko, V M Antonov, E L Boyarintsev, A V Melekhov, V G Posukh, I F Shaikhislamov**

Institute of Laser Physics (ILP), Siberian Branch of Russian Academy of Sciences (SB RAS), Novosibirsk, Av. Lavrentyeva 13/3, 630090, Russia

E-mail: ki1z@mail.ru

**Abstract**. Problem of the global and even catastrophic modification of the Earth's magnetosphere (into Artificial one) by impulsive and huge plasma ejecta, was proposed for the first time during our study of possible after-effects of high-energy explosions against asteroids at near-Earth space. Later, a similar problem of extreme compression of the Earth's magnetopause from its usual $R_{mp} \approx 10R_E$ up to new stand-off distance $R_m^* \sim 3R_E$, by plasma of giant Coronal Mass Ejections (CME, with effective energy $E_0 \sim 10^{28}$ J), was considered for its simulations by Laser-Produced Plasma (LPP) at KI-1 facility of ILP, that were done initially without "Solar Wind" (in AMEX experiment). Here we present the first results of the "full" laboratory simulations of the CME-problem with the up-stream impact of LPP (with $E_0 \sim 1$ kJ) onto classical terrella-model of "stationary" magnetopause (with $R_{mp} \approx 17$ cm), formed near compact dipole in a flow of background $H^+$-plasma, imitated Solar Wind. As a result, we have observed for the first time a two-fold compression of magnetopause size, accompanied by very strong and near expected value of dipole magnetic field's compression up to the factor $7 \div 8 \approx (R_{mp}/R_m^*)^3$ inside of magnetopause. Our data allow to predict a global CME-effect at $E_0 \sim 10^{29}$J.

## 1. KI-1 Program and basic researches of Plasma-Dipole interaction for laboratory simulations

More than 50 years ago in many plasma laboratories had started various experiments with magnetic dipoles to create a model of the Earth's magnetosphere, so called "terrella". But up to now, none of them has not well confirmed possibilities to reproduce in a full measure (and to study) a whole physical processes [1], even in more simple case of front MagnetoPause (MP) with the main Chapman-Ferraro current system, which forms a magnetic cavity around dipole in Solar Wind (SW) . Moreover, in the past were absent a technical opportunities to simulate a non-stationary phenomena in MP-region, caused by strong shock-disturbances of SW, most powerful of which are a natural giant Coronal Mass Ejections (CME) from the Sun [2,3] as well as a artificial anti-asteroid explosions at near-Earth space [4]. Namely for simulation of both given cases of the global SW-disturbances, a general scientific Program was formulated near 40 years ago [5] during creation of the facility [6,15] "Kosmic Investigations - 1" (KI-1) of ILP, based on the using of high-power $CO_2$-lasers for generation of Laser-Produced Plasma (LPP) to imitate CME (and its shocks) or space explosions. From the very beginning of the realisation of KI-1 Program a study of the basic collisionless interaction processes between LPP and magnetic fields (or magnetized Background Plasma - BP), including dipole one, became the main goal of all investigations [6]. As a result, at their final stage a new and very

Journal of Physics: Conference Series **688** (2016) 012129 doi:10.1088/1742-6596/688/1/012129

important PIC-modeling [7], experimental [8] and theoretical [7,9] data on the Plasma-Dipole (PD) interaction was obtained, for the conditions required to simulate MP structure correctly. According to this data, the observed "anomalous" penetration of plasma (up to $R_n$ ~ Stoermer radius [8,10]) inside of classical MHD-scale of the MP at $R_m$ and opposite outward shift of cavity boundary up to $R_x$ are caused mainly by Hall-effects according to new model [7,9] at presence of observed Large Larmor Flute Instability [11-13]. Both non-MHD "$R_n$" and "$R_x$-effects" are determined by the low-value D <1, as shown for latter one at figure 1, where $D = R_m/\Lambda_i$ is a general physical criterion of the problem [14] for ion inertia scale $\Lambda_i = C/\omega_{pi}$. So, for the correct simulation of magnetic cavity boundary ($\gamma \approx 1$) we need to supply $D \geq 1$, while to get a plasma front $R_n$ rather close to $R_m$ one needs a value $D \geq 2\div3$ [7,9].

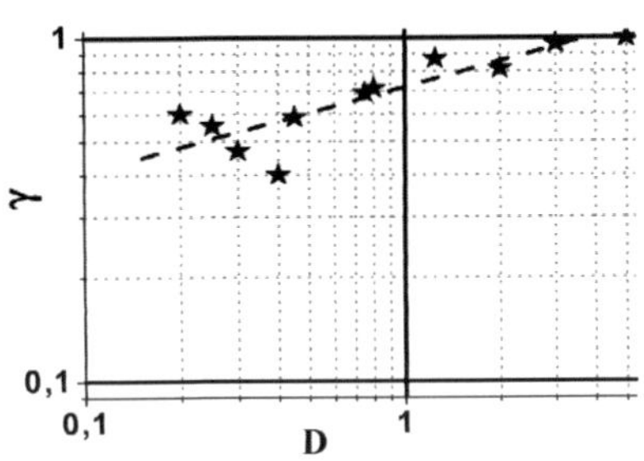


**Figure 1.** Combined data (*) of experiments and 3D/PIC-simulations of ILP [7-9] (with another calculation data by the same PIC-code [12] of Kyushu University) on the stand-off size $R_x$ of magnetic cavity around dipole, in comparison with its Hall-MHD model [9] – dashed line. Here 3D-model dimensionless dependence (for $D \leq 1$), like $\gamma = (R_x/R_m)^{-1} \approx \alpha D^{1/4}$, is determined by the main criterion D [14]. We had used the data of quasi-stationary overflowing of dipole by plasma stream (with MP at $R_m = R_{mp}$) and dipole in a flow of exploding plasmas, when $R_m = R_m^*$.

In a review papers [1,9] a lot of KI-1 experimental data on mainly steady (and some LPP) plasma overflowing of various dipoles μ are presented for regimes of different D up to $D \approx 5$, when a both BP' front and MP-position are very close to MHD scale $R_{mp} = (\mu^2/4\pi n_* m_* V_*^2)^{1/6}$, determined by density $n_*$, ion mass $m_*$ and velocity $V_*$ of SW. For the case of exploding plasmas, a very useful approximation $R_m^* \approx 0{,}75 R_0/æ^{1/6}$ for MHD MP-position was suggested [4,11], based on the value of another, energetic criterion $æ = 3E_0R_0^3/\mu^2$ of problem, for the distance $R_0$ between a dipole and point of plasma ejection. Given relation for $R_m^*$ was tested in comparison with a simple MHD-model [4,11] and 3D/PIC-simulations [11-13] of PD-interaction in vacuum as well as was verified in a series [11-13] of Artificial Magnetosphere Experiments (AMEX) with LPP in a wide range of $20 \leq æ \leq 5000$.

## 2. A "full" simulative HERMEX experiment at KI-1 with Background and Laser Plasmas

The characteristics of new simulative scheme HERMEX (HERmean Magnetosphere EXperiment) shown at figure 2, are most suitable to study the non-stationary CME-related processes in the Mercury, more simple magnetosphere: without ionosphere and almost non-conducting planetary surface, that means a small influence of the additional current systems, like a field-aligned (Birkeland or FAC) one. This new scheme is similar to our previous one [15], when during a KEVL [11] experiment with high-density BP, the surface of dipole (with small moment $\mu \sim 10^5$ G*cm$^3$) had been hit by LPP blob. Here,

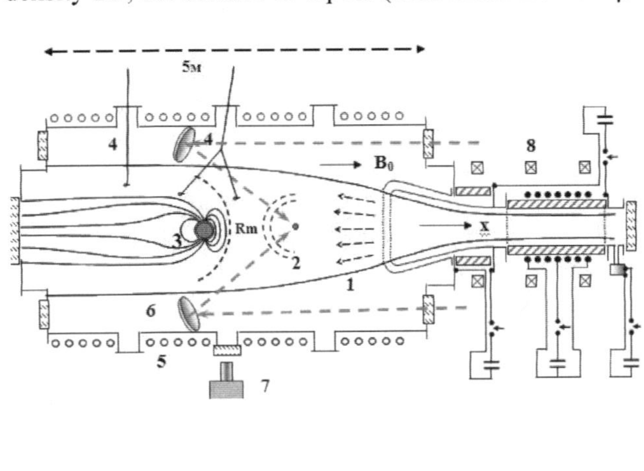


**Figure 2.** Experimental scheme of **HERMEX** with Background Plasma **(1)** and Laser-Produced Plasma **(2)** at distance $R_0$ = 46 cm from dipole **(3)**, in high-vacuum chamber (∅ 120 cm), supplied by: Langmuir and magnetic probes **(4)**, coil system(**5**) for uniform magnetic field $B_0$, focusing elements **(6)** for $CO_2$-Laser beams (2*100J / 100 ns), Gated Optical Imager-GOI/100 ns **(7)** and θ-pinch **(8)** to generate flow of Background Plasma.

we had used a new compact dipole with the moment up to $10^6$ G*cm$^3$, which allows to supply all nessessary conditions (see Table) to simulate magnetopause of the Earth and its super-compression due to using of enhanced effective LPP-energy, i.e. $E_0=(dE_0/d\Omega)*4\pi$ (and effective number $N_{e0} \sim 10^{19}$ for electrons), while $E_k=(dE_0/d\Omega)*<\Omega>$. Laser spot was ~ 3 cm$^2$ at polyethylene target of convex form.

| **Parameters** | **Notations** | **Values** |
|---|---|---|
| Density of Background Plasma - BP (its diameter) | $n_*$ (∅) | $1.5\cdot10^{13}$ cм$^{-3}$(∅ ~70-80 cm) |
| Velocity of BP (and ions) | $V_*$ ($H^+$) | 50-60 km/s ($H^+$) |
| Size of MagnetoPause (MP) in the BP | $R_{mp}$ | 17-18 cm |
| Magnetic moment of dipole and radius $R_d$ | $\mu$ | $5*10^5$ G*cm$^3$, $R_d$ = 2,7 cm |
| Uniform magnetic field in chamber (along to $V_*$ of BP) | $B_0$ | 20 G |
| Kinetic energy of Laser Plasma-LPP (Effective Energy) | $E_k$ ($E_0$) | ~ 60 J (700 J) |
| Expansion velocity of LPP (and its solid angle <Ω>) | $V_0$ (of front) | ≈ 210 km/s (<Ω> ~1 Sr) |
| Size of MP in the stream of LPP (in vacuum) | $R_m^*$ | 8 cm |
| Hall parameter D for plasmas BP and LPP | $D = R_m/(C/\omega_{pi})$ | $D_* \approx 3$ and $D \approx 1,5$ |
| Alfven-Mach number for LPP | $M_{a0} = V_0/C_{A*}$ | ~ 20 >> 1 |
| Energetic criterion LPP ($H^+/C^{+4}$ ions,<m/z>≈2.5a.e.m.) | $æ = 3E_0R_0^3/\mu^2$ | ~ $10^4$ >> 1 |
| Ratio of velocities of LPP to BP | $V_0/V_*$ | ~ 4 (~10) |
| Ratio of the initial MP-size to its compressed one | $\nu = R_{mp}/R_m^*$ | ~ 2 |
| *For the case of CME impact onto Earth' MP: $M_{ao}$~70 >>1, $V_0/V_*$ ~7(~10), ν ~ 2÷3; æ ~$10^{23}$ >>1, $D_*$~D~1000* | | |

## 3. Electromagnetic Global Effect of CME and main results of HERMEX experiment

The fast and global changes ΔB of geomagnetic field (SSC) during MP-compression by CMEs, probably are the most dangerous of their after-effects [2-4,16], leading for example to well-known damages of high-power transformers and other grids. This problem is characterized by the value ΔB/Δt which could reach a few 1000 nT/min for giant CME, according to our estimations [11], which is comparable with the local ΔB-changes of auroral electrojets [17]. Magnetic field $B_{CF}$ of Chapman-Ferraro current system (at semi-spherical MP) is a quasi-uniform [12] with the level $B_{CF}\approx\mu/R_m^3$ from which its independence from dipole μ followed, for both MHD-expression $R_{mp}$ and $R_m^*$ for MP-position. Data of GOI (Figure 3**a,b**) show an effect of two-fold compression of MP by LPP in BP up to the size of $R_m^* \approx 8$ cm for the given case ($\mu = 5*10^5$ G*cm$^3$) of LPP in vacuum (**c**), since a scale of LPP interaction with BP [4,11] is $R_*=(3N_{e0}<m/z>/4\pi n_* m_*)^{1/3} \approx 75$ cm and larger than $R_0 = 46$ cm, as corresponds to the value of specific criterion for BP-problem $\xi = R_*/R_0 \sim 2$ also in space (at $E_0 \geq 10^{27}$ J). GOI-data give dependence of MP-size for BP+LPP cases rather close to LPP (in vacuum, Figure 4).

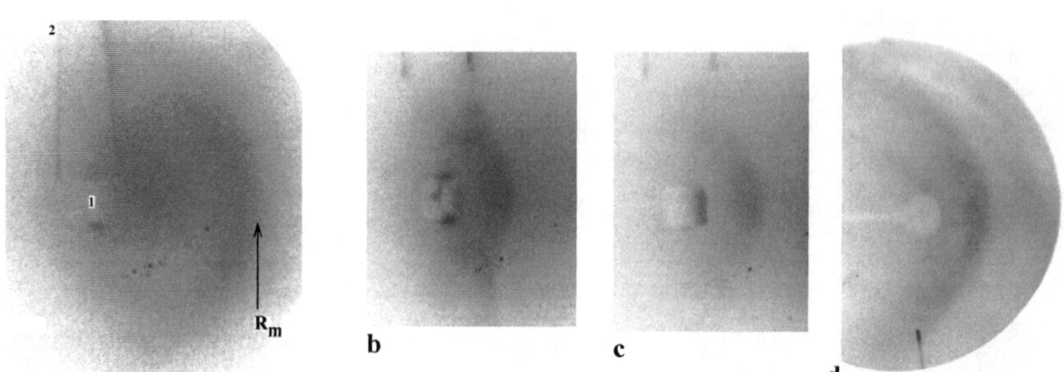


**Figure 3.** Meridian shape of MP formed by Background Plasma (BP):1–dipole (∅5,5 cм); 2–probes

The same for joint action of both BP and LPP at t=2,2 μs.

Meridian (**c**) and equatorial (**d**) shapes of MP formed by Laser-Produced Plasma (LPP) in vacuum (t=2,2 μs and t=2,9 μs).

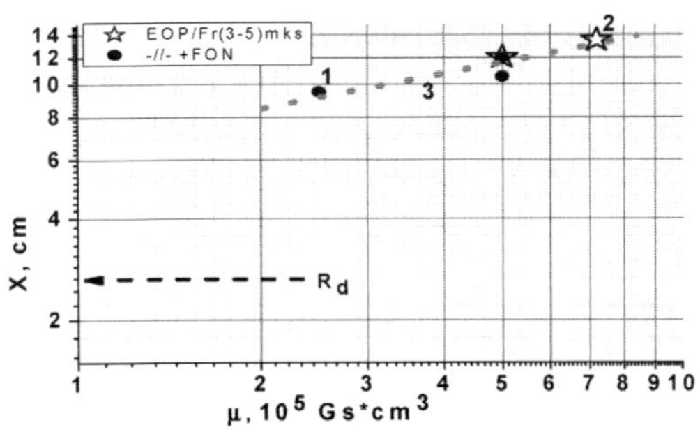


**Figure 4.** GOI data on MP-position depending upon dipole μ for regimes: **1**–Laser Plasma with Background one; **2**–Laser Plasma only; **3**– scaling $R_x \propto \mu^{1/3}$ from $R_m^*$ (left point at t≈3 μs, others t≈4,8÷5 μs). $R_d$ - dipole rad.

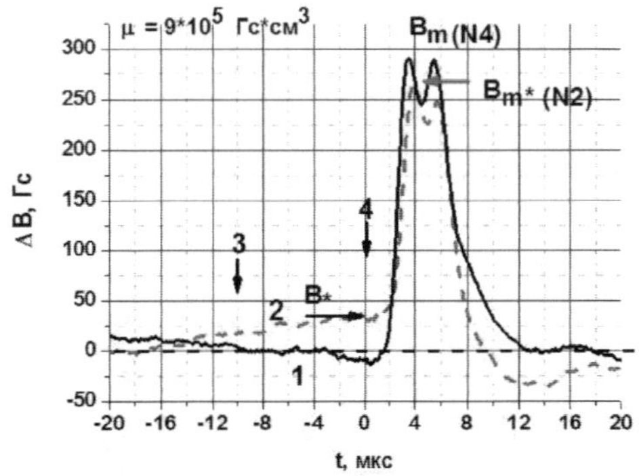


**Figure 5.** Compression of magnetic field near dipole by Laser Plasma: **1** – $B_m$ in vacuum; **2**– $B_{m*}$ with Background Plasma (its $\Delta B = B_*$ starts at point **3**); **4**–Laser starts.

Typical examples of magnetic field compression ΔB is shown at Figure 5 and for both values of ΔB (from BP and LPP) were are almost independent from the μ-value (as was expected). Indeed the both corresponding quantities $\Delta B = B_*$ and $B_{m*}$ are slightly increase, when μ decrease from initial (at Figure 5), as $B_*/B_{m*} \approx 35/267$; 40/283 and 47/310 (in G) for μ = 9; 5 and 2,5 (in $10^5$ G*cm$^3$). Although BP did not affect onto MP, it seems that at presence of background, some more complicated plasma (and field) structures are formed near dipole surface (Figure 3b), like a FAC with negative ΔB at Figure 5 (t=12 μs). Therefore, in the case of "Superflare" [18] with $E_0 \sim 10^{29}$ J, not only a super-compression of magnetopause up to $R_m^* \approx 2R_E$ could occur, but also extreme FAC near polar regions could be generate.

* This work was supported by the ILP Project II.8.1.4 of SB RAS on Fundamental Researches, Integrated Project 113 of SB RAS, Program N 22 Presidium RAS and in Part by CRDF-Project 14864.

## References

[1] Antonov V M, Boyarintsev E L, Zakharov Yu P, Melekhov A V, Posukh V G, Ponomarenko A G, Shaikhislamov I F 2013 in *Plasma Heliogeophysica*, 20 p. – In Press
[2] Ponomarenko A G, Zakharov Yu P, et al. 2006 *IAU Symposium "Solar Activity..."* **233**, 389
[3] Manchester W B, Ridley A J, Gombosi T I et al. 2006 *Adv. Space Res.* **38** 253 and 263
[4] Zakharov Yu, Nikitin S, Ponomarenko A, et al. 1997 *Annals New-York Academy Sc.* **822** 583 and Zakharov Yu, Nikitin S, Ponomarenko A. 1996 *Computational Technologies* **1**, 36 (In Russ.) in https://www.researchgate.net/profile/Yuri_Zakharov/publications/4?sorting=published
[5] Gevorkyan L, Orishich A, Ponomarenko A, Soloukhin R, et al. 1976 in *Physical Gasdynamics*, **6** Novosibirsk, Inst. Pure and Applied Mechanics AN USSR 192 (in Russ.)
[6] Zakharov Yu P 2003 *IEEE Trans. Plasma Sci.* **31** 1243
[7] Zakharov Yu P, Antonov V M, Boyarintsev E L, Vchivkov K V, Shaikhislamov I F, Melekhov A V, Posukh V G, Ponomarenko A G, et al. 2009 *Int. Conf. on Space Science and Commun.* 96
[8] Zakharov Yu P, Antonov V M, Shaikhislamov I F, et al. 2011 *Contrib. Plasma Phys.* **51** 18
[9] Shaikhislamov I F, Antonov V M, Zakharov Yu P, et al. 2013 *Adv. Space Res.* **52** 422
[10] Fujita K 2004 *J. Space Technol. Sci.* **20** 26
[11] Ponomarenko A G, Zakharov Yu P, Antonov V M, et al. 2007 *IEEE Trans. Plasma Sci.* **35** 813
[12] Zakharov Yu P, Antonov V M, Boyarintsev E L et al. 2008 *J. Phys.: Conf. Ser.* **112** 042011
[13] Zakharov Yu P, Ponomarenko A G, Horton W, et al. 2009 *Astrophys. Space Sc.* **322** 151
[14] Omidi N, Blanco-Cano X, Russell C T, Karimabadi H 2004 *Adv. Space Res.* **33** 1996
[15] Ponomarenko AG, Zakharov Yu P, et al. 2008 *Plasma Phys. Contr. Fus.* **50** 074015
[16] Kuznetsov V D, Makhutov N A 2012 *Herald Russian Academy Sciences* **82** 36
[17] Kappenman J G, in *Space Storm and Space weather Hazards*, NATO Sci. Ser., **38**, 335 (2001)
[18] Schaefer B E, et al. 2000 *Astrophys. J.* **529** 1026; Maehara H, et al. 2012 *Nature* **485** 478

# Charge Accretion Rate and Injection Radius of Ionized-Induced Injections in Laser Wakefield Accelerators

**Ming Zeng, Min Chen and Zheng-Ming Sheng**
Key Laboratory for Laser Plasmas (Ministry of Education), Department of Physics and Astronomy, Shanghai Jiao Tong University, Shanghai 200240, China
E-mail: zmsheng@sjtu.edu.cn

**Abstract.** Ionization-induced injection has recently been proved to be a stable injection method with several advantages in laser wakefield accelerators. However, the controlling of this injection process aiming at producing high uality electron beams is still challenging. In this paper, we examine the ionization injection processes and estimate the injection rate with two-dimensional particle-in-cell simulations. The injection rate is shown to increase linearly with the high-Z gas density as long as its ratio is smaller than some threshold in the mix gases. It is also shown that by changing the transverse mode of the driving lasers one can control the injection rate.

## 1. Introductions

In laser wakefield accelerators (LWFA), one of the most important issues to obtain high beam qualities is optimizing the injection mechanism. Injection means placing the electrons into the accelerating and focusing phase with proper velocities that they can gain kinetic energy from the wake continuously. The injection schemes can be either pre-accelerating the electrons to the phase velocity of the wake such as the colliding pulse injection scheme [1, 2, 3], or slowing down the wake so that the electrons have enough time to catch up with the accelerating phase before deceleration such as the density ramping or transition injection scheme [4, 5, 6, 7]. A recent proposed injection scheme, called ionization-induced injections, generates free electrons inside a wake at appropriate phases [8, 9]. This idea is relatively easily operated in experiments due to its simple experimental configuration. Usually the ionization-induced injection is to utilize the higher ionization threshold of the inner shell of some high- gas (such as nitrogen, oxygen or argon) mixed with a low- gas (usually hydrogen or helium) to control the starting phase of the ionized electrons. Although the ionization-induced injection has been experimentally demonstrated by several groups [10, 11, 12], the detailed studies on the effect of the mixing ratio and injection length have not been conducted yet.

In this paper, we study the injection dynamics by two-dimensional (2D) particle-in-cell (PIC) simulations and find the relation between the gas mixing ratio and the injection rate. The transverse radius of the injection section is proved to be a few microns, and can be changed by using different transverse laser modes. At the same time, the betatron radiation can be enhanced when an appropriate laser mode is used.

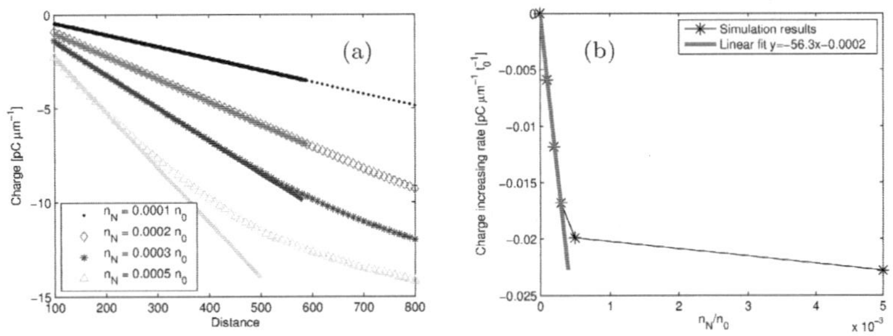


**Figure 1.** (Color) (a) Injected charge vs. laser propagating distance from 2D PIC simulations. Only electrons in the first bucket are considered. The different symbols represent cases for different nitrogen atom densities as shown in the legend. The solid lines are linear fittings for data with propagation distance from 100 to 300 $\mu$m. $n_0 = 2.829 \times 10^{19}$ cm$^{-3}$ is the density normalization unit in the simulations. Because the simulations are in 2D slab geometry, the charge unit is pico-Coulomb per $\mu$m. (b) Beam charge increasing rate vs. nitrogen atom density. Charge increasing rate is in pico-Coulomb per $\mu$m per simulation unit time $t_0 = 3.33$ fs.

## 2. Charge accretion rate in ionization injection process

One of the most important issues in ionization-induced injection is to choose the mixing ratio of the injection provider and the background gas. The injection provider is a gas that has the electrons with higher ionization threshold, such as nitrogen, oxygen or argon. The background is another gas that has low  number and provides the majority of the background plasma electrons which cannot be trapped, such as hydrogen or helium. Some reported experiments chose the mixing ratio smaller than 1:9 to reduce the defocusing effect owning to the preionization [12]. In this section, we study the relation between the mixing ratio and the injected charge by PIC simulations and find out a way to estimate the growth rate of the injected beam charge.

In the following 2D simulations, the injection provider is chosen to be nitrogen and the background is helium. The simulation box size is $50\times100$ $\mu$m$^2$, the cell size is $0.015625\times0.25$ $\mu$m$^2$ and the time step interval is 0.033 fs. The helium density is kept to be $0.1n_0$, where $n_0 = 2.829 \times 10^{19}$ cm$^{-3}$ is the density unit in the simulations. The nitrogen atom density varies from 0.0001 to $0.005n_0$. The ADK ionization model is used in the simulation code.

The wake driver is a laser beam with wavelength of 800 nm, pulse duration of 33fs in FWHM, waist of $W_0 = 7.48$ $\mu$m, and normalized vector potential of $a_0 = 2.8$, which corresponds to

$$P\ [\mathrm{GW}] = 21.5 \times \left(\frac{a_0 W_0}{\lambda_0}\right)^2 = 14.7.\ [\mathrm{TW}] \tag{1}$$

The laser pulse satisfies the so called transverse matching condition[13]($k_\mathrm{p} = \sqrt{0.2}$ $\mu$m$^{-1}$ in our cases) for self-guided propagation $k_\mathrm{p}W_0 = 2\sqrt{a_0}$, which means that the laser can stably propagate for a distance much longer than the Rayleigh length. The mixed gas has a profile of a half infinite flat-top with a small up-ramp at the beginning, and the laser is focused at the up-ramp.

We measure the injected beam charge vs. time in each simulations. The charge variations with different nitrogen density cases are shown in Fig. 1 (a). One may see that the charge linearly increased with laser propagation distance in the first few microns. We fit the data in this linear region and obtain the beam charge increasing rate. After obtaining enough slops, one can get the charge increasing rate vs. nitrogen atom density as shown in Fig. 1 (b). We can see in low $n_\mathrm{N}$ limit the beam charge increasing rate is linear with $n_\mathrm{N}$, while there is a saturation in larger $n_\mathrm{N}$.

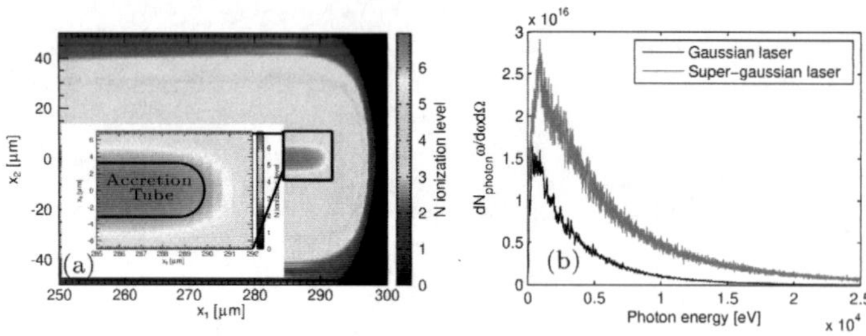


**Figure 2.** (Color) (a) Nitrogen ionization level plot after laser propagates a few hundred microns. The nitrogen inner shell includes the 6th and 7th electrons. Applying the analysis in Sec. 2, we can define a "Accretion Tube" inside which the electrons ionized from the inner shell of nitrogen atoms are injected. The Accretion Tube encircles the region of ionization level of 6 or above as one can see in the figure. (b)Forward betatron radiation spectra for the cases of using Gaussian and super-Gaussian driver lasers. The laser parameters are the same: $a_0 = 2.8$, $W_0 = 7.48$ $\mu$m, and the measurements were taken after the injection saturation, which means that the injected charge is almost the same in these two cases.

The physical modelling is straightforward. Assume in the first few hundred microns in the gas target, the laser keeps its shape and amplitude so that the ionization-induced injection is relatively stable. And in low $n_{\rm N}$ limit, only the inner shell electrons within few microns away from the central region of nitrogen are ionized and injected as shown in Fig. 2 (a). We call this few microns' section the "Accretion Tube", inside which the ionized 6th and 7th electrons of nitrogen atoms are trapped in the wake. Thus the increasing rate of the injected beam charge, or "Accretion Rate" $R_{\rm A}$ can be written as

$$R_{\rm A} = \pi r_\sigma^2 \cdot v_{\rm g} \cdot 2n_{\rm N} \cdot e \tag{2}$$

in 3D geometry, or

$$R_{\rm A} = 2r_\sigma \cdot v_{\rm g} \cdot 2n_{\rm N} \cdot e \tag{3}$$

in 2D slab geometry, where $r_\sigma$ is the radius of the accretion tube, $v_{\rm g}$ is the laser group velocity and $e$ is the electron charge. Notice that in 3D, $R_{\rm A}$ is defined as charge gain per unit time, while in 2D slab geometry $R_{\rm A}$ is actually charge gain per unit time per unit length (in the 3rd direction). By taking $v_{\rm g} \approx c$ and applying our simulation parameters, one obtain

$$R_{\rm A}[\mathrm{pC} \cdot \mu\mathrm{m}^{-1} \cdot \mathrm{t}_0^{-1}] = -18.1 \times \mathrm{r}_\sigma[\mu\mathrm{m}] \cdot \frac{\mathrm{n_N}}{\mathrm{n_0}} \tag{4}$$

where $t_0 = 1$ $\mu$m$/c = 3.33$ fs is the time unit in the simulations. Finally, by comparing with the linear fitting in Fig. 1 (b) $R_{\rm A}[\mathrm{pC} \cdot \mu\mathrm{m}^{-1} \cdot t_0^{-1}] = -56.3 \times \frac{n_{\rm N}}{n_0}$, one obtains the accretion radius $r_\sigma = 3.1$ $\mu$m, which is consistent with Fig. 2 (a).

In high $n_{\rm N}$ limit, the accretion rate is shown to be saturated. This may because of the beam loading effect [14] or ionization etching effect on the laser pulse.

## 3. Charge accretion rate and betatron radiation with super-Gaussian lasers

In this section, we use the same parameters as those in Sec. 2, but change the laser transverse mode to super Gaussian, i. e. $E_{\rm laser} \propto \exp[-(r/W_0)^4]$. The total laser energy changes due to the mode change is small and neglected. We repeat the simulations shown in Sec. 2, i. e. a few simulations with $n_{\rm N}$ varying from 0.0001 to 0.0005 $n_0$, calculate the injected charge increasing rate $R_{\rm A}$, and find the linear relation between $n_{\rm N}$ and $R_{\rm A}$, $R_{\rm A}[\mathrm{pC} \cdot \mu\mathrm{m}^{-1} \cdot t_0^{-1}] = -72.5 \times \frac{n_{\rm N}}{n_0}$.

Compared with Eq. 4, the accretion radius is obtained as $r_\sigma = 4.0$ $\mu$m in the super-Gaussian mode driven case, which is larger than the Gaussian mode case. In a real experiment, the laser beam is commonly neither a pure Gaussian mode nor a pure super Gaussian mode. This makes the precise prediction of the accretion radius difficult. However, the approach described above can be applied in real experiments to measure the accretion radius.

Measuring the betatron radiation is another possible way to diagnose the injected beam radius. We use the "particle tracking" technique of the PIC simulations to obtain the trajectories of the injected particles, and calculate the far-field radiation spectra plotted in Fig. 2 (b). From simulation results we know that the total injected charge is almost the same in Gaussian or super-Gaussian driver case. This is because the measurements were taken after injection saturation. However, the radiation of the super-Gaussian laser driven case is about twice of the Gaussian driver case. There are two reasons. The first is that the super-Gaussian case has larger charge increasing rate, so that a larger number of electrons are injected earlier and thus gain higher energy compared with the Gaussian case. The second is that in the super-Gaussian laser driven case, the off-axial laser field is slightly higher, so that there are more inner-shell electrons ionized and injected off-axially compared with the Gaussian case, thus the averaged transverse displacement of the injected electrons is larger. As measured from the simulations, the emittance after injection saturation is 1.05 mm $\cdot$ mrad for the gaussian driver case, and 2.13 mm $\cdot$ mrad for the super-gaussian driver case. These explains why the betatron radiation of the later is about twice of the former.

## 4. Summary

We present a method to measure the injection rate of the ionization-induced injection in laser wakefield accelerators. The injection rate is proportional to the density of the injection provider (nitrogen in our case) when the density is small. It is also proportional to the injection area which can be controlled by using different laser modes. By using super-Gaussian transverse modes, one can obtain increased injection section areas and amplitudes of the betatron motions, thus enhanced betatron radiations.

## Acknowledgments

The authors would like to acknowledge the OSIRIS Consortium, consisting of UCLA and IST (Lisbon, Portugal) for the use of OSIRIS and the vis D framework. This work is supported by the National Basic Research Program of China (Grant No. 2013CBA01504) and the National Science Foundation of China (Grants No. 11121504, 11129503, 11075105 and 11205101). Simulations were performed on the Hoffman2 at UCLA, Hopper at NERSC, and Supercomputer Π at Shanghai Jiao Tong University.

## References

[1] Esarey E *et al.*, 1997 *Phys. Rev. Lett.* **79**(14) 2682 2685.
[2] Faure J *et al.* 2006 *Nature* **444** 737 739.
[3] Rechatin C *et al.* 2009 *Phys. Rev. Lett.* **102**(16) 164801
[4] Geddes C G R *et al.* 2008 *Phys. Rev. Lett.* **100**(21) 215004.
[5] Gonsalves A *et al.* 2011 *Nature Physics* **7** 862 866.
[6] Bulanov S *et al.* 1998 *Phys. Rev. E* **58**(5) R5257 R5260.
[7] Li F Y *et al.* 2013 *Phys. Rev. Lett.* **110**(13) 135002.
[8] Chen M *et al.* 2006 *J. Appl. Phys.* **99** 056109, Chen M *et al.* 2012 *Phys. Plasmas* **19** 033101.
[9] Oz E *et al.* 2007 *Phys. Rev. Lett.* **98**(8) 084801, Zeng M *et al.* 2013 *Phys. Plasmas* **21** 030701.
[10] Clayton C E *et al.* 2010 *Phys. Rev. Lett.* **105**(10) 105003.
[11] McGuffey C *et al.* 2010 *Phys. Rev. Lett.* **104**(2) 025004.
[12] Pak A *et al.* 2010 *Phys. Rev. Lett.* **104**(2) 025003.
[13] Lu W *et al.* 2007 *Phys. Rev. ST Accel. Beams* **10**(6) 061301.
[14] Tzoufras M *et al.* 2008 *Phys. Rev. Lett.* **101**(14) 145002.

# Measurement of preheat in aluminium target in indirect drive using the SGIII prototype facilities

**C Zhang[1,2], Z B Wang[3], H Liu[3], X S Peng[3], F Wang[3], Y K Ding[3] and J Zheng[1,2]**

[1] CAS Key Lab of Basic Plasma Physics, Hefei, Anhui 230026, P. R, China
[2] Department of modern physics, University of Science and Technology of China, Hefei, Anhui 230026, P. R. China
[3] Research Center of Laser Fusion, China Academy of Engineering Physics, Mianyang ,Sichuan 621900, P. R. China

E-mail: zcph@mail.ustc.edu.cn

**Abstract**. The velocity interferometer system for any reflector (VISAR) is used to demonstrate preheat effect in aluminium in indirect drive. The rear surface motion prior to shock front was observed and compared with a multi-group calculation. By properly adjusting the hard x-ray portion of the radiation source, the calculated rear surface motion fits well with the experimental results, which gives us confidence to predict the preheated temperature of the sample by hard x-rays. Further, the effect of hohlraum geometry is compared and discussed experimentally. The result suggests gas-filled hohlraum or hohlraum with low Z substrates should be considered to further reduce preheating.

## 1. Introduction

The study of sample preheating is an inescapable issue during the equation of state measurement. For the indirect drive scheme, the hohlraum provides a radiation source with adequate spatial uniformity and compresses samples into high energy density states with strong shocks [1]. However, some high energy photons like M-band x-rays are generated in the hohlraum [2]. These photons would penetrate deep into the target and heat the sample before shocks arrival. This preheating may play a role of violation and degenerate the diagnostic accuracy [3]. As a result, the experimental diagnostics of sample preheating and a way to reduce it is of great importance.

In this article, the preheating of aluminum is investigated by using the VISAR system. The rear surface velocity of the sample is accurately measured. The results are verified by using multi-group hydrodynamic simulation [4], which gives a relatively good description for the preheating phenomenon. Besides, a special designed hohlraum is tested to control the preheating for the sample. The effect of hohlraum geometry on the sample preheating is experimentally compared and discussed. The comprehensive analysis suggests gas fill hohlraum or hohlraum with low Z substrate is needed to prevent plasma from convergence and to further reduce preheating.

## 2. Experimental setup

The configuration of the experiment is illustrated in figure 1 (a). It is performed in the SGIII prototype facility [5], which has eight Nd: glass beams operating at 1.053 μm. In our experiment, we use the 3ω mode. The continuous phase plate (CPP) technique is applied to smooth out each beam. The laser focal spot has a full-width-half-maximum (FWHM) around 0.6 mm. The pulse width is 3 ns with the total energy of 5.2 kJ, resulting in the laser spot intensity of $1 \times 10^{14}$ W/cm$^2$.

A gold cylinder hohlraum with the size of 1.5 mm in diameter and 1.2 mm in length is placed in the center of the vacuum chamber. The hohlraum has a laser entrance hole (LEH) in the upper surface of 0.9 mm and a diagnostic hole (DH) in the bottom of 0.7 mm × 0.4 mm. The eight laser beams are simultaneously delivered into the hohlraum from the LEH. The radiation temperature of the source is measured by using a multichannel soft x-ray spectrometer (SXS) [6] from the 20° azimuth angle with respect to the hohlraum axis. A set of flat-response x-ray detector (FXRD) [7] array is equipped to verify the total flux diagnostics.

A planar aluminum with a thickness of 35μm is attached to the DH, which is the preheating sample in our experiment. The observed rear surface is defined as the critical density for the probe beam, which has a wavelength of 532 nm. It is the location where reflection occurs. Through a reflective mirror assembled on the target with designed angle, the observation direction of VISAR is perpendicular to the aluminum surface and it has a view of the whole aluminum sample. The preheating effect of the sample is mainly diagnosed through this VISAR system. The probe beam is reflected from the aluminum surface and transmitted into an interferometer system. Then the interference fringe is imaged onto the slit of an optical streak camera. The position of the image is adjusted so that its central part is observed to ensure the feasibility of one dimensional approximation in our experiment. The system enables a spatial resolution of 10 μm and a temporal resolution of 20ps. The rear surface expansion due to preheat is obtained from the fringe shift. The sensitivities of the two channels are 2.35 km/s/fringe and 3.04 km/s/fringe, respectively. The detection threshold is 1/10 fringes and the minimum detectable velocity is around 0.2 km/s. Since aluminum is widely used as a standard material in EOS measurements, the study could help us to specify the effect of preheating on EOS diagnostics.

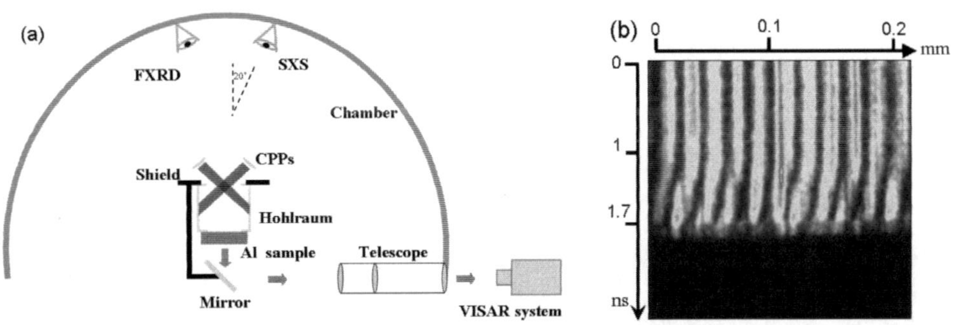


**Figure 1 (a):** Experimental setup is illustrated. **(b):** Image from VISAR measurement.

## 3. Experimental results and discussion

In our experiment, the contribution to preheating from hot electrons can be ignored because of the low intensity and the uniform laser spot. The laser plasmas interaction (LPI) processes are not effectively stimulated. Thus, the preheating mainly contributes from x-ray radiation transport.

### *3.1. Rear surface motion*

The typical VISAR result is shown in figure 1 (b). The horizontal axis provides the spatial direction and the vertical one is the time axis. Time zero stands for the arrival of the main pulse. No movement is observed of the interference fringe before 1 ns due to low preheating level. However, between $t = 1$ ns to $t = 1.7$ ns, apparent shift occurs, indicting the motion of the sample's rear surface. After $t = 1.7$ ns, the shock breaks out from the rear surface. The released plasma becomes a good absorber for the probe beam, resulting in a sudden disappearance of the fringe.The velocity of the rear surface is deduced accordingly in figure 2 (a) and matches the simulation result from MULTI1D well. In our simulation, we choose Thermos database [8] as the opacity tables and divide the photons into 35 groups. A factor of 0.82 is multiplied to the radiation temperature obtain by SXS in order to match the shock break-out time. This factor may represent the transport efficiency from the LEH to the DH. The spectrum we use in our simulation is constructed by a Planckian spectrum plus two Gaussian bumps centered at energy $E_1 = 1.4$ keV and $E_2 = 3.5$ keV, respectively. The former bump just locates near aluminum K-edge, which has a higher penetrability. The temporal evolution of the two bumps is illustrated in figure 2 (b). The fractions remain small before $t = 1$ ns, corresponding to a low preheating level in figure 1 (b) and figure 2 (a). After then both fractions increase notably, which probably results from the motion and convergence of gold plasma in the center of the hohlraum. With higher electron density and temperature, the emission of the piled up plasma tends to be stronger and more energy is converted into hard x-rays which shall enhance preheating.

According to the simulation, the highest preheating temperature of aluminum ahead of the shock is about 1800 K in our experiment, which is below aluminum's boiling point of over 2700 K. As a result, the state of the sample is still liquid. Further more, the reflectivity maintains high value before shocks. This phenomenon indicates the reflective surface should be in metallic state rather than plasma state, which is a strong absorber for the probe beam. Based on the above two facts, it is deduced the fringe shifts should be due to the rear surface motion instead of other factors, such as plasma expansion.

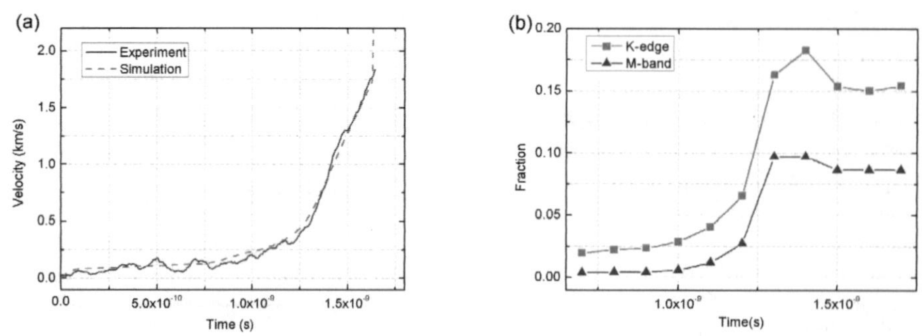


**Figure 2 (a)**: the experimental result (blue curve) compared with simulation (dashed red). **(b)**: the temporal evolution for the fractions of aluminum K-edge (red) and gold M-band (blue).

### 3.2. *Effect of hohlraum geometry*

In order to suppress preheating in the hohlraum, a special hohlraum is designed to shield the laser spot from the sample, as shown in figure 3 (a) and figure 3 (b). The special designed hohlraum contains a double-cylinder structure. The laser beams just lay within the turning corner of the two cylinders. This special geometric design serves to prevent the sample from "seeing" the laser spot directly.

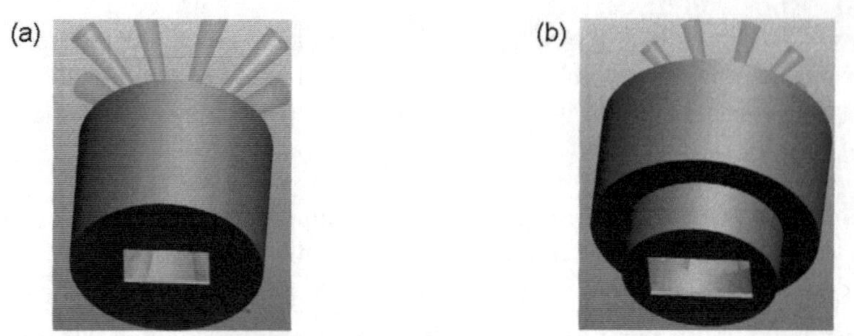


**Figure 3 (a)**: the typical shape of a hohlraum design. **(b)**: the special designed hohlraum, which contains a double-cylinder structures.

The resultant rear surface velocity and the predicted preheating temperature for different hohlraum designs are given in figure 4 (a) and figure 4 (b), respectively. The rise of velocity for the typical hohlraum is apparently much faster, which implies a stronger preheating level. The predicted temperature of the rear surface for the typical hohlraum is approximately twice the magnitude of the modified ones at the same time. However, sample preheat would last a longer time due to smaller source intensity and continuous plasma motion. The final preheat temperature is of the same order as the normal hohlraum. This result suggests additional improvements, such as low Z substrates in the hohlraum, are needed to further suppress preheating.

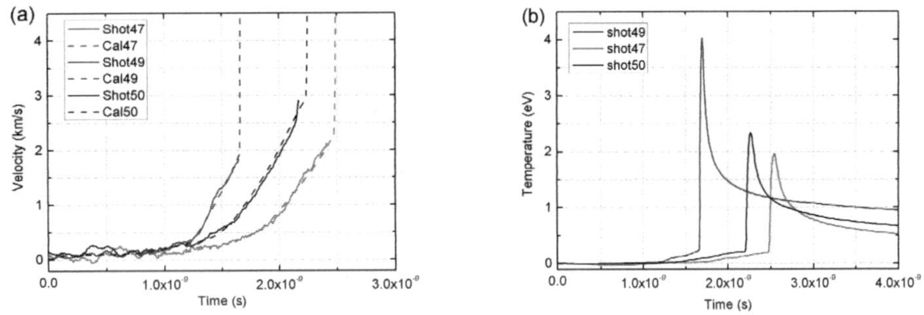


**Figure 4 (a)**: the rear surface motion of different hohlraum designs. **(b)**: the predicted temperature evolution for each shot. The typical hohlraum result is blue, while the special designed hohlraum is black and red.

## 4. Summary

We have investigated the preheating effect in indirect drive scheme in SGIII prototype facility. The rear surface velocity is measured using high sensitivity VISAR system. The phenomenon is reproduced by multi-group hydrodynamic simulation, which gives a good prediction for preheating studies. Besides, a suitable hohlraum design could reduce preheating to a lower level and current result implies there is still space to improve hohlraum designs.

## References

[1] Löwer Th, Sigel R, Eidmann K, Földes I B, Hüller S, Massen J, Tsakiris G D, Witkowski S, Preuss W, Nishimura H, Shiraga H, Kato Y, Nakai S and Endo T 1994 *Phys. Rev. Lett.* **72** 3186

[2] McDonald J W, Suter L J, Landen O L, Foster J M, Celeste J R, Holder J P, Dewald E L, Schneider M B, Hinkel D E, Kauffman R L, Atherton L J, Bonanno R E, Dixit S N, Eder D C, Haynam C A, Kalantar D H, Koniges A E, Lee F D, MacGowan B J, Manes K R, Munro D H, Murray J R, Shaw M J, Stevenson R M, Parham T G, Van Wonterghem B M, Wallace R J, Wegner P J, Whitman P K, Young B K, Hammel B A and Moses E I 2006 *Phys. Plasmas.* **13** 032703

[3] Rothman S D, Evans A M, Horsfield C J, Graham P and Thomas B R 2002 *Phys. Plasmas.* **9** 1721

[4] Ramis R, Schmalz R and Meyer-Ter-Vehn J 1988 *Comp. Phys. Commu.* **49** 475

[5] Li P, Jing F, Wu D S, Zhao R C, Li H, Lin H H and Su J Q 2012 *Proc. SPIE* **8433** 843317

[6] Song T M, Yang J M and Yi R Q 2010 *High Power Laser and Particle Beams* **22**, 2017

[7] Li Z C, Jiang X H, Liu S Y, Huang T X, Zheng J, Yang J M, Li S W, Guo L, Zhao X F, Du H B, Song T M, Yi R Q, Liu Y G, Jiang S E and Ding Y K 2010 *Rev. Sci. Instrum.* **81** 073504

[8] Ivanov E M, Rozanov V B and Vergunova G A 2003 *Proc. of SPIE* **12** 79

Journal of Physics: Conference Series **688** (2016) 012132 doi:10.1088/1742-6596/688/1/012132

# Quantitative K$\alpha$ line spectroscopy for energy transport in ultra-intense laser plasma interaction

**Z. Zhang[1], H. Nishimura[1], S. Fujioka[1], Y. Arikawa[1], M. Nakai[1], H. Chen[2], J. Park[2], G. J. Williams[2], T. Ozaki[3], H. Shiraga[1], S. Kojima[1], T. Johzaki[5], A. Sunahara[6], N. Miyanaga[1], J. Kawanaka[1], Y. Nakata[1], T. Jitsuno[1], and H. Azechi[1]**

K$\alpha$ [1] Institute of Laser Engineering, Osaka University, 2-6 Yamada-oka, Suita, Osaka 565-0871, Japan

[2]Lawrence Livermore National Laboratory, Livermore, CA 94550, USA

[3] Institute for Laser Technology, 2-6 Yamada-oka, Suita, Osaka 565-0871, Japan

[4] The Graduate School of Engineering, Hiroshima University, 1-4-1 Kagamiyama, Higashi Hiroshima 739-8527, Japan

[5] Institute for Laser Technology, 2-6 Yamadaoka, Suita, Osaka 565-0871, Japan

E-mail: zhang-z@ile.osaka-u.ac.jp

**Abstract.** Absolute K$\alpha$ line spectroscopy is proposed for studying laser-plasma interactions taking place in the cone-guided fast ignition targets. X-ray spectra ranging from 20 to 100 keV were quantitatively measured with a Laue spectrometer. The absolute sensitivities of the Laue spectrometer system were calibrated using pre-characterized laser-produced x-ray sources and radioisotopes. The integrated reflectivity for the crystal is in good agreement with predictions by an open code for x-ray diffraction. The energy transfer efficiency from incident laser beams to hot electrons, as the energy transfer agency, is derived as a consequence of this work. The absolute yield of Au and Ta K$\alpha$ lines were measured in the fast ignition experimental campaign performed at Institute of Laser Engineering, Osaka University. Applying the hot electron spectrum information from the electron spectrometer, an energy transfer efficiency of the incident LFEX [1], a kJ-class PW laser, to hot electrons was derived for a planar and cone-guided geometry.

## 1. Introduction

Fast ignition is recognized as a promising pathway to efficient thermonuclear fusion in laser-driven inertial confinement fusion. A cone-guided CD-shell has been used as a base-line target for the fast ignition experiment [1, 2]. It has long been expected to provide more quantitative information about the hot electron generation and transport in the cone than those derived only with x-ray imaging and neutron detection. In this research, we propose an absolute K$\alpha$ line spectroscopy dedicated for quantitative measurement of hot electron generation and transportation in high-Z targets. This diagnostic provides local information about the hot electrons propagating through specific materials composing the cone-guided target.

In this study, Au and Ta were chosen as tracers since they are representative higher-Z materials which are available for the guide cone, which provide stronger stopping for the MeV-hot electrons and the ratio of the number of hot electrons to the number of K$\alpha$ photons produced are better matching with MeV-hot electrons than lower-Z tracers such as Cu.

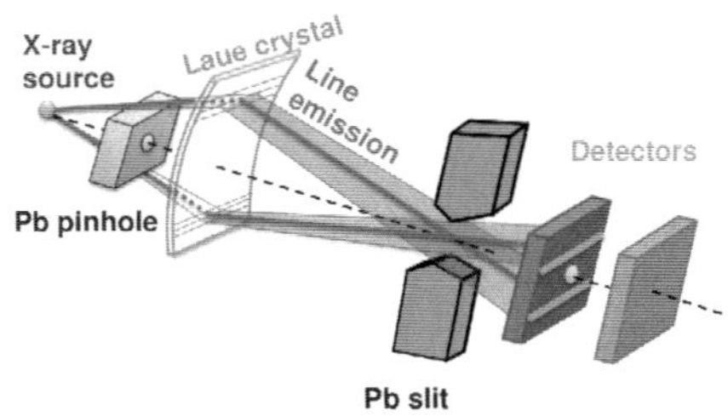


**Figure 1.** The Laue spectrometer system.

## 2. High energy K$\alpha$ x-ray spectrometer

A Laue spectrometer was developed to cover the high energy K$\alpha$ lines from Mo (K$\alpha_1$: 17.48 keV, K$\alpha_2$: 17.37 keV) to Au (K$\alpha_1$: 68.80 keV, K$\alpha_2$: 66.99 keV)[3]. As shown in Fig. 1, the spectrometer consists of a cylindrically curved quartz (10-11) plate and a detector. The detector can either be an imaging plate (IP) from Fujifilm Inc. or a charge coupled device (CCD: Andor Model DH420-FO) with a fiber-optic plate coated with a CsI phosphor of 100 $\mu$m thickness. The quartz plate is bent such that the diffracted x-rays are focused at an intermediate slit. X-ray components propagating in a straightforward manner are shielded from the detector directly with a lead pinhole plate located in front of the crystal and a pair of lead shields located at the intermediate x-ray focus. To avoid influence of hard x-rays from plasma on output signal, whole body of the spectrometer and the detector are covered with lead shielding. This Cauchois geometry effectively discriminates $0^{th}$ order component, stray x-rays and fluorescence from spectrometer components such as filters [4, 5]. By varying the distances from the crystal to the source and detector, this spectrometer can cover the energy range of either 10-60 keV or 22-100 keV.

## 3. Hot electron trajectory

The hot electron trajectory inside the solid target is tracked with Monte-Carlo simulation. A 3-dimensional code PHITS was applied. The code uses the continuous slowing down approximation (CSDA), which means that the electrons change their direction of motion due to elastic scattering, and lose their energy between two scattering points continuously. An example is shown in Fig. 2. The black square in the center indicates a 1 mm Ta cube, which is surrounded by low density air with a pressure of $10^{-5}$ Torr. The hot electron beam was irradiated on the Ta cube from the left side. An initial divergence of 45° and a Gaussian spatial profile were assumed. The energy spectrum was set based on the electron spectrometer (ESM) measurement in experiments. From Fig. 2, it is clearly seen that most of the electrons with energy less than 0.1 MeV were reflected at the front surface. The penetration depth increased along with electron energy; and only part of the electrons with energy higher than 1 MeV can propagate through the Ta cube and escape from the back surface.

## 4. Laser transfer efficiency

The laser transfer efficiency, $\eta_{TE}$, is defined as the fraction of energy deposited into hot electrons. The number of K$\alpha$ photons $N_{K\alpha}$ generated from hot electron with number $N_h$ and temperature $T_h$ can be estimated by the following model:

$$N_{K\alpha} = N_h \frac{n_A\, \omega_{K\alpha}}{4\,\pi\, T_h} \int_0^{\infty} dE\, \sigma_{K\alpha}(E) \times \int_0^{d} dx\, f_h(E_0, x) \exp(-\frac{x}{\lambda_{mfp}\cos(\theta)}) \quad , \tag{1}$$

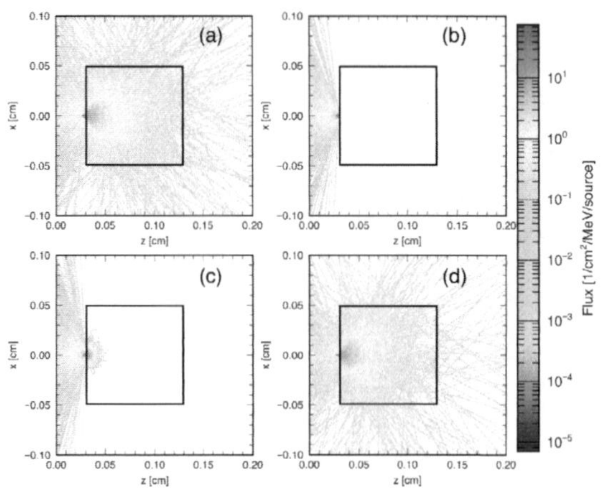


**Figure 2.** Electron trajectory in the Ta cube: (a) all electrons, (b) electron with energy less than 0.1 MeV, (c) electron with energy from 0.1 to 1 MeV, (d) electron with energy more than 1 MeV. In this simulation, an initial hot electron temperature of 5 MeV is assumed, as $f_h = \exp(-\frac{E}{5})$.

where $\sigma_{K\alpha}$, $\omega_{K\alpha}$, and $n_A$ are, respectively, the cross section for K-shell ionization, the K$\alpha$ fluorescence yield, and the atomic number density. The term $\exp(-\frac{x}{\lambda_{mfp}\cos(\theta)})$ describes the reabsorption of K$\alpha$ photons during the propagation through the target material, where $\theta$ is the angle between the spectrometer and target normal. $f_h(E_0, x)$ describes the energy spectrum for the hot electron propagating inside the Ta cube with a depth $x$, where the information was achieved with PHITS simulations. Considering the absolute K$\alpha$ photon number $N_{K\alpha}$ measured by the Laue spectrometer, the transfer efficiency $\eta_{TE}$ can be estimated by comparing the experimental measurement and simulation results.

The transfer efficiency from LFEX laser to target has been estimated in the cases of planar and cone-guided geometry. In the case of planar target, an Au plate was placed at the target chamber center and the LFEX laser with maximum energy of 1.8 kJ was focused on the front surface with an incident angle of 10°. The K$\alpha$ from Au was recorded by the Laue spectrometer and an ESM was located from the back surface for the hot electron temperature. The $\eta_{TE}$ as a function of laser intensity is shown in Fig. 3. As a general trend, the $\eta_{TE}$ is increased by increasing the laser intensity.

The cone-guided configuration is shown in Fig. 4. A cone is attached with a hemi-CH shell, which was irradiated by three beams of Gekko-XII laser [1]. A dense plasma surrounding the tip of the cone was produced to mimic the condition of the fast ignition Au cone+CD shell target [1]. The Ta cube was attached as the K$\alpha$ tracer after the hemi-CH shell. Four types of cone were used, as shown in Fig. 4 (b)-(e): (b) is the standard Au cone with 7 $\mu$m thickness; (c) is

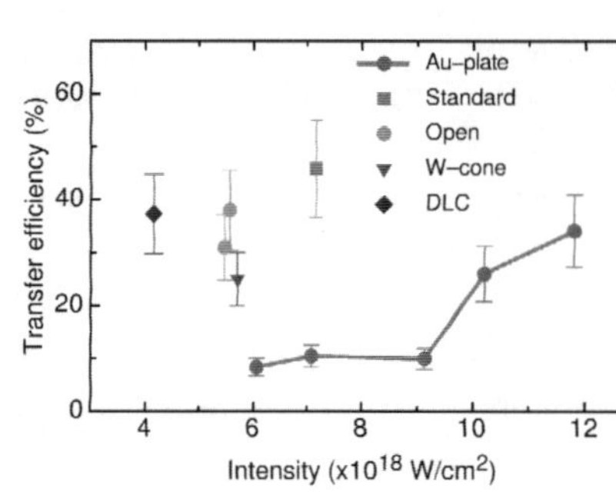


**Figure 3.** The $\eta_{TE}$ as a function of laser intensity for the planar and cone-guided targets.

the open-cone without the tip; (d) is the W-cone with double Au layers; and (e) is the diamond like carbon (DLC) cone. The estimated $\eta_{TE}$ was shown in Fig. 3.

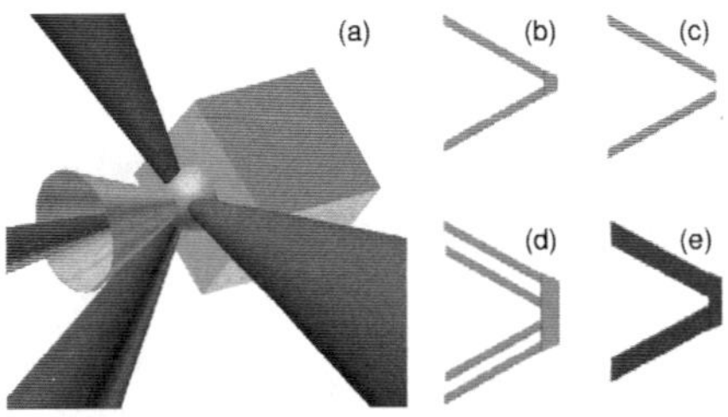


**Figure 4.** (a). The configuration of a cone+CH hemi shell+Ta cube target; (b). Standard Au cone; (c). Open Au cone; (d). W-cone; (d). DLC cone.

## 5. Conclusion

K$\alpha$ line spectroscopy, particularly for hard-x-ray region, has been proposed for quantitative measurement of cone-guided fast ignition experiment. The Au and Ta K$\alpha$ lines were observed and energy transfer efficiency was provided as a preliminary study. Compared with the planar geometry, the LFEX laser transfer efficiency is significantly enhanced with a guiding cone. In near future, absolute measurement of hard x-ray continuum will be made together with that of the K$\alpha$ line to improve the accuracy of energy transfer measurements in the cone-guided fast ignition targets.

## Acknowledgments

The authors would like to thank the Gekko-XII and LFEX laser operation crew, the target fabrication group, the plasma diagnostics group, and the computer operation staffs for their great contribution to this work. The work by LLNL staffs were performed under the auspices of the U.S. DOE by LLNL under Contract DE-AC52-07NA27344.

## References

[1] Shiraga H *et al* 2012 *High Energy Density Phys.* **8** 227
[2] Kodama R *et al* 2001 *Nature* **412** 798
[3] Zhang Z *et al* 2012 *Rev. Sci. Instrum.* **83** 053502
[4] Cauchois Y 1932 *J. Phys. et le Radium* **3** 320
[5] Seely J F *et al* 2007 *High Energy Density Phys.* **3** 263